Bio and Synthetic Based Polymer Composite Materials

Bio and Synthetic Based Polymer Composite Materials

Editors

Emin Bayraktar
S.M. Sapuan
R. A. Ilyas

MDPI • Basel • Beijing • Wuhan • Barcelona • Belgrade • Manchester • Tokyo • Cluj • Tianjin

Editors
Emin Bayraktar
Department of Mechanical and Manufacturing
ISAE-Supmeca-Paris
Saint-Ouen
Malaysia

S.M. Sapuan
Department of Mechanical and Manufacturing Engineering, Faculty of Engineering
Universiti Putra Malaysia
Serdang
Malaysia

R. A. Ilyas
Faculty of Chemical and Energy Engineering
Universiti Teknologi Malaysia
Skudai
Malaysia

Editorial Office
MDPI
St. Alban-Anlage 66
4052 Basel, Switzerland

This is a reprint of articles from the Special Issue published online in the open access journal *Polymers* (ISSN 2073-4360) (available at: www.mdpi.com/journal/polymers/special_issues/recycling_of_bio_and_synthetic_polymer_based_materials).

For citation purposes, cite each article independently as indicated on the article page online and as indicated below:

LastName, A.A.; LastName, B.B.; LastName, C.C. Article Title. *Journal Name* **Year**, *Volume Number*, Page Range.

ISBN 978-3-0365-5240-8 (Hbk)
ISBN 978-3-0365-5239-2 (PDF)

Contents

About the Editors

Emin Bayraktar

Emin Bayraktar (Prof. Emeritus, Habil., Dr (Ph.D.), DSc—Doctor of Science) is an academic and research staff member in Mechanical and Manufacturing Engineering at SUPMECA/Paris, France. His research areas include manufacturing techniques for new materials (basic composites—hybrid), the metal formation of thin sheets (design + test + FEM), static and dynamic behavior and optimization of materials (experimental and FEM—utilization and design of composite-based metallic and non-metallic, powder metallurgy, and energetic material aeronautical applications), metallic-based and non-metallic materials, powder metallurgy and metallurgy of steels, welding, and heat treatment, as well as the processing of new composites, sintering techniques, sinter–forging, thixoforming, etc. He has authored more than 200 publications in the International Journals and International Conference Proceedings and has also authored more than 90 research reports (European = Steel Committee projects, Test + Simulation). He has already advised 32 Ph.D. and 120 MSc theses and is currently advising seven. He is a Fellow of WAMME (World Academy of Science in Materials and Manufacturing Engineering), an Editorial Board—Member of JAMME (International Journal of Achievement in Materials and Manufacturing Engineering), an Advisory board member of AMPT—2009 (Advanced Materials Processing technologies), and a member of APCMP—2008 and APCMP—2010. He was a Visiting Professor at Nanyang Technology University, Singapore, in 2012; Xi'an Northwestern Technical University, Aeronautical Engineering, in 2016; and University of Campinas, UNICAMP-Brazil, in 2013 until 2023. He is a recipient of the Silesian University Prix pour "FREDERIK STAUB Golden Medal-2009"by the Academy of WAMME, "World Academy of Science", Poland, materials science section, and a recipient of the William Johnson International Gold Medal,2014, AMPT academic association.

S.M. Sapuan

S.M. Sapuan is an A-Grade Professor of composite materials at the Department of Mechanical and Manufacturing, Universiti Putra Malaysia (UPM), and a Head of Laboratory of Biocomposite Technology, INTROP, UPM. He has a BEng in Mechanical Engineering from the University of Newcastle, Australia, an MSc in Engineering Design from Loughborough University, UK, and a PhD in Material Engineering from De Montfort University, UK. He is a Professional Engineer, a Society of Automotive Engineers Fellow, an Academy of Science Malaysia Fellow, a Plastic and Rubber and Institute Malaysia Fellow, a Malaysian Scientific Association Fellow, an International Biographical Association Fellow, and an Institute of Material Malaysia Fellow. He is an Honorary Member and immediate past Vice President of the Asian Polymer Association based in IIT Delhi and the Founding Chairman and Honorary Member of Society of Sugar Palm Development and Industry, Malaysia. He is the co-editor-in-chief of Functional Composites and Structures, and a member of editorial boards of more than two dozen journals. To date, he has produced more than 1800 publications, including over 860 journal papers, 50 books, and 175 chapters in book. He has delivered over 50 plenary and keynote lectures, and over 150 invited lectures. He has organized 30 journal Special Issues as a guest editor, presented over 650 technical articles in conferences and seminars, reviewed over 1300 journal papers, and has eight patents. He has successfully supervised 91 PhD and 70 MSc students and 15 postdoctoral researchers. His current h-index is 93, and his number of citations is 31,647 (Google Scholar). He received nine Outstanding Researcher Awards from UPM, ISESCO Science Award (Gold Medal), the Plastic and Rubber Institute Malaysia Fellowship Award, and the Forest Research

Institute Malaysia First Prize Publication Award. He also received the Khwarizimi International Award and the SEARCA Regional Professorial Chair award.

R. A. Ilyas

R.A. Ilyas is a senior lecturer in the School of Chemical and Energy Engineering, Faculty of Engineering, Universiti Teknologi Malaysia, Malaysia. He received his PhD degree in the field of Biocomposite Technology & Design at the Institute of Tropical Forestry and Forest Products (INTROP) UPM. R.A. Ilyas was the recipient of an MVP Doctor of Philosophy Gold Medal Award UPM, 2019, for Best PhD Thesis and Top Student Award, INTROP, UPM. He was awarded an Outstanding Reviewer recognition by {Carbohydrate Polymers}, Elsevier, United Kingdom; Best Paper Award (11th AUN/SEED-Net Regional Conference on Energy Engineering); National Book Award 2018; Best Paper Award in various international conferences; and Top Cited Article 2020-2021 Journal {Polymer Composite}, Wiley, 2022. R.A. Ilyas was also listed and awarded among the world's top 2% scientists (subject-wise) in terms of citation impact during the single calendar years f=if 2019 and 2020 by Stanford University, US, and was given a PERINTIS Publication Award in 2021 and 2022 by Persatuan Saintis Muslim Malaysia; Emerging Scholar Award by Automotive and Autonomous Systems 2021, Belgium; Young Scientists Network - Academy of Sciences Malaysia (YSN-ASM) 2021; UTM Young Research Award 2021; UTM Publication Award 2021; and UTM Highly Cited Researcher Award 2021. His main research interests are (1) polymer engineering (biodegradable polymers, biopolymers, polymer composites, polymer gels) and (2) naterial engineering (natural-fiber-reinforced polymer composites, biocomposites, cellulose materials, nano-composites). To date, he has authored or co-authored more than 404 publications (published/accepted) in 164 journals indexed in JCR/Scopus, 2 non-index journals, 15 books, 104 book chapters, 78 conference proceedings/seminars, 4 research bulletins, and 10 conference papers (abstract published in the book of abstracts) and has been guest editor of 17 Special Issues of journals and editor/co-editor of 10 conferences/seminar proceedings on subjects related to green materials.

Preface to "Bio and Synthetic Based Polymer Composite Materials"

For decades, synthetic fibers have been the leading commodity in the composites industry. However, synthetic fibers have many disadvantages, as they are non-biodegradable. Since synthetic fibers have many shortcomings, researchers have had growing interest in producing polymers that incorporate natural fibers. Natural fibers are becoming more common as a viable option due to the harmful environmental and health consequences of synthetic fibers. Concerns about the environment and the rising greenhouse effect, as well as increasing interest in the use of sustainable materials, have motivated researchers to investigate biocomposite materials. In today's manufacturing environment, natural fiber composites are playing a prominent role in many vital applications, such as in fuselages and propellers in the aerospace industry, racing car bodies, wings of wind turbines, bicycle frames, automobile interiors, seat cushions, and door panels. The great interest in natural fiber composites is due to their high performance, biodegradability, nonabrasive light weight, and low cost. Moreover, the widespread adoption of natural fibers and biopolymers as green materials is being motivated by the rapid depletion of petroleum supplies, as well as by a growing recognition of global environmental issues associated with the use of traditional plastics. The successful application of biopolymers and the promise of alternative pathways with a reduced carbon footprint arising from the use of green materials bodes well for the future design and development of ever more sophisticated green materials.

Emin Bayraktar, S.M. Sapuan, and R. A. Ilyas
Editors

polymers

Editorial

Bio and Synthetic Based Polymer Composite Materials

R. A. Ilyas [1,2,3,*], S. M. Sapuan [4] and Emin Bayraktar [5]

1 School of Chemical and Energy Engineering, Faculty of Engineering, Universiti Teknologi Malaysia (UTM), Johor Bahru 81310, Malaysia
2 Centre for Advanced Composite Materials (CACM), Universiti Teknologi Malaysia (UTM), Johor Bahru 81310, Malaysia
3 Institute of Tropical Forest and Forest Products (INTROP), Universiti Putra Malaysia, Serdang 43400, Malaysia
4 Advanced Engineering Materials and Composites, Department of Mechanical and Manufacturing Engineering, Faculty of Engineering, Universiti Putra Malaysia, Serdang 43400, Malaysia
5 School of Mechanical and Manufacturing Engineering, ISAE-SUPMECA Institute of Mechanics of Paris, 93400 Saint-Ouen, France
* Correspondence: ahmadilyas@utm.my

Citation: Ilyas, R.A.; Sapuan, S.M.; Bayraktar, E. Bio and Synthetic Based Polymer Composite Materials. *Polymers* **2022**, *14*, 3778. https://doi.org/10.3390/polym14183778

Received: 31 August 2022
Accepted: 2 September 2022
Published: 9 September 2022

Bio and Synthetic Based Polymer Composite Materials is a newly opened Special Issue of *Polymers*, which aims to publish original and review papers on new scientific and applied research and make contributions to the findings and understanding of the reinforcing effects of various bio and synthetic-based polymers on the performance of polymer composites. This Special Issue also covers the fibre-reinforced polymer composites' fundamentals, characterisation, and applications.

It is recognised that synthetic-based polymer or petroleum-based plastics have great barrier and thermomechanical properties, as well as a low production cost and require lightweight materials, which produce good performance in terms of overall criteria [1]. The environmental impact of petroleum-based plastic materials, which are non-biodegradable and the increasing need for more sustainable green materials, especially for packaging and plastics in particular, have become an issue of concern. These phenomena are ever-growing global concerns. Thus, in order to overcome this problem, solutions to reduce and in some cases to replace those petroleum-based plastic materials are prioritised in research efforts. One of the current focuses is replacing synthetic-based polymer with bio-based polymer, also known as biopolymers. In recent years, the development of biopolymers based on constituents obtained from natural resources has gained much attention [2,3]. The exploitation of biopolymers to engineer advanced biocomposites and hybrid composite materials is the focus of increasing scientific activity, explained by the growing environmental concerns and interest in the novel features and multiple functionalities of these macromolecules. Biopolymers such as thermoplastic starch (TPS), chitosan, polyhydroxyalkanoates (PHA), cellulose, lignin, chitin, polyhydroxybutyrate (PHB), and poly lactic acid (PLA) have been pursued as alternative solutions. The most widely used is PLA, which is mainly used in packaging applications. It is used for films or thermoformed or injected packages for relative short-term and mild temperature contact conditions, such as fresh salads and beverage drinks, because of its low resistance to temperature. One major limitation commonly referred to is its high price and commercial shortage, as compared to conventional plastics. Thermoplastic starch (TPS) has also been used for replacing petroleum-based plastics [4,5]. However, TPS has lower mechanical properties which make it unsuitable to be used in packaging applications [6]. Thus, one of the ways to overcome this problem is by reinforcing TPS with fiber, which can improve its mechanical properties.

On the other hand, synthetic fibers have been the leading commodity in the composites industry. However, synthetic fibers possess many disadvantages as they are non-biodegradable. Since synthetic fibers have many shortcomings, researchers have expressed growing interest in producing polymers that incorporate natural fibers [7–9]. Natural fibers are becoming more common as a viable option due to the harmful environmental and health consequences of synthetic fibers [10]. Concerns about the environment, the rising greenhouse effect and increasing interest in the use of sustainable materials has motivated researchers to investigate biocomposite materials. Today, fibre-reinforced polymers are used in several applications including in packaging [3,11,12]; electrical and electronic appliances [13,14]; crossarm structures [15]; foam structures [16]; as energy storage [17,18]; in automotives [19]; in filter, coating, and bone tissue engineering; in drug delivery [20]; human prosthetics [21]; and more. The continuous development and appearance on the market of new high-performance reinforcing fibers in polymer composites have constituted a strong challenge for researchers to design and adapt new functional composites for several applications [22–28]. Such natural fibers are comprised of various lignocellulosic plant fiber, cellulose, nanocrystalline cellulose [29–32], nanofibrillated cellulose [33,34], bacterial nanocellulose [35,36], and lignin nanoparticles [37]. The great interest in natural fiber composites is due to their high performance, biodegradability, non-abrasiveness, light weight, and low cost. Moreover, the widespread adoption of natural fibers and biopolymers as green materials is motivated by the rapid depletion of petroleum supplies, as well as by a growing recognition of global environmental issues associated with the use of traditional plastics. The successful application of biopolymers and the promise of alternative pathways with a reduced carbon footprint arising from the use of bio-based materials bodes well for the future design and development of ever more sophisticated green materials.

Thus, in this Special Issue, we aim to capture the cutting edge of the state of the art in research pertaining to bio and synthetic-based polymer composite materials and their advanced applications. Contributions to the processing of bio and synthetic polymers, the use of diverse polymer sources, the reinforcement of fiber materials with polymers, and applications of these polymers composites constitute the backbone of this Special Issue.

Author Contributions: Conceptualization, R.A.I., S.M.S. and E.B.; writing—original draft preparation, R.A.I., S.M.S. and E.B.; writing—review and editing, R.A.I., S.M.S. and E.B.; project administration, R.A.I., S.M.S. and E.B. All authors have read and agreed to the published version of the manuscript.

Funding: The authors would like to express gratitude for the financial support received from the Universiti Teknologi Malaysia for the project "The impact of Malaysian bamboos' chemical and fibre characteristics on their pulp and paper properties", grant number PY/2022/02318—Q.J130000.3851.21H99. The research has been carried out under the programme Research Excellence Consortium (JPT (BPKI) 1000/016/018/25 (57)) provided by the Ministry of Higher Education Malaysia (MOHE).

Conflicts of Interest: The authors declare no conflict of interest.

References

1. Haris, N.I.N.; Ilyas, R.A.; Hassan, M.Z.; Sapuan, S.M.; Afdzaluddin, A.; Jamaludin, K.R.; Zaki, S.A.; Ramlie, F. Dynamic Mechanical Properties and Thermal Properties of Longitudinal Basalt/Woven Glass Fiber Reinforced Unsaturated Polyester Hybrid Composites. *Polymers* **2021**, *13*, 3343. [CrossRef] [PubMed]
2. Tarique, J.; Sapuan, S.; Khalina, A.; Ilyas, R.; Zainudin, E. Thermal, flammability, and antimicrobial properties of arrowroot (Maranta arundinacea) fiber reinforced arrowroot starch biopolymer composites for food packaging applications. *Int. J. Biol. Macromol.* **2022**, *213*, 1–10. [CrossRef] [PubMed]
3. Nazrin, A.; Sapuan, S.M.; Zuhri, M.Y.M.; Tawakkal, I.S.M.A.; Ilyas, R.A. Flammability and physical stability of sugar palm crystalline nanocellulose reinforced thermoplastic sugar palm starch/poly(lactic acid) blend bionanocomposites. *Nanotechnol. Rev.* **2021**, *11*, 86–95. [CrossRef]
4. Diyana, Z.; Jumaidin, R.; Selamat, M.; Ghazali, I.; Julmohammad, N.; Huda, N.; Ilyas, R. Physical Properties of Thermoplastic Starch Derived from Natural Resources and Its Blends: A Review. *Polymers* **2021**, *13*, 1396. [CrossRef]
5. Abotbina, W.; Sapuan, S.M.; Sultan, M.T.H.; Alkbir, M.F.M.; Ilyas, R.A. Development and Characterization of Cornstarch-Based Bioplastics Packaging Film Using a Combination of Different Plasticizers. *Polymers* **2021**, *13*, 3487. [CrossRef]
6. Mohammed, A.A.B.A.; Omran, A.A.B.; Hasan, Z.; Ilyas, R.A.; Sapuan, S.M. Wheat Biocomposite Extraction, Structure, Properties and Characterization: A Review. *Polymers* **2021**, *13*, 3624. [CrossRef]

7. Nurazzi, N.M.; Khalina, A.; Chandrasekar, M.; Aisyah, H.; Rafiqah, S.A.; Ilyas, R.; Hanafee, Z. Effect of fiber orientation and fiber loading on the mechanical and thermal properties of sugar palm yarn fiber reinforced unsaturated polyester resin composites. *Polimery* **2020**, *65*, 115–124. [CrossRef]
8. Izwan, S.M.; Sapuan, S.; Zuhri, M.; Mohamed, A. Thermal Stability and Dynamic Mechanical Analysis of Benzoylation Treated Sugar Palm/Kenaf Fiber Reinforced Polypropylene Hybrid Composites. *Polymers* **2021**, *13*, 2961. [CrossRef]
9. Suriani, M.; Zainudin, H.; Ilyas, R.; Petrů, M.; Sapuan, S.; Ruzaidi, C.; Mustapha, R. Kenaf Fiber/Pet Yarn Reinforced Epoxy Hybrid Polymer Composites: Morphological, Tensile, and Flammability Properties. *Polymers* **2021**, *13*, 1532. [CrossRef]
10. Norrrahim, M.N.F.; Huzaifah, M.R.M.; Farid, M.A.A.; Shazleen, S.S.; Misenan, M.S.M.; Yasim-Anuar, T.A.T.; Naveen, J.; Nurazzi, N.M.; Rani, M.S.A.; Hakimi, M.I.; et al. Greener Pretreatment Approaches for the Valorisation of Natural Fibre Biomass into Bioproducts. *Polymers* **2021**, *13*, 2971. [CrossRef]
11. Rozilah, A.; Jaafar, C.N.A.; Sapuan, S.M.; Zainol, I.; Ilyas, R.A. The Effects of Silver Nanoparticles Compositions on the Mechanical, Physiochemical, Antibacterial, and Morphology Properties of Sugar Palm Starch Biocomposites for Antibacterial Coating. *Polymers* **2020**, *12*, 2605. [CrossRef] [PubMed]
12. Alias, A.H.; Norizan, M.N.; Sabaruddin, F.A.; Asyraf, M.R.M.; Norrrahim, M.N.F.; Ilyas, A.R.; Kuzmin, A.M.; Rayung, M.; Shazleen, S.S.; Nazrin, A.; et al. Hybridization of MMT/Lignocellulosic Fiber Reinforced Polymer Nanocomposites for Structural Applications: A Review. *Coatings* **2021**, *11*, 1355. [CrossRef]
13. Azman, M.A.; Asyraf, M.R.M.; Khalina, A.; Petrů, M.; Ruzaidi, C.M.; Sapuan, S.M.; Wan Nik, W.B.; Ishak, M.R.; Ilyas, R.A.; Suriani, M.J. Natural Fiber Reinforced Composite Material for Product Design: A Short Review. *Polymers* **2021**, *13*, 1917. [CrossRef]
14. Sharma, S.; Sudhakara, P.; Omran, A.A.B.; Singh, J.; Ilyas, R.A. Recent Trends and Developments in Conducting Polymer Nanocomposites for Multifunctional Applications. *Polymers* **2021**, *13*, 2898. [CrossRef] [PubMed]
15. Asyraf, M.R.M.; Ishak, M.R.; Sapuan, S.M.; Yidris, N.; Ilyas, R.A.; Rafidah, M.; Razman, M.R. Evaluation of Design and Simulation of Creep Test Rig for Full-Scale Crossarm Structure. *Adv. Civ. Eng.* **2020**, *2020*, 6980918. [CrossRef]
16. Amelia, J.J.N.; Zuhri, M.Y.M.; Leman, Z.; Zahari, N.I.; As'Arry, A.; Ilyas, R.A. Quasi-Static Compression Properties of Bamboo and PVC Tube Reinforced Polymer Foam Structures. *Polymers* **2021**, *13*, 3603. [CrossRef]
17. Nurazzi, N.M.; Sabaruddin, F.A.; Harussani, M.M.; Kamarudin, S.H.; Rayung, M.; Asyraf, M.R.M.; Aisyah, H.A.; Norrrahim, M.N.F.; Ilyas, R.A.; Abdullah, N.; et al. Mechanical Performance and Applications of CNTs Reinforced Polymer Composites—A Review. *Nanomaterials* **2021**, *11*, 2186. [CrossRef]
18. Jha, K.; Tyagi, Y.K.; Kumar, R.; Sharma, S.; Huzaifah, M.R.M.; Li, C.; Ilyas, R.A.; Dwivedi, S.P.; Saxena, A.; Pramanik, A. Assessment of Dimensional Stability, Biodegradability, and Fracture Energy of Bio-Composites Reinforced with Novel Pine Cone. *Polymers* **2021**, *13*, 3260. [CrossRef]
19. Nurazzi, N.M.; Asyraf, M.R.M.; Rayung, M.; Norrrahim, M.N.F.; Shazleen, S.S.; Rani, M.S.A.; Shafi, A.R.; Aisyah, H.A.; Radzi, M.H.M.; Sabaruddin, F.A.; et al. Thermogravimetric Analysis Properties of Cellulosic Natural Fiber Polymer Composites: A Review on Influence of Chemical Treatments. *Polymers* **2021**, *13*, 2710. [CrossRef] [PubMed]
20. Sharma, S.; Sudhakara, P.; Singh, J.; Ilyas, R.A.; Asyraf, M.R.M.; Razman, M.R. Critical Review of Biodegradable and Bioactive Polymer Composites for Bone Tissue Engineering and Drug Delivery Applications. *Polymers* **2021**, *13*, 2623. [CrossRef]
21. Khare, J.M.; Dahiya, S.; Gangil, B.; Ranakoti, L.; Sharma, S.; Huzaifah, M.R.M.; Ilyas, R.A.; Dwivedi, S.P.; Chattopadhyaya, S.; Kilinc, H.C.; et al. Comparative Analysis of Erosive Wear Behaviour of Epoxy, Polyester and Vinyl Esters Based Thermosetting Polymer Composites for Human Prosthetic Applications Using Taguchi Design. *Polymers* **2021**, *13*, 3607. [CrossRef] [PubMed]
22. Ilyas, R.A.; Zuhri, M.Y.M.; Aisyah, H.A.; Asyraf, M.R.M.; Hassan, S.A.; Zainudin, E.S.; Sapuan, S.M.; Sharma, S.; Bangar, S.P.; Jumaidin, R.; et al. Natural Fiber-Reinforced Polylactic Acid, Polylactic Acid Blends and Their Composites for Advanced Applications. *Polymers* **2022**, *14*, 202. [CrossRef] [PubMed]
23. Norfarhana, A.; Ilyas, R.; Ngadi, N. A review of nanocellulose adsorptive membrane as multifunctional wastewater treatment. *Carbohydr. Polym.* **2022**, *291*, 119563. [CrossRef]
24. Ilyas, R.A.; Zuhri, M.Y.M.; Norrrahim, M.N.F.; Misenan, M.S.M.; Jenol, M.A.; Samsudin, S.A.; Nurazzi, N.M.; Asyraf, M.R.M.; Supian, A.B.M.; Bangar, S.P.; et al. Natural Fiber-Reinforced Polycaprolactone Green and Hybrid Biocomposites for Various Advanced Applications. *Polymers* **2022**, *14*, 182. [CrossRef] [PubMed]
25. Ilyas, R.A.; Aisyah, H.A.; Nordin, A.H.; Ngadi, N.; Zuhri, M.Y.M.; Asyraf, M.R.M.; Sapuan, S.M.; Zainudin, E.S.; Sharma, S.; Abral, H.; et al. Natural-Fiber-Reinforced Chitosan, Chitosan Blends and Their Nanocomposites for Various Advanced Applications. *Polymers* **2022**, *14*, 874. [CrossRef] [PubMed]
26. Haris, N.I.N.; Hassan, M.Z.; Ilyas, R.; Suhot, M.A.; Sapuan, S.; Dolah, R.; Mohammad, R.; Asyraf, M. Dynamic mechanical properties of natural fiber reinforced hybrid polymer composites: A review. *J. Mater. Res. Technol.* **2022**, *19*, 167–182. [CrossRef]
27. Ilyas, R.; Sapuan, S.; Asyraf, M.; Dayana, D.; Amelia, J.; Rani, M.; Norrrahim, M.; Nurazzi, N.; Aisyah, H.; Sharma, S.; et al. Polymer Composites Filled with Metal Derivatives: A Review of Flame Retardants. *Polymers* **2021**, *13*, 1701. [CrossRef]
28. Suriani, M.J.; Ilyas, R.A.; Zuhri, M.Y.M.; Khalina, A.; Sultan, M.T.H.; Sapuan, S.M.; Ruzaidi, C.M.; Wan, F.N.; Zulkifli, F.; Harussani, M.M.; et al. Critical Review of Natural Fiber Reinforced Hybrid Composites: Processing, Properties, Applications and Cost. *Polymers* **2021**, *13*, 3514. [CrossRef]
29. Ilyas, R.; Sapuan, S.; Atikah, M.; Asyraf, M.; Rafiqah, S.A.; Aisyah, H.; Nurazzi, N.M.; Norrrahim, M. Effect of hydrolysis time on the morphological, physical, chemical, and thermal behavior of sugar palm nanocrystalline cellulose (*Arenga pinnata (Wurmb.) Merr*). *Text. Res. J.* **2020**, *91*, 152–167. [CrossRef]

30. Abral, H.; Ariksa, J.; Mahardika, M.; Handayani, D.; Aminah, I.; Sandrawati, N.; Pratama, A.B.; Fajri, N.; Sapuan, S.; Ilyas, R. Transparent and antimicrobial cellulose film from ginger nanofiber. *Food Hydrocoll.* **2019**, *98*, 105266. [CrossRef]
31. Sabaruddin, F.A.; Tahir, P.; Sapuan, S.M.; Ilyas, R.A.; Lee, S.H.; Abdan, K.; Mazlan, N.; Roseley, A.S.M.; Hps, A.K. The Effects of Unbleached and Bleached Nanocellulose on the Thermal and Flammability of Polypropylene-Reinforced Kenaf Core Hybrid Polymer Bionanocomposites. *Polymers* **2020**, *13*, 116. [CrossRef]
32. Norrrahim, M.N.F.; Kasim, N.A.M.; Knight, V.F.; Ong, K.K.; Noor, S.A.M.; Halim, N.A.; Shah, N.A.A.; Jamal, S.H.; Janudin, N.; Misenan, M.S.M.; et al. Emerging Developments Regarding Nanocellulose-Based Membrane Filtration Material against Microbes. *Polymers* **2021**, *13*, 3249. [CrossRef] [PubMed]
33. Ilyas, R.; Sapuan, S.; Ishak, M.; Zainudin, E. Sugar palm nanofibrillated cellulose (*Arenga pinnata (Wurmb.) Merr*): Effect of cycles on their yield, physic-chemical, morphological and thermal behavior. *Int. J. Biol. Macromol.* **2018**, *123*, 379–388. [CrossRef] [PubMed]
34. Syafri, E.; Jamaluddin; Sari, N.H.; Mahardika, M.; Amanda, P.; Ilyas, R.A. Isolation and characterization of cellulose nanofibers from *Agave gigantea* by chemical-mechanical treatment. *Int. J. Biol. Macromol.* **2021**, *200*, 25–33. [CrossRef] [PubMed]
35. Abral, H.; Chairani, M.K.; Rizki, M.D.; Mahardika, M.; Handayani, D.; Sugiarti, E.; Muslimin, A.N.; Sapuan, S.; Ilyas, R. Characterization of compressed bacterial cellulose nanopaper film after exposure to dry and humid conditions. *J. Mater. Res. Technol.* **2021**, *11*, 896–904. [CrossRef]
36. Kadier, A.; Ilyas, R.A.; Huzaifah, M.R.M.; Harihastuti, N.; Sapuan, S.M.; Harussani, M.M.; Azlin, M.N.M.; Yuliasni, R.; Ibrahim, R.; Atikah, M.S.N.; et al. Use of Industrial Wastes as Sustainable Nutrient Sources for Bacterial Cellulose (BC) Production: Mechanism, Advances, and Future Perspectives. *Polymers* **2021**, *13*, 3365. [CrossRef]
37. Trevisan, H.; Rezende, C.A. Pure, stable and highly antioxidant lignin nanoparticles from elephant grass. *Ind. Crop. Prod.* **2020**, *145*, 112105. [CrossRef]

Short Biography of Authors

R. A. Ilyas is a senior lecturer at the School of Chemical and Energy Engineering, in the Faculty of Engineering, at the Universiti Teknologi Malaysia (UTM), Malaysia. He received his Diploma in Forestry at Universiti Putra Malaysia, Bintulu Campus (UPMKB), Sarawak, Malaysia, from May 2009 to April 2012. In 2012, he was awarded the Public Service Department (JPA) scholarship to pursue his bachelor's degree (BSc) in Chemical Engineering at Universiti Putra Malaysia (UPM). Upon completing his BSc programme in 2016, he was awarded the Graduate Research Fellowship (GRF) by the Universiti Putra Malaysia (UPM) to undertake a PhD degree in the field of Biocomposite Technology and Design at Institute of Tropical Forestry and Forest Products (INTROP) UPM. R.A. Ilyas was the recipient of the MVP Doctor of Philosophy Gold Medal Award UPM 2019, for the Best Ph.D. Thesis and Top Student Award, INTROP, UPM. He was awarded Outstanding Reviewer by Carbohydrate Polymers, Elsevier United Kingdom; Best Paper Award (11th AUN/SEED-Net Regional Conference on Energy Engineering); National Book Award 2018, Best Paper Award (Seminar Enau Kebangsaan 2019, Persatuan Pembangunan dan Industri Enau Malaysia); Top Cited Article 2020-2021, Journal Polymer Composite, Wiley, 2022. R.A. Ilyas was also listed among the World's Top 2% Scientists (Subject-Wise) citation impact during the single calendar year between 2019 and 2020 by Stanford University, US, and awarded the PERINTIS Publication Award 2021 and 2022 by Persatuan Saintis Muslim Malaysia, the Emerging Scholar Award by Automotive and Autonomous Systems 2021, Belgium, Young Scientists Network—Academy of Sciences Malaysia (YSN-ASM) 2021, the UTM Young Research Award 2021, UTM Publication Award 2021, and the UTM Highly Cited Researcher Award 2021. His main research interests are: (1) polymer engineering (biodegradable polymers, biopolymers, polymer composites, and polymer gels) and (2) material engineering (natural fibre-reinforced polymer composites, biocomposites, cellulose materials, and nano-composites). To date, he has authored or co-authored more than 404 publications (published/accepted): 164 Journals Indexed in JCR/Scopus, 2 non-index Journal, 15 books, 104 book chapters, 78 conference proceedings/seminars, 4 research bulletins, 10 conference papers (abstract published in book of abstract), 17 Guest Editor of Journal Special Issues and 10 Editor/Co-Editor of Conference/Seminar Proceedings on green materials related subjects.

S. M. Sapuan is an "A" Grade Professor of composite materials at the Department of Mechanical and Manufacturing, Universiti Putra Malaysia (UPM) and a Head of Laboratory of Biocomposite Technology, INTROP, UPM. He has a BEng in Mechanical Engineering from the University of Newcastle, Australia, an MSc in Engineering Design from Loughborough University, UK, and a PhD in Material Engineering from De Montfort University, UK. He is a Professional Engineer, a Society of Automotive Engineers Fellow, an Academy of Science Malaysia Fellow, a Plastic and Rubber and Institute Malaysia Fellow, a Malaysian Scientific Association Fellow, an International Biographical Association Fellow, and an Institute of Material Malaysia Fellow. He is an Honorary Member and immediate past Vice President of the Asian Polymer Association based in IIT Delhi and the Founding Chairman and Honorary Member of Society of Sugar Palm Development and Industry, Malaysia. He is the co-editor-in-chief of Functional Composites and Structures, and a member of editorial boards of more than two dozen journals. To date he has produced more than 1800 publications including over 860 journal papers, 50 books, and 175 chapters in book. He has delivered over 50 plenary and keynote lectures, and over 150 invited lectures. He organized 30 journal special issues as a guest editor, presented over 650 technical articles in conferences and seminars, reviewed over 1300 journal papers, and has eight patents. He successfully supervised 91 PhD and 70 MSc students and 15 postdoctoral researchers. His current h-index is 93 and the number of citations is 31,647 (Google Scholar). He received nine Outstanding Researcher Awards from UPM, ISESCO Science Award (Gold Medal), Plastic and Rubber Institute Malaysia Fellowship Award and Forest Research Institute Malaysia First Prize Publication Award. He also received the Khwarizimi International Award, the SEARCA Regional Professorial Chair award, the Kuala Lumpur Royal Rotary Gold Medal Research Award, and two National Book Awards. He received the Endeavour Research Promotion Award by TMU/IEEE India; the Citation of Excellence Award, Emerald, UK; Malaysia's Research Star Award, Elsevier/Ministry of Education Malaysia; the Publons Peer Review Award, Publons, USA; the Professor of Eminence Award from Aligarh Muslim University, India; the Top Research Scientists' Malaysia Award, Academy of Science Malaysia; the Gold in Invention and Innovation Awards; the Malaysia Technology Expo; PERINTIS Publication Award, PERINTIS, Malaysia. He was listed among the World Top 2% Scientists by Stanford University, USA. He is the finalist of IET Achievements Award, IET, UK and the 2021 SAE Subir Chowdhury Medal of Quality Leadership, SAE, USA.

Emin Bayraktar (Prof. Emeritus, Habil., Dr (Ph.D.), DSc—Doctor of Science) is an academic and research staff member in Mechanical and Manufacturing engineering at SUPMECA/Paris, France. His research areas include manufacturing techniques of new materials (basic composites—hybrid), the metal forming of thin sheets (design + test + FEM), static and dynamic behavior and optimization of materials (experimental and FEM—utilization and design of composite-based metallic and non-metallic, powder metallurgy, and energetic material aeronautical applications), metallic-based and non-metallic materials, powder metallurgy and metallurgy of steels, welding, and heat treatment, as well as the processing of new composites, sintering techniques, sinter–forging, thixoforming, etc. He has authored more than 200 publications in the International Journals and International Conference Proceedings and has also authored more than 90 research reports (European = Steel Committee projects, Test + Simulation). He already advised 32 Ph.D. and 120 MSc theses and is currently advising seven. He is a Fellow of WAMME (World Academy of Science in Materials and Manufacturing Engineering), an Editorial Board—Member of JAMME (*International Journal of Achievement in Materials and Manufacturing Engineering*), an Advisory board member of AMPT—2009 (*Advanced Materials Processing technologies*), and a member of APCMP—2008 and APCMP—2010. He was a Visiting Professor at Nanyang Technology University, Singapore in 2012; Xi'an Northwestern Technical University, Aeronautical Engineering, in 2016; University of Campinas, UNICAMP-Brazil in 2013 until 2023. He is a recipient of the Silesian University Prix pour "FREDERIK STAUB Golden Medal-2009" by the Academy of WAMME, "World Academy of Science"—Poland, materials science section, and a recipient of the William Johnson International Gold Medal—2014, AMPT academic association.

polymers

MDPI

Article

Development and Characterization of Cornstarch-Based Bioplastics Packaging Film Using a Combination of Different Plasticizers

Walid Abotbina [1], S. M. Sapuan [1,2,*], M. T. H. Sultan [3], M. F. M. Alkbir [4,5] and R. A. Ilyas [6,7]

1 Advanced Engineering Materials and Composites Research Centre, Department of Mechanical and Manufacturing Engineering, Universiti Putra Malaysia (UPM), Serdang 43400, Selangor, Malaysia; walidhmeed@gmail.com
2 Laboratory of Biocomposite Technology, Institute of Tropical Forest and Forest Products, Universiti Putra Malaysia (UPM), Serdang 43400, Selangor, Malaysia
3 Department of Aerospace Engineering, Universiti Putra Malaysia (UPM), Serdang 43400, Selangor, Malaysia; thariq@upm.edu.my
4 Advanced Facilities Engineering Technology Research Cluster, Malaysian Institute of Industrial Technology (MITEC), University Kuala Lumpur, Persiaran Sinaran Ilmu, Bandar Seri Alam, Johor Bahru 81750, Johor, Malaysia; m.f.m.alkbir@gmail.com
5 Facilities Maintenance Engineering Section, Malaysian Institute of Industrial Technology, Universiti Kuala Lumpur (UniKL), Persiaran Sinaran Ilmu, Bandar Seri Alam, Johor Bahru 81750, Johor, Malaysia
6 School of Chemical and Energy Engineering, Faculty of Engineering, Universiti Teknologi Malaysia (UTM), Johor Bahru 81310, Johor, Malaysia; ahmadilyas@utm.my
7 Centre for Advanced Composite Materials (CACM), Universiti Teknologi Malaysia (UTM), Johor Bahru 81310, Johor, Malaysia
* Correspondence: sapuan@upm.edu.my; Tel.: +60-19-3863191

Citation: Abotbina, W.; Sapuan, S.M.; Sultan, M.T.H.; Alkbir, M.F.M.; Ilyas, R.A. Development and Characterization of Cornstarch-Based Bioplastics Packaging Film Using a Combination of Different Plasticizers. *Polymers* **2021**, *14*, 3487. https://doi.org/10.3390/polym13203487

Academic Editor: Marina Patricia Arrieta Dillon

Received: 19 August 2021
Accepted: 30 September 2021
Published: 11 October 2021

Abstract: This work aims to develop cornstarch (CS) based films using fructose (F), glycerol (G), and their combination (FG) as plasticizers with different ratios for food packaging applications. The findings showed that F-plasticized film had the lowest moisture content, highest crystallinity among all films, and exhibited the highest tensile strength and thermostability. In contrast, G-plasticized films showed the lowest density and water absorption with less crystallinity compared to the control and the other plasticized film. In addition, SEM results indicated that FG-plasticized films had a relatively smoother and more coherent surface among the tested films. The findings have also shown that varying the concentration of the plasticizers significantly affected the different properties of the plasticized films. Therefore, the selection of a suitable plasticizer at an appropriate concentration may significantly optimize film properties to promote the utilization of CS films for food packaging applications.

Keywords: cornstarch; plasticizer; fructose; glycerol; film; properties

1. Introduction

Currently, petroleum-based plastics, which are characterized by a long polymer chain, are widely used for different applications on a large scale due to their diverse mechanical properties and low cost [1–5]. The global production of plastics was estimated to be up to 311 million tons in 2014 [6,7], reached up to 381 million tons by 2015, and is predicted to increase four-fold by 2050 [6]. Despite the considerable contribution of petroleum-based plastics to the global economy, which exceeded billions of dollars, the non-degradability of these materials represents a huge challenge for the ecosystem, leading to many environmental crises [8–11]. It was reported that plastic waste represented approximately 79% of the waste that was discharged in landfills, recording more than 300 million tons of waste in 2015; moreover, only 12% of this waste was incinerable, and only 9% of it was recyclable [12,13].

Being recalcitrant or not biodegradable makes petroleum-based plastics one of the most critical issues to the environment currently. Thus, there is an urgent necessity to substitute other eco-friendly alternatives for these materials [14–17]. In this context, biodegradable plastics, which are naturally made from biopolymers or synthesized bio-based polymers, seem a promising alternative to replace or at least to reduce the extensive use of conventional plastics and their harmful waste [18–23]. Additionally, biopolymers derived from renewable resources such as plants and animals can play an essential role in overcoming the challenges related to the depletion of oil reserves along with the environmental issues related to the increased use of petroleum-based plastics. These biopolymers can be natural fibers, cellulose, polysaccharides, proteins, lipopolysaccharides, polyhydroxyalkanoates, or glycolipids, which are suitable in environmental applications [24–29].

Due to its ability as a linking matrix between fillers, starch is the most used biopolymer for the fabrication of biofilms with high performance and biodegradability [30–36]. Additionally, it is more attractive for the industrial sector because of its availability and cost-effectiveness [37]. Corn starch represents more than 85% of global starch, making the corn plant the primary source of commercial starches globally. Other plants such as wheat, rice, and potato are also important sources of native starch; however, they contribute to a minor proportion of global starch production [38]. Semi-crystalline starch represents around 70% of the mass of the corn granule, and the rest is carbohydrate, protein, oil, and ash [39–41]. In recent years, starch-based materials have received increased attention in different packaging applications pushed by the rising concerns about global warming [42–44]. Although success has been achieved in the biopolymers market, especially in the environmental aspects, this market still faces some challenges to substitute petroleum-based plastics. This is mainly due to the sensitive nature of biopolymer-based films to high moisture [45,46], and their poor mechanical properties [47–49]. Due to their low molecular weight and non-volatility nature, plasticizers have been widely used to improve the workability and durability of polymers since the early 1800s [50]. The primary role of these plasticizers is to weaken the attraction of hydrogen bonds in starch network amylose and amylopectin; plasticizers can also increase the mobility rate of the polymer macromolecular chain, which reduces the glass transition temperature, in turn, and enhances the flexibility and stiffness of the plasticized corn starch films [42,51–54]. Plasticizers do not only improve the physical properties of biopolymers but can also effectively enhance the processing characteristics; as a property modifier, plasticizers can decrease the second order transition temperature and the elasticity modulus, which improves the cold flexibility. Moreover, the softening effect of plasticizers helps in lowering the required processing temperature, and providing better flow properties [55]. Various types of plasticizers have been reported in the literature as adding materials for biopolymers fabrication, including fructose, glucose, and sucrose [56,57], urea [57,58], glycerol [56,59], tri-ethanolamine and glycol [57,60], as well as sorbitol and xylitol [61–63]. Ibrahim et al. (2019) investigated the effect of different plasticizers on corn-starch film properties, and found that film plasticized with 25% fructose produced the best set of features and achieved the highest mechanical performance [54]. In the same context, Mali et al. (2006) studied the effect of glycerol concentration (0, 20, and 40%) on corn starch and revealed that the tensile strength was reduced while the elongation at break increased with the increase in glycerol concentration [64]. In another study, Zhang et al. (2006) reported that monosaccharides (mannose, glucose, and fructose) resulted in stronger starch films (higher tensile strength). whereas polyols (glycerol and sorbitol) exhibited higher water vapor permeability [65]. Thus, the combination among different plasticizers seems attractive for optimizing the properties of CS films in different aspects. To our knowledge, no prior studies have examined the effect of combining fructose and glycerol plasticizers on the properties of CS-based film. Thus, this present study aims to investigate the potential effects of using glycerol, fructose, and a combination of glycerol and fructose at different concentrations as plasticizers on the properties of corn-starch-based bioplastics. Moreover, CS-based polymers are sensitive to humidity and have a low moisture barrier, which has limited their wider use in food applications. Because of that, special attention will be paid to optimizing the moisture barrier criteria of CS film for food packaging applications.

2. Materials and Methods

2.1. Materials

Corn starch was purchased from Thye Huat Chan Sdn. Bhd (Sungai Buloh, SGR, Malaysia), and the glycerol and fructose plasticizers were purchased from Evergreen Engineering & Resources Sdn. Bhd (Semenyih, SGR, Malaysia). The commercial starch was graded to 0.25 mm size in a sieve machine Matest A060-01 (Arcore, MB, Italy) to prepare starch powder for the characterization.

2.2. Preparation of Cornstarch Biopolymers

The CS-based films were prepared by the application of the solution casting technique, as depicted in Figure 1. First, both types of plasticizers were introduced into a beaker containing 180 mL of distilled water. The mixture was then heated at a temperature of 85 °C for 20 min using a water bath to prepare a homogeneous solution. After that, 10 g of corn powder was separately added into the prepared solution at different plasticizer concentrations (0, 30, 45, and 60% *w*/*w*). The solution was placed again in the water bath for 20 min at the same temperature, and the slurry was left to cool before casting on a thermal platform. The casting dishes were weighed at 45 g to ensure uniformity of film thickness. The blend was the dried for 15 h using an air circulation oven at a temperature of 65 °C. The dehydrated films were collected from the casting plates and maintained at room temperature for one week inside plastic bags before characterization. The different samples of plasticized films were coded as follows: F30%, F45%, F60% for fructose, G30%, G45%, G60% for glycerol, FG30%, FG45%, and FG60% for fructose/glycerol in a ratio of 1:1 (*v*/*v*), and CS for the control corn starch film (non-plasticized).

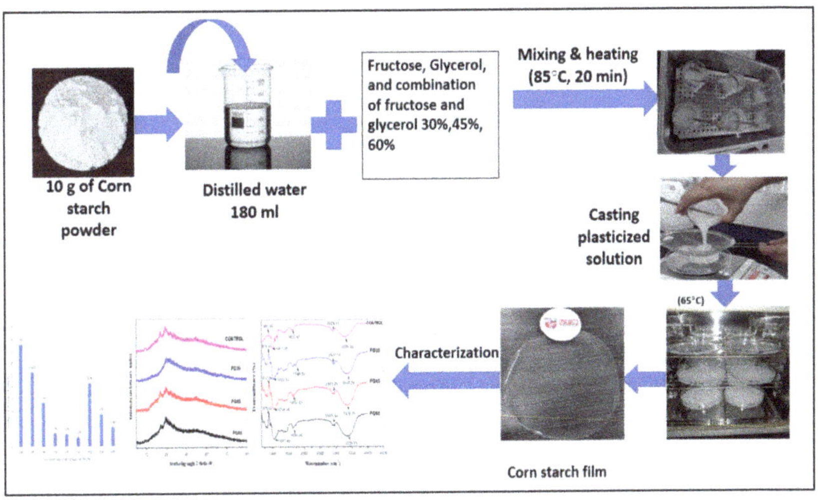

Figure 1. Flow chart of film preparation.

2.3. Physical Properties

2.3.1. Film Density

The density of prepared films was measured using a densimeter (Mettler-Toledo (M) Sdn. Bhd., Selangor, Malaysia). Xylene was selected as a dipping solvent in this work instead of distilled water to avoid the absorption of water by the hydrophilic samples. The low density of xylene prevents the floating of prepared films. Dehydration of samples was conducted for seven days using desiccators equipped with SiO_2 as the drying agent.

The initial dry mass (m) of each sample was measured, and the biofilm was immersed in the liquid to note the volume of displaced liquid (v). The density measurement (ρ) can be calculated using Equation (1) as follows:

$$\rho = m/v = \text{g/m}^3 \tag{1}$$

2.3.2. Film Moisture Content (MC)

To determine the content of moisture in the studied films, three samples of each film were dried for 24 h in an oven at a temperature of 105 °C. The samples were weighed before (W_1, gram) and after drying (W_2, gram) and the calculation of moisture content was performed using Equation (2) [66]:

$$MC\ (\%) = \left(\frac{W_1 - W_2}{W_1}\right) \times 100 \tag{2}$$

2.3.3. Film Thickness

The thickness of each film sample was measured using a digital micrometer (Mitutoyo Co., Kawasaki, Japan) with an accuracy of 0.001 mm. For more reliable results, the thickness measurement of each sample was replicated five times at different areas of the film, and the mean value of the films' thickness was calculated.

2.3.4. Film Solubility

To determine the film solubility, three samples (2 cm diameter) were collected from each film and dehydrated for 24 h using an oven at a temperature of 105 °C to measure the initial dry matter of each film (W_i). Each sample was then immersed for 24 h in a beaker containing 30 mL of distilled water at a temperature of 23 ± 2 °C with periodic stirring. After that, the insoluble portion of the film sample was separated from the solution and dried for 24 h at 105 °C. The mass of the dried insoluble sample (W_0) was used to calculate the fraction of soluble matter, which represents the solubility of the samples in water using Equation (3) as follows:

$$\text{Solubility}\ (\%) = \left(\frac{W_i - W_0}{W_i}\right) \times 100 \tag{3}$$

2.3.5. Water Absorption

The water absorption of a material is measured by the volume of water retained by 1 g of the dehydrated material. Water absorption of the films was evaluated by a similar method reported by Ibrahim et al. [41] with little modification. One gram of the film was introduced in a pre-weighed centrifugal tube ($M_{initial}$), then immersed in 25 mL of distilled water and centrifuged for 25 min at 3000 rpm. The supernatant was then removed, and the residue was dried for 30 min at 50 °C in an air circulation oven before being reweighed (M_{final}). These steps were repeated several times until a constant mass of the sample was reached. The water absorption (WA) percentage was then calculated using Equation (4) (Ibrahim et al., 2020):

$$\text{WA}\ (\%) = \left(\frac{M_{final} - M_{initial}}{M_{initial}}\right) \times 100 \tag{4}$$

2.4. Structural Properties

2.4.1. Fourier Transform Infrared Spectroscopy (FTIR)

The Fourier transform infrared (FTIR) spectroscopy analysis was conducted to test possible changes of functional groups in the films. The analysis was performed for each sample in an IR spectrometer Nicolet 6700 AEM- Thermo Nicolet Corporation (Madison, WI, USA) at 4 cm^{-1} resolution in 4000 to 500 cm^{-1} range with 42 total scans.

2.4.2. X-ray Diffraction (XRD)

All of the samples were analyzed with an X-ray diffractometer 2500 (Rigaku, Tokyo, Japan) with a speed of scattering at 0.02(θ) s^{-1} over a range of angles between 5 and 60° (2θ). The operating current and voltage were fixed at 35 mA and 40 kV, respectively. The outcomes from the XRD test include relative crystallinity (Xc), crystalline area (Ic), and amorphous area (Io). Equation (5) defines relative crystallinity as a ratio between crystalline and amorphous space.

$$Xc\ (\%) = ((Ic - Io)/Ic) \times 100 \tag{5}$$

2.4.3. Scanning Electron Microscopy (SEM)

Before conducting scanning electron microscopy (SEM), the samples were covered with a layer of gold with an argon plasma metallizer (sputter coater KK575X) (Edwards Limited, Crawley, UK) to prevent unnecessary charging. The fractured sample surface was then examined under SEM, model Hitachi S-3400N, with a 10 kV acceleration voltage.

2.5. Thermogravimetric Analysis (TGA)

A thermal gravimetric analysis (TGA) was performed using an analyzer Q500 V20.13 Build 39 (TA Instruments, Hüllhorst, Germany). A 10 mm sample of the film was introduced in platinum crucibles under a nitrogen atmosphere and heated starting from room temperature to a temperature of 450 °C at a rate of 10 °C/min. This test of thermal analysis assessed the thermal stability of each sample by following the mass loss over time as a function of temperature.

2.6. Tensile Properties

Tensile properties, including tensile strength, elongation at break, and Young's modulus, were measured using a universal tensile machine (5 kN INSTRON, Instron, Norwood, MA, USA) to assess the mechanical behavior of the different CS film samples. The tensile machine clamps were attached to a film strip (70 to 10 mm) that was pulled at 2 mm/min crosshead speed, with an effective grip distance of 30 mm. The machine was connected to computer software (Bluehill 3), which provided the mean of each parameter using five replicates of the tested sample.

2.7. Statistical Analyses

The statistical analyses of the findings were performed using Microsoft Excel 365, and the obtained data were plotted using Origin® 8.5 software for the graphical presentation of the results.

3. Results and Discussion

3.1. Physical Properties

3.1.1. Film Moisture Content (MC) and Density

As shown in Table 1, the results of film average moisture content using different plasticizers at different concentrations showed that the G-plasticized film had the highest moisture content followed by FG-plasticized film, whereas the lowest moisture content was observed for F-plasticized film. Moreover, a significant decrease in the moisture content was observed with the increase of F-plasticizer concentration from 30 to 60%, whereas increasing the concentration of glycerol plasticizer in the F-plasticized film from 30 to 60% significantly increased the moisture content from 11.06 to 16.97%. Similar to the G-plasticizer, the increase in the concentration of combined GF-plasticizer significantly raised the moisture content, where the moisture contents of 10.64 and 13.63% were observed for FG-plasticizer concentration of 30 and 60%, respectively. However, the observed increase in the moisture content by adding GF-plasticizer was less than the observed increase in the G-plasticized film.

Table 1. Moisture content and density of biofilm at different concentrations of plasticizers.

Sample	Control	Fructose			Glycerol			Combination		
		F30	F45	F60	G30	G45	G60	FG30	FG45	FG60
MC (%)	10.15	6.11	5.40	5.32	11.06	15.19	16.97	10.64	13.31	13.63
Density (g/cm^3)	1.69	1.49	1.46	1.4	1.40	1.36	1.34	1.45	1.41	1.39

The low moisture content of F-plasticized films, when compared to glycerol-containing films, might be related to the high similarity in the molecular structure between fructose and glucose units of the polymer, which strengthened the interactions between fructose molecules and the intermolecular chains in the film [41,53]. This resulted in lower probability of the interaction between the fructose and water molecules. Meanwhile, glycerol molecules consisting of hydroxyl groups were characterized by a high water affinity, which made forming hydrogen bonds and retaining water within the matrix easier in the G-plasticized films [67,68]. Thus, fructose acted as a water-resistant agent, whereas glycerol was considered as a water-holding agent [53].

The findings presented in Table 1 show that all the plasticized biofilms had a lower density than the control biofilm: the density ranged between 1.34 and 1.49 g/cm^3 for the plasticized films compared to a density of 1.69 g/cm^3 observed for the control biofilm. Although the results did not show a significant difference among the selected plasticizers as the density values were too close, the lowest density was noted for G-plasticizer at a concentration of 60%. The same findings were reported by Sahari et al. (2012), who stated that density values did not demonstrate much significant difference between the various plasticizer types [69].

Moreover, a slight decrease in the density was observed when the concentration was increased from 20 to 60% for all plasticizers. Tarique et al. (2011) reported that this decrease in the density might be due to the association between increasing plasticizer content and the thickness of the biofilm and its volume [70].

3.1.2. Film Thickness and Solubility

Figure 2 shows the thickness of different plasticized films at different plasticizer concentrations. Regardless of the type of plasticizer used, the results show that the increase of plasticizer concentration from 30 to 60% was associated with the increase in the thickness of plasticized films. Sanyang et al. (2015) reported the same findings and indicated the effect of plasticizers in the deformation of intermolecular polymer chain matrix, which provided more free volume increasing film thickness. In addition, the thickness results from varying plasticizer types showed that the thicknesses of different plasticized films were very close, although the molar mass of fructose (180 g/mole) is almost two-fold that of the molar mass of glycerol (92 g/mole). Hence, there was no significant influence of plasticizer molar mass on film thickness. This result disagreed with the findings reported in previous studies concluding that the thickness of the plasticized film was significantly related to the molar mass of the plasticizer used [42,53].

The water solubility of the film is also a critical property that should be considered during the characterization of biopolymers, especially for food packaging applications where water insolubility and water-resistance are sometimes required [71]. Similar to the thickness results, the same findings were observed for the solubility test of the plasticized films, where increasing the plasticizer concentration from 30 to 60% significantly increased the solubility for all types of plasticizers, as shown in Figure 3. The possible explanation of these findings is the hydrophilicity of these plasticizers as polyols contributed to weakening the bonds between polymer molecules and enlarging the free space volume in the chains. This facilitated the diffusion of water into the polymer matrix and, in turn, increased the films' solubility [42]. The findings also indicated relatively close solubility in the films in terms of the type of plasticizer, as shown in Figure 3, where solubility ranged from 39.42 to 47.98%, 35.79 to 50.09%, and from 37.99 to 51.26% were observed for G-plasticized films,

G- plasticized films, and FG-plasticized films, respectively. These similar results might be attributed to the high affinity of both glycerol and fructose to water molecules.

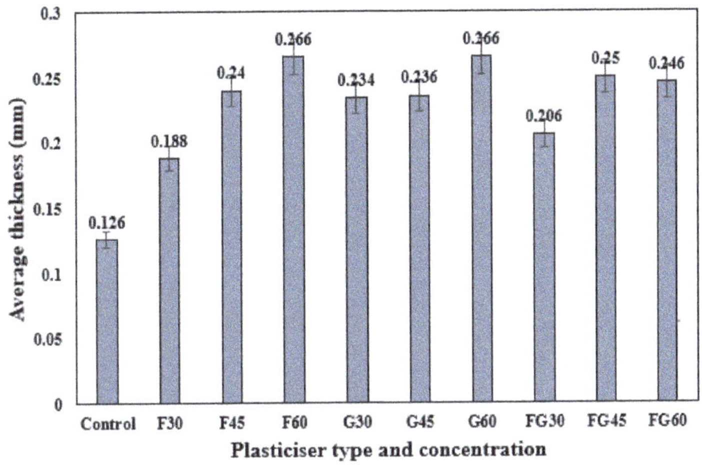

Figure 2. Thickness of corn starch films with various plasticizer types at different concentrations.

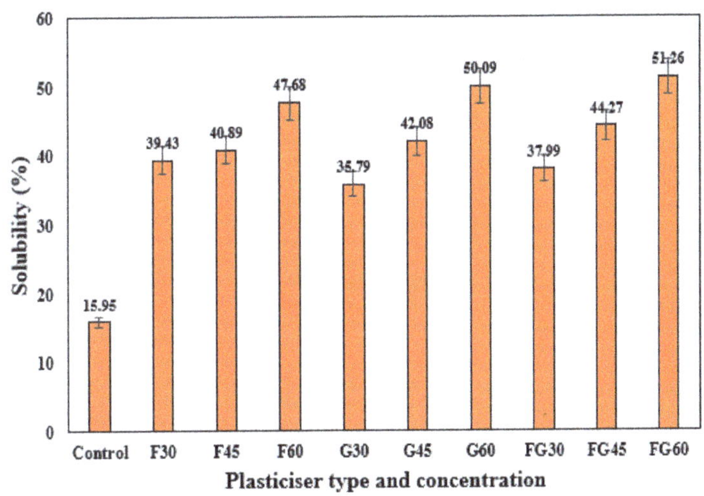

Figure 3. Solubility of corn starch films with various plasticizer types at different concentrations.

3.1.3. Water Absorption (WA)

Water absorption ability is a critical property for starch films due to the significant role that water plays in a plasticizer. Thus, plasticized films with higher water content are characterized by higher flexibility [42,54]. In this study, the duration of the biofilms immersion in water was set at 120 min as it was reported that the plasticized samples start to dissolve in water at 140 min [72]. Figure 4 shows the results of investigating the water absorption of the plasticized films at different plasticizer concentrations. The findings show that all the films, including the control film, reached the saturation point at 40 min after the immersion, where the amount of water absorbed after this point was negligible. The highest water absorption was observed for the control sample at approximately 194.3%. Among the three studied plasticizers, F-plasticized film had the highest water absorption

(187.87%), followed by FG-plasticized film, and G-plasticized film, with water absorption of 106.23 and 98.82%, respectively, at 30% of plasticizer concentration. Moreover, the findings indicated that the increase in the concentration of the plasticizers for the three types of plasticized films led to a decline in water absorption. For an increase from 30 to 60% in the plasticizer concentration, a decrease in water absorption from 187.87 to 74.10% was observed in F-plasticizer; reductions in water absorption from 98.82 to 50.58% and from 106.23 to 50.90% were recorded for FG-plasticized film and G-plasticized film, respectively. Thus, G-plasticized and FG-plasticized films possessed better water resistance than G-plasticized films. In turn, they can provide a more palatable texture and longer shelf life to high moisture products. This is understood due to the high hydrophobicity of glycerol; soluble plasticizers may block the micro-voids in the matrix of the film, causing a decrease in water absorption. At the same time, hydrophobic plasticizers can cause the formation of different phases in the produced film, which decreases the flexibility or forms discontinued areas in the matrix of the films [55].

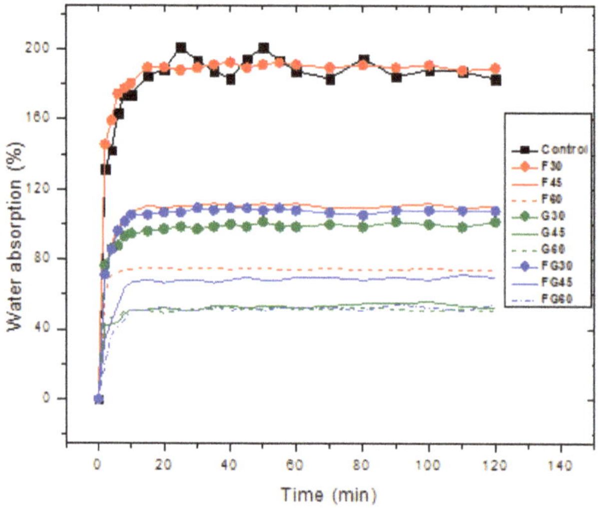

Figure 4. Water absorption for biofilms using different plasticizers at varying concentrations.

3.2. Structural Properties

3.2.1. Scanning Electron Microscopy (SEM)

The SEM images of the control samples exhibited a uniform and relatively smooth surface with the appearance of some non-dissolved granules related to the morphological structure of corn starch [54]. The FG- plasticized film with 30% concentration showed the smoothest surface with the absence of undissolved particles and pores (see Supplementary Materials). A relatively smooth and coherent surface without any pores was observed for the G-plasticized film at a concentration of 60%, with the presence of some impurities and agglomerated starch covering the surface. The other plasticized films, including F30, F45, F60, G30, G45, FG45, and FG60, showed less consistent surfaces with the presence of pores and microcracks in large areas of the surfaces. Although it has been reported by Edhirej et al. (2016) that F-plasticized films were evidenced to be rather smooth, coherent, and more homogeneous, the findings of the current study showed that all F-plasticized films were not smooth or homogeneous. This might be due to some differences during the preparation of the biofilm. Overall, the addition of plasticizer into CS films at appropriate concentrations led to the total dissolving of starch molecules, enhancing the coherence and integrity in the surface structure of the film [73]. The homogeneity of the matrix in the films is a good indicator of their structural integrity [74]; therefore, using a higher proportion of plasticizer produced weak and incoherent films that were hard to peel off the casting container. In contrast, preparing the film with a low concentration of plasticizer produced a film that appeared to be brittle, sticky, and difficult to remove from the casting container.

3.2.2. Spectroscopy Analysis of the Film Using FTIR

To identify the changes that occurred in the chemical structures of the films by adding different plasticizers at different concentrations, FTIR analysis was conducted to investigate the intermolecular rearrangement of polysaccharide chain orientation, as shown in Figure 5 [75]. The control film showed peaks at 890.23, 1019.91, 1653.21, 2942.45, and 3358.15 cm^{-1}. The peak observed around 3358 cm^{-1} was associated with the stretching of the O–H groups, whereas the bands identified at 2942.45 cm^{-1} were attributed to C–H stretching. The peak at 1653.21 cm^{-1} was assigned to the bending mode of the absorbed water [76]. The characteristic peaks at 1019.91 cm^{-1} were associated with the C–O bond stretching of the C–O–C groups in the corn starch anhydroglucose ring, whereas the vibrational modes of the D-glucopyranosyl ring were around 890.23 [77,78]. Although Yin et al. [6] reported that changes in the characteristic spectral bands indicate the chemical interactions between two or more substances physically blended [79], the FTIR results in this study showed that all the plasticized films had similar spectra to the FTIR spectrum observed for the control film. These findings were in accordance with the previous study conducted by Hazrati et al. [20], who reported that the similarity observed in FTIR spectra of the control film and plasticized films was due to the fact that starch and the used plasticizers have the same functional groups, as all of them are polyols [78].

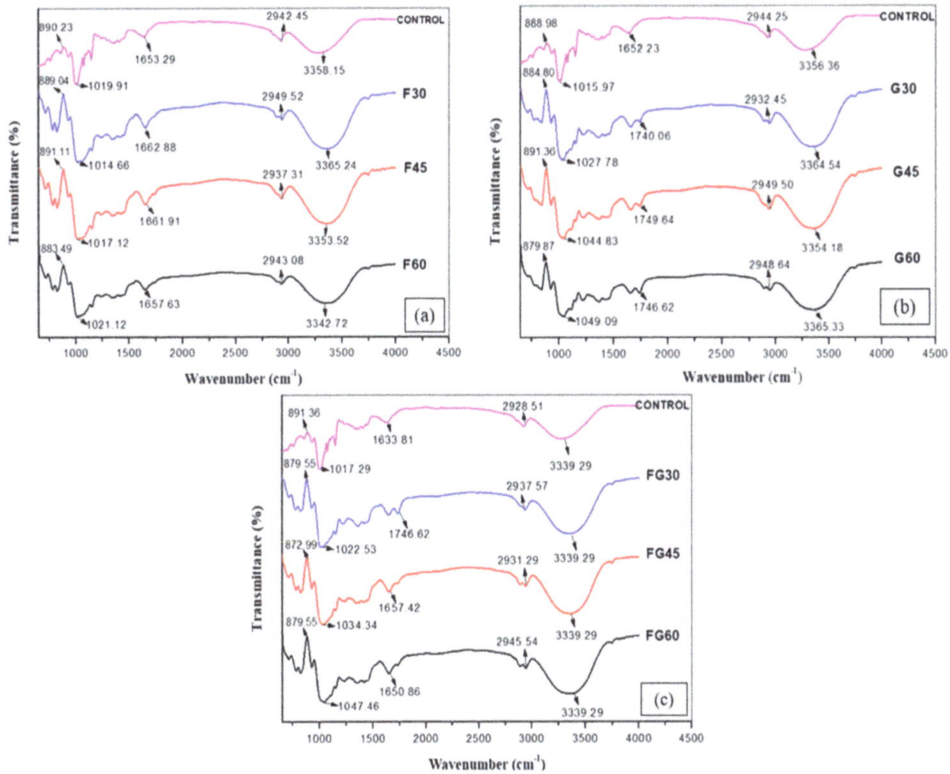

Figure 5. FTIR spectra of CS films with various plasticizers at different concentrations; (**a**) F-plasticized film, (**b**) G-plasticized film, (**c**) FG-plasticized film.

3.2.3. X-ray Diffraction (XRD)

The X-ray Diffraction analysis showed that the majority of corn starch crystals were gelatinized and retrograded during the preparation of the film. The XRD analysis of the control CS-film showed four peaks at the points 15, 17.4, 19.8, and 21.8°; these findings were in agreement with the results reported by Hazrol et al. [53]. The observed peaks reflected that the control CS-film had the B-type diffraction pattern [78]. The F-plasticized film showed the same diffraction peaks observed for the control film for all the concentrations; however, a decrease in intensity of the peak at 19.8° was observed by increasing the concentration of the F-plasticizer from 30 to 60%. The addition of G-plasticizer at a concentration of 30% to the control film led to the immersion of the peak at 17.4° that appeared again when G-plasticizer was added at concentrations of 45% and 60%; meanwhile, the intensity of the peak at 19.8° increased. The FG-plasticized films at 45% and 60% of plasticizer concentrations showed a similar pattern of diffraction to the control film; however, both peaks at 17.4° and 21.8° were immersed at 25% FG-plasticized film, resulting in a large peak at 19.8° (see Supplementary Materials).

The crystallinity degree of the different samples was calculated based on the XRD pattern and are presented in Table 2. For all plasticizers, a significant increase in the crystallinity index was noticed when the plasticizer concentration was increased from 30 to 60%. Moreover, the F-plasticized film showed a higher crystallinity degree than the G- and FG-plasticized films. According to [80–82], there is a significant relationship between the increase in crystallinity degree of starch-based films and the decrease in the moisture content, which was compatible with the findings of this study, in which the low moisture content and the higher crystallinity degree were observed for F-plasticized films.

Table 2. Crystallinity Index of samples using different plasticizers.

Sample	Crystallinity Index (%)
Control	21.3
F-30	33.4
F-45	36.5
F-60	37.2
G-30	20.8
G-45	19.7
G-60	19.3
FG-30	21.1
FG-45	20.6
FG-60	20.2

3.3. Thermogravimetric Analysis (TGA)

The thermostability of cornstarch-based biofilms was assessed using thermogravimetric analysis as represented in Figure 6 and Table 3 to determine the decomposition temperatures of each material and the fraction of material residues at the maximum degradation rate. For all the samples, three stages of degradation were recorded. The initial weight loss was observed at a temperature of below 100 °C due to the removal of a small amount of water via evaporation and dehydroxylation processes [83–85]. Additional heating at the range of temperature 150–200 °C led to the second loss in the film weight; the volatilization of fructose molecules and residuals was the driving process of the weight loss at this stage [86]. As illustrated in Figure 6, the weight loss of the films was limited in this stage. At the third stage of degradation, the mass loss was primarily caused by the decomposition of the water-soluble amylopectin in the films [41]. Overall, the final degradation rate of F-plasticized films ranged between 65% and 67.3%, which was very similar to the weight loss of the control sample, which recorded 65.2% as the degradation rate. The G-plasticized films showed the highest mass loss, which ranged between 87.6% and 90.5%, increasing with the concentration of plasticizers from 30 to 60%; followed by the FG-plasticized films, where the final loss of weight was estimated between 78.7% and

84.3% from the total mass. However, these results indicated that increasing the plasticizer concentration resulted in increasing the final mass loss for G-plasticized films, whereas increasing the plasticizer concentration in FG-plasticized films was associated with the decrease in the degradation rate.

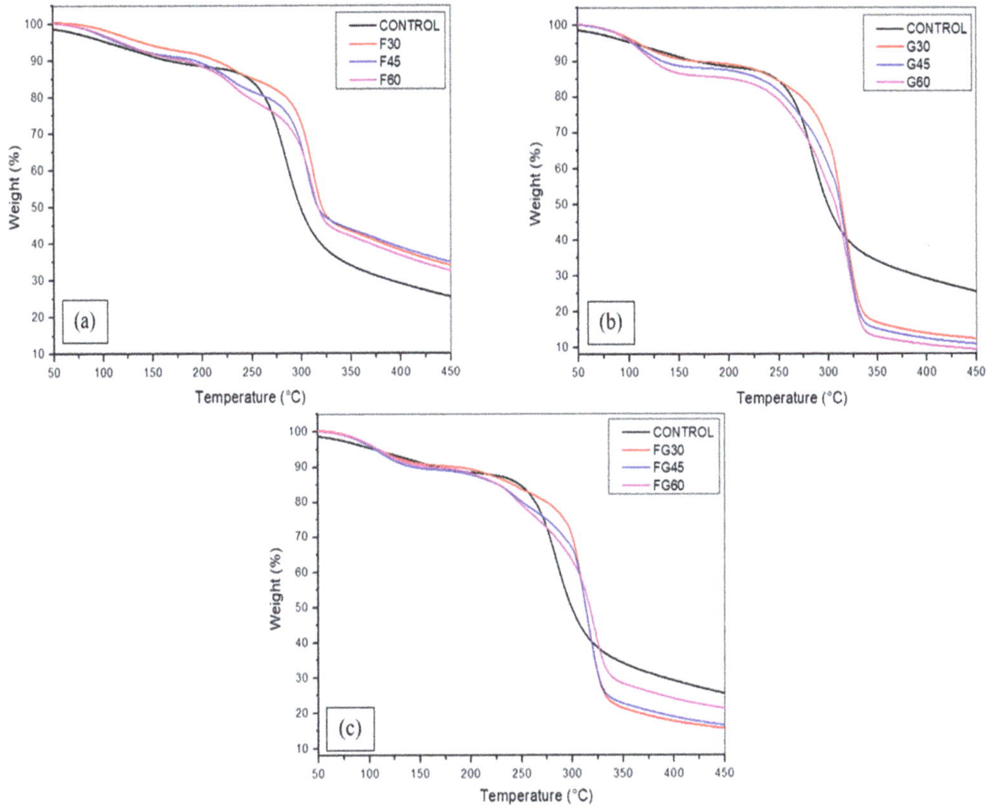

Figure 6. Thermogravimetric analysis of corn starch films with various plasticizers type at different concentrations: (**a**) F-plasticized film; (**b**) G-plasticized film; and (**c**) FG-plasticized film.

Table 3. Weight loss of all samples at different stages of degradation.

Temperature Range	Weight Loss (%)									
	C	F30	F45	F60	G30	G45	G60	FG30	FG45	FG60
20–150 °C	7.7	8.0	7.8	5.7	9.2	11.2	13.3	9.0	10.3	9.7
150–200 °C	3.8	3.5	2.7	2.9	6.3	7.4	7.9	6.3	9.1	10.8
200–500 °C	53.7	55.8	54.5	57.3	72.1	70.4	69.3	69.0	63.9	58.2
Total loss (%)	65.2	67.3	65.0	65.9	87.6	89	90.5	84.3	83.3	78.7

3.4. Tensile Properties of Films

Tensile analysis of the different plasticized films was conducted to evaluate the tensile strength, extension at the break, and Young's modulus at different concentrations. Figure 7 demonstrates the effect of various plasticizers at different concentrations on the tensile strength of CS films. The F-plasticized film showed the highest tensile strength of 17.15 MPa for a concentration of 30%, followed by the FG plasticizer, which showed a tensile strength of 10.66 MPa at a plasticized concentration of 30%, whereas the highest tensile of G-plasticized film was 2.24 MPa. These results were consistent with the study conducted by [76], which reported the greater tensile strength of F-plasticized CS film compared to other plasticizers. For all types of plasticizers, the increase in their concentration significantly decreased the tensile strength. The increase in plasticizer concentration from 30 to 60% led to a decrease from 17.15 to 7.44 MPa in the tensile strength of F-plasticized film, a decrease from 2.24 to 1.52 MPa in the tensile strength of G-plasticized film, and a decrease from 10.66 to 3.28 MPa in the tensile strength of FG-plasticized film. Several works reported the decrease of tensile strength in the starch-based films as a response to the increase of plasticizers concentration [42,54,87]. The addition of plasticizers increased the formation of hydrogen bonds between the starch molecules and the plasticizers, causing the weakness of intramolecular interactions among the molecules of starch chains [78].

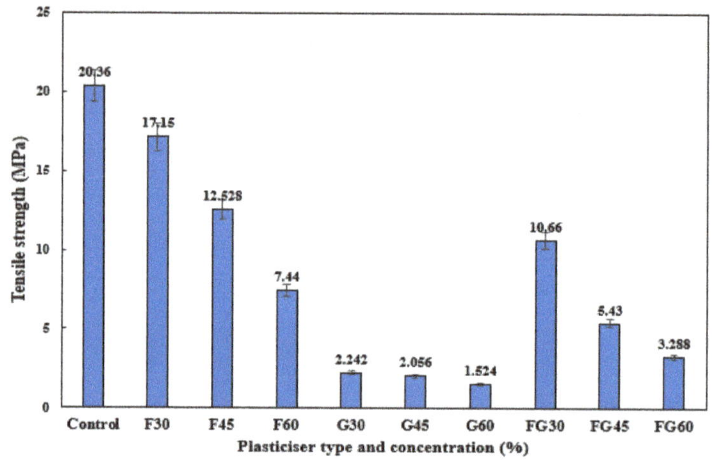

Figure 7. Tensile strength analysis of corn starch film with various plasticizers type at different concentrations: F30–F60: F-plasticized film; G30–G60: G-plasticized film; and FG30–FG60: FG-plasticized film.

Young's modulus analysis was conducted to assess the stiffness of the films, where a high Young's modulus reflects a high degree of stiffness of the material. As shown in Figure 8, F-plasticized films showed the highest Young's modulus, followed by the FG-plasticized films and G-plasticized films, respectively. Among all the tested plasticizers, a decrease in tensile modulus was observed when plasticizer concentration was increased from 30 to 60% in all films, indicating that increasing plasticizer content made the films less rigid. The decrease of stiffness with increasing plasticizer content in hydrophilic films has been reported previously [88]. This behavior could be related to the structural modifications of the starch network when plasticizers were incorporated and the matrix of the films became less dense [76].

For elongation at break, the F-plasticized films and FG-plasticized films showed higher elongations at break compared to G-plasticized films, as shown in Figure 9. However, the increase in plasticizer concentration resulted in increasing the elongation at break at F-plasticized films, whereas a decrease in elongation at break was observed in FG-plasticized films when the plasticizer concentration was increased from 30 to 60%. At the same time, the effect of plasticizer concentration on the elongation at break was not significant for G-plasticized films, where elongation of break of 33.02% and 33.87% was recorded at glycerol plasticizer concentrations of 30% and 60%, respectively. These findings were in agreement with the reported results in previous studies, which indicated the decrease of elongation break as a result of increasing plasticized concentration for F-plasticized film. Meanwhile, the opposite effect was observed when other types of plasticizers were used [76].

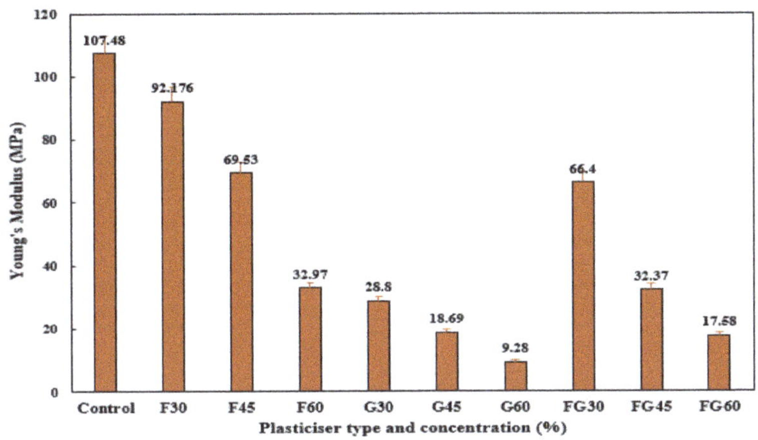

Figure 8. Young's modulus of corn starch film with various plasticizer types at different concentrations: F30–F60: F-plasticized film; G30–G60: G-plasticized film; and FG30–FG60: FG-plasticized film.

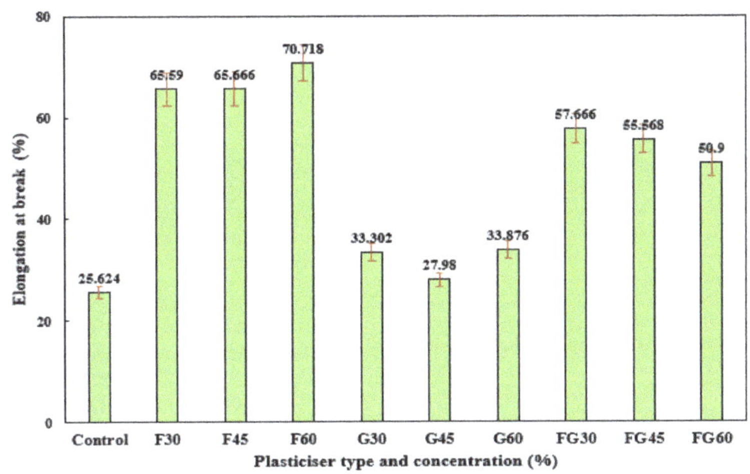

Figure 9. Extension at break of corn starch film with various plasticizer types at different concentrations: F30–F60: F-plasticized film; G30–G60: G-plasticized film; and FG30–FG60: FG-plasticized film.

3.5. Potential of the Plasticized CS Film for Food Packaging Considering Water-Resistant Ability

As stated earlier, the hydrophobicity of CS films resulted in high moisture content and high-water absorption as challenging factors that limit their broader use in food applications. Therefore, one of the aims of this study is to reduce the moisture content and water absorption of CS film by the incorporation of different plasticizers. Among the tested films, F-60 plasticized film exhibited the lowest moisture content, in which the incorporation of fructose plasticizer at a concentration of 60% reduced the moisture content by 47.58% compared to the control film. Additionally, G60 and FG60 biofilms showed the lowest water absorption recording 48.2% and 48.8%, respectively, compared to 74% as water absorption observed for F-60 biofilm. Moreover, the comparison of the properties of the prepared films with previous studies on CS-based films using different plastics showed the high quality of the films as water-resistant material (Table 4).

Table 4. Comparison of F-plasticized film properties with previous studies.

CS Film Type	Moisture Content (%)	Water Absorption (%)	Mechanical Properties			Ref.
			Tensile Strength (MPa)	Tensile Modulus (MPa)	Elongation at Break (%)	
Control	10.15	194.3	20.36	107.48	25.62	This study
Fructose	5.32 (F60)	74.1(F60)	7.44 (F60)	32.97 (F60)	70.718 (F60)	This study
Glycerol	11.06 (G30)	48.2 (G60)	2.24 (G30), 1.52 (G60)	28.8 (G30), 9.28 (G60)	33.30 (G30), 33.88 (G60)	This study
Fructose/Glycerol	10.64 (FG30)	48.8(FG60)	10.66 (FG30), 3.29 (FG60)	66.4 (FG30), 17.58 (FG60)	57.67 (FG30), 50.90 (FG60)	This study
Sorbitol	9.25–10.04	147	13.62 (S30)	495.97 (S30)	NA	[53]
Glycerol	14.7–16.55	112	2.53 (G30)	19.43 (G30)	NA	[53]
Sorbitol/glycerol	9.11–14.99	135	5.74 (SG30)	47.17 (SG30)	NA	[53]
Urea	21.05–27.86	NA	0.62 (U25)	1.67 (U25)	NA	[54]
Formamide	NA	96.09	3.42	NA	105.21	[89]
Ethylene Glycol	NA	92.24	3.8	NA	98.22	[89]

There is no doubt that the ability of packaging materials to prevent or minimize moisture transfer between the food and the surrounding environment is a crucial property for effective food packaging [90]; however, there was a trade-off between the mechanical properties and water resistance properties of the biofilms in the present study, reflected in high water resistance biofilms with low mechanical properties and vice-versa. In this regard, FG30 biofilm, which exhibited a moderate water resistance (WC = 10.64%, WA = 103.4%), thermal resistance, as well as tensile strength (TS = 10.66 MPa) seems to be the best plasticizer as a compromise to meet the requirements of food packaging applications.

4. Conclusions

In this work, the physical, structural, thermal, and mechanical properties of plasticized corn starch films were investigated to highlight the effect of plasticizer type and concentration on these properties. Incorporation of the selected plasticizers, including fructose, glycerol, and the combination of fructose/glycerol, significantly enhanced the properties of CS starch films. Moreover, the findings indicated that the addition of a specific plasticizer might enhance some properties over others. In this context, CS films containing fructose exhibited low moisture content, higher crystallinity, higher thermostability, and higher tensile strength; the G-plasticized films demonstrated the lowest density and water absorption. Thus, it seems that combining both plasticizers in FG-plasticized films effectively led to the development of corn-starch films with intermediate properties. Over the different tested plasticizers, the data indicated that the incorporation ratio of plasticizers significantly influenced the different properties of the films, which reflected the importance of determining the suitable concentration of plasticizers to optimize the quality of the film to meet the requirements of food packaging applications. In this regard, the FG30 plasticized CS film seems a promising biopolymer for food packaging applications, providing a low moisture content with a reduced water absorption capacity and good mechanical and thermal properties. Moreover, further efforts are required to investigate the chemical mechanisms of the plasticizers that affect the physical, thermal, and mechanical properties of CS starch films, which contribute to developing new techniques and methods for enhancing these properties.

Supplementary Materials: The following are available online at https://www.mdpi.com/article/10.3390/polym13203487/s1, Figure S1: Scanning electron micrograph of CS films with various plasticizers and concentrations, Figure S2: XRD analysis of corn starch film with various plasticizers type at different concentrations; (a) F-plasticized film, (b) G-plasticized film, (c) FG-plasticized film.

Author Contributions: Conceptualization, S.M.S. and W.A.; methodology, S.M.S., W.A. and R.A.I.; software, W.A.; validation, S.M.S. and R.A.I.; formal analysis, W.A.; investigation, W.A.; resources, S.M.S.; data curation, W.A.; writing—original draft preparation, W.A.; writing—review and editing, S.M.S., R.A.I., M.T.H.S. and M.F.M.A.; visualization, W.A.; supervision, S.M.S., R.A.I., M.T.H.S. and M.F.M.A.; project administration, S.M.S., R.A.I., M.T.H.S. and M.F.M.A.; funding acquisition, S.M.S. All authors have read and agreed to the published version of the manuscript.

Funding: The authors would like to appreciate University Putra Malaysia for financial support through through Geran Putra Berimpak (UPM/800-3/3/1/GPB/2019/9679800) and the grant Inisiatif Putra Siswazah (GP-IPS/2021/9697100).

Institutional Review Board Statement: Not applicable.

Informed Consent Statement: Not applicable.

Data Availability Statement: Not applicable.

Conflicts of Interest: The authors declare no conflict of interest.

References

1. Ilyas, R.A.; Sapuan, S.M. Biopolymers and Biocomposites: Chemistry and Technology. *Curr. Anal. Chem.* **2020**, *16*, 500–503. [CrossRef]
2. Baihaqi, N.M.Z.N.; Khalina, A.; Nurazzi, N.M.; Aisyah, H.A.; Sapuan, S.M.; Ilyas, R.A. Effect of fiber content and their hybridization on bending and torsional strength of hybrid epoxy composites reinforced with carbon and sugar palm fibers. *Polimery* **2021**, *66*, 36–43. [CrossRef]
3. Ilyas, R.A.; Sapuan, S.M.; Harussani, M.M.; Hakimi, M.Y.A.Y.; Haziq, M.Z.M.; Atikah, M.S.N.; Asyraf, M.R.M.; Ishak, M.R.; Razman, M.R.; Nurazzi, N.M.; et al. Polylactic Acid (PLA) Biocomposite: Processing, Additive Manufacturing and Advanced Applications. *Polymers* **2021**, *13*, 1326. [CrossRef] [PubMed]
4. Supian, A.B.M.; Sapuan, S.M.; Jawaid, M.; Zuhri, M.Y.M.; Ilyas, R.A.; Syamsir, A. Crashworthiness Response of Filament Wound Kenaf/Glass Fibre-reinforced Epoxy Composite Tubes with Influence of Stacking Sequence under Intermediate-Velocity Impact Load. *Fibers Polym.* **2021**, 1–12. [CrossRef]
5. Harussani, M.M.; Sapuan, S.M.; Rashid, U.; Khalina, A.; Ilyas, R.A. Pyrolysis of polypropylene plastic waste into carbonaceous char: Priority of plastic waste management amidst COVID-19 pandemic. *Sci. Total Environ.* **2021**, *803*, 149911. [CrossRef]
6. Iram, D.; Riaz, R.; Iqbal, R.K. Usage of potential micro-organisms for degradation of plastics. *Open J. Environ. Biol.* **2019**, *4*, 7–15.
7. Abotbina, W.; Sapuan, S.M.; Sulaiman, S.; Ilyas, R.A. Review of corn starch biopolymer. In *Proceedings of the 7th Postgraduate Seminar on Natural Fibre Reinforced Polymer Composites 2020*; Institute of Tropical Forestry and Forest Product (INTROP), Universiti Putra Malaysia: Serdang, Selangor, Malaysia, 2020; pp. 37–40.
8. Ilyas, M.; Ahmad, W.; Khan, H.; Yousaf, S.; Khan, K.; Nazir, S. Plastic waste as a significant threat to environment–a systematic literature review. *Rev. Environ. Health* **2018**, *33*, 383–406. [CrossRef]
9. Ilyas, R.A.; Sapuan, S.M.; Ibrahim, R.; Abral, H.; Ishak, M.R.; Zainudin, E.S.; Atiqah, A.; Atikah, M.S.N.; Syafri, E.; Asrofi, M.; et al. Thermal, Biodegradability and Water Barrier Properties of Bio-Nanocomposites Based on Plasticised Sugar Palm Starch and Nanofibrillated Celluloses from Sugar Palm Fibres. *J. Biobased Mater. Bioenergy* **2020**, *14*, 234–248. [CrossRef]
10. Ilyas, R.A.; Sapuan, S.M.; Asyraf, M.R.M.; Dayana, D.A.Z.N.; Amelia, J.J.N.; Rani, M.S.A.; Norrrahim, M.N.F.; Nurazzi, N.M.; Aisyah, H.A.; Sharma, S.; et al. Polymer Composites Filled with Metal Derivatives: A Review of Flame Retardants. *Polymers* **2021**, *13*, 1701. [CrossRef] [PubMed]
11. Mohd Nurazzi, N.; Asyraf, M.R.M.; Khalina, A.; Abdullah, N.; Sabaruddin, F.A.; Kamarudin, S.H.; Ahmad, S.; Mahat, A.M.; Lee, C.L.; Aisyah, H.A.; et al. Fabrication, Functionalization and Application of Carbon Nanotube-Reinforced Polymer Composite: An Overview. *Polymers* **2021**, *13*, 1047. [CrossRef] [PubMed]
12. Geyer, R.; Jambeck, J.R.; Law, K.L. Production, use, and fate of all plastics ever made. *Sci. Adv.* **2017**, *3*, e1700782. [CrossRef]
13. Jang, Y.-C.; Lee, G.; Kwon, Y.; Lim, J.; Jeong, J. Recycling and management practices of plastic packaging waste towards a circular economy in South Korea. *Resour. Conserv. Recycl.* **2020**, *158*, 104798. [CrossRef]
14. Tarique, J.; Sapuan, S.M.; Khalina, A.; Sherwani, S.F.K.; Yusuf, J.; Ilyas, R.A. Recent developments in sustainable arrowroot (*Maranta arundinacea* Linn) starch biopolymers, fibres, biopolymer composites and their potential industrial applications: A review. *J. Mater. Res. Technol.* **2021**, *13*, 1191–1219. [CrossRef]
15. Nurazzi, N.M.; Asyraf, M.R.M.; Khalina, A.; Abdullah, N.; Aisyah, H.A.; Rafiqah, S.A.; Sabaruddin, F.A.; Kamarudin, S.H.; Norrrahim, M.N.F.; Ilyas, R.A.; et al. A Review on Natural Fiber Reinforced Polymer Composite for Bullet Proof and Ballistic Applications. *Polymers* **2021**, *13*, 646. [CrossRef] [PubMed]
16. Suriani, M.J.; Rapi, H.Z.; Ilyas, R.A.; Petrů, M.; Sapuan, S.M. Delamination and Manufacturing Defects in Natural Fiber-Reinforced Hybrid Composite: A Review. *Polymers* **2021**, *13*, 1323. [CrossRef] [PubMed]
17. Nurazzi, N.M.; Sabaruddin, F.A.; Harussani, M.M.; Kamarudin, S.H.; Rayung, M.; Asyraf, M.R.M.; Aisyah, H.A.; Norrrahim, M.N.F.; Ilyas, R.A.; Abdullah, N.; et al. Mechanical Performance and Applications of CNTs Reinforced Polymer Composites—A Review. *Nanomaterials* **2021**, *11*, 2186. [CrossRef] [PubMed]

18. George, A.; Sanjay, M.R.; Srisuk, R.; Parameswaranpillai, J.; Siengchin, S. A comprehensive review on chemical properties and applications of biopolymers and their composites. *Int. J. Biol. Macromol.* **2020**, *154*, 329–338. [CrossRef]
19. Hazrati, K.Z.; Sapuan, S.M.; Zuhri, M.Y.M.; Jumaidin, R. Recent and potential developments in Dioscorea Hispida biopolymer composites: A review. In *Proceedings of the 7th Postgraduate Seminar on Natural Fibre Reinforced Polymer Composites 2020*; Institute of Tropical Forestry and Forest Product (INTROP), Universiti Putra Malaysia: Serdang, Selangor, Malaysia, 2020; pp. 14–17.
20. Hazrati, K.Z.; Sapuan, S.M.; Zuhri, M.Y.M.; Jumaidin, R. Extraction and Characterization of Potential Biodegradable Materials Based on Dioscorea hispida Tubers. *Polymers* **2021**, *13*, 584. [CrossRef]
21. Kashirina, A.; Yao, Y.; Liu, Y.; Leng, J. Biopolymers as bone substitutes: A review. *Biomater. Sci.* **2019**, *7*, 3961–3983. [CrossRef]
22. Rostami, M.; Yousefi, M.; Khezerlou, A.; Mohammadi, M.A.; Jafari, S.M. Application of different biopolymers for nanoencapsulation of antioxidants via electrohydrodynamic processes. *Food Hydrocoll.* **2019**, *97*, 105170. [CrossRef]
23. Bernaerts, T.M.M.; Gheysen, L.; Foubert, I.; Hendrickx, M.E.; Van Loey, A.M. The potential of microalgae and their biopolymers as structuring ingredients in food: A review. *Biotechnol. Adv.* **2019**, *37*, 107419. [CrossRef]
24. Harussani, M.M.; Sapuan, S.M.; Rashid, U.; Khalina, A. Development and characterization of polypropylene waste from personal protective equipment (Ppe)-derived char-filled sugar palm starch biocomposite briquettes. *Polymers* **2021**, *13*, 1707. [CrossRef]
25. Kalia, S.; Avérous, L. *Biopolymers: Biomedical and Environmental Applications*; John Wiley & Sons: Hoboken, NJ, USA, 2011; Volume 70, ISBN 9780470639238.
26. Omran, A.A.B.; Mohammed, A.A.B.A.; Sapuan, S.M.; Ilyas, R.A.; Asyraf, M.R.M.; Koloor, S.S.R.; Petrů, M. Micro- and Nanocellulose in Polymer Composite Materials: A Review. *Polymers* **2021**, *13*, 231. [CrossRef]
27. Syafiq, R.; Sapuan, S.M.; Zuhri, M.Y.M.; Ilyas, R.A.; Nazrin, A.; Sherwani, S.F.K.; Khalina, A. Antimicrobial activities of starch-based biopolymers and biocomposites incorporated with plant essential oils: A review. *Polymers* **2020**, *12*, 2403. [CrossRef] [PubMed]
28. Abral, H.; Chairani, M.K.; Rizki, M.D.; Mahardika, M.; Handayani, D.; Sugiarti, E.; Muslimin, A.N.; Sapuan, S.M.; Ilyas, R.A. Characterization of compressed bacterial cellulose nanopaper film after exposure to dry and humid conditions. *J. Mater. Res. Technol.* **2021**, *11*, 896–904. [CrossRef]
29. Norrrahim, M.N.F.; Huzaifah, M.R.M.; Farid, M.A.A.; Shazleen, S.S.; Misenan, M.S.M.; Yasim-Anuar, T.A.T.; Naveen, J.; Nurazzi, N.M.; Rani, M.S.A.; Hakimi, M.I.; et al. Greener Pretreatment Approaches for the Valorisation of Natural Fibre Biomass into Bioproducts. *Polymers* **2021**, *13*, 2971. [CrossRef] [PubMed]
30. Ilyas, R.A.; Sapuan, S.M.; Atikah, M.S.N.; Asyraf, M.R.M.; Rafiqah, S.A.; Aisyah, H.A.; Nurazzi, N.M.; Norrrahim, M.N.F. Effect of hydrolysis time on the morphological, physical, chemical, and thermal behavior of sugar palm nanocrystalline cellulose (*Arenga pinnata* (*Wurmb.*) *Merr*). *Text. Res. J.* **2021**, *91*, 152–167. [CrossRef]
31. Chan, J.X.; Wong, J.F.; Petrů, M.; Hassan, A.; Nirmal, U.; Othman, N.; Ilyas, R.A. Effect of Nanofillers on Tribological Properties of Polymer Nanocomposites: A Review on Recent Development. *Polymers* **2021**, *13*, 2867. [CrossRef] [PubMed]
32. Jumaidin, R.; Diah, N.A.; Ilyas, R.A.; Alamjuri, R.H.; Yusof, F.A.M. Processing and Characterisation of Banana Leaf Fibre Reinforced Thermoplastic Cassava Starch Composites. *Polymers* **2021**, *13*, 1420. [CrossRef] [PubMed]
33. Diyana, Z.N.; Jumaidin, R.; Selamat, M.Z.; Ghazali, I.; Julmohammad, N.; Huda, N.; Ilyas, R.A. Physical Properties of Thermoplastic Starch Derived from Natural Resources and Its Blends: A Review. *Polymers* **2021**, *13*, 1396. [CrossRef] [PubMed]
34. Ilyas, R.A.; Sapuan, S.M.; Ishak, M.R.; Zainudin, E.S. Development and characterization of sugar palm nanocrystalline cellulose reinforced sugar palm starch bionanocomposites. *Carbohydr. Polym.* **2018**, *202*, 186–202. [CrossRef]
35. Kumari, N.; Bangar, S.P.; Petrů, M.; Ilyas, R.A.; Singh, A.; Kumar, P. Development and Characterization of Fenugreek Protein-Based Edible Film. *Foods* **2021**, *10*, 1976. [CrossRef]
36. Punia Bangar, S.; Nehra, M.; Siroha, A.K.; Petrů, M.; Ilyas, R.A.; Devi, U.; Devi, P. Development and Characterization of Physical Modified Pearl Millet Starch-Based Films. *Foods* **2021**, *10*, 1609. [CrossRef] [PubMed]
37. Ghanbarzadeh, B.; Almasi, H.; Entezami, A.A. Improving the barrier and mechanical properties of corn starch-based edible films: Effect of citric acid and carboxymethyl cellulose. *Ind. Crop. Prod.* **2011**, *33*, 229–235. [CrossRef]
38. Waterschoot, J.; Gomand, S.V.; Fierens, E.; Delcour, J.A. Production, structure, physicochemical and functional properties of maize, cassava, wheat, potato and rice starches. *Starch/Staerke* **2015**, *67*, 14–29. [CrossRef]
39. Bertoft, E. Understanding starch structure: Recent progress. *Agronomy* **2017**, *7*, 56. [CrossRef]
40. McAloon, A.; Taylor, F.; Yee, W.; Ibsen, K.; Wooley, R. *Determining the Cost of Producing Ethanol from Corn Starch and Lignocellulosic Feedstocks*; No. NREL/TP-580-28893; National Renewable Energy Laboratory: Golden, CO, USA, 2000.
41. Ibrahim, M.I.J.; Sapuan, S.M.; Zainudin, E.S.; Zuhri, M.Y.M. Extraction, chemical composition, and characterization of potential lignocellulosic biomasses and polymers from corn plant parts. *BioResources* **2019**, *14*, 6485–6500. [CrossRef]
42. Sanyang, M.L.; Sapuan, S.M.; Jawaid, M.; Ishak, M.R.; Sahari, J. Effect of plasticizer type and concentration on physical properties of biodegradable films based on sugar palm (*Arenga pinnata*) starch for food packaging. *J. Food Sci. Technol.* **2016**, *53*, 326–336. [CrossRef]
43. Zentou, H.; Rosli, N.S.; Wen, C.H.; Abdul Azeez, K.; Gomes, C. The viability of biofuels in developing countries: Successes, failures, and challenges. *Iran. J. Chem. Chem. Eng.* **2019**, *38*, 173–182.
44. Zentou, H.; Zainal Abidin, Z.; Yunus, R.; Awang Biak, D.R.; Abdullah Issa, M.; Yahaya Pudza, M. A new model of alcoholic fermentation under a byproduct inhibitory effect. *ACS Omega* **2021**, *6*, 4137–4146. [CrossRef]

45. Shahabi-Ghahfarrokhi, I.; Goudarzi, V.; Babaei-Ghazvini, A. Production of starch based biopolymer by green photochemical reaction at different UV region as a food packaging material: Physicochemical characterization. *Int. J. Biol. Macromol.* **2019**, *122*, 201–209. [CrossRef] [PubMed]
46. Özeren, H.D.; Olsson, R.T.; Nilsson, F.; Hedenqvist, M.S. Prediction of plasticization in a real biopolymer system (starch) using molecular dynamics simulations. *Mater. Des.* **2020**, *187*, 108387. [CrossRef]
47. Yu, F.; Prashantha, K.; Soulestin, J.; Lacrampe, M.-F.; Krawczak, P. Plasticized-starch/poly (ethylene oxide) blends prepared by extrusion. *Carbohydr. Polym.* **2013**, *91*, 253–261. [CrossRef] [PubMed]
48. Wang, X.-L.; Yang, K.-K.; Wang, Y.-Z. Properties of starch blends with biodegradable polymers. *J. Macromol. Sci. Part C Polym. Rev.* **2003**, *43*, 385–409. [CrossRef]
49. Pelissari, F.M.; Yamashita, F.; Garcia, M.A.; Martino, M.N.; Zaritzky, N.E.; Grossmann, M.V.E. Constrained mixture design applied to the development of cassava starch–chitosan blown films. *J. Food Eng.* **2012**, *108*, 262–267. [CrossRef]
50. Tyagi, V.; Bhattacharya, B. Role of plasticizers in bioplastics. *MOJ Food Process. Technol.* **2019**, *7*, 128–130.
51. Santana, R.F.; Bonomo, R.C.F.; Gandolfi, O.R.R.; Rodrigues, L.B.; Santos, L.S.; dos Santos Pires, A.C.; de Oliveira, C.P.; da Costa Ilhéu Fontan, R.; Veloso, C.M. Characterization of starch-based bioplastics from jackfruit seed plasticized with glycerol. *J. Food Sci. Technol.* **2018**, *55*, 278–286. [CrossRef]
52. Hafiza, M.N.; Isa, M.I.N. Correlation between structural, ion transport and ionic conductivity of plasticized 2-hydroxyethyl cellulose based solid biopolymer electrolyte. *J. Membr. Sci.* **2020**, *597*, 117176. [CrossRef]
53. Hazrol, M.D.; Sapuan, S.M.; Zainudin, E.S.; Zuhri, M.Y.M.; Abdul Wahab, N.I. Corn starch (*Zea mays*) biopolymer plastic reaction in combination with sorbitol and glycerol. *Polymers* **2021**, *13*, 242. [CrossRef]
54. Ibrahim, M.I.J.; Sapuan, S.M.; Zainudin, E.S.; Zuhri, M.Y.M. Physical, thermal, morphological, and tensile properties of cornstarch-based films as affected by different plasticizers. *Int. J. Food Prop.* **2019**, *22*, 925–941. [CrossRef]
55. Vieira, M.G.A.; da Silva, M.A.; dos Santos, L.O.; Beppu, M.M. Natural-based plasticizers and biopolymer films: A review. *Eur. Polym. J.* **2011**, *47*, 254–263. [CrossRef]
56. Versino, F.; López, O.V.; García, M.A. Sustainable use of cassava (Manihot esculenta) roots as raw material for biocomposites development. *Ind. Crop. Prod.* **2015**, *65*, 79–89. [CrossRef]
57. Cao, N.; Yang, X.; Fu, Y. Effects of various plasticizers on mechanical and water vapor barrier properties of gelatin films. *Food Hydrocoll.* **2009**, *23*, 729–735. [CrossRef]
58. Wang, J.; Cheng, F.; Zhu, P. Structure and properties of urea-plasticized starch films with different urea contents. *Carbohydr. Polym.* **2014**, *101*, 1109–1115. [CrossRef] [PubMed]
59. Jouki, M.; Khazaei, N.; Ghasemlou, M.; HadiNezhad, M. Effect of glycerol concentration on edible film production from cress seed carbohydrate gum. *Carbohydr. Polym.* **2013**, *96*, 39–46. [CrossRef]
60. Audic, J.-L.; Chaufer, B. Influence of plasticizers and crosslinking on the properties of biodegradable films made from sodium caseinate. *Eur. Polym. J.* **2005**, *41*, 1934–1942. [CrossRef]
61. Müller, C.M.O.; Yamashita, F.; Laurindo, J.B. Evaluation of the effects of glycerol and sorbitol concentration and water activity on the water barrier properties of cassava starch films through a solubility approach. *Carbohydr. Polym.* **2008**, *72*, 82–87. [CrossRef]
62. Mantzari, G.; Raphaelides, S.N.; Exarhopoulos, S. Effect of sorbitol addition on the physicochemical characteristics of starch–fatty acid systems. *Carbohydr. Polym.* **2010**, *79*, 154–163. [CrossRef]
63. Fishman, M.L.; Coffin, D.R.; Konstance, R.P.; Onwulata, C.I. Extrusion of pectin/starch blends plasticized with glycerol. *Carbohydr. Polym.* **2000**, *41*, 317–325. [CrossRef]
64. Mali, S.; Grossmann, M.V.E.; García, M.A.; Martino, M.N.; Zaritzky, N.E. Effects of controlled storage on thermal, mechanical and barrier properties of plasticized films from different starch sources. *J. Food Eng.* **2006**, *75*, 453–460. [CrossRef]
65. Zhang, Y.; Han, J.H. Mechanical and Thermal Characteristics of Pea Starch Films Plasticized with Monosaccharides and Polyols. *J. Food Sci.* **2006**, *71*, E109–E118. [CrossRef]
66. Jumaidin, R.; Sapuan, S.M.; Jawaid, M.; Ishak, M.R.; Sahari, J. Thermal, mechanical, and physical properties of seaweed/sugar palm fibre reinforced thermoplastic sugar palm Starch/Agar hybrid composites. *Int. J. Biol. Macromol.* **2017**, *97*, 606–615. [CrossRef]
67. Ibrahim, M.I.J.; Sapuan, S.M.; Zainudin, E.S.; Zuhri, M.Y.M. Preparation and characterization of cornhusk/sugar palm fiber reinforced Cornstarch-based hybrid composites. *J. Mater. Res. Technol.* **2020**, *9*, 200–211. [CrossRef]
68. Cerqueira, M.A.; Souza, B.W.S.; Teixeira, J.A.; Vicente, A.A. Effect of glycerol and corn oil on physicochemical properties of polysaccharide films—A comparative study. *Food Hydrocoll.* **2012**, *27*, 175–184. [CrossRef]
69. Sahari, J.; Sapuan, S.M.; Ismarrubie, Z.N.; Rahman, M.Z.A. Physical and chemical properties of different morphological parts of sugar palm fibres. *Fibres Text. East. Eur.* **2012**, *2*, 21–24.
70. Tarique, J.; Sapuan, S.M.; Khalina, A. Effect of glycerol plasticizer loading on the physical, mechanical, thermal, and barrier properties of arrowroot (*Maranta arundinacea*) starch biopolymers. *Sci. Rep.* **2021**, *11*, 13900. [CrossRef]
71. Basiak, E.; Lenart, A.; Debeaufort, F. How glycerol and water contents affect the structural and functional properties of starch-based edible films. *Polymers* **2018**, *10*, 412. [CrossRef]
72. Ibrahim, M.I.J.; Sapuan, S.M.; Zainudin, E.S.; Zuhri, M.Y.M.; Edhirej, A. Processing and characterization of cornstalk/sugar palm fiber reinforced cornstarch biopolymer hybrid composites. In *Advanced Processing, Properties, and Applications of Starch and Other Bio-Based Polymers*; Elsevier: Amsterdam, The Netherlands, 2020; pp. 35–46.

73. Dai, H.; Yu, J.; Geng, F.; Ma, X. Preparation and properties of starch-based film using N-(2-hydroxyethyl) formamide as a new plasticizer. *Polym. Plast. Technol. Eng.* **2009**, *48*, 866–870. [CrossRef]
74. Galdeano, M.C.; Mali, S.; Grossmann, M.V.E.; Yamashita, F.; García, M.A. Effects of plasticizers on the properties of oat starch films. *Mater. Sci. Eng. C* **2009**, *29*, 532–538. [CrossRef]
75. Syafiq, R.M.O.; Sapuan, S.M.; Zuhri, M.R.M. Effect of cinnamon essential oil on morphological, flammability and thermal properties of nanocellulose fibre–reinforced starch biopolymer composites. *Nanotechnol. Rev.* **2020**, *9*, 1147–1159. [CrossRef]
76. Edhirej, A.; Sapuan, S.M.; Jawaid, M.; Zahari, N.I. Effect of various plasticizers and concentration on the physical, thermal, mechanical, and structural properties of cassava-starch-based films. *Starch Stärke* **2017**, *69*, 1500366. [CrossRef]
77. Nordin, N.; Othman, S.H.; Rashid, S.A.; Basha, R.K. Effects of glycerol and thymol on physical, mechanical, and thermal properties of corn starch films. *Food Hydrocoll.* **2020**, *106*, 105884. [CrossRef]
78. Hazrati, K.Z.; Sapuan, S.M.; Zuhri, M.Y.M.; Jumaidin, R. Effect of plasticizers on physical, thermal, and tensile properties of thermoplastic films based on Dioscorea hispida starch. *Int. J. Biol. Macromol.* **2021**. [CrossRef]
79. Yin, Y.J.; Yao, K.D.; Cheng, G.X.; Ma, J.B. Properties of polyelectrolyte complex films of chitosan and gelatin. *Polym. Int.* **1999**, *48*, 429–432. [CrossRef]
80. Hu, G.; Chen, J.; Gao, J. Preparation and characteristics of oxidized potato starch films. *Carbohydr. Polym.* **2009**, *76*, 291–298. [CrossRef]
81. Famá, L.; Rojas, A.M.; Goyanes, S.; Gerschenson, L. Mechanical properties of tapioca-starch edible films containing sorbates. *LWT Food Sci. Technol.* **2005**, *38*, 631–639. [CrossRef]
82. Gutiérrez, T.J.; Tapia, M.S.; Pérez, E.; Famá, L. Structural and mechanical properties of edible films made from native and modified cush-cush yam and cassava starch. *Food Hydrocoll.* **2015**, *45*, 211–217. [CrossRef]
83. Suppakul, P.; Chalernsook, B.; Ratisuthawat, B.; Prapasitthi, S.; Munchukangwan, N. Empirical modeling of moisture sorption characteristics and mechanical and barrier properties of cassava flour film and their relation to plasticizing–antiplasticizing effects. *LWT Food Sci. Technol.* **2013**, *50*, 290–297. [CrossRef]
84. Dang, K.M.; Yoksan, R. Development of thermoplastic starch blown film by incorporating plasticized chitosan. *Carbohydr. Polym.* **2015**, *115*, 575–581. [CrossRef]
85. Huang, Q.; Zhao, J.; Liu, M.; Chen, J.; Zhu, X.; Wu, T.; Tian, J.; Wen, Y.; Zhang, X.; Wei, Y. Preparation of polyethylene polyamine@ tannic acid encapsulated MgAl-layered double hydroxide for the efficient removal of copper (II) ions from aqueous solution. *J. Taiwan Inst. Chem. Eng.* **2018**, *82*, 92–101. [CrossRef]
86. Vega, D.; Villar, M.A.; Failla, M.D.; Vallés, E.M. Thermogravimetric analysis of starch-based biodegradable blends. *Polym. Bull.* **1996**, *37*, 229–235. [CrossRef]
87. Jantrawut, P.; Chaiwarit, T.; Jantanasakulwong, K.; Brachais, C.H.; Chambin, O. Effect of plasticizer type on tensile property and in vitro indomethacin release of thin films based on low-methoxyl pectin. *Polymers* **2017**, *9*, 289. [CrossRef]
88. Mali, S.; Grossmann, M.V.E.; Garcia, M.A.; Martino, M.N.; Zaritzky, N.E. Microstructural characterization of yam starch films. *Carbohydr. Polym.* **2002**, *50*, 379–386. [CrossRef]
89. Zuo, Y.; Gu, J.; Tan, H.; Zhang, Y. Thermoplastic starch prepared with different plasticizers: Relation between degree of plasticization and properties. *J. Wuhan Univ. Technol. Sci. Ed.* **2015**, *30*, 423–428. [CrossRef]
90. Dai, L.; Qiu, C.; Xiong, L.; Sun, Q. Characterisation of corn starch-based films reinforced with taro starch nanoparticles. *Food Chem.* **2015**, *174*, 82–88. [CrossRef] [PubMed]

polymers

MDPI

Review

Emerging Developments Regarding Nanocellulose-Based Membrane Filtration Material against Microbes

Mohd Nor Faiz Norrahim [1], Noor Azilah Mohd Kasim [1,2,*], Victor Feizal Knight [1,*], Keat Khim Ong [1,2], Siti Aminah Mohd Noor [1,2], Norhana Abdul Halim [3], Noor Aisyah Ahmad Shah [2], Siti Hasnawati Jamal [2], Nurjahirah Janudin [1], Muhammad Syukri Mohamad Misenan [4], Muhammad Zamharir Ahmad [5], Mohd Hanif Yaacob [6] and Wan Md Zin Wan Yunus [7,*]

1 Research Centre for Chemical Defence, Universiti Pertahanan Nasional Malaysia, Kem Perdana Sungai Besi, Kuala Lumpur 57000, Malaysia; faiznorrrahim@gmail.com (M.N.F.N.); ongkhim@upnm.edu.my (K.K.O.); s.aminah@upnm.edu.my (S.A.M.N.); nurjahirahjanudin@upnm.edu.my (N.J.)

2 Department of Chemistry and Biology, Centre for Defence Foundation Studies, Universiti Pertahanan Nasional Malaysia, Kem Perdana Sungai Besi, Kuala Lumpur 57000, Malaysia; aisyah@upnm.edu.my (N.A.A.S.); hasnawati@upnm.edu.my (S.H.J.)

3 Department of Physics, Centre for Defence Foundation Studies, Universiti Pertahanan Nasional Malaysia, Kem Perdana Sungai Besi, Kuala Lumpur 57000, Malaysia; norhana@upnm.edu.my

4 Department of Chemistry, College of Arts and Science, Yildiz Technical University, Davutpasa Campus, Esenler, Istanbul 34220, Turkey; syukrimisenan@gmail.com

5 Biotechnology and Nanotechnology Research Centre, Malaysia Agricultural Research and Development Institute, Persiaran MARDI-UPM, Serdang 43400, Selangor, Malaysia; zamharir@mardi.gov.my

6 Wireless and Photonics Network Research Centre (WiPNET), Universiti Putra Malaysia, Serdang 43400, Selangor, Malaysia; hanif@upm.edu.my

7 Research Centre for Tropicalisation, Universiti Pertahanan Nasional Malaysia, Kem Perdana Sungai Besi, Kuala Lumpur 57000, Malaysia

* Correspondence: azilah@upnm.edu.my (N.A.M.K.); victor.feizal@upnm.edu.my (V.F.K.); wanmdzin@upnm.edu.my (W.M.Z.W.Y.)

Citation: Norrrahim, M.N.F.; Mohd Kasim, N.A.; Knight, V.F.; Ong, K.K.; Mohd Noor, S.A.; Abdul Halim, N.; Ahmad Shah, N.A.; Jamal, S.H.; Janudin, N.; Misenan, M.S.M.; et al. Emerging Developments Regarding Nanocellulose-Based Membrane Filtration Material against Microbes. *Polymers* **2021**, *13*, 3249. https://doi.org/10.3390/polym13193249

Academic Editors: Emin Bayraktar, S. M. Sapuan and R. A. Ilyas

Received: 31 August 2021
Accepted: 22 September 2021
Published: 24 September 2021

Abstract: The wide availability and diversity of dangerous microbes poses a considerable problem for health professionals and in the development of new healthcare products. Numerous studies have been conducted to develop membrane filters that have antibacterial properties to solve this problem. Without proper protective filter equipment, healthcare providers, essential workers, and the general public are exposed to the risk of infection. A combination of nanotechnology and biosorption is expected to offer a new and greener approach to improve the usefulness of polysaccharides as an advanced membrane filtration material. Nanocellulose is among the emerging materials of this century and several studies have proven its use in filtering microbes. Its high specific surface area enables the adsorption of various microbial species, and its innate porosity can separate various molecules and retain microbial objects. Besides this, the presence of an abundant OH groups in nanocellulose grants its unique surface modification, which can increase its filtration efficiency through the formation of affinity interactions toward microbes. In this review, an update of the most relevant uses of nanocellulose as a new class of membrane filters against microbes is outlined. Key advancements in surface modifications of nanocellulose to enhance its rejection mechanism are also critically discussed. To the best of our knowledge, this is the first review focusing on the development of nanocellulose as a membrane filter against microbes.

Keywords: nanocellulose; membrane filter; microbes; surface functionalization

1. Introduction

Throughout the evolutionary process, among the significant issues faced by society today are the protection of natural resources and the implementation of eco-friendly approaches to sustaining a high quality of life. Environmental pollution is a worldwide concern and the majority of pollutants have long-term negative impacts on humans. Focusing on microbial pollution, the most common bulk transportation media for particulate contaminants are air and water. Microbes are microscopic living organisms that can be found everywhere, including in water, soil and air, but they are too small to be seen with the naked eye. These microbes are commonly viruses, bacteria, and fungi and may involve microscopic parasites. Certain microbes are harmful to our health, while others are beneficial. Table 1 shows several types of infectious diseases caused by microbes.

Table 1. Several infectious diseases caused by microbes.

Infectious Disease	Microbe That Causes the Disease	Type of Microbe	Reference
Coronavirus (COVID-19)	Severe acute respiratory syndrome coronavirus 2 (SARS-CoV-2)	Virus	[1]
Cold	Rhinovirus	Virus	[2]
Chickenpox	*Varicella zoster*	Virus	[3]
German measles	Rubella	Virus	[4]
Whooping cough	*Bordatella pertussis*	Bacteria	[5]
Bubonic plague	*Yersinia pestis*	Bacteria	[6]
TB (Tuberculosis)	*Mycobacterium tuberculosis*	Bacteria	[7]
Malaria	*Plasmodium falciparum*	Protozoa	[8]
Tinea barbae (dermatophyte infection)	*Trichophyton rubrum*	Fungus	[9]
Athletes' foot	*Trichophyton mentagrophytes*	Fungus	[10]

Microbial contamination in water can be dangerous to health, causing serious waterborne disease outbreaks, such as gastroenteritis, cholera, giardiasis and cryptosporidiosis. The most common bacteria involved in these outbreaks are *Shigella dysenteriae, Vibrio cholera, Legionella* sp., *Escherichia coli,* and *Campylobacter jejuni* [11]. Whereas giardiasis and cryptosporidiosis are gastrointestinal diseases caused by microscopic parasites (protozoa), namely *Giardia duodenalis* and *Cryptosporidium sp.,* respectively. When invading the gastrointestinal tract, these microbes can cause local reactions to their presence and may even cause systemic effects from toxins they secrete (only certain microbes secrete toxins). Some microbes may invade the bloodstream, where they can cause sepsis [12]. Annually, it is estimated that approximately 485,000 people die from diarrheal disease as a result of drinking contaminated water [13]. Hence, microbially contaminated wastewater must be treated before it is discharged into water bodies or water courses.

As mentioned previously, microbes can also be transmitted through the air. According to the World Health Organization (WHO), airborne transmission differs from droplet transmission as it refers to the presence of microbes within droplet nuclei that are typically less than 5 μm in diameter and can circulate in the air for significant periods and be transmitted to others over distances more than 1 m [14]. Whereas droplet transmission occurs when a person is in close contact (within 1 m) with a symptomatic patient with respiratory symptoms such as coughing or sneezing and is thus at risk of exposure to potentially infective respiratory droplets (typically >5–10 μm in diameter). Nowadays, the threat of the newly discovered infectious coronavirus disease (COVID-19) is worrying, as this pandemic outbreak has already killed millions of people worldwide. The outbreak is exacerbated by the occurrence of frequent mutations, which makes it difficult to rapidly produce omnipotent vaccines [1]. Therefore, an effective, robust, and inexpensive air-borne virus removal membrane filter is an urgent need to provide a means to prevent virus spread in hospitals, transportation hubs, schools, and other venues with high social traffic turn-over in order to minimize the risks arising from the COVID-19 pandemic.

Microbe removal can be done through a variety of methods, such as, filtration (either depth filtration or surface screening), partitioning and fractionation (centrifugation), and chromatography (ion-exchange, affinity, gel permeation) [15]. Of these different techniques, filtration is a desirable choice, as it is non-destructive and non-interfering, implying that it will not threaten the quality of biological samples or induce immune reactions. Membrane filters have been made from a variety of synthetic and semi-synthetic polymers, designed to achieve a desired filtration pore size. The membrane filter is also an effective and widely used method for detecting microbiological pollution in collection samples. It requires less planning than certain other conventional methods and is one of the few methods that allows for microorganism separation and subsequent determination. Microbes cannot be retained by the normal membrane filter because the membrane pores are too large. Therefore, it is critical to have a more effective material for microbe filtration, and there are studies that have led to the discovery of new filtering media made from cellulose with efficient filtration capability. The ultimate objective would be to be able to effectively and securely filter microbes from the environment at an affordable cost.

Current filter materials which are typically non-biodegradable and non-renewable have also received much attention among scientists. These membrane filters are primarily made from polymers which include proprietary non-ionic polymers, polytetrafluoroethylene (PTFE), polypropylene (PP), polysulphone (PS), polyvinylidene fluoride (PVDF) and polyethylene (PE) [16,17]. These non-biodegradable polymers when being disposed of after use are known to be harmful to the environment [18]. Figure 1 shows an example of used surgical masks incorporating PP that have been discarded and not properly disposed of thereby causing problems of littering both on land and at sea as well as in waterways. Therefore, scientists urgently need to find better solutions to this problem. It is important to have a more efficient renewable and biodegradable material, of which nanocellulose is a prime candidate.

Figure 1. Used surgical masks that were not properly disposed.

Cellulose is a versatile industrial product because of its abundance, renewability, biodegradability and ability to be readily chemically modified [19–21]. With the development of nanocellulosic materials, since 1977 there has been extensive research into their use in many fields, such as in biocomposites, bioadsorbents, textiles, biomedicals, military, automotives, sensors, energy, packaging, as well as membrane filtration [22–31]. Generally, nanocellulose can be classified into three types, which are cellulose nanofiber (CNF), cellulose nanocrystals (CNC), and bacterial nanocellulose (BNC) [1,32,33]. The production of nanocellulose can be accomplished in two ways, including up-down or down-up techniques [34]. Up-down techniques can be used to synthesize CNF and CNC. The most popular way to make CNC is to use inorganic acids to acid hydrolyse pure cellulose. To reduce energy consumption, CNF is made via mechanical disintegration with high shear forces, maybe in combination with chemical or enzymatic pre-treatment [35]. On the other hand, bacterial nanocellulose (BNC) is made using a down-up technique, in which cellulose-producing bacteria such as *Acetobacter xylinum* is used [36].

According to Hassan et al. (2020) [37], two approaches that utilize nanocellulose for filtration have been explored; namely, the first approach which incorporated them into other polymer matrices to enhance the effectiveness of prepared membranes. This was done by dissolving the polymer in suitable solvents and ensuring that the nanocellulose materials were well dispersed within the polymer solution before being film cast. Meanwhile, the second approach was one first introduced by Ma et al. (2011) [38], in which the authors developed membranes from a nanocellulose layer with adequate porosity laid over polymeric supports without having to dissolve the cellulose matrix and using the film casting method to produce porosity. The latter approach was found to be more desirable and intriguing.

Nanocellulose-based membrane filters have been found to be effective at removing microbes in previous research [39–41]. When compared to synthetic polymers or plastic membranes, a significant benefit of the nanocellulose-based membrane filters is that they are entirely made from natural resources, making their disposal much easier, as they were made up of predominantly biodegradable materials [42].

This review is intended to provide a comprehensive overview of the recent advances in the development of membrane filters for microbial removal, which are made up either entirely from nanocellulose or utilizing a modified approach, incorporating nanocellulose. This review will include (1) a description of the types of membrane filters and the rejection mechanism they use; (2) details of the nanocellulose used in the production of membrane filters; (3) a suite of antimicrobial technologies used for nanocellulose functionalization. This manuscript provides knowledge and direction for scientists to stimulate future research in this area.

2. Types and Rejection Mechanisms of Membrane Filters

The role of membranes alone in the removal of pathogens is discussed here. Membrane filters can be categorized by the size of the pollutant they are able to reject (see Figure 2), namely: reverse osmosis (0.1–1 nm), nanofiltration (NF) (1–2 nm), ultrafiltration (UF) (2–100 nm) and microfiltration (MF) (100 nm–10 µm). The two most important features of a membrane are its permeability and selectivity. To enhance the productivity of membrane separation processes, it is always necessary to develop membranes with high permeability and high selectivity [43].

Referring to Figure 3, it can be seen that there are various pore sizes of nanocellulose-based membrane filters that are available, depending on their origin. The different pore sizes of nanocellulose membrane filters fulfil various filtration modes. Thus, nanocellulose membrane filters obtained from electrospun CNF usually have a pore size of more than 100 nm (Figure 3a). Thus, making this membrane filter more suitable as an MF. Meanwhile, CNF (Figure 3b) and CNC (Figure 3c,d), which are obtained through other methods usually have a smaller pore size, which ranges from 5 nm to 100 nm. Thus, this makes them suitable for use in NF and UF. The pore size of nanocellulose depends on several factors, such as the cellulose origin, the isolation process concentration and processing conditions as well as pretreatments administered to the cellulose. Moreover, CNF has more advantages to be used as a membrane filter. This is because CNF led to a higher strength and modulus compared to CNC, due to CNF's larger aspect ratio and fibre entanglement, but lower strain-at-failure because of their relatively large fibre agglomerates [44]. Table 2 summarized the properties of CNF films from different sources.

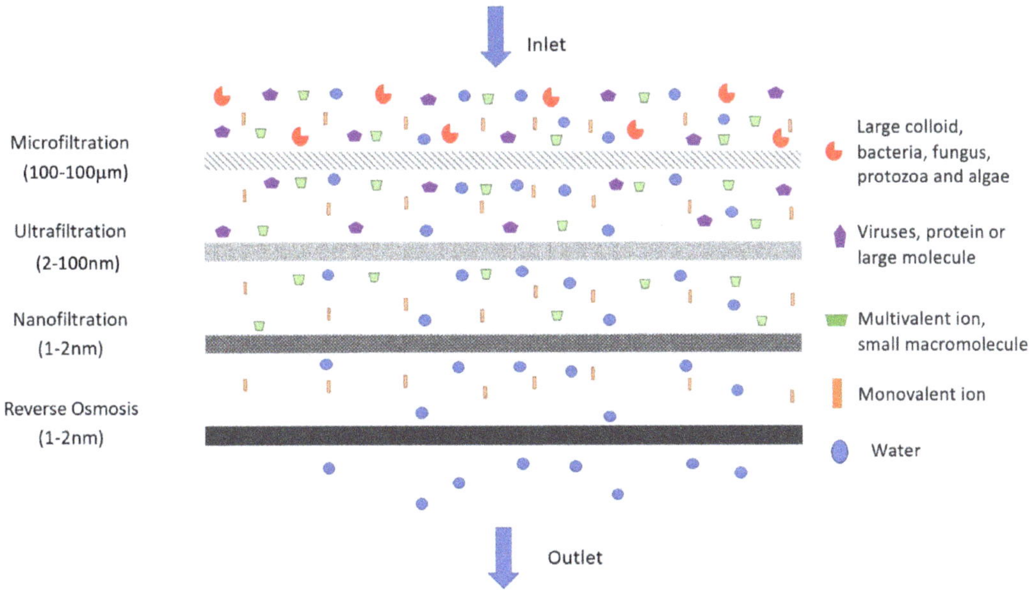

Figure 2. Comparison of the size and type of contaminants rejected by membrane filtration techniques.

Table 2. Representative studies on the properties of CNF films from different raw materials.

Source of CNF	Modulus (GPa)	Tensile Strength (MPa)	Strain to Failure (%)	Reference
Pulp	10.4–13.7	129–214	3.3–10.1	[45]
Kraft pulp	17	250	2–6	[46]
Wood	6.2–6.9	222–233	7.0–7.6	[47]
Wood	13	223	-	[48]

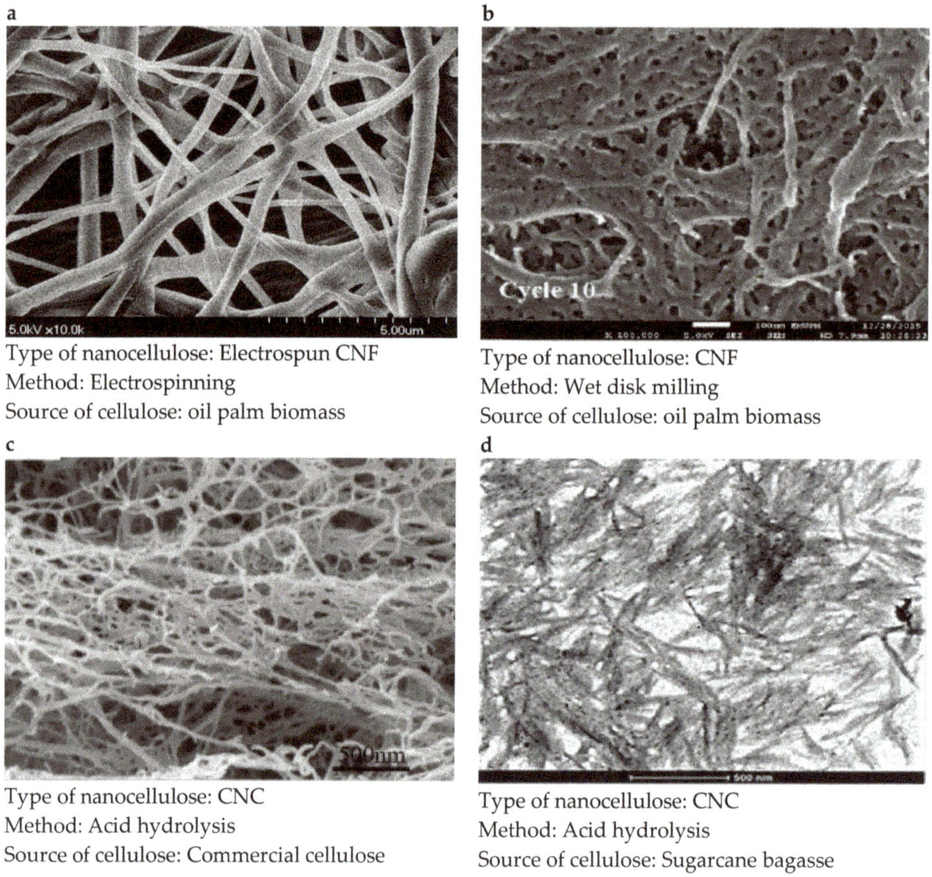

Figure 3. The morphology of nanocellulose produced from various methods. (**a**) electrospinning, (**b**) wet disk milling, (**c**,**d**) acid hydrolysis. This figure is adapted with permission from the [49–52].

Figure 2 illustrates the membrane filtration spectrum, which operates by utilizing the size exclusion method in rejecting or inhibiting the pathogenic microorganisms. Wu et al. (2019) [53] described it as an established, reliable, and robust method, considering its ability to physically remove various types of infective microorganisms, including virus. Of note, the other method that utilized the affinity principle could also be used for filtration.

There are two types of membrane filters with different pore sizes, commonly used in the retention of microorganisms. The first one is the MF membrane, which has a pore size of 0.1–10 μm, while the other is the UF membrane, which has a smaller pore size, ranging from 5 to 100 nm. Both types of membrane filters are applicable to the removal of protozoa and algae (i.e., size range between 3 to 14 μm) [54]. In addition, Francy et al. (2012) [55] outlined that tertiary disinfection is not necessary, as the pore size of both MF and UF membranes are too nominal when compared to the size of coliform bacteria, suggesting the total removal of bacteria by size exclusion of the membranes. However, a particular concern was raised regarding the removal of virus via direct membrane filtration considering its smaller size compared to bacteria.

Size exclusion is a widely used technique in the chromatography method, which separates molecules, depending on their sizes or "hydrodynamic volume" in solution. Filtration takes place through a gel composed of spherical beads with a particular size distribution of pores. When molecules of various sizes are incorporated or omitted from the pores inside the matrix, separation occurs. Small molecules diffuse into the pores, slowing their flow through the column, whereas large molecules (or having the greatest hydrodynamic volume) do not penetrate the pores and are eluted in the column's void volume [56]. As a result, molecules segregate according to their size as they move down the column and are eluted in decreasing order of molecular weight (MW).

There are a few criteria that influence the effectiveness of the size exclusion technique, particularly the pH and ionic strength of the load buffers, which have a major impact on the retention of diverse specimens. In neutral membranes, the sieving behaviour of charged pollutants is different from that of neutral pollutants. Due to electrostatic interactions with ions in solution, charged pollutants have a double layer of electrical charge on their surface. A solution entering a pore will compress this electrical double layer if the pollutants and pore sizes are of the same magnitude. This is not energetically advantageous, resulting in a reduction in the sieving of the charged pollutants. Interestingly, when an ionic solution encounters a charged membrane, the Donnan model gives a well-known classical description of the electrochemical equilibrium that occurs. It ignores ion size effects while accounting for electrostatic interactions, since the theory regards ions as point charges [57]. When neutral pollutants are applied to a charged membrane, similar effects occur. The presence of neutral pollutants causes a compression of the electrical double layer, associated with the pore wall. With similarly charged pores and pollutants, this impact is amplified much further. Due to charge repulsion, membranes with charged ligands (with identical charge) have poorer sieving of specimens [58].

On the other hand, membrane filters utilize the affinity principle, known to use adsorption to remove pollutants based on the electrostatic interaction between functional groups of the membranes with the pollutant. This type of membrane filter includes composite or hybrid filter structures that consist of a porous substrate with either nanocellulose moieties attached to their surface or impregnated within. It is interesting to note that size exclusion and affinity regime approaches have been explored in many studies on membrane filters utilizing the nanocellulose.

Adsorption is an exothermic surface-based process in which molecules of a substance in a certain state aggregate on an adsorbent surface. The adsorbate is the substance that is adsorbed on the adsorbent. Desorption, on the other hand, is the release of adsorbed molecules from the adsorbent's surface, which is the reverse of adsorption. Adsorption of molecules to the adsorbent surface can take place in two ways: "physical adsorption," also known as "physisorption," and "chemical sorption," also known as "chemisorption." This is determined by the interactions of the molecules with the surface. Weak forces, such as electrostatic interactions and Van der Waals forces, are involved in physical adsorption. Chemical adsorption results in the formation of strong chemical bonds, such as covalent bonds, between the surface and the adsorbed molecules. A monomolecular layer (monolayer) is developed on the adsorbent surface during chemical adsorption, whereas a thick multilayer is created during physical adsorption on the adsorbent surface [59].

2.1. Fabrication of Nanocellulose Membrane

Numerous nanocellulose membrane production processes have been devised, taking into account the unique features of nanocellulose fibres and the membrane casting suspension, as shown in Figure 4 and summarized as follows.

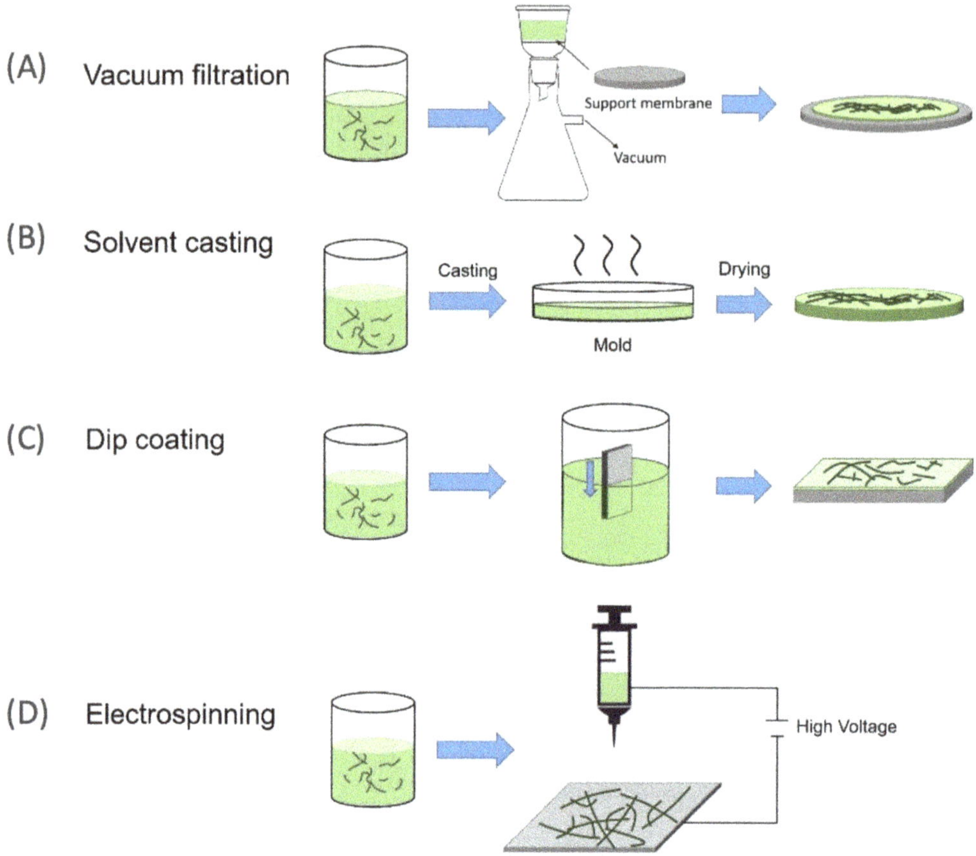

Figure 4. Various methods used for the fabrication of the nanocellulose membrane filter. Reproduced with permission from [60].

2.1.1. Vacuum Filtration

Vacuum filtration, followed by optional hot-pressing, is a rapid, easy, and accessible procedure for producing layered structures of nanocellulose membrane filters. The amount and concentration of nanocellulose suspensions could be used to alter the thickness and pore size of the resulting membranes [61].

2.1.2. Casting Evaporation and Coating Self-Standing

Normally, self-standing membranes are made by evaporating a dilute nanocellulose suspension in a petri dish. In general, to avoid agglomeration, the nanocellulose dispersion should be diluted to a low concentration (depending on surface chemistry and fibril diameter, but usually less than 1 wt%) [62,63].

2.1.3. Electrospinning

Electrospun membranes have a smaller base weight, a greater effective surface area, and a higher effective porosity, with pores that are continually interconnected. Nanocellulose could change the membrane surface charge density, increase total effective surface area, and improve functional group density by incorporating it into electrospinning membranes.

Furthermore, the nanocellulose content of this multi-layered nanofibrous system could alter the mean pore size and pore size distribution, and hence the separation performance [60].

3. Attributes of Nanocellulose Membrane Filtration of Microbes

In filtration systems, diameter, length, cross-section shape, internal structure (cellular or solid), and strength properties, which include tensile strength, stretch or elongation, and stiffness, are the most important physical characteristics of fibres for use in filter media. Fibre qualities that optimize the bulk, air permeability, and pore size of the filter media are ideal. The purpose of the filter design is to optimize bulk and air permeability to allow for breathing while reducing pore size to improve filtration efficiency [64].

Generally, polypropylene (PP) is used to produce the mask accords with the technical standards. The pore diameter of polypropylene is 25 mm. They were treated with dimethyl-dioctadecylammonium bromide to impart a positive electrical charge capable of attracting bacteria. Bacterial or viral filtration efficiency was almost 100% for the PP mask [65].

Normal membrane filters usually have pores that are too large to retain microbes. The advances in nanotechnology have made nanocellulose the more suitable material for the filtration of microbes. Nanocellulose with pore dimensions measuring between 1 and 100 nm offers certain unique characteristics, which include high strength, chemical inertness, hydrophilic surface chemistry and high surface area, thereby making it a promising material for use as a high-performance membrane filter that can effectively remove microbes from either air or liquids [22,49,66–72]. Moreover, membrane filter constructed entirely of CNF has recently been discovered to be capable of filtering even the tiniest viruses with up to 99.9980–99.9995% effectiveness.

The comparison between the PP membrane and nanocellulose membrane is tabulated in Table 3. On the other hand, Table 4 summarizes the importance of several special properties of nanocellulose which are related to the application as a membrane filtration material against microbes.

Table 3. Comparison of PP and nanocellulose membrane.

Characteristic	PP	Nanocellulose
Fibre length (nm)	-	400
Diameter (nm)	25,000	1–100
Efficiency against pathogens	~100%	99.9980–99.9995%
Tensile modulus (GPa)	1.5	145
Tensile strength (GPa)	0.02	7.5
Poison's ratio	0.4	0.3

As described earlier, nanocellulose can be classified into three types (CNF, CNC, and BNC) according to their manner of origin. These three types are as shown in Table 5, below. Essentially cellulose is a molecule that consists of β-1, 4-glucose, and three active hydroxyls at the C2, C3, and C6 sites of the glucopyranose ring, and its configuration provides sufficient sites for several surface functionalization's. These sites may undergo oxidation, esterification, and etherification to enable these variable functional groups that may include aldehyde groups and quaternary ammonium. The details of these surface functions will be explained in the following section.

Table 4. Certain interesting properties of nanocellulose related to membrane filtration materials.

Property	Advantages	Reference
Nanoporosity	Good virus filtration using size-exclusion method. Typically, the pore size of nanocellulose is below 100 nm.	[15]
Surface functionalization	Functionalization nanocellulose with several compounds to make it cationic charged causes an increase in its binding affinity towards viruses.	[73]
High specific surface area	Provides a large surface area for functionalization. Thereby increasing interaction efficiency.	[74]
Renewable	Nanocellulose can be easily sourced from plant bio-waste. Its use can eliminate the use of other non-renewable polymers as mentioned in the Introduction section.	[49,66]
Biodegradability	An important aspect to save the environment. It is biodegradable in landfills. Hence, current environment issues from used and discarded surgical masks can be reduced or even eliminated.	[75]
High mechanical strength	High strength membrane filters can be fabricated using it.	[76]
Stability in water	It can reduce biofouling of membrane filters. This is important for application in membrane filters for wastewater.	[77]

Each type of nanocellulose has different dimensions and properties, mainly due to their different methods of preparation/fabrication [78]. There are various methods used to extract cellulose nanoparticles which then result in these particles having different crystallinities, surface chemistries and mechanical properties [79]. These production methods range from top-down enzymatic/chemical/physical methods aimed at isolating them from wood and forest or agricultural residues to bottom-up production of cellulose nanofibrils from glucose by bacterial action [80].

Table 5. Types of nanocellulose according to their sources, treatments and dimensions [60,81–83].

Nanocellulose	Abbreviation	Sources	Main Treatments	Dimensions
Cellulose nanofiber 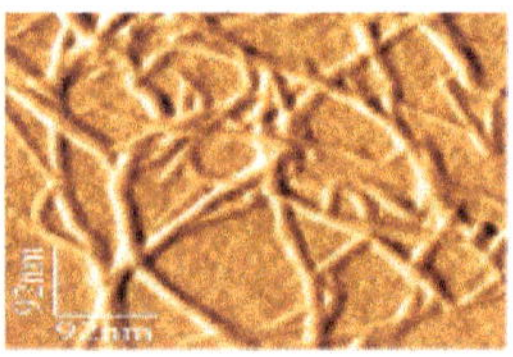	CNF	Plants	Mechanical fibrillation	Diameter: 5–50 nm Length: Several µm
Cellulose nanocrystals nanowhiskers/nanorods 	CNC	Plants	Acid hydrolysis	Diameter: 2–20 nm Length: 100 nm to several µm
Bacterial nanocellulose/biocellulose 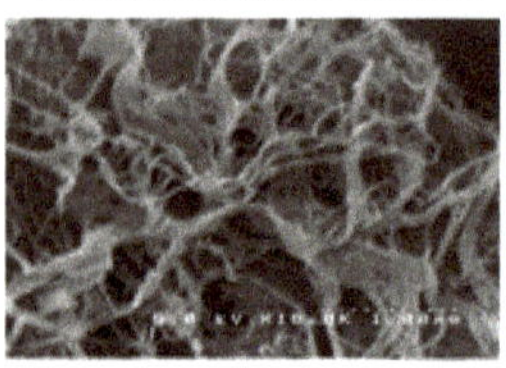	BNC	Microorganisms	Polymerization and crystallization	Diameter: 2–4 nm Length: 100µm

4. Modifications on Nanocellulose to Improve Filter Efficiency

The surface functionalization of nanocellulose is a key step to promoting an increase in the efficiency of a membrane filter. This is an important step when the membrane filter operates using the affinity regime. This can be done using different strategies of surface functionalization that will involve the chemistry of hydroxyl function [84]. Referring to Figure 5, a search using keywords 'functionalization of nanocellulose' was performed by lens.org, (https://www.lens.org/; accessed on 15 February 2021), a problem-solving non-profit social enterprise website utilizing linked open knowledge artefacts and metadata. It was found that manuscripts focusing on the functionalization of nanocellulose have been increasing from 2011 until now. This showed that the area of research interest has grown among scientists in this decade.

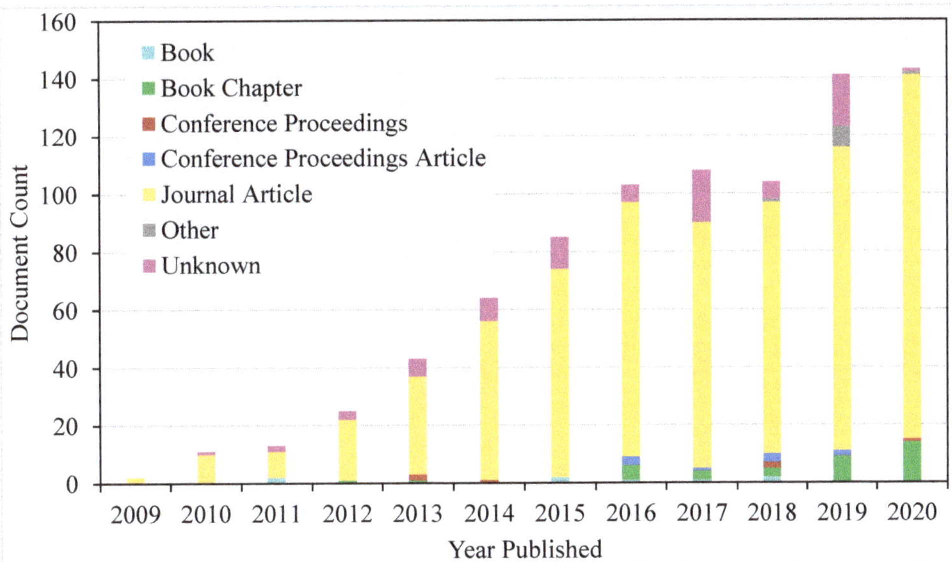

Figure 5. A chart of published manuscripts focusing on the functionalization of nanocellulose from Lens.org.

The surface functionalization is also important to increase the performance and lifetimes of the nanocellulose-based membrane filter. Nanocellulose is a fibrous water-loving polymer, due to the presence of an abundant number of OH groups. Thus, it can cause the membrane filter to swell and weaken in the presence of water. However, surface functionalization can overcome this problem by enhancing the hydrophobicity of nanocellulose [71,72] Besides that, surface functionalization also can improve the mechanical strength of nanocellulose-based membrane filters [85].

Surface functionalization on nanocellulose can be made through several processes, such as oxidation, esterification, and etherification, which eventually results in the introduction of new functional groups on the material. Other than this, previous studies have showed that nanocellulose can also be subjected to modification by adding compounds such as aldehyde, quaternary compounds (both cationic and anionic forms), activated carbon, citric acid, antibiotics, and nanomaterials. For instance, the aldehyde groups are grafted onto nanocellulose through the oxidation process, using oxidants such as periodate sodium and 2,2,6,6-teramethylpiperidinyloxy (TEMPO), which results in the TEMPO positioning on the surface of the nanocellulose under aqueous conditions, while the hydroxyl group located at the C6 position of the nanocellulose can be converted to carboxyl and aldehyde functional groups.

Besides this, low toxicity and environmentally friendly quaternary compounds, such as poly(N,N-dimethylaminoethyl methacrylate), amines, anionic polyelectrolytes, and polyglutamic acid could be used to quaternize nanocellulose to improve its efficiency as a membrane filter, as these quaternary compounds have the ability to form electrostatic affinity towards microbes. This quaternization process can be performed using grinding and high-pressure homogenization processes. Table 6 illustrates several examples of functionalized nanocellulose using quaternary compounds for virus removal applications. Most viruses and certain microbial species have polar charges on their surface; thus, modification on the surface charge of the nanocellulose improves the electrostatic interaction properties of the material, which consequently results in high efficiency filtration.

The main challenge with chemical functionalization to nanocellulose is to select an appropriate time for the functionalization to occur. Surface functionalization can be carried out during the preparation step or post-production of nanocellulose [86]. This process can be greatly affecting the final properties of functionalized nanocellulose such as crystallinity, yield, dimensions and morphology, surface chemistry, physicochemical, and thermal properties. If the wrong selection of functionalization approach, the 3D crystal network can be interrupted during modification, thus the mechanical properties of nanocellulose could deteriorate, which could consequently limit the applications of modified nanocellulose. In some approaches, nanocellulose is modified during the production step [87], while in other strategies the nanocellulose was produced, first followed by the modifications [88–90].

For example, Henschen (2019) [91] functionalized the nanocellulose with polyelectrolytes. It was found that the suspended nanocellulose is too small to be easily recovered if added to different solutions and it has a tendency to aggregate in polyelectrolyte solutions. Therefore, to adsorb polyelectrolytes onto nanocellulose, the functionalization is preferably done after the production of nanocellulose.

In the following section, several interesting findings concerning quaternized nanocellulose for microbial removal will be discussed.

Table 6. Functionalized nanocellulose with quaternary compounds for virus removal applications.

Functional Group	Chemical Structure
a. Functional group: aminoethyl methacrylate or poly(*N*-(2-aminoethylmethacrylamide)	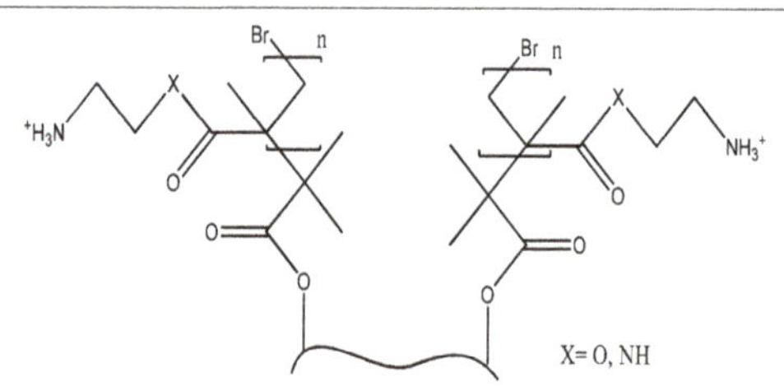
b. Functional group: 2,3-epoxypropyl trimethylammonium chloride	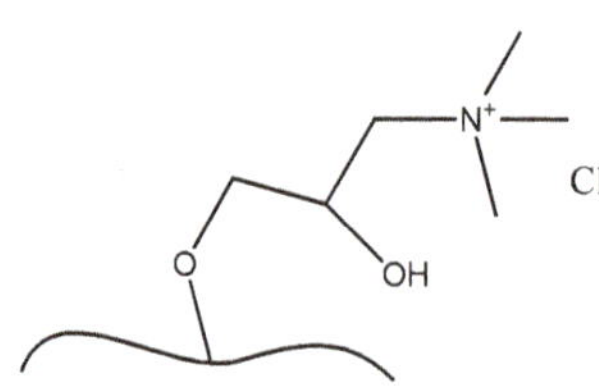

Table 6. *Cont.*

Functional Group	Chemical Structure
c. Imidazolium	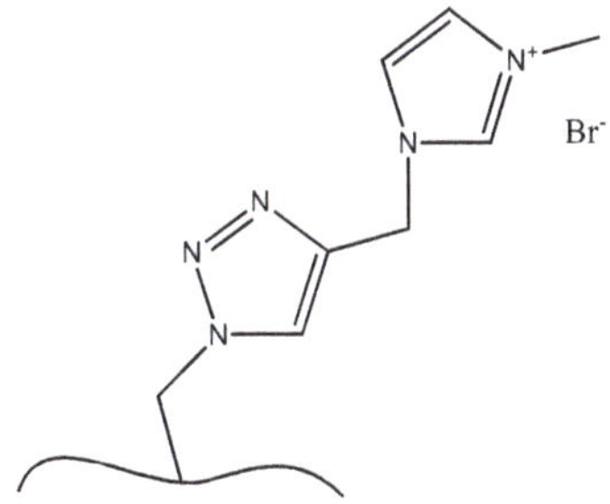
d. Pyridinium	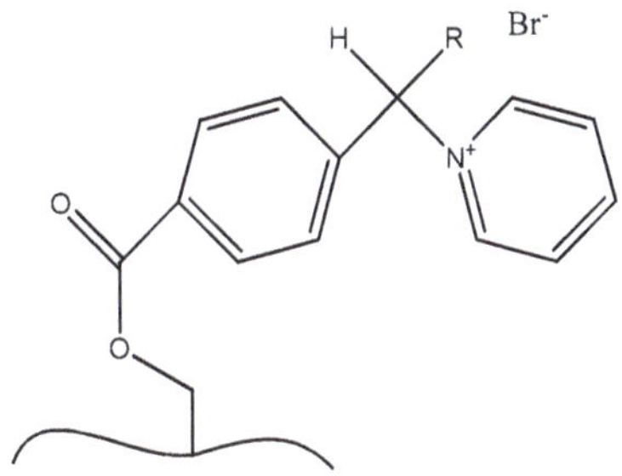
e. e-vinylpyidine	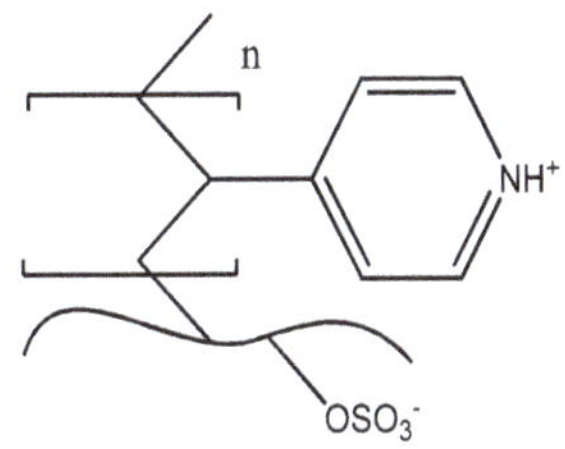
	 = nanocellulose

5. Recent Developments on Nanocellulose as a Filtration Material against Microbes

In this section, several developments concerning nanocellulose-based membrane filters capable of removing microbes will be reviewed. An important aspect of the modification of nanocellulose materials is to increase the binding affinity of the materials towards microbes. There are a number of studies that focused on the filtration removal of viruses and bacteria; however, very limited studies have been conducted, which concern other types of microbes, such as fungi, algae and protozoa.

5.1. Viruses

The ensnarement of viruses is one of the most crucial steps in biopharmaceutical and clinical processes and applications [92]. Of the various types of microbes, virus is among the smallest and most difficult to deal with, as compared to other microbes.

Exploration of nanocellulose as a filtration material against several types of viruses has received much research attention. As mentioned earlier, several viruses, including COVID-19, are airborne viruses that can be dispersed and spread through human nasal or saliva secretions from an infected person. Therefore, in order to minimize infection risks from viruses, an efficient, robust and affordable air-borne virus removal filter is an urgent requirement. Multiple research articles were recently published with regard to this type of air filter.

Several factors, such as filter thickness, pore size, number of layers, size of the virus, the charge on the filter surface, its ionic strength and surface chemistry are usually influenced by the efficiency of the air filtration process [15]. Generally, the use of size-exclusion type filtration has several benefits, such as flexibility and ease of use since it provides virus removal predictability through its physical properties, allows for the filtration of viral markers, enabling easy validation of the filtration process, and does not use toxic or mutagenic chemicals for viral inactivation [15,93,94].

Gustafsson et al. (2018) [95] evaluated membrane filter made from nanocellulose in a mille-feuille arrangement of varying thicknesses using a simulated wastewater matrix to explore its ability to remove viruses for drinking water purification applications. The filtrations of various samples of simulated wastewater with its total suspended solid content being 30 nm latex particles as surrogate waste material and 28 nm ΦX174 bacteriophages as the viral contamination. The authors examined the performance of these membrane filters at a pressure of 1 and 3 bar with varying thicknesses of 9 and 29 μm. The data they obtained demonstrate that a membrane filter made from 100% nanocellulose has the capacity to efficiently remove even the smallest of viruses, with up to 99.9980–99.9995%.

Manukyan et al. (2019) [96] fabricated nanocellulose-based mille-feuille type membrane filter for use in upstream applications for serum-free growth media filtration and it was designed to remove ΦX174 bacteriophages. The filter performance was evaluated based on its ability to filter small–medium-sized viruses using varying thicknesses of the fabricated membrane filter (i.e., 11 and 33 μm), as well as by varying the operating pressures (i.e., 1 and 3 bar). Based on their results, the 33 μm thick filter showed more stability and had better virus removal as compared to the 11 μm thick filter, although their flux was nominally lower. The findings of this study suggest that the nanocellulose membrane filter would be a viable alternative for the filtration of large volumes of cell culture media in upstream bioprocessing.

Asper et al. (2015) [97], in their study, used a membrane filter composed of 100% CNF to remove xenotropic murine leukaemia virus (xMuLV). It was found that the particle retention properties of the nanocellulose membrane filter were verified following the filtration of 100 nm latex beads, as shown in Figure 6. The results of this filtration of xMuLV suggested that the nanocellulose membrane filter was useful for removal of endogenous rodent retroviruses and retrovirus-like particles during the production of recombinant proteins.

Metreveli et al. (2014) [15] reported that one of the most challenging tasks for designing the virus removal membrane filter is tailoring the membrane upper pore size cut-off so that the filter retains viruses with a particle size of between 12 and 300 nm, while allowing for the unhindered passage of proteins which typically range between 4 and 12 nm in size. Therefore, high porosity of the nanocellulose-based filters is required to circumvent the problem of low permeance. In their study, the developed nanocellulose membrane filter, sized at 70 μm with a total porosity of 35%, was able to remove Swine Influenza A Virus (SIV), which had a particle size of 80–120 nm. The latex beads and SIV particles are observed as stacked structures on the surface of the porous membrane filter. They also found that the proteins pass unhindered through the membrane filter. Therefore, the pore size distribution presented in their work is promising for virus filtration applications, especially for large viruses $\geq$50 nm.

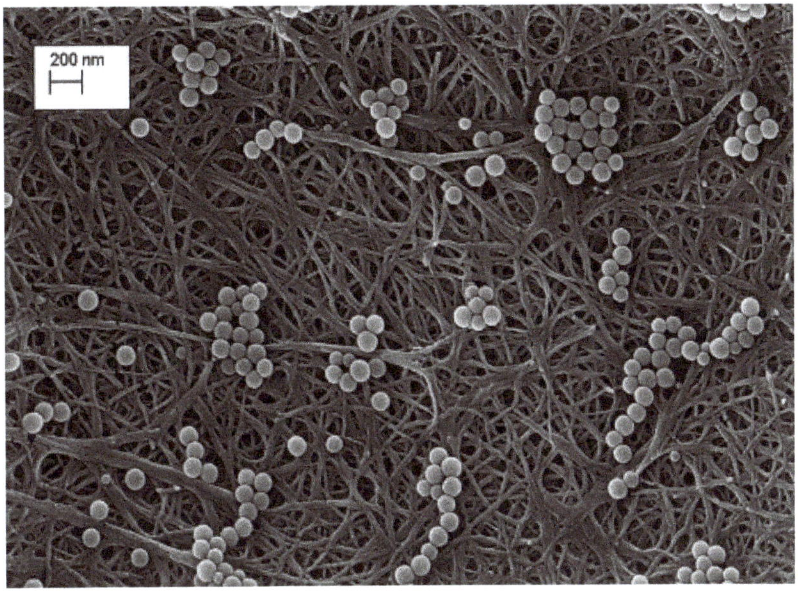

Figure 6. SEM images of 100 nm latex beads retained on the nanocellulose membrane filter. Reproduced with permission from [97].

Besides this, Mautner et al. (2021) [98] also produced BNC membrane filters with high porosity for optimized permeance and rejection of nm-pollutants. The BNC was treated with organic liquids (alcohol, ketone, ether) before being further processed into the membrane filter. The treated BNC membrane filter has a porosity of 67%, which is higher than the untreated BNC membrane filter (33%). It also exhibits 40 times higher permeance, caused by a lower membrane density. Despite their higher porosity, the developed membrane filter also still has pore sizes of 15–20 nm, which is similar to the untreated BNC membrane filter. Thus, the developed membrane filter enables the removal of viruses by a size-exclusion mechanism at high permeance.

The strength of the nanocellulose is also important in designing a good membrane filter to remove viruses via a size exclusion mechanism. A study by Quellmalz and Mihranyan, (2015) [85] found that the citric acid cross-linked nanocellulose-based membrane filter has better mechanical performance than the untreated nanocellulose. It was observed that the untreated nanocellulose membrane filter was readily cracked at pressure gradients above 15 kPa, which could be limiting for its industrial application. The improved strength of the cross-linked nanocellulose membrane filter enables increasing the pressure gradient applied for filtration without compromising the integrity of the filter. It is concluded that citric acid cross-linking of nanocellulose is beneficial to be used in several industrial applications for removing viruses.

Previous studies on the surface modification of nanocellulose have led to the improvement of filtration efficiency against viruses. Electrostatic interaction between nanocellulose and viruses is improved dramatically with the incorporation of quaternary compounds as discussed in Section 4, above. For instance, viruses such as coronavirus have a negatively charged surface and would interact with the cationic or anionic charge of nanocellulose-quaternary compounds [99]. Figure 7 shows a schematic diagram of the coronavirus structure with proteins embedded in its bilayer membrane and negatively charged lipid head groups protruding to the outer side of the membrane.

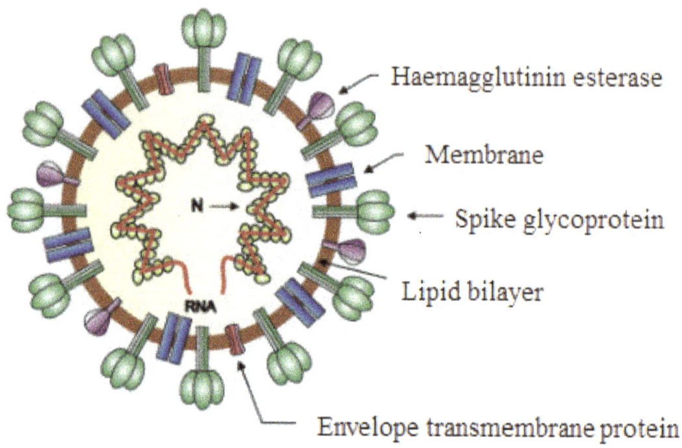

Figure 7. A structure design of coronavirus particle. Reproduced with permission from [100].

The entrapment of the virus onto nanocellulose matrix is due to the presence of electrostatic force attraction between the negatively charged virus particle and the positively charged nanocellulose membrane. Several studies have demonstrates successful results in filtering negatively charged viruses using cationic nanocellulose. For example, Mi et al. (2020) [101] developed a filtration setup using modified CNC with a positively charged guanidine group to adsorb porcine parvovirus and Sindbis virus and to completely filter out those viruses from water. It is interesting to point out that this filtration system has exceeded the Environment Protection Agency (EPA) virus removal standard requirement for portable water. In addition to the electrostatic interaction between the virus and guanidine group, Meingast and Heldt (2020) [102] outlined that the complete virus removal from water was also due to the protonated guanidine groups on the cationic CNC forming ionic and hydrogen bonds with the proteins and lipids on the virus surface.

Other than that, Rosilo et al. (2014) [103] in their study observed a very high affinity binding between the cationic CNC (known as CNC-g-P(QDMAEMA)s) mixture and cowpea chlorotic mottle virus (CCMV) and norovirus-like particles in water dispersions. Of note, this cationic CNC mixture was prepared by surface-initiated atom-transfer radical polymerization of poly(N,N-dimethylamino ethyl methacrylate) and its subsequent quaternization of the polymer pendant amino groups.

In addition, the anionic CCMVs could also be removed using functionalized lignin with a quaternary amine. In their study, they found that the CCMVs would form agglomerated complexes with cationic lignin [104]. Therefore, suggesting its potential use as material in the development of membrane filter for the removal of CCMVs.

Besides that, Sun et al. (2020) [105] reported that covalent modification on CNF (i.e., functionalization of nanocellulose) using polyglutamic acid (PGS) and mesoporous silica nanoparticles (MSNs) resulted in the successful filtration of EV71 virus and Sindbis virus. This is particularly due to the interaction between two exposed positively charged amino acids (His10 and Lys14) and the negatively charged MSNs on the modified CNF [105].

Table 7 summarizes the development of nanocellulose-based membrane filtration material for virus removal that have been discovered/explored so far. In addition to the guanidine groups, lignin, nanoparticles, and citric acid, nanocellulose could also be functionalized with several other compounds, such as small organic molecules, porphyrin dendrimers and others polymers in order to make it positively or negatively charged [73]. However, it is important to note that not all of these examples have been tested as a filter to remove viruses. It can be summarized that several present studies have shown the capability of nanocellulose as a filtration material for virus removal. Separation by size exclusion and adsorption mechanism are the most common approaches. Factors such as pore size distribution, porosity, thickness, strength and surface functionalization of nanocellulose can greatly influence the filtration efficiency.

Table 7. Nanocellulose developed filtration material for virus removal.

Microbes	Type of Nanocellulose	Functionalization	Findings	Reference
A/swine/Sweden/9706/2010 (H1N2)—Swine influenza	BNC	Not applicable	• The newly developed non-woven, μm thick membrane filter consisting of crystalline BNC able to remove virus particles solely based on the size-exclusion principle, with a log 10 reduction value ≥6.3, thereby matching the performance of industrial synthetic polymer virus removal filters currently in use.	[15]
Xenotropic murine	BNC	Not applicable	• The developed BNC membrane filter could remove the endogenous rodent retroviruses and retrovirus-like particles.	[97]
MS2 viruses	BNC	Not applicable	• This study highlights the efficiency of the nanocellulose-based membrane filter in removing/filtering out the ΦX174 bacteriophage with value of 5−6 log virus clearance (28 nm; pI 6.6).	[53]
ColiphagesΦX174	BNC	Not applicable	• The nanocellulose-based membrane filter exhibited 5−6 log virus clearance of MS2 viruses (27 nm; pI 3.9). This study also showed the possibility of producing cost-efficient viral removal filters (i.e., manufacturing process).	[53]
Parvoviruses	BNC	Not applicable	• The developed filter was the first non-woven, wet-laid membrane filter composed of 100% native cellulose. This study showed that the non-enveloped parvoviruses could be eliminated using this filter.	[106]

Table 7. *Cont.*

Microbes	Type of Nanocellulose	Functionalization	Findings	Reference
EV71	CNF	Polyglutamic acid and mesoporous silica nanoparticles	• This study showed that the modified microfibers could strongly adsorb the epitope of the EV71 capsid which is useful for virus removal.	[105]
Sindbis virus	CNC	Guanidine	• Functionalization of guanidine on CNC resulted in over 4 log removal value against the Sindbis virus.	[101]
Porcine parvo virus	CNC	Guanidine	• Authors also revealed that functionalization of guanidine on CNC managed to remove the Porcine parvo virus with over 4 log removal value.	[101]

5.2. Bacteria

The development of nanocellulose as a filtration material against bacteria also been widely discovered. Generally, the diameter of most waterborne bacteria is larger than 0.2 µm [38]. Thereby, it would be easy for nanocellulose-based membrane filters to entrap most bacteria species using the size-exclusion mechanism. Moreover, as discussed in Section 4 earlier, the modification of nanocellulose by surface functionalization can also be performed to increase the removal efficiency of bacteria. In this review, we highlight several findings concerning bacterial removal using nanocellulose based membrane filters.

Wang et al. (2013) [107] demonstrated that a multi-layered nanofibrous microfiltration system with high flux, low-pressure drops and high retention capability against bacteria (*Brevundimonas diminuta* and *Escherichia coli*) was possible by impregnating ultrafine CNF into an electrospun polyacrylonitrile (PAN) nanofibrous scaffold supported by a poly (ethylene terephthalate) (PET) non-woven substrate. The CNF was functionalized prior to impregnation with carboxylate and aldehyde groups using TEMPO oxidation. It was observed that this CNF-based microfiltration membrane exhibited full retention capability against those bacteria.

Otoni et al. (2019) [108] developed a cationic CNF compound using Girard's reagent T (GRT) and shaped it into foam using several protocols, such as cryo-templating to remove the ubiquitous human pathogen *Escherichia coli*. The porosity of this foam, which is associated directly with its surface area and pore size plays a significant role in the removal of *Escherichia coli*. The cryogel foams produced by this method had porosities of *circa* 98% and were established to be able to achieve an approximately 85% higher anti *Escherichia coli* activity when compared to sample foams made up of unmodified CNF. The cationic CNF using GRT demonstrated good potential for both air and liquid filtration, with excellent absorbency through functional coating. Access to safe drinking water in high- and low-income countries has become one of the biggest challenges in the world as natural resources become scarcer.

Gouda et al. (2014) [109] invented a modified electrospun CNF containing silver nanoparticles (AgNPs) as a water disinfecting system for water purification systems. The AgNP content, physical characterization, surface morphology and antimicrobial efficacy of the developed membrane filter were then studied. AgNP, which belongs to the group of biocidal nanoparticles, has antimicrobial properties and is commonly used due to its size quantization effect. This can cause a shift in the reactivity of metals in the nanoscale. The developed membrane filter had an excellent ability to remove bacteria, including *Escherichia coli*, *Salmonella typhi*, and *Staphylococcus aureus* with a percentage filtration of more than 91% in contaminated water samples.

Ottenhall et al. (2018) [110] developed a CNF-based membrane filter, modified with polyelectrolyte multilayers to produce multilayer cationic polyvinyl amine (PVAm) and anionic polyacrylic acid (PAA). The authors successfully modified the CNF with cationic polyelectrolyte PVAm, together with the anionic polyelectrolyte PAA in either single layers or multilayers (3 or 5 layers) using a water-based process at room temperature. Based on filtration analysis, the functionalized CNF-based membrane filters with several layers were physically able to remove more than 99.9% of *Escherichia coli* from water. The three-layer membrane filter could remove more than 97% of cultivatable bacteria from natural water samples, which was a remarkable performance, as compared with the simple processing technique using plain nanocellulose filters.

Table 8 summarizes the effectiveness of nanocellulose-based membrane filters that have been functionalized with bioactive compounds for the removal of bacteria. It can be concluded that bacterial separation by size exclusion mechanism is easier as compared to the virus. This is because the size of bacteria is usually larger as compared to a virus. The surface functionalization on nanocellulose is capable of introducing anti-bacterial properties to the developed filtration material. However, limited studies were reported for the removal of other bacteria species using nanocellulose-based filtration material.

Table 8. Nanocellulose developed filtration material for bacterial removal.

Microbes	Type of Nanocellulose	Functionalization	Findings	Reference
Escherichia coli	CNC	Silver nanoparticles	• It possesses high adsorption capacity and is reusable. Beneficial in total removal of *Escherichia coli*.	[40]
Bacillus subtilis and *Escherichia coli*	CNF	ZnO and CeO_2	• It has high anti-bacterial activity, MIC_{50} against Bacillus *subtilis* (10.6 μg mL^{-1}) and *Escherichia coli* (10.3 μg mL^{-1}).	[111]
Escherichia coli	BNC	Not applicable	• The significance of Brownian motion caused by microorganisms captured with BNC-based membrane filter through theoretical modelling and filtration experiments was investigated by the authors. • It was found that the BNC-based filter was capable of filtering the bacteria.	[112]
Escherichia coli, *Staphylococcus aureus*	CNF	Activated carbon	• The two-layer AC/OCNF/CNF membrane able to remove *Escherichia coli* bacteria up to ~96–99% and inhibits the growth of *Escherichia coli* and *Staphylococcus aureus* on the upper CNF surface	[41]

Table 8. *Cont.*

Microbes	Type of Nanocellulose	Functionalization	Findings	Reference
Escherichia coli	BNC	Silver nanoparticle	• Higher amount of silver nanoparticles loaded onto the BNC membrane surface could increase the inhibition zone hence highlighting its good antimicrobial property against *Escherichia coli.*	[113]
Escherichia coli, Staphylococcus aureus, Pseudomonas aeruginosa	BNC	Silver nanoparticle	• BNC-silver nanoparticle membrane showed strong antimicrobial activity against Gram positive (*Staphylococcus aureus*) and Gram-negative (*Escherichia coli* and *Pseudomonas aeruginosa*) bacteria.	[114]
Escherichia coli, Staphylococcus aureus	BNC	Silver nanoparticle	• The developed Ag/BNC membrane exhibited good property as antimicrobial agent against *Escherichia coli* and *Staphylococcus aureus* as the antibacterial ratio against *Escherichia coli* and *Staphylococcus aureus* reached 99.4% and 98.4%, respectively.	[115]
Escherichia coli	CNF	Polyethersulfone (PES) membranes	• TEMPO oxidized-CNF coating is effective against *Escherichia coli.* The effectiveness was attributed to the pH reduction effect induced by carboxyl groups	[116]

5.3. Other Types of Microbes

Nanocellulose would also be able to act as a removal agent for other types of microbes which are larger in size than bacteria, such as fungi, algae and protozoa. However, it is noteworthy that there is still a lack of studies regarding this matter. To the best of our knowledge, there are no available reports on the development of a nanocellulose-based membrane filter for the removal of fungi.

Algae is also a major contributor to microbial contamination in water resources and their presence could change the taste or odour of water. For example, blue–green algae and coloured flagellates (especially the *Chrysophyta* and *Euglenophyta* genera of algae) are the best-known algae that cause contamination in water resources. Furthermore, green algae may also be a significant contamination factor as well [117]. Hence, the potential of nanocellulose should be explored by scientists to define their role as a membrane filtration material suitable for removing algae and protozoa from the contaminated water efficiently. Algae and protozoa are known to have a larger size than viruses and bacteria, thus the removal of these microbes could be effectively carried out using the size-exclusion mechanism.

However, similar to viruses and bacteria, the nanocellulose needs to be modified with other compounds such as metal nanoparticles, enzymes and proteins in order to increase its filtration efficiency [118]. Studies have shown that different charges between the cellular membrane of algae and protozoa do play a dominant role in the adsorption/retention of these microbes on a filtration membrane's surface (i.e., through the electrostatic interaction principle) [110,119].

A previous study carried out by Ge et al. (2016) [120] discovered the sustainability and the most efficient approach in harvesting algae using a modified CNC. The modification was made by introducing a 1-(3-aminopropyl)-imidazole (APIm) structure as a reversible coagulant. As shown in Figure 8, the coagulation process occurs when the positively charged CNC–APIm interacts with the negatively charged *Chlorella vulgaris* in the presence of carbon dioxide (carbonated water). Their findings are in agreement with the works of Qiu et al. (2019) [121], in which it was found that harvesting efficiency could reach up to 85% with only 0.2 g CNC–APIm mass ratio, 5 s of CO_2 sparging time, and a 50 mL/min flow rate. This signifies that the CNC–APIm complex could be an alternative to current conventional coagulants for harvesting algae in industrial applications.

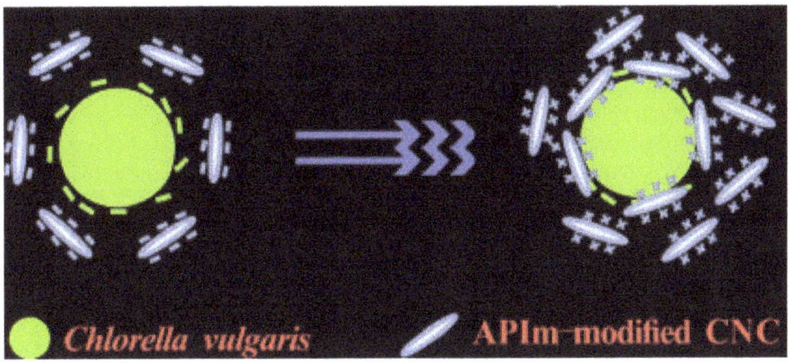

Figure 8. Illustration on the electrostatic attraction between *Chlorella vulgaris* and APIm-modified CNC. This figure was adapted from [120].

Algae harvesting is important for biodiesel industry and many studies have been carried out to increase its sustainability on a global scale. For example, the capability of CNF and CNC in harvesting algae was investigated by Yu et al. (2016) [122]. In their study, they discovered that the CNF did not require any surface modification to harvest the algae, as it played a role as an algae flocculant via its network geometry, something that the CNC (even cationic modified CNC) could not do. By referring to Figure 9, flocculation of algae did not happen when CNC was used, as the freely moving algae cannot be entrapped by the nanoparticle structure formation of CNC. However, this study only focuses on the flocculation capability of CNF and CNC, which could lead to a further study on the filtration efficiency of both materials for algae harvesting. This finding could indirectly point to the development of a nanocellulose-based membrane filter for algae removal in the future.

Overall, nanocellulose has shown its capability to filter algae. The functionalization is also important to improve filtration efficiency. However, the development of nanocellulose as a filtration material of fungus and protozoa is still limited. It is important to further investigate the capability of nanocellulose to filter these microbes. Moreover, several other factors which could influence the filtration efficiency, as discussed before, can also be considered for future studies.

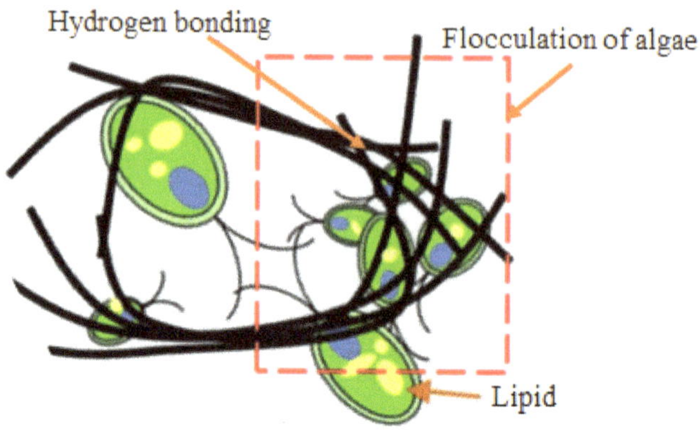

Figure 9. Schematic design of CNF induced microalgal flocculation. This figure was adapted from [122].

6. Challenges and Future Recommendations

This review has shown several undoubtedly interesting features of nanocellulose which is useful in filtering viruses, bacteria, fungi, algae, and protozoa through the mechanism of membrane filtration. The properties and characteristics of nanocellulose as a filtration material is very promising and is an exciting area for current and future research. Several recent developments in the application of nanocellulose as a membrane filtration material were discussed here. It is interesting to note that the functionalization of nanocellulose with a variety of functional groups is among the important key factors of success to enhance the removal of microbes from air and water.

Even though different works on the nanocellulose as a membrane filter material has shown several effective findings, improvements in this area are still needed. There are several other types of microbial species which have not been explored. The use of nanocellulose-based membrane filters as a means of eradicating the COVID-19 virus has also not been explored. Moreover, research on the use of nanocellulose-based membrane filter materials to remove fungus, protozoa and algae is still very limited. Therefore, more works concerning those microbes remains an urgent need.

The functionalization of nanocellulose is a very important step to obtaining the improved performance of membrane filtration material. In this review, several compounds have been shown to be capable of being incorporated with nanocellulose. However, their side effects towards the environment as a result of this functionalization of nanocellulose is also an important consideration. For example, the functionalization of nanocellulose with TEMPO can be harmful to the environment. This is because the synthesis of TEMPO can generate chemical by-products which are toxic to aquatic life when released as waste effluent into the environment [123]. In the future, research should also be focused on this concern to determine the actual feasibility and sustainability of these developed functionalized nanocellulose products towards the environment.

In addition, further research on generating hybrid nanostructures on the surface of nanocellulose to enable interaction with multiple species of microbes is urgently needed, which will pave the way towards the development of new materials capable of eliminating various kind of microbes at once.

This review has identified several difficulties concerning the use of nanocellulose as a membrane filter material. One of them is particularly related to the high production cost especially in large industrial scale. Furthermore, high energy consumption during the production of nanocellulose is still a concern and cries for a more reliable and reproducible production technique to pave way to using nanocellulose as a commercially viable membrane filter material. To the best of our knowledge, there is much progress in research studies focusing on reducing the energy consumption and production cost of nanocellulose and have been attempted in numerous pilot-scale productions worldwide. Other than that, issue regarding the biodegradability of nanocellulose as adsorbent need to be evaluated by considering several factors such as the types of water and presence of certain microbes that may cause cellulose degradation.

All in all, nanocellulose is a good alternative for a membrane filter material and is expected to be fully utilized in numerous industries in the near future, considering the solutions for the outlined challenges and difficulties that have been met.

Author Contributions: Conceptualization, M.N.F.N., N.A.M.K., and V.F.K.; validation, K.K.O., and W.M.Z.W.Y.; writing—original draft preparation, M.N.F.N., N.A.M.K., V.F.K., N.J., and M.S.M.M.; supervision, S.A.M.N., N.A.H., N.A.A.S., and S.H.J.; project administration, M.Z.A. and M.H.Y.; funding acquisition, N.A.M.K. All authors have read and agreed to the published version of the manuscript.

Funding: This research was funded by Ministry of Higher Education, Malaysia through CHEMDEF research grant (UPNM/2018/CHEMDEF/ST/4).

Institutional Review Board Statement: Not applicable.

Informed Consent Statement: Not applicable.

Data Availability Statement: Not applicable.

Acknowledgments: The authors would like to highly acknowledge financial support (UPNM/2018/CHEMDEF/ST/4) from the Ministry of Higher Education Malaysia. The authors also gratefully acknowledge the support of Universiti Pertahanan Nasional Malaysia (National Defence University of Malaysia) in the preparation of this review.

Conflicts of Interest: The authors declare no conflict of interest in the preparation of this review.

References

1. Norrrahim, M.N.F.; Norizan, M.N.; Jenol, M.A.; Farid, M.A.A.; Janudin, N.; Ujang, F.A.; Yasim-Anuar, T.A.T.; Najmuddin, S.U.F.S.; Ilyas, R.A. Emerging Development on Nanocellulose as Antimicrobial Material: An Overview. *Mater. Adv.* **2021**, *2*, 3538–3551. [CrossRef]
2. Atkinson, S.K.; Sadofsky, L.R.; Morice, A.H. How does rhinovirus cause the common cold cough? *BMJ Open Respir. Res.* **2016**, *3*, 1–11. [CrossRef] [PubMed]
3. Lamont, R.F.; Sobel, J.D.; Carrington, D.; Mazaki-Tovi, S.; Kusanovic, J.P.; Vaisbuch, E.; Romero, R. Varicella-zoster virus (chickenpox) infection in pregnancy. *BJOG Int. J. Obstet. Gynaecol.* **2011**, *118*, 1155–1162. [CrossRef] [PubMed]
4. Leung, A.K.; Hon, K.L.; Leong, K.F. Rubella (German measles) revisited. *Hong Kong Med. J.* **2019**, *25*, 134–141. [CrossRef]
5. Hegerle, N.; Guiso, N. Epidemiology of whooping cough & typing of Bordetella pertussis. *Future Microbiol.* **2013**, *8*, 1391–1403. [PubMed]
6. Spyrou, M.A.; Tukhbatova, R.I.; Wang, C.C.; Valtueña, A.A.; Lankapalli, A.K.; Kondrashin, V.V.; Tsybin, V.A.; Khokhlov, A.; Kühnert, D.; Herbig, A.; et al. Analysis of 3800-year-old Yersinia pestis genomes suggests Bronze Age origin for bubonic plague. *Nat. Commun.* **2018**, *9*, 1–10. [CrossRef]
7. Gagneux, S. Ecology and evolution of Mycobacterium tuberculosis. *Nat. Rev. Microbiol.* **2018**, *16*, 202–213. [CrossRef]
8. Bhatt, S.; Weiss, D.J.; Cameron, E.; Bisanzio, D.; Mappin, B.; Dalrymple, U.; Battle, K.E.; Moyes, C.L.; Henry, A.; Eckhoff, P.A.; et al. The effect of malaria control on Plasmodium falciparum in Africa between 2000 and 2015. *Nature* **2015**, *526*, 207–211. [CrossRef] [PubMed]
9. Furlan, K.C.; Kakizaki, P.; Chartuni, J.C.N.; Valente, N.Y.S. Sycosiform tinea barbae caused by trichophyton rubrum and its association with autoinoculation. *Bras. Dermatol.* **2017**, *92*, 160–161. [CrossRef] [PubMed]
10. Makola, N.F.; Magongwa, N.M.; Matsaung, B.; Schellack, G.; Schellack, N. Managing athlete's foot. *S. Afr. Fam. Pract.* **2018**, *60*, 37–41. [CrossRef]
11. Fahimirad, S.; Fahimirad, Z.; Sillanpää, M. Efficient removal of water bacteria and viruses using electrospun nanofibers. *Sci. Total Environ.* **2020**, *751*, 141673. [CrossRef]
12. Sussman, E.M. Escherichia coli: Mechanisms of Virulence. *J. R. Soc. Med.* **1997**, *90*, 466.
13. World Health Organization Drinking-Water. Available online: https://www.who.int/news-room/fact-sheets/detail/drinking-water (accessed on 20 September 2021).
14. World Health Organization. Modes of Transmission of Virus Causing COVID-19: Implications for IPC Precaution Recommendations. Available online: https://www.who.int/news-room/commentaries/detail/modes-of-transmission-of-virus-causing-covid-19-implications-for-ipc-precaution-recommendations (accessed on 20 September 2021).
15. Metreveli, G.; Wågberg, L.; Emmoth, E.; Belák, S.; Strømme, M.; Mihranyan, A. A Size-Exclusion Nanocellulose Filter Paper for Virus Removal. *Adv. Healthc. Mater.* **2014**, *3*, 1546–1550. [CrossRef]

16. Hai, F.I.; Riley, T.; Shawkat, S.; Magram, S.F.; Yamamoto, K. Removal of pathogens by membrane bioreactors: A review of the mechanisms, influencing factors and reduction in chemical disinfectant dosing. *Water* **2014**, *6*, 3603–3630. [CrossRef]
17. Das, O.; Neisiany, R.E.; Capezza, A.J.; Hedenqvist, M.S.; Försth, M.; Xu, Q.; Jiang, L.; Ji, D.; Ramakrishna, S. The need for fully bio-based facemasks to counter coronavirus outbreaks: A perspective. *Sci. Total Environ.* **2020**, *736*, 1–7. [CrossRef]
18. Ilyas, R.; Sapuan, S.; Norrrahim, M.N.F.; Yasim-Anuar, T.A.T.; Kadier, A.; Kalil, M.S.; Atikah, M.; Ibrahim, R.; Asrofi, M.; Abral, H.; et al. Nanocellulose/starch biopolymer nanocomposites: Processing, manufacturing, and applications. In *Advanced Processing, Properties, and Applications of Starch and Other Bio-Based Polymers*; Elsevier: Amsterdam, The Netherlands, 2020; pp. 65–88.
19. Norrrahim, M.N.F.; Huzaifah, M.R.M.; Farid, M.A.A.; Shazleen, S.S.; Misenan, M.S.M.; Yasim-Anuar, T.A.T.; Naveen, J.; Nurazzi, N.M.; Rani, M.S.A.; Hakimi, M.I.; et al. Greener Pretreatment Approaches for the Valorisation of Natural Fibre Biomass into Bioproducts. *Polymers* **2021**, *13*, 2971. [CrossRef] [PubMed]
20. Norrrahim, M.N.F.; Ilyas, R.A.; Nurazzi, N.M.; Rani, M.S.A.; Atikah, M.S.N.; Shazleen, S.S. Chemical Pretreatment of Lignocellulosic Biomass for the Production of Bioproducts: An Overview. *Appl. Sci. Eng. Prog.* **2021**, *13*, 2971.
21. Lee, C.H.; Lee, S.H.; Padzil, F.N.M.; Ainun, Z.M.A.; Norrrahim, M.N.F.; Chin, K.L. Biocomposites and Nanocomposites. In *Composite Materials*; CRC Press: Boca Raton, FL, USA, 2021; pp. 29–60.
22. Yasim-Anuar, T.A.T.; Ariffin, H.; Norrrahim, M.N.F.; Hassan, M.A.; Andou, Y.; Tsukegi, T.; Nishida, H. Well-Dispersed Cellulose Nanofiber in Low Density Polyethylene Nanocomposite by Liquid-Assisted Extrusion. *Polymers* **2020**, *12*, 927. [CrossRef]
23. Norrrahim, M.N.F.; Ariffin, H.; Yasim-Anuar, T.A.T.; Hassan, M.A.; Ibrahim, N.A.; Yunus, W.M.Z.W.; Nishida, H. Performance Evaluation of Cellulose Nanofiber with Residual Hemicellulose as a Nanofiller in Polypropylene-Based Nanocomposite. *Polymers* **2021**, *13*, 1064. [CrossRef] [PubMed]
24. Misenan, M.S.M.; Ali, E.S.; Khiar, A.S.A. Conductivity, dielectric and modulus study of chitosan-methyl cellulose-BMIMTFSI polymer electrolyte doped with cellulose nano crystal. In *AIP Conference Proceedings*; AIP Publishing LLC: Melville, NY, USA, 2018; Volume 1972. [CrossRef]
25. Norrrahim, M.N.F.; Yasim-Anuar, T.A.T.; Jenol, M.A.; Mohd Nurazzi, N.; Ilyas, R.A.; Sapuan, S. Performance Evaluation of Cellulose Nanofiber Reinforced Polypropylene Biocomposites for Automotive Applications. In *Biocomposite and Synthetic Composites for Automotive Applications*; Woodhead Publishing Series: Amsterdam, The Netherlands, 2020; pp. 119–215.
26. Sharip, N.S.; Yasim-Anuar, T.A.T.; Norrrahim, M.N.F.; Shazleen, S.S.; Nurazzi, N.M.; Sapuan, S.M.; Ilyas, R.A. A review on nanocellulose composites in biomedical application. In *Composites in Biomedical Applications*; CRC Press: Boca Raton, FL, USA, 2020; pp. 161–190.
27. Ilyas, R.A.; Sapuan, S.M.; Harussani, M.M.; Hakimi, M.Y.A.Y.; Haziq, M.Z.M.; Atikah, M.S.N.; Asyraf, M.R.M.; Ishak, M.R.; Razman, M.R.; Nurazzi, N.M.; et al. Polylactic Acid (PLA) Biocomposite: Processing, Additive Manufacturing and Advanced Applications. *Polymers* **2021**, *13*, 1326. [CrossRef]
28. Ilyas, R.; Sapuan, S.; Atikah, M.; Asyraf, M.; Rafiqah, S.A.; Aisyah, H.; Nurazzi, N.M.; Norrrahim, M. Effect of hydrolysis time on the morphological, physical, chemical, and thermal behavior of sugar palm nanocrystalline cellulose (*Arenga pinnata* (*Wurmb.*) *Merr*). *Text. Res. J.* **2021**, *91*, 152–167. [CrossRef]
29. Ilyas, R.A.; Sapuan, S.M.; Mohd Nurazzi, N.; Norrrahim, M.N.F.; Ibrahim, R.; Atikah, M.S.N.; Huzaifah, M.R.M.; Radzi, A.M.; Izwan, S.; Noor Azammi, A.M.; et al. Macro to Nanocscale Natural Fibre Composites for Automotive Components: Research, Development, and Application. In *Biocomposite and Synthetic Composites for Automotive Applications*; Woodhead Publishing Series: Amsterdam, The Netherlands, 2020; pp. 51–105.
30. Norrrahim, M.N.F.; Yasim-Anuar, T.A.T.; Sapuan, S.M.; Ilyas, R.A.; Hakimi, M.I.; Najmuddin, S.U.F.S.; Jenol, M.A. Nanocellulose Reinforced Polypropylene and Polyethylene Composite for Packaging Application. In *Bio-Based Packaging: Material Environmental and Economic Aspects*; Sapuan, S.M., Ilyas, R.A., Eds.; John Wiley & Sons Ltd.: Hoboken, NJ, USA, 2021; pp. 133–150. [CrossRef]
31. Norrrahim, M.N.F.; Kasim, N.A.M.; Knight, V.F.; Misenan, M.S.M.; Janudin, N.; Shah, N.A.A.; Kasim, N.; Yusoff, W.Y.W.; Noor, S.A.M.; Jamal, S.H.; et al. Nanocellulose: A bioadsorbent for chemical contaminant remediation. *RSC Adv.* **2021**, *11*, 7347–7368. [CrossRef]
32. Norrrahim, M.N.F.; Kasim, N.A.M.; Knight, V.F.; Ujang, F.A.; Janudin, N.; Razak, M.A.I.A.; Shah, N.A.A.; Noor, S.A.M.; Jamal, S.H.; Ong, K.K.; et al. Nanocellulose: The Next Super Versatile Material for the Military. *Mater. Adv.* **2021**, *2*, 1485–1506. [CrossRef]
33. Norrrahim, M.N.F.; Kasim, N.A.M.; Knight, V.F.; Halim, N.A.; Shah, N.A.A.; Noor, S.A.M.; Jamal, S.H.; Ong, K.K.; Yunus, W.M.Z.W.; Farid, M.A.A.; et al. Performance Evaluation of Cellulose Nanofiber Reinforced Polymer Composites. *Funct. Compos. Struct.* **2021**, *3*. [CrossRef]
34. Misenan, M.S.M.; Janudin, N.; Idayu, M.A.; Norrrahim, M.N.F.; Jamal, S.H.; Wan Yusoff, W.Y.; Kasim, N.; Yunus, W.M.D.Z.W.; Ernest, V.F.K.V.; Kasim, N.A.M. Cellulose Nanofiber as Potential Absorbent Material for Chloride Ion. *Solid State Phenom.* **2021**, *317*, 263–269. [CrossRef]
35. Ariffin, H.; Tengku Yasim-Anuar, T.A.; Norrrahim, M.N.F.; Hassan, M.A. Synthesis of Cellulose Nanofiber from Oil Palm Biomass by High Pressure Homogenization and Wet Disk Milling. In *Nanocellulose: Synthesis, Structure, Properties and Applications*; World Scientific: Singapore, 2021; pp. 51–64.
36. Jasmania, L. Preparation of nanocellulose and its potential application. *Int. J. Nanomater. Nanotechnol. Nanomed.* **2018**, *4*, 14–21. [CrossRef]
37. Hassan, M.L.; Fadel, S.M.; Abouzeid, R.E.; Elseoud, W.S.A.; Hassan, E.A.; Berglund, L.; Oksman, K. Water purification ultrafiltration membranes using nanofibers from unbleached and bleached rice straw. *Sci. Rep.* **2020**, *10*, 11278. [CrossRef]

38. Ma, H.; Burger, C.; Hsiao, B.S.; Chu, B. Ultra-fine cellulose nanofibers: New nano-scale materials for water purification. *J. Mater. Chem.* **2011**, *21*, 7507–7510. [CrossRef]
39. Thakur, V.K.; Voicu, S.I. Recent Advances in Cellulose and Chitosan Based Membranes for Water Purification: A Concise Review. *Carbohydr. Polym.* **2016**, *146*, 148–165. [CrossRef]
40. Suman; Kardam, A.; Gera, M.; Jain, V.K. A novel reusable nanocomposite for complete removal of dyes, heavy metals and microbial load from water based on nanocellulose and silver nano-embedded pebbles. *Environ. Technol.* **2014**, *36*, 706–714. [CrossRef] [PubMed]
41. Hassan, M.; Abou-Zeid, R.; Hassan, E.; Berglund, L.; Aitomäki, Y.; Oksman, K. Membranes Based on Cellulose Nanofibers and Activated Carbon for Removal of Escherichia coli Bacteria from Water. *Polymers* **2017**, *9*, 335. [CrossRef]
42. Norrrahim, M.N.F.; Kasim, N.A.M.; Knight, V.F.; Ong, K.K.; Noor, S.A.M.; Jamal, S.H.; Shah, N.A.A.; Halim, N.A.; Ilyas, R.A.; Yunus, W.M.Z.W. Cationic Nanocellulose as Promising Candidate for Filtration Material of COVID-19: A Perspective. *Appl. Sci. Eng. Prog.* **2021**. [CrossRef]
43. Du, N.; Park, H.B.; Dal-Cin, M.M.; Guiver, M.D. Advances in high permeability polymeric membrane materials for CO_2 separations. *Energy Environ. Sci.* **2012**, *5*, 7306–7322. [CrossRef]
44. Xu, X.; Liu, F.; Jiang, L.; Zhu, J.Y.; Haagenson, D.; Wiesenborn, D.P. Cellulose nanocrystals vs. Cellulose nanofibrils: A comparative study on their microstructures and effects as polymer reinforcing agents. *ACS Appl. Mater. Interfaces* **2013**, *5*, 2999–3009. [CrossRef]
45. Henriksson, M.; Berglund, L.A.; Isaksson, P.; Lindström, T.; Nishino, T. Cellulose nanopaper structures of high toughness. *Biomacromolecules* **2008**, *9*, 1579–1585. [CrossRef]
46. Syverud, K.; Stenius, P. Strength and barrier properties of MFC films. *Cellulose* **2009**, *16*, 75–85. [CrossRef]
47. Klemm, D.; Kramer, F.; Moritz, S.; Lindström, T.; Ankerfors, M.; Gray, D.; Dorris, A. Nanocelluloses: A new family of nature-based materials. *Angew. Chem.-Int. Ed.* **2011**, *50*, 5438–5466. [CrossRef] [PubMed]
48. Nogi, M.; Iwamoto, S.; Nakagaito, A.N.; Yano, H. Optically Transparent Nanofiber Paper. *Adv. Mater.* **2009**, *21*, 1595–1598. [CrossRef]
49. Norrrahim, M.N.F.; Ariffin, H.; Yasim-Anuar, T.A.T.; Ghaemi, F.; Hassan, M.A.; Ibrahim, N.A.; Ngee, J.L.H.; Yunus, W.M.Z.W. Superheated steam pretreatment of cellulose affects its electrospinnability for microfibrillated cellulose production. *Cellulose* **2018**, *25*, 3853–3859. [CrossRef]
50. Norrrahim, M.N.F. Superheated Steam Pretreatment of Oil Palm Biomass for Improving Nanofibrillation of Cellulose and Performance of Polypropylene/Cellulose Nanofiber Composites. Ph.D. Thesis, Universiti Putra Malaysia, Selangor, Malaysia, 2018.
51. Zhang, T.; Zhang, Y.; Jiang, H.; Wang, X. Adjusting the pore size of nano-cellulose aerogel by heat treatment in the gel stage. *Mater. Res. Express* **2019**, *6*, 075027. [CrossRef]
52. Kumar, A.; Negi, Y.S.; Choudhary, V.; Bhardwaj, N.K. Characterization of cellulose nanocrystals produced by acid-hydrolysis from sugarcane bagasse as agro-waste. *J. Mater. Phys. Chem.* **2014**, *2*, 1–8. [CrossRef]
53. Wu, L.; Manukyan, L.; Mantas, A.; Mihranyan, A. Nanocellulose-Based Nanoporous Filter Paper for Virus Removal Filtration of Human Intravenous Immunoglobulin. *ACS Appl. Nano Mater.* **2019**, *2*, 6352–6359. [CrossRef]
54. Ottoson, J.; Hansen, A.; Westrell, T.; Johansen, K.; Norder, H.; Stenström, T.A. Removal of noro-and enteroviruses, giardia cysts, cryptosporidium oocysts, and fecal indicators at four secondary wastewater treatment plants in Sweden. *Water Environ. Res.* **2006**, *78*, 828–834. [CrossRef] [PubMed]
55. Francy, D.S.; Stelzer, E.A.; Bushon, R.N.; Brady, A.M.; Williston, A.G.; Riddell, K.R.; Borchardt, M.A.; Spencer, S.K.; Gellner, T.M. Comparative effectiveness of membrane bioreactors, conventional secondary treatment, and chlorine and UV disinfection to remove microorganisms from municipal wastewaters. *Water Res.* **2012**, *46*, 4164–4178. [CrossRef]
56. Held, D.; Kilz, P. Size-exclusion chromatography as a useful tool for the assessment of polymer quality and determination of macromolecular properties. *Chem. Teach. Int.* **2021**, *3*, 77–103. [CrossRef]
57. Fornasiero, F.; Park, H.G.; Holt, J.K.; Stadermann, M.; Grigoropoulos, C.P.; Noy, A.; Bakajin, O. Mechanism of ion exclusion by sub-2nm carbon nanotube membranes. *Mater. Res. Soc. Symp. Proc.* **2008**, *1106*, 8–14. [CrossRef]
58. Harinarayan, C.J.; Mueller, A.; Ljunglöf, R.; Fahrner, J.; Van Alstine, R. An Exclusion Mechanism in Ion Exchange Chromatography. *Biotechnol. Bioeng.* **2006**, *95*, 775–787. [CrossRef] [PubMed]
59. Lombardo, S.; Thielemans, W. *Thermodynamics of Adsorption on Nanocellulose Surfaces*; Springer: Dordrecht, The Netherlands, 2019; Volume 26.
60. Dai, Z.; Ottesen, V.; Deng, J.; Helberg, R.M.L.; Deng, L. A brief review of nanocellulose based hybrid membranes for CO2 separation. *Fibers* **2019**, *7*, 40. [CrossRef]
61. Cheng, Q.; Ye, D.; Chang, C.; Zhang, L. Facile fabrication of superhydrophilic membranes consisted of fibrous tunicate cellulose nanocrystals for highly efficient oil/water separation. *J. Membr. Sci.* **2017**, *525*, 1–8. [CrossRef]
62. Ansaloni, L.; Salas-Gay, J.; Ligi, S.; Baschetti, M.G. Nanocellulose-based membranes for CO2 capture. *J. Membr. Sci.* **2017**, *522*, 216–225. [CrossRef]
63. Dai, Z.; Deng, J.; Yu, Q.; Helberg, R.M.L.; Janakiram, S.; Ansaloni, L.; Deng, L. Fabrication and Evaluation of Bio-Based Nanocomposite TFC Hollow Fiber Membranes for Enhanced CO2 Capture. *ACS Appl. Mater. Interfaces* **2019**, *11*, 10874–10882. [CrossRef] [PubMed]

64. Hutten, I.M. CHAPTER 1-Introduction to Nonwoven Filter Media. In *Handbook of Nonwoven Filter Media*; Hutten, I.M.B.T.-H., Ed.; Butterworth-Heinemann: Oxford, UK, 2007; pp. 1–28. ISBN 978-1-85617-441-1.
65. Good, T.; Hospital, S. Evaluation of the Efficiency of Medical Masks and the Creation of New Medical Masks. *J. Int. Med. Res.* **2007**, *35*, 213–223.
66. Voisin, H.; Bergström, L.; Liu, P.; Mathew, A. Nanocellulose-Based Materials for Water Purification. *Nanomaterials* **2017**, *7*, 57. [CrossRef] [PubMed]
67. Ariffin, H.; Norrrahim, M.N.F.; Yasim-Anuar, T.A.T.; Nishida, H.; Hassan, M.A.; Ibrahim, N.A.; Yunus, W.M.Z.W. Oil Palm Biomass Cellulose-Fabricated Polylactic Acid Composites for Packaging Applications. In *Bionanocomposites for Packaging Applications*; Springer: Berlin/Heidelberg, Germany, 2018; pp. 95–105. ISBN 978-331-967-3-196.
68. Norrrahim, M.N.F.; Ariffin, H.; Yasim-Anuar, T.A.T.; Hassan, M.A.; Nishida, H.; Tsukegi, T. One-pot nanofibrillation of cellulose and nanocomposite production in a twin-screw extruder. *IOP Conf. Ser. Mater. Sci. Eng.* **2018**, *368*, 1–9. [CrossRef]
69. Halib, N.; Perrone, F.; Cemazar, M.; Dapas, B.; Farra, R.; Abrami, M.; Chiarappa, G.; Forte, G.; Zanconati, F.; Pozzato, G.; et al. Potential applications of nanocellulose-containing materials in the biomedical field. *Materials* **2017**, *10*, 977. [CrossRef]
70. Bacakova, L.; Pajorova, J.; Bacakova, M.; Skogberg, A.; Kallio, P.; Kolarova, K.; Svorcik, V. Versatile application of nanocellulose: From industry to skin tissue engineering and wound healing. *Nanomaterials* **2019**, *9*, 164. [CrossRef]
71. Fareez, I.M.; Jasni, A.H.; Norrrahim, M.N.F. Nanofibrillated Cellulose Based Bio-phenolic Composites. In *Phenolic Polymers Based Composite Materials*; Springer: Singapore, 2020; pp. 139–151.
72. Norrrahim, M.N.F.; Ariffin, H.; Hassan, M.A.; Ibrahim, N.A.; Yunus, W.M.Z.W.; Nishida, H. Utilisation of superheated steam in oil palm biomass pretreatment process for reduced chemical use and enhanced cellulose nanofibre production. *Int. J. Nanotechnol.* **2019**, *16*, 668–679. [CrossRef]
73. Sunasee, R.; Hemraz, U.D. Synthetic strategies for the fabrication of cationic surface-modified cellulose nanocrystals. *Fibers* **2018**, *6*, 15. [CrossRef]
74. Lin, N.; Dufresne, A. Nanoscale Preparation, properties and applications of polysaccharide nanocrystals in advanced functional nanomaterials: A review. *Nanoscale* **2012**, *4*, 3274–3294. [CrossRef]
75. Mahfoudhi, N.; Boufi, S. Nanocellulose as a novel nanostructured adsorbent for environmental remediation: A review. *Cellulose* **2017**, *24*, 1171–1197. [CrossRef]
76. Rusli, R.; Eichhorn, S.J. Determination of the stiffness of cellulose nanowhiskers and the fiber- matrix interface in a nanocomposite using Raman spectroscopy Determination of the stiffness of cellulose nanowhiskers and the fiber-matrix interface in a nanocomposite using Raman spe. *Appl. Phys. Lett.* **2008**, *93*, 1–3. [CrossRef]
77. Nguyen, T.; Roddick, F.A.; Fan, L. Biofouling of Water Treatment Membranes: A Review of the Underlying Causes, Monitoring Techniques and Control Measures. *Membranes* **2012**, *2*, 804–840. [CrossRef] [PubMed]
78. Khalil, H.P.S.A.; Davoudpour, Y.; Islam, N.; Mustapha, A.; Sudesh, K.; Dungani, R.; Jawaid, M. Production and modification of nanofibrillated cellulose using various mechanical processes: A review. *Carbohydr. Polym.* **2014**, *99*, 649–665. [CrossRef]
79. Abitbol, T.; Rivkin, A.; Cao, Y.; Nevo, Y.; Abraham, E.; Ben-Shalom, T.; Lapidot, S.; Shoseyov, O. Nanocellulose, a tiny fiber with huge applications. *Curr. Opin. Biotechnol.* **2016**, *39*, 76–88. [CrossRef]
80. Kongruang, S. Bacterial Cellulose Production by *Acetobacter xylinum* Strains from Agricultural Waste Products. *Appl. Biochem. Biotechnol.* **2008**, *148*, 245–256. [CrossRef]
81. Ahmad, I.; Thomas, S.; Dufresne, A.; Huang, J.; Lin, N. Advances in cellulose nanomaterials. *Cellulose* **2018**, *25*, 2151–2189. [CrossRef]
82. Islam, N.; Rahman, F. Production and modification of nanofibrillated cellulose composites and potential applications. In *Green Composites for Automotive Applications*; Elsevier Ltd: Amsterdam, The Netherlands, 2019; pp. 115–141. ISBN 978-008-102-1-774.
83. Abol-fotouh, D.; Hassan, M.A.; Shokry, H.; Roig, A.; Azab, M.S.; Kashyout, A.E.B. Bacterial nanocellulose from agro-industrial wastes: Low-cost and enhanced production by Komagataeibacter saccharivorans. *Sci. Rep.* **2020**, *10*, 3491. [CrossRef] [PubMed]
84. Islam, M.T.; Alam, M.M.; Zoccola, M. Review on modification of nanocellulose for application in composites. *Int. J. Innov. Res. Sci.* **2013**, *2*, 5444–5451.
85. Quellmalz, A.; Mihranyan, A. Citric Acid Cross-Linked Nanocellulose-Based Paper for Size-Exclusion Nanofiltration. *ACS Biomater. Sci. Eng.* **2015**, *1*, 271–276. [CrossRef]
86. Wei, L.; Agarwal, U.P.; Hirth, K.C.; Matuana, L.M.; Sabo, R.C.; Stark, N.M. Chemical modification of nanocellulose with canola oil fatty acid methyl ester. *Carbohydr. Polym.* **2017**, *169*, 108–116. [CrossRef]
87. Radhakrishnan, A.; Nahi, J.; Beena, B. Synthesis and characterization of multi-carboxyl functionalized nanocellulose/graphene oxide-zinc oxide composite for the adsorption of uranium (VI) from aqueous solutions: Kinetic and equilibrium profiles. *Mater. Today Proc.* **2020**, *41*, 557–563. [CrossRef]
88. Li, L.; Yu, C.; Yu, C.; Chen, Q.; Yu, S. Nanocellulose as template to prepare rough-hydroxy rich hollow silicon mesoporous nanospheres (R-nCHMSNs) for drug delivery. *Int. J. Biol. Macromol.* **2021**, *180*, 432–438. [CrossRef] [PubMed]
89. Wu, M.; Zhou, Z.; Yang, J.; Zhang, M.; Cai, F.; Lu, P. ZnO nanoparticles stabilized oregano essential oil Pickering emulsion for functional cellulose nanofibrils packaging films with antimicrobial and antioxidant activity. *Int. J. Biol. Macromol.* **2021**, *190*, 433–440. [CrossRef]

90. Fan, W.; Wang, J.; Zhang, Z.; Li, Z. Synergistic enhancement of UV-resistance and electrical conductivity of waterborne polyurethane composite with combination of functionalized 2D graphene oxide and 1D nanocellulose. *Prog. Org. Coat.* **2021**, *151*, 106017. [CrossRef]
91. Henschen, J. Bio-Based Preparation of Nanocellulose and Functionalization Using Polyelectrolytes. Ph.D. Thesis, KTH Royal Institute of Technology, Stockholm, Sweden, 2019.
92. Lebrun, G.J.L. Cellulose-based virus-retentive filters: A review. *Rev. Environ. Sci. Biotechnol.* **2017**, *16*, 455–489. [CrossRef]
93. Burnouf, T.; Radosevich, M. Nanofiltration of plasma-derived biopharmaceutical products. *Haemophilia* **2003**, *19*, 24–37. [CrossRef] [PubMed]
94. Kalbfuss, B.; Wolff, M.; Morenweiser, R.; Reichl, U. Purification of cell culture-derived human influenza A virus by size-exclusion and anion-exchange chromatography. *Biotechnol. Bioeng.* **2007**, *96*, 932–944. [CrossRef]
95. Gustafsson, O.; Manukyan, L.; Mihranyan, A. High-Performance Virus Removal Filter Paper for Drinking Water Purification. *Glob. Chall.* **2018**, *2*, 31. [CrossRef] [PubMed]
96. Manukyan, L.; Li, P.; Gustafsson, S.; Mihranyan, A. Growth media filtration using nanocellulose-based virus removal filter for upstream biopharmaceutical processing. *J. Membr. Sci.* **2019**, *572*, 464–474. [CrossRef]
97. Asper, M.; Hanrieder, T.; Quellmalz, A.; Mihranyan, A. Removal of xenotropic murine leukemia virus by nanocellulose based filter paper. *Biologicals* **2015**, *43*, 452–456. [CrossRef] [PubMed]
98. Mautner, A.; Bismarck, A. Bacterial nanocellulose papers with high porosity for optimized permeance and rejection of nm-sized pollutants. *Carbohydr. Polym.* **2021**, *251*, 117130. [CrossRef]
99. Leung, W.W.F.; Sun, Q. Electrostatic charged nanofiber filter for filtering airborne novel coronavirus (COVID-19) and nano-aerosols. *Sep. Purif. Technol.* **2020**, *250*, 116886. [CrossRef]
100. Artika, I.M.; Dewantari, A.K.; Wiyatno, A. Molecular biology of coronaviruses: Current knowledge. *Heliyon* **2020**, *6*, e04743. [CrossRef] [PubMed]
101. Mi, X.; Albukhari, S.M.; Heldt, C.L.; Heiden, P.A. Virus and chlorine adsorption onto guanidine modified cellulose nanofibers using covalent and hydrogen bonding. *Carbohydr. Res.* **2020**, *498*, 108153. [CrossRef] [PubMed]
102. Meingast, C.; Heldt, C.L. Arginine-enveloped virus inactivation and potential mechanisms. *Biotechnol. Prog.* **2020**, *30*, 108–112. [CrossRef] [PubMed]
103. Rosilo, H.; McKee, J.R.; Kontturi, E.; Koho, T.; Hytönen, V.P.; Ikkala, O.; Kostiainen, M.A. Cationic polymer brush-modified cellulose nanocrystals for high-affinity virus binding. *Nanoscale* **2014**, *6*, 11871–11881. [CrossRef]
104. Rivière, G.N.; Korpi, A.; Sipponen, M.H.; Zou, T.; Kostiainen, M.A.; Österberg, M. Agglomeration of Viruses by Cationic Lignin Particles for Facilitated Water Purification. *ACS Sustain. Chem. Eng.* **2020**, *8*, 4167–4177. [CrossRef]
105. Sun, M.; Wang, H.; Li, X. Modification of cellulose microfibers by polyglutamic acid and mesoporous silica nanoparticles for Enterovirus 71 adsorption. *Mater. Lett.* **2020**, *277*, 128320. [CrossRef]
106. Gustafsson, S.; Lordat, P.; Hanrieder, T.; Asper, M.; Schaefer, O.; Mihranyan, A. Mille-feuille paper: A novel type of filter architecture for advanced virus separation applications. *Mater. Horiz.* **2016**, *3*, 320–327. [CrossRef]
107. Wang, R.; Guan, S.; Sato, A.; Wang, X.; Wang, Z.; Yang, R.; Hsiao, B.S.; Chu, B. Nanofibrous microfiltration membranes capable of removing bacteria, viruses and heavy metal ions. *J. Membr. Sci.* **2013**, *446*, 376–382. [CrossRef]
108. Otoni, C.G.; Figueiredo, J.S.L.; Capeletti, L.B.; Cardoso, M.B.; Bernardes, J.S.; Loh, W. Tailoring the Antimicrobial Response of Cationic Nanocellulose-Based Foams through Cryo-Templating. *ACS Appl. Bio Mater.* **2019**, *2*, 1975–1986. [CrossRef]
109. Gouda, M.; Hebeish, A.A.; Al-Omair, M.A. Development of silver-containing nanocellulosics for effective water disinfection. *Cellulose* **2014**, *21*, 1965–1974. [CrossRef]
110. Ottenhall, A.; Henschen, J.; Illergård, J.; Ek, M. Cellulose-based water purification using paper filters modified with polyelectrolyte multilayers to remove bacteria from water through electrostatic interactions. *Environ. Sci. Water Res. Technol.* **2018**, *4*, 2070–2079. [CrossRef]
111. Nath, B.K.; Chaliha, C.; Kalita, E.; Kalita, M.C. Synthesis and characterization of ZnO:CeO_2:nanocellulose:PANI bionanocomposite. A bimodal agent for arsenic adsorption and antibacterial action. *Carbohydr. Polym.* **2016**, *148*, 397–405. [CrossRef] [PubMed]
112. Gustafsson, O.; Gustafsson, S.; Manukyan, L.; Mihranyan, A. Significance of Brownian motion for nanoparticle and virus capture in nanocellulose-based filter paper. *Membranes* **2018**, *8*, 90. [CrossRef] [PubMed]
113. Zelal, I.; Ali, U.; Nadir, D. Filtration and Antibacterial Properties of Bacterial Cellulose Membranes for Textile Wastewater Treatment. *AVvicenna J. Environ. Health Eng.* **2018**, *5*, 106–114.
114. Barud, H.S.; Regiani, T.; Marques, R.F.C.; Lustri, W.R.; Messaddeq, Y.; Ribeiro, S.J.L. Antimicrobial Bacterial Cellulose-Silver Nanoparticles Composite Membranes. *J. Nanomater.* **2011**, *2011*, 1–8. [CrossRef]
115. Zhang, X.; Fang, Y.; Chen, W. Preparation of Silver/Bacterial Cellulose Composite Membrane and Study on Its Antimicrobial Activity. *Synth. React. Inorg. Met.-Org. Nano-Met. Chem.* **2013**, *43*, 907–913. [CrossRef]
116. Aguilar-Sanchez, A.; Jalvo, B.; Mautner, A.; Nameer, S.; Pöhler, T.; Tammelin, T.; Mathew, A.P. Waterborne nanocellulose coatings for improving the antifouling and antibacterial properties of polyethersulfone membranes. *J. Membr. Sci.* **2020**, *620*, 118842. [CrossRef]
117. Sen, B.; Tahir, M.; Sonmez, F.; Turan Kocer, M.A.; Canpolat, O. Relationship of Algae to Water Pollution and Waste Water Treatment. In *Water Treatment*; IntechOpen: London, UK, 2013; pp. 335–354.

118. Gopakumar, D.A.; Arumughan, V.; Pasquini, D.; Leu, S.Y.; Abdul Khalil, H.P.S.; Thomas, S. *Nanocellulose-Based Membranes for Water Purification*; Elsevier Inc.: Amsterdam, The Netherlands, 2018; ISBN 9780128139271.
119. Kim, T.-I.; Park, H.; Han, M. Development of algae removal method based on positively charged bubbles. *KSCE J. Civ. Eng.* **2017**, *21*, 2567–2572. [CrossRef]
120. Ge, S.; Champagne, P.; Wang, H.; Jessop, P.G.; Cunningham, M.F. Microalgae Recovery from Water for Biofuel Production Using CO2-Switchable Crystalline Nanocellulose. *Environ. Sci. Technol.* **2016**, *50*, 7896–7903. [CrossRef] [PubMed]
121. Qiu, S.; Wang, L.; Champagne, P.; Cao, G.; Chen, Z.; Wang, S.; Ge, S. Effects of crystalline nanocellulose on wastewater-cultivated microalgal separation and biomass composition. *Appl. Energy* **2019**, *239*, 207–217. [CrossRef]
122. Yu, S., II; Min, S.K.; Shin, H.S. Nanocellulose size regulates microalgal flocculation and lipid metabolism. *Sci. Rep.* **2016**, *6*, 1–9. [CrossRef]
123. Patankar, S.C.; Renneckar, S. Greener synthesis of nanofibrillated cellulose using magnetically separable TEMPO nanocatalyst. *Green Chem.* **2017**, *19*, 4792–4797. [CrossRef]

polymers

MDPI

Article

Compression Behaviour of Bio-Inspired Honeycomb Reinforced Starfish Shape Structures Using 3D Printing Technology

S. A. S. A. Saufi [1], M. Y. M. Zuhri [1,2,*], M. Lalegani Dezaki [3,4], S. M. Sapuan [1,2], R. A. Ilyas [5,6], A. As'arry [3], M. K. A. Ariffin [3] and M. Bodaghi [4]

1 Advanced Engineering Materials and Composites Research Centre, Department of Mechanical and Manufacturing Engineering, Faculty of Engineering, Universiti Putra Malaysia, Serdang 43400, Malaysia; syedsaufi@gmail.com (S.A.S.A.S.); sapuan@upm.edu.my (S.M.S.)
2 Laboratory of Biocomposite Technology, Institute of Tropical Forestry and Forest Product (INTROP), Universiti Putra Malaysia, Serdang 43400, Malaysia
3 Department of Mechanical and Manufacturing Engineering, Faculty of Engineering, Universiti Putra Malaysia, Serdang 43400, Malaysia; lalegani.mr@gmail.com (M.L.D.); zizan@upm.edu.my (A.A.); khairol@upm.edu.my (M.K.A.A.)
4 Department of Engineering, School of Science and Technology, Nottingham Trent University, Nottingham NG11 8NS, UK; mahdi.bodaghi@ntu.ac.uk
5 School of Chemical and Energy Engineering, Faculty of Engineering, Universiti Teknologi Malaysia, UTM, Johor Bahru 81310, Malaysia; ahmadilyas@utm.my
6 Centre for Advanced Composite Materials (CACM), Universiti Teknologi Malaysia, UTM, Johor Bahru 81310, Malaysia
* Correspondence: zuhri@upm.edu.my

Citation: Saufi, S.A.S.A.; Zuhri, M.Y.M.; Dezaki, M.L.; Sapuan, S.M.; Ilyas, R.A.; As'arry, A.; Ariffin, M.K.A.; Bodaghi, M. Compression Behaviour of Bio-Inspired Honeycomb Reinforced Starfish Shape Structures Using 3D Printing Technology. *Polymers* **2021**, *13*, 4388. https://doi.org/10.3390/polym13244388

Academic Editor: Chin-San Wu

Received: 24 September 2021
Accepted: 19 October 2021
Published: 14 December 2021

Abstract: The bio-inspired structure (e.g., honeycomb) has been studied for its ability to absorb energy and its high strength. The cell size and wall thickness are the main elements that alter the structural ability to withstand load and pressure. Moreover, adding a secondary structure can increase the compressive strength and energy absorption (EA) capability. In this study, the bio-inspired structures are fabricated by fused deposition modelling (FDM) technology using polylactic acid (PLA) material. Samples are printed in the shape of a honeycomb structure, and a starfish shape is used as its reinforcement. Hence, this study focuses on the compression strength and EA of different cell sizes of 20 and 30 mm with different wall thicknesses ranging from 1.5 to 2.5 mm. Subsequently, the deformation and failure of the structures are determined under the compression loading. It is found that the smaller cell size with smaller wall thickness offered a crush efficiency of 69% as compared to their larger cell size with thicker wall thickness counterparts. It is observed that for a 20 mm cell size, the EA and maximum peak load increase, respectively, when the wall thickness increases. It can be concluded that the compression strength and EA capability increase gradually as the cell size and wall thickness increase.

Keywords: 3D printing; bio-inspired structure; energy absorption; fused deposition modelling; honeycomb structure

1. Introduction

A sandwich structure is a combination of the core structure and joint parts with layers of face sheets [1]. The honeycomb structure is one of the most common bio-inspired structures that has been studied and optimised. Different types of honeycomb structures are differentiated by their geometry shape, such as square, hexagonal, tetrahedral, pyramidal and pentagonal [2,3]. Sandwich structures have been widely used in various fields, such as aerospace and construction, due to their high strength and stiffness [4,5]. Moreover, sandwich structures can be used in 4D printing due to their specific features, such as absorbing energy [6]. These structures are well known for their excellent ability to absorb energy impacts. The hexagonal shape structure can provide superior energy absorption (EA) compared to other forms of sandwich structures under compression loading [7].

The EA of honeycomb structures can be altered to increase their crashworthiness capability. These studies showed the bio-inspired structure can increase EA [8,9]. Wang et al. [10] stated that the performance of the honeycomb structure could be enhanced by introducing a secondary structure and increasing the value of the structure's stiffness. Another finding from Shan et al. [11] suggested that the wall thickness and cell size of the core structure could influence the failure behaviour of the sandwich structure. The multilayer hexagonal shape of the honeycomb can absorb about 31% to 60% of energy compared to the rectangular form [12]. A comparison study proved that the hexagonal shape honeycomb performed better stiffness than a foam sandwich due to the buckling and their cell size. Hence, controlling the crashworthiness of the honeycomb structure is more accessible [13]. Variation of cell size, material and wall thickness have different results on EA. By optimising these parameters, researchers can increase the EA based on the experiment conducted by Tao et al. [14]. They suggested that by changing the wall thickness, the structure had a different result on their crushing strength and EA. Aside from that, the plastic deformation near the cell wall could improve the EA by increasing the material at the intersecting area [15].

In addition, the EA capability of the cellular structure depends on the unit cell geometry, relative density, the properties of base material and the loading force [16,17]. Xu et al. [18] studied the EA capability between the standard hexagonal shape with a combination of hybrid structures consisting of auxetic and hexagonal honeycomb cells under in-plane compression. The hybrid structure improved by 38% compared to the standard hexagonal honeycomb structure. Moreover, Wang et al. [19] worked on the inhomogeneity honeycomb structure. One of the factors that could influence the structure failure during the compression is the regularity and adhesive failure of cell size. Almost all failures of honeycomb structures during compression occur near the joint, where the cells are connected. It creates an over-stiff area with a higher stress concentration [20]. It is shown that the capability of reinforcement in honeycomb structures had a higher significant result in terms of strength compared to the typical geometry structures. Moreover, the elastic modulus and the EA enhanced by 26% and 73%, respectively [21]. Another study showed the integration of reinforced structure not only increased the EA but also increased the peak load. Sun et al. [22] investigated a combination of sandwich structures reinforced with a grid at higher peak loads.

According to Tuo et al. [23], a honeycomb structure integrated with a plate made of basalt fibre, compressed under an edgewise position, proved that the structure had a larger shear modulus of elasticity and good plastic deformation ability. Lei et al. [24] examined the edgewise compression to investigate the behaviour of reinforced fibreglass honeycombs columns. The results showed that the load decreased with an increase of the columns' length. Similarly, Kara et al. [25] stated that less deflection occurs when a higher load is exerted, and they suggested reinforcing the honeycomb structure for major improvement. Another study on the comparison of edgewise and flatwise testing positions found that the former position for sandwich beam exhibited higher stiffness and strength bearing capacity compared to the latter position [26]. There are three phases that a structure undergoes during the compression, which are linear region, plateau region and densification. The phases are defined through a compression test in a stress-strain curve or load-displacement curve. The structure starts to fail when a continuous load presses the sample over the elastic region.

Meanwhile, recent developments and rapid usage of additive manufacturing (AM), mostly on fused deposition modelling (FDM) technology, has led to more possibilities to fabricate and design complex products [27,28]. Three-dimensional printing technology is capable of producing a wide range of structures, from simple to complex shapes [29]. The nature of the FDM process is to fabricate products layer by layer [30,31]. A barrage of materials from polymer to composite can be used in this process, which is in the shape of filament [32]. Polylactic acid (PLA) is a friendly biodegradable material that is made of starch and is converted into dextrose, a fermentable sugar by enzymatic hydrolysis, a process where the micro-organisms break the starch into smaller pieces called lactic acid [33,34]. In a study performed by Rebelo et al. [35], the EA of a sacrificial honeycomb cladding was analysed by using PLA material. The findings concluded that the dynamic force and specific energy absorption (SEA) were directly proportional to the relative density, which controls the buckling of the interior cell wall and the deformity of the structure for the top and bottom layer.

Chen et al. [36] studied a hierarchical honeycomb under large compressive deformation. Due to the complexity of their specimens, the products were made by a 3D printer and were tested under compression loading. To validate the experiment result, samples were analysed in FEA, and the results were similar. Meanwhile, Kucewic et al. [37] worked on the deformation behaviour, and failure of 3D printed cellular structures and compared their experimental findings with the numerical analysis. Moreover, Sahu et al. [38] studied the effect of the cell size of high-density polyethylene (HDPE) sandwich honeycomb structures made with the 3D printing process. The results indicated that a smaller cell size inhabited a larger SEA. As the cell size increased, the crushing force efficiency (CFE) value is decreased, while opposite results were found with a constant cell size with a different wall thickness. In addition, a study on hierarchical honeycomb under a larger compressive deformation is conducted, where the honeycomb wall is replaced by a triangular lattice. The honeycomb showed an improvement in stiffness and EA [36].

This study aims to improve the compression properties of the current honeycomb structure with reinforcement of a bio-inspired starfish shape. Aside from that, the main objectives of this study are to investigate the energy capability of bio-inspired structures, such as their EA, SEA and CFE, fabricated by using FDM technology, as well as their failure deformation under quasi-static loading conditions, and it focuses on the effect of cell size and cell wall thickness.

2. Material and Methods

2.1. Material

The material used is polylactic acid (PLA) supplied by PolyLiteTM Polymaker, (Shanghai, China) with the same properties: density ranging from 1.17 to 1.24 g/cm^3 and the melting temperature between 190 and 230 °C. The materials come in white and black in colour.

2.2. Methods

The bio-inspired structures are based on the combination of honeycomb and starfish structures (see Figure 1). The internal structure of the honeycomb is reinforced with the inspired structure of starfish. It is modified by adding 6 branches of the starfish for better stability and symmetry. The design of the combined structure is drawn using Solidworks® software (Version 2010) before converting into the standard tessellation language (STL) format for the 3D printing process. Figure 2 illustrates the complete honeycomb reinforced starfish structure. The curve design of the branch is used to reduce the load at the centre and distribute the load to the edge, as suggested by Wang et al. [39]. Each branch has the same angle with the value of 60°, (a) is the radius of the starfish shape, (b) thickness of the starfish shape (which is equivalent to the honeycomb wall thickness) and (c) is the honeycomb cell size. Prior to conducting testing under quasi-static loading, three sets of samples for each size and thickness are prepared, which is similar to Khan [40].

Figure 3 shows the samples with different cell sizes, where all have the same width of 15 mm. Here, the controlled parameter is the cell size, these being 20, 25 and 30 mm. Each size has 3 different wall thicknesses, which are 1.5, 2.0 and 2.5 mm. Table 1 presents the physical properties of the 3D printed structures. The model of the FDM machine used is Ultimaker 2+ (Utrecht, The Netherlands). The nozzle diameter is 0.25 mm, and the nozzle temperatures are between 190 and 230 °C, respectively. The layer thickness is 60 microns with a built-up speed of 60 mm^3/s. The build orientation is set to 0° due to the optimum angle in the printing process. The final 3D printed samples are shown in Figure 4.

Figure 1. Starfish shape [41].

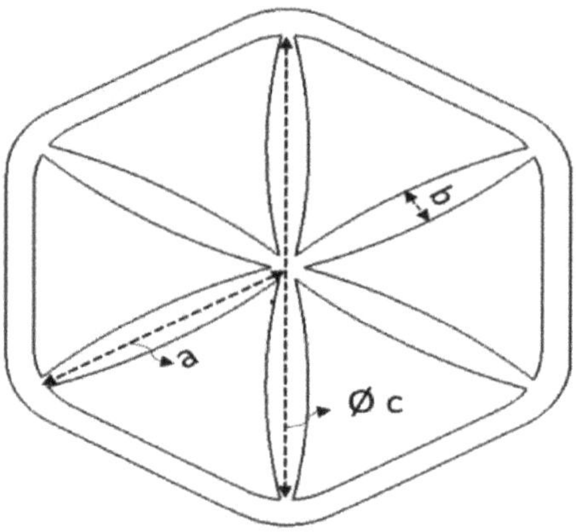

Figure 2. Illustration of honeycomb reinforced starfish structure.

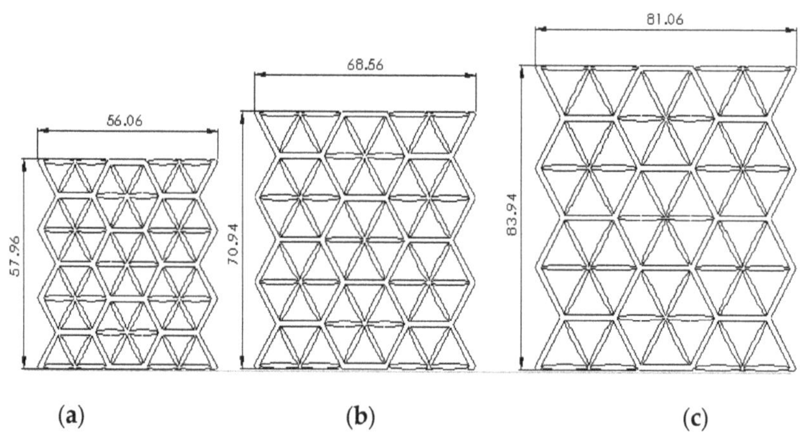

Figure 3. Cell size of (**a**) 20, (**b**) 25 and (**c**) 30 mm.

Table 1. Physical properties of the 3D printed samples.

Cell Size (mm)	Sample No.	Height (mm)	Length (mm)	Width (mm)	Wall Thickness (mm)	Weight (g)
20	S1	57.96	56.06	15	1.5	23
	S2	57.96	56.06			23
	S3	57.96	56.06			23
	S4	59.96	58.08		2.0	31
	S5	59.96	58.08			31
	S6	59.96	58.08			31
	S7	61.96	60.11		2.5	38
	S8	61.96	60.11			39
	S9	61.96	60.11			38
25	S10	70.94	68.56		1.5	28
	S11	70.94	68.56			29
	S12	70.94	68.56			29
	S13	72.94	70.58		2.0	39
	S14	72.94	70.58			39
	S15	72.94	70.58			39
	S16	74.94	72.61		2.5	49
	S17	74.94	72.61			48
	S18	74.94	72.61			48
30	S19	83.94	81.06		1.5	34
	S20	83.94	81.06			34
	S21	83.94	81.06			34
	S22	85.94	83.08		2.0	46
	S23	85.94	83.08			47
	S24	85.94	83.08			47
	S25	87.94	85.11		2.5	57
	S26	87.94	85.11			57
	S27	87.94	85.11			59

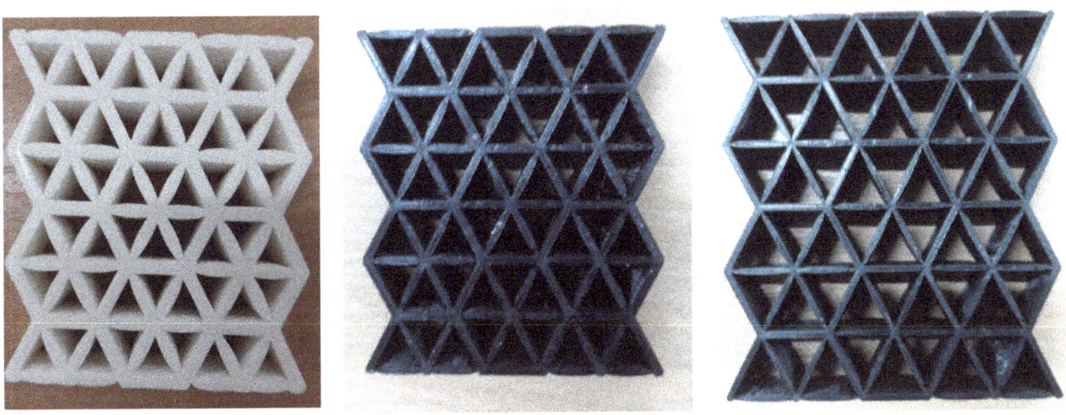

(a) (b) (c)

Figure 4. The final 3D printed specimens for (**a**) cell size 20 mm, thickness 2.5 mm, (**b**) cell size 25 mm, thickness 2 mm and (**c**) cell size 30 mm, thickness 1.5 mm.

The ASTM standard D7336/D7336M recommends three criteria of energy absorption. The capability of the structure to resist deformation from an external force depends on its energy absorption. The energy absorption capability of the cellular structure depends on the unit cell geometry, relative density, the properties of base material and the loading force [17]. Energy absorption can be obtained by determining the area under the load-displacement curve. It is derived from the mathematical formula as follows:

$$EA = \int_{0}^{d} F(x)dx \tag{1}$$

where the F(x) is the function of displacement x and d is the deformation. Total energy absorption is the accumulative load-displacement curve from zero to the maximum deformation. Specific energy absorption (SEA) is derived from the value of energy absorption per mass.

$$SEA = \frac{EA}{m} \tag{2}$$

Crushing force efficiency (CFE) analysis is the performance of the structure during the compression process. It is the ratio of the average load in the plateau region to the load at the maximum peak load. As explained from the definition in ASTM D7336, the mathematical expression of CFE is obtained as follows:

$$CFE = \frac{F_{ave}}{F_{max}} \tag{3}$$

where F_{ave} is the average load in the plateau region and F_{max} is the maximum peak load.

Following this, an edgewise compression test is performed following the ASTM C364 standard. An INSTRON 3382 universal testing machine is used to conduct the compression test with a load cell of 100 kN and crosshead displacement of 1 mm/min. The orientation of the sample is facing perpendicular to the force exerted, as shown in Figure 5. This method is similar to Ivañez et al. [16]. The four stages of the structure are photographed at every 5 mm of displacement to determine its failure deformation.

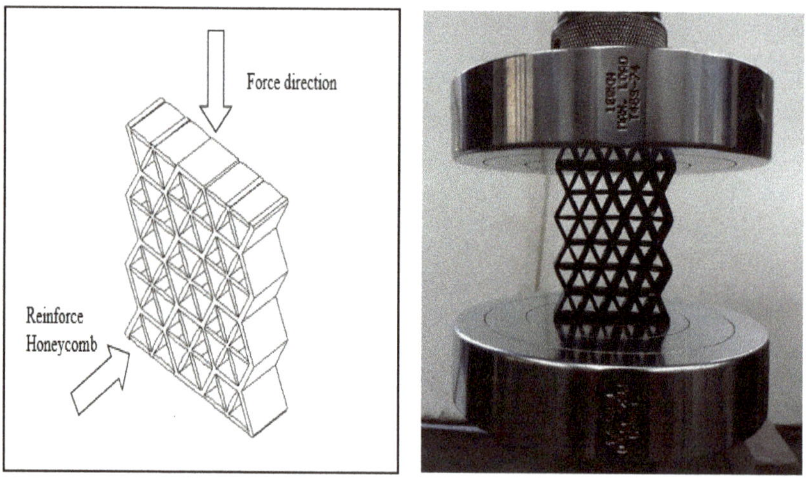

(a) (b)

Figure 5. (**a**) Schematic of sample orientation during test. (**b**) Sample orientation during test.

3. Results and Discussion

The results obtained from the compression test are plotted in the form of a load-displacement curve. The maximum peak means the maximum value of loading that the structure can resist. The value of the stress determines the strength capability of the structure. Here, the maximum peak value of the tested samples is determined from the highest value stress in the linear region. The CFE is taken from the load-displacement curve. It is calculated from the ratio between average forces in the plateau region and the maximum peak value. Meanwhile, the EA value is calculated based on the area under the load-displacement curve. The failure deformation and the reaction of the structure during the experiment are also investigated. In this study, the maximum peak value is discussed in two ways. First, the comparison under variable cell sizes with constant wall thickness and secondly constant cell size with variable wall thicknesses. This work focuses on investigating the effects of cell size with variable wall thicknesses and vice versa. For constant wall thickness, the sample with the highest maximum peak load is used to compare with the other results.

3.1. Effects of Cell Size on the Structure

Figure 6 presents the experimental result for a batch of samples with a 20 mm cell size and a 2 mm wall thickness. It is observed that the structure failed in three stages, these being uniform compression and elastic instability (A), plastic deformation (B) and plastic instability (C). In section A, a linear region is observed where the load increases proportionally with an increase in displacement. This is where the reinforced starfish shape structure provides some additional support to the main honeycomb core. At the end of this section (maximum load), the sample started to collapse and deformed due to the breakage of the starfish shape joint from the honeycomb core (as shown in Table 2). As the loading continued, the structure started to lose its stability and the load decreased gradually due to the less interaction of the reinforced starfish shape to its main structure. Thus, this leads to the plateau region loading under section B, which shows a constant loading compared to sections A and C, where the main structure (honeycomb) takes its role to support the overall structure with minimal support from the secondary support (starfish shape). In section C, the structure is fully collapsing, where the region is called densification, where the structure is no longer able to withstand any force.

However, the sample with a 25 mm cell size and wall thickness of 2 mm showed different behaviour compared to other samples. Here, the sample is likely to behave as two different structures. At the first stage, the loading started to decrease after a displacement of 3 mm, as shown in Figure 7a. Following this, the loading starts to increase again. This indicates that the structure has entered the second stage of the compression loading, where it starts from zero again before continuing with linear and plateau regions and finally the densification region (see Figure 7b). At the first stage, the failure is due to the breakage of the link between the structure at both sides of the structure, which causes the decrease of the compression loading. As the load continued, the bottom side of the collapsed branch between the honeycomb structure started to overlap each other and formed a 'new' structure. In contrast, most of the structural centre parts remain in their original form.

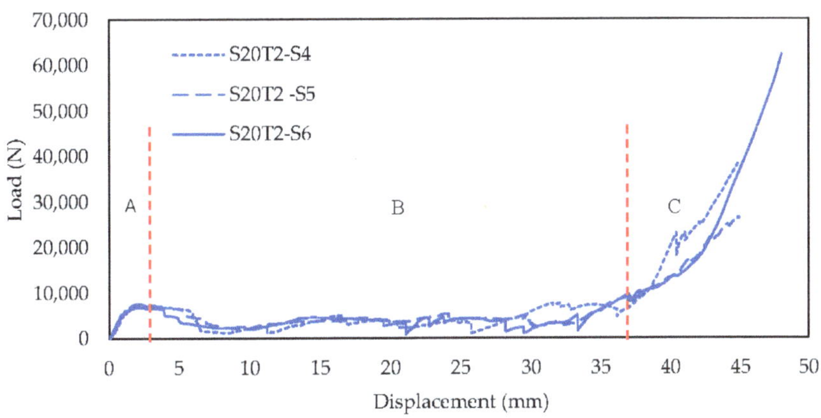

Figure 6. The load-displacement curve for samples with 20 mm cell size with 2 mm wall thickness.

Table 2. Comparison of structure failure deformation under different cell sizes.

Stage	20 mm Cell Size 2.0 mm Wall Thickness	25 mm Cell Size 2.0 mm Wall Thickness
I (Initial stage)		
II (Max peak)		
III (Plateau region)		
IV (Densification)		

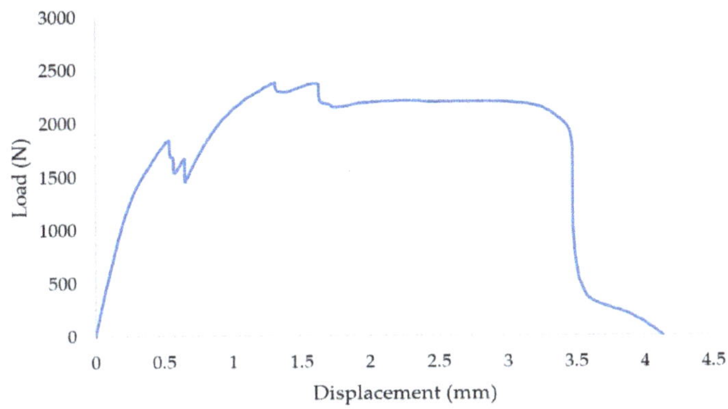

(a)

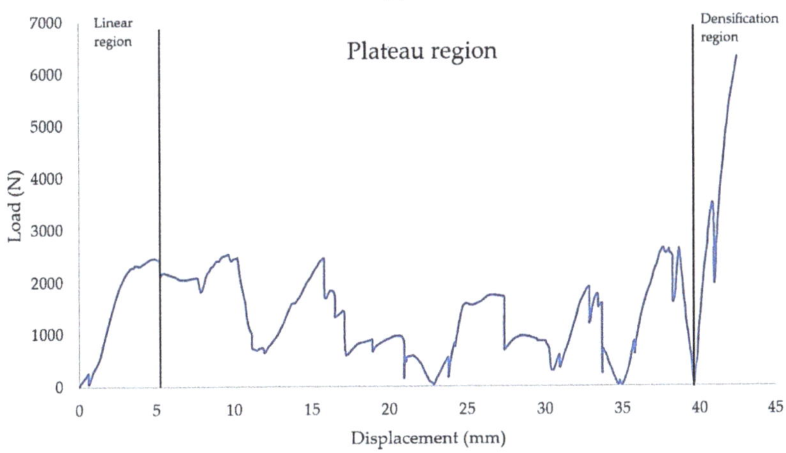

(b)

Figure 7. Load-displacement curve of a sample with a 25 mm cell size, 2 mm wall thickness, showing the (**a**) initial stage and (**b**) second stage of the failure.

During the plateau region, the side structure started to break away from the honeycomb structure. Hence, the structure lost support from the edge side. As the load pressed, the side of the structure continued to move further from the main structure. Simultaneously, in the middle of the structure, the branch (see red circle in Table 2) started to break. Meanwhile, at 30 mm of displacement (Plateau region), the wall thickness at the centre structure split and became thinner. The split part started to fold at 40 mm displacement (densification) as the structure height compressed. The link that connected to the honeycomb started to break from the joining, and a major loss happened for support. Moreover, at this stage, inhomogeneity of cells occurred. This resulted in adhesive failure between the cell sizes [19]. A failure is also recorded at the middle of the core structure, where the starfish shape broke from the honeycomb structure. In the densification region, only the centre of the structure formed a solid structure.

Table 2 shows a comparison of structure deformation between the cell size of 20 mm with 2 mm wall thickness and the cell size of 25 mm with 2 mm wall thickness. From the observation, the behaviour of both structures at initial and maximum peak values showed no differences. The geometry for both structures showed no sign of rupture, buckling or break. Due to the continuous load, in the collapsed structure, a solid form structure started to develop mainly on the bottom side. For the cell size 25 mm with 2 mm wall thickness, an internal tear started to develop during the densification region. Slicing the link of the structure in half and a portion of the structure broke from the main structure. At the densification area, a 20 mm cell size with 2 mm wall thickness showed a larger solid area compared to the 25 mm cell size with 2 mm wall thickness. It is observed that the structure of 20 mm cell size with 2 mm wall thickness started to reach its densification point at the displacement of 40 mm. On the other hand, the 25 mm cell size with 2 mm thickness started its densification at the displacement of 45 mm. The split that occurred in the structures is due to the kink-band failure [42]. The main reason might be due to the shortened infill during the process, as stated by Jerez-Mesa et al. [43]. They stated that the infill density could influence the fatigue lifespan of the PLA material. It is also found that the type of infill can be contributed to the strength of the printed structure [44]. The collapse sequences were almost similar for all samples except the cell size of 25 mm with 2 mm wall thickness. Increasing cell size reduced the maximum peak value [15]. Figure 8 gives information about the wall thicknesses of 1.5 to 2.5 mm. The maximum peak value decreased gradually as the cell size increased. Smaller cell size exhibited a larger peak value compared to the larger cell size, regardless of the wall thickness size.

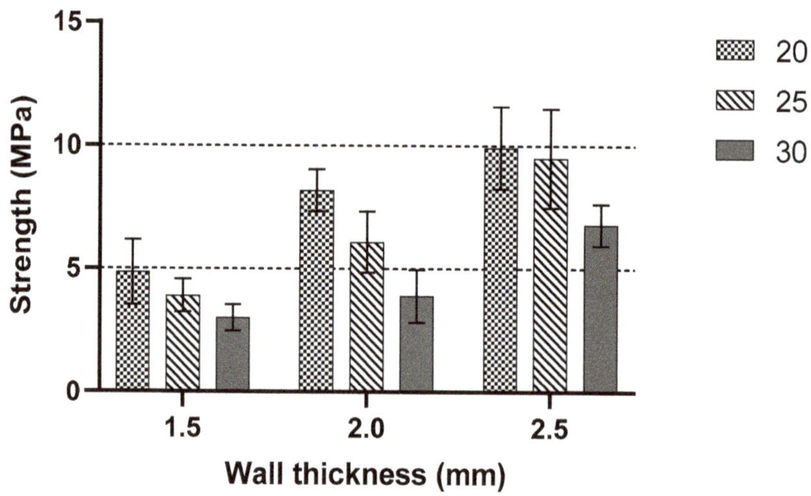

Figure 8. Comparison between constant wall thicknesses with variable cell sizes.

3.2. *Effects of Wall Thickness on the Structure*

Figure 9 illustrates a load-displacement graph for 20 mm cell size with 2.5 mm wall thickness. Table 3 shows the collapse behaviour in the plateau region. The structure started to collapse after reaching the maximum peak value. The load pressed the structure, and samples started to expand to the edge side. At 5 mm displacement, the link between the cell size at the top upper side started to fracture and detach from the main structure; however, the centre of the part still maintained its shape (see stage I). After it reached 10 mm of displacement, the edge side of the structure started to move to the outer part of the structure. It can be seen the fracture that broken section parts move further away. At the same time, the starfish structure inside the honeycomb had a failure breaking and created an almost solid shape at the top centre and the edge sides.

As the displacement reached 15 mm, there was no significant shape-changing at the bottom side, while the centre part of the structure became thick. A portion in the top right side of the structure breaks from its main structure. As shown in the graph, the load decreases at 20 mm displacement until 25 mm displacement. The load increases at 30 mm displacement. Then, the centre of the part becomes more solidified, even though some sections from the left of the sample move away from the main structure. The left side of the structure collapsed from the main structure. The structure started to become almost solid at 40 mm displacement and indicated that the structure entered the densification region. Papka and Kyriakides [45] concluded their finding by defining the elastic region. The relationship of load and displacement is quite uniform, and the behaviour of the geometry is fairly stiff. As the load reached the highest peak, a sharp fall happened in the load-displacement chart subsequently.

From the analysis of the maximum peak value for the three cell sizes with a different wall thickness, a comparison between 20, 25 and 30 mm cell sizes for the highest peak value is shown in Figure 10. The cell size with the largest wall thickness had the highest maximum value compared to the smaller cell size. The wall thickness of the structure affected the peak load value. This bar chart shows a comparison between samples with various cell sizes and wall thicknesses. The experimental data showed an average of 5.28 MPa stress for 1.5 mm thickness with 20 mm cell size. Moreover, all samples showed an almost identical value accordingly. As the wall thickness increased to 2 mm, the average peak value increased with an average of 8.13 MPa for 20 mm cell size. The 25 mm cell size sample showed an almost similar pattern to the 20 mm cell size. Except for 2.0 mm thickness, one of the samples showed a large different record with 3.06 MPa of peak value compared to the other two samples, 6.30 and 5.16 MPa, respectively. Overall, the maximum peak value increased as the cell size increased. The 25 mm cell size trend showed that an increase in wall thickness caused the peak value to change proportionally. For a cell size of 30 mm, the same trend is recorded as the other two cell sizes. As the wall thickness increased, the value of the maximum peak value increased gradually. However, from the obtained data, the values of 30 mm were smaller than 20 and 25 mm, which were similar to the Shan et al. [11]. The peak stress decreased with an increase in sample size. The highest stress value among samples is recorded around 6.67 MPa for the 30 mm cell size, 8.39 MPa for the 20 mm cell size and 8.79 MPa for the 25 mm cell size.

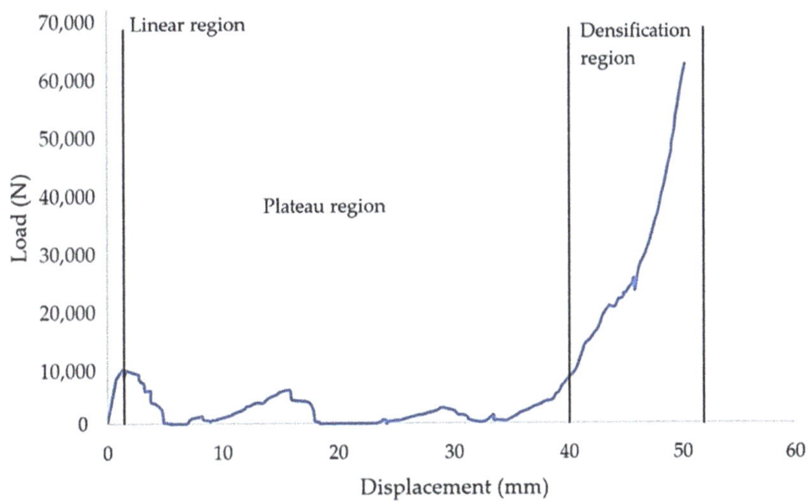

Figure 9. Cell size 20 mm, wall thickness 2.5 mm load-displacement mean curve.

Table 3. Failure sequence of the structure with 20 mm cell size and 2.5 mm wall thickness.

Stage	20 mm Cell Size and 2.5 mm Wall Thickness
I (Initial stage)	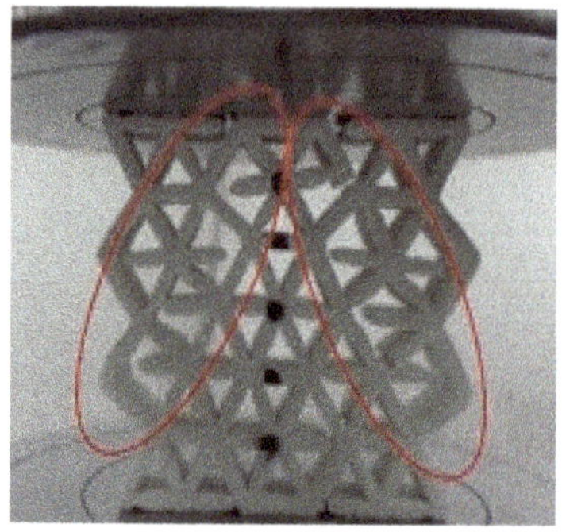
II (Max peak)	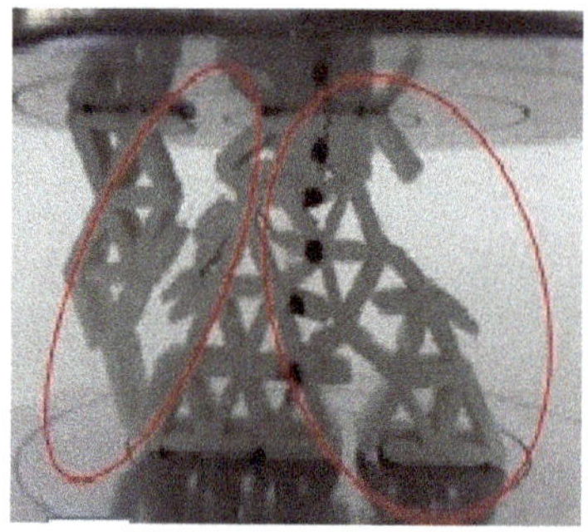
III (Plateau region)	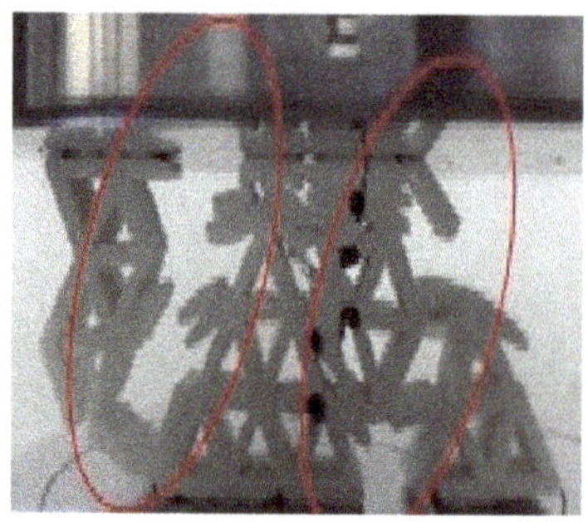
IV (Densification)	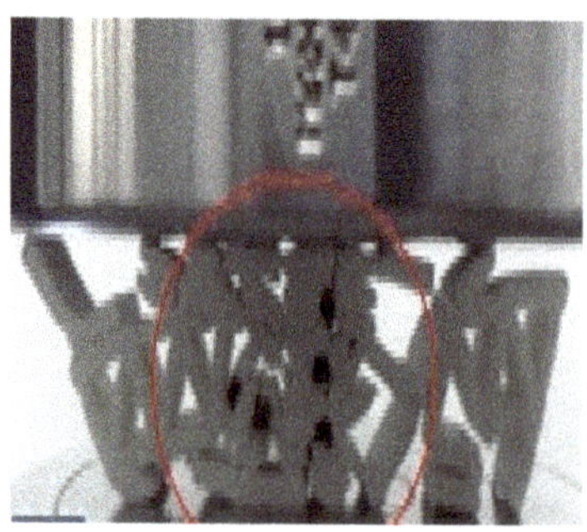

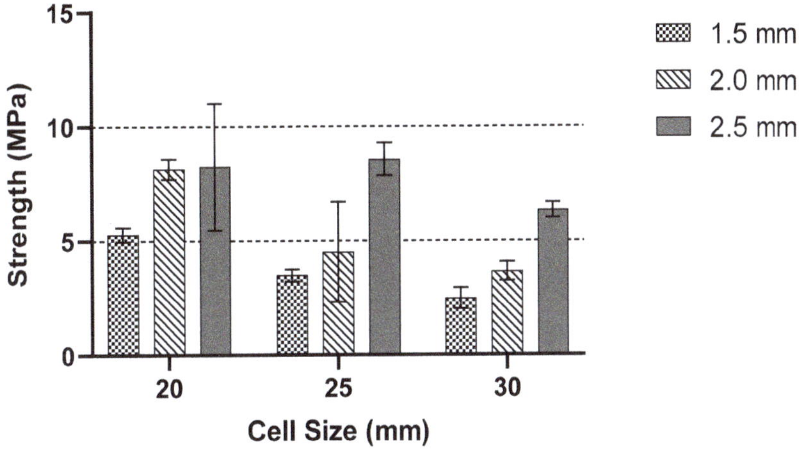

Figure 10. Comparison between constant cell sizes with variable thickness.

3.3. Effects of Bio-Inspired Structure on Energy Efficiency and Energy Absorption

Figure 11 shows the CFE values of cell sizes 20, 25 and 30 mm, respectively. Based on the experimental result, the 20 mm cell size with a lower wall thickness exhibited a higher ratio average force in the plateau region to peak force, which is near to 0.7, compared to 2.5 mm wall thickness with a ratio of 0.28. The peak load at 2.5 mm is much higher than 1.5 mm, but the average load in the plateau region in 1.5 mm is higher compared to 2.5 mm wall thickness. Different behaviour is recorded in the 25 mm cell size, where 2.5 mm wall thickness recorded a higher peak load compared to 20 mm cell size. However, the CFE values were still lower than the 1.5 mm wall thickness. The CFE value decreased as the wall thickness increased, similar to the findings by Ivañez et al. [16]. As the cell size decreased, the CFE value decreased as well. The highest CFE occurred when the cell size was 20 mm with 1.5 mm wall thickness, which was the smallest sample.

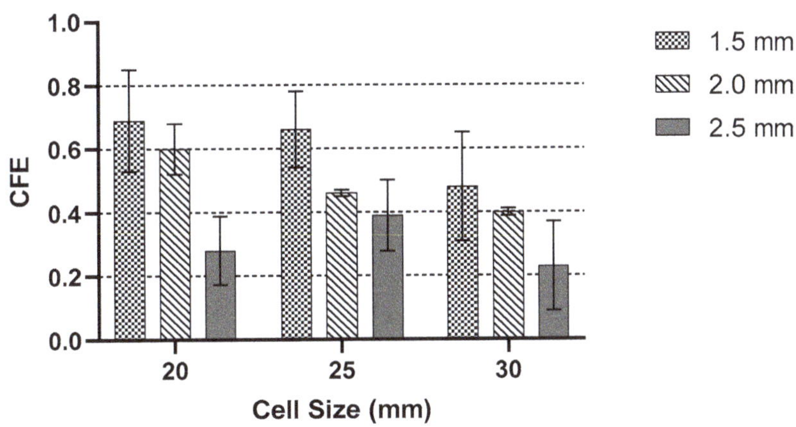

Figure 11. Comparison of CFE with different cell sizes and wall thickness.

The result for the 30 mm cell size showed a similar pattern to the 20 mm and 25 mm cell sizes. Figures 12 and 13 show the results for EA and SEA for all specimens, respectively. From the given result, the highest average EA and SEA occurred where the wall thickness was 2 mm with a 20 mm cell size. The highest value was recorded at 348.18 kJ of EA and

8356.21 kJ/Kg for the SEA. Furthermore, the difference between all three samples with 2.0 mm wall thickness was quite large. Overall, the given values for both EA and SEA were lower than the 20 mm cell size. Observing the highest value of both EA and SEA for the 25 mm cell size in this study showed that the wall thickness is an important factor. As an example, a major improvement with the value of 71% can be seen in the sample with 20 mm cell size and a 2.0 mm thickness compared to those with 30 mm cell size and 1.5 mm thickness.

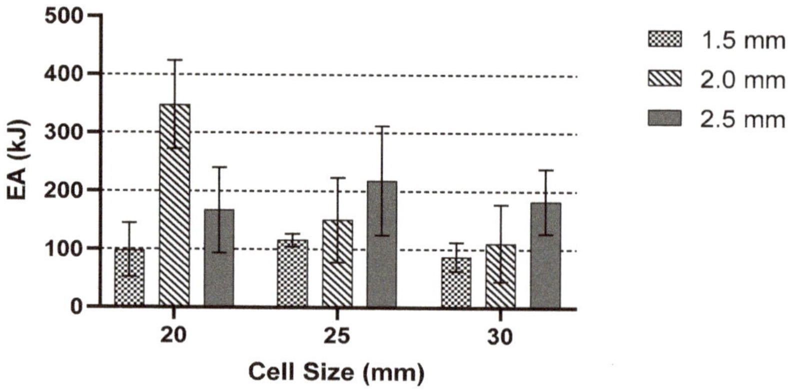

Figure 12. The comparison of EA for the same cell size with different wall thicknesses.

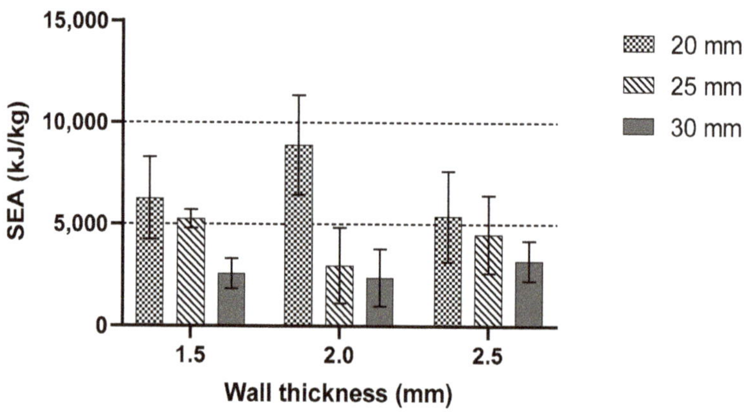

Figure 13. The SEA comparison of the same wall thickness with different cell sizes.

Figure 12 summarises a comparison of EA in terms of cell size, of which the highest values of EA are selected. In brief, a larger cell size generated a smaller EA. From the graphical chart, the behaviour between 25 and 30 mm cell sizes is similar. As the wall thickness increased, the amount of EA decreased. In addition, Figure 13 shows a comparison between the SEA values in terms of wall thickness, which present a similar pattern as in Figure 11. Based on the analysis for maximum peak force, CFE, EA and SEA, the 20 mm cell size with a 2 mm wall thickness showed a remarkable result compared to the rest of the samples. Moreover, similar experimental setups and testing parameters from previous studies are used. The results from Xiang and Du [46] show the SEA and EA of different thicknesses and cell sizes of bio-inspired structures. The results, after adding the bio-reinforcement, increased by 35.97%. Thomas and Tiwari [15] studied the behaviour of aluminium honeycomb reinforced structure, where the sample sizes were 10 mm in cell size with a 0.1 mm wall thickness. The samples were strengthened by a secondary reinforcement structure. An EA produced by an aluminium reinforcement structure had a higher value than ABS honeycomb structure from the study by Kucewicz et al. [20]. It is shown that an additional secondary structure as a reinforced structure could increase the EA capability of the common honeycomb structure. Another study is conducted on a novel bio-inspired honeycomb sandwich panel (BHSP) based on a woodpecker's beak. They found out the SEA of this novel structure is higher by 125% and 63.7% compared to the conventional sandwich panel. They recorded different values for each sample with different thicknesses. The same results happened in this study compared to the recorded experiments.

4. Conclusions

A range of compression testing was conducted to analyse the strength and EA capability of a starfish-reinforced honeycomb shape structure. The effect of cell size and wall thickness was investigated. To achieve a higher peak load, a larger cell size and wall thickness offered the optimum criteria. In this case, a comparison between the wall thicknesses presented a uniform pattern. As the wall thickness increases, the peak load increases, regardless of the core size. On the other hand, as the cell size increases, the peak load will decrease. The compression behaviour of the smallest wall thickness gives better performance than the larger wall thickness. As the wall thickness increases, the efficiency decreases (regardless of the cell size). In addition, the results indicated that a higher CFE value was generated from a smaller value of energy absorption. In brief, as the wall thickness increases, the CFE value will decrease, yet the EA is increased. For future works, a study on the structural joint or node can be further investigated to increase its performance, as most of the structure failed at the interlink between cell size. Their capability under dynamic loading conditions should also be considered.

Author Contributions: Project administration, M.Y.M.Z. and S.A.S.A.S.; Visualisation, S.A.S.A.S. and M.L.D.; Writing—original draft, S.A.S.A.S., M.Y.M.Z. and M.L.D.; Writing—review and editing, S.M.S., R.A.I., A.A., M.K.A.A. and M.B. All authors have read and agreed to the published version of the manuscript.

Funding: This work was supported by the Universiti Putra Malaysia through the Putra Grant (GP-IPS/2018/9663200).

Acknowledgments: Gratitude is also expressed to Mohd Hafizul Hashim and Muhammad Wildan Ilyas Mohamed Ghazali for their assistance in conducting this experimental work.

Conflicts of Interest: The authors declare no conflict of interest.

References

1. Birman, V.; Kardomateas, G.A. Review of current trends in research and applications of sandwich structures. *Compos. Part B Eng.* **2018**, *142*, 221–240. [CrossRef]
2. Thomas, T.; Tiwari, G. Crushing behavior of honeycomb structure: A review. *Int. J. Crashworthiness* **2019**, *24*, 555–579. [CrossRef]
3. Lalegani Dezaki, M.; Ariffin, M.K.A.M.; Appalanaidoo, D.; Wahid, Z.; Rage, A.M. 3D printed object's strength-to-weight ratio analysis for M3 liquid material. *Adv. Mater. Process. Technol.* **2020**, 1–15. [CrossRef]
4. Sairajan, K.K.; Aglietti, G.S.; Mani, K.M. A review of multifunctional structure technology for aerospace applications. *Acta Astronaut.* **2016**, *120*, 30–42. [CrossRef]
5. Sujiatanti, S.H.; Zubaydi, A.; Budipriyanto, A. Finite Element Analysis of Ship Deck Sandwich Panel. *Appl. Mech. Mater.* **2018**, *874*, 134–139. [CrossRef]
6. Bodaghi, M.; Serjouei, A.; Zolfagharian, A.; Fotouhi, M.; Rahman, H.; Durand, D. Reversible energy absorbing meta-sandwiches by FDM 4D printing. *Int. J. Mech. Sci.* **2020**, *173*, 105451. [CrossRef]
7. Sang, L.; Han, S.; Peng, X.; Jian, X.; Wang, J. Development of 3D-printed basalt fiber reinforced thermoplastic honeycombs with enhanced compressive mechanical properties. *Compos. Part A Appl. Sci. Manuf.* **2019**, *125*, 105518. [CrossRef]
8. Audibert, C.; Chaves-Jacob, J.; Linares, J.-M.; Lopez, Q.-A. Bio-inspired method based on bone architecture to optimize the structure of mechanical workspieces. *Mater. Des.* **2018**, *160*, 708–717. [CrossRef]
9. Tasdemirci, A.; Akbulut, E.F.; Guzel, E.; Tuzgel, F.; Yucesoy, A.; Sahin, S.; Guden, M. Crushing behavior and energy absorption performance of a bio-inspired metallic structure: Experimental and numerical study. *Thin-Walled Struct.* **2018**, *131*, 547–555. [CrossRef]
10. Wang, Z.; Zhang, Y.; Jiefu, L. Comparison between five typical reinforced honeycomb structures. In Proceedings of the 5th International Conference on Advanced Engineering Materials and Technology, Guangzhou, China, 22–23 August 2015.

11. Shan, J.; Xu, S.; Zhou, L.; Wang, D.; Liu, Y.; Zhang, M.; Wang, P. Dynamic fracture of aramid paper honeycomb subjected to impact loading. *Compos. Struct.* **2019**, *223*, 110962. [CrossRef]
12. Ali, M.; Ohioma, E.; Kraft, F.; Alam, K. Theoretical, numerical, and experimental study of dynamic axial crushing of thin walled pentagon and cross-shape tubes. *Thin-Walled Struct.* **2015**, *94*, 253–272. [CrossRef]
13. Crupi, V.; Epasto, G.; Guglielmino, E. Comparison of aluminium sandwiches for lightweight ship structures: Honeycomb vs. foam. *Mar. Struct.* **2013**, *30*, 74–96. [CrossRef]
14. Tao, Y.; Duan, S.; Wen, W.; Pei, Y.; Fang, D. Enhanced out-of-plane crushing strength and energy absorption of in-plane graded honeycombs. *Compos. Part B Eng.* **2017**, *118*, 33–40. [CrossRef]
15. Thomas, T.; Tiwari, G. Energy absorption and in-plane crushing behavior of aluminium reinforced honeycomb. *Vacuum* **2019**, *166*, 364–369. [CrossRef]
16. Ivañez, I.; Fernandez-Cañadas, L.M.; Sanchez-Saez, S. Compressive deformation and energy-absorption capability of aluminium honeycomb core. *Compos. Struct.* **2017**, *174*, 123–133. [CrossRef]
17. Zhang, L.; Feih, S.; Daynes, S.; Chang, S.; Wang, M.Y.; Wei, J.; Lu, W.F. Energy absorption characteristics of metallic triply periodic minimal surface sheet structures under compressive loading. *Addit. Manuf.* **2018**, *23*, 505–515. [CrossRef]
18. Xu, M.; Xu, Z.; Zhang, Z.; Lei, H.; Bai, Y.; Fang, D. Mechanical properties and energy absorption capability of AuxHex structure under in-plane compression: Theoretical and experimental studies. *Int. J. Mech. Sci.* **2019**, *159*, 43–57. [CrossRef]
19. Wang, Z.; Li, Z.; Zhou, W.; Hui, D. On the influence of structural defects for honeycomb structure. *Compos. Part B Eng.* **2018**, *142*, 183–192. [CrossRef]
20. Kucewicz, M.; Baranowski, P.; Małachowski, J. A method of failure modeling for 3D printed cellular structures. *Mater. Des.* **2019**, *174*, 107802. [CrossRef]
21. Zhang, Y.; Yan, L.; Zhang, W.; Su, P.; Han, B.; Guo, S. Metallic tube-reinforced aluminum honeycombs: Compressive and bending performances. *Compos. Part B Eng.* **2019**, *171*, 192–203. [CrossRef]
22. Sun, Z.; Shi, S.; Guo, X.; Hu, X.; Chen, H. On compressive properties of composite sandwich structures with grid reinforced honeycomb core. *Compos. Part B Eng.* **2016**, *94*, 245–252. [CrossRef]
23. Tuo, W.; Wei, P.; Chen, J.; Okabe, Y.; Zhang, X.; Xu, M. Experimental study of the edgewise compressive mechanical properties of biomimetic fully integrated honeycomb plates. *J. Sandw. Struct. Mater.* **2017**, *21*, 2735–2750. [CrossRef]
24. Lei, H.; Yao, K.; Wen, W.; Zhou, H.; Fang, D. Experimental and numerical investigation on the crushing behavior of sandwich composite under edgewise compression loading. *Compos. Part B Eng.* **2016**, *94*, 34–44. [CrossRef]
25. Kara, E.; Crupi, V.; Epasto, G.; Guglielmino, E.; Aykul, H. Flexural behaviour of glass fiber reinforced aluminium honeycomb sandwiches in flatwise and edgewise positions. In Proceedings of the 16th European Conference on Composite Materials, ECCM 2014, Seville, Spain, 22–26 June 2014.
26. Zhang, F.; Liu, W.; Fang, H.; Jia, Z. Flexural behavior of composite sandwich beams with different kinds of GFRP ribs in flatwise and edgewise positions. *Compos. Part B Eng.* **2019**, *156*, 229–239. [CrossRef]
27. Meng, L.; Liang, H.; Yu, H.; Yang, J.; Li, F.; Wang, Z.; Zeng, X. The energy absorption and bearing capacity of light-weight bio-inspired structures produced by selective laser melting. *J. Mech. Behav. Biomed. Mater.* **2019**, *93*, 170–182. [CrossRef]
28. Lalegani Dezaki, M.; Mohd Ariffin, M.K.A.; Hatami, S. An overview of fused deposition modelling (FDM): Research, development and process optimisation. *Rapid Prototyp. J.* **2021**, *27*, 562–582. [CrossRef]
29. Paolini, A.; Kollmannsberger, S.; Rank, E. Additive manufacturing in construction: A review on processes, applications, and digital planning methods. *Addit. Manuf.* **2019**, *30*, 100894. [CrossRef]
30. Liu, Z.; Wang, Y.; Wu, B.; Cui, C.; Guo, Y.; Yan, C. A critical review of fused deposition modeling 3D printing technology in manufacturing polylactic acid parts. *Int. J. Adv. Manuf. Technol.* **2019**, *102*, 2877–2889. [CrossRef]
31. Płaczek, M.; Ariffin, M.K.A.; Baharudin, B.T.H.T.; Lalegani Dezaki, M. The Effects of 3D Printing Structural Modelling on Compression Properties for Material Jetting and FDM Process. In *Experiments and Simulations in Advanced Manufacturing*; Kyratsis, P., Davim, J.P., Eds.; Springer International Publishing: Cham, Switzerland, 2021; pp. 171–194.
32. Bourell, D.; Kruth, J.P.; Leu, M.; Levy, G.; Rosen, D.; Beese, A.M.; Clare, A.J.C.A. Materials for additive manufacturing. *CIRP Ann.* **2017**, *66*, 659–681. [CrossRef]
33. Avinc, O.; Khoddami, A. Overview of Poly(lactic acid) (PLA) Fibre. *Fibre Chem.* **2009**, *41*, 391–401. [CrossRef]
34. Singh, S.K.; Ahmed, S.U.; Pandey, A. Metabolic engineering approaches for lactic acid production. *Process Biochem.* **2006**, *41*, 991–1000. [CrossRef]
35. Rebelo, H.B.; Lecompte, D.; Cismasiu, C.; Jonet, A.; Belkassem, B.; Maazoun, A. Experimental and numerical investigation on 3D printed PLA sacrificial honeycomb cladding. *Int. J. Impact Eng.* **2019**, *131*, 162–173. [CrossRef]
36. Chen, Y.; Li, T.; Jia, Z.; Scarpa, F.; Yao, C.-W.; Wang, L. 3D printed hierarchical honeycombs with shape integrity under large compressive deformations. *Mater. Des.* **2018**, *137*, 226–234. [CrossRef]
37. Kucewicz, M.; Baranowski, P.; Małachowski, J.; Popławski, A.; Płatek, P. Modelling, and characterization of 3D printed cellular structures. *Mater. Des.* **2018**, *142*, 177–189. [CrossRef]
38. Sahu, S.K.; Badgayan, N.D.; Samanta, S.; Sahu, D.; Sreekanth, P.S.R. Influence of cell size on out of plane stiffness and in-plane compliance character of the sandwich beam made with tunable PCTPE nylon honeycomb core and hybrid polymer nanocomposite skin. *Int. J. Mech. Sci.* **2018**, *148*, 284–292. [CrossRef]

39. Wang, D.; Abdalla, M.M.; Zhang, W. Buckling optimization design of curved stiffeners for grid-stiffened composite structures. *Compos. Struct.* **2017**, *159*, 656–666. [CrossRef]
40. Khan, M.K. Compressive and lamination strength of honeycomb sandwich panels with strain energy calculation from ASTM standards. *Proc. Inst. Mech. Eng. Part G J. Aerosp. Eng.* **2006**, *220*, 375–386. [CrossRef]
41. Mah, C. *Starfish: Biology & Ecology of the Asteroidea*; Lawrence, J., Ed.; JHU Press: Baltimore, MD, USA, 2013; Volume 53, pp. 871–873. [CrossRef]
42. Yuan, Y.; Yao, X.; Niu, K.; Liu, B.; Wuyun, Q. Compressive failure of fiber reinforced polymer composites by imperfection. *Compos. Part A Appl. Sci. Manuf.* **2019**, *118*, 106–116. [CrossRef]
43. Jerez-Mesa, R.; Travieso-Rodriguez, J.A.; Llumà-Fuentes, J.; Gomez-Gras, G.; Puig, D. Fatigue lifespan study of PLA parts obtained by additive manufacturing. *Procedia Manuf.* **2017**, *13*, 872–879. [CrossRef]
44. Lubombo, C.; Huneault, M.A. Effect of infill patterns on the mechanical performance of lightweight 3D-printed cellular PLA parts. *Mater. Today Commun.* **2018**, *17*, 214–228. [CrossRef]
45. Papka, S.D.; Kyriakides, S. In-plane compressive response and crushing of honeycomb. *J. Mech. Phys. Solids* **1994**, *42*, 1499–1532. [CrossRef]
46. Xiang, J.; Du, J. Energy absorption characteristics of bio-inspired honeycomb structure under axial impact loading. *Mater. Sci. Eng. A* **2017**, *696*, 283–289. [CrossRef]

polymers MDPI

Review

Use of Industrial Wastes as Sustainable Nutrient Sources for Bacterial Cellulose (BC) Production: Mechanism, Advances, and Future Perspectives

Abudukeremu Kadier [1], R. A. Ilyas [2,3,*], M. R. M. Huzaifah [4,*], Nani Harihastuti [5], S. M. Sapuan [6,7], M. M. Harussani [6], M. N. M. Azlin [7,8], Rustiana Yuliasni [5], R. Ibrahim [9], M. S. N. Atikah [10], Junying Wang [1], K. Chandrasekhar [11], M Amirul Islam [12], Shubham Sharma [13], Sneh Punia [14], Aruliah Rajasekar [15,*], M. R. M. Asyraf [16] and M. R. Ishak [16]

1 Laboratory of Environmental Science and Technology, The Xinjiang Technical Institute of Physics and Chemistry, Key Laboratory of Functional Materials and Devices for Special Environments, Chinese Academy of Sciences, Urumqi 830011, China; abudukeremu@ms.xjb.ac.cn (A.K.); wangjunying19@mails.ucas.ac.cn (J.W.)
2 School of Chemical and Energy Engineering, Faculty of Engineering, Universiti Teknologi Malaysia (UTM), Johor Bahru 81310, Johor, Malaysia
3 Centre for Advanced Composite Materials (CACM), Universiti Teknologi Malaysia (UTM), Johor Bahru 81310, Johor, Malaysia
4 Faculty of Agricultural Science and Forestry, Bintulu Campus, Universiti Putra Malaysia, Bintulu 97000, Sarawak, Malaysia
5 Centre of Industrial Pollution Prevention Technology, The Ministry of Industry, Jawa Tengah 50136, Indonesia; nanisoeharto@yahoo.com (N.H.); rustianay@yahoo.com (R.Y.)
6 Advanced Engineering Materials and Composites Research Centre (AEMC), Department of Mechanical and Manufacturing Engineering, Faculty of Engineering, Universiti Putra Malaysia (UPM), Serdang 43400, Selangor, Malaysia; sapuan@upm.edu.my (S.M.S.); mmharussani17@gmail.com (M.M.H.)
7 Laboratory of Technology Biocomposite, Institute of Tropical Forestry and Forest Products (INTROP), Universiti Putra Malaysia (UPM), Serdang 43400, Selangor, Malaysia; mohdazlin@uitm.edu.my
8 Department of Textile Technology, School of Industrial Technology, Universiti Teknologi MARA, Universiti Teknologi Mara Negeri Sembilan, Kuala Pilah 72000, Negeri Sembilan, Malaysia
9 Innovation & Commercialization Division, Forest Research Institute Malaysia, Kepong 52109, Selangor Darul Ehsan, Malaysia; rushdan@frim.gov.my
10 Department of Chemical and Environmental Engineering Engineering, Faculty of Engineering, Universiti Putra Malaysia (UPM), Serdang 43400, Selangor, Malaysia; sitinuratikah_asper7@yahoo.com
11 School of Civil and Environmental Engineering, Yonsei University, Seoul 03722, Korea; chanduibt@gmail.com
12 Laboratory for Quantum Semiconductors and Photon-Based BioNanotechnology, Department of Electrical and Computer Engineering, Faculty of Engineering, Université de Sherbrooke, Sherbrooke, QC J1K 2R1, Canada; amirul.geb@gmail.com
13 Department of Mechanical Engineering, IK Gujral Punjab Technical University, Jalandhar 144001, India; shubham543sharma@gmail.com
14 Department of Food, Nutrition and Packaging Sciences, Clemson University, Clemson, SC 29634, USA; snehpunia69@gmail.com
15 Environmental Molecular Microbiology Research Laboratory, Department of Biotechnology, Thiruvalluvar University, Serkkadu, Vellore 632115, India
16 Department of Aerospace Engineering, Universiti Putra Malaysia (UPM), Serdang 43400, Selangor, Malaysia; asyrafriz96@gmail.com (M.R.M.A.); mohdridzwan@upm.edu.my (M.R.I.)
* Correspondence: ahmadilyas@utm.my (R.A.I.); muhammadhuzaifah@upm.edu.my (M.R.M.H.); rajasekargood@gmail.com (A.R.)

Citation: Kadier, A.; Ilyas, R.A.; Huzaifah, M.R.M.; Harihastuti, N.; Sapuan, S.M.; Harussani, M.M.; Azlin, M.N.M.; Yuliasni, R.; Ibrahim, R.; Atikah, M.S.N.; et al. Use of Industrial Wastes as Sustainable Nutrient Sources for Bacterial Cellulose (BC) Production: Mechanism, Advances, and Future Perspectives. *Polymers* **2021**, *13*, 3365. https://doi.org/10.3390/polym13193365

Academic Editors: Andrea Maio and Debora Puglia

Received: 1 September 2021
Accepted: 22 September 2021
Published: 30 September 2021

Abstract: A novel nanomaterial, bacterial cellulose (BC), has become noteworthy recently due to its better physicochemical properties and biodegradability, which are desirable for various applications. Since cost is a significant limitation in the production of cellulose, current efforts are focused on the use of industrial waste as a cost-effective substrate for the synthesis of BC or microbial cellulose. The utilization of industrial wastes and byproduct streams as fermentation media could improve the cost-competitiveness of BC production. This paper examines the feasibility of using typical wastes generated by industry sectors as sources of nutrients (carbon and nitrogen) for the commercial-scale production of BC. Numerous preliminary findings in the literature data have revealed the potential to yield a high concentration of BC from various industrial wastes. These findings indicated the need to optimize culture conditions, aiming for improved large-scale production of BC from waste streams.

Keywords: bacterial cellulose (BC); biopolymer; industrial waste; microbial cellulose; carbon source; nitrogen source

1. Introduction

As a novel nanomaterial, Bacterial Cellulose (BC) has continued to draw scholarly interests since it was first discovered due to its unique properties, such as a high degree of purity, biodegradability, biocompatibility, and ease of polymerisation [1,2], making BC a material with a wide range of applications including skin and bone tissue engineering, barrier technology, and electrical, electrochemical, and sensing applications [3–8]. Despite offering many beneficial properties, its expensive production cost bounds its industrial-scale application. Conventionally, producers utilize fructose and glycerol as conventional carbon sources, however, the costs of these materials are remarkably high. A growing research body studies methods of minimizing the BC production cost. However, it has ended up with unconvincing and inadequate findings [9]. Recent research on reducing the production costs has emphasized utilizing waste products for sources of carbon or nitrogen. At present, active research to investigate the cost-effectiveness of BC synthesis from different waste products is ongoing and needs to be elaborated.

Nevertheless, the literature analysis compiles crucial developments in the field and, hence, enables the assessment of the future practicability of this manufacturing of BC for various applications [10–12]. The feasibility of using waste in BC production is examined in this paper through an extensive literature review to strengthen the current phase of knowledge and analyse discernible trends and gaps in inexperience. Many industrial wastes are rich in carbon and nitrogen content; hence, utilizing them as substrates may yield high microbial cellulose concentrations with the optimization of culture conditions.

2. Overview of Bacterial Cellulose (BC) and Its Applications

Bacterial cellulose (BC), commonly known as biocellulose, which is the purest form of cellulose, continues to receive widespread focus due to its superior physicochemical properties compared to plant cellulose, in which impurities such as hemicellulose and lignin are often found [13–21]. Some of the superior physicochemical properties of BC include high tensile strength, crystallinity, and water holding capacity (WHC), as well as a slow water release rate (WRR) and remarkable moldability into three-dimensional structures [22]. The water molecules are bonded through hydrogen bonds within the complex structure of BC. The unbonded free water molecules will penetrate and exit the BC molecular structure, as shown in Figure 1 [23].

Bi et al. [24] characterized the BC synthesized from different strains in agitated culture. The macrostructure morphology of BC varied depending on the different culture methods [25]. The research used isolated bacteria, namely *Komagataeibacter nataicola Y19* (BC-1) and *Gluconacetobacter entanii* ACCC10215 (BC-2). Both bacteria were fermented in Hestrin–Schramm medium. The BC morphology result depicts that both samples have different sizes and shapes, as shown in Figure 2. Figure 2a,b illustrates the optical image of BC-1 and BC2, while Figure 2c,d shows the morphology of BC samples. The BC-1 shows the flocky asterisk-like and solid sphere-like for BC-2. In addition, based on Pang et al. [7], BC is useful as a natural renewable polymer in many fields due to its versatility and numerous notable properties such as biocompatibility, chirality, structure-forming potential, hydrophilicity, high crystallinity, high purity, a high degree of polymerization, high porosity, large specific area, favourable permeability, flexibility, hygroscopicity, and biodegradability. BC is produced as extrusions of glucose chains from the bacterial body via small pores present on their cell envelope. These extrusions then form microfibrils that further aggregate into web-shaped cellulose ribbon networks with many empty spaces between the fibres. The well-separated non-fibrils of BC create an expanded surface area and highly porous matrix. The basic fibril structure contains a β-1→4 glucan chain with the molecular formula $(C_6H_{10}O_5)_n$ and is held together by hydrogen bonds. These microfibrils are approximately 100-fold smaller than the fibrils of vegetal cellulose [26]. Until recently, much research was done on the production of BC and its modification and applications in various fields. As displayed in Figure 3, the number of BC publications has increased since 2000 from 81 to 819 publications.

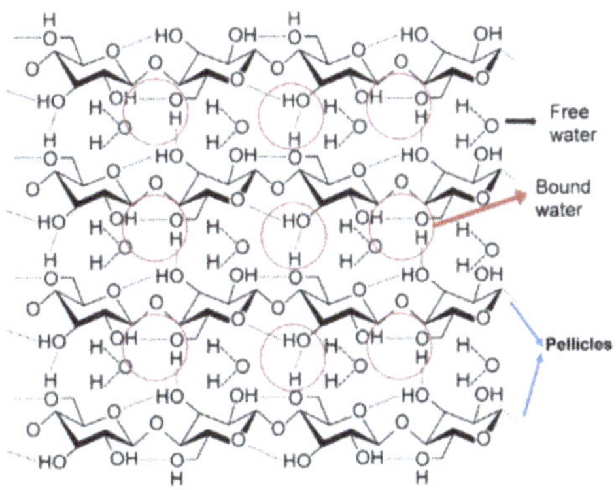

Figure 1. Schematic of the molecular structure of bacterial cellulose and its bound and free water [23].

Figure 2. Optical images (**a**,**b**), scanning electron microscope (SEM) images (**c**) of BC samples and ultrastructural transmission electron microscopy (TEM) images (**d**) of BC samples [24,27,28].

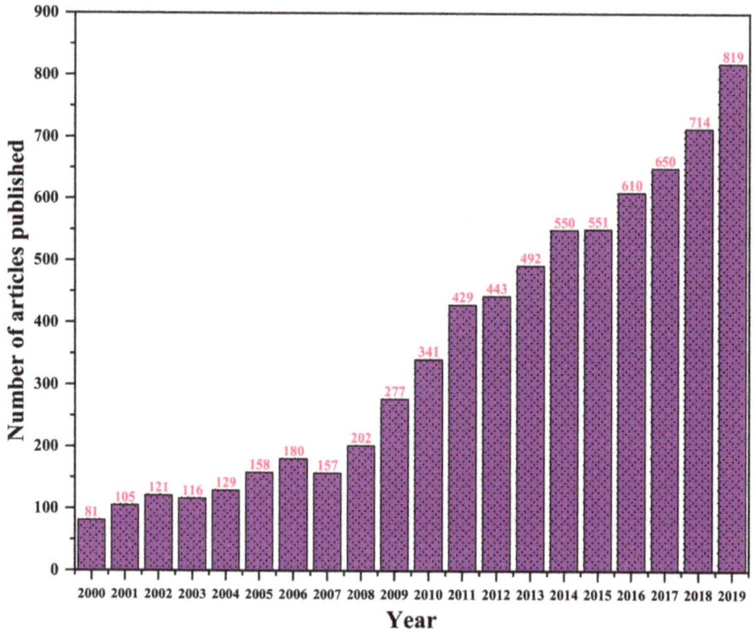

Figure 3. Number of publications on bacterial cellulose since 2000–2020 (Scopus search engine system, the search term "bacterial cellulose").

The unique macro-physical and outstanding thermal and mechanical properties of BC make it an ideal material to be applied in various fields of applications (Figure 4). BC possesses good thermal stability and low or no chronic inflammatory response, which has attracted huge attention for BC as a novel functional material in applications such as nonwoven fabric-like products and paper [29]. BC is also used as a binder in advanced paper technology due to its nano-sized structure, a property that significantly improves the durability and strength of pulp when reinforced into paper [30]. One of the main reasons it is being used in biomedicine is its excellent biocompatibility [14]. In addition, the weight-average degree of polymerization (DPw) of BC is high, such as the DPw of BC produce by *Acetobacter xylinum* BPR2001, which remained in the range of 14,000 of 16,000 [7,31]. BC possesses nanofibrillar and ultrafine structured material with an excellent combination of properties such as high flexibility and tensile strength (Young modulus of 114 GPa) [32], as well as high crystallinity (84–89%) [33]. Therefore, due to its outstanding mechanical properties, BC nanocomposites had been fabricated by reinforcing it with other polymers to be developed in various applications, including paper [29], treating tympanic membrane perforation [34,35], shielding film [36], food packaging films [37], audio speaker diaphragms [38], and so on. Development of BC for paper products had been actively conducted by Ajinomoto Corporation along with Mitsubishi Paper Mills in Japan since 1995 (JP patent 63295793) [39].

Due to the high porosity combined with a large specific area of three-dimensional structure, research on BC has opened up opportunities for it to be used as a photocatalyst [40], electronic sensing platform [41], and biosensing material [42,43] (Figure 5). BC has also been used widely in biomedical applications such as wound-dressing [44–47] (applied on the wounded torso, hand, and face) and cell culture [48–51] because of its excellent flexibility, high mechanical strength at wet state, water holding capacity, very low risk for irritation due to its ultra-high purity, hygroscopicity, liquid/gasses permeability, and ease of wound inspection due to its transparency. Biopolymer such as polylactic acid (PLA), starch, polyhydroxyalkanoate (PHA) [52–56], and synthetic polymer such as polyvinyl alcohol (PVA) and unsaturated polyester (UP) [57,58] are potential polymers to be reinforced with BC. The outstanding properties of BC such as biodegradability, good controllability during BC production, and possessing net-like morphology that is almost similar to human collagen as a biomimetic feature makes it favoured in the medical field and has been widely utilized in controlled drug delivery [59], medical pads [41], artificial skin [7,60], cartilage [61] and bone [62,63], bone tissue engineering scaffolds [64–66], hormones [72], and nerve guides for spinal cord injury treatment [73]. vascular grafting [67,68], blood vessel tubes [69,70], dental implant [71], proteins and hormones [72], and nerve guides for spinal cord injury treatment [73].

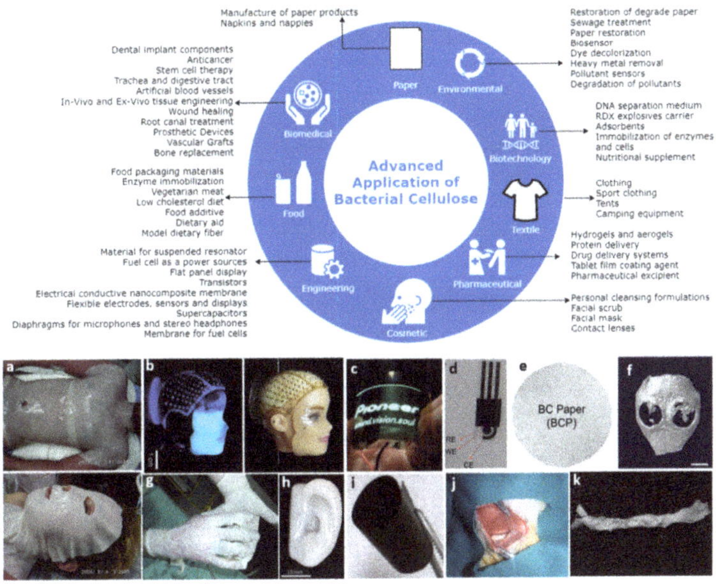

Figure 4. Advanced application of bacterial cellulose (BC). (**a**) A never-dried microbial cellulose membrane shows remarkable conformability to the various body contours, maintains a moist environment, and significantly reduces pain [74]. (**b**) A doll face was scanned, and a 4.5 wt % Flink containing A. xylinum was deposited onto the face using a custom-built 3D printer. In situ cellulose growth leads to the formation of a cellulose-reinforced hydrogel that, after removal of all biological residues, can serve as a skin transplant [75]. (**c**) Luminescence of an organic light-emitting diode deposited onto a flexible, low-CTE, and optically transparent cellulose nanocomposite [76]. (**d**) Screen-printed electrodes made on BC substrate [77]. (**e**) BC paper [78]. (**f**) Bone regeneration efficacy of the scaffolds [79]. (**g**) Microbial cellulose dressing applied on a wounded hand [80]. (**h**) 3D Bioprinting Human Chondrocytes with nanocellulose–alginate bioink [81]. (**i**) Flexible freestanding nanocellulose paper-based Si Anodes for Lithium-ion batteries [82]. (**j**) Cellulose acetate/poly lactic acid coaxial wet-electrospun scaffold containing citalopram-loaded gelatin nanocarriers for neural tissue [83]. (**k**) Artificial Bacterial cellulose ligament or tendons [84].

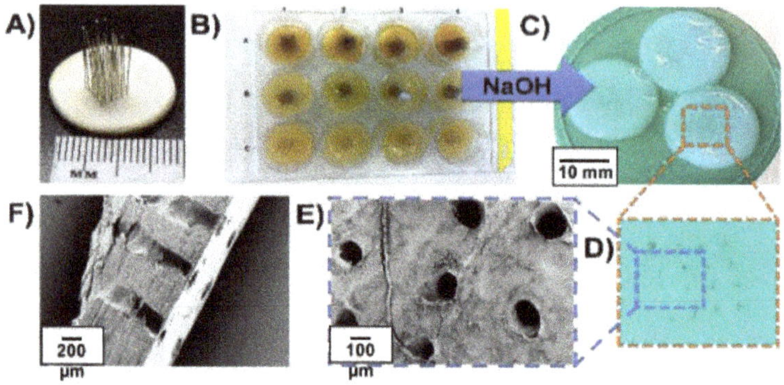

Figure 5. Experimental images of (**A**) clay-needle template with needles at the centre, (**B**) growing BC scaffolds with aid from clay-needle templates in static cultures, and (**C**) clean BC pellicles with channels to be an effective hydrogel-like material for different tissue engineering applications. (**D**) Enlargement of the channel area in (**C**). The channel diameter was approximately 250 µm and the inter-distance approximately 1 mm. (**E**) Scanning electron images (SEM) of channeled area in (**C**). (**F**) Cross-section of channels [85].

BC possess large surface areas and have the capability to absorb liquids. Hence, a small amount of BC can be utilized for producing coating, thickening, and binding agents, especially in the food industry. Remarkably, in 1992, BC was categorized as "generally recognized as safe" (GRAS) by the USA Food and Drug Administration (FDA) and, hence, is suitable to be used in food industry applications [86]. The largest industrial-scale production of BC that has been produced so far is led by Cetus Co. (Emeryville, CA, USA) and Weyerhaeuser Co. (Tacoma, Washington, DC, USA). Both companies develop a Cellulon, a bulking agent with a wide range of applications such as in coating, binding, and thickening applications [87]. Besides that, BC also can be used in the oil and gas recovery sector, cosmetics, adhesives, paints, and mining. High-end audio speaker systems had been fabricated by Sony Corporation using BC. This might be due to its good acoustic properties [88].

Food packing, battery separator, transparent coating or film, pharmaceutical industries, adsorbent, cosmetics, water treatment, ethanol production, biomaterials, artificial blood vessels, electric conductors or magnetic materials, and scaffolds for tissue engineering are examples of uses of BC in industrial and medical areas [10,90]. This can be observed in Figure 6. Besides that, BC has been utilized in biomedical applications such as scaffolds and ex-situ and in situ modified through different processes [91]. The culture conditions are modified with additives or reinforcement materials via in situ modification, whereas the modification of ex-situ is performed after BC harvest. Incorporation of the additive materials can be added into a growing BC microfibril for the preparation of BC composites with required properties. A biocomposite is a material composed of two or more distinct constituent materials (one being naturally derived) which are combined to yield a new material with improved performance over single constituent materials [92–95]. This modification type can be employed in a static method for the purpose of control of properties, shape, and structure of modified BC. This application is mostly applied in bone tissue engineering, in which, in order to produce BC scaffolds with microporous structure, paraffin wax microspheres were added into culture medium via an in situ modification technique [96].

Gonçalves-Pimentel et al. [97] conducted experiments on BC as a support for the growth of retinal pigment epithelium, showing that all surface-modified BC substrates showed similar permeation coefficients with solutes of up to 300 kDa. Surface modification of BC greatly improved the proliferation and adhesion of retinal pigment epithelium cells. All samples showed. Insignificant stress–strain behaviour was observed in all samples, of which acetylated BC showed the highest elastic modulus; however, after some period, it exhibited a slightly smaller tensile strength and elongation at break as compared to control BC [98]. A study conducted by Buruaga-Ramiro [78] on the suitability of BC matrices to prepare enzymatically active nanocomposites shows improvement in durability, reusability, and thermal stability of BC/lipase nanocomposites (Figure 7). Besides that, the enzyme immobilised onto BC/lipase nanocomposites paper retained 60% of its activity after 48 h at 60 °C. The results attained suggest that BC/lipase nanocomposites are promising biomaterials for the development of green biotechnological devices with potential applications to be used as part of biosensor devices with applications in many fields such as food quality control, environmental monitoring, and clinical diagnosis.

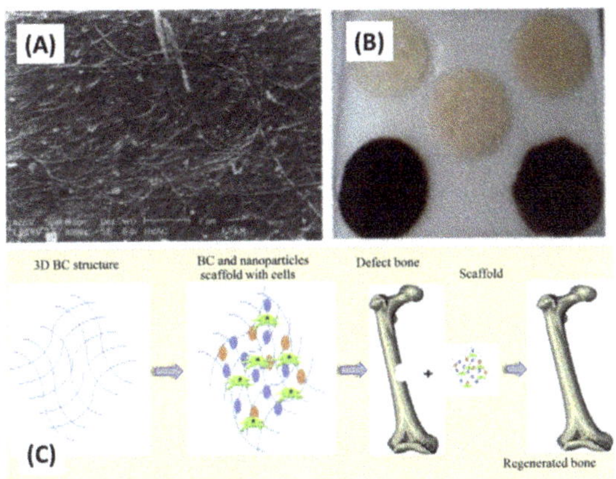

Figure 6. (**A**) Network structure of ribbon-shaped fibrils of BC, (**B**) natural biomaterial of BC, and (**C**) 3D-shaped BC for bone tissue engineering [87,89].

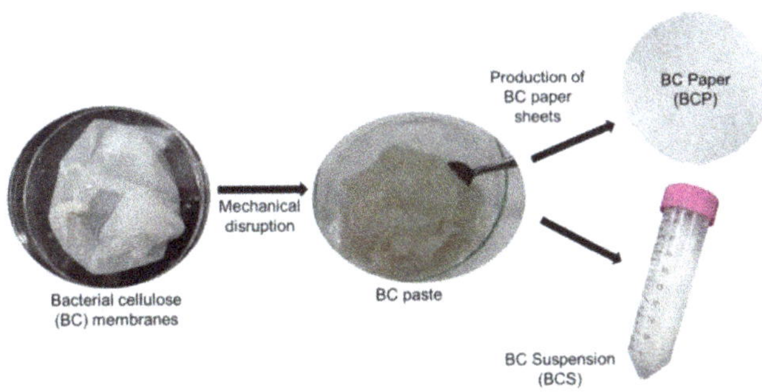

Figure 7. Schematic diagram to explain the approach for bacterial cellulose matrices production [78].

The effect of BC on disintegrability in composting conditions of plasticized polyhydroxybutyrate (PHB) nanocomposites [99] shows that the compounds with BC and plasticizer presented a similar behaviour to that of control plasticized PHB. This might be due to the low dispersion and low interfacial adhesion of BC in the matrix. However, the crystallinity of PHB nanocomposites was increased. Another study conducted by Zhang et al. [100] on the reinforcement of BC with polyvinyl alcohol (PVA) coated with Bichar-Nanosilver (C-Ag) antibacterial composite membranes, shows that the BC was homogeneously mixed into the PVA gel and that the C-Ag particles were uniformly immobilized in the PVA/BC hybrid composites membrane. These hybrid composites show excellent antibacterial activity against *Escherichia coli* and good reusability to be used as drinking water treatment applications. Hamedi et al. [101] conducted experiments on double-network antibacterial hydrogel based on aminated BC and schizophyllan (SPG) biopolymer nanocomposites. A novel hydrogel composed of BC/SPG biopolymers shows an improvement in antibacterial, swelling, and mechanical properties. MTT assay displayed that amine-grafted BC/SPG stimulated the proliferation of normal human fibroblast cells. They concluded that this novel nanocomposite can be utilized in diverse areas such as anti-wrinkle dressing masks, wound healing, and absorption biomaterial for water treatment applications.

A hybrid of BCNCs–AgNPs/alginate–MoO_3NPs was effectively developed for H_2S gas sensors [102]. In this study, BC was produced by *Gluconacetobacter xylinus* strain under static culture. The bionanocomposites film was successfully fabricated using a solution casting method and has the ability to detect H_2S gas emission. Through the shift in the oxidation number of MoO_3NPs, the colour of the film was changed. Once activated by AgNPs, MoO_3NPs were readily reduced to a coloured sub-oxide by atomic hydrogen that was produced and received from the reaction of H_2S gas [102].

Cazón et al. [3] conducted a study on BC reinforced polyvinyl alcohol (PVOH) composite film with eco-friendly UV-protective properties. The addition of PVOH shows improvement in mechanical and transparency properties and reduced the water vapour permeability of composite films. Thus, they concluded that these novel composite films have huge advantages to be utilized in the food industry to prevent oxidation of proteins, lipids, and vitamins, as well as the degradation of antioxidants in foods. Besides that, it can be s substitute novel material to antioxidants to increase food shelf-life as well as to maintain the quality of food products [3]. AgNP produced using CUR:HPβCD (cAgNP) reinforced with BC-based hydrogels for wound dressing applications has been developed by Gupta et al. [103]. The composites show high cytocompatibility between cAgNp and BC with high moisture content and a good level of transparency. These hydrogels-based composites also showed broad-spectrum antimicrobial activity along with antioxidant properties.

In terms of electrical applications, there are a few applications of BC, as bioelectrical devices are hard to fabricate. However, several previous works have been conducted. According to Di Pasquale et al. [104], electrodes of the sensor are made of BC that has been treated with ionic solutions and coated with conducting polymers. The mechano-electric transduction properties of the composite are used to create a generating sensor. The device, which is placed in a cantilever arrangement, is used to detect anchor acceleration. On the other hand, Di Pasquale and co-researchers are testing an all-organic Bacterial Cellulose-Conducting Polymer (BC)-PEDOT:PSS composite soaked with Ionic Liquids (ILs) as a mass sensor [104]. As a result of the applied deformation, the sensor functions as a vibrating transducer in a cantilever arrangement, producing a voltage signal. The effect of the additional mass on the system's frequency response is used to estimate the value of the measurand. The sensing system is made of low-cost, flexible, and environmentally friendly components that may be used to create smart ubiquitous sensing systems in the future.

Wang et al. [105] created a novel wound care system that uses an aligned bacterial cellulose (BC)/gelatin membrane in combination with EF stimulation to direct cell migration and improve wound healing. The produced BC/gelatin membranes had a well-aligned fibre structure, a strong mechanical property, a high thermal stability, good light transmittance, foldability, and surface roughness, and great biocompatibility. Especially, the 40% stretched BC/gelatin membrane promoted the adhesion, orientation, and migration of NIH3T3 cells in vitro. For further increase in electrical conductivity and cell survival of polyaniline (PANI) coated BC nanocomposites, BC fibres are chemically functionalized with a poly(4-vinylaniline) (PVAN) interlayer [106]. PVAN was discovered to have increased PANI yield by promoting the creation of a uniform PANI layer with nanofiber- and nanorod-like supramolecular structures. These new electrically conductive BC/PVAN/PANI nanocomposites have the potential to enable a wide range of biomedical applications, including bioelectronic interfaces and the manufacturing of biosensors. Table 1 displays BC and its biocomposites yielded in static and agitation/shaking culture bioreactor and their various applications.

Table 1. Fabrication of BC and BC-based biocomposites under static and agitation culture methods, their properties, and applications.

Bacterial Cellulose and Bacterial Cellulose-Based Biocomposites	Applications	Structure and Properties	References
Fabrication of BC and BC-based composites under static culture methods			
BC	BC mask	Fast healing, high moisture donation, and high conformability	Saxena et al. [107]
	Blood vessel; Vascular grafts	Excellent mechanical properties, thin layers, dense	Putra et al. [108]
	Implant material for auricular cartilage regeneration and for ear cartilage replacement	Compatible mechanical strength and patient-specific shapes	Nimeskern et al. [109]
	Potential meniscus implant	High compression strain and mechanical strength	Bodin et al. [110]
	Replacement of blood vessels and diseased arteries	High water holding capacity and mechanical strength	Charpentier et al. [111]
	Artificial blood vessels for microsurgery	The smooth inner surface, moldability, and high mechanical properties	Klemm et al. [112]
	Artificial cornea and eye bioengineering Retinal pigment epithelium (RPE)	High elastic modulus, tensile strength and elongation at break, high initial cell adhesion, porous, permeable up to 300 kDa, and dimensionally stable	Padra et al. [98]
BC/polycaprolactone biocomposites	Tissue substitutes in rabbits' cornea	Signs of the moderate inflammatory process, protected ocular surface and remained stable in corneal tissue during the 45-day follow-up	Sepúlveda et al. [113]
BC/polycaprolactone (PCL) biocomposites	Biodegradable food packaging	Good transparency of the BC/PCL, smooth surface morphology	Barud et al. [114]
BC/benzoyltrifluoroacetone	Phosphors and UV to Visible energy converting devices	Improvement of the luminescence characteristics	Caiut et al. [115]
BC/ AgNPs/ lginate-molybdenum trioxide nanoparticles (MoO_3NPs)	Hydrogen sulfide (H_2S) gas sensor	Successfully detected H_2S gas	Sukhavattanakul et al. [102]
BC/chitosan biocomposites	Wound dressing	The improved proliferation and fibroblast adhesion	Kim et al. [116]
BC/Lipase nanocomposites	Bioactive paper for developing a simple, handheld, and disposable devices; industrials bio- processes of detergents and food industry and biomedicine	Specific activity was higher for BC/ Lipase suspension (4.2 U/mg), improved thermal stability, reusability, and durability	Buruaga-Ramiro [78]
BC/ SOD (Procel-Super) and poviargol (Procel-PA) biocomposites	Skin regeneration scaffold; Membranes for skin tissue regeneration	Highly transparency, antibacterial activity	Legeza et al. [117]
BC/ PVOH	The food industry, food packaging	Improved mechanical properties; UV-light barrier properties; Reduce WVP and porosity value	Cazón et al. [3]
BC/ PHB	Food packaging applications	low dispersion of BC in the matrix; increased crystallinity of the incubated samples; low interfacial adhesion	Seoane et al. [99]

Table 1. *Cont.*

Bacterial Cellulose and Bacterial Cellulose-Based Biocomposites	Applications	Structure and Properties	References
BC/ciprofloxacin biocomposites	Contact lens for better tissue regeneration, enhance the recovery of ocular burns, replacement for antibiotics eye drops, wound dressing after eye surgery or protection against bacteria.	No mutagenicity, genotoxicity and cytotoxicity effects	Messaddeq et al. [118]
BC/ polyvinyl alcohol coated biochar nanosilver biocomposites	Drinking water treatment	BC was uniformly mixed into the PVA gel; PVA/BC/C-Ag composite membranes exhibited excellent antibacterial activity; good reusability	Zhang et al. [100]
BC/polycaprolactone biocomposites	Tissue substitutes in rabbit cornea	High transparency and mechanical properties	Sepúlveda et al. [113]
BC/polyvinyl alcohol biocomposites	BC gloves	Skin cell support and fabrication of optimal moist condition	Osorio et al. [119]
BC/ cAgNP	Wound dressing	High cytocompatibility; high moisture content and; good level of transparency; broad-spectrum antimicrobial activity along with antioxidant properties	Gupta et al. [103]
Fabrication of BC and BC-based composites under agitation/shaking culture method			
BC	Sewage treatment; Immobilized reaction; Adsorption of Pb^{2+} bio-separation and bovine serum albumin	Porous and loose structure, BC adhering to each other; diameter of composites with a size range of 0.5–6 mm	Zhu, Jia, Yang, et al. [120]
	The production of immobilized glucoamylase was supported by BC spheres for industrial applications usage	BC spheres were produced with various range of sizes such as 0.5–1.5, 2–3, and 4–5 mm; Large functional group, as well as great surface area to connect with enzymes, resulted to the higher activity of small spheres.	Wu & Li, [121]
	For good viability and adhesion on the surface of the material	Sphere formation was affected by temperature; solid structure formed; diameter of composites with a size range of 2–8 mm formed	Hu et al. [122]
	Fermentation	IR: 6.52–3.85; Crystallinity: 81.43–84.35 %; Flocky asterisk-like; diameter of composites with a size range of 5–10 mm,	Bi et al. [24]
	Food, healthcare, and materials applications	Diameter is less than 1–8 mm at 150 rpm; Form solid structure however the central region is not layered; Layer spacing 10 μm (150 rpm) and 20 μm (125 rpm)	Hu & Catchmark [123]
	Good production yield	Thinner microfibrils structure; IR: 4.48; crystallinity: 84%; large and unique spheres; diameter of composites with a size range of 5–10 mm	Czaja et al. [124]
	High-efficiency lipase-immobilization system for large-scale industrial hydrolysis of fats and oils Suitable for enzymatic immobilization.	High hydrolytic activity; High operational activity; Lipase immobilized BC sphere; Size of diameter between 3–9 mm.	Cai et al. [125]

Table 1. *Cont.*

Bacterial Cellulose and Bacterial Cellulose-Based Biocomposites	Applications	Structure and Properties	References
	Pectin and xyloglucan can be used to enhance cellulose growth and cellulose assembly.	Xyloglucan: Layered structure, densely packed cellulose bundles with the layered structure were formed; Central core is not obviously seen; diameter of composites with a size range of 4–5 mm; aster-likePectin: Densely packed cellulose bundles with layered structure; diameter of composites with a size range of 5–6 mm; aster-like Xylan: Pore structure of cellulose bundles with a few tails formed on the surface of sphere; diameter of composites with a size range of 7–8 mm; layered structure Arabinogalactan: Cellulose linkage between layered structure; diameter of composites with a size range of 4–6 mm; Sphere	Gu & Catchmark [126]
BC/ schizophyllan (SPG) biopolymers	Anti-wrinkle dressing masks, wound healing and absorption materials	Mechanical, swelling and antibacterial properties were improved; moderate antibacterial activity	Hamedi et al. [101]
BC/CNT biocomposites	–	BC: IR index 2.23, crystallinity 67.2%; snow like structuredBC/CNT composites: IR index 2.56; crystallinity 76.2%, the diameter of composites with a size range of 2–5 mm, rice-like structured	Yan et al. [127]
BC/Fe_3O_4 biocomposites	Elution: $Mn^{2+} > Pb^{2+} > Cr^{3+}$ Superparamagnetic Adsorption: $Pb^{2+} > Mn^{2+} > Cr^{3+}$	Iron(II,III) oxide (Fe_3O_4) particles with a size of 15 nm were distributed uniformly in spheres The diameter of composites with a size range of 3–5 mm	Zhu, Jia, Wan, et al. [128]
BC/GO biocomposites	Superabsorbent for water environmental protection	Superior absorption capacity; Interconnected structure with a honeycomb-like surface pattern; diameter of composites with a size range of 3–7 mm	Hu [129]

3. Principal Pathways of Cellulose Production

Biopolymer cellulose can be produced using four distinguishing methods, including cellulose extraction, cellulose biosynthesis, enzymatic synthesis, and chemosynthesis. The most well-known method is cellulose extraction from plants, including the elimination of lignin and hemicelluloses using alkali or acid treatments. According to Klemm et al. [112], there are two main sources in cellulose production including plants and microorganisms, as shown in Figure 1. Extensive research has been conducted on the extraction of cellulose fibre from various plant fibre, i.e., sugar palm fibre [130–138], water hyacinth [139], ginger fibre [140,141], kenaf [142], sugarcane [143,144], lemongrass [145], cassava, corn, oat, palm oil fibre, and others [146–148]. Next is cellulose biosynthesis by using different types of microorganisms; (i) ***bacteria*** (gram-negative: *Alcaligenes* [149], *Salmonella, Enterobacter, Pseudomonas* [150], *Gluconacetobacter xylinus* [151], *Agrobacterium* [152], *Komagataeibacter Medellinensis* [153], *Aerobacter, Achromobacter insuavis* [154], *Rhizobium leguminosarum* [155], *Acetobacter* spp. [156], *Acetobacter xylinum* [157], *Zoogloea* [97], and gram-positive: *Sarcina ventriculi* [158], *Leifsonia sp* [159], *Rhizosphere bacterium, Bacillus subtilis* [70,160]); (ii) ***fungi*** (*Aspergillus ornatus* [161], *Penicillium* sp. [162,163], *Aspergillus terreus* MS105 [164], *Aspergillus terreus* M1 [165], *Aspergillus niger, Rhizopus* sp. [166,167], *Aspergillus niger* [168], *Trichoderma longibrachiatum* [169], *Beauveria Bassiana* [170], *Ascomycota* [171,172], or *Basidiomycota* [173]); (iii) ***algae*** (*Gelidium elegans* [174], *Posidonia oceanica* [175], *Aegagropila Linnaei* [176], *Komagataeibacter hansenii* [177], *Cladophora glomerata* [178]). However, extracellular synthesized cellulose as fibres is not achievable in some microorganisms. From the scientific viewpoint, the first enzymatic in vitro synthesis was initiated from cellobiosyl fluoride [179,180], and the earliest chemosynthesis was started from glucose via ring-opening polymerization of benzylated and pivaloylated derivatives [181]. These principle paths are schematically described in Figure 8 [112].

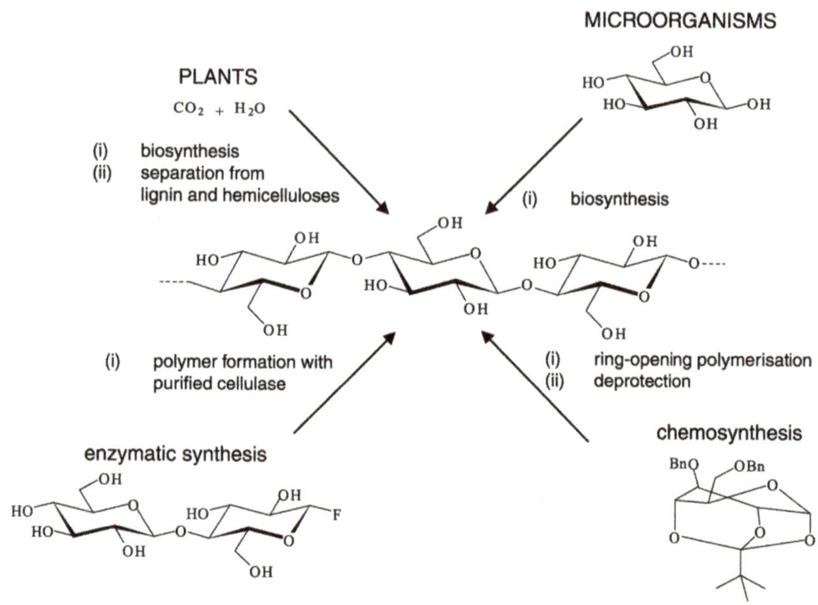

Figure 8. Major pathways to the cellulose [112].

Different modified methods or additives have been applied to enhance BC production. BC gained from bioreactors have been characterized and analysed for structure, shape, and properties of BC, thermogravimetric analysis, density, porosity, yield, water holding capacity, Fourier transform infrared, purity, zeta potential, degree of polymerization, surface area, chemical structure, pore size and distribution, degree of crystallinity, and microstructure, as well as macroscopic morphology [86,182].

4. Fundamentals of Bacterial Cellulose (BC) Production Process

Numerous aerobic and non-pathogenic bacteria yield BC from the genera *Gluconacetobacter*, *Sarcina*, *Rhizobium*, and *Agrobacterium* either in synthetic or non-synthetic media [22]. However, these bacteria are non-photosynthetic; therefore, they need glucose or organic substrate synthesized by the photosynthetic organism to accumulate their cellulose [183]. BC production comprises fermentation in static or agitated conditions. Among the cultivation media, the most frequently used cultivation medium is a chemically defined medium known as the Hestrin–Schramm (HS) medium [22]. This medium involves somewhat expensive additional components, such as peptone, yeast extract, citric acid, glucose, and disodium phosphate, resulting in costly production. According to Abol-Fotouh et al. [184], thermal-acidic pre-treatment was proposed to enhance the characteristics of molasses, boost its (glucose-fructose) content per volume, and remove the majority of contaminants that might stifle microbial development or reduce product output, as shown in Figure 9 [185,186]. The function of thermal acidic pre-treatment of molasses in virtually complete breakdown of the contained sucrose to its original constituents, glucose and fructose, was clarified by Bae and Shoda [187].

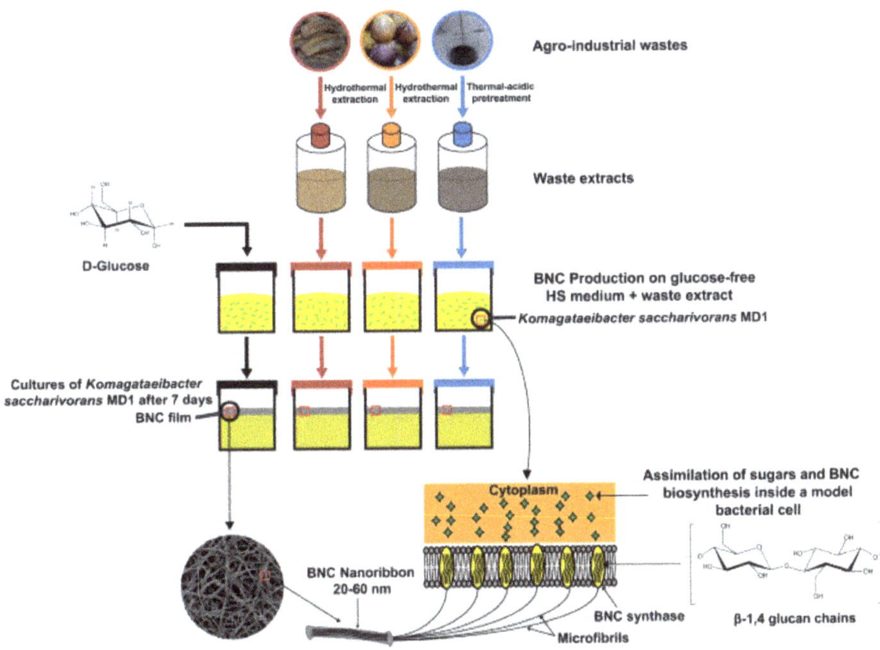

Figure 9. Schematic illustrations of pre-treatment of wastes for BC biosynthesis [184].

Alteration of growth conditions; temperature, pH, and sources of carbon and their concentrations influenced both the quality and quantity of BC yielded. In addition, different cultivation pathways led to the production of BC with different properties and structures [22]. Figure 10 illustrates the mechanism of bacterial cellulose synthesis from *G. oxydans* on the surface cell of cellulose [188].

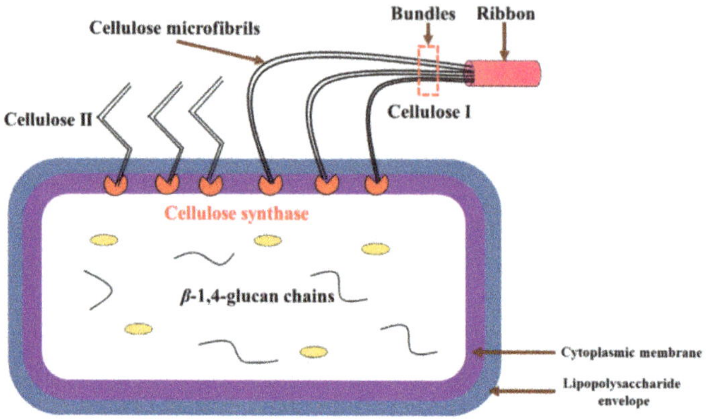

Figure 10. Representation of cellulose chains formation in microbial cells, and formation of micro- and macro fibrils, bundles, and ribbons [188].

The BC obtained after the fermentation process has yielded good properties for several applications. Stable and efficient bacteria strains will influence the effectiveness of bacterial cellulose (BC) production. The hydroxyl groups in the BC structure have enabled direct modification by introducing other polymers into the BC network [189]. However, some modifications can be done on BC by combining other materials into the polymeric system for a broader range of applications [188]. The modification process can be divided into two main groups, which are in situ and ex-situ modifications. An inadequate supply of oxygen causes bacteria to be inactive, which is a significant constraint in static production environments. Agitated conditions result in higher yields; however, the BC formation mechanism remains uncertain under different conditions [190].

More comprehensive applications of BC depend on practical considerations regarding production costs and scale-up capability. Recently, many studies have focused on cheap nutrient sources, diverse strains of cellulose-producing microorganisms, and supplementary components to produce value-effective BC [26]. Many waste products from different fields, such as whey, industrial waste, wastewater, and agro-industrial waste, have been examined as alternative substrates for the enhanced production of BC. Various additives or modified methods have been used to improve the production of BC. The BC harvested from other bioreactors has been characterized in terms of structure and properties such as macroscopic morphology, microstructure, degree of crystallinity, chemical structure, polymerization degree, purity, water holding capacity, porosity, and thermogravimetric ability [191]. Table 2 shows the BC production specifications, modifications, and advantages of different reactors.

Table 2. Various modifications, product specifications, and advantages of different reactors for BC production.

Modification	Production Specification and Advantages	BC Production	References
Enriched Oxygen Bioreactors			
Bubble column (controlled pH) Aeration rate:1.0 vvm (30 L/min)	**Attributes:** Low concentrated solution state culture; Low shear stress; Low mechanical properties: 17.15 to 11.66 MPa; Low crystallinity: 86 to 79.6%, Low degree of polymerization and molecular weight **Advantages:** High oxygen transfer rate	0.07–0.09 g/L/h	Choi et al. [192]
High oxygen concentration	**Attribute:** After 30 h the production decreased **Advantages:** Higher productivity; High oxygen transfer rate; Low power requirement.	0.20 g/L/h	Chao et al. [193]
Internal loop airlift with controlled pH/ fresh and glucose medium	**Attribute:** The highest concentration: 10.4 g/L at 60–70 g/L fructose **Advantages:** Formed a unique ellipse; Low mechanical strength; High hydrodynamic characteristic; High volumetric oxygen transfer	0.22 g/L/h	Chao et al. [194]
Internal loop airlift with enriched oxygen	**Advantages:** Unique ellipse was formed; High hydrodynamic characteristic; High volumetric oxygen transfer	0.116 g/L/h	Chao et al. [195]
Shaking flask with controlled pH/ Hestrin & Schramm medium	**Attribute:** A membrane-type BC was produced **Advantages:** Varying the net plates number would result in high Young's modulus and water holding capacity	-	Wu and Li [121]
Rotating disc bioreactors			
A rotating disk bioreactor	**Attribute:** A consistent product **Advantages:** Produced strong and intact cellulosic matrix, BC pore size of 10–15 μm; High tensile strength	-	Mormino & Bungay [196] Zahan et al. [197]
Rotating disk bioreactor supported by plastic composites	**Attribute:** A semi-continuous process **Advantages:** Low mechanical property (Young's modulus of 372.5 MPa); Low crystallinity: 66.9%; similar thermostability and water content with BC produced by static culture	0.01 g/L/day	Lin et al. [198]

Table 2. *Cont.*

Modification	Production Specification and Advantages	BC Production	References
Rotating disk bioreactor with different additions supported by plastic composites	**Attribute:** A semi-continuous process **Advantages:** Similar strain but lower stress for carboxymethylcellulose and avicel, respectively; High water retention properties of 98.6–99%; Disc rotation speed and oxygen concentration improved the fermentation process; Fructose concentration was decreased from 50 to 10 g/L; No re-inoculation	0.64 g/slice with 0.8% carboxymethylcellulose and avicel	Lin et al. [199]
Rotating magnetic field	**Advantages:** Yield BC with an altered degree of porosity and microstructure; Increased biochemical properties; Positive impact on the growth of bacteria; Increased water retention by 26% as compared to the control sample; high density with tangled and long fibres	-	Fijałkowski et al. [200–203]
Other bioreactors for BC production			
Spin filter supporting bioreactor	**Advantages:** BC concentration was increased from 5.65 to 11.52 g/L/140 h; An abundance of Cel + cells were converted into Cel- mutants	0.02 to 0.06 g/L/h	Jung et al. [204]
Fed-batch principle	**Advantages:** The gradient of a graph in the load-displacement diagram: (aerosol bioreactor = 34.7 N/10 mm, usual surface culture = 8.9 N/10 mm); High tensile strength: 114 N; High-quality cellulose; the degree of polymerization of BC is 5200; Best time interval: 6 h; BC layer or slices (3–4 cm); Culture box: low cost	-	Hornung et al. [205]
Biofilm reactor	**Advantages:** High crystallinity: 93% with a crystal size of 5.2 nm; high biomass density; Water retention ability up to 95 %; better thermal performance	7.05 g/L	Cheng et al. [206]
Biofilm reactor with additives	**Advantages:** Continuous BC production; High biomass density; High Young's modulus and tensile strength; High crystallinity: 80% with a crystal size of 4.2 nm; potential application of BC paper sheets	13 g/L	Cheng et al. [207]

5. Industrial Waste Streams as Feedstock for the Production of Bacterial Cellulose

Industrial-scale applications of BC manufacturing encountered some drawbacks such as high culture medium-cost as well as low yield production. In the fermentation process, the cost of the medium for the cultivation of BC accounts for 50–65% of the total production. Thus, according to Velásquez-Riaño et al. [208] and Vazquez et al. [190], the establishment of a cost-effective culture medium for optimum product yield is important in order to enhance the process of fermentation. Some hard work has been done to minimize the production cost of BC, such as using the low-cost medium for BC cultivation and accessible and renewable sources of the nutrient. Over the last twenty years, significant global energy, environmental, and economic concerns have set the prominence of accessible and sustainable utilization of various industrial wastes such as agro-industrial waste products, wastewater treatment plants of dairy industries, brewery and beverages industries waste, waste from textile mills, waste from the micro-algae industry, etc. The innovations in clean and green technology techniques, as well as biotechnological methods, have equipped scientists and researchers with platforms for renewable natural sources consumption, i.e., using industrial waste to produce BC. The utilization of these industrial wastes for BC production helps prevent disposal and environmental pollution, aid in waste management, and, hence, reduce industrial waste disposal costs. From this novel approach, the production of BC from the industrial wastes can be categorized into six individual industrial wastes as illustrated in Figure 11: (1) brewery and beverages industries wastes; (2) agro-industrial wastes; (3) lignocellulosic biorefineries, pulp mills, and sugar industries wastes; (4) textile mills; (5) micro-algae industry wastes; (6) biodiesel industry wastes.

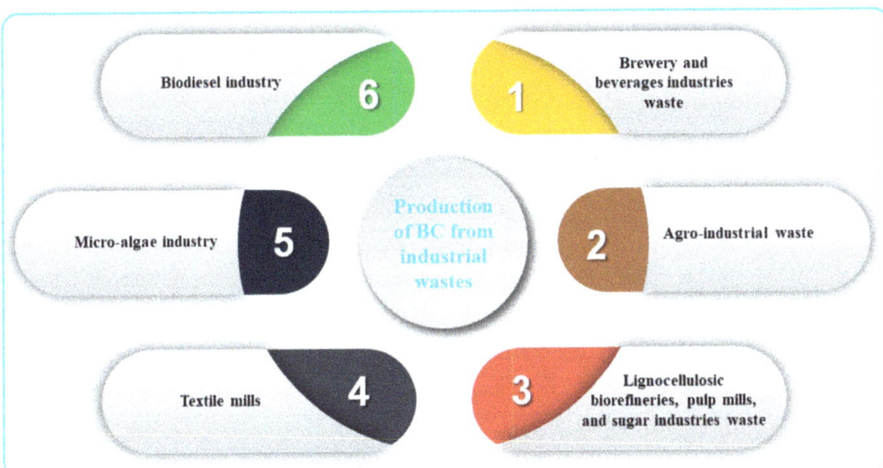

Figure 11. Schematic overview of Bacterial Cellulose (BC) production from different industrial wastes.

The examples are a small number of possible limitations of more lignocellulosic, sugar, brewery, and other industrial wastes as media without any additional nutrient source or as nitrogen and carbon sources with an additional nutrient source for the production of BC. Among all industrial waste, agro-industrial wastes are seen as highly potent and can be extensively utilized for producing BC. This might be due to the higher BC productivity and large-scale accessibility. Besides that, municipal waste is becoming a progressively more prominent source of biomass waste with high organic content, especially carbon, as a result of fast urbanization around the world, especially in developing countries [209,210]. The potential of upscale production of BC at a large scale or industrial scale from all the low-cost industries waste media are elaborated here; specifically, those that do not require complicated or complex supplementation, detoxification, and pre-treatments. Currently, the production of BC from industrial waste media has been observed to have comparable yield, physical, physico-chemical, crystallinity, and mechanical properties compared to standard media.

Industrial waste is a rich source of carbon for the bacterial synthesis of cellulose. In the past few decades, the urge to achieve 'zero waste' in the industrial sector has led many researchers to utilize industrial waste and byproducts as potential nutrient sources for microbial cultivation. Of many such industries, wastes and byproducts from the food, agriculture, and brewery industries are the most commonly utilised and can be a rich source of carbon [86]. All confectionery products are made of varying amounts of sugar and sugar substitutes. Since waste from confectionaries is rich in carbohydrates, this suggests that it can yield substantial amounts of carbon [211].

In terms of efficacy of BC produced by different utilized wastes, numerous researchers have extensively studied this issue. Kongruang [212] mentioned that agro-industries waste is richer in proteins, carbohydrates, and trace elements. Thus, it resulted in a higher BC productivity. Furthermore, Goelzer et al. [213] had stated that brewery and beverage industries waste mainly affects BC production, and similar influence can be seen from other different wastes such as wastewater sugar industries [214,215], lignocellulosic biorefineries wastes [216–218], and micro-algae biomass industries waste [177]. The results showed that utilizing pre-treated orange peel medium produced seven times more BC than using standard (HS) medium. However, structural research revealed that BC made from various wastes had a thicker and denser pack of nanofibrils, but FTIR spectra revealed no significant differences [214]. Plus, according to Fan et al. [216], in comparison to BC produced from HS medium, BC created from waste medium had no significant variations in microstructure, features, FTIR peaks, crystallinity index, or color parameter. These observations are in line with results discovered by Qi et al. [218], who found that the BC samples obtained from these hydrolysates had similar physico-chemical structural characteristics (microscopic morphology, functional groups, and crystallinity), but had a high water holding capacity and low mechanical strength. Furthermore, the physico-chemical characteristics of BC generated in various media were similar. However, when compared to HS medium, the viscosity of BC formed from molasses medium is low [217]. As a result, it may be inferred that agricultural wastes from all over the world can be used as a low-cost, readily available, and abundant feedstock for BC production.

Furthermore, fiber/textile industry waste derived hydrolysate was used as growth medium for BC production and the results showed 83% higher yield (10.8 g/L) and 79% higher tensile strength (0.070 MPa) of BC as compared to the production by glucose-based HS medium [215]. Cotton-based textile wastes were treated with the ionic liquid 1-allyl-3-methylimidazolium chloride before being hydrolyzed with enzymes. This resulted in a decreasing sugar concentration of 17 g/L in the hydrolysate. Because the natural sources used in the fiber/textile sector are often high in cellulose content, the wastes generated can be used to produce a variety of value-added products such as BC after detoxification and hydrolysis treatments [219,220].

Research by Costa et al. [26] revealed that industrial debris waste, namely sugarcane molasses, corn steep liquor (CSL), and jeans laundry effluent are rich sources of carbon and nitrogen that maintain BC production using *Gluconacetobacter hansenii* [26]. The wastes yielded a high concentration of microbial cellulose; however, the main limitation of the study was a substantial deformation observed in the product. According to Gao et al. [221], the cracks in the polymer might be due to the existence of crazing at the tip of the crack during tensile testing. Besides, BC microfibrils and nanofibrils sustained the cracks until rupture occurred [222]. The further analysis exposed that tensile testing caused the fibres to deform, leading to the formation of nodes [223,224].

In another study, Bıyık & Çoban [225] studied the potential of a bacterial strain isolated from a wine sample called *Acetobacter pasteurianus* for cellulose production using industrial waste and examined its performance with different carbon and nitrogen sources. The results showed that the presence of glucose and yeast extract in the media manufactured the highest quantity of microbial cellulose of 0.45 g/L. Among the industrial wastes (CSL, molasses, and whey), molasses produced the highest amount of BC (0.31 g/L). Further analysis of the structural properties of cellulose using Thin Layer Chromatography (TLC), Scanning Electron Microscopy (SEM), and Carbon-13 NMR revealed similarities in the structural characteristics of the BC with plant cellulose, indicated by the presence of non-branched polymer with D-glucopyranose units bonded with β-1, 4 bonds. Moreover, Voon et al. [226] used *Beijerinkia fluminensis* WAUPM53 and *Gluconacetobacter xylinus* 0416 bacteria to produce BC in sago byproducts (SBM), coconut water (CWM), and the standard Hestrin–Schramm mediums (HSM). The highest BC production was recorded in HSM followed by SBM and CWM for about 0.52 g/L, 0.47 g/L, and 0.45 g/L, respectively [226].

5.1. Brewery and Beverages Industries Wastes

There is a rising interest in the production of brewery and beverage industries because of the increasing user demand worldwide. According to Uzuner et al. [227], the beverage industry has become one of the biggest food processing industries. This industry can be categorized into two main groups that are non-alcoholic (i.e., whey, tea, cordial, coffee, apple, lassi, carbonated soft drink, etc.) and alcoholic drinks (i.e., whiskey, wine, beer, etc.). Carbonated soft drinks are consumed the most compared to other drinks, which are consumed at a rate of 48.8 gal/person, followed by bottled water, coffee, and beer with a value of 29.1, 24.6, and 21.8 gal/person, respectively [228]. This industry produces a large volume of waste per day and becomes a concern for management, spurring an effort to reduce the cost of disposal. These wastes are rich in nutrients; thus, they can be used for the biological treatment to produce BC for cost-effective and efficient waste management. Table 3 shows the evaluation of several waste or byproducts generated from the brewery and beverage industries to be used for BC production.

Whey is known to be rich in various nutritional components; hence, a growing literature body examines the feasibility of utilizing waste products as low-cost substrates for improved BC production [229]. Specifically, whey protein functions as an excellent source of nutrients. Revin et al. [230] examined the utilization of the dairy and alcohol industries acidic wastes, stillage (TS) and cheese whey, for the economical manufacturing of BC with *Gluconacetobacter sucrofermentans*. The findings revealed that, in three days of cultivation, the bacterial strain in whey produced up to 5.45 g/L of B and C structural properties analysis showed similarities between the synthesised cellulose with plant cellulose, despite morphological differences associated with crystallinity. The findings also indicated that acidic byproducts of dairy industries, such as wheat stillage and whey, are potential affordable sources of nitrogen and carbon for BC production.

Thin stillage (TS), a liquid byproduct produced after microbial fermentation of carbohydrates by yeast, contains various organic compounds. Hence, it is a potential source of nitrogen and carbon for BC synthesis. TS quantification via NMR methods showed that whey TS is rich in nutrition, containing high concentrations of lactic acid (7.41 g/L), dextrin (11.65 g/L), ethanol (1.31 g/L), acetic acid (2.72 g/L), and glycerol (7.87 g/L) [231]. TS wastewater from rice wine distilleries demonstrated the capability of producing BC with a concentration of 6.26 g/L in a seven days period of *Gluconacetobacter xylinus* cultivation [232]. From the study, it is confirmed that low-cost production of BC using TS as a substitute for HS medium is possible and the best alternative. Furthermore, the research revealed a facile and more practical approach for wastewater disposal. There have been efforts to enhance BC formation under static conditions by evaluating BC formation using *Gluconacetobacter xylinus* and a combination of whey and fruits as a culture medium by Jozala et al. [233]. The findings were in good correspondence to results achieved in other studies using *Gluconacetobacter sucrofermentans* [151].

5.2. *Agro-Industry Waste*

Several studies on the practicability of using different sources of agro-industry waste in BC production are reported. For instance, using *Komagataeibacter hansenii* for BC manufacturing from sisal juice as the substrate [234]. The researchers evaluated the effects of various variables on the potential of production, including the sugar concentration, pH, duration of cultivation, and nitrogen supplementation. From the findings, the best BC yield achieved from sisal waste was 3.38 g/L, which was yielded after 10 days of cultivation at a pH of 5. The study recommended that sisal waste is a precious resource for BC production; however, concerns arise regarding the ease of availability of sisal waste for large-scale manufacturing.

In a related study, Castro et al. [235] characterized the structural properties of BC obtained from agrochemical wastes of sugarcane and pineapple using *Gluconacetobacter swingsii*. HS medium was used as the reference standard for the comparisons. The results revealed that pineapple peel juice produced BC of higher quality than the reference standard, with values of 2.8 g/L and 2.1 g/L, respectively. The findings were parallel with other studies that concluded that utilizing agro-industry waste in general, and pineapple and sugarcane substrates in particular, are feasible for BC production. Whereas, when HS medium was utilized, some structural similarities were observed using SEM, while ATR-FR-IR spectra displayed chemical similarities in the microfibrils.

Zhao et al. [236] evaluated the potential of using yeast lees from fermentation vessels during fruit production using *Glucoacetobacter xylinum* for BC production. From the findings, yeast residue was identified as a potential substrate for economic BC production. However, for optimum production, modifications to the medium component and culture conditions of the bacterial strain are necessary. This is particularly important, given that the BC yield decreased with loading volume into cultivation vessels, which could have been associated with a reduced concentration of oxygen in the media [237].

For the yeast lees, researchers determined that mango pulp could supply essential substrate during BC production. Mango and guava purees displayed similar results due to the significant increase in water vapor permeability of the product [238]. Additional alterations in the produced BC included enhanced elongation and tensile strength reduction. Several studies suggested the addition of hydrophobic compounds [239] as a method to improve water resistance through cross-linking mechanisms [240].

Recent attempts to manufacture BC by *Gluconacetobacter xylinus* using pulp mill and lignocellulosic biorefinery waste fibre sludge displayed the potential to generate close to 11 g/L cellulose [241]. Producing high-quality BC at a low cost by utilizing sequential fermentation of residual streams from pulp mills and biorefinery processes is the most crucial contribution of this paper. The findings are in good agreement with the results achieved in other studies using various substrates.

There have been efforts to evaluate the possibility of utilizing other agricultural wastes for carbon sources in BC production, including corn products, coffee cherry husk (CCH), date fruits, and banana peel. CCH waste is an abundant agro-industrial waste. This method, using CCH as a substrate to produce BC achieved up to 8.2 g/L of BC using 8% of CCH extract combined with steep corn liquor under optimized conditions [242]. The findings were parallel with research evidence that steep corn liquor is rich in nutrition, which supplied organic content during BC production, such as carbon and nitrogen [243,244].

Banana peel is another potential agricultural waste being studied for a carbon source in BC production using *Acetobacter xylinum* [245]. The concentration of BC produced was 19.46 g/L of the product in a period of 15 days and a temperature of 30 °C. Similar results were achieved with coconut water and pineapple juice as substrates for the same bacteria [245]. The date is a fruit with a potential carbon source for BC production. Date trees are grown mainly in tropical and arid areas of North Africa and Southwest Asia. Date syrup consists of essential nutrients that are sufficient for the growth of numerous microorganisms [246]. However, date processing is accompanied by massive loss and wastage, which can be converted to useful byproducts. Lotfiman et al. [247] assessed the viability of producing BC from date syrup using *A. Xylinum*. The researchers examined sugar content in the waste sample using HPLC and tested different concentrations of the fruit syrup at different culture times. BC production was attained with 3% (*w*/*v*) date in the medium cultivated for a duration of eight days. Alteration of the HS medium resulted in an increase in BC yield of up to 68%. It was determined that date waste is a potential source of carbon. Other cellulosic non-food wastes have also been utilized for BC formation with reduced BC yields, such as olive mill residues that produced 0.81 g/L of BC [248]. This BC yield was lower compared to date syrup. The findings indicated that agricultural waste could be used as a potential carbon source substitute compared to non-food sources.

Besides the substrates explained above, date industry waste is another possible substrate for the economic production of BC. One such byproduct of the industry is date syrup (DS), which is rich in carbohydrates [249]. A study utilized low-quality DS with very little commercial value; Moosavi-Nasab and Yousefi [246] found that BC production displayed a steady increment up to day 14 compared to sucrose, for which the production of BC remained almost constant. At the end of the cultivations, cellulose yield from DS (4.35 g/L) were more than two-folds that of sucrose (1.69 g/L). This is associated with the DS consisting of reducing sugars in abundance compared to sucrose, a disaccharide [246]. The same substrate was studied by Lotfiman et al. (2018) for the investigation of BC production by *A. xylinum*. Results of their study showed that *A. xylinum* produced up to 5.8 g/L of BC that was 68% higher compared to that of the standard HS medium [247].

Coffee cherry husk (CCH), a byproduct that is present in abundance from coffee cherry processing, is seen as a potential substrate for BC production [242]. Results of research attempts for BC production from CCH showed that the production capacity of up to 8.2 g/L was attained using 8% CCH extract combined with steep corn liquor under optimized conditions.

Besides the examples listed, banana peel is also being studied as a potential substrate for the economic production of BC. *Acetobacter xylinum* generated 19.46 g/L of BC in a cultivation period of 15 days at 30 °C [245]. Similar results were achieved when coconut water and pineapple juice were used as a substrate for the same microorganism [250].

5.3. Wastewater Sugar Industries, Pulp Mills and Lignocellulosic Biorefineries Wastes

Zhao et al. [236] studied the use of fermented wastewater as a substrate that showed a BC yield of 1.177 g/L, which much lower than HS medium (1.757 g/L). This could be ascribed to a low concentration of nitrogen and carbon contents in the substrate. However, BC yield from wastewater was sufficient to maintain large-scale commercial applications, with the low costs of the production process taken into consideration. These results are in good agreement with the existing evidence, which indicated that the ideal system for cellulose biosynthesis does not exist, even with gram-negative bacterium such as *Gluconacetobacter xylinus* that is able to secrete large volumes of cellulose as microfibrils from different waste products [251]. Other cellulosic wastes from non-food wastes are also being studied and have resulted in reduced BC yields [248].

In another related study, Li et al. [252] determined that jujube processing industry wastewater could provide an inexpensive raw material for BC production using *Gluconacetobacter xylinum* [252]. The experiments exhibited the potential to yield 2.25 g/L of BC in hydrolysate with acid treatment. However, the setup involves the usage of special filters of between 3 and 14 nm to produce nanostructures. This study showed the possibility of improving BC yield by adjusting the level of crystallinity and manipulating ammonium citrate concentration. Further research revealed that candied jujube consists of various nutritional compounds, such as amino acids, saccharides, and vitamins, which make an ideal substrate for BC synthesis of considerable quantities [253]. However, the crystallinity of the microbial cellulose is an important factor to be taken into consideration when utilizing jujube for BC, which was altered significantly in different cultivation media as a result of the effect of fibre size distribution. This effect is also observed during BC production using other waste feedstocks [254].

5.4. Textile Industries Waste

The growth of industrialization worldwide is affected by the increasing number of the world population. This phenomenon has resulted in an increase in the utilization of fabric and textiles to produce clothes and other materials related to textile-based products [90]. The growing demand for these products has led to the generation of tons of waste produced by the textile industries and consumers. Natural fibre resources such as cotton are commonly used in the textile industry to produce the fabric. According to Estur [255], world textile fibre consumption is projected to expand at an annual average rate of 4% to reach 70 million tons in 2010 and by 2.8% per year to reach 87 million tons in 2020. Used cotton fabric is not recycled because it does not provide a satisfactory level of use. It is usually dumped at the landfilled or garbage collection station or disposed of by incineration.

These waste cotton textiles have the potential to be used as an effective alternative to producing high-value products at low cost through enzymatic hydrolysis and microbial conversion processes. In addition, these wastes also have the potential to reduce environmental problems and save natural resources. Kuo et al. [256] conducted an experiment on enzymatic saccharification of dissolution pretreated cellulosic waste fabrics for BC production by *Gluconacetobacter xylinus*, which has shown that the BC produced from discoloured hydrolysate (1.88 g/L) by *G. xylinus* in static cultivation of seven days was about 20% higher compared to that in the coloured hydrolosate (1.59 g/L). This might be attributed to the fact that the coloured reducing sugars that were removed by chitosan adsorption prevent the fermentation activity of *Gluconacetobacter xylinus* for BC production.

Guo et al. [257] showed that BC could be successfully produced using waste dyed cotton fabrics cellulose through pretreatment with the ionic liquid (IL) 1-allyl-3-methyl-imidazolium chloride ([AMIM]Cl) with *Gluconacetobacter xylinus* followed by the production of enzymes with *Trichoderma reesei* via enzymatic saccharification. They found that the BNC yield obtained from the purple bed sheet (14.2 g/L) by *Gluconacetobacter xylinus* in static cultivation of 10 days was higher compared to that in the red bed sheet (13.7 g/L) and green bed sheet (14.1 g/L) [257]. Moreover, according to Guo et al. [257], this is due to the supplementation of calcium ions during treatment of $Ca(OH)_2$ detoxification as well as the removal of dyes from the enzymatic hydrolysates.

Previous studies by Hong et al. [215] reported on production of high-quality carbon sources for BC from cotton-based waste textiles by *Gluconacetobacter xylinus*. These fabrics were pre-treated with the ([AMIM]Cl) followed by enzymatic hydrolysis. The results show that the yield and tensile strength of BC are 83% (10.8 g/L) and 79% (0.07 MPa) higher compared to a culture grown on a glucose-based medium [215]. The studies describing the use of textile mills waste for BC production are displayed in Table 3.

5.5. Biodiesel Industry Waste

BC is well known as a natural biomaterial with a broad range of applications. However, high-cost production in terms of raw materials, as well as low yields, have limited the industrial and commercial applications of BC. Hence, the usage of low cost-alternative raw materials as fermentation media would enhance BC production's cost-competitiveness. The worldwide biodiesel production was more than 2.8 billion liters in 2018, and it has increased by 933% over the last 20 years [258,259]. It is estimated that crude glycerol is generated as a 10% (*w*/*w*) byproduct from transesterification of triglycerides with alcohol, most frequently methanol, which is equivalent to 0.28 billion liters.

In one of the studies, Tsouko et al. [259] investigated the feasibility of using fermentation media obtained from the confectionery industry and sunflower-based biodiesel industries waste streams and found that confectionary industry waste provides rich sources for carbon and nitrogen required for highly efficient BC production [259]. Batch fermentations using *Komagataeibacter sucrofermentans* (DSM) in synthetic media yielded BC concentrations of up to 13.3 g/L. The experimental results showed similar yields using both waste streams. The findings determined the significance of *Komagataeibacter sucrofermentans* DSM strain for high concentrations of BC production from the confectionery and biodiesel industry wastes. More importantly, the findings of this study on the water holding capacity (WHC) of the BC were parallel with the existing literature [260–263].

Previous studies reported the production and characterization of BC produced from non-detoxified crude glycerol as an alternative medium by *Gluconacetobacter xylinus* strain [264]. The highest BC production is 12.31 g/L. However, increasing crude glycerol has resulted in decreased BC production. This phenomenon might be due to the impurities in crude glycerol that might affect the activity of the cell. Besides that, from the research conducted by Soemphol et al. [264], it was shown that production of BC could improve by the addition of pineapple peel extract (PPE) into crude glycerol without any supplementation, and the optimal BC production was seen at acidic pH. The usage of these wastes or byproducts from biodiesel industries will not only produce value-added materials, it will also reduce environmental pollution and non-renewable energy consumption. The studies describing the use of biodiesel wastes for the production of BC are displayed in Table 3.

5.6. Micro-Algae Biomass Industries

Bioactive compounds such as carotenoids, polyunsaturated fatty acids, protein, vitamins, and minerals can be found in various commercial forms of micro-algal biomass (i.e., capsule, tablet, oil, liquid, flour, or powder forms). They play essential roles in numerous applications such as cosmetic products, pharmaceutical chemicals, feed product for animals (for fish, shellfish, poultry, and cattle) or functional food (i.e., supplements, dye, oil-derivatives, pastas, dairy products, and dessert) or with favorable outcomes upon human health, including antiviral, antimicrobial, anti-inflammatory, and antioxidant effects, as well as prevention of hypertension, diabetes, anaemia, constipation, and gastric ulcers [265]. Starch is one of the valuable constituents of microalgae biomass. Low-cost starch biomass products can be yielded from outdoor photobioreactors of *Chlorella cultures* microalgae [177]. Besides that, there are several studies that have been conducted to increase the starch content of algal biomass under different conditions (i.e., light intensities, nitrogen starvation, and sulphur). Freshwater algae *Chlorella vulgaris* can produce low-cost starch in large quantities (Dragone et al., 2011). This starch can be utilized as a promising alternative carbon source medium for the production of BC. Several studies reported the use of a byproduct of the micro-algae medium as a carbon source for BC. Several byproducts of micro-algae industries have been evaluated for BC production, as stated in Table 3.

Uzyol & Saçan [177] produced BC with *Komagataeibacter hansenii* using algae-based glucose, and showed that the BC production yields were 1.202 g/L and 1.104 g/L from glucose and algae-based glucose, respectively. The morphological structure of algae-based BC was observed to be similar to those of glucose-based BC. Another study, conducted on the production of green BC by utilizing renewable resources of algae with corn steep liquor [266], shows that the maximum BC production is 4.86 g/L. Therefore, based on the literature review, it can be summarized that integrating the metabolic components in algal biomass (i.e., corn steep liquor, glucose, yeast, starch, peptone, etc.) in the production of BC with the biorefinery concept would bring economic and environmental benefits, including the achievement of large scale production at low cost, and protecting the environment.

Table 3. Industrial wastes utilized as sustainable feedstock for the production of bacterial cellulose (BC).

Microorganism	Production Mode	BC Production	Time	Industrial Waste	Additional Nutrients	References
Beverages/Brewery						
Waste as carbon source with additional nutrients						
Komagataeibacter xylinus CICC No.10529	Static	5.7 g/L	8 days	Citrus peel and pomace enzymolysis medium	Yeast extract, ethanol and peptone	Fan et al. [216]
Gluconacetobacter xylinus NRRL B-42	Static	8.00 g/L	14 days	Grape bagasse	Corn steep liquor	Vazquez et al. [190]
Gluconacetobacter xylinus NRRL B-42	Static	7.20 g/L	14 days	Grape bagasse	Diammonium phosphate	
Gluconacetobacter xylinus ATCC®® 10788™	Static	0.35 g/L	3 days	Makgeolli sludge filtrate	Modified HS (MHS) medium	Hyun et al. [267]
Gluconacetobacter xylinus ATCC®® 10788™	Static	1.2 g/L	3 days	Makgeolli sludge filtrate	Mixed modified HS (MMHS)	
Gluconacetobacter xylinus BCRC 12334	Static	0.90 g/L	7 days	Thin stillage (TS) wastewater	50% TS	Wu & Liu [232]
Gluconacetobacter xylinus BCRC 12334	Static	6.26 g/L	7 days	Thin stillage (TS) wastewater	50/50 TS-HS	
Gluconacetobacter oboediens	Shaking	10.8 g/L	72 h	Distillery effluent	Sucrose (carbon source) and corn steep liquor (nitrogen source)	Jahan et al. [268]
Gluconacetobacter hansenii PJK KCTC 10505BP	Static	13.95 g/L	336 h	Untreated WBFB	1% Glucose	Ha et al. [269]
Gluconacetobacter hansenii PJK KCTC 10505BP	Shaking	1.50 g/L	168 h	Untreated WBFB	1% Glucose	
Gluconacetobacter hansenii PJK KCTC 10505BP	Static	7.37 g/L	336 h	Autolyzed WBFB	Glucose	
Gluconacetobacter hansenii PJK KCTC 10505BP	Static	3.64 g/L	336 h	Hydrolysed WBFB	1% Glucose	
Waste as a complex medium without any additional nutrients						
Komagataeibacter saccharivorans strain BC1 (K. saccharivorans strain BC1)	Static	1.24 g/L	8 days	UB breweries limited, Baikampady, Mangalore, India	-	Gayathri et al. [270]
Gluconacetobacter xylinus BCRC 12334	Static	3.10 g/L	7 days	Thin stillage (TS) wastewater	-	Wu & Liu [232]
Gluconacetobacter xylinus NRRL B-42	Static	4.20 g/L	14 days	Grape bagasse	-	Vazquez et al. [190]
Gluconacetobacter xylinus ATCC®® 10788™	Static	0.30 g/L	3 days	Makgeolli sludge filtrate	-	Hyun et al. [267]
Gluconacetobacter medellinensis ID13488	Static	1.5 g/L	14 days	Fresh apple peel/ sugar cane ratio (*w*/*w*) (1/2.3)	-	Urbina et al. [271]
Gluconacetobacter medellinensis ID13488	Static	1.4 g/L	14 days	Apple residue (AR)/ sugar cane (SC) ratio (*w*/*w*) (1/2.3)	-	

Table 3. *Cont.*

Microorganism	Production Mode	BC Production	Time	Industrial Waste	Additional Nutrients	References
Gluconacetobacter medellinensis ID13488	Static	2.0 g/L	14 days	AR/SC ratio (*w*/*w*) (0.5/2.8)	-	
Gluconacetobacter medellinensis ID13488	Static	1.2 g/L	14 days	AR/SC ratio (*w*/*w*) (2/1.3)	-	
Gluconacetobacter medellinensis ID13488	Static	2.5 g/L	14 days	AR/SC ratio (*w*/*w*) (1.5/2.3)	-	
Gluconoacetobacter xylinum ATCC 23768	Static	2.9 g/L	9 days	Black strap molasses	-	Khattak et al. [272]
Gluconoacetobacter xylinum ATCC 23768	Shaking	3.05 g/L	9 days	Black strap molasses	-	
Gluconoacetobacter xylinum ATCC 23768	Static	1.70 g/L	9 days	Brewery molasses	-	
Gluconoacetobacter xylinum ATCC 23768	Shaking	1.75 g/L	9 days	Brewery molasses	-	
Gluconacetobacter oboediens	Shaking	8.5 g/L	72 h	Crude effluent	-	Jahan et al. [268,273]
Acetobacter xylinum NRRL B-42	Static	6.7 g/L	21 days	Grape pomace extract/corn steep liquor	-	Cerrutti et al. [274]
Gluconacetobacter hansenii PJK KCTC 10505BP	Static	8.46 g/L	336 h	Untreated Waste from beer fermentation broth (WBFB)	-	Ha et al. [270]
Gluconacetobacter hansenii PJK KCTC 10505BP	Static	2.00 g/L	336 h	Autolyzed WBFB	-	
Gluconacetobacter hansenii PJK KCTC 10505BP	Static	2.82 g/L	336 h	Hydrolysed WBFB	-	
Gluconacetobacter sucrofermentans B-11267	Shaking	2.40 g/L	3 days	Hestrin and Schramm (HS) medium	-	Revin et al. [230]
Gluconacetobacter sucrofermentans B-11267	Shaking	6.19 g/L	3 days	Thin stillage	-	
Gluconacetobacter sucrofermentans B-11267	Shaking	5.50 g/L	3 days	Cheese whey	-	
Gluconacetobacter sucrofermentans B-11267	Shaking	6.19 g/L	3 days	Thin stillage pH 3.95	-	
Gluconacetobacter sucrofermentans B-11267	Shaking	5.40 g/L	3 days	Thin stillage pH 5	-	
Gluconacetobacter sucrofermentans B-11267	Shaking	3.50 g/L	3 days	Thin stillage pH 6	-	
Gluconacetobacter xylinus	Static	2.90 g/L	4 days	Acid hydrolysate of waste oleaginous yeast biomass	-	Luo et al. [275]
Gluconacetobacter hansenii CGMCC 3917	Static	3.89 g/L	14 days	Waste beer yeast treated with ultrasonication treatment	-	Lin et al. [237]
Gluconacetobacter hansenii CGMCC 3917	Static	2.40 g/L	14 days	Waste beer yeast treated with NaOH treatment	-	
Gluconacetobacter hansenii CGMCC 3917	Static	2.00 g/L	14 days	Waste beer yeast treated with high speed homogenizer treatment	-	

Table 3. *Cont.*

Microorganism	Production Mode	BC Production	Time	Industrial Waste	Additional Nutrients	References
Gluconacetobacter hansenii CGMCC 3917	Static	1.50 g/L	14 days	Waste beer yeast treated with microwaves treatment	-	
Gluconacetobacter hansenii CGMCC 3917	Static	1.20 g/L	14 days	Waste beer yeast treated with untreatment	-	
Gluconacetobacterxylinus BC-11 K.	Static	1.18 g/L	10 days	Wastewater after pullulan polysaccharide fermentation	-	Zhao et al. [236]
Agro industrial waste						
Waste as nitrogen source						
Gluconacetobacter swingsii	Static	2.8 g/L	13 days	Pineapple peel juice	Glucose, fructose and sucrose	Castro et al. [235]
Waste as carbon source with additional nutrients						
Gluconacetobacter swingsii	Static	-	13 days	Sugar cane juice	Glucose, fructose and sucrose	Castro et al. [235]
Gluconacetobacter xylinum bacterium (ATCC 700178)	Shaking	10.6 g/L	7 days	Wheat straw	Corn steep liquor (CSL)	Goyat [266]
Gluconacetobacterxylinus	Static	1.8 g/L	9 days	Carob and haricot bean (CHb) medium	Citric acid	Bilgi et al. [276,277]
Komagataeibacter rhaeticus	Static	6.0 g/L	7 days	HS medium and Cashew tree exudates (HSCTE)	HS medium	Pacheco et al. [278]
Komagataeibacter rhaeticus	Static	6.0 g/L	7 days	HS medium and Cashew tree exudates (HSCG)	HS medium	
Acetobacter aceti ATCC 23770	Shaking and static	2.12 g/L	8 days	Cheap agricultural product konjac powder	Yeast extract and tryptone	Hong & Qiu [279]
Gluconacetobacter hansenii UAC09	Static	8.2 g/L	14 days	Coffee cherry husk (CCH)	8% corn steep liquor (CSL)	Rani & Appaiah [242]
Gluconacetobacter hansenii UAC09	Static	6.5 g/L	14 days	Coffee cherry husk (CCH)	0.2% Urea	
Gluconacetobacter hansenii UAC09	Static	6.9 g/L	14 days	Coffee cherry husk (CCH)	Ethyl alcohol (EA) + Acetic acid (AA)	
Gluconacetobacter hansenii UAC09	Static	7.5 g/L	14 days	Coffee cherry husk (CCH)	8% CSL + EA + AA	
Gluconacetobacter hansenii UAC09	Static	6.6 g/L	14 days	Coffee cherry husk (CCH)	0.2% urea + EA + AA	

Table 3. *Cont.*

Microorganism	Production Mode	BC Production	Time	Industrial Waste	Additional Nutrients	References
Acetobacter xylinus ATCC 23770	Static	8.3 g/L	7 days	Enzymatic hydrolysate of wheat straw	Other components are same as of HS medium	Chen et al. [280]
Acetobacter xylinum 0416 MARDI	Static	4.0 g/L	8 days	Extracted date syrup (DSH-2%)	Other components are same as of HS medium	Lotfiman et al. [247]
Acetobacter xylinum 0416 MARDI	Static	5.8 g/L	8 days	Extracted date syrup (DSH-3%)	Other components are same as of HS medium	
Acetobacter xylinum 0416 MARDI	Static	4.5 g/L	8 days	Extracted date syrup (DSH-5%)	Other components are same as of HS medium	
Gluconacetobacter sacchari	Static	0.1 g/L	96 h	Grape skins aqueous extract, cheese whey, crude glycerol and sulfite pulping liquor	Organic or inorganic nitrogen	Carreira et al. [281]
Acinetobacter sp. BAN1	Static	0.3 g/L	15 days	Pineapple juice medium (PIJM)	Other components are same as that of HS medium	Adebayo-Tayo et al. [282]
Acinetobacter sp. BAN1	Static	6.4 g/L	15 days	Pawpaw juice medium (PAJM)	Other components are same as that of HS medium	
Acinetobacter sp. BAN1	Static	0.6 g/L	15 days	Watermelon juice medium (WMJM)	Other components are same as that of HS medium	
Acetobacter pasteurianus PW1	Static	0.1 g/L	15 days	Pineapple juice medium (PIJM)	Other components are same as that of HS medium	
Acetobacter pasteurianus PW1	Static	7.7 g/L	15 days	Pawpaw juice medium (PAJM)	Other components are same as that of HS medium	
Acetobacter pasteurianus PW1	Static	0.4 g/L	15 days	Watermelon juice medium (WMJM)	Other components are same as that of HS medium	
Gluconoacetobacter xylinus BCRC 12334	Static	3.40 g/L	8 days	Orange peel fluid and orange peel hydrolysate	Acetate buffer, peptone and yeast extract	Kuo et al. [214]
Beijerinkia fluminensis WAUPM53	Static	0.47 g/L	14 days	Sago byproduct	Other components are same as of HS medium	Voon et al. [226]
Gluconacetobacter xylinus 0416	Static	1.55 g/L	14 days	Sago byproduct	Other components are same as of HS medium	
Acetobacter xylinum NBRC 13693	Static	4.1 g/L	14 days	Pineapple	Disodium hydrogen phosphate buffer	Kurosumi et al. [283]
Acetobacter xylinum NBRC 13693	Static	3.95 g/L	14 days	Apple	Disodium hydrogen phosphate buffer	
Acetobacter xylinum NBRC 13693	Static	5.9 g/L	14 days	Orange	Disodium hydrogen phosphate buffer	
Acetobacter xylinum NBRC 13693	Static	3.5 g/L	14 days	Japanese pear	Disodium hydrogen phosphate buffer	
Acetobacter xylinum NBRC 13693	Static	1.8 g/L	14 days	Grape	Disodium hydrogen phosphate buffer	

Table 3. *Cont.*

Microorganism	Production Mode	BC Production	Time	Industrial Waste	Additional Nutrients	References
Acetobacter xylinum NBRC 13693	Static	0.5 g/L	14 days	Pineapple	Sugar reagent (glucose, fructose and sucrose)	
Acetobacter xylinum NBRC 13693	Static	0.2 g/L	14 days	Apple	Sugar reagent (glucose, fructose and sucrose)	
Acetobacter xylinum NBRC 13693	Static	1.85 g/L	14 days	Orange	Sugar reagent (glucose, fructose and sucrose)	
Acetobacter xylinum NBRC 13693	Static	0.5 g/L	14 days	Japanese pear	Sugar reagent (glucose, fructose and sucrose)	
Acetobacter xylinum NBRC 13693	Static	0.4 g/L	14 days	Grape	Sugar reagent (glucose, fructose and sucrose)	
Gluconacetobacter sacchari	Static	1.7 g/L	96 h	Dry olive mill residue (DOR100) Water extraction at 100 °C	Nitrogen	Gomes et al. [248]
Gluconacetobacter sacchari	Static	1.4 g/L	96 h	Dry olive mill residue (DOR100) Water extraction at 100 °C	Phosphorus	
Komagataeibacter hansenii MCM B-967	Static	125 g/L	7 days	Pineapple and watermelon peels	Sucrose, ammonium sulfate and cycloheximide	Kumbhar et al. [284]
Acetobacter xylinum DSMZ2004	Static	8.6 g/L	48 h	Poor quality apple residues in combination with glycerol	Apple glucose equivalents, glycerol, ammonium sulfate and citric acid	Casarica et al. [285]
Acetobacter xylinum BCRC 14182 (purchased)	Static	-	3–7 days	Coconut-water	Sugar	Lin et al. [286]
Waste as complex medium without any additional nutrients						
Komagataeibacter hansenii GA2016	Static	2.06 BC/100 g peel	21 days	Lemon peels (LBC)	-	Güzel & Akpınar [287]
Komagataeibacter hansenii GA2016	Static	3.92 BC/100 g peel	21 days	Mandarin peels (MBC)	-	
Komagataeibacter hansenii GA2016	Static	2.33 BC/100 g peel	21 days	Orange peels (OBC)	-	
Komagataeibacter hansenii GA2016	Static	2.68 BC/100 g peel	21 days	Grapefruit peels (GBC)	-	
Komagataeibacter xylinus	Static	2.90 g/L	10 days	Discarded waste durian shell	-	Luo, Huang et al. [275]
Gluconacetobacter xylinus CH001	Static	2.67 g/L	10 days	Discarded waste durian shell	-	Luo, Huang, et al.[288]

Table 3. *Cont.*

Microorganism	Production Mode	BC Production	Time	Industrial Waste	Additional Nutrients	References
Gluconacetobacter medellinensis	Static	3.24 g/L	7 days	Sugar cane juice and pineapple residues	-	Algar et al. [289]
Gluconacetobacter medellinensis	Dynamic	0.82 g/L	7 days	Sugar cane juice and pineapple residues	-	
Acinetobacter sp. BAN1	Static	0.4–0.6 g/L	15 days	Pineapple waste medium (PIWAM)	-	Adebayo-Tayo et al. [290]
Acinetobacter sp. BAN1	Static	0.2–1.1 g/L	15 days	Pawpaw waste medium (PAWAM)	-	
Acetobacter pasteurianus PW1	Static	0.2–1.0 g/L	15 days	Pawpaw waste medium (PAWAM)	-	
Acetobacter pasteurianus PW1	Static	0.1–3.9 g/L	15 days	Pineapple waste medium (PIWAM)	-	
Komagataeibacter rhaeticus iGEM	Static	–	10 days	Fermented tea	-	Florea et al. [291]
Gluconacetobactersacchari	-	1.28 g/L	-	Industrial residues from olive oil production	-	Gomes et al. [248]
Gluconacetobacter persimmonis GH-2	Static	5.75 g/L	14 days	Watermelon + HS medium	-	Hungund et al. [292]
Gluconacetobacter persimmonis GH-2	Static	5.98 g/L	14 days	Orange juice + HS medium	-	
Gluconacetobacter persimmonis GH-2	Static	6.18 g/L	14 days	Muskmelon + HS medium	-	
Gluconacetobacter persimmonis GH-2	Static	8.08 g/L	14 days	Coconut water +HS medium	-	
Acetobacterxylinum	Static	19.46 g/L	15 days	Banana peel	-	Hungund et al. [245]
Gluconacetobacter xylinus ATCC 53582	Static	60 g/L	96 h	Rotten fruit culture	-	Jozala et al. [293]
Gluconacetobacter xylinus CGMCC 2955	Static	2.25 g/L	114 h	Waste water of candied jujube hydrolysate	-	Li et al. [252]
Acetobacter xylinum 0416	Rotary disc reactor	28.30 g/L	4 days	Pineapple waste medium	-	Zahan et al. [197]
Komagataeibacter rhaeticus	Static	2.8 g/L	7 days	Cashew tree exudates (CTE)	-	Pacheco et al. [278]
Komagataeibacter rhaeticus	Static	2.3 g/L	7 days	Cashew gum (CG)	-	
Gluconacetobacter hansenii UAC09	Static	5.6 g/L	14 days	Coffee cherry husk (CCH)	-	Rani & Appaiah [242]

Table 3. *Cont.*

Microorganism	Production Mode	BC Production	Time	Industrial Waste	Additional Nutrients	References
Gluconacetobacter sacchari	Static	0.81 g/L	96 h	Dry olive mill residue (DOR40) Water extraction at 40 °C	-	Gomes et al. [248]
Gluconacetobacter sacchari	Static	0.85 g/L	96 h	Dry olive mill residue (DOR100) Water extraction at 100 °C	-	
Sugar industries, pulp mills and lignocellulosic biorefineries wastes						
Waste as carbon source with additional nutrients						
Komagatacibacter xylinus PTCC 1734	Static	7.02 g/L	10 days	Vinasse	Other components are same as of HS medium	Barshan et al. [294]
Acetobacter xylinum BPR2001	Rotary shaker	3.01 g/L	70 h	Molasses	Corn steep liquor	Bae & Shoda [187]
Acetobacter xylinum BPR2001	Rotary shaker	5.30 g/L	70 h	H_2SO_4 heat treated molasses	Corn steep liquor	
Gluconacetobacter xylinus	Static	5.9 g/L	14 days	Cane molasses	Corn steep liquor and diammonium phosphate	Vazquez et al. [190]
Acetobacter sp. V6	Agitated	3.12 g/L	168 h	Molasses and corn steep liquor	Acetic acid	Jung et al. [204]
Acetobacter xylinum ATCC 10245	Static	223% as compared to 100% in HS medium	7 days	Sugar cane molasses	Carbohydrates, minerals, vitamins and amino acids	Premjet et al. [295]
Komagataeibacter rhaeticus	Static	3.90 g/L	120 h	Sugarcane molasses (SCM) 10 g/L of SCM	40 g/L of glucose	Machado et al. [296]
Komagataeibacter rhaeticus	Static	4.01 g/L	120 h	20 g/L of SCM	30 g/L of glucose	
Komagataeibacter rhaeticus	Static	3.7 g/L	120 h	30 g/L of SCM	20 g/L of glucose	
Komagataeibacter rhaeticus	Static	3.50 g/L	120 h	40 g/L of SCM	10 g/L of glucose	
Gluconacetobacter xylinus ATCC 23770	Static	11 g/L	7 days	Waste fiber sludge sulfate	Yeast extract and tryptone	Cavka et al. [241]
Gluconacetobacter xylinus ATCC 23770	Static	10 g/L	7 days	Waste fiber sludge sulfite	Yeast extract and tryptone	
Acetobacter xylinum ATCC 10245	Static	20.6 %	7 days	Softwood purified water-soluble (SPWS)	Other components are same as of HS medium	Uraki et al. [297]
Acetobacter xylinum ATCC 10245	Static	33 %	7 days	Hardwood purified water-soluble (HPWS)	Other components are same as of HS medium	

Table 3. *Cont.*

Microorganism	Production Mode	BC Production	Time	Industrial Waste	Additional Nutrients	References
Acetobacter xylinum ATCC 53582	Static	5.4 %	7 days	Softwood purified water-soluble (SPWS)	Other components are same as of HS medium	
Acetobacter xylinum ATCC 53582	Static	8.9 %	7 days	Hardwood purified water-soluble (HPWS)	Other components are same as of HS medium	
Waste as carbon source without any additional nutrients						
Acetobacter xylinus 23769		0.15 g/L		Hot water extract	-	Erbas Kiziltas et al. [298]
Gluconoacetobacter xylinum ATCC 23768	Shaking	2.51 g/L	10 days	Scum of sugarcane jaggery or gur (JS)	-	Khattak, Khan, Ul-Islam, Wahid, et al. [299]
Gluconoacetobacter xylinum ATCC 23768	Static	2.13 g/L	10 days	Scum of sugarcane jaggery or gur (JS)	-	
Komagataeibacter europaeus SGP37	Static	6.30 g/L	16 days	Sweet lime pulp waste	-	Dubey et al. [300]
G. persimmonis GH-2	Static	5.75 g/L	14 days	Molasses + HS medium	-	Hungund et al. [292]
G. intermedius SNT-1	Static	12.6 g/L	10 days	Molasses pretreated with hea	-	Tyagi et al. [301]
Gluconacetobacter xylinus (PTCC, 1734)	Static	4.35 g/L	336 h	Date syrup	-	Moosavi-Nasab [246]
Komagataeibacter rhaeticus	Static	1.90 g/L	120 h	50 g/L of SCM	-	Machado et al. [296]
Gluconaceter xylinus CH001	Static	0.66 g/L	5 days	Lipid fermentation wastewater	-	Huang et al. [302]
Gluconaceterxylinus	Static	1·34 g/L	7 days	Acetone-butanol-ethanol(ABE) fermentation wastewater	-	Huang et al. [303]
Gluconaceterxylinus BC-11	Static	1.177 g/L	10 days	Wastewater after pullulan polysaccharide fermentation	-	Zhao et al. [236]
Acetobacter xylinum 23769	Static	0.15 g/L	672 h	Wood hot water extract	-	Erbas Kiziltas et al. [298]
Textile mills						
Waste as carbon source with additional nutrients						
Gluconacetobacter xylinus ATCC 23770	Static	10.8	14 days	Cotton-based waste textiles	Glucose, yeast extract and peptone	Hong et al. [215]

Table 3. *Cont.*

Microorganism	Production Mode	BC Production	Time	Industrial Waste	Additional Nutrients	References
Gluconacetobacter xylinus	Static	14.2 g/L	10 days	Waste dyed cotton fabrics hydrolysate - Purple bed sheet (PBS)	Peptone and yeast extract	Guo et al. [257]
Gluconacetobacter xylinus	Static	13.7 g/L	10 days	Waste dyed cotton fabrics hydrolysate- rose -Red bed sheet (RRBS)	Peptone and yeast extract	
Gluconacetobacter xylinus	Static	14.1 g/L	10 days	Waste dyed cotton fabrics hydrolysate- green bed sheet (GBS)	Peptone and yeast extract	
Gluconacetobacter xylinus	Static	1.59 g/L	7 days	Coloured hydrolysate	Peptone and yeast extract	Kuo et al. [256]
Gluconacetobacter xylinus	Static	1.88 g/L	7 days	Discoloured hydrolysate	Peptone and yeast extract	Kuo et al. [256]
Biodiesel industry						
Waste as carbon source with additional nutrients						
Gluconaceter xylinus BNKC19	Static	12.31 g/L	7 days	Non-detoxified crude glycerol	Pineapple and in combination with HS medium components	Soemphol et al. [264]
Gluconacetobacter xylinus DSM 46604	Agitated	2.87 g/L	5 days	20 g/L glycerol	Yeast extract, ammonium sulphate, potassium hydrogen orthophosphate and magnesium sulphate	Adnan [304]
Gluconacetobacter xylinus DSM 46604	Agitated	2.87 g/L	5 days	50 g/L glucose	Yeast extract, ammonium sulphate, potassium hydrogen orthophosphate and magnesium sulphate	Adnan [304]
Gluconacetobacter xylinus	Static	10 g/L	14 days	Glycerol from biodiesel	Diammonium phosphate and corn steep liquor	Vazquez et al. [190]
Gluconacetobacter intermedius NEDO-01	Static	3.4 g/L	4 days	Waste glycerol	Carboxymethyl Cellulose	Kose et al. [305]
Komagataeibacter sucrofermentans DSM 15973	Shaking	3.2 g/L	15 days	Crude glycerol from biodiesel	Yeast extract and peptone	Tsouko et al. [259]
Komagataeibacter sucrofermentans DSM 15973	Shaking	13.3 g/L	15 days	Crude glycerol from biodiesel	Sunflower meal hydrolysates	Tsouko et al. [259]
Komagataeibacter sucrofermentans DSM 15973	Shaking	13 g/L	15 days	Crude glycerol from biodiesel	Flour-rich hydrolysates	Tsouko et al. [259]
Waste as carbon source without additional nutrients						
Gluconacetobacter xylinus	Static	3.5 g/L	14 days	Glycerol from biodiesel	-	Vazquez et al. [190]

Table 3. *Cont.*

Microorganism	Production Mode	BC Production	Time	Industrial Waste	Additional Nutrients	References
Micro-algae industry						
Waste as carbon source with additional nutrients						
Gluconacetobacter xylinum bacterium (ATCC 700178)	Shaking	4.86 g/L	7 days	Algae	Corn steep liquor (CSL)	Goyat [266]
Gluconacetobacter xylinus (ATCC #700178)	Static	77%	7 days	*Chlorella vulgaris*	Glucose/yeast extract	Chen et al. [306]
Gluconacetobacter xylinus (ATCC #700178)	Static	94%	7 days	*Scenedesmus obliqnus*	Glucose/yeast extract	
Gluconacetobacter xylinus (ATCC #700178)	Static	85%	7 days	*Chlamydomonas reinhardtii*	Glucose/yeast extract	
Komagataeibacter hansenii DSMZ	Static	1.104 g/L	7 days	Algae (*Chlorella vulgaris*) algae based glucose	Meat extract, peptone, NaCl and ethanol	Uzyol & Saçan [177]
Waste as carbon source without additional nutrients						
Komagataeibacter saccharivorans	Static	85.1%	14 days	Algae (*Chlamydomonas debaryana*) (BEA0067)	-	Nóbrega et al. [307]

6. Future Perspectives

Different industrial sectors produce a large amount of waste on a daily basis. More brewery, sugar, lignocellulosic, and other industrial wastes could be valorized as complex media without additional nutrient sources or as carbon and nitrogen sources with additional nutrient sources for BC production. Due to their large-scale availability and increased BC productivity, agro-industrial wastes can be widely utilised for BC production. Due to increased urbanization around the world, particularly in economically developing nations, municipal waste is anticipated to become an increasingly major source of waste biomass with higher organic content. All of the low-cost waste media of the industries discussed here, especially those that do not require complex pre-treatments, detoxification, or supplementing, have a lot of potential for upscale production of BC on an industrial scale. When compared to regular media, BC created from waste media has similar physico-chemical characteristics and a higher yield.

Because these wastes are available in huge quantities, waste producers may be able to sell them to commercial BC producers or academic institutions. The BC that was obtained could be used as a raw material by a variety of biomedical enterprises for commercial purposes as well as by scientists for study. Since BC derived from some agro-wastes might be colored and absorb unwanted compounds, proper purification is required. These facts may justify limiting the use of BC in industries with stringent regulatory standards, such as biomedicine, pharmaceutics, cosmetics, or the food industry. In terms of the environment, eliminating these industrial outputs will allow for proper waste management, lowering the environmental and health risks associated with these wastes. This will be a realistic option for dealing with pollution issues.

7. Conclusions

Bacterial cellulose (BC) is considered a desirable biomaterial for various applications across many fields due to its unique structural features and desirable properties. This review mainly discusses the technical and economic feasibility of producing microbial cellulose from industrial wastes from agro-industry, textile, biodiesel, micro-algae biomass, wastewater sugar, and lignocellulosic biorefineries, breweries, and beverages. The overarching conclusion is that most industrial wastes have the potential to produce high concentrations of BC. The production of high concentrations of BC can be obtained by optimizing bacterial culture conditions, such as temperature and pH. More importantly, the findings demonstrate that the produced microbial cellulose would have desirable chemical, physical, and mechanical properties, which suit various advanced applications. This review shows that the production of BC from industrial waste is successful. The future of using industrial wastes for BC production seems promising, since the source of nutrients in BC production from industrial wastes has reduced the production cost. Moreover, tonnes of industrial waste are generated daily, and using some of these wastes in BC production can mitigate waste disposal problems. The high yield and low production cost of BC is the main challenge that needs to be contemplated. A lot of progress can be made by developing new fermentation methods, new bioreactor design, and using a cheaper waste media that aims to increase the yield of BC at a lower cost. The BC has been used in various industries in manufacturing products as well as advanced applications. Products such as BC masks, BC gloves, paper, biodegradable food packaging, and wound dressing have been on the market. More advanced BC applications have shown promising results, such as never-dried microbial cellulose membranes, skin transplants, optically transparent cellulose nanocomposites, and artificial bacterial cellulose ligaments. Overall, large-scale commercial production and demand of microbial cellulose using waste as a carbon and energy source can lower the biomaterial production cost and help eliminate or reduce the economic and environmental burden of industrial waste.

Author Contributions: Conceptualization, A.K. and R.A.I.; methodology, A.K. and R.A.I.; investigation, A.K. and R.A.I.; writing—original draft preparation, A.K., R.A.I., J.W., M.R.M.H., N.H., R.Y., S.M.S., M.M.H. and M.N.M.A.; writing—review and editing, R.I., M.S.N.A., K.C., M.A.I., S.S., S.P., A.R., M.R.M.A. and M.R.I.; visualization, A.K. and R.A.I.; supervision, A.K. and R.A.I.; project administration, A.K. and R.A.I.; funding acquisition, M.R.M.H. All authors have read and agreed to the published version of the manuscript.

Funding: The authors would like express gratitude for the financial support received from Universiti Teknologi Malaysia, project CRG 30.3: Retardant coating using graphene/bamboo aerogel mixtures on SAR robotics system, grant number PY/2020/03495—R.J130000.7351.4B534. The research has been carried out under program Research Excellent Consortium (JPT)(BKI)1000/016/018/25(57) provided by Ministry of Higher Education Malaysia (MOHE).

Institutional Review Board Statement: Not applicable.

Informed Consent Statement: Not applicable.

Acknowledgments: The authors would like express gratitude for the financial support received from Universiti Teknologi Malaysia, project CRG 30.3: Retardant coating using graphene/bamboo aerogel mixtures on SAR robotics system, grant number PY/2020/03495—R.J130000.7351.4B534". The research has been carried out under program Research Excellence Consortium (JPT (BPKI) 1000/016/018/25 (57)) provided by Ministry of Higher Education Malaysia (MOHE). The author also would like to thank Universiti Teknologi Malaysia (UTM) for work and facilities support. The article processing fee of this manuscript is funded by Research Management Centre (RMC), Univesiti Putra Malaysia.

Conflicts of Interest: The authors declare that they have no conflict of interest.

References

1. Abral, H.; Pratama, A.B.; Handayani, D.; Mahardika, M.; Aminah, I.; Sandrawati, N.; Sugiarti, E.; Muslimin, A.N.; Sapuan, S.M.; Ilyas, R.A. Antimicrobial Edible Film Prepared from Bacterial Cellulose Nanofibers/Starch/Chitosan for a Food Packaging Alternative. *Int. J. Polym. Sci.* **2021**, *2021*, 1–11. [CrossRef]
2. Abral, H.; Chairani, M.K.; Rizki, M.D.; Mahardika, M.; Handayani, D.; Sugiarti, E.; Muslimin, A.N.; Sapuan, S.M.; Ilyas, R.A. Characterization of compressed bacterial cellulose nanopaper film after exposure to dry and humid conditions. *J. Mater. Res. Technol.* **2021**, *11*, 896–904. [CrossRef]
3. Cazón, P.; Velazquez, G.; Vázquez, M. Characterization of mechanical and barrier properties of bacterial cellulose, glycerol and polyvinyl alcohol (PVOH) composite films with eco-friendly UV-protective properties. *Food Hydrocoll.* **2020**, *99*, 105323. [CrossRef]
4. Lin, C.-M.; Chang, Y.-C.; Cheng, L.-C.; Liu, C.-H.; Chang, S.C.; Hsien, T.-Y.; Wang, D.-M.; Hsieh, H.-J. Preparation of graphene-embedded hydroxypropyl cellulose/chitosan/polyethylene oxide nanofiber membranes as wound dressings with enhanced antibacterial properties. *Cellulose* **2020**, *27*, 2651–2667. [CrossRef]
5. Kamiński, K.; Jarosz, M.; Grudzień, J.; Pawlik, J.; Zastawnik, F.; Pandyra, P.; Kołodziejczyk, A.M. Hydrogel bacterial cellulose: A path to improved materials for new eco-friendly textiles. *Cellulose* **2020**, *27*, 5353–5365. [CrossRef]
6. Galdino, C.J.S.; Maia, A.D.; Meira, H.M.; Souza, T.C.; Amorim, J.D.P.; Almeida, F.C.G.; Costa, A.F.S.; Sarubbo, L.A. Use of a bacterial cellulose filter for the removal of oil from wastewater. *Process Biochem.* **2020**, *91*, 288–296. [CrossRef]
7. Pang, M.; Huang, Y.; Meng, F.; Zhuang, Y.; Liu, H.; Du, M.; Ma, Q.; Wang, Q.; Chen, Z.; Chen, L.; et al. Application of bacterial cellulose in skin and bone tissue engineering. *Eur. Polym. J.* **2020**, *122*, 109365. [CrossRef]
8. Norizan, M.N.; Harussani, M.M.; Demon, S.Z.N.; Halim, N.A.; Samsuri, A.; Mohamad, I.S.; Knight, V.F.; Abdullah, N. Carbon nanotubes: Functionalisation and their application in chemical sensors. *RSC Adv.* **2020**, 43704–43732. [CrossRef]
9. Esa, F.; Tasirin, S.M.; Rahman, N.A. Overview of Bacterial Cellulose Production and Application. *Agric. Agric. Sci. Procedia* **2014**, *2*, 113–119. [CrossRef]
10. Andriani, D.; Apriyana, A.Y.; Karina, M. The optimization of bacterial cellulose production and its applications: A review. *Cellulose* **2020**, *27*, 6747–6766. [CrossRef]
11. Ilyas, R.A.; Sapuan, S.M.; Harussani, M.M.; Atikah, M.S.N.; Ibrahim, R.; Asyraf, M.R.M.; Radzi, A.M.; Nadlene, R.; Kian, L.K.; Mali, S. Development and characterization of roselle nanocellulose and its potential in reinforced nanocomposites. In *Roselle*; Elsevier: Amsterdam, The Netherlands, 2021; pp. 285–317.
12. Omran, A.A.B.; Mohammed, A.A.B.A.; Sapuan, S.M.; Ilyas, R.A.; Asyraf, M.R.M.; Koloor, S.S.R.; Petrů, M. Micro- and Nanocellulose in Polymer Composite Materials: A Review. *Polymers* **2021**, *13*, 231. [CrossRef] [PubMed]
13. Sari, N.H.; Pruncu, C.I.; Sapuan, S.M.; Ilyas, R.A.; Catur, A.D.; Suteja, S.; Sutaryono, Y.A.; Pullen, G. The effect of water immersion and fibre content on properties of corn husk fibres reinforced thermoset polyester composite. *Polym. Test.* **2020**, *91*, 106751. [CrossRef]

14. Picheth, G.F.; Pirich, C.L.; Sierakowski, M.R.; Woehl, M.A.; Sakakibara, C.N.; de Souza, C.F.; Martin, A.A.; da Silva, R.; de Freitas, R.A. Bacterial cellulose in biomedical applications: A review. *Int. J. Biol. Macromol.* **2017**, *104*, 97–106. [CrossRef] [PubMed]
15. Aisyah, H.A.; Paridah, M.T.; Sapuan, S.M.; Ilyas, R.A.; Khalina, A.; Nurazzi, N.M.; Lee, S.H.; Lee, C.H. A Comprehensive Review on Advanced Sustainable Woven Natural Fibre Polymer Composites. *Polymers* **2021**, *13*, 471. [CrossRef] [PubMed]
16. Norizan, M.N.; Moklis, M.H.; Alias, A.H.; Rushdan, A.I.; Norrrahim, M.N.F.; Abdan, K.; Abdullah, N. Treatments of Natural Fibre as Reinforcement in Polymer Composites-Short Review. *Funct. Compos. Struct.* **2021**, *3*, 024002. [CrossRef]
17. Norrrahim, M.N.F.; Huzaifah, M.R.M.; Farid, M.A.A.; Shazleen, S.S.; Misenan, M.S.M.; Yasim-Anuar, T.A.T.; Naveen, J.; Nurazzi, N.M.; Rani, M.S.A.; Hakimi, M.I.; et al. Greener Pretreatment Approaches for the Valorisation of Natural Fibre Biomass into Bioproducts. *Polymers* **2021**, *13*, 2971. [CrossRef] [PubMed]
18. Azman, M.A.; Asyraf, M.R.M.; Khalina, A.; Petrů, M.; Ruzaidi, C.M.; Sapuan, S.M.; Wan Nik, W.B.; Ishak, M.R.; Ilyas, R.A.; Suriani, M.J. Natural Fiber Reinforced Composite Material for Product Design: A Short Review. *Polymers* **2021**, *13*, 1917. [CrossRef]
19. Nurazzi, N.M.; Asyraf, M.R.M.; Khalina, A.; Abdullah, N.; Aisyah, H.A.; Rafiqah, S.A.; Sabaruddin, F.A.; Kamarudin, S.H.; Norrrahim, M.N.F.; Ilyas, R.A.; et al. A Review on Natural Fiber Reinforced Polymer Composite for Bullet Proof and Ballistic Applications. *Polymers* **2021**, *13*, 646. [CrossRef]
20. Alsubari, S.; Zuhri, M.Y.M.; Sapuan, S.M.; Ishak, M.R.; Ilyas, R.A.; Asyraf, M.R.M. Potential of natural fiber reinforced polymer composites in sandwich structures: A review on its mechanical properties. *Polymers* **2021**, *13*, 423. [CrossRef] [PubMed]
21. Ayu, R.S.; Khalina, A.; Harmaen, A.S.; Zaman, K.; Isma, T.; Liu, Q.; Ilyas, R.A.; Lee, C.H. Characterization Study of Empty Fruit Bunch (EFB) Fibers Reinforcement in Poly(Butylene) Succinate (PBS)/Starch/Glycerol Composite Sheet. *Polymers* **2020**, *12*, 1571. [CrossRef]
22. Ullah, H.; Santos, H.A.; Khan, T. Applications of bacterial cellulose in food, cosmetics and drug delivery. *Cellulose* **2016**, *23*, 2291–2314. [CrossRef]
23. Rebelo, A.R.; Archer, A.J.; Chen, X.; Liu, C.; Yang, G.; Liu, Y. Dehydration of bacterial cellulose and the water content effects on its viscoelastic and electrochemical properties. *Sci. Technol. Adv. Mater.* **2018**, *19*, 203–211. [CrossRef] [PubMed]
24. Bi, J.-C.; Liu, S.-X.; Li, C.-F.; Li, J.; Liu, L.-X.; Deng, J.; Yang, Y.-C. Morphology and structure characterization of bacterial celluloses produced by different strains in agitated culture. *J. Appl. Microbiol.* **2014**, *117*, 1305–1311. [CrossRef] [PubMed]
25. Evans, B.R.; O'Neill, H.M. Effect of surface attachment on synthesis of bacterial cellulose. *Appl. Biochem. Biotechnol. Part A Enzym. Eng. Biotechnol.* **2005**, *121*, 439–450. [CrossRef]
26. Costa, A.F.S.; Almeida, F.C.G.; Vinhas, G.M.; Sarubbo, L.A. Production of bacterial cellulose by Gluconacetobacter hansenii using corn steep liquor as nutrient sources. *Front. Microbiol.* **2017**, *8*, 1–12. [CrossRef]
27. Torres, F.G.; Commeaux, S.; Troncoso, O.P. Biocompatibility of bacterial cellulose based biomaterials. *J. Funct. Biomater.* **2012**, *3*, 864–878. [CrossRef]
28. Boisset, C.; Fraschini, C.; Schülein, M.; Henrissat, B.; Chanzy, H. Imaging the enzymatic digestion of bacterial cellulose ribbons reveals the endo character of the cellobiohydrolase Cel6A from *Humicola insolens* and its mode of synergy with cellobiohydrolase Cel7A. *Appl. Environ. Microbiol.* **2000**, *66*, 1444–1452. [CrossRef]
29. Basta, A.H.; El-Saied, H. Performance of improved bacterial cellulose application in the production of functional paper. *J. Appl. Microbiol.* **2009**, *107*, 2098–2107. [CrossRef]
30. El-Saied, H.; Basta, A.H.; Gobran, R.H. Research Progress in Friendly Environmental Technology for the Production of Cellulose Products (Bacterial Cellulose and Its Application). *Polym. Plast. Technol. Eng.* **2004**, *43*, 797–820. [CrossRef]
31. Tahara, N.; Tabuchi, M.; Watanabe, K.; Yano, H.; Morinaga, Y.; Yoshinaga, F. Degree of Polymerization of Cellulose from *Acetobacter xylinum* BPR2001 Decreased by Cellulase Produced by the Strain. *Biosci. Biotechnol. Biochem.* **1997**, *61*, 1862–1865. [CrossRef]
32. Grande, C.J.; Torres, F.G.; Gomez, C.M.; Troncoso, O.P.; Canet-Ferrer, J.; Martínez-Pastor, J. Development of self-assembled bacterial cellulose-starch nanocomposites. *Mater. Sci. Eng. C* **2009**, *29*, 1098–1104. [CrossRef]
33. Gorgieva, S.; Trček, J. Bacterial cellulose: Production, modification and perspectives in biomedical applications. *Nanomaterials* **2019**, *9*, 1352. [CrossRef]
34. Lou, Z.C. A better design is needed for clinical studies of chronic tympanic membrane perforations using biological materials. *Eur. Arch. Oto Rhino-Laryngol.* **2016**, *273*, 4045–4046. [CrossRef]
35. Biskin, S.; Damar, M.; Oktem, S.N.; Sakalli, E.; Erdem, D.; Pakir, O. A new graft material for myringoplasty: Bacterial cellulose. *Eur. Arch. Oto-Rhino-Laryngol.* **2016**, *273*, 3561–3565. [CrossRef]
36. Lang, N.; Merkel, E.; Fuchs, F.; Schumann, D.; Klemm, D.; Kramer, F.; Mayer-Wagner, S.; Schroeder, C.; Freudenthal, F.; Netz, H.; et al. Bacterial nanocellulose as a new patch material for closure of ventricular septal defects in a pig model. *Eur. J. Cardio-Thorac. Surg.* **2014**, *47*, 1013–1021. [CrossRef]
37. Azeredo, H.M.C.; Rosa, M.F.; Mattoso, L.H.C. Nanocellulose in bio-based food packaging applications. *Ind. Crops Prod.* **2017**, *97*, 664–671. [CrossRef]
38. Gellner, P.E.L.; Dang, A.E.; Jay, H. Acoustic Diaphragm and Method for Producing Same. US Patent US005274199A, 28 December 1993.
39. Hioki, N.; Hori, Y.; Watanabe, K.; Morinaga, Y.; Yoshinaga, F.; Hibino, Y.; Ogura, T. Bacterial cellulose; as a new material for papermaking. *Jpn. TAPPI J.* **1995**, *49*, 718–723. [CrossRef]

40. Yang, J.; Yu, J.; Fan, J.; Sun, D.; Tang, W.; Yang, X. Biotemplated preparation of CdS nanoparticles/bacterial cellulose hybrid nanofibers for photocatalysis application. *J. Hazard. Mater.* **2011**, *189*, 377–383. [CrossRef] [PubMed]
41. Benito-González, I.; López-Rubio, A.; Gómez-Mascaraque, L.G.; Martínez-Sanz, M. PLA coating improves the performance of renewable adsorbent pads based on cellulosic aerogels from aquatic waste biomass. *Chem. Eng. J.* **2020**, *390*, 124607. [CrossRef]
42. Morales-Narváez, E.; Golmohammadi, H.; Naghdi, T.; Yousefi, H.; Kostiv, U.; Horák, D.; Pourreza, N.; Merkoçi, A. Nanopaper as an Optical Sensing Platform. *ACS Nano* **2015**, *9*, 7296–7305. [CrossRef]
43. Yuen, J.D.; Shriver-Lake, L.C.; Walper, S.A.; Zabetakis, D.; Breger, J.C.; Stenger, D.A. Microbial nanocellulose printed circuit boards for medical sensing. *Sensors* **2020**, *20*, 2047. [CrossRef]
44. Portela, R.; Leal, C.R.; Almeida, P.L.; Sobral, R.G. Bacterial cellulose: A versatile biopolymer for wound dressing applications. *Microb. Biotechnol.* **2019**, *12*, 586–610. [CrossRef]
45. Zhang, X.; Peng, X.; Zhang, S.W. Biodegradable medical polymers. In *Science and Principles of Biodegradable and Bioresorbable Medical Polymers*; Zhang, X., Ed.; Elsevier: Duxford, UK, 2017; pp. 1–33.
46. Czaja, W.; Krystynowicz, A.; Kawecki, M.; Wysota, K.; Sakiel, S.; Wróblewski, P.; Glik, J.; Nowak, M.; Bielecki, S. Biomedical Applications of Microbial Cellulose in Burn Wound Recovery. In *Cellulose: Molecular and Structural Biology*; Springer: Dordrecht, The Netherlands, 2007; pp. 307–321. [CrossRef]
47. Sulaeva, I.; Hettegger, H.; Bergen, A.; Rohrer, C.; Kostic, M.; Konnerth, J.; Rosenau, T.; Potthast, A. Fabrication of bacterial cellulose-based wound dressings with improved performance by impregnation with alginate. *Mater. Sci. Eng. C* **2020**, *110*, 110619. [CrossRef]
48. Blanco Parte, F.G.; Santoso, S.P.; Chou, C.C.; Verma, V.; Wang, H.T.; Ismadji, S.; Cheng, K.C. Current progress on the production, modification, and applications of bacterial cellulose. *Crit. Rev. Biotechnol.* **2020**, *40*, 397–414. [CrossRef]
49. Żur, J.; Piński, A.; Michalska, J.; Hupert-Kocurek, K.; Nowak, A.; Wojcieszyńska, D.; Guzik, U. A whole-cell immobilization system on bacterial cellulose for the paracetamol-degrading Pseudomonas moorei KB4 strain. *Int. Biodeterior. Biodegrad.* **2020**, *149*, 104919. [CrossRef]
50. Zhang, S.; He, H.; Guan, S.; Cai, B.; Li, Q.; Rong, S. Bacterial cellulose-alginate composite beads as yarrowia lipolytica cell carriers for lactone production. *Molecules* **2020**, *25*, 928. [CrossRef] [PubMed]
51. Feil, G.; Horres, R.; Schulte, J.; Mack, A.F.; Petzoldt, S.; Arnold, C.; Meng, C.; Jost, L.; Boxleitner, J.; Kiessling-Wolf, N.; et al. Bacterial cellulose shifts transcriptome and proteome of cultured endothelial cells towards native differentiation. *Mol. Cell. Proteom.* **2017**, *16*, 1563–1577. [CrossRef]
52. Chan, J.X.; Wong, J.F.; Petrů, M.; Hassan, A.; Nirmal, U.; Othman, N.; Ilyas, R.A. Effect of Nanofillers on Tribological Properties of Polymer Nanocomposites: A Review on Recent Development. *Polymers* **2021**, *13*, 2867. [CrossRef] [PubMed]
53. Punia Bangar, S.; Nehra, M.; Siroha, A.K.; Petrů, M.; Ilyas, R.A.; Devi, U.; Devi, P. Development and Characterization of Physical Modified Pearl Millet Starch-Based Films. *Foods* **2021**, *10*, 1609. [CrossRef] [PubMed]
54. Kumari, N.; Bangar, S.P.; Petrů, M.; Ilyas, R.A.; Singh, A.; Kumar, P. Development and Characterization of Fenugreek Protein-Based Edible Film. *Foods* **2021**, *10*, 1976. [CrossRef] [PubMed]
55. Ilyas, R.A.; Sapuan, S.M.; Harussani, M.M.; Hakimi, M.Y.A.Y.; Haziq, M.Z.M.; Atikah, M.S.N.; Asyraf, M.R.M.; Ishak, M.R.; Razman, M.R.; Nurazzi, N.M.; et al. Polylactic Acid (PLA) Biocomposite: Processing, Additive Manufacturing and Advanced Applications. *Polymers* **2021**, *13*, 1326. [CrossRef] [PubMed]
56. Ilyas, R.A.; Sapuan, S.M. Biopolymers and Biocomposites: Chemistry and Technology. *Curr. Anal. Chem.* **2020**, *16*, 500–503. [CrossRef]
57. Abral, H.; Atmajaya, A.; Mahardika, M.; Hafizulhaq, F.; Kadriadi; Handayani, D.; Sapuan, S.M.; Ilyas, R.A. Effect of ultrasonication duration of polyvinyl alcohol (PVA) gel on characterizations of PVA film. *J. Mater. Res. Technol.* **2020**, *9*, 2477–2486. [CrossRef]
58. Sapuan, S.M.; Aulia, H.S.; Ilyas, R.A.; Atiqah, A.; Dele-Afolabi, T.T.; Nurazzi, M.N.; Supian, A.B.M.; Atikah, M.S.N. Mechanical properties of longitudinal basalt/woven-glass-fiber-reinforced unsaturated polyester-resin hybrid composites. *Polymers* **2020**, *12*, 2211. [CrossRef] [PubMed]
59. Pötzinger, Y.; Kralisch, D.; Fischer, D. Bacterial nanocellulose: The future of controlled drug delivery? *Ther. Deliv.* **2017**, *8*, 753–761. [CrossRef]
60. Unal, S.; Gunduz, O.; Uzun, M. Tissue Engineering Applications of Bacterial Cellulose Based Nanofibers. In *Green Nanomaterials. Advanced Structured Materials*; Ahmed, S., Ali, W., Eds.; Springer: Singapore, 2020; pp. 319–346.
61. Wang, K.; Ma, Q.; Zhang, Y.M.; Han, G.T.; Qu, C.X.; Wang, S.D. Preparation of bacterial cellulose/silk fibroin double-network hydrogel with high mechanical strength and biocompatibility for artificial cartilage. *Cellulose* **2020**, *27*, 1845–1852. [CrossRef]
62. Klinthoopthamrong, N.; Chaikiawkeaw, D.; Phoolcharoen, W.; Rattanapisit, K.; Kaewpungsup, P.; Pavasant, P.; Hoven, V.P. Bacterial cellulose membrane conjugated with plant-derived osteopontin: Preparation and its potential for bone tissue regeneration. *Int. J. Biol. Macromol.* **2020**, *149*, 51–59. [CrossRef]
63. Junka, A.; Bartoszewicz, M.; Dziadas, M.; Szymczyk, P.; Dydak, K.; Żywicka, A.; Owczarek, A.; Bil-Lula, I.; Czajkowska, J.; Fijałkowski, K. Application of bacterial cellulose experimental dressings saturated with gentamycin for management of bone biofilm in vitro and ex vivo. *J. Biomed. Mater. Res. Part B Appl. Biomater.* **2020**, *108*, 30–37. [CrossRef]
64. Svensson, A.; Nicklasson, E.; Harrah, T.; Panilaitis, B.; Kaplan, D.L.; Brittberg, M.; Gatenholm, P. Bacterial cellulose as a potential scaffold for tissue engineering of cartilage. *Biomaterials* **2005**, *26*, 419–431. [CrossRef]

65. Codreanu, A.; Balta, C.; Herman, H.; Cotoraci, C.; Mihali, C.V.; Zurbau, N.; Zaharia, C.; Rapa, M.; Stanescu, P.; Radu, I.C.; et al. Bacterial cellulose-modified polyhydroxyalkanoates scaffolds promotes bone formation in critical size calvarial defects in mice. *Materials* **2020**, *13*, 1433. [CrossRef]
66. Zhang, W.; Wang, X.C.; Li, X.Y.; Zhang, L.L.; Jiang, F. A 3D porous microsphere with multistage structure and component based on bacterial cellulose and collagen for bone tissue engineering. *Carbohydr. Polym.* **2020**, *236*, 116043. [CrossRef] [PubMed]
67. Lina, F.; Chandra, P.; Adrianna, M.; Wankei, W. Bacterial cellulose production using a novel microbe. *Front. Bioeng. Biotechnol.* **2016**, *4*, 1–2. [CrossRef]
68. Bodea, I.M.; Cătunescu, G.M.; Stroe, T.F.; Dîrlea, S.A.; Beteg, F.I. Applications of bacterial-synthesized cellulose in veterinary medicine—A review. *Acta Vet. Brno* **2019**, *88*, 451–471. [CrossRef]
69. Lin, N.; Dufresne, A. Nanocellulose in biomedicine: Current status and future prospect. *Eur. Polym. J.* **2014**, *59*, 302–325. [CrossRef]
70. Kim, Y.-K.; Lee, S.-C.; Cho, Y.-Y.; Oh, H.-J.; Ko, Y.H. Isolation of Cellulolytic Bacillus subtilis Strains from Agricultural Environments. *ISRN Microbiol.* **2012**, *2012*, 1–9. [CrossRef]
71. Inoue, B.S.; Streit, S.; dos Santos Schneider, A.L.; Meier, M.M. Bioactive bacterial cellulose membrane with prolonged release of chlorhexidine for dental medical application. *Int. J. Biol. Macromol.* **2020**, *148*, 1098–1108. [CrossRef]
72. Li, N.; Yang, L.; Pan, C.; Saw, P.E.; Ren, M.; Lan, B.; Wu, J.; Wang, X.; Zeng, T.; Zhou, L.; et al. Naturally-occurring bacterial cellulose-hyperbranched cationic polysaccharide derivative/MMP-9 siRNA composite dressing for wound healing enhancement in diabetic rats. *Acta Biomater.* **2020**, *102*, 298–314. [CrossRef]
73. Stumpf, T.R.; Tang, L.; Kirkwood, K.; Yang, X.; Zhang, J.; Cao, X. Production and evaluation of biosynthesized cellulose tubes as promising nerve guides for spinal cord injury treatment. *J. Biomed. Mater. Res.-Part A* **2020**, *108*, 1380–1389. [CrossRef]
74. Czaja, W.K.; Young, D.J.; Kawecki, M.; Brown, R.M. The Future Prospects of Microbial Cellulose in Biomedical Applications. *Biomacromolecules* **2007**, *8*, 1–12. [CrossRef]
75. Schaffner, M.; Rühs, P.A.; Coulter, F.; Kilcher, S.; Studart, A.R. 3D printing of bacteria into functional complex materials. *Sci. Adv.* **2017**, *3*, eaao6804. [CrossRef]
76. Okahisa, Y.; Yoshida, A.; Miyaguchi, S.; Yano, H. Optically transparent wood-cellulose nanocomposite as a base substrate for flexible organic light-emitting diode displays. *Compos. Sci. Technol.* **2009**, *69*, 1958–1961. [CrossRef]
77. Gomes, N.O.; Carrilho, E.; Antonio, S.; Machado, S.; Sgobbi, F. Bacterial cellulose-based electrochemical sensing platform: A smart material for miniaturized biosensors. *Electrochim. Acta* **2020**, *349*, 136341. [CrossRef]
78. Buruaga-Ramiro, C.; Valenzuela, S.V.; Valls, C.; Roncero, M.B.; Pastor, F.I.J.; Díaz, P.; Martínez, J. Bacterial cellulose matrices to develop enzymatically active paper. *Cellulose* **2020**, *27*, 3413–3426. [CrossRef]
79. Park, S.; Park, J.; Jo, I.; Cho, S.P.; Sung, D.; Ryu, S.; Park, M.; Min, K.A.; Kim, J.; Hong, S.; et al. In situ hybridization of carbon nanotubes with bacterial cellulose for three-dimensional hybrid bioscaffolds. *Biomaterials* **2015**, *58*, 93–102. [CrossRef] [PubMed]
80. Czaja, W.; Krystynowicz, A.; Bielecki, S.; Brown, J.R.M. Microbial cellulose—The natural power to heal wounds. *Biomaterials* **2006**, *27*, 145–151. [CrossRef] [PubMed]
81. Markstedt, K.; Mantas, A.; Tournier, I.; Martínez Ávila, H.; Hägg, D.; Gatenholm, P. 3D Bioprinting Human Chondrocytes with Nanocellulose–Alginate Bioink for Cartilage Tissue Engineering Applications. *Biomacromolecules* **2015**, *16*, 1489–1496. [CrossRef] [PubMed]
82. Zhang, X.; Liu, D.; Yang, L.; Zhou, L.; You, T. Self-assembled three-dimensional graphene-based materials for dye adsorption and catalysis. *J. Mater. Chem. A* **2015**, *3*, 10031–10037. [CrossRef]
83. Naseri-Nosar, M.; Salehi, M.; Hojjati-Emami, S. Cellulose acetate/poly lactic acid coaxial wet-electrospun scaffold containing citalopram-loaded gelatin nanocarriers for neural tissue engineering applications. *Int. J. Biol. Macromol.* **2017**, *103*, 701–708. [CrossRef]
84. Mathew, A.P.; Oksman, K.; Pierron, D.; Harmand, M.F. Biocompatible Fibrous Networks of Cellulose Nanofibres and Collagen Crosslinked Using Genipin: Potential as Artificial Ligament/Tendons. *Macromol. Biosci.* **2013**, *13*, 289–298. [CrossRef]
85. Sämfors, S.; Karlsson, K.; Sundberg, J.; Markstedt, K.; Gatenholm, P. Biofabrication of bacterial nanocellulose scaffolds with complex vascular structure. *Biofabrication* **2019**, *11*, 45010. [CrossRef]
86. Hussain, Z.; Sajjad, W.; Khan, T.; Wahid, F. Production of bacterial cellulose from industrial wastes: A review. *Cellulose* **2019**, *26*, 2895–2911. [CrossRef]
87. Vandamme, E.J.; De Baets, S.; Vanbaelen, A.; Joris, K.; De Wulf, P. Improved production of bacterial cellulose and its application potential. *Polym. Degrad. Stab.* **1998**, *59*, 93–99. [CrossRef]
88. Bajpai, P. *Biobased Polymers: Properties and Applications in Packaging*, 1st ed.; Elsevier Inc.: Amsterdam, The Netherlands, 2019; ISBN 9780128184042.
89. Torgbo, S.; Sukyai, P. Bacterial cellulose-based scaffold materials for bone tissue engineering. *Appl. Mater. Today* **2018**, *11*, 34–49. [CrossRef]
90. Mohd Nurazzi, N.; Asyraf, M.R.M.; Khalina, A.; Abdullah, N.; Sabaruddin, F.A.; Kamarudin, S.H.; Ahmad, S.; Mahat, A.M.; Lee, C.L.; Aisyah, H.A.; et al. Fabrication, Functionalization, and Application of Carbon Nanotube-Reinforced Polymer Composite: An Overview. *Polymers* **2021**, *13*, 1047. [CrossRef] [PubMed]
91. Stumpf, T.R.; Yang, X.; Zhang, J.; Cao, X. In situ and ex situ modifications of bacterial cellulose for applications in tissue engineering. *Mater. Sci. Eng. C* **2018**, *82*, 372–383. [CrossRef]

92. Asyraf, M.R.M.; Ishak, M.R.; Sapuan, S.M.; Yidris, N.; Ilyas, R.A.; Rafidah, M.; Razman, M.R. Potential Application of Green Composites for Cross Arm Component in Transmission Tower: A Brief Review. *Int. J. Polym. Sci.* **2020**, *2020*, 1–15. [CrossRef]
93. Asyraf, M.R.M.; Rafidah, M.; Ishak, M.R.; Sapuan, S.M.; Ilyas, R.A.; Razman, M.R. Integration of TRIZ, Morphological Chart and ANP method for development of FRP composite portable fire extinguisher. *Polym. Compos.* **2020**, *41*, 2917–2932. [CrossRef]
94. Nazrin, A.; Sapuan, S.M.; Zuhri, M.Y.M.; Tawakkal, I.S.M.A.; Ilyas, R.A. Water barrier and mechanical properties of sugar palm crystalline nanocellulose reinforced thermoplastic sugar palm starch (TPS)/poly(lactic acid) (PLA) blend bionanocomposites. *Nanotechnol. Rev.* **2021**, *10*, 431–442. [CrossRef]
95. Tarique, J.; Sapuan, S.M.; Khalina, A.; Sherwani, S.F.K.; Yusuf, J.; Ilyas, R.A. Recent developments in sustainable arrowroot (*Maranta arundinacea* Linn) starch biopolymers, fibres, biopolymer composites and their potential industrial applications: A review. *J. Mater. Res. Technol.* **2021**, *13*, 1191–1219. [CrossRef]
96. Zaborowska, M.; Bodin, A.; Bäckdahl, H.; Popp, J.; Goldstein, A.; Gatenholm, P. Microporous bacterial cellulose as a potential scaffold for bone regeneration. *Acta Biomater.* **2010**, *6*, 2540–2547. [CrossRef]
97. Gonçalves-Pimentel, C.; Moreno, G.M.M.; Trindade, B.S.; Isaac, A.R.; Rodrigues, C.G.; Savariradjane, M.; de Albuquerque, A.V.; de Andrade Aguiar, J.L.; Andrade-da-Costa, B.L.d.S. Cellulose exopolysaccharide from sugarcane molasses as a suitable substrate for 2D and 3D neuron and astrocyte primary cultures. *J. Mater. Sci. Mater. Med.* **2018**, *29*, 139. [CrossRef] [PubMed]
98. Padra, J.; Silva, P.; Sencadas, V. Bacterial Cellulose as a Support for the Growth of Retinal Pigment Epithelium. *Biomacromolecules* **2015**, *16*, 1341–1351. [CrossRef]
99. Seoane, I.T.; Manfredi, L.B.; Cyras, V.P.; Torre, L.; Fortunati, E.; Puglia, D. Effect of Cellulose Nanocrystals and Bacterial Cellulose on disintegrability in composting conditions of Plasticized PHB Nanocomposites. *Polymers* **2017**, *9*, 561. [CrossRef] [PubMed]
100. Zhang, L.; Zheng, S.; Hu, Z.; Zhong, L.; Wang, Y.; Zhang, X.; Xue, J. Preparation of Polyvinyl Alcohol/Bacterial-Cellulose-Coated Biochar–Nanosilver Antibacterial Composite Membranes. *Appl. Sci.* **2020**, *10*, 752. [CrossRef]
101. Hamedi, S.; Shojaosadati, S.A.; Najafi, V.; Alizadeh, V. A novel double-network antibacterial hydrogel based on aminated bacterial cellulose and schizophyllan. *Carbohydr. Polym.* **2020**, *229*, 115383. [CrossRef]
102. Sukhavattanakul, P.; Manuspiya, H. Fabrication of hybrid thin film based on bacterial cellulose nanocrystals and metal nanoparticles with hydrogen sulfide gas sensor ability. *Carbohydr. Polym.* **2020**, *230*, 115566. [CrossRef]
103. Gupta, A.; Briffa, S.M.; Swingler, S.; Gibson, H.; Kannappan, V.; Adamus, G.; Kowalczuk, M.; Martin, C.; Radecka, I. Synthesis of Silver Nanoparticles Using Curcumin-Cyclodextrins Loaded into Bacterial Cellulose-Based Hydrogels for Wound Dressing Applications. *Biomacromolecules* **2020**, *21*, 1802–1811. [CrossRef]
104. Di Pasquale, G.; Graziani, S.; Pollicino, A.; Trigona, C. Performance characterization of a biodegradable deformation sensor based on bacterial cellulose. *IEEE Trans. Instrum. Meas.* **2019**, *69*, 2561–2569. [CrossRef]
105. Wang, L.; Mao, L.; Qi, F.; Li, X.; Ullah, M.W.; Zhao, M.; Shi, Z.; Yang, G. Synergistic effect of highly aligned bacterial cellulose/gelatin membranes and electrical stimulation on directional cell migration for accelerated wound healing. *Chem. Eng. J.* **2021**, *424*, 130563. [CrossRef]
106. Rebelo, A.; Liu, Y.; Liu, C.; Schäfer, K.-H.; Saumer, M.; Yang, G. Poly (4-vinylaniline)/polyaniline bilayer functionalized bacterial cellulose membranes as bioelectronics interfaces. *Carbohydr. Polym.* **2019**, *204*, 190–201. [CrossRef]
107. Saxena, I.M.; Brown, R.M. A Perspective on the Assembly of Cellulose-Synthesizing Complexes: Possible Role of KORRIGAN and Microtubules in Cellulose Synthesis in Plants. In *Cellulose: Molecular and Structural Biology*; Springer: Dordrecht, The Netherlands, 2007; pp. 169–181.
108. Putra, A.; Kakugo, A.; Furukawa, H.; Gong, J.P.; Osada, Y. Tubular bacterial cellulose gel with oriented fibrils on the curved surface. *Polymer* **2008**, *49*, 1885–1891. [CrossRef]
109. Nimeskern, L.; Martínez Ávila, H.; Sundberg, J.; Gatenholm, P.; Müller, R.; Stok, K.S. Mechanical evaluation of bacterial nanocellulose as an implant material for ear cartilage replacement. *J. Mech. Behav. Biomed. Mater.* **2013**, *22*, 12–21. [CrossRef]
110. Bodin, A.; Concaro, S.; Brittberg, M.; Gatenholm, P. Bacterial cellulose as a potential meniscus implant. *J. Tissue Eng. Regen. Med.* **2007**, *1*, 406–408. [CrossRef]
111. Charpentier, P.A.; Maguire, A.; Wan, W. Surface modification of polyester to produce a bacterial cellulose-based vascular prosthetic device. *Appl. Surf. Sci.* **2006**, *252*, 6360–6367. [CrossRef]
112. Klemm, D.; Schumann, D.; Udhardt, U.; Marsch, S. Bacterial synthesized cellulose—Artificial blood vessels for microsurgery. *Prog. Polym. Sci.* **2001**, *26*, 1561–1603. [CrossRef]
113. Sepúlveda, R.V.; Valente, F.L.; Reis, E.C.C.; Araújo, F.R.; Eleotério, R.B.; Queiroz, P.V.S.; Borges, A.P.B. Bacterial cellulose and bacterial cellulose/polycaprolactone composite as tissue substitutes in rabbits' cornea. *Pesqui. Veterinária Bras.* **2016**, *36*, 986–992. [CrossRef]
114. Barud, H.S.; Ribeiro, S.J.L.; Carone, C.L.P.; Ligabue, R.; Einloft, S.; Queiroz, P.V.S.; Borges, A.P.B.; Jahno, V.D. Optically transparent membrane based on bacterial cellulose/polycaprolactone. *Polimeros* **2013**, *23*, 135–138. [CrossRef]
115. Caiut, J.M.A.; Barud, H.S.; Santos, M.V.; Oliveira, U.L.; Menezes, J.F.S.; Messaddeq, Y.; Ribeiro, S.J.L. Luminescent multifunctional biocellulose membranes. In Proceedings of the Nanostructured Thin Films IV, San Diego, CA, USA, 21–25 August 2011; Volume 8104, p. 81040. [CrossRef]
116. Kim, J.; Cai, Z.; Lee, H.S.; Choi, G.S.; Lee, D.H.; Jo, C. Preparation and characterization of a Bacterial cellulose/Chitosan composite for potential biomedical application. *J. Polym. Res.* **2011**, *18*, 739–744. [CrossRef]

117. Legeza, V.I.; Galenko-Yaroshevskii, V.P.; Zinov'ev, E.V.; Paramonov, B.A.; Kreichman, G.S.; Turkovskii, I.I.; Gumenyuk, E.S.; Karnovich, A.G.; Khripunov, A.K. Effects of new wound dressings on healing of thermal burns of the skin in acute radiation disease. *Bull. Exp. Biol. Med.* **2004**, *138*, 311–315. [CrossRef] [PubMed]
118. Messaddeq, Y.; Ribeiro, S.J.L.; Thomazini, W. Contact Lens for Therapy, Method and Apparatus for Their Production and Use. Brazil Patent BR PI0603704-6, 2008.
119. Osorio, M.; Velásquez-Cock, J.; Restrepo, L.M.; Zuluaga, R.; Gañán, P.; Rojas, O.J.; Ortiz-Trujillo, I.; Castro, C. Bioactive 3D-Shaped Wound Dressings Synthesized from Bacterial Cellulose: Effect on Cell Adhesion of Polyvinyl Alcohol Integrated In Situ. *Int. J. Polym. Sci.* **2017**, *2017*, 1–10. [CrossRef]
120. Zhu, H.; Jia, S.; Yang, H.; Jia, Y.; Yan, L.; Li, J. Preparation and Application of Bacterial Cellulose Sphere: A Novel Biomaterial. *Biotechnol. Biotechnol. Equip.* **2011**, *25*, 2233–2236. [CrossRef]
121. Wu, S.-C.; Li, M.-H. Production of bacterial cellulose membranes in a modified airlift bioreactor by *Gluconacetobacter xylinus*. *J. Biosci. Bioeng.* **2015**, *120*, 444–449. [CrossRef] [PubMed]
122. Hu, Y.; Catchmark, J.M.; Vogler, E.A. Factors Impacting the Formation of Sphere-Like Bacterial Cellulose Particles and Their Biocompatibility for Human Osteoblast Growth. *Biomacromolecules* **2013**, *14*, 3444–3452. [CrossRef] [PubMed]
123. Hu, Y.; Catchmark, J.M. Formation and Characterization of Spherelike Bacterial Cellulose Particles Produced by *Acetobacter xylinum* JCM 9730 Strain. *Biomacromolecules* **2010**, *11*, 1727–1734. [CrossRef] [PubMed]
124. Czaja, W.; Romanovicz, D.; malcolm Brown, R. Structural investigations of microbial cellulose produced in stationary and agitated culture. *Cellulose* **2004**, *11*, 403–411. [CrossRef]
125. Cai, Q.; Hu, C.; Yang, N.; Wang, Q.; Wang, J.; Pan, H.; Hu, Y.; Ruan, C. Enhanced activity and stability of industrial lipases immobilized onto spherelike bacterial cellulose. *Int. J. Biol. Macromol.* **2018**, *109*, 1174–1181. [CrossRef] [PubMed]
126. Gu, J.; Catchmark, J.M. Impact of hemicelluloses and pectin on sphere-like bacterial cellulose assembly. *Carbohydr. Polym.* **2012**, *88*, 547–557. [CrossRef]
127. Yan, Z.; Chen, S.; Wang, H.; Wang, B.; Jiang, J. Biosynthesis of bacterial cellulose/multi-walled carbon nanotubes in agitated culture. *Carbohydr. Polym.* **2008**, *74*, 659–665. [CrossRef]
128. Zhu, H.; Jia, S.; Wan, T.; Jia, Y.; Yang, H.; Li, J.; Yan, L.; Zhong, C. Biosynthesis of spherical Fe3O4/bacterial cellulose nanocomposites as adsorbents for heavy metal ions. *Carbohydr. Polym.* **2011**, *86*, 1558–1564. [CrossRef]
129. Hu, D. Study on Structural Modulation and Compounding with Graphene of Bacterial Cellulose for Adsorption of Organics. Master's Thesis, Tianjin University, Tianjin, China, 2014.
130. Ilyas, R.A.; Sapuan, S.M.; Ishak, M.R.; Zainudin, E.S. Development and characterization of sugar palm nanocrystalline cellulose reinforced sugar palm starch bionanocomposites. *Carbohydr. Polym.* **2018**, *202*, 186–202. [CrossRef]
131. Ilyas, R.A.; Sapuan, S.M.; Ishak, M.R.; Zainudin, E.S. Sugar palm nanofibrillated cellulose (*Arenga pinnata* (*Wurmb.*) *Merr*): Effect of cycles on their yield, physic-chemical, morphological and thermal behavior. *Int. J. Biol. Macromol.* **2019**, *123*, 379–388. [CrossRef] [PubMed]
132. Ilyas, R.A.; Sapuan, S.M.; Ibrahim, R.; Abral, H.; Ishak, M.R.; Zainudin, E.S.; Atikah, M.S.N.; Mohd Nurazzi, N.; Atiqah, A.; Ansari, M.N.M.; et al. Effect of sugar palm nanofibrillated celluloseconcentrations on morphological, mechanical andphysical properties of biodegradable films basedon agro-waste sugar palm (*Arenga pinnata* (*Wurmb.*) *Merr*) starch. *J. Mater. Res. Technol.* **2019**, *8*, 4819–4830. [CrossRef]
133. Hazrol, M.D.; Sapuan, S.M.; Ilyas, R.A.; Othman, M.L.; Sherwani, S.F.K. Electrical properties of sugar palm nanocrystalline cellulose reinforced sugar palm starch nanocomposites. *Polimery* **2020**, *65*, 363–370. [CrossRef]
134. Ilyas, R.A.; Sapuan, S.M.; Ibrahim, R.; Abral, H.; Ishak, M.R.; Zainudin, E.S.; Asrofi, M.; Atikah, M.S.N.; Huzaifah, M.R.M.; Radzi, A.M.; et al. Sugar palm (*Arenga pinnata* (*Wurmb.*) *Merr*) cellulosic fibre hierarchy: A comprehensive approach from macro to nano scale. *J. Mater. Res. Technol.* **2019**, *8*, 2753–2766. [CrossRef]
135. Ilyas, R.A.; Sapuan, S.M.; Atiqah, A.; Ibrahim, R.; Abral, H.; Ishak, M.R.; Zainudin, E.S.; Nurazzi, N.M.; Atikah, M.S.N.; Ansari, M.N.M.; et al. Sugar palm (*Arenga pinnata* [*Wurmb.*] *Merr*) starch films containing sugar palm nanofibrillated cellulose as reinforcement: Water barrier properties. *Polym. Compos.* **2019**, *41*, 459–467. [CrossRef]
136. Ilyas, R.A.; Sapuan, S.M.; Atikah, M.S.N.; Asyraf, M.R.M.; Rafiqah, S.A.; Aisyah, H.A.; Nurazzi, N.M.; Norrrahim, M.N.F. Effect of hydrolysis time on the morphological, physical, chemical, and thermal behavior of sugar palm nanocrystalline cellulose (*Arenga pinnata* (*Wurmb.*) *Merr*). *Text. Res. J.* **2021**, *91*, 152–167. [CrossRef]
137. Rozilah, A.; Jaafar, C.N.A.; Sapuan, S.M.; Zainol, I.; Ilyas, R.A. The Effects of Silver Nanoparticles Compositions on the Mechanical, Physiochemical, Antibacterial, and Morphology Properties of Sugar Palm Starch Biocomposites for Antibacterial Coating. *Polymers* **2020**, *12*, 2605. [CrossRef]
138. Syafiq, R.; Sapuan, S.M.; Zuhri, M.Y.M.; Ilyas, R.A.; Nazrin, A.; Sherwani, S.F.K.; Khalina, A. Antimicrobial Activities of Starch-Based Biopolymers and Biocomposites Incorporated with Plant Essential Oils: A Review. *Polymers* **2020**, *12*, 2403. [CrossRef] [PubMed]
139. Syafri, E.; Sudirman; Mashadi; Yulianti, E.; Deswita; Asrofi, M.; Abral, H.; Sapuan, S.M.; Ilyas, R.A.; Fudholi, A. Effect of sonication time on the thermal stability, moisture absorption, and biodegradation of water hyacinth (*Eichhornia crassipes*) nanocellulose-filled bengkuang (*Pachyrhizus erosus*) starch biocomposites. *J. Mater. Res. Technol.* **2019**, *8*, 6223–6231. [CrossRef]
140. Abral, H.; Ariksa, J.; Mahardika, M.; Handayani, D.; Aminah, I.; Sandrawati, N.; Pratama, A.B.; Fajri, N.; Sapuan, S.M.; Ilyas, R.A. Transparent and antimicrobial cellulose film from ginger nanofiber. *Food Hydrocoll.* **2020**, *98*, 105266. [CrossRef]

141. Abral, H.; Ariksa, J.; Mahardika, M.; Handayani, D.; Aminah, I.; Sandrawati, N.; Sapuan, S.M.; Ilyas, R.A. Highly transparent and antimicrobial PVA based bionanocomposites reinforced by ginger nanofiber. *Polym. Test.* **2019**, *81*, 106186. [CrossRef]
142. Sabaruddin, F.A.; Paridah, M.T.; Sapuan, S.M.; Ilyas, R.A.; Lee, S.H.; Abdan, K.; Mazlan, N.; Roseley, A.S.M.; Abdul Khalil, H.P.S. The effects of unbleached and bleached nanocellulose on the thermal and flammability of polypropylene-reinforced kenaf core hybrid polymer bionanocomposites. *Polymers* **2020**, *13*, 116. [CrossRef]
143. Asrofi, M.; Sapuan, S.M.; Ilyas, R.A.; Ramesh, M. Characteristic of composite bioplastics from tapioca starch and sugarcane bagasse fiber: Effect of time duration of ultrasonication (Bath-Type). *Mater. Today Proc.* **2020**, *46*, 1626–1630. [CrossRef]
144. Asrofi, M.; Sujito; Syafri, E.; Sapuan, S.M.; Ilyas, R.A. Improvement of Biocomposite Properties Based Tapioca Starch and Sugarcane Bagasse Cellulose Nanofibers. *Key Eng. Mater.* **2020**, *849*, 96–101. [CrossRef]
145. Kamaruddin, Z.H.; Jumaidin, R.; Selamat, M.Z.; Ilyas, R.A. Characteristics and Properties of Lemongrass (Cymbopogan Citratus): A Comprehensive Review. *J. Nat. Fibers* **2021**, 1–18. [CrossRef]
146. Wahab, M.; Sapuan, S.M.; Harussani, M.M.; Zuhri, M.Y.M.; Saleh, A.A. Conceptual Design of Glass/Renewable Natural Fibre-Reinforced Polymer Hybrid Composite Motorcycle Side Cover. *J. Renew. Mater.* **2021**, *9*, 1973–1989. [CrossRef]
147. Ilyas, R.A.; Sapuan, S.M.; Sanyang, M.L.; Ishak, M.R.; Zainudin, E.S. Nanocrystalline cellulose as reinforcement for polymeric matrix nanocomposites and its potential applications: A Review. *Curr. Anal. Chem.* **2018**, *14*, 203–225. [CrossRef]
148. Harussani, M.M.; Sapuan, S.M.; Rashid, U.; Khalina, A. Development and Characterization of Polypropylene Waste from Personal Protective Equipment (PPE)-Derived Char-Filled Sugar Palm Starch Biocomposite Briquettes. *Polymers* **2021**, *13*, 1707. [CrossRef] [PubMed]
149. Jung, H.I.; Jeong, J.H.; Lee, O.M.; Park, G.T.; Kim, K.K.; Park, H.C.; Lee, S.M.; Kim, Y.G.; Son, H.J. Influence of glycerol on production and structural-physical properties of cellulose from *Acetobacter* sp. V6 cultured in shake flasks. *Bioresour. Technol.* **2010**, *101*, 3602–3608. [CrossRef]
150. Rangaswamy, B.E.; Vanitha, K.P.; Hungund, B.S. Microbial Cellulose Production from Bacteria Isolated from Rotten Fruit. *Int. J. Polym. Sci.* **2015**, *2015*, 280784. [CrossRef]
151. Singh, O.; Panesar, P.S.; Chopra, H.K. Response surface optimization for cellulose production from agro industrial waste by using new bacterial isolate *Gluconacetobacter xylinus* C18. *Food Sci. Biotechnol.* **2017**, *26*, 1019–1028. [CrossRef] [PubMed]
152. Fan, C.; Xu, X.; Song, L.; Guan, W.; Li, J.; Liu, B.; Shi, P.; Zhang, W. The use of *Agrobacterium*-mediated insertional mutagenesis sequencing to identify novel genes of *Humicola insolens* involved in cellulase production. *3 Biotech* **2018**, *8*, 153. [CrossRef]
153. Molina-Ramírez, C.; Castro, M.; Osorio, M.; Torres-Taborda, M.; Gómez, B.; Zuluaga, R.; Gómez, C.; Gañán, P.; Rojas, O.J.; Castro, C. Effect of different carbon sources on bacterial nanocellulose production and structure using the low pH resistant strain *Komagataeibacter medellinensis*. *Materials* **2017**, *10*, 639. [CrossRef]
154. Reese, C. Characterization of WssF; a Putative Acetyltransferase from *Achromobacter insuavis* and *Pseudomonas fluorescens*. Master's Thesis, Wilfrid Laurier University, Waterloo, ON, Canada, 2019.
155. Ahmed, S.A.; Kazim, A.R.; Hassan, H.M. Increasing Cellulose Production from *Rhizobium leguminosarum* bv. *viciae*. *J. Al-Nahrain Univ.* **2017**, *20*, 120–125. [CrossRef]
156. Anusuya, R.S.; Anandham, R.; Kumutha, K.; Gayathry, G.; Mageshwaran, V. Characterization and optimization of bacterial cellulose produced by *Acetobacter* spp. *J. Environ. Biol.* **2020**, *41*, 207–215. [CrossRef]
157. Sun, B.; Zi, Q.; Chen, C.; Zhang, H.; Gu, Y.; Liang, G.; Sun, D. Study of specific metabolic pattern of *Acetobacter xylinum* NuST4.2 and bacterial cellulose production improvement. *Cellul. Chem. Technol.* **2018**, *52*, 795–801.
158. Canale-Parola, E.; Wolfe, R.S. Synthesis of cellulose by *Sarcina ventriculi*. *Biochim. Biophys. Acta-Gen. Subj.* **1964**, *82*, 403–405. [CrossRef]
159. Rastogi, A.; Banerjee, R. Production and characterization of cellulose from *Leifsonia* sp. *Process Biochem.* **2019**, *85*, 35–42. [CrossRef]
160. Sreena, C.P.; Sebastian, D. Augmented cellulase production by *Bacillus subtilis* strain MU S1 using different statistical experimental designs. *J. Genet. Eng. Biotechnol.* **2018**, *16*, 9–16. [CrossRef] [PubMed]
161. Toor, Y.; Ilyas, U. Optimization of cellulase production by *Aspergillus ornatus* by the solid state fermentation of *Cicer arietinum*. *Am. J. Res.* **2014**, *2*, 125–141.
162. Picart, P.; Diaz, P.; Pastor, F.I.J. Cellulases from two *Penicillium* sp. strains isolated from subtropical forest soil: Production and characterization. *Lett. Appl. Microbiol.* **2007**, *45*, 108–113. [CrossRef]
163. Prasanna, H.N.; Ramanjaneyulu, G.; Rajasekhar Reddy, B. Optimization of cellulase production by *Penicillium* sp. *3 Biotech* **2016**, *6*, 162. [CrossRef]
164. Sohail, M.; Ahmad, A.; Khan, S.A. Production of cellulase from *Aspergillus terreus* MS105 on crude and commercially purified substrates. *3 Biotech* **2016**, *6*, 103. [CrossRef] [PubMed]
165. Gao, J.; Weng, H.; Zhu, D.; Yuan, M.; Guan, F.; Xi, Y. Production and characterization of cellulolytic enzymes from the thermoacidophilic fungal *Aspergillus terreus* M11 under solid-state cultivation of corn stover. *Bioresour. Technol.* **2008**, *99*, 7623–7629. [CrossRef] [PubMed]
166. Santos, T.C.D.; Abreu Filho, G.; Brito, A.R.D.; Pires, A.J.V.; Bonomo, R.C.F.; Franco, M. Production and characterization of cellulolytic enzymes by *Aspergillus niger* and *Rhizopus* sp. by solid state fermentation of prickly pear. *Rev. Caatinga* **2016**, *29*, 222–233. [CrossRef]
167. Lee, C.K.; Darah, I.; Ibrahim, C.O. Production and Optimization of Cellulase Enzyme Using *Aspergillus niger* USM AI 1 and Comparison with *Trichoderma reesei* via Solid State Fermentation System. *Biotechnol. Res. Int.* **2011**, *2011*, 1–6. [CrossRef]

168. Qurat-Ul-Ain; Baig, S.; Saleem, M. Production and characterization of cellulases of *Aspergillus niger* by using rice husk and saw dust as substrates. *Pak. J. Bot.* **2012**, *44*, 377–382.
169. Pachauri, P.; Aranganathan, V.; More, S.; Sullia, S.B.; Deshmukh, S. Purification and characterization of cellulase from a novel isolate of *Trichoderma longibrachiatum*. *Biofuels* **2020**, *11*, 85–91. [CrossRef]
170. Petlamul, W.; Sripornngam, T.; Buakwan, N.; Buakaew, S.; Mahamad, K. The Capability of Beauveria Bassiana for Cellulase Enzyme Production. In Proceedings of the 7th International Conference on Bioscience, Biochemistry and Bioinformatics–ICBBB '17, Bangkok Thailand, 21–23 January 2017; ACM Press: New York, NY, USA, 2017; pp. 62–66.
171. Hernández, C.; Milagres, A.M.F.; Vázquez-Marrufo, G.; Muñoz-Páez, K.M.; García-Pérez, J.A.; Alarcón, E. An ascomycota coculture in batch bioreactor is better than polycultures for cellulase production. *Folia Microbiol.* **2018**, *63*, 467–478. [CrossRef] [PubMed]
172. Schuerg, T.; Gabriel, R.; Baecker, N.; Baker, S.E.; Singer, S.W. *Thermoascus aurantiacus* is an Intriguing Host for the Industrial Production of Cellulases. *Curr. Biotechnol.* **2017**, *6*, 89–97. [CrossRef]
173. Baldrian, P.; Valášková, V. Degradation of cellulose by basidiomycetous fungi. *FEMS Microbiol. Rev.* **2008**, *32*, 501–521. [CrossRef] [PubMed]
174. Chen, Y.W.; Lee, H.V.; Juan, J.C.; Phang, S.-M. Production of new cellulose nanomaterial from red algae marine biomass *Gelidium elegans*. *Carbohydr. Polym.* **2016**, *151*, 1210–1219. [CrossRef] [PubMed]
175. Tarchoun, A.F.; Trache, D.; Klapötke, T.M. Microcrystalline cellulose from *Posidonia oceanica* brown algae: Extraction and characterization. *Int. J. Biol. Macromol.* **2019**, *138*, 837–845. [CrossRef] [PubMed]
176. Sebeia, N.; Jabli, M.; Ghith, A.; Elghoul, Y.; Alminderej, F.M. Production of cellulose from *Aegagropila Linnaei* macro-algae: Chemical modification, characterization and application for the bio-sorptionof cationic and anionic dyes from water. *Int. J. Biol. Macromol.* **2019**, *135*, 152–162. [CrossRef] [PubMed]
177. Uzyol, H.K.; Saçan, M.T. Bacterial cellulose production by *Komagataeibacter hansenii* using algae-based glucose. *Environ. Sci. Pollut. Res.* **2017**, *24*, 11154–11162. [CrossRef] [PubMed]
178. Xiang, Z.; Gao, W.; Chen, L.; Lan, W.; Zhu, J.Y.; Runge, T. A comparison of cellulose nanofibrils produced from *Cladophora glomerata* algae and bleached eucalyptus pulp. *Cellulose* **2016**, *23*, 493–503. [CrossRef]
179. Kobayashi, S.; Kashiwa, K.; Shimada, J.; Kawasaki, T.; Shoda, S. Enzymatic polymerization: The first in vitro synthesis of cellulose via nonbiosynthetic path catalyzed by cellulase. *Makromol. Chem. Macromol. Symp.* 55. [CrossRef]
180. Kobayashi, S.; Kashiwa, K.; Kawasaki, T.; Shoda, S. Novel method for polysaccharide synthesis using an enzyme: The first in vitro synthesis of cellulose via a nonbiosynthetic path utilizing cellulase as catalyst. *J. Am. Chem. Soc.* **1991**, *113*, 3079–3084. [CrossRef]
181. Nakatsubo, F.; Kamitakahara, H.; Hori, M. Cationic Ring-Opening Polymerization of 3,6-Di-O-benzyl-α-d-glucose 1,2,4-Orthopivalate and the First Chemical Synthesis of Cellulose. *J. Am. Chem. Soc.* **1996**, *118*, 1677–1681. [CrossRef]
182. Wu, H.; Williams, G.R.; Wu, J.; Wu, J.; Niu, S.; Li, H.; Wang, H.; Zhu, L. Regenerated chitin fibers reinforced with bacterial cellulose nanocrystals as suture biomaterials. *Carbohydr. Polym.* **2018**, *180*, 304–313. [CrossRef]
183. Keshk, S.M. Bacterial Cellulose Production and its Industrial Applications. *J. Bioprocess. Biotech.* **2014**, *4*, 2. [CrossRef]
184. Abol-Fotouh, D.; Hassan, M.A.; Shokry, H.; Roig, A.; Azab, M.S.; Kashyout, A.E.-H.B. Bacterial nanocellulose from agro-industrial wastes: Low-cost and enhanced production by *Komagataeibacter saccharivorans* MD1. *Sci. Rep.* **2020**, *10*, 1–14.
185. Bae, S.O.; Sugano, Y.; Ohi, K.; Shoda, M. Features of bacterial cellulose synthesis in a mutant generated by disruption of the diguanylate cyclase 1 gene of *Acetobacter xylinum* BPR 2001. *Appl. Microbiol. Biotechnol.* **2004**, *65*, 315–322. [CrossRef]
186. Cakar, F.; Özer, I.; Aytekin, A.Ö.; Şahin, F. Improvement production of bacterial cellulose by semi-continuous process in molasses medium. *Carbohydr. Polym.* **2014**, *106*, 7–13. [CrossRef] [PubMed]
187. Bae, S.O.; Shoda, M. Production of bacterial cellulose by *Acetobacter xylinum* BPR2001 using molasses medium in a jar fermentor. *Appl. Microbiol. Biotechnol.* **2005**, *67*, 45–51. [CrossRef] [PubMed]
188. Poddar, M.K.; Dikshit, P.K. Recent development in bacterial cellulose production and synthesis of cellulose based conductive polymer nanocomposites. *Nano Sel.* **2021**, *2*, 1605–1628. [CrossRef]
189. Shi, Z.; Li, Y.; Chen, X.; Han, H.; Yang, G. Double network bacterial cellulose hydrogel to build a biology-device interface. *Nanoscale* **2014**, *6*, 970–977. [CrossRef]
190. Vazquez, A.; Foresti, M.L.; Cerrutti, P.; Galvagno, M. Bacterial Cellulose from Simple and Low Cost Production Media by *Gluconacetobacter xylinus*. *J. Polym. Environ.* **2013**, *21*, 545–554. [CrossRef]
191. Wang, J.; Tavakoli, J.; Tang, Y. Bacterial cellulose production, properties and applications with different culture methods—A review. *Carbohydr. Polym.* **2019**, *219*, 63–76. [CrossRef]
192. Choi, C.N.; Song, H.J.; Kim, M.J.; Chang, M.H.; Kim, S.J. Properties of bacterial cellulose produced in a pilot-scale spherical type bubble column bioreactor. *Korean J. Chem. Eng.* **2009**, *26*, 136–140. [CrossRef]
193. Chao, Y.P.; Sugano, Y.; Kouda, T.; Yoshinaga, F.; Shoda, M. Production of bacterial cellulose by *Acetobacter xylinum* with an air-lift reactor. *Biotechnol. Tech.* **1997**, *11*, 829–832. [CrossRef]
194. Chao, Y.; Sugano, Y.; Shoda, M. Bacterial cellulose production under oxygen-enriched air at different fructose concentrations in a 50-L, internal-loop airlift reactor. *Appl. Microbiol. Biotechnol.* **2001**, *55*, 673–679. [CrossRef]
195. Chao, Y.; Ishida, T.; Sugano, Y.; Shoda, M. Bacterial cellulose production by *Acetobacter xylinum* in a 50-L internal-loop airlift reactor. *Biotechnol. Bioeng.* **2000**, *68*, 345–352. [CrossRef]

196. Mormino, R.; Bungay, H. Composites of bacterial cellulose and paper made with a rotating disk bioreactor. *Appl. Microbiol. Biotechnol.* **2003**, *62*, 503–506. [CrossRef]
197. Zahan, K.A.; Pa'e, N.; Muhamad, I.I. An evaluation of fermentation period and discs rotation speed of rotary discs reactor for bacterial cellulose production. *Sains Malays.* **2016**, *45*, 393–400.
198. Lin, S.-P.; Hsieh, S.-C.; Chen, K.-I.; Demirci, A.; Cheng, K.-C. Semi-continuous bacterial cellulose production in a rotating disk bioreactor and its materials properties analysis. *Cellulose* **2014**, *21*, 835–844. [CrossRef]
199. Lin, S.-P.; Liu, C.-T.; Hsu, K.-D.; Hung, Y.-T.; Shih, T.-Y.; Cheng, K.-C. Production of bacterial cellulose with various additives in a PCS rotating disk bioreactor and its material property analysis. *Cellulose* **2016**, *23*, 367–377. [CrossRef]
200. Fijałkowski, K.; Żywicka, A.; Drozd, R.; Junka, A.F.; Peitler, D.; Kordas, M.; Konopacki, M.; Szymczyk, P.; Rakoczy, R. Increased water content in bacterial cellulose synthesized under rotating magnetic fields. *Electromagn. Biol. Med.* **2017**, *36*, 192–201. [CrossRef]
201. Fija, K.; Anna, Ż.; Junka, A.F.; Kordas, M.; Rakoczy, R.; Fijałkowski, K.; Drozd, R.; Żywicka, A.; Junka, A.F.; Kordas, M.; et al. Biochemical and cellular properties of *Gluconacetobacter xylinus* cultures exposed to different modes of rotating magnetic field. *Pol. J. Chem. Technol.* **2017**, *19*, 107–114. [CrossRef]
202. Fijałkowski, K.; Rakoczy, R.; Żywicka, A.; Drozd, R.; Zielińska, B.; Wenelska, K.; Cendrowski, K.; Peitler, D.; Kordas, M.; Konopacki, M.; et al. Time Dependent Influence of Rotating Magnetic Field on Bacterial Cellulose. *Int. J. Polym. Sci.* **2016**, *2016*, 7536397. [CrossRef]
203. Fijałkowski, K.; Żywicka, A.; Drozd, R.; Niemczyk, A.; Junka, A.F.; Peitler, D.; Kordas, M.; Konopacki, M.; Szymczyk, P.; El Fray, M.; et al. Modification of bacterial cellulose through exposure to the rotating magnetic field. *Carbohydr. Polym.* **2015**, *133*, 52–60. [CrossRef]
204. Jung, J.Y.; Khan, T.; Park, J.K.; Chang, H.N. Production of bacterial cellulose by *Gluconacetobacter hansenii* using a novel bioreactor equipped with a spin filter. *Korean J. Chem. Eng.* **2007**, *24*, 265–271. [CrossRef]
205. Hornung, M.; Ludwig, M.; Schmauder, H.P. Optimizing the Production of Bacterial Cellulose in Surface Culture: A Novel Aerosol Bioreactor Working on a Fed Batch Principle (Part 3). *Eng. Life Sci.* **2007**, *7*, 35–41. [CrossRef]
206. Cheng, K.-C.; Catchmark, J.M.; Demirci, A. Enhanced production of bacterial cellulose by using a biofilm reactor and its material property analysis. *J. Biol. Eng.* **2009**, *3*, 12. [CrossRef] [PubMed]
207. Cheng, K.-C.; Catchmark, J.M.; Demirci, A. Effects of CMC Addition on Bacterial Cellulose Production in a Biofilm Reactor and Its Paper Sheets Analysis. *Biomacromolecules* **2011**, *12*, 730–736. [CrossRef] [PubMed]
208. Velásquez-Riaño, M.; Bojacá, V. Production of bacterial cellulose from alternative low-cost substrates. *Cellulose* **2017**, *24*, 2677–2698. [CrossRef]
209. Harussani, M.M.; Salit, M.S.; Rashid, U.; Abdan, K. Plastic Waste Conversion into Electrical, Thermal and Fuel Energy via Incinerations and Pyrolysis amidst COVID-19 Pandemic. In Proceedings of the AIUE Proceedings of the 2nd Energy and Human Habitat Conference, Cape Town, South Africa, 26–27 July 2021.
210. Harussani, M.M.; Sapuan, S.M.; Khalina, A.; Rashid, U.; Tarique, J. Slow pyrolysis of disinfected COVID-19 non-woven polypropylene (PP) waste. In Proceedings of the International Symposium on Applied Sciences and Engineering ISASE2021, Erzurum, Turkey, 7–9 April 2021; Office of International Affairs, Atatürk University: Erzurum, Turkey, 2021; pp. 310–312.
211. Miah, J.H.; Griffiths, A.; McNeill, R.; Halvorson, S.; Schenker, U.; Espinoza-Orias, N.D.; Morse, S.; Yang, A.; Sadhukhan, J. Environmental management of confectionery products: Life cycle impacts and improvement strategies. *J. Clean. Prod.* **2018**, *177*, 732–751. [CrossRef]
212. Kongruang, S. Bacterial cellulose production by *Acetobacter xylinum* strains from agricultural waste products. In *Biotechnology for Fuels and Chemicals*; Springer: Berlin/Heidelberg, Germany, 2007; pp. 763–774.
213. Goelzer, F.D.E.; Faria-Tischer, P.C.S.; Vitorino, J.C.; Sierakowski, M.-R.; Tischer, C.A. Production and characterization of nanospheres of bacterial cellulose from *Acetobacter xylinum* from processed rice bark. *Mater. Sci. Eng. C* **2009**, *29*, 546–551. [CrossRef]
214. Kuo, C.-H.H.; Huang, C.-Y.Y.; Shieh, C.-J.J.; Wang, H.-M.M.D.; Tseng, C.-Y.Y. Hydrolysis of Orange Peel with Cellulase and Pectinase to Produce Bacterial Cellulose using *Gluconacetobacter xylinus*. *Waste Biomass Valorization* **2019**, *10*, 85–93. [CrossRef]
215. Hong, F.; Guo, X.; Zhang, S.; Han, S.F.; Yang, G.; Jönsson, L.J. Bacterial cellulose production from cotton-based waste textiles: Enzymatic saccharification enhanced by ionic liquid pretreatment. *Bioresour. Technol.* **2012**, *104*, 503–508. [CrossRef]
216. Fan, X.; Gao, Y.; He, W.; Hu, H.; Tian, M.; Wang, K.; Pan, S. Production of nano bacterial cellulose from beverage industrial waste of citrus peel and pomace using *Komagataeibacter xylinus*. *Carbohydr. Polym.* **2016**, *151*, 1068–1072. [CrossRef] [PubMed]
217. Keshk, S.; Sameshima, K. The utilization of sugar cane molasses with/without the presence of lignosulfonate for the production of bacterial cellulose. *Appl. Microbiol. Biotechnol.* **2006**, *72*, 291–296. [CrossRef]
218. Qi, G.; Luo, M.; Huang, C.; Guo, H.; Chen, X.; Xiong, L.; Wang, B.; Lin, X.; Peng, F.; Chen, X. Comparison of bacterial cellulose production by *Gluconacetobacter xylinus* on bagasse acid and enzymatic hydrolysates. *J. Appl. Polym. Sci.* **2017**, *134*, 45066. [CrossRef]
219. Pensupa, N.; Leu, S.-Y.; Hu, Y.; Du, C.; Liu, H.; Jing, H.; Wang, H.; Lin, C.S.K. Recent trends in sustainable textile waste recycling methods: Current situation and future prospects. In *Chemistry and Chemical Technologies in Waste Valorization*; Springer: Berlin/Heidelberg, Germany, 2017; pp. 189–228.
220. Wang, Y. Fiber and textile waste utilization. *Waste Biomass Valorization* **2010**, *1*, 135–143. [CrossRef]

221. Gao, X.; Shi, Z.; Liu, C.; Yang, G.; Sevostianov, I.; Silberschmidt, V.V. Inelastic behaviour of bacterial cellulose hydrogel: In aqua cyclic tests. *Polym. Test.* **2015**, *44*, 82–92. [CrossRef]
222. Almeida, L.R.; Martins, A.R.; Fernandes, E.M.; Oliveira, M.B.; Correlo, V.M.; Pashkuleva, I.; Marques, A.P.; Ribeiro, A.S.; Durães, N.F.; Silva, C.J.; et al. New biotextiles for tissue engineering: Development, characterization and in vitro cellular viability. *Acta Biomater.* **2013**, *9*, 8167–8181. [CrossRef] [PubMed]
223. Kowalska-Ludwicka, K.; Cala, J.; Grobelski, B.; Sygut, D.; Jesionek-Kupnicka, D.; Kolodziejczyk, M.; Bielecki, S.; Pasieka, Z. Modified bacterial cellulose tubes for regeneration of damaged peripheral nerves. *Arch. Med. Sci.* **2013**, *9*, 527–534. [CrossRef] [PubMed]
224. Huang, Y.; Zhu, C.; Yang, J.; Nie, Y.; Chen, C.; Sun, D. Recent advances in bacterial cellulose. *Cellulose* **2014**, *21*, 1–30. [CrossRef]
225. Bıyık, H.; Çoban, E.P. Evaluation of different carbon, nitrogen sources and industrial wastes for bacterial cellulose production. *Eur. J. Biotechnol. Biosci.* **2017**, *5*, 74–80. [CrossRef]
226. Voon, W.W.Y.; Muhialdin, B.J.; Yusof, N.L.; Rukayadi, Y.; Meor Hussin, A.S. Bio-cellulose Production by *Beijerinckia fluminensis* WAUPM53 and *Gluconacetobacter xylinus* 0416 in Sago By-product Medium. *Appl. Biochem. Biotechnol.* **2019**, *187*, 211–220. [CrossRef]
227. Uzuner, S.; Cekmecelioglu, D. Enzymes in the beverage industry. In *Enzymes in Food Biotechnology*; Elsevier Inc.: Amsterdam, The Netherlands, 2018; ISBN 9780128132807.
228. Chandrasekaran, M. *Enzymes in Food and Beverage Processing*; CRC Press: Boca Raton, FL, USA, 2015; pp. 1–518. [CrossRef]
229. Jain, S. Development of Low Cost Nutritional Beverage from Whey. *IOSR J. Environ. Sci. Toxicol. Food Technol.* **2013**, *5*, 73–88. [CrossRef]
230. Revin, V.; Liyaskina, E.; Nazarkina, M.; Bogatyreva, A.; Shchankin, M. Cost-effective production of bacterial cellulose using acidic food industry by-products. *Braz. J. Microbiol.* **2018**, *49*, 151–159. [CrossRef]
231. Ratanapariyanuch, K.; Shen, J.; Jia, Y.; Tyler, R.T.; Shim, Y.Y.; Reaney, M.J.T. Rapid NMR method for the quantification of organic compounds in thin stillage. *J. Agric. Food Chem.* **2011**, *59*, 10454–10460. [CrossRef] [PubMed]
232. Wu, J.M.; Liu, R.H. Cost-effective production of bacterial cellulose in static cultures using distillery wastewater. *J. Biosci. Bioeng.* **2013**, *115*, 284–290. [CrossRef] [PubMed]
233. Jozala, A.F.; Pértile, R.A.N.; dos Santos, C.A.; de Carvalho Santos-Ebinuma, V.; Seckler, M.M.; Gama, F.M.; Pessoa, A. Bacterial cellulose production by *Gluconacetobacter xylinus* by employing alternative culture media. *Appl. Microbiol. Biotechnol.* **2015**, *99*, 1181–1190. [CrossRef] [PubMed]
234. Lima, H.L.S.; Nascimento, E.S.; Andrade, F.K.; Brígida, A.I.S.; Borges, M.F.; Cassales, A.R.; Muniz, C.R.; Souza Filho, M.D.S.M.; Morais, J.P.S.; Rosa, M.D.F. Bacterial cellulose production by *Komagataeibacter hansenii* ATCC 23769 using sisal juice—An agroindustry waste. *Braz. J. Chem. Eng.* **2017**, *34*, 671–680. [CrossRef]
235. Castro, C.; Zuluaga, R.; Putaux, J.L.; Caro, G.; Mondragon, I.; Gañán, P. Structural characterization of bacterial cellulose produced by *Gluconacetobacter swingsii* sp. from Colombian agroindustrial wastes. *Carbohydr. Polym.* **2011**, *84*, 96–102. [CrossRef]
236. Zhao, H.; Xia, J.; Wang, J.; Yan, X.; Wang, C.; Lei, T.; Xian, M.; Zhang, H. Production of bacterial cellulose using polysaccharide fermentation wastewater as inexpensive nutrient sources. *Biotechnol. Biotechnol. Equip.* **2018**, *32*, 350–356. [CrossRef]
237. Lin, D.; Lopez-Sanchez, P.; Li, R.; Li, Z. Production of bacterial cellulose by *Gluconacetobacter hansenii* CGMCC 3917 using only waste beer yeast as nutrient source. *Bioresour. Technol.* **2014**, *151*, 113–119. [CrossRef]
238. Viana, R.M.; Sá, N.M.S.M.; Barros, M.O.; Borges, M.d.F.; Azeredo, H.M.C. Nanofibrillated bacterial cellulose and pectin edible films added with fruit purees. *Carbohydr. Polym.* **2018**, *196*, 27–32. [CrossRef]
239. Manrich, A.; Moreira, F.K.V.; Otoni, C.G.; Lorevice, M.V.; Martins, M.A.; Mattoso, L.H.C. Hydrophobic edible films made up of tomato cutin and pectin. *Carbohydr. Polym.* **2017**, *164*, 83–91. [CrossRef]
240. Azeredo, H.M.C.; Kontou-Vrettou, C.; Moates, G.K.; Wellner, N.; Cross, K.; Pereira, P.H.F.; Waldron, K.W. Wheat straw hemicellulose films as affected by citric acid. *Food Hydrocoll.* **2015**, *50*, 1–6. [CrossRef]
241. Cavka, A.; Guo, X.; Tang, S.-J.J.; Winestrand, S.; Jönsson, L.J.; Hong, F. Production of bacterial cellulose and enzyme from waste fiber sludge. *Biotechnol. Biofuels* **2013**, *6*, 1–10. [CrossRef]
242. Rani, M.U.; Appaiah, K.A.A. Production of bacterial cellulose by *Gluconacetobacter hansenii* UAC09 using coffee cherry husk. *J. Food Sci. Technol.* **2013**, *50*, 755–762. [CrossRef]
243. Xiao, X.; Hou, Y.; Liu, Y.; Liu, Y.; Zhao, H.; Dong, L.; Du, J.; Wang, Y.; Bai, G.; Luo, G. Classification and analysis of corn steep liquor by UPLC/Q-TOF MS and HPLC. *Talanta* **2013**, *107*, 344–348. [CrossRef]
244. Joshi, S.; Goyal, S.; Reddy, M.S. Corn steep liquor as a nutritional source for biocementation and its impact on concrete structural properties. *J. Ind. Microbiol. Biotechnol.* **2018**, *45*, 657–667. [CrossRef]
245. Khami, S.; Khamwichit, W.; Suwannahong, K.; Sanongraj, W. Characteristics of Bacterial Cellulose Production from Agricultural Wastes. *Adv. Mater. Res.* **2014**, *931-932*, 693–697. [CrossRef]
246. Moosavi-Nasab, M.; Yousefi, A. Biotechnological production of cellulose by *Gluconacetobacter xylinus* from agricultural waste. *Iran. J. Biotechnol.* **2011**, *9*, 94–101.
247. Lotfiman, S.; Awang Biak, D.R.; Ti, T.B.; Kamarudin, S.; Nikbin, S. Influence of Date Syrup as a Carbon Source on Bacterial Cellulose Production by *Acetobacter xylinum* 0416. *Adv. Polym. Technol.* **2018**, *37*, 1085–1091. [CrossRef]
248. Gomes, F.P.; Silva, N.H.C.S.; Trovatti, E.; Serafim, L.S.; Duarte, M.F.; Silvestre, A.J.D.; Neto, C.P.; Freire, C.S.R. Production of bacterial cellulose by *Gluconacetobacter sacchari* using dry olive mill residue. *Biomass Bioenergy* **2013**, *55*, 205–211. [CrossRef]

249. Al-Farsi, M.; Alasalvar, C.; Al-Abid, M.; Al-Shoaily, K.; Al-Amry, M.; Al-Rawahy, F. Compositional and functional characteristics of dates, syrups, and their by-products. *Food Chem.* **2007**, *104*, 943–947. [CrossRef]
250. Lestari, P.; Elfrida, N.; Suryani, A.; Suryadi, Y. Study on the Production of Bacterial Cellulose from *Acetobacter xylinum* Using Agro-Waste. *Jordan J. Biol. Sci.* **2014**, *7*, 75–80. [CrossRef]
251. Gülsah, K.; Gülnur, K.; Mikhael, B.; Céline, P.-B.; Mualla, Ö. Potential of polyhydroxyalkanoate (PHA) polymers family as substitutes of petroleum based polymers for packaging applications and solutions brought by their composites to form barrier materials. *Pure Appl. Chem.* **2017**, *89*, 1841. [CrossRef]
252. Li, Z.; Wang, L.; Hua, J.; Jia, S.; Zhang, J.; Liu, H. Production of nano bacterial cellulose from waste water of candied jujube-processing industry using *Acetobacter xylinum*. *Carbohydr. Polym.* **2015**, *120*, 115–119. [CrossRef] [PubMed]
253. Lin, Q.B.; Che, L.L.; Guo, J.; Wang, R.Z. Use of 4-chloro-3, 5-dinitrobenzotrifluoride (CNBF) Derivatization and Ultrahigh-performance Liquid Chromatography Tandem Mass Spectrometry for the Determination of 20 Free Amino Acids in Chinese Jujube Date. *Food Anal. Methods* **2014**, *7*, 571–579. [CrossRef]
254. Zhang, Y.; Wang, X.; Zhao, G.; Wang, Y. Influence of oxidized starch on the properties of thermoplastic starch. *Carbohydr. Polym.* **2013**, *96*, 358–364. [CrossRef]
255. Estur, G. (Ed.) *Cotton Exporter's Guide*, 1st ed.; International Trade Centre UNCTAD/WTO; University of California: Berkeley, CA, USA, 2007.
256. Kuo, C.H.; Lin, P.J.; Lee, C.K. Enzymatic saccharification of dissolution pretreated waste cellulosic fabrics for bacterial cellulose production by *Gluconacetobacter xylinus*. *J. Chem. Technol. Biotechnol.* **2010**, *85*, 1346–1352. [CrossRef]
257. Guo, X.; Chen, L.; Tang, J.; Jönsson, L.J.; Hong, F.F. Production of bacterial nanocellulose and enzyme from [AMIM]Cl-pretreated waste cotton fabrics: Effects of dyes on enzymatic saccharification and nanocellulose production. *J. Chem. Technol. Biotechnol.* **2016**, *91*, 1413–1421. [CrossRef]
258. Shao, Y.; Yashiro, T.; Okubo, K.; Fujii, T. Effect of cellulose nano fiber (CNF) on fatigue performance of carbon fiber fabric composites. *Compos. Part A Appl. Sci. Manuf.* **2015**, *76*, 244–254. [CrossRef]
259. Tsouko, E.; Kourmentza, C.; Ladakis, D.; Kopsahelis, N.; Mandala, I.; Papanikolaou, S.; Paloukis, F.; Alves, V.; Koutinas, A. Bacterial cellulose production from industrial waste and by-product streams. *Int. J. Mol. Sci.* **2015**, *16*, 14832–14849. [CrossRef]
260. Schrecker, S.T.; Gostomski, P.A. Determining the water holding capacity of microbial cellulose. *Biotechnol. Lett.* **2005**, *27*, 1435–1438. [CrossRef] [PubMed]
261. Moosavi-nasab, M.; Yousefi, A.R. Investigation of Physicochemical Properties of the Bacterial. *World Acad. Sci. Eng. Technol.* **2010**, *44*, 1258–1263.
262. Gayathry, G.; Gopalaswamy, G. Production and characterisation of microbial cellulosic fibre from *Acetobacter xylinum*. *Indian J. Fibre Text. Res.* **2014**, *39*, 93–96.
263. Wang, B.; Qi, G.X.; Huang, C.; Yang, X.Y.; Zhang, H.R.; Luo, J.; Chen, X.F.; Xiong, L.; Chen, X.D. Preparation of Bacterial Cellulose/Inorganic Gel of Bentonite Composite by In Situ Modification. *Indian J. Microbiol.* **2016**, *56*, 72–79. [CrossRef] [PubMed]
264. Soemphol, W.; Hongsachart, P.; Tanamool, V. Production and characterization of bacterial cellulose produced from agricultural by-product by *Gluconacetobacter* strains. *Mater. Today Proc.* **2018**, *5*, 11159–11168. [CrossRef]
265. Camacho, F.; Macedo, A.; Malcata, F. Potential industrial applications and commercialization of microalgae in the functional food and feed industries: A short review. *Mar. Drugs* **2019**, *17*, 312. [CrossRef] [PubMed]
266. Goyat, M. Production of Green Bacterial Cellulose Nanofibers by Utilizing Renewable Resources of Algae in Comparison with Agricultural Residue. Ph.D. Thesis, Ryerson University, Toronto, ON, Canada, 2016.
267. Hyun, J.Y.; Mahanty, B.; Kim, C.G. Utilization of Makgeolli Sludge Filtrate (MSF) as Low-Cost Substrate for Bacterial Cellulose Production by *Gluconacetobacter xylinus*. *Appl. Biochem. Biotechnol.* **2014**, *172*, 3748–3760. [CrossRef] [PubMed]
268. Jahan, F.; Kumar, V.; Saxena, R.K.K. Distillery effluent as a potential medium for bacterial cellulose production: A biopolymer of great commercial importance. *Bioresour. Technol.* **2018**, *250*, 922–926. [CrossRef]
269. Ha, J.H.; Shehzad, O.; Khan, S.; Lee, S.Y.; Park, J.W.; Khan, T.; Park, J.K. Production of bacterial cellulose by a static cultivation using the waste from beer culture broth. *Korean J. Chem. Eng.* **2008**, *25*, 812–815. [CrossRef]
270. Gayathri, G.; Srinikethan, G. Bacterial Cellulose production by *K. saccharivorans* BC1 strain using crude distillery effluent as cheap and cost effective nutrient medium. *Int. J. Biol. Macromol.* **2019**, *138*, 950–957. [CrossRef]
271. Urbina, L.; Hernández-Arriaga, A.M.; Eceiza, A.; Gabilondo, N.; Corcuera, M.A.; Prieto, M.A.; Retegi, A. By-products of the cider production: An alternative source of nutrients to produce bacterial cellulose. *Cellulose* **2017**, *24*, 2071–2082. [CrossRef]
272. Khattak, W.A.; Khan, T.; Ul-Islam, M.; Ullah, M.W.; Khan, S.; Wahid, F.; Park, J.K. Production, characterization and biological features of bacterial cellulose from scum obtained during preparation of sugarcane jaggery (gur). *J. Food Sci. Technol.* **2015**, *52*, 8343–8349. [CrossRef] [PubMed]
273. Jahan, F.; Kumar, V.; Rawat, G.; Saxena, R.K. Production of microbial cellulose by a bacterium isolated from fruit. *Appl. Biochem. Biotechnol.* **2012**, *167*, 1157–1171.
274. Cerrutti, P.; Roldán, P.; García, R.M.; Galvagno, M.A.; Vázquez, A.; Foresti, M.L. Production of bacterial nanocellulose from wine industry residues: Importance of fermentation time on pellicle characteristics. *J. Appl. Polym. Sci.* **2016**, *133*, 1–9. [CrossRef]
275. Luo, M.-T.; Huang, C.; Chen, X.-F.; Huang, Q.-L.; Qi, G.-X.; Tian, L.-L.; Xiong, L.; Li, H.-L.; Chen, X.-D. Efficient bioconversion from acid hydrolysate of waste oleaginous yeast biomass after microbial oil extraction to bacterial cellulose by *Komagataeibacter xylinus*. *Prep. Biochem. Biotechnol.* **2017**, *47*, 1025–1031. [CrossRef] [PubMed]

276. Bilgi, E.; Bayir, E.; Sendemir-Urkmez, A.; Hames, E.E. Statistical optimization of bacterial cellulose production by gluconacetobacter xylinus using alternative raw materials. In Proceedings of the International Biomedical Engineering Congress, Toronto, Canada, 7–12 June 2015; p. 49.
277. Bilgi, E.; Bayir, E.; Sendemir-Urkmez, A.; Hames, E.E. Optimization of bacterial cellulose production by *Gluconacetobacter xylinus* using carob and haricot bean. *Int. J. Biol. Macromol.* **2016**, *90*, 2–10. [CrossRef] [PubMed]
278. Pacheco, G.; Nogueira, C.R.; Meneguin, A.B.; Trovatti, E.; Silva, M.C.C.; Machado, R.T.A.; Ribeiro, S.J.L.; da Silva Filho, E.C.; Barud, H.d.S. Development and characterization of bacterial cellulose produced by cashew tree residues as alternative carbon source. *Ind. Crops Prod.* **2017**, *107*, 13–19. [CrossRef]
279. Hong, F.; Qiu, K. An alternative carbon source from konjac powder for enhancing production of bacterial cellulose in static cultures by a model strain *Acetobacter aceti* subsp. *xylinus ATCC 23770. Carbohydr. Polym.* **2008**, *72*, 545–549. [CrossRef]
280. Chen, L.; Hong, F.; Yang, X.; Han, S. Biotransformation of wheat straw to bacterial cellulose and its mechanism. *Bioresour. Technol.* **2013**, *135*, 464–468. [CrossRef]
281. Carreira, P.; Mendes, J.A.S.; Trovatti, E.; Serafim, L.S.; Freire, C.S.R.; Silvestre, A.J.D.; Neto, C.P. Utilization of residues from agro-forest industries in the production of high value bacterial cellulose. *Bioresour. Technol.* **2011**, *102*, 7354–7360. [CrossRef]
282. Adebayo-Tayo, B.; Akintunde, M.; Sanusi, J. Effect of Different Fruit Juice Media on Bacterial Cellulose Production by *Acinetobacter* sp. BAN1 and *Acetobacter pasteurianus* PW1. *J. Adv. Biol. Biotechnol.* **2017**, *14*, 1–9. [CrossRef]
283. Kurosumi, A.; Sasaki, C.; Yamashita, Y.; Nakamura, Y. Utilization of various fruit juices as carbon source for production of bacterial cellulose by *Acetobacter xylinum* NBRC 13693. *Carbohydr. Polym.* **2009**, *76*, 333–335. [CrossRef]
284. Kumbhar, J.V.; Rajwade, J.M.; Paknikar, K.M. Fruit peels support higher yield and superior quality bacterial cellulose production. *Appl. Microbiol. Biotechnol.* **2015**, *99*, 6677–6691. [CrossRef] [PubMed]
285. Casarica, A.; Campeanu, G.; Moscovici, M.; Ghiorghita, A.; Manea, V. Improvement of bacterial cellulose production by *Aceobacter xyilinum* dsmz-2004 on poor quality horticultural substrates using the taguchi method for media optimization. Part I. *Cellul. Chem. Technol.* **2013**, *47*, 61–68.
286. Lin, W.C.; Lien, C.C.; Yeh, H.J.; Yu, C.M.; Hsu, S.H. Bacterial cellulose and bacterial cellulose-chitosan membranes for wound dressing applications. *Carbohydr. Polym.* **2013**, *94*, 603–611. [CrossRef] [PubMed]
287. Güzel, M.; Akpınar, Ö. Production and Characterization of Bacterial Cellulose from Citrus Peels. *Waste Biomass Valorization* **2019**, *10*, 2165–2175. [CrossRef]
288. Luo, M.-T.; Zhao, C.; Huang, C.; Chen, X.-F.; Huang, Q.-L.; Qi, G.-X.; Tian, L.-L.; Xiong, L.; Li, H.-L.; Chen, X.-D. Efficient Using Durian Shell Hydrolysate as Low-Cost Substrate for Bacterial Cellulose Production by *Gluconacetobacter xylinus*. *Indian J. Microbiol.* **2017**, *57*, 393–399. [CrossRef] [PubMed]
289. Algar, I.; Fernandes, S.C.M.; Mondragon, G.; Castro, C.; Garcia-Astrain, C.; Gabilondo, N.; Retegi, A.; Eceiza, A. Pineapple agroindustrial residues for the production of high value bacterial cellulose with different morphologies. *J. Appl. Polym. Sci.* **2015**, *132*, 41237. [CrossRef]
290. Adebayo-Tayo, B.C.; Akintunde, M.O.; Alao, S.O. Comparative effect of agrowastes on bacterial cellulose production by *Acinetobacter* sp. BAN1 and *Acetobacter pasteurianus* PW1. *Turk. J. Agri. Nat. Sci.* **2014**, *4*, 145–154.
291. Florea, M.; Hagemann, H.; Santosa, G.; Abbott, J.; Micklem, C.N.; Spencer-Milnes, X.; de Arroyo Garcia, L.; Paschou, D.; Lazenbatt, C.; Kong, D.; et al. Engineering control of bacterial cellulose production using a genetic toolkit and a new cellulose-producing strain. *Proc. Natl. Acad. Sci. USA* **2016**, *113*, E3431–E3440. [CrossRef]
292. Hungund, B.; Prabhu, S.; Shetty, C.; Acharya, S.; Prabhu, V.; Gupta, S.G. Production of bacterial cellulose from *Gluconacetobacter persimmonis* GH-2 using dual and cheaper carbon sources. *J. Microb. Biochem. Technol.* **2013**, *5*, 31–33. [CrossRef]
293. Jozala, A.F.; de Lencastre-Novaes, L.C.; Lopes, A.M.; de Carvalho Santos-Ebinuma, V.; Mazzola, P.G.; Pessoa, A., Jr.; Grotto, D.; Gerenutti, M.; Chaud, M.V. Bacterial nanocellulose production and application: A 10-year overview. *Appl. Microbiol. Biotechnol.* **2016**, *100*, 2063–2072. [CrossRef]
294. Barshan, S.; Rezazadeh-Bari, M.; Almasi, H.; Amiri, S. Optimization and characterization of bacterial cellulose produced by *Komagatacibacter xylinus* PTCC 1734 using vinasse as a cheap cultivation medium. *Int. J. Biol. Macromol.* **2019**, *136*, 1188–1195. [CrossRef]
295. Premjet, S.; Premjet, D.; Ohtani, Y. The Effect of Ingredients of Sugar Cane Molasses on Bacterial Cellulose Production by *Acetobacter xylinum* ATCC 10245. *Fiber* **2007**, *63*, 193–199. [CrossRef]
296. Machado, R.T.A.; Meneguin, A.B.; Sábio, R.M.; Franco, D.F.; Antonio, S.G.; Gutierrez, J.; Tercjak, A.; Berretta, A.A.; Ribeiro, S.J.L.; Lazarini, S.C.; et al. *Komagataeibacter rhaeticus* grown in sugarcane molasses-supplemented culture medium as a strategy for enhancing bacterial cellulose production. *Ind. Crops Prod.* **2018**, *122*, 637–646. [CrossRef]
297. Uraki, Y.; Morito, M.; Kishimoto, T.; Sano, Y. Bacterial Cellulose Production Using Monosaccharides Derived from Hemicelluloses in Water-Soluble Fraction of Waste Liquor from Atmospheric Acetic Acid Pulping. *Holzforschung* **2002**, *56*, 341–347. [CrossRef]
298. Erbas Kiziltas, E.; Kiziltas, A.; Gardner, D.J. Synthesis of bacterial cellulose using hot water extracted wood sugars. *Carbohydr. Polym.* **2015**, *124*, 131–138. [CrossRef]
299. Khattak, W.A.; Khan, T.; Ul-Islam, M.; Wahid, F.; Park, J.K. Production, Characterization and Physico-mechanical Properties of Bacterial Cellulose from Industrial Wastes. *J. Polym. Environ.* **2015**, *23*, 45–53. [CrossRef]

300. Dubey, S.; Singh, J.; Singh, R.P. Biotransformation of sweet lime pulp waste into high-quality nanocellulose with an excellent productivity using *Komagataeibacter europaeus* SGP37 under static intermittent fed-batch cultivation. *Bioresour. Technol.* **2018**, *247*, 73–80. [CrossRef] [PubMed]
301. Tyagi, N.; Suresh, S. Production of cellulose from sugarcane molasses using *Gluconacetobacter intermedius* SNT-1: Optimization & characterization. *J. Clean. Prod.* **2016**, *112*, 71–80. [CrossRef]
302. Huang, C.; Guo, H.J.; Xiong, L.; Wang, B.; Shi, S.L.; Chen, X.F.; Lin, X.Q.; Wang, C.; Luo, J.; Chen, X.D. Using wastewater after lipid fermentation as substrate for bacterial cellulose production by *Gluconacetobacter xylinus*. *Carbohydr. Polym.* **2016**, *136*, 198–202. [CrossRef]
303. Huang, C.; Yang, X.-Y.; Xiong, L.; Guo, H.-J.; Luo, J.; Wang, B.; Zhang, H.-R.; Lin, X.-Q.; Chen, X.-D. Evaluating the possibility of using acetone-butanol-ethanol (ABE) fermentation wastewater for bacterial cellulose production by *Gluconacetobacter xylinus*. *Lett. Appl. Microbiol.* **2015**, *60*, 491–496. [CrossRef] [PubMed]
304. Adnan, A.B. Production of Bacterial Cellulose Using Low-Cost Media. Ph.D. Dissertation, University of Waikato, Hamilton, New Zealand, 2015.
305. Kose, R.; Sunagawa, N.; Yoshida, M.; Tajima, K. One-step production of nanofibrillated bacterial cellulose (NFBC) from waste glycerol using *Gluconacetobacter intermedius* NEDO-01. *Cellulose* **2013**, *20*, 2971–2979. [CrossRef]
306. Chen, Q.; Fan, Q.; Zhang, Z.; Mei, Y.; Wang, H. Effective in situ harvest of microalgae with bacterial cellulose produced by *Gluconacetobacter xylinus*. *Algal Res.* **2018**, *35*, 349–354. [CrossRef]
307. Nóbrega, V.; Faria, M.; Quintana, A.; Kaufmann, M.; Ferreira, A.; Cordeiro, N. From a basic microalga and an acetic acid bacterium cellulose producer to a living symbiotic biofilm. *Materials* **2019**, *12*, 2275. [CrossRef]

polymers MDPI

Review

Greener Pretreatment Approaches for the Valorisation of Natural Fibre Biomass into Bioproducts

Mohd Nor Faiz Norrrahim [1,2,*], Muhammad Roslim Muhammad Huzaifah [3,*], Mohammed Abdillah Ahmad Farid [2,*], Siti Shazra Shazleen [4], Muhammad Syukri Mohamad Misenan [5], Tengku Arisyah Tengku Yasim-Anuar [6], Jesuarockiam Naveen [7], Norizan Mohd Nurazzi [8,*], Mohd Saiful Asmal Rani [9], Mohd Idham Hakimi [2], Rushdan Ahmad Ilyas [10,*] and Mohd Azwan Jenol [2]

1 Research Center for Chemical Defence, Universiti Pertahanan Nasional Malaysia, Kem Sungai Besi, Kuala Lumpur 57000, Malaysia
2 Department of Bioprocess Technology, Faculty of Biotechnology and Biomolecular Sciences, Universiti Putra Malaysia UPM, Serdang 43400, Malaysia; idhamhakimi@ymail.com (M.I.H.); azwan.jenol@gmail.com (M.A.J.)
3 Faculty of Agricultural Science and Forestry, Bintulu Campus, Universiti Putra Malaysia, Bintulu Sarawak 97000, Malaysia
4 Laboratory of Biopolymer and Derivatives, Institute of Tropical Forestry and Forest Products (INTROP), Universiti Putra Malaysia UPM, Serdang 43400, Malaysia; shazra.shazleen@yahoo.com
5 Department of Chemistry, College of Arts and Science, Yildiz Technical University, Davutpasa Campus, Esenler, Istanbul 34220, Turkey; syukrimisenan@gmail.com
6 Nextgreen Pulp & Paper Sdn. Bhd., Menara LGB, Jalan Wan Kadir, Taman Tun Dr. Ismail, Kuala Lumpur 60000, Malaysia; tengkuarisyah@gmail.com
7 School of Mechanical Engineering, Vellore Institute of Technology, Vellore 632014, India; gandhi.naveen66@gmail.com
8 Center for Defence Foundation Studies, Universiti Pertahanan Nasional Malaysia, Kem Perdana Sungai Besi, Kuala Lumpur 57000, Malaysia
9 School of Materials and Mineral Resources Engineering, Engineering Campus, Universiti Sains Malaysia, Nibong Tebal 14300, Malaysia; saifulasmal@gmail.com
10 School of Chemical and Energy Engineering, Faculty of Engineering, Universiti Teknologi Malaysia UTM, Johor Bahru 81310, Malaysia
* Correspondence: faiznorrrahim@gmail.com (M.N.F.N.); muhammadhuzaifah@upm.edu.my (M.R.M.H.); abdillah.upm@gmail.com (M.A.A.F.); mohd.nurazzi@gmail.com (N.M.N.); ahmadilyas@utm.my (R.A.I.)

Citation: Norrrahim, M.N.F.; Huzaifah, M.R.M.; Farid, M.A.A.; Shazleen, S.S.; Misenan, M.S.M.; Yasim-Anuar, T.A.T.; Naveen, J.; Nurazzi, N.M.; Rani, M.S.A.; Hakimi, M.I.; et al. Greener Pretreatment Approaches for the Valorisation of Natural Fibre Biomass into Bioproducts. *Polymers* **2021**, *13*, 2971. https://doi.org/10.3390/polym13172971

Academic Editor: Emin Bayraktar

Received: 8 August 2021
Accepted: 27 August 2021
Published: 31 August 2021

Abstract: The utilization of lignocellulosic biomass in various applications has a promising potential as advanced technology progresses due to its renowned advantages as cheap and abundant feedstock. The main drawback in the utilization of this type of biomass is the essential requirement for the pretreatment process. The most common pretreatment process applied is chemical pretreatment. However, it is a non-eco-friendly process. Therefore, this review aims to bring into light several greener pretreatment processes as an alternative approach for the current chemical pretreatment. The main processes for each physical and biological pretreatment process are reviewed and highlighted. Additionally, recent advances in the effect of different non-chemical pretreatment approaches for the natural fibres are also critically discussed with a focus on bioproducts conversion.

Keywords: non-chemical pretreatment; lignocellulosic biomass; bioproducts

1. Introduction

Many industries currently produce many tons of agro-industrial wastes. However, direct utilization of lignocellulosic biomass as a feedstock for bioproducts is challenging due to their complex structure (as represented in Figure 1). A variety of useful components, including sugars, protein, lipids, cellulose, and lignin, are present in natural fibres. The major issue that limits their utilization is, however, the tight bonding within their components [1]. Cross-linking of polysaccharides and lignin occurs through ester and ether bonds, while microfibrils produced by cellulose, hemicellulose, and lignin aid in the stability of plant cell wall structure [2,3]. These strong cross-linking connections exist between the components of the plant cell wall that act as a barrier to its disintegration.

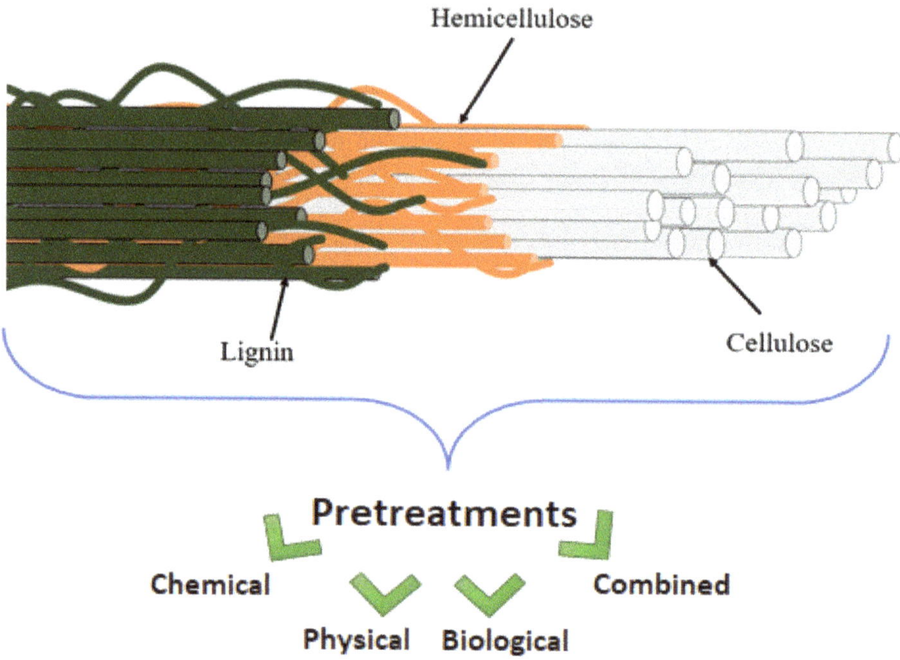

Figure 1. Overview of the complex structure of natural fibers and pretreatments.

Pretreatment helps to fractionate biomass prior to further processes, making it simpler to handle in the process [4–6]. It enables biomass hydrolysis and makes building blocks for biobased products, fuels, and chemicals. It is often the initial stage of the biorefining process and enables the following steps such as enzymatic hydrolysis and fermentation to be carried out more quickly, effectively, and economically [7]. The pretreatment method used is entirely dependent on the targeted application. Numerous pretreatment methods are mainly developed to effectively separate these interconnected components in order to get the most advantages from the lignocellulosic biomass's constituents.

Pretreatment of natural fibres is not as straightforward as it may seem. In fact, it is the second most expensive procedure after the installation of a power generator. Hydrogen bond disruption, cross-linked matrix disruption, as well as increased porosity and surface area, are the three objectives that a good pretreatment technique accomplishes in crystalline cellulose. Additionally, the result of pretreatment varies attributed to the different ratios of cell wall components [8]. More criteria to take into consideration for efficient and economically feasible pretreatment process include less chemical usage, prevention of hemicellulose and cellulose from denaturation, minimum energy demand, low price, and the capacity to reduce size.

Biomass recalcitrance is a term used for the ability of natural fibres to resist chemical and biological degradation. While there are many components involved in the recalcitrance of lignocellulosic biomass, the crystalline structure of cellulose, the degree of lignification, accessible surface area (porosity), the structural heterogeneity, and complexity of cell-wall are primary causes [9,10]. As a consequence of breaking the resistant structure of lignocellulose, it causes lignin sheath, hemicellulose, and crystallinity to all be degraded, as well as casuing a decrease in cellulose's degree of polymerization [11].

Depending on the types of natural fibres employed, the preference for the pretreatment method varies according to the composition of cellulose, hemicellulose, and lignin. Figure 2 depicts the general differences between the many common approaches which come under the four categories of physical, chemical, biological, and combination pretreatment [4]. While some of these methods have successfully transitioned from a research platform to an industrial stage, there are many hurdles, and one of the greatest is the requirement for highly toxic waste generation and high-energy inputs. From here, a serious issue that must be addressed is the lack of green and cost-effective solutions. Nevertheless, it has only lately garnered significant attention as a potential solution to the problem by focusing on the employment of non-chemical pretreatment. This could be reflected by the increment in article publications that reviewed lignocellulosic fibre pretreatment via individual greener approach as highlighted in Table 1 indicating that this topic is increasingly well-known owing to environmental concerns. The development of technology that maximises the use of raw resources, reduces waste, and avoids the use of poisonous and hazardous compounds is critical to accomplishing this objective. However, a review of all greener pretreatment approaches for lignocellulosic biomass is missing in the current literature.

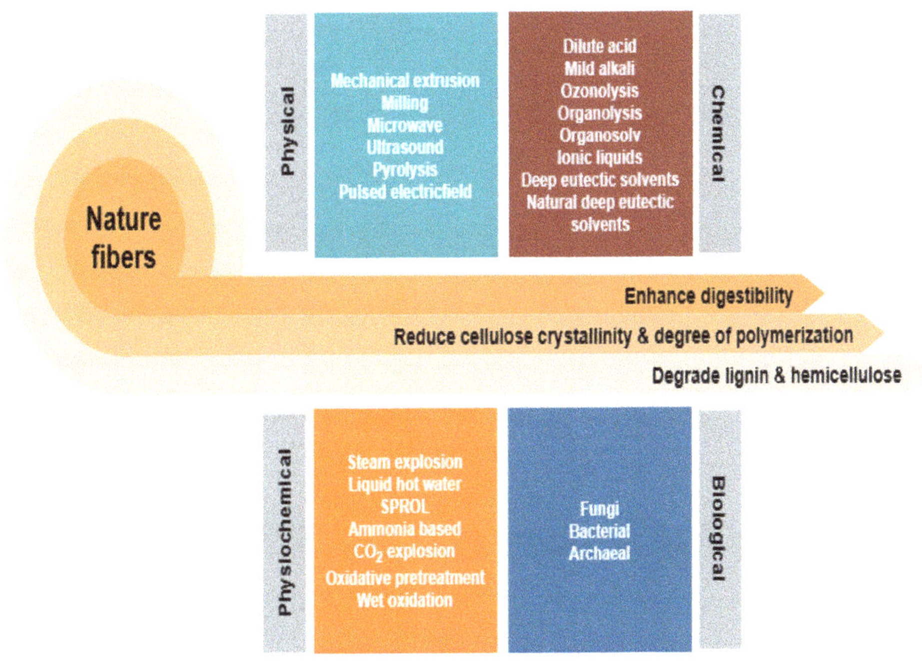

Figure 2. Different pretreatments, which fall into four main categories: physical, chemical, biological, and combination have been used to improve lignocellulosic fractionation for natural fibres.

Table 1. Recent review articles related to greener pretreatment approaches for lignocellulosic biomass.

No.	Title	Highlights of Review	Ref.
1.	Enzymatic pretreatment of lignocellulosic biomass for enhanced biomethane production-A review	• Reviews the anaerobic digestion process, challenges in degrading lignocellulosic materials, the current status of research to improve the biogas rate and yield from the anaerobic digestion of lignocellulosic biomass via enzymatic pretreatment, and the future trend in research for the reduction of enzymatic pretreatment cost	[12]
2.	A review on the environment-friendly emerging techniques for pretreatment of lignocellulosic biomass: Mechanistic insight and advancement	• Discusses the important aspects of the emerging pretreatment techniques of lignocellulosic biomass including the advancements, and the mechanistic insight for large scale of commercial implementation in a lignocellulosic biorefinery.	[13]
3.	Recent Insights into Lignocellulosic Biomass Pyrolysis: A Critical Review on Pretreatment, Characterization, and Products Upgrading	• Provides an outline of the pyrolysis process including physical and chemical pretreatment of biomass, pyrolysis mechanism, and process products upgrading.	[14]
4.	Recent advances in the pretreatment of lignocellulosic biomass for biofuels and value-added products	• Briefly presents recent findings on the chemical pretreatment for the conversion of lignocellulosic materials into fuel and value-added products.	[15]
5.	Emerging technologies for the pretreatment of lignocellulosic biomass	• Reviews the application of emerging technologies in chemical and mechanical pretreatment.	[16]

Hence, the green pretreatment approaches for lignocellulosic biomass such as physical, biological, and combination methods, as well as their impact on the separation of the complex components of different lignocellulosic sources, are reviewed in more detail in the next sections.

2. Physical Pretreatment

The physical pretreatment allows increasing the specific surface area of the fibres via mechanical comminution. It also contributes to reduce the crystallinity of the natural fibres and enhance their digestibility. The physical pretreatment usually does not affect the chemical composition of natural fibres. Physical pretreatment can be conducted by using milling, extrusion and ultrasound. Physical pretreatment is often an essential step prior to or following chemical or biochemical processing. However, the information on the mechanism of how physical pretreatment modifies the structures of the fibre is still limited.

There are some drawbacks of physical pretreatment that need to be considered. Physical pretreatment lacks the ability to remove the lignin and hemicellulose which limits the enzymes' access to cellulose. Besides that, physical pretreatment requires high energy consumption which limits its large-scale implementation and environmental safety concerns.

2.1. Mechanical Extrusion

Mechanical extrusion is one of the most conventional methods of pretreatment [17]. In this pretreatment, the fibres are subjected to a heating process (>300 °C) under shear mixing. Due to the combined effects of high temperatures that are maintained in the barrel and the shearing force generated by the rotating screw blades, the amorphous and crystalline cellulose matrix in the biomass residues is disrupted. Besides that, extrusion requires a significant amount of high energy, making it a cost-intensive method and difficult to scale up for industrial purposes [17].

Temperature and screw speed of extrusion are the main important factors. Karunanithy and Muthukumarappan [18] studied the effect of these factors on the pretreatment of corn cobs. When pretreatment was carried out at different temperatures (25, 50, 75, 100, and 125 °C) and different screw speeds (25, 50, 75, 100, and 125 rpm), maximum concentration sugars were obtained at 75 rpm and 125 °C using cellulase and β-glucosidase in the ratio of 1:4, which were nearly 2.0 times higher than the controls.

2.2. Milling

Mechanical milling is used to reduce the crystallinity of cellulose. It can reduce the size of fibre up to 0.2 mm. However, studies found that further reduction of biomass particles below 0.4 mm has no significant effect on the rate and yield of hydrolysis [17]. The type of milling and milling duration are important factors that influence the milling process. These factors can greatly affect the specific surface area, the final degree of polymerization, and a net reduction in cellulose crystallinity.

Wet disk milling has been a popular mechanical pretreatment due to its low energy consumption as compared to other milling processes. Disk milling enhances cellulose hydrolysis by producing fibres and more effective as compared to hammer milling which produces finer bundles [19]. Hideno et al. [20] compared the effect of wet disk milling and conventional ball milling pretreatment method over rice straw. The optimal conditions obtained were 60 min of milling time in case of dry ball milling while 10 repeated milling operations were required in case of wet disk milling.

2.3. Ultrasound

Ultrasound is relatively a new technique used for the pretreatment of fibres [17]. Ultrasound waves affect the physical, chemical, and morphological properties of fibres. Ultrasound treatment leads to the formation of small cavitation bubbles. These bubbles can rupture the cellulose and hemicellulose fractions. The ultrasonic field is influenced by ultrasonic frequency and duration, reactor geometry, and types of solvent used. Besides that, fibres characteristics and reactor configuration also influence the pretreatment [21].

The power and duration of ultrasound are important to be optimised depending on the fibres and slurry characteristics. This is important to meet the pretreatment objectives. Duration of ultrasound pretreatment has maximum effect on pretreatment of fibres. Besides that, a higher ultrasound power level has an adverse effect on the pretreatment. It can lead to the formation of bubbles near the tip of the ultrasound transducer which hinders the transfer of energy to the liquid medium [22].

3. Biological Pretreatment

Retting is a biological process in which enzymatic activity removes non-cellulosic components connected to the fibre bundle, resulting in detached cellulosic fibres. The dew retting uses anaerobic bacteria fermentation and fungal colonization to produce enzymes that hydrolyse fibre-binding components on fibre bundles. *Clostridium* sp. is an anaerobic bacterium commonly found in lakes, rivers, and ponds. Plant stems were cut and equally scattered in the fields during the dew retting process, where bacteria, sunlight, atmospheric air and dew caused the disintegration of stem cellular tissues and sticky compounds that encircled the fibres [23]. For the dew retting procedure to enhance fungal colonization, locations with a warm day and heavy might dew are recommended.

Bleuze et al. [24] investigated the flax fibre's modifications during the dew retting process. Microbial colonization can be affected the chemical compositions of cell walls. After seven days, fungal hyphae and parenchyma were found on the epidermis and around fibre bundles, respectively. After the retting process (42 days), signs of parenchyma deterioration and fibre bundle decohesion revealed microbial infestation at the stem's inner core.

Fila et al. [25] found 23 different varieties of dew-retting agent fungi in Southern Europe. All *Aspergillus* and *Penicillium* strains yield high-quality retted flax fibres, according to the researchers. Besides that, under field conditions, Repeckien and Jankauskiene [26] investigated the effects of fungal complexes on flax dew-retting acceleration. *Cladosporium* species variations with high colonization rates (25–29%) have been identified as a good fungus for fibre separation. Most fungi survived on flax fed with fungal complex N-3, which contained six different fungal strains.

On a commercial scale, Jankauskiene et al. [27] optimised the dew retting method. Two fungal combinations were created and put to straw after the swath was pulled and returned. Furthermore, after spraying *Cladosporium herbarum* suspension during fibre harvesting, extremely high fibre separation was found.

Bacterial and Fungi Interaction

Fungi colonization is thought to be the most important enzymatic active mechanism for dew retting. Recent research has focused on the interplay of the bacterial and fungal communities during dew retting. The association between the chemical contents of hemp fibres and microbial population fluctuation during the retting process was investigated by Liu et al. [28]. In the first seven days, fungal colonization was discovered with very little bacteria. After 20 days, there was a gradually risen in bacterial attachments on the fibre surface, with fewer fungal hyphae. The area with the highest bacterial concentration was found to severely deteriorate. The phylogenetic tree for the bacterial and fungal population in dew-retting hemp fibres is shown in Figure 3. While Table 2 shows ultrastructural changes in hemp stems and fibres as a result of microbial activity during the retting process.

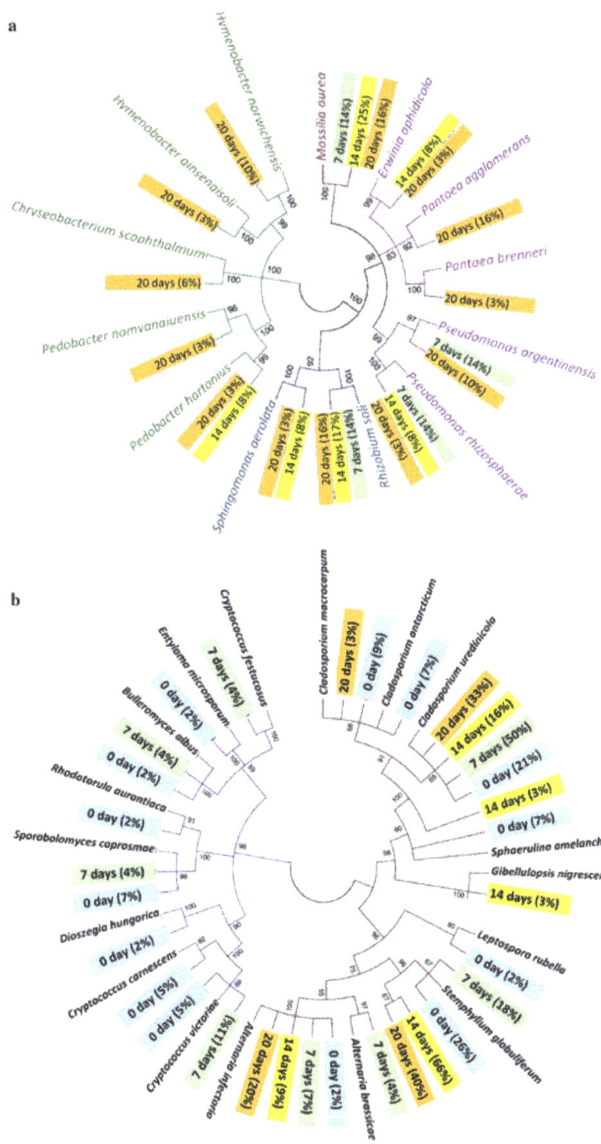

Figure 3. The phylogenetic tree of the (**a**) bacterial and (**b**) fungus communities found in hemp fibre samples. The color of the branches indicates the type of proteobacteria present, while the color of the tag indicates the number of bacteria/fungi present on different days [28].

Table 2. Highlights of ultrastructural changes on hemp stems and fibres associated with microbial activity during the retting process [29].

Retting Period	0 Days	7 Days	14–20 Days	After 50 Days
Changes in the hemp stem's and fibre's ultrastructure	(i) Stem with a well-preserved layered structure (ii) Un-collapsed, unbroken cells with their original cell geometry (iii) Living cells with cytoplasm (iv) Cuticle and trichomes are unharmed on the clear surface. (v) Chloroplasts in abundance in the upper epidermis	(i) The structure as a whole is in good condition. (ii) Fungal growth on the outside of the stems and inside the stems (iii) With damaged epidermis and parenchyma, cellular architecture is less stable.	(i) Cuticle has seriously deteriorated. (ii) Changes in cellular anatomy, as well as significant loss of live cells (iii) Fibre bundles were isolated from each other and the epidermis. (iv) Thick-walled cells populate seldom; parenchyma degrades completely, although chlorenchyma suffers less harm. (v) Bast fibres with sporadic moderate attacks (vi) Fungi colonisation and decay morphology were both affected by fibre morphology.	(i) The structure of hemp was severely harmed and dissolved. (ii) The epidermis and cambium were heavily invaded by dominating bacteria. (iii) In the bast regions, the parenchyma cells have been destroyed, and the structural integrity has been lost. (iv) All cell types, including fibre cells, have hyphae inside their lumina. (v) BFIs are more intense inside the stem. (vi) Anatomy and ultrastructure have been severely harmed. (vii) Bast fibres with a thick wall and degradation properties (viii) Effects on the ultrastructure of the fibre wall. • CML loosening/degradation, resulting in delamination and defibration • The S3 layer is loosening and decaying • Delamination within the S2 trans wall and intra wall cracks in the S2 layer have a noticeable effect • S2 materials have been removed directly (e.g., S2 thinning, broken S2, and disintegration into nanosized cellulose fibrillar structures)
The dynamics and activity of microbes	Fungi (i) Rarely seen Bacteria (ii) Not observed Fungi	Fungi (i) Mycelia with sparse growth (ii) Less variety (iii) Outside of the cortical layers, colonisation occurs largely in live cells. (iv) Trichomes near to the surface trichomes have dense colonisation. (v) Dependence on readily available food (vi) Damage to cell walls is reduced. Bacteria (i) Less abundant	Fungi (i) Extensive and plentiful (ii) Mycelia densely covering the cuticle (iii) diverse population (iv) a large number of spores (v) Interactions and activities that are intense Bacteria (i) Abundant (ii) Diverse population iii) Over the cuticle, colonies (iv) Associated with hyphae and fungal spores (v) After 20 days, there are more noticeable activity (vi) Cuticle has severely deteriorated	Fungi (i) Less abundant on the outside of the stem (ii) Mycelia on the surface is dead, but there are active hyphae inside the stem (iii) Mycelia, an invading bacteria's sole source of nourishment, showed bacterial mycophagy (i.e., extracellular and endocellular biotrophic and extracellular necrotrophic activities). Bacteria (i) Highly abundant inside and outside the stems (ii) Highly dominant and diverse role. (iii) Visible as dense overlay representing (a) Biofilms (b) Morphologically different colonies (c) Randomly scattered cells (iv) Showed strong BFIs (v) Using fungal highways, bacterial movement occurs over and inside the hemp stem. (vi) Cutinolytic and cellulolytic activities were improved.

4. Combination Pretreatment

It can be noticed from the green pretreatment techniques applied to pretreat the lignocellulosic biomass reviewed in the previous section that, while each pretreatment method makes a significant contribution, no single pretreatment approach yields efficient results without its own inherent limitations. Therefore, the combined pretreatment strategies could minimise the drawbacks while still achieving the intended result.

4.1. Physiochemical Pretreatment

Physiochemical pretreatment could be achieved by temperature elevation and irradiation in the processing of lignocellulosic material. Physiochemical pretreatment by steam such as superheated steam, hydrothermal and steam explosion is the most common pretreatments applied on natural fibre for several purposes. Physiochemical pretreatment is usually applied to remove the hemicellulose and lignin from the natural fibres [30].

4.1.1. Superheated Steam

Pretreatment of fibres by superheated steam is gaining interest recently, as this pretreatment is considered as an environmentally friendly technique to remove hemicellulose. This could be a great alternative to chemical pretreatment in order to isolate the cellulose. Superheated steam is believed as the most economical pretreatment as compared to the other physical pretreatments as discussed before.

Superheated steam is unsaturated (dry) steam generated by the addition of heat to saturated (wet) steam [31]. It has several advantages such as improved energy efficiency, higher drying rate, being conducted at atmospheric pressure and reduced environmental impact when condensate is reused [32,33]. Saturated steam cannot be superheated when it is in contact with water which is also heated, and condensation of superheated steam cannot occur without being reduced to the temperature of saturated steam. It has a high heat transfer coefficient, enabling rapid and uniform heating. Drying rates with superheated steam are faster than those with conventional hot air. Steam in a dried state or superheated steam is assumed to behave like a perfect gas. Although superheated steam is considered a perfect gas, it possesses properties like those of gases namely pressure, volume, temperature, internal energy, enthalpy and entropy. The pressure, volume, and temperature of steam as a vapour is not connected by any simple relationship such as is expressed by the characteristic equation for a perfect gas. Figure 4 shows the schematic diagram of superheated steam pretreatment. The saturated steam was generated in the boiler. The saturated steam produced was further heated by a super-heater to produce superheated steam. Then, the superheated steam was subjected to the fibres.

Superheated steam has been managed to alter the chemical composition of natural fibres. It has been proven that superheated steam pretreatment managed to remove high amount of hemicellulose from the lignocellulose fibres [34–40]. According to Warid et al. [40], superheated steam pretreatment on oil palm biomass at higher temperature and shorter time managed to remove a high amount of hemicellulose while maintaining the cellulose composition as compared to the method reported by Norrrahim et al. [39]. It was found that oil palm mesocarp fibre pretreated at 260 °C/30 min managed to remove hemicellulose of 68%, while cellulose degradation is maintained below 5%. Besides that, superheated steam was also able to remove silica bodies from the fibres where the presence of silica bodies increases the difficulty in grinding the fibre and causes abrasive wear and screw damage [32].

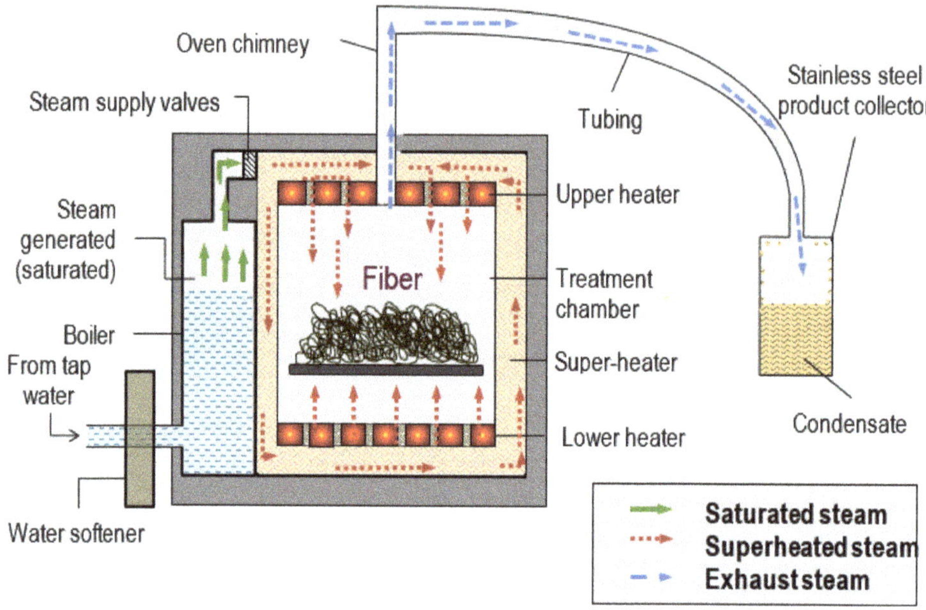

Figure 4. Schematic design of superheated steam pretreatment. Reprinted with permission from ref. [31]. 2018 Universiti Putra Malaysia.

4.1.2. Hydrothermal

Hydrothermal treatment is another pretreatment that has been proven to effectively remove impurities such as hemicellulose, lignin, and silica from lignocellulosic biomass. This treatment is being widely used in industry, owing to its low cost of production, high effectiveness in removing impurities without affecting the cellulose structure, disorganizing hydrogen bonds, swelling of the lignocellulosic biomass, as well as minimum requirements of preparation and handling [41,42]. In contrast to the superheated steam system that uses steam as the main mechanism, hydrothermal pretreatment only relies on water that will be subjected to a high temperature during the whole processing [43]. This treatment is also considered as an autohydrolysis of lignocellulosic linkages, with the presence of hydronium ions (H_3O^+) generated from water and acetic groups released from hemicellulose. The hydronium ions (H_3O^+) will act as a catalyst to break down and loosen the lignocellulosic structure [41,44]. This then will improve the effectiveness of further treatments such as enzymatic hydrolysis for biosugar production [43] and anaerobic digestion for biomethane production [45].

Numerous studies have reported the effectiveness of this treatment in reducing impurities, especially at a very high temperature. Zhang et al. [46] studied the effects of different hydrothermal temperatures which were 170, 190, and 210 °C at 20 min pretreatment time on corn stover. This study reported a drastic reduction in hemicellulose with an increase in hydrothermal temperature. In fact, no content of hemicellulose was detected and almost 125% of lignin was removed after hydrothermal treatment at 210 °C. Similarly, Phuttaro et al. [47] also reported the same trend of results, in which no amount of hemicellulose was detected in Napier grass after pretreatet at 200 °C for 15 min. Both studies agreed that hydrothermal pretreatment plays a significant effect in improving the enzymatic hydrolysis yield afterwards. Meanwhile, Lee and Park [42] reported that sunflower biomass treated with hydrothermal pretreatment at 160–220 °C for 30 min demonstrated a reduction of hemicellulose and lignin up to 25 and 15%, respectively. This then led to higher methane yield (213.87–289.47 mL g^{-1}) and biodegradability (43–63%) than the non-hydrothermally treated biomass. All of these reviews highlighted that despite of using a simple mechanism, hydrothermal can still efficiently removed impurities and improve the chemical and physical properties of lignocellulosic biomass prior to further treatments.

4.1.3. Steam Explosion

Steam explosion involves the use of high pressure and heat to pretreat lignocellulosic biomass. The biomass will be subjected to heat ranging from 160–280 °C and high pressure ranging from 0.2–5 MPa, depending on biomass source, duration, and other conditions [48,49]. Before the discovery of superheated steam and hydrothermal treatment, the steam explosion was widely applied in the industry due to its low energy consumption and chemical usage [49]. Figure 5 shows an example of the steam explosion process. Theoretically, biomass needs to be subjected to high temperature and pressure in a close reactor. The water contained in the biomass will then be evaporated and expanded, led to hydrolysis to a certain extent. Explosive decompression will then occur by promptly reducing the pressure to the atmospheric level [50,51].

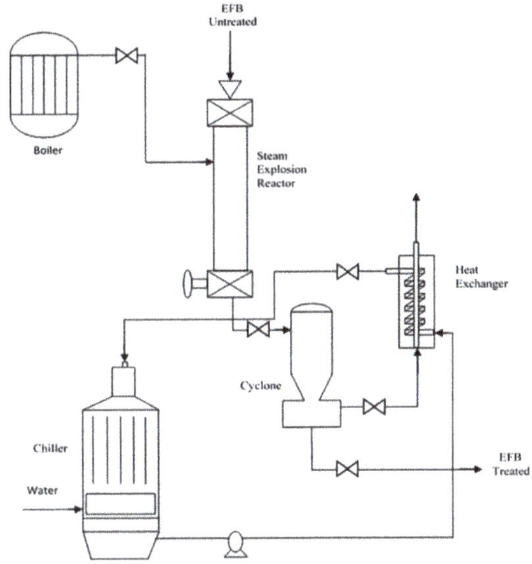

Figure 5. Schematic diagram of steam explosion process. Page: 11 Reprinted with permission from ref. [51]. 2016 Elsevier.

Steam explosion treatment helps to reduce the particle size of biomass, disrupt the structure of lignocellulosic biomass by removing amorphous structures such as hemicellulose and other impurities, and reduce cellulose crystallinity [52]. Similar to hydrothermal, the steam explosion also carried out auto-hydrolysis. During processing, acetic acids and other organic acids will be formed, and this will assist in the breakdown of ester and ether bonds in the cellulose-hemicellulose-lignin matrix. For steam explosion, reaction temperature, pressure, and processing duration are considered as the key factors.

Numerous studies have reported the effectiveness of this treatment in reducing impurities and enhancing the effectiveness of further treatments. For example, Abraham et al. [52] discovered that the sudden pressure drop due to explosion has pre-defibrillated the raw banana, jute, and pineapple leaf fibre biomass after pretreated for 1 h, which then eases and enhances the efficiency of fibrillation process by acid hydrolysis for the production of nanocellulose. Meanwhile, Medina et al. [51] discovered an application of steam explosion pretreatment for empty fruit bunches. The heating time was around 2 min and the reaction time was controlled after the temperature was reached. It was found that the application of steam explosion helped to enhance the production of glucans to 34.69%, reduce the amount of hemicellulose to 68.11%, and increase enzymatic digestibility to 33%. This was all due to steam explosion pretreatment, which helped in increasing the fibre porosity of empty fruit bunches. Marques et al. [53] also highlighted that the oil palm mesocarp fibre which has been treated to the steam explosion has higher purity, thermal stability, and crystallinity than the non-treated biomass. The reaction time was between 3 to 17 min. The cellulose pulp yield was increased by 47%. In addition, high-quality lignin was obtained as a co-product of steam explosion pretreatment, which can potentially be used for other purposes such as in the development of resin.

4.2. Biological-Chemical Pretreatment

In recent years, a more often used combined pretreatment method is physical and chemical combined pretreatment, while biological and chemical combined pretreatment has yet to be thoroughly researched. Combining microbial and chemical pretreatments, for instance, is seen as a cost-effective technique for reducing pretreatment times, minimizing chemical usage and hence secondary pollution [54]. Table 3 listed different biological-chemical pretreatment approaches to pretreat lignocellulosic biomass. Till now, the biological-alkaline pretreatment for lignocellulosic biomass has been the most widely researched.

Table 3. Previous research on biological-chemical pretreatment approaches to pretreat lignocellulosic biomass. Data retrieved from Ref. [54].

Substrate	Conditions		Component's Degradation (%)		
	1st Step	2nd Step	Lignin	Hemicellulose	Cellulose
		Biological—alkaline pretreatment			
Corn stalks	*Irpex lacteus* (28 °C, 15 d)	0.25 M NaOH solution (75 °C, 2 h)	80	51.37	6.62
Populus tomentosa	*Trametes velutina* D10149 (28 °C, 28 d)	70% (v/v) ethanol aqueous solution containing 1%(w/v) NaOH (75 °C, 3 h)	23.08	22.22	18.91
Willow sawdust	*Leiotrametes menziesii* (27 °C, 30 d)	1% (w/v) NaOH (80 °C, 24 h)	59.8	68.1	51.2
	Abortiporus biennis (27 °C, 30 d)		54.2	51.8	29.1
		Biological—acid pretreatment			
Populus tomentosa	*Trametes velutina* D1014 (28 °C, 56 d)	1% sulphuric acid (140 °C, 1 h)	23.82	75.96	(+) 18.74
Oil palm empty fruit bunches	*Pleurotus floridanus* LIPIMC996 (31 °C, 28 d)	Ball milled at 29.6/s for 4 min. Phosphoric acid treatment (50 °C, 5 h)	(+) 8.29	60.63	(+) 37.52
Olive tree biomass	*Irpex lacteus* (Fr.238 617/93) (30 °C, 28 d)	2% w/v H_2SO_4 (130 °C, 1.5 h)	(+) 105.82	75.29	(+) 62.95
		Biological—oxidative pretreatment			
Corn Straw	*Echinodontium taxodii* (25 °C, 15 d)	0.0016% NaOH and 3% H_2O_2 (25 °C, 16 h)	52.00	23.64	(+) 45.45
Hemp chips	*Pleurotus eryngii* (28 °C, 21 d)	3% NaOH and 3% (v/v) H_2O_2 (40 °C, 24 h)	55.7	23.2	25.1
		Biological—organosolv pretreatment			
Sugarcane straw	*Ceriporiopsis subvermispora* (27 °C, 15 d)	Acetosolv pulping (Acetic acid with 0.3% w/w HCl) (120 °C, 5 h	86.8	93.8	32.1
Pinus radiata	*Gloeophyllum trabeum* (27 °C, 28 d)	60% ethanol in water solvent (200 °C, 1 h)	74.26	80.74	-
		Biological—liquid hot water (LHW) pretreatment			
Soybean	Liquid Hot water (170 °C, 3 min, 400 rpm, 110 psi, solid to liquid ratio of 1:10)	*Ceriporiopsis subvermispora* (28 °C, 18 d)	36.69	41.34	0.84
Corn stover			41.99	42.91	7.09
Wheat straw	Hot water extraction (HWE) (85 °C, 10 min, solid to liquid ratio of 1:20)	*Ceriporiopsis subvermispora* (28 °C, 18 d)	24.87	13.19	1.86
Corn stover			30.09	28.14	4.96
Soybean			0.09	0.09	0.09
		Biological—steam explosion pretreatment			
Beech woodmeal	*Phanerochaete chrysosporium* (37 °C, 28 d)	Steam explosion (215 °C, 6.5 min)	42.00	-	-
Sawtooth oak, corn and bran	*Lentinula edodes* (120 d)	Steam explosion (214 °C, 5 min, 20 atm)	17.1	80.43	(+) 5.19

(+): represents the increment in fibre content.

5. The Influence of Pretreatment of Natural Fibre on Several Applications

Pretreatment of lignocellulosic materials has long been known for its advantages. It has been applied for various applications such as biocomposites, adsorbent, paper, packaging, military, biosugars, biomedical, bioenergy and more [55–68]. In Table 4, the purposes of the non-chemical pretreatment strategies and their benefits and drawbacks are summarised [69]. Since there are so many pretreatment-related applications, discussing each pretreatment technique in depth becomes very challenging. For certain applications, pretreatment techniques applied on natural fibre are summarised in the following sections.

Table 4. Purposes of the pretreatment strategies and their advantages and disadvantages.

Pretreatments	Preferred Natural Fibres	Purposes	Advantages	Disadvantages
Physical	Hardwoods and agricultural residues	Enhance the digestibility of lignocellulosic biomass by increase the available specific surface area, and reduce both the degree of polymerisation and cellulose crystallinity	(1) No recycling cost (2) No chemical usage (3) Increase biogas, bioethanol and biohydrogen yields	(1) Excessive size reduction decreases biofuel production (2) Formation of fermentation inhibitors at high temperature (3) Incomplete digestion of lignin-carbohydrate matrix (4) The need to wash the hydrolysate decreases sugar yield (5) High energy requirement
Biological	Hardwoods, softwoods, and agricultural residues	Leverage the action of fungi capable of producing enzymes that can degrade lignin, hemicellulose, and polyphenols	(1) The depolymerisation is very selective and efficient (2) Low-capital cost (3) Low energy requirement (4) No chemicals requirement (5) Mild process conditions	(1) The rate of biological pretreatment is too slow for industrial purposes (10–14 days) (2) Require careful growth conditions and a large amount of space (3) A fraction of carbohydrate is consumed by the microbes, thus reduces the sugar yield

5.1. *Influence of Physical Pretreatment on Applications*

Physical pretreatment is responsible for the changes in specific surface area, particle sizes, crystallinity index, or polymerization degree of biomass. The physical pretreatment avoids the use of chemicals, thus reducing the generation of waste and inhibitors for subsequent reactions. The management of biomass after harvesting, storage, and transportation is made easier by a higher bulk density [70]. Reduced particle size and increased specific surface area facilitate the following process by establishing a phase barrier between lignocellulosic material and chemicals and eliminating heat transfer limitation [71]. Mechanical, microwave or ultrasound pretreatments are the most common techniques carried out to improve the efficiency of the main steps in biomass processing.

It has been discovered that milling leads to higher production of biogas, bioethanol, and biohydrogen. Given the high energy requirements of industrial milling and the increasing energy demand, it seems doubtful that milling will be economically viable [69]. While most studies demonstrated that milling after chemical pretreatment reduces the amount of energy used and the cost of solid-liquid separation, the amount of mixing in pretreatment slurries and fermentation inhibitors are avoided [72]. Thus, understanding the characteristics of the feedstock is critical for making the best choice of technique and equipment for mechanical processing, and this should guarantee an adequate cost-effectiveness balance [73].

De la Rubia et al. [74] discovered that the excessive reduction in biomass may lower biofuel generation and impede methane synthesis during anaerobic digestion by the formation of inhibitory volatile fatty acids (VFA). When coupled with other pretreatment techniques, size reduction is more successful. The greatest biogas generation from rice straw was achieved via a combination of milling, grinding, and heating treatment. Milling is beneficial since it eliminates inhibitors of fermentation such as furfural and hydroxyl methyl furfural [75]. There have also been suggestions for other types of physical pretreatment, including the use of gamma rays to break the ß-1,4 glycosidic linkages, which results in a higher surface area and a lower crystallinity [76]. Ball milling pretreatment gave the lowest particle size compared to mashing or chipping but resulted in a lower hydrolysis rate [77]. On a wide scale, this technique will certainly be extremely costly, and it will raise significant environmental and safety issues. The use of a twin-screw extruder for methane production may reduce 50% of hemicellulose content. This concomitantly increases the fraction of soluble chemicals, e.g., carbohydrates, proteins, lipids, minerals, and vitamins, and rapidly converted to 15–21% more biogas by methanogenic microorganisms [78]. Table 5 summarises the applications that used physical pretreatments and their yield improvement and product properties.

Table 5. A summary of physical pretreatments applied for numerous applications.

Bioproducts	Natural Fibres	Pretreatments	Conditions	Yield Improvement/Product Properties	References
Biohydrogen	Corn stover	Steam explosion	1.5 Mpa and 198 °C for 1.5 min	51.9 L H_2 kg^{-1} TS *	[79]
	Rice straw	Hydrothermal	pH 7.0, 210 °C, 15.4 °C min^{-1}, and 20% TS	28.0 mL H_2 g^{-1} VS *	[80]
Biomethane	Sugarcane bagasse	Hydrolysis	178.6 °C, 43.4 min, and solid to liquid ratio of 0.24	1.56 Nm^3 CH_4 kg^{-1} TOC *	[81]
	Wheat straw	Microwave irradiation	260 °C, 33 bars, 3 min	28%	[82]
	Pennisetum hybrid			12%	[83]
	Blend of maize, ryegrass, and rice straw	Extrusion	Exit slit opened at 60%	11.5–13.4%	[84]
	Hay	Steam explosion	220 °C for 15 min	16%	[85]
	Vine trimming shoot	Extrusion	200 g h^{-1} feed rate	51–58% hemicellulose reduction, 15.7–21.4% CH_4 increased	[78]
Biosugar	Wheat straw	Supercritical CO_2 & steam explosion	A steam explosion at 200 °C for 15 min and supercritical CO_2 of 12 MPa at 190 °C for 60 min	36.5%	[86]
	Poplar wood chips	Mechanical pulping & steam	Disc clearance set 0.5–0.1 mm for mechanical pulping and steam pretreatment at 210 °C for 5 min	76%	[87]
	Poplar wood	Steam explosion	180 °C and 18 min	94%	[88]
	Cane bagasse	Hydrothermal	200 °C	4 mg xylose ml^{-1} *	[89]
	Pinewood		240 °C and 10 min	32% **	[90]
	Rapeseed meal		260 °C and 10 min	51 g glucose kg^{-1} *	[91]
Nanocellulose	Poplar wood	Steam explosion	2 MPa for 180 s	13.2%	[92]
	Cotton	High-pressure homogenization	80 MPa for 30 HPH	10–20 nm in diameter, reduced thermal stability, and crystallinity	[93]
	Sugarcane bagasse			10–20 nm in diameter, reduced thermal stability, and crystallinity	[94]
	Oil palm biomass	Superheated steam	260 °C for 30 min	<100 nm diameter, 27% crystallinity reduced	[35]

* The highest yield obtained. ** In carbohydrate.

5.2. Influence of Biological Pretreatment on Applications

The majority of pretreatment methods involve costly instruments or equipment that consumes a lot of energy, depending on the process. Biomass conversion in particular requires a large amount of energy for physical and thermochemical operations. Biological treatment with different kinds of rot fungus is being recommended more than ever as a low-energy delignification technique. The pretreatment is renowned for working with fungal and enzyme-assisted processes to break down the barrier that has formed within the cell wall, allowing for more abundant lignocellulosic components to be utilised in the activities of cellulase enzymes, hence increase their digestibility and yield. For instance, a pretreatment may enhance the enzymatic hydrolysis rate by 3–10-fold [1].

Additionally, any pretreatment should prevent carbohydrate degradation or loss, as well as the production of by-products that are detrimental to future hydrolysis and fermentation. The presence of white-rot fungus allows the organism to delignify, without compensating for the carbohydrate content, resulting in enhanced 30–35% cellulose conversion to sugar [95] and an additional 10–96% methane production [96,97]. In contrast to thermochemical techniques, chemical pretreatment suffers from silica scaling that prohibits the recovery of alkaline chemicals, due to the high silica concentration of many agricultural feedstocks, such as rice and wheat straw. The economic feasibility of scaling up biological pretreatment is higher since it does not need a large initial capital investment due to the lack of or reduced use of chemicals and heat, as well as the absence of a necessity for feedstock size reduction [19]. A further disadvantage of the thermochemical method is that it often produces low-molecular-mass molecules with high pretreatment severities, which may act as an inhibitor to the primary process [98]. As a result, it needs a detoxification step after the thermochemical reaction, which adds to the cost [99].

Another possibility of biological pretreatment is the potential to produce a variety of value-added co-products or intermediates, including enzymes, reducing sugars, furfural, ethanol, protein and amino acids, carbohydrates, lipids, organic acids, phenols, activated carbon, degradable plastic composites, cosmetics, adsorbents, resins, medicines, foods and feeds, methane, pesticides, promoters, secondary metabolites, surfactants, fertiliser, and other miscellaneous products [100–104]. Despite many successful attempts, economic separation and co-products recovery have remained a problem. Nonetheless, the diversity of the product allows for a wide range of markets, which means that market saturation is less of a concern [97].

Despite the advantage of requiring no additional nutrients, the usual fungal breakdown process needs a lengthy incubation period of up to 14–56 days [105]. Carbohydrates also gradually degrade over this period, which results, even with selective lignin-degrading fungus, in a reduced sugar yield, therefore making fungal biological pretreatment is impracticable for use in industrial production. The use of enzymes rather than fungus may overcome these significant drawbacks, including less carbohydrate consumption, shorter treatment time, and better yield [106]. However, only a limited number of enzyme treatments are as efficacious in pulping as fungal treatments, since solid wood enzymes cannot penetrate effectively and need high pressure to get better results [97]. Table 6 summarises the applications that used biological pretreatments and their yield improvement and product properties.

Table 6. A summary of biological pretreatments applied for numerous applications.

Bioproducts	Natural Fibres	Type of Microbes/Enzymes	Hydrolysis Conditions	Yield Improvement/Product Properties	References
Biohydrogen	Corn stover	*Clostridium cellulolyticum* and hydrogen fermentation bacteria	20 mL of medium, 5% (*v*/*v*) inoculum, 10 g L^{-1} carbon source, at 37 °C for 96 hrs	40.3 L H_2 kg^{-1} TS *	[79]
Bioethanol	Corn stover	*Ceriporiopsis subvermispora*	28 °C for 18 days	57.8% yield increased	[107]
	Corn stover	*Ceriporiopsis subvermispora*	28 °C for 35 days	66.6% yield increased	[107]
	Potato and cassava peel	*Gloeophyllum sepiarium* and *Pleurotus ostreatus*	28 °C for 7 days	26% yield increased	[108]
	Straw	*Neosartorya fischeri–Myceliophthora thermophila* and *Aeromonas hydrophila–Pseudomonas poae*	30–55 °C for 6 days	7-fold yield increased	[109]
	Corn stover	*Irpex lacteus*	28 °C for 42 days	66.9% yield increased	[110]
	Corn stalks		28 °C for 28 days	82% yield increased	[111]
Biomethane	Wheat straw	*Trametes versicolor*	Laccase at 500 U/L, 25 °C for 6 days	10–18% yield increase	[96]
	Cassava	Yeast and cellulolytic bacteria	100 mL of PCS medium, at 55 °C for 12 h	96.6% yield increased	[112]
	Microalgae	Enzyme mix (cellulase, glucohydrolase and xylanase)	1% enzyme mix, 37 °C for 24 hrs	15% yield increased	[113]
	Sawdust	*Methanobrevibacter thaueri* MB-1, *Methanosarcina acetivorans* MB-2, and *Methanococcus voltae* MB 3.	60 °C for 6 days	92.2% yield increased	[114]
Biosugar	Corn stover	*Ceriporiopsis subvermispora*	28 °C for 5–7 days	57–67% yield increase	[107]
	Silver grass	*Bacillus, Pseudomonas, Exiguobacterium*, and *Aeromonas*	37 °C for 7 days	2.2-fold yield increased	[115]
	Sugarcane bagasse	*Ceriporiopsis submervispora*	27 °C for 60 days	47% yield increased	[116]
	Sawdust	*Pleurotus pulmonarius*	28 °C for 30 days	94.8% yield increased	[117]
	Paddy straw	*Pleurotus florida*	25–29 °C for 28 days	75.3% yield increased	[118]
	Rice straw	*Pholiota adiposa*	25 °C for 120 h	716 mg g^{-1} *	[119]
		Pholiota adipose and *Armillaria gemina*	27 °C for 45 days	74.2% yield increased	[120]
	Populus tomentiglandulosa	*Armillaria gemina* SKU2114	30 °C for 48 h	62% yield increased	[121]
Nanocellulose	Eucalyptus	Endoglucanase and cellobiohydrolase	7 pH, 50 °C for 48 h	20 nm diameter, >500 nm length	[122]
	Wood fibre	Endoglucanase	4.8 pH, 50 °C for 2 h	5–30 nm diameter	[123]
	Orange residues	ß-glucosidase	4 pH, 50 °C for 48 h	180 nm diameter, 1.3 mm length	[124]
	Sugarcane bagasse	ß-glucosidase and endoglucanase	5 pH, 50 °C for 24 h	14–18 nm diameter, 195–250 nm length	[125]
	Maple pulp	Cellic CTec 2 and Cellic HTec 2 (commercial enzymes)	4.8 pH, 50 °C for 72 h	5–10 nm diameter, 1 µm length	[126]
	Cotton linters	Cellulase	5 pH, 55 °C for 24 h	35 nm diameter, 0.3 mm length	[127]

* The highest yield obtained.

6. Challenges and Future Recommendations

There is currently an issue with agro-industrial waste disposal across the world. Therefore, it is vital to continuously explore for alternatives to manage the problem effectively. A review of recent advances in the effect of different pretreatment techniques on the conversion of natural fibres to bioproducts has been discussed. It can be inferred that the downstream application has a profound effect on the selection and optimization of a feasible pretreatment technique. Among all of these, non-chemical approaches for natural fibres pretreatment are gaining popularity since they are more advantageous and greener than chemical pretreatment due to their chemical-free processability, cost-effectiveness, and sustainability. This is due to the fact that an ideal natural fibres pretreatment should have minimum or no solvent costs and also the capacity to process at high solids loadings with shorter treatment times and minimal inhibitor formation.

In fact, each pretreatment has its own set of limitations or shortcomings, and no specific technique can be used to pretreat all types of biomasses. Hence, a thorough understanding of the relationship between biomass structure and pretreatment is needed. Each pretreatment has a substantial impact on fibre properties. The selection of pretreatment is determined by the widespread application of natural fibre materials. Several factors such as type of fibre, crystallinity, molecular weight and other properties may influence in selecting the most effective pretreatment method. Additionally, operating conditions, such as temperature, time, etc must be taken into consideration during pretreatments as they have a direct influence on the fibre properties.

As mentioned previously, each pretreatment method has its benefits and shortcomings depending on the source of biomass, the processes employed, and the desired end product. Nevertheless, many previous studies have been conducted on a small scale, yet there is a significant disparity between laboratory preliminary findings, pilot-scale outcomes, and, eventually, industrial-scale results. Hence, further research is needed to address these issues and provide a feasible pretreatment approach for large-scale biorefinery operations.

Besides that, utilization of by-products derived from the pretreatment is also important to be investigated. For example, the SHS pretreatment had partially degraded the lignocellulosic structure of the biomass into smaller compounds such as acetic acid, formic, levulinic and succinic. These compounds were found useful to be used as antimicrobial agents. This indicates the possibility of having lignocellulosic components degradation products as byproducts during SHS pretreatment [33]. However, to the best of our knowledge, lack of reports was focused on the other type of pretreatments. Moreover, it is also important to ensure that there is no consequence generation of contaminants could be derived during the pretreatment process of fibres.

Even though most of the non-chemical pretreatment as discussed here are known to be more environmentally friendly, improvements are still needed. This is due to the fact that some of the techniques such as milling, SHS, and hydrothermal pretreatment require high production cost, especially at industrial levels. The high energy consumption and long processing time related to the pretreatment of fibres is still an issue that hampers the industrial applicability of some of these pretreatments. However, to the best of our knowledge, progress has been accomplished in this area, and numerous studies to overcome this issue have now been conducted worldwide.

Author Contributions: Conceptualization, M.N.F.N., M.R.M.H., M.A.A.F., S.S.S., M.S.M.M. and T.A.T.Y.-A.; validation, J.N. and N.M.N.; writing—original draft preparation, M.N.F.N., M.R.M.H., M.A.A.F., S.S.S., M.S.M.M., T.A.T.Y.-A. and M.I.H.; supervision, R.A.I., and M.A.J.; project administration, M.S.A.R. and M.I.H.; funding acquisition, M.R.M.H. All authors have read and agreed to the published version of the manuscript.

Funding: This research received no external funding and The APC was funded by Universiti Putra Malaysia (UPM).

Institutional Review Board Statement: Not applicable.

Informed Consent Statement: Not applicable.

Data Availability Statement: Not applicable.

Acknowledgments: The authors gratefully acknowledge the technical and financial support from the Universiti Putra Malaysia (UPM).

Conflicts of Interest: There are no conflict of interest to declare.

References

1. Aftab, M.N.; Iqbal, I.; Riaz, F.; Karadag, A.; Tabatabaei, M. Different Pretreatment Methods of Lignocellulosic Biomass for Use in Biofuel Production. In *Biomass for Bioenergy-Recent Trends and Future Challenges*; IntechOpen: London, UK, 2019.
2. Xiao, C.; Bolton, R.; Pan, W. Lignin from rice straw Kraft pulping: Effects on soil aggregation and chemical properties. *Bioresour. Technol.* **2007**, *98*, 1482–1488. [CrossRef]
3. Himmel, M.E.; Ding, S.-Y.; Johnson, D.K.; Adney, W.S.; Nimlos, M.R.; Brady, J.W.; Foust, T.D. Biomass Recalcitrance: Engineering Plants and Enzymes for Biofuels Production. *Science* **2007**, *315*, 804–807. [CrossRef]
4. Tu, W.-C.; Hallett, J.P. Recent advances in the pretreatment of lignocellulosic biomass. *Curr. Opin. Green Sustain. Chem.* **2019**, *20*, 11–17. [CrossRef]
5. Nurazzi, N.M. Treatments of natural fiber as reinforcement in polymer composites—A short review. *Funct. Compos. Struct.* **2021**, *3*, 2. [CrossRef]
6. Norrrahim, M.N.F.; Ilyas, R.A.; Nurazzi, N.M.; Rani, M.S.A.; Atikah, M.S.N.; Shazleen, S.S. Chemical Pretreatment of Lignocellulosic Biomass for the Production of Bioproducts: An Overview. *Appl. Sci. Eng. Prog.* **2021**. [CrossRef]
7. Williams, C.L.; Emerson, R.M.; Tumuluru, J.S. Biomass Compositional Analysis for Conversion to Renewable Fuels and Chemicals. In *Biomass Volume Estimation and Valorization for Energy*; IntechOpen Limited: London, UK, 2017. [CrossRef]
8. Mood, S.H.; Golfeshan, A.H.; Tabatabaei, M.; Jouzani, G.S.; Najafi, G.; Gholami, M.; Ardjmand, M. Lignocellulosic biomass to bioethanol, a comprehensive review with a focus on pretreatment. *Renew. Sustain. Energy Rev.* **2013**, *27*, 77–93. [CrossRef]
9. Guerriero, G.; Hausman, J.F.; Strauss, J.; Ertan, H.; Siddiqui, K.S. Lignocellulosic biomass: Biosynthesis, degradation, and industrial utilization. *Eng. Life Sci.* **2016**, *16*, 1–16. [CrossRef]
10. Barakat, A.; De Vries, H.; Rouau, X. Dry fractionation process as an important step in current and future lignocellulose biorefineries: A review. *Bioresour. Technol.* **2013**, *134*, 362–373. [CrossRef]
11. Chen, H.; Liu, J.; Chang, X.; Chen, D.; Xue, Y.; Liu, P.; Lin, H.; Han, S. A review on the pretreatment of lignocellulose for high-value chemicals. *Fuel Process. Technol.* **2017**, *160*, 196–206. [CrossRef]
12. Koupaie, E.h.; Dahadha, S.; Lakeh, A.A.B.; Azizi, A.; Elbeshbishy, E. Enzymatic pretreatment of lignocellulosic biomass for enhanced biomethane production-A review. *J. Environ. Manag.* **2019**, *233*, 774–784. [CrossRef]
13. Haldar, D.; Purkait, M.K. A review on the environment-friendly emerging techniques for pretreatment of lignocellulosic biomass: Mechanistic insight and advancements. *Chemosphere* **2021**, *264*, 128523. [CrossRef]
14. Zadeh, Z.E.; Abdulkhani, A.; Aboelazayem, O.; Saha, B. Recent Insights into Lignocellulosic Biomass Pyrolysis: A Critical Review on Pretreatment, Characterization, and Products Upgrading. *Processes* **2020**, *8*, 799. [CrossRef]
15. Mahmood, H.; Moniruzzaman, M.; Iqbal, T.; Khan, M.J. Recent advances in the pretreatment of lignocellulosic biomass for biofuels and value-added products. *Curr. Opin. Green Sustain. Chem.* **2019**, *20*, 18–24. [CrossRef]
16. Hassan, S.; Williams, G.A.; Jaiswal, A.K. Emerging technologies for the pretreatment of lignocellulosic biomass. *Bioresour. Technol.* **2018**, *262*, 310–318. [CrossRef]
17. Kumar, A.K.; Sharma, S. Recent updates on different methods of pretreatment of lignocellulosic feedstocks: A review. *Bioresour. Bioprocess.* **2017**, *4*, 1–19. [CrossRef]
18. Karunanithy, C.T.; Muthukumarappan, K. Influence of Extruder Temperature and Screw Speed on Pretreatment of Corn Stover while Varying Enzymes and Their Ratios. *Appl. Biochem. Biotechnol.* **2010**, *162*, 264–279. [CrossRef]
19. Zhu, J.; Wang, G.; Pan, X.; Gleisner, R. Specific surface to evaluate the efficiencies of milling and pretreatment of wood for enzymatic saccharification. *Chem. Eng. Sci.* **2009**, *64*, 474–485. [CrossRef]
20. Hideno, A.; Inoue, H.; Tsukahara, K.; Fujimoto, S.; Minowa, T.; Inoue, S.; Endo, T.; Sawayama, S. Wet disk milling pretreatment without sulfuric acid for enzymatic hydrolysis of rice straw. *Bioresour. Technol.* **2009**, *100*, 2706–2711. [CrossRef]
21. Bussemaker, M.J.; Zhang, D. Effect of Ultrasound on Lignocellulosic Biomass as a Pretreatment for Biorefinery and Biofuel Applications. *Ind. Eng. Chem. Res.* **2013**, *52*, 3563–3580. [CrossRef]
22. Gogate, P.R.; Sutkar, V.S.; Pandit, A.B. Sonochemical reactors: Important design and scale up considerations with a special emphasis on heterogeneous systems. *Chem. Eng. J.* **2011**, *166*, 1066–1082. [CrossRef]
23. Sanjay, M.R.; Siengchin, S.; Parameswaranpillai, J.; Jawaid, M.; Pruncu, C.I.; Khan, A. A comprehensive review of techniques for natural fibers as reinforcement in composites: Preparation, processing and characterization. *Carbohydr. Polym.* **2019**, *207*, 108–121. [CrossRef]
24. Bleuze, L.; Lashermes, G.; Alavoine, G.; Recous, S.; Chabbert, B. Tracking the dynamics of hemp dew retting under controlled environmental conditions. *Ind. Crop. Prod.* **2018**, *123*, 55–63. [CrossRef]
25. Fila, G.; Manici, L.M.; Caputo, F. In vitro evaluation of dew-retting of flax by fungi from southern Europe. *Ann. Appl. Biol.* **2001**, *138*, 343–351. [CrossRef]

26. Repečkiene, J.; Jankauskiene, Z. Application of fungal complexes to improve flax dew-retting. *Biomed. Moksl.* **2009**, *83*, 63–71.
27. Jankauskiene, Z.; Lugauskas, A.; Repeckiene, J. New Methods for the Improvement of Flax Dew Retting. *J. Nat. Fibers* **2007**, *3*, 59–68. [CrossRef]
28. Liu, M.; Ale, M.T.; Kołaczkowski, B.; Fernando, D.; Daniel, G.; Meyer, A.S.; Thygesen, A. Comparison of traditional field retting and Phlebia radiata Cel 26 retting of hemp fibres for fibre-reinforced composites. *AMB Express* **2017**, *7*, 1–15. [CrossRef]
29. Fernando, D.; Thygesen, A.; Meyer, A.S.; Daniel, G. Elucidating field retting mechanisms of hemp fibres for biocomposites: Effects of microbial actions and interactions on the cellular micro-morphology and ultrastructure of hemp stems and bast fibres. *BioResources* **2019**, *14*, 4047–4084. [CrossRef]
30. Farid, M.A.A.; Hassan, M.A.; Roslan, A.M.; Ariffin, H.; Norrrahim, M.N.F.; Othman, M.R.; Yoshihito, S. Improving the decolorization of glycerol by adsorption using activated carbon derived from oil palm biomass. *Environ. Sci. Pollut. Res.* **2021**, *28*, 27976–27987. [CrossRef]
31. Norrrahim, M.N.F. Superheated Steam Pretreatment of Oil Palm Biomass for Improving Nanofibrillation of Cellulose and Performance of Polypropylene/Cellulose Nanofiber Composites. Doctoral Thesis, Universiti Putra Malaysia, Selangor, Malaysia, 2018.
32. Nordin, N.I.A.A.; Ariffin, H.; Andou, Y.; Hassan, M.A.; Shirai, Y.; Nishida, H.; Yunus, W.M.Z.W.; Karuppuchamy, S.; Ibrahim, N.A. Modification of Oil Palm Mesocarp Fiber Characteristics Using Superheated Steam Treatment. *Molecules* **2013**, *18*, 9132–9146. [CrossRef]
33. Sharip, N.S.; Ariffin, H.; Hassan, M.A.; Nishida, H.; Shirai, Y. Characterization and application of bioactive compounds in oil palm mesocarp fiber superheated steam condensate as an antifungal agent. *RSC Adv.* **2016**, *6*, 84672–84683. [CrossRef]
34. Megashah, L.N.; Ariffin, H.; Zakaria, M.R.; Hassan, M.A.; Andou, Y.; Padzil, F.N.M. Modification of cellulose degree of polymerization by superheated steam treatment for versatile properties of cellulose nanofibril film. *Cellulose* **2020**, *27*, 7417–7429. [CrossRef]
35. Bahrin, E.K.; Baharuddin, A.S.; Ibrahim, M.F.; Razak, M.N.A.; Sulaiman, A.; Aziz, S.A.; Hassan, M.A.; Shirai, Y.; Nishida, H. Physicochemical property changes and enzymatic hydrolysis enhancement of oil palm empty fruit bunches treated with superheated steam. *BioResources* **2012**, *7*, 1784–1801. [CrossRef]
36. Norrrahim, M.N.F.; Ariffin, H.; Hassan, M.A.; Ibrahim, N.A.; Yunus, W.M.Z.W.; Nishida, H. Utilisation of superheated steam in oil palm biomass pretreatment process for reduced chemical use and enhanced cellulose nanofibre production. *Int. J. Nanotechnol.* **2019**, *16*, 668. [CrossRef]
37. Norrrahim, M.; Ariffin, H.; Yasim-Anuar, T.; Hassan, M.; Ibrahim, N.; Yunus, W.; Nishida, H. Performance Evaluation of Cellulose Nanofiber with Residual Hemicellulose as a Nanofiller in Polypropylene-Based Nanocomposite. *Polymers* **2021**, *13*, 1064. [CrossRef]
38. Zakaria, M.R.; Norrrahim, M.N.F.; Hirata, S.; Hassan, M.A. Hydrothermal and wet disk milling pretreatment for high conversion of biosugars from oil palm mesocarp fiber. *Bioresour. Technol.* **2015**, *181*, 263–269. [CrossRef] [PubMed]
39. Norrrahim, M.N.F.; Ariffin, H.; Yasim-Anuar, T.A.T.; Ghaemi, F.; Hassan, M.A.; Ibrahim, N.A.; Ngee, J.L.H.; Yunus, W.M.Z.W. Superheated steam pretreatment of cellulose affects its electrospinnability for microfibrillated cellulose production. *Cellulose* **2018**, *25*, 3853–3859. [CrossRef]
40. Warid, M.N.M.; Ariffin, H.; Hassan, M.A.; Shirai, Y. Optimization of Superheated Steam Treatment to Improve Surface Modification of Oil Palm Biomass Fiber. *Bioresources* **2016**, *11*, 5780–5796. [CrossRef]
41. Lei, H.; Cybulska, I.; Julson, J. Hydrothermal Pretreatment of Lignocellulosic Biomass and Kinetics. *J. Sustain. Bioenergy Syst.* **2013**, *3*, 250–259. [CrossRef]
42. Lee, J.; Park, K.Y. Impact of hydrothermal pretreatment on anaerobic digestion efficiency for lignocellulosic biomass: Influence of pretreatment temperature on the formation of biomass-degrading byproducts. *Chemosphere* **2020**, *256*, 127116. [CrossRef]
43. Zakaria, M.R.; Hirata, S.; Hassan, M.A. Hydrothermal pretreatment enhanced enzymatic hydrolysis and glucose production from oil palm biomass. *Bioresour. Technol.* **2015**, *176*, 142–148. [CrossRef]
44. Rasmussen, H.; Sørensen, H.R.; Meyer, A.S. Formation of degradation compounds from lignocellulosic biomass in the biorefinery: Sugar reaction mechanisms. *Carbohydr. Res.* **2014**, *385*, 45–57. [CrossRef]
45. Bianco, F.; Şenol, H.; Papirio, S. Enhanced lignocellulosic component removal and biomethane potential from chestnut shell by a combined hydrothermal–alkaline pretreatment. *Sci. Total. Environ.* **2021**, *762*, 144178. [CrossRef]
46. Zhang, H.; Li, J.; Huang, G.; Yang, Z.; Han, L. Understanding the synergistic effect and the main factors influencing the enzymatic hydrolyzability of corn stover at low enzyme loading by hydrothermal and/or ultrafine grinding pretreatment. *Bioresour. Technol.* **2018**, *264*, 327–334. [CrossRef] [PubMed]
47. Phuttaro, C.; Sawatdeenarunat, C.; Surendra, K.; Boonsawang, P.; Chaiprapat, S.; Khanal, S.K. Anaerobic digestion of hydrothermally-pretreated lignocellulosic biomass: Influence of pretreatment temperatures, inhibitors and soluble organics on methane yield. *Bioresour. Technol.* **2019**, *284*, 128–138. [CrossRef]
48. Megashah, L.N. Development of Efficient Processing Method for the Production of Cellulose Nanofibrils from Oil Palm Biomass. Doctoral Thesis, Universiti Putra Malaysia, Selangor, Malaysia, 2020.
49. Sarker, T.R.; Pattnaik, F.; Nanda, S.; Dalai, A.K.; Meda, V.; Naik, S. Hydrothermal pretreatment technologies for lignocellulosic biomass: A review of steam explosion and subcritical water hydrolysis. *Chemosphere* **2021**, *284*, 131372. [CrossRef]

50. Marques, F.P.; Soares, A.K.L.; Lomonaco, D.; e Silva, L.M.A.; Santaella, S.T.; Rosa, M.D.F.; Leitão, R.C. Steam explosion pretreatment improves acetic acid organosolv delignification of oil palm mesocarp fibers and sugarcane bagasse. *Int. J. Biol. Macromol.* **2021**, *175*, 304–312. [CrossRef]
51. Medina, J.D.C.; Woiciechowski, A.; Filho, A.Z.; Nigam, P.S.; Ramos, L.P.; Soccol, C.R. Steam explosion pretreatment of oil palm empty fruit bunches (EFB) using autocatalytic hydrolysis: A biorefinery approach. *Bioresour. Technol.* **2016**, *199*, 173–180. [CrossRef]
52. Abraham, E.; Deepa, B.; Pothan, L.A.; Jacob, M.; Thomas, S.; Cvelbar, U.; Anandjiwala, R. Extraction of nanocellulose fibrils from lignocellulosic fibres: A novel approach. *Carbohydr. Polym.* **2011**, *86*, 1468–1475. [CrossRef]
53. Marques, F.P.; Silva, L.M.A.; Lomonaco, D.; Rosa, M.D.F.; Leitão, R.C. Steam explosion pretreatment to obtain eco-friendly building blocks from oil palm mesocarp fiber. *Ind. Crop. Prod.* **2020**, *143*, 111907. [CrossRef]
54. Meenakshisundaram, S.; Fayeulle, A.; Leonard, E.; Ceballos, C.; Pauss, A. Fiber degradation and carbohydrate production by combined biological and chemical/physicochemical pretreatment methods of lignocellulosic biomass—A review. *Bioresour. Technol.* **2021**, *331*, 125053. [CrossRef]
55. Ariffin, H.; Norrrahim, M.N.F.; Yasim-Anuar, T.A.T.; Nishida, H.; Hassan, M.A.; Ibrahim, N.A.; Yunus, W.M.Z.W. Oil Palm Biomass Cellulose-Fabricated Polylactic Acid Composites for Packaging Applications. In *Bionanocomposites for Packaging Applications*; Springer Science and Business Media LLC.: Berlin, Germany, 2017; pp. 95–105.
56. Ilyas, R.; Sapuan, S.; Nurazzi, N.M.; Norrrahim, M.N.F.; Ibrahim, R.; Atikah, M.; Huzaifah, M.; Radzi, A.; Izwan, S.; Azammi, A.N.; et al. Macro to nanoscale natural fiber composites for automotive components: Research, development, and application. In *Biocomposite and Synthetic Composites for Automotive Applications*; Elsevier BV: Amsterdam, The Netherlands, 2021; pp. 51–105.
57. Ilyas, R.; Sapuan, S.; Harussani, M.; Hakimi, M.; Haziq, M.; Atikah, M.; Asyraf, M.; Ishak, M.; Razman, M.; Nurazzi, N.; et al. Polylactic Acid (PLA) Biocomposite: Processing, Additive Manufacturing and Advanced Applications. *Polymers* **2021**, *13*, 1326. [CrossRef]
58. Sharip, N.S.; Yasim-Anuar, T.A.T.; Norrrahim, M.N.F.; Shazleen, S.S.; Nurazzi, N.M.; Sapuan, S.M.; Ilyas, R.A. A Review on Nanocellulose Composites in Biomedical Application. In *Composites in Biomedical Applications*; CRC Press: Boca Raton, FL, USA, 2020; pp. 161–190.
59. Norrrahim, M.N.F.; Kasim, N.A.M.; Knight, V.F.; Ujang, F.A.; Janudin, N.; Razak, M.A.I.A.; Shah, N.A.A.; Noor, S.A.M.; Jamal, S.H.; Ong, K.K.; et al. Nanocellulose: The next super versatile material for the military. *Mater. Adv.* **2021**, *2*, 1485–1506. [CrossRef]
60. Norrrahim, M.N.F.; Kasim, N.A.M.; Knight, V.F.; Misenan, M.S.M.; Janudin, N.; Shah, N.A.A.; Kasim, N.; Yusoff, W.Y.W.; Noor, S.A.M.; Jamal, S.H.; et al. Nanocellulose: A bioadsorbent for chemical contaminant remediation. *RSC Adv.* **2021**, *11*, 7347–7368. [CrossRef]
61. Yasim-Anuar, T.A.T. Well-Dispersed Cellulose Nanofiber in Low Density Polyethylene Nanocomposite by Liquid-Assisted Extrusion. *Polymers* **2020**, *12*, 927. [CrossRef] [PubMed]
62. Norrrahim, M.N.F.; Nurazzi, N.M.; Jenol, M.A.; Farid, M.A.A.; Janudin, N.; Ujang, F.A.; Yasim-Anuar, T.A.T.; Najmuddin, S.U.F.S.; Ilyas, R.A. Emerging development of nanocellulose as an antimicrobial material: An overview. *Mater. Adv.* **2021**, *2*, 3538–3551. [CrossRef]
63. Nurazzi, N.M.; Asyraf, M.R.M.; Rayung, M.; Norrrahim, M.N.F.; Shazleen, S.S.; Rani, M.S.A.; Shafi, A.R.; Aisyah, H.A.; Radzi, M.H.M.; Sabaruddin, F.A.; et al. Thermogravimetric Analysis Properties of Cellulosic Natural Fiber Polymer Composites: A Review on Influence of Chemical Treatments. *Polymers* **2021**, *13*, 2710. [CrossRef]
64. Norrrahim, M.N.F. Cationic Nanocellulose as Promising Candidate for Filtration Material of COVID-19: A Perspective. *Appl. Sci. Eng. Prog.* **2021**. [CrossRef]
65. Norrrahim, M.N.F.; Yasim-Anuar, T.A.T.; Sapuan, S.; Ilyas, R.; Hakimi, M.I.; Najmuddin, S.U.F.S.; Jenol, M.A. Nanocellulose Reinforced Polypropylene and Polyethylene Composite for Packaging Application. In *Bio-based Packaging*; Wiley: Hoboken, NJ, USA, 2021; pp. 133–150.
66. Lee, C.H.; Lee, S.H.; Padzil, F.N.M.; Ainun, Z.M.A.; Norrrahim, M.N.F.; Chin, K.L. Biocomposites and Nanocomposites. In *Composite Materials*; Informa UK Limited: London, UK, 2021; pp. 29–60.
67. Nurazzi, N.; Asyraf, M.; Athiyah, S.F.; Shazleen, S.; Rafiqah, S.; Harussani, M.; Kamarudin, S.; Razman, M.; Rahmah, M.; Zainudin, E.; et al. A Review on Mechanical Performance of Hybrid Natural Fiber Polymer Composites for Structural Applications. *Polymers* **2021**, *13*, 2170. [CrossRef]
68. Nurazzi, N.; Asyraf, M.; Khalina, A.; Abdullah, N.; Aisyah, H.; Rafiqah, S.; Sabaruddin, F.; Kamarudin, S.; Norrrahim, M.; Ilyas, R.; et al. A Review on Natural Fiber Reinforced Polymer Composite for Bullet Proof and Ballistic Applications. *Polymers* **2021**, *13*, 646. [CrossRef]
69. Agbor, V.B.; Cicek, N.; Sparling, R.; Berlin, A.; Levin, D.B. Biomass pretreatment: Fundamentals toward application. *Biotechnol. Adv.* **2011**, *29*, 675–685. [CrossRef]
70. Moset, V.; Xavier, C.D.A.N.; Feng, L.; Wahid, R.; Møller, H. Combined low thermal alkali addition and mechanical pre-treatment to improve biogas yield from wheat straw. *J. Clean. Prod.* **2018**, *172*, 1391–1398. [CrossRef]
71. Khan, A.S.; Man, Z.; Bustam, M.A.; Kait, C.F.; Khan, M.I.; Muhammad, N.; Nasrullah, A.; Ullah, Z.; Ahmad, P. Impact of Ball-Milling Pretreatment on Pyrolysis Behavior and Kinetics of Crystalline Cellulose. *Waste Biomass-Valorization* **2016**, *7*, 571–581. [CrossRef]

72. Zhu, J.Y.; Pan, X.; Zalesny, R.S. Pretreatment of woody biomass for biofuel production: Energy efficiency, technologies, and recalcitrance. *Appl. Microbiol. Biotechnol.* **2010**, *87*, 847–857. [CrossRef] [PubMed]
73. Jędrzejczyk, M.; Soszka, E.; Czapnik, M.; Ruppert, A.M.; Grams, J. Physical and Chemical Pretreatment of Lignocellulosic Biomass. In *Second and Third Generation of Feedstocks: The Evolution of Biofuels*; Elsevier: Amsterdam, The Netherlands, 2019; pp. 143–196. [CrossRef]
74. De la Rubia, M.A.; Fernández-Cegrí, V.; Raposo, F.; Borja, R. Influence of particle size and chemical composition on the performance and kinetics of anaerobic digestion process of sunflower oil cake in batch mode. *Biochem. Eng. J.* **2011**, *58–59*, 162–167. [CrossRef]
75. Ramos, L.P. The chemistry involved in the steam treatment of lignocellulosic materials. *Química Nova* **2003**, *26*, 863–871. [CrossRef]
76. Takács, E.; Wojnárovits, L.; Földváry, C.; Hargittai, P.; Borsa, J.; Sajó, I. Effect of combined gamma-irradiation and alkali treatment on cotton-cellulose. *Radiat. Phys. Chem.* **2000**, *57*, 399–403. [CrossRef]
77. Naimi, L.J.; Sokhansanj, S. Data-based equation to predict power and energy input for grinding wheat straw, corn stover, switchgrass, miscanthus, and canola straw. *Fuel Process. Technol.* **2018**, *173*, 81–88. [CrossRef]
78. Pérez-Rodríguez, N.; García-Bernet, D.; Domínguez, J.M. Faster methane production after sequential extrusion and enzymatic hydrolysis of vine trimming shoots. *Environ. Chem. Lett.* **2018**, *16*, 295–299. [CrossRef]
79. Zhang, S.-C.; Lai, Q.-H.; Lu, Y.; Liu, Z.-D.; Wang, T.; Zhang, C.; Xing, X.-H. Enhanced biohydrogen production from corn stover by the combination of Clostridium cellulolyticum and hydrogen fermentation bacteria. *J. Biosci. Bioeng.* **2016**, *122*, 482–487. [CrossRef]
80. He, L.; Huang, H.; Lei, Z.; Liu, C.; Zhang, Z. Enhanced hydrogen production from anaerobic fermentation of rice straw pretreated by hydrothermal technology. *Bioresour. Technol.* **2014**, *171*, 145–151. [CrossRef]
81. Baeta, B.; Lima, D.R.S.; Adarme, O.F.H.; Gurgel, L.; de Aquino, S.F. Optimization of sugarcane bagasse autohydrolysis for methane production from hemicellulose hydrolyzates in a biorefinery concept. *Bioresour. Technol.* **2016**, *200*, 137–146. [CrossRef]
82. Jackowiak, D.; Bassard, D.; Pauss, A.; Ribeiro, T. Optimisation of a microwave pretreatment of wheat straw for methane production. *Bioresour. Technol.* **2011**, *102*, 6750–6756. [CrossRef]
83. Li, L.; Kong, X.; Yang, F.; Li, D.; Yuan, Z.; Sun, Y. Biogas Production Potential and Kinetics of Microwave and Conventional Thermal Pretreatment of Grass. *Appl. Biochem. Biotechnol.* **2012**, *166*, 1183–1191. [CrossRef] [PubMed]
84. Simona, M.; Gianfranco, A.; Jody, G.; Paolo, B. Energetic assessment of extrusion as pre-treatment to improve the anaerobic digestion of agricultural ligno-cellulosic biomasses. In Proceedings of the 15th International Conference Ramiran 2013, Versailles, France, 3–5 June 2013; Volume 1.
85. Bauer, A.; Lizasoain, J.; Theuretzbacher, F.; Agger, J.W.; Rincón, M.; Menardo, S.; Saylor, M.K.; Enguídanos, R.; Nielsen, P.J.; Potthast, A.; et al. Steam explosion pretreatment for enhancing biogas production of late harvested hay. *Bioresour. Technol.* **2014**, *166*, 403–410. [CrossRef]
86. Alinia, R.; Zabihi, S.; Esmaeilzadeh, F.; Kalajahi, J.F. Pretreatment of wheat straw by supercritical CO2 and its enzymatic hydrolysis for sugar production. *Biosyst. Eng.* **2010**, *107*, 61–66. [CrossRef]
87. Chandra, R.P.; Chu, Q.; Hu, J.; Zhong, N.; Lin, M.; Lee, J.-S.; Saddler, J. The influence of lignin on steam pretreatment and mechanical pulping of poplar to achieve high sugar recovery and ease of enzymatic hydrolysis. *Bioresour. Technol.* **2016**, *199*, 135–141. [CrossRef]
88. DeMartini, J.D.; Foston, M.; Meng, X.; Jung, S.; Kumar, R.; Ragauskas, A.J.; Wyman, C.E. How chip size impacts steam pretreatment effectiveness for biological conversion of poplar wood into fermentable sugars. *Biotechnol. Biofuels* **2015**, *8*, 1–16. [CrossRef] [PubMed]
89. Walch, E.; Zemann, A.; Schinner, F.; Bonn, G.; Bobleter, O. Enzymatic saccharification of hemicellulose obtained from hydrothermally pretreated sugar cane bagasse and beech bark. *Bioresour. Technol.* **1992**, *39*, 173–177. [CrossRef]
90. Ståhl, M.; Nieminen, K.; Sixta, H. Hydrothermolysis of pine wood. *Biomass-Bioenergy* **2018**, *109*, 100–113. [CrossRef]
91. Pińkowska, H.; Krzywonos, M.; Wolak, P. Valorization of Rapeseed Meal by Hydrothermal Treatment—Effect of Reaction Parameters on Low Molecular Products Distribution. *Cellul. Chem. Technol.* **2019**, *53*, 755–765. [CrossRef]
92. Zhang, Y.; Chen, J.; Zhang, L.; Zhan, P.; Liu, N.; Wu, Z. Preparation of nanocellulose from steam exploded poplar wood by enzymolysis assisted sonication Preparation of nanocellulose from steam exploded poplar wood by enzymolysis assisted sonication. *Mater. Res. Express* **2020**, *7*, 035010. [CrossRef]
93. Wang, Y.; Wei, X.; Li, J.; Wang, Q.; Wang, F.; Kong, L. Homogeneous Isolation of Nanocellulose from Cotton Cellulose by High Pressure Homogenization. *J. Mater. Sci. Chem. Eng.* **2013**, *1*, 49–52. [CrossRef]
94. Li, J.; Wei, X.; Wang, Q.; Chen, J.; Chang, G.; Kong, L.; Su, J.; Liu, Y. Homogeneous isolation of nanocellulose from sugarcane bagasse by high pressure homogenization. *Carbohydr. Polym.* **2012**, *90*, 1609–1613. [CrossRef]
95. Hatakka, A. Pretreatment of wheat straw by white-rot fungi for enzymic saccharification of cellulose. *Appl. Microbiol. Biotechnol.* **1983**, *18*, 350–357. [CrossRef]
96. Çağrı, A.; Ince, O.; Bozan, M.; Ozbayram, G.; Ince, B. Biological pretreatment with Trametes versicolor to enhance methane production from lignocellulosic biomass: A metagenomic approach. *Ind. Crop. Prod.* **2019**, *140*. [CrossRef]
97. Chen, S.; Zhang, X.; Singh, D.; Yu, H.; Yang, X. Biological pretreatment of lignocellulosics: Potential, progress and challenges. *Biofuels* **2010**, *1*, 177–199. [CrossRef]

98. Tanahashi, M. Characterization and degradation mechanisms of wood components by steam explosion and utilization of exploded wood. In *Bulletin of the Wood Research Institute Kyoto University*; Kyoto University Research Information Repository: Kyoto, Japan, 1990; Volume 77, pp. 49–117.
99. Gregg, D.; Saddler, J.N. A Techno-Economic Assessment of the Pretreatment and Fractionation Steps of a Biomass-to-Ethanol Process. In *Seventeenth Symposium on Biotechnology for Fuels and Chemicals*; Humana Press: Totowa, NJ, USA, 1996; Volume 57, pp. 711–727. [CrossRef]
100. De Souza, C.G.M.; Tychanowicz, G.K.; De Souza, D.F.; Peralta, R.M. Production of laccase isoforms byPleurotus pulmonarius in response to presence of phenolic and aromatic compounds. *J. Basic Microbiol.* **2004**, *44*, 129–136. [CrossRef]
101. Galbe, M.; Zacchi, G. A review of the production of ethanol from softwood. *Appl. Microbiol. Biotechnol.* **2002**, *59*, 618–628. [CrossRef]
102. Mtui, G.Y.S. Trends in industrial and environmental biotechnology research in Tanzania. *Afr. J. Biotechnol.* **2007**, *6*, 2860–2867. [CrossRef]
103. Ubalua, A.O. Cassava wastes: Treatment options and value addition alternatives. *Afr. J. Biotechnol.* **2007**, *6*, 2065–2073. [CrossRef]
104. Demirbaş, A. Utilization of Urban and Pulping Wastes to Produce Synthetic Fuel via Pyrolysis. *Energy Sources* **2002**, *24*, 205–213. [CrossRef]
105. Guerra, A.; Mendonça, R.; Ferraz, A. Molecular weight distribution of wood components extracted from Pinus taeda biotreated by Ceriporiopsis subvermispora. *Enzym. Microb. Technol.* **2003**, *33*, 12–18. [CrossRef]
106. Maijala, P.; Kleen, M.; Westin, C.; Poppius-Levlin, K.; Herranen, K.; Lehto, J.; Reponen, P.; Mäentausta, O.; Mettälä, A.; Hatakka, A. Biomechanical pulping of softwood with enzymes and white-rot fungus Physisporinus rivulosus. *Enzym. Microb. Technol.* **2008**, *43*, 169–177. [CrossRef]
107. Wan, C.; Li, Y. Microbial pretreatment of corn stover with Ceriporiopsis subvermispora for enzymatic hydrolysis and ethanol production. *Bioresour. Technol.* **2010**, *101*, 6398–6403. [CrossRef]
108. Oyeleke, S.B.; Dauda, B.E.N.; Oyewole, O.A.; Okoliegbe, I.N.; Ojebode, T. Production of bioethanol from cassava and sweet potato peels. *Ad. Environ. Biol* **2011**, *5*, 3729–3733.
109. Taha, M.; Shahsavari, E.; Al-Hothaly, K.; Mouradov, A.; Smith, A.; Ball, A.; Adetutu, E.M. Enhanced Biological Straw Saccharification Through Coculturing of Lignocellulose-Degrading Microorganisms. *Appl. Biochem. Biotechnol.* **2015**, *175*, 3709–3728. [CrossRef]
110. Song, L.; Yu, H.; Ma, F.; Zhang, X. Biological Pretreatment under Non-sterile Conditions for Enzymatic Hydrolysis of Corn Stover. *Bioresources* **2013**, *8*, 3802–3816. [CrossRef]
111. Du, W.; Yu, H.; Song, L.; Zhang, J.; Weng, C.; Ma, F.; Zhang, X. The promoting effect of byproducts from Irpex lacteus on subsequent enzymatic hydrolysis of bio-pretreated cornstalks. *Biotechnol. Biofuels* **2011**, *4*, 37. [CrossRef]
112. Zhang, Q.; He, J.; Tian, M.; Mao, Z.; Tang, L.; Zhang, J.; Zhang, H. Enhancement of methane production from cassava residues by biological pretreatment using a constructed microbial consortium. *Bioresour. Technol.* **2011**, *102*, 8899–8906. [CrossRef] [PubMed]
113. Passos, F.; Hom-Diaz, A.; Blanquez, P.; Vicent, T.; Ferrer, I. Improving biogas production from microalgae by enzymatic pretreatment. *Bioresour. Technol.* **2016**, *199*, 347–351. [CrossRef]
114. Ali, S.S.; Al-Tohamy, R.; Manni, A.; Luz, F.C.; Elsamahy, T.; Sun, J. Enhanced digestion of bio-pretreated sawdust using a novel bacterial consortium: Microbial community structure and methane-producing pathways. *Fuel* **2019**, *254*, 115604. [CrossRef]
115. Guo, H.; Zhao, Y.; Chen, X.; Shao, Q.; Qin, W. Pretreatment of Miscanthus with biomass-degrading bacteria for increasing delignification and enzymatic hydrolysability. *Microb. Biotechnol.* **2019**, *12*, 787–798. [CrossRef]
116. Machado, A.D.S.; Ferraz, A. Biological pretreatment of sugarcane bagasse with basidiomycetes producing varied patterns of biodegradation. *Bioresour. Technol.* **2017**, *225*, 17–22. [CrossRef]
117. Castoldi, R.; Bracht, A.; de Morais, G.R.; Baesso, M.L.; Corrêa, R.C.G.; Peralta, R.; Moreira, R.; Polizeli, M.D.L.; de Souza, C.G.M.; Peralta, R.M. Biological pretreatment of Eucalyptus grandis sawdust with white-rot fungi: Study of degradation patterns and saccharification kinetics. *Chem. Eng. J.* **2014**, *258*, 240–246. [CrossRef]
118. Kumar, M.N.; Ravikumar, R.; Sankar, M.K.; Thenmozhi, S. New insight into the effect of fungal mycelia present in the bio-pretreated paddy straw on their enzymatic saccharification and optimization of process parameters. *Bioresour. Technol.* **2018**, *267*, 291–302. [CrossRef]
119. Jagtap, S.; Dhiman, S.; Kim, T.-S.; Li, J.; Kang, Y.C.; Lee, J.-K. Characterization of a β-1,4-glucosidase from a newly isolated strain of Pholiota adiposa and its application to the hydrolysis of biomass. *Biomass-Bioenergy* **2013**, *54*, 181–190. [CrossRef]
120. Dhiman, S.S.; Haw, J.-R.; Kalyani, D.; Kalia, V.C.; Kang, Y.C.; Lee, J.-K. Simultaneous pretreatment and saccharification: Green technology for enhanced sugar yields from biomass using a fungal consortium. *Bioresour. Technol.* **2015**, *179*, 50–57. [CrossRef]
121. Jagtap, S.; Dhiman, S.; Kim, T.-S.; Li, J.; Lee, J.-K.; Kang, Y.C. Enzymatic hydrolysis of aspen biomass into fermentable sugars by using lignocellulases from Armillaria gemina. *Bioresour. Technol.* **2013**, *133*, 307–314. [CrossRef] [PubMed]
122. Zhu, J.Y.; Sabo, R.; Luo, X. Integrated production of nano-fibrillated cellulose and cellulosic biofuel (ethanol) by enzymatic fractionation of wood fibers. *Green Chem.* **2011**, *13*, 1339–1344. [CrossRef]
123. Henriksson, M.; Berglund, L.; Lindström, T. An environmentally friendly method for enzyme-assisted preparation of microfibrillated cellulose (MFC) nanofibers. *Eur. Polym. J.* **2007**, *43*, 3434–3441. [CrossRef]
124. Tsukamoto, J.; Durán, N.; Tasic, L. Nanocellulose and Bioethanol Production from Orange Waste using Isolated Microorganisms. *J. Braz. Chem. Soc.* **2013**, *24*, 1537–1543. [CrossRef]

125. de Camargo, L.A.; Pereira, S.C.; Correa, A.C.; Farinas, C.S.; Marconcini, J.M.; Mattoso, L.H.C. Feasibility of Manufacturing Cellulose Nanocrystals from the Solid Residues of Second-Generation Ethanol Production from Sugarcane Bagasse. *BioEnergy Res.* **2016**, *9*, 894–906. [CrossRef]
126. Martelli-Tosi, M.; Torricillas, M.D.S.; Martins, M.A.; De Assis, O.B.G.; Tapia-Blácido, D.R. Using Commercial Enzymes to Produce Cellulose Nanofibers from Soybean Straw. *J. Nanomater.* **2016**, *2016*, 1–10. [CrossRef]
127. Beltramino, F.; Roncero, M.B.; Vidal, T.; Torres, A.L.; Valls, C. Increasing yield of nanocrystalline cellulose preparation process by a cellulase pretreatment. *Bioresour. Technol.* **2015**, *192*, 574–581. [CrossRef] [PubMed]

polymers

MDPI

Review

Critical Review of Natural Fiber Reinforced Hybrid Composites: Processing, Properties, Applications and Cost

M. J. Suriani [1,*], R. A. Ilyas [2,3,*], M. Y. M. Zuhri [4,5,*], A. Khalina [5,6], M. T. H. Sultan [5,7], S. M. Sapuan [4,5], C. M. Ruzaidi [1], F. Nik Wan [1], F. Zulkifli [1], M. M. Harussani [4], M. A. Azman [1], F. S. M. Radzi [1] and Shubham Sharma [8]

1 Faculty of Ocean Engineering Technology and Informatics, Universiti Malaysia Terengganu, Kuala Nerus 21030, Terengganu, Malaysia; ruzaidi@umt.edu.my (C.M.R.); wnfatihahnw@gmail.com (F.N.W.); fakhratulz@umt.edu.my (F.Z.); asyrafazman23@yahoo.com (M.A.A.); fathinsakinah96@gmail.com (F.S.M.R.)
2 School of Chemical and Energy Engineering, Faculty of Engineering, Universiti Teknologi Malaysia (UTM), Johor Bahru 81310, Johor, Malaysia
3 Centre for Advanced Composite Materials (CACM), Universiti Teknologi Malaysia (UTM), Johor Bahru 81310, Johor, Malaysia
4 Advanced Engineering Materials and Composites Research Centre (AEMC), Department of Mechanical and Manufacturing Engineering, Faculty of Engineering, Universiti Putra Malaysia, Serdang 43400, Selangor, Malaysia; sapuan@upm.edu.my (S.M.S.); mmharussani17@gmail.com (M.M.H.)
5 Laboratory of Biocomposite Technology, Institute of Tropical Forestry and Forest Products (INTROP), Universiti Putra Malaysia, Serdang 43400, Selangor, Malaysia; khalina@upm.edu.my (A.K.); thariq@upm.edu.my (M.T.H.S.)
6 Department of Biological and Agricultural Engineering, Faculty of Engineering, Universiti Putra Malaysia, Serdang 43400, Selangor, Malaysia
7 Department of Aerospace Engineering, Faculty of Engineering, Universiti Putra Malaysia, Serdang 43400, Selangor, Malaysia
8 Department of Mechanical Engineering, IK Gujral Punjab Technical University, Main Campus-Kapurthala, Punjab 144603, India; shubham543sharma@gmail.com
* Correspondence: surianimatjusoh@umt.edu.my (M.J.S.); ahmadilyas@utm.my (R.A.I.); zuhri@upm.edu.my (M.Y.M.Z.)

Citation: Suriani, M.J.; Ilyas, R.A.; Zuhri, M.Y.M.; Khalina, A.; Sultan, M.T.H.; Sapuan, S.M.; Ruzaidi, C.M.; Wan, F.N.; Zulkifli, F.; Harussani, M.M.; et al. Critical Review of Natural Fiber Reinforced Hybrid Composites: Processing, Properties, Applications and Cost. *Polymers* **2021**, *13*, 3514. https://doi.org/10.3390/polym13203514

Academic Editor: Emin Bayraktar

Received: 15 September 2021
Accepted: 7 October 2021
Published: 13 October 2021

Abstract: Increasing scientific interest has occurred concerning the utilization of natural fiber-enhanced hybrid composites that incorporate one or more types of natural enhancement. Annual natural fiber production is estimated to be 1,783,965 $\times$ 10^3 tons/year. Extensive studies have been conducted in the domains of natural/synthetic as well as natural/natural hybrid composites. As synthetic fibers have better rigidity and strength than natural fibers, natural/synthetic hybrid composites have superior qualities via hybridization compared to natural composites in fibers. In general, natural fiber compounds have lower characteristics, limiting the use of natural composites reinforced by fiber. Significant effort was spent in enhancing the mechanical characteristics of this group of materials to increase their strengths and applications, especially via the hybridization process, by manipulating the characteristics of fiber-reinforced composite materials. Current studies concentrate on enhancing the understanding of natural fiber-matrix adhesion, enhancing processing methods, and natural fiber compatibility. The optimal and resilient conceptions have also been addressed due to the inherently more significant variabilities. Moreover, much research has tackled natural fiber reinforced hybrid composite costs. In addition, this review article aims to offer a review of the variables that lead to the mechanical and structural failure of natural fiber reinforced polymer composites, as well as an overview of the details and costings of the composites.

Keywords: natural fiber; hybrid composite; cellulose; costing; processing; fiber-matrix adhesion

1. Introduction

Composite materials are produced from a combination of two or more elements that are easily distinguishable to enhance the characteristics of the individual element [1,2].

Newly invented materials may be favored for various reasons, e.g., they are stronger, more lightweight, and less costly than existing materials [3–5]. In general, the individual materials making up composites are known as constituents. Principally, most composites comprise two constituent materials—reinforcement and matrix—however, in general, the composites may contain not only two components, matrix and reinforcement, but also other types of components: fillers, compatibilizers, coupling agents, pigments, lubricants, surfactants, solvents, etc. Only the simplest textile-based composites, also called textolites, contain two constituents, polymer matrix and reinforcement such as natural, synthetic or hybrid fibers, or fabrics. The reinforcement is significantly stiffer and stronger than the matrix, contributing to the composites' superior characteristics [6,7]. The main functions of the polymer matrix in textile-based composites are to bind reinforcements (fabric, fibers, or nanofibers) and maintain the integrity of the composite.

Composites' reinforcements can be fabrics elements, fibers, or nanofibers [8–10]. Fiber is defined as one very long axis with two other axes that frequently are either circular or near-circular. The fibers have a pronounced axial orientation. As is known, Young's modulus and tensile strength of fiber in the longitudinal direction (along the fiber axis) is usually an order of magnitude higher than in the lateral direction of the fiber. Nanofibers have an ideal form; however, they are smaller in diameter and length compared to fibers [11]. Figure 1 shows the type of reinforcements in composites.

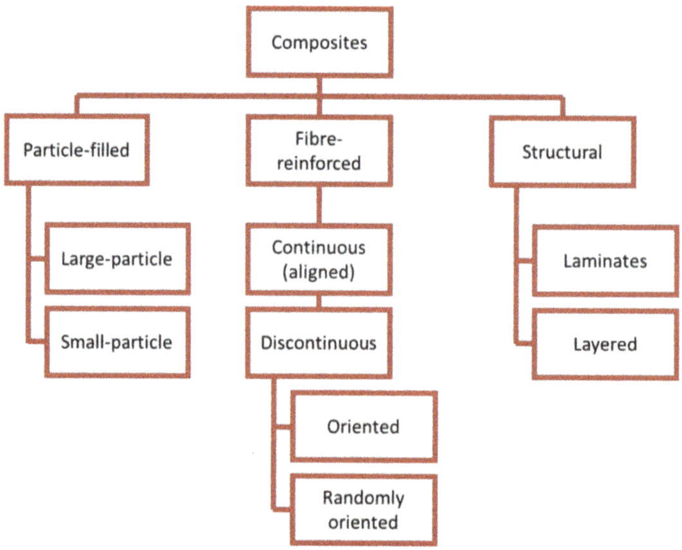

Figure 1. Types of reinforcements in composites. Redrawn with permission from Saba and Jawaid [12].

The matrix might be thermoset, thermoplastic, or biopolymer. Polyvinyl chloride (PVC) is the most common thermoplastic matrix used in natural fiber composites. Besides this, due to the limited compatibility of nonpolar hydrophobic polyethylene and polypropylene with polar and hydrophilic natural cellulose fibers, these thermoplastics typically are not employed as a matrix for natural cellulose fibers. Meanwhile, polyester, epoxy, and phenolics are the most often used thermoset polymers [13,14].

Nowadays, the availability of bio-based polymer matrices on the market is comparatively meager; however, it has been speedily expanding, thanks to a huge number of industrial investigations and continuous research into the advantages of these materials, as well as their practicability in actual applications. The preliminary data on the essential characteristics of composites attained from eco-based matrices were reported in [15,16].

2. Natural Fiber (NF)

Fiber is the continuous filaments hair-like material of elongated piece that is similar to a thread, while fibers is a group of fiber that can be coiled into rope, filaments, or thread [17–19]. They are useful as of the composite materials' element and were also formed into sheets to produce felt or paper. Fibers are categorized into three groups: (1) natural fiber and (2) man-made, and (3) synthetic fiber. Natural fibers are either animal, plant or mineral-based that are extracted from nature without compromising the environment. Commonly used animal-based natural fibers including feather, wool, silk, hair, etc. Examples of plant-based natural fibers are banana, jute, hemp, bamboo, flax, sisal, etc., which are broadly applied to manufacture natural fiber reinforced polymer (NFRP) composites [20]. The classifications of natural fibers are shown in Figure 2, and the annual natural fiber production is presented in Table 1.

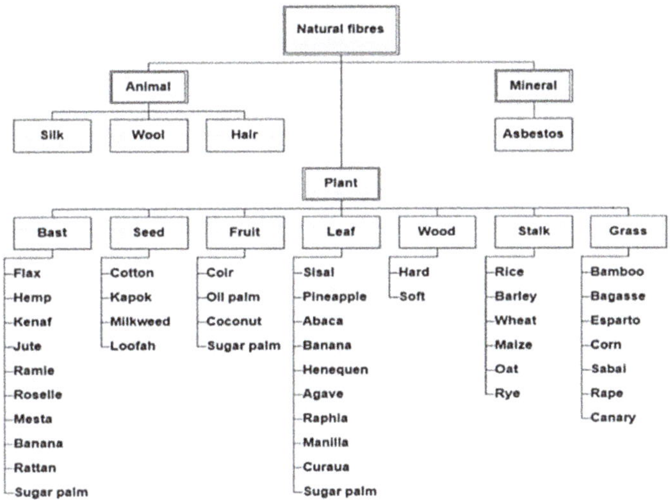

Figure 2. Classifications of natural fibers. Redrawn with permission from Jawaid and Khalil [12].

Over the past few years, natural fibers have become eminent reinforcing fibers in polymer-matrix composites (PMC). They offer rapidly increasing and abundant characteristics, allowing them to be obtained at a small cost. Numerous attempts in terms of studies have been performed worldwide to prove the eligibility of natural fiber-based composites to substitute the synthetic as newly engineered fibers. Due to the growing need for renewable, cost-effective, and environmentally friendly materials, the use of natural fibers as composite materials' reinforcement has proliferated over the years.

When compared to glass fiber composites, natural fiber composites are more advantageous for being more lightweight, biodegradable in nature, ease of machinability, zero toxicity, cheap cost, convenience, and non-abrasive nature [13,21,22]. Many natural fiber composites are reputable and have been satisfactorily proven in research. The number of natural fiber composites has been investigated previously in terms of their physical and mechanical properties, such as arrowroot, hemp, sisal, coir, jute, kenaf, date, pine cone, vakka and bamboo [23–27].

Table 1. Annual natural fibers production. Extracted with permission from [28].

Natural Fiber	Source	World Production ($\times 10^3$ Tons)
Abaca	Leaf	70
Banana	Stem	200
Bamboo	Stem	10,000
Broom	Stem	Abundant
Coir	Fruit	100
Cotton lint	Seed	18,500
Elephant	Stem	Abundant
Flax	Stem	810
Hemp	Stem	215
Jute	Stem	2500
Kenaf	Stem	770
Linseed	Fruit	Abundant
Oil Palm Fruit	Fruit	Abundant
Ramie	Stem	100
Rice Husk	Grain	Abundant
Roselle	Stem	250
Sisal	Leaf	380
Sun hemp	Stem	70
Wood	Stem	1,750,000

3. Recent Development of Natural Fiber Reinforced Hybrid Composites

It is known that natural fibers possess some limitations compared with those common fibers such as glass and carbon, where it is having more inferior mechanical properties and a higher water absorption [29]. Therefore, an introduction of hybrid biocomposites consists of two or more fibers in one matrix is seen as a solution to enhance the natural fiber-reinforced polymer composites' properties. Zahra et al. [30] stated that hybridizing one natural fiber with another natural fiber/synthetic fiber in one matrix will improve it's thermal and mechanical than the individual fiber composites [31]. This has shown that hybrid composites are more reliable for various applications, besides being more environmentally friendly.

The hybridization of natural fiber-based reinforced polymer composites can be done through a combination of natural–natural fibers, natural–synthetic fibers, natural fiber with carbonaceous materials, and natural fiber with metal [32]. Due to their varied properties and considerations of interfacial adhesion, hybrid natural fiber composite materials are facing difficulties in fabrication. Composites are manufactured in a variety of ways, such as through basic mixing and open or closed molding. Many factors can affect the interactions between the fiber and matrix, for example, and could be mild owing to the existing van der Waals forces, hydrogen bonding, and weak electrostatic interactions. In addition, a good interaction could also exist due to the chemical interactions between those materials. Therefore, studies on hybrid natural fiber composites keep increasing in order to discover the ability of hybrids to be a possible alternative, replacing various petroleum-based products. Some examples of the studies of hybrid natural fiber composites are presented in Table 2.

Table 2. Natural fiber reinforced hybrid composites.

Natural Fiber	Matrix	Hybrids	Process	Ref.
Sugar palm fiber (SPF)				
Sugar palm fiber	Unsaturated polyester	Woven glass	Compression molding	[28]
Sugar palm fiber	Unsaturated polyester	Glass fiber	Hand lay-up	[33]
Sugar palm fiber	Thermoplastic polyurethane	Glass fiber	Melt compounding	[34]
Sugar palm yarn fiber	Epoxy	Carbon fiber	Hand lay-up	[35]
Benzoyl treated sugar palm fiber	Epoxy	Glass fiber	Hand lay-up	[36,37]
Sugar palm fiber	Thermoplastic sugar palm starch/agar	Seaweed fiber	Hydraulic thermo-press	[38]
Sugar palm fiber	Thermoplastic polyurethane	Roselle fiber	Hot press	[39]
Sugar palm fiber	Cornstarch	Cornhusk	Solution casting	[40]
Sugar palm yarn fiber	Unsaturated polyester	Glass fiber	Sheet molding	[41]
Sugar palm fiber	Cassava starch	Cassava fiber	Casting	[42]
Sugar palm fiber	Polypropylene	Kenaf fiber	Compression molding	[43]
Sugar palm fiber	Cornstarch	Cornstalk fiber	Solution casting	[44]
Sugar palm fiber	Epoxy	Ramie fiber	Compression molding	[45]
Sugar palm fiber	Vinyl ester	Roselle fiber	Hand lay-up	[46]
Sugar palm fiber	Polypropylene	Glass fiber	Film stacking and hot compression	[47]
Kenaf fiber (KF)				
Kenaf fiber	Epoxy	Glass fiber	Sheet molding	[48]
Kenaf fiber	unsaturated polyester (UPE)	Glass fiber	Sheet molding	[49]
Kenaf fiber	Epoxy	Silica	Hand lay-up	[50,51]
Kenaf fiber	Epoxy	Bamboo fiber/nanoclay	Hand lay-up	[52]
Kenaf fiber	Epoxy	Oil palm/montmorillonite	Hand lay-up	[53]
Kenaf fiber	Polypropylene-grafted maleic anhydride (PP-g-MA)	Graphene nanosheets	Hot press	[54]
Kenaf core	Polypropylene	Bleached nanocellulose	Melt mixing compounding	[55]
Kenaf fiber	Epoxy	Glass fiber	Filament winding	[56]
Kenaf fiber	Epoxy	Pet yarn	Cold press	[57]
Kenaf fiber	Polyethylene terephthalate	Glass fiber	Compression molding	[58]
Kenaf fiber	Epoxy	Kevlar	Hand lay-up	[59]
Kenaf fiber	Polyester	Banana fiber	Hand lay-up	[60]
Kenaf fiber	Indian almond fiber	Kenaf fiber	Hand lay-up	[61]
Kenaf fiber	Epoxy	Glass/waste tea leaf fiber	Compression molding	[62]
Kenaf fiber	Epoxy	Oil palm fiber	Hand lay-up	[63]
Woven kenaf fiber	Polypropylene	Glass fiber	Hot press molding	[64]
Kenaf fiber	Polypropylene	E-glass	Hot compression molding	[65]
Kenaf fiber	Epoxy	Bamboo fiber	Hand lay-up	[66]
Kenaf fiber	Polypropylene	Wood flour	Injection molding	[67]
	Polylactic acid (PLA)	Kenaf	Fused Deposition Modeling (FDM)	[68]
Oil palm empty fruit bunches fiber (OPEFB)				
Oil palm empty fruit bunches fiber	Epoxy	MgO_2 pet yarn	Compression molding	[69]
Oil palm empty fruit bunches fiber	Polyester resin	MgO_2 pet yarn	Compression molding	[69]
Oil palm empty fruit bunches fiber	Epoxy	Woven kenaf fabric	Hand lay-up	[70]
Oil palm empty fruit bunches fiber	Polypropylene (PP) matrix		Injection molding	[71]
Oil palm empty fruit bunches fiber	Phenolic formaldehyde (PF) resin	Sugarcane bagasse (SCB) fiber	Hand lay-up	[72]
Oil palm empty fruit bunches fiber	Resin	Gamma-irradiated kevlar	Hand lay-up	[73]
Oil palm empty fruit bunches fiber	Recycled polypropylene (RPP)	Glass fiber	Extrusion and injection molding	[74]
Oil palm fibers	Polyester resin	Chopped strand mat (CSM) glass fibers	Hybrid laminates	[75]
Oil palm empty fruit bunches fiber	Polypropylene	Glass fiber	Hot pressing	[76]

Table 2. *Cont.*

Natural Fiber	Matrix	Hybrids	Process	Ref.
		Pineapple leaf fibers (PALF)		
Pineapple leaf fibers	Carbon hybrid laminate		Vacuum infusion	[77]
Pineapple leaf fibers	Polylactic acid (PLA)	Alkali treated coir	Compression molding	[78]
Pineapple leaf fibers	Vinyl ester	Glass fiber	Automated spray-up	[79]
Pineapple leaf fibers	Polyester	Banana/glass fiber	Hand lay-up	[80]
Silane treated pineapple leaf fiber	Phenolic hybrid	Kenaf fiber	Hydraulic pressure hot press	[81]
Pineapple leaf fibers	Polyester	Sisal fiber	Injection molding	[82]
		Bamboo fiber (BF)		
Long bamboo fiber	Epoxy		Compression molding	[83]
Short bamboo fiber	POlypropylene	Glass fiber	Injection molding	[84]
Bamboo fiber	Maleic anhydride grafted polypropylene (MAPP)	Glass fiber	Injection molding	[85]
Bamboo fiber	Polypropylene	Glass fiber	Compression molding	[86,87]
Bamboo fiber	Epoxy	Ceramic fillers	Compression molding	[88]
Bamboo fiber	Epoxy polymer	Jute fiber	Hand lay-up	[89]
Bamboo fiber	Epoxy	Flax fiber mat	Hand lay-up	[90]
Bamboo fiber	Epoxy	Sisal fiber	Hand lay-up	[91]
Bamboo fiber	Epoxy	Cotton yarn	Compression molding	[92]
Bamboo leaf fiber ash	Aluminium metal matrix	Rice husk ash	Hand lay-up	[93]
Bamboo fiber	Epoxy	Kenaf fiber	Hand lay-up	[66]
		Jute fiber (JF)		
Alkali treated jute fiber	Vinyl ester resin		Compression molding	[94]
Jute fiber	Epoxy	Carbon fiber	Hand lay-up	[95]
Jute fiber	Epoxy polymer	Bamboo fiber	Hand lay-up	[89]
Woven jute	Polyester	Glass fabric	Hand lay-up	[96]
Woven jute	Vinyl ester	Ramie fiber	Hand lay-up	[97]
Jute fiber	Epoxy resin	Glass fiber	Resin infusion	[98]
Jute fiber	Polyester	Glass fiber	Injection molding	[99]
Jute fiber		Hemp/Flax fiber	Hand lay-up	[100]
Jute fiber	Polyester	Cotton woven fabric	Hand lay-up	[101]
Jute fiber	Polyester	Woven fabric basalt fiber	Compression molding	[102]
		Hemp fiber (HF)		
Alkaline-treated hemp fiber	Polyester resin	Carbon fiber	Hand lay-up	[103]
Hemp fiber mat	Green epoxy	Sisal fiber	Hand lay-up method and hot press	[104]
Hemp fiber	Unsaturated polyester	Soybean oil/nanoclay	Compression molding	[105]
Hemp fiber	Polylactic acid	Sisal fiber	Injection molding	[106]
Hemp fiber	HDPE	Basalt fiber	Injection molding	[107]
Interwoven hemp fiber	PET		Vacuum infusion	[108]
		Flax fiber (FF)		
Flax fiber	Epoxy	Hemp fiber	Hand lay-up	[100]
Flax fiber	Epoxy	Jute/hemp fiber	Hand lay-up	[100]
Flax fiber	Vinyl ester	Glass fiber	Resin infusion	[109]
Short flax fiber	Polypropylene		Injection molding	[110]
Flax fiber	Polypropylene	Kenaf/hemp fiber	Compression molding	[111]
Flax fiber	Polylactic acid	Kenaf/hemp fiber	Compression molding	[111]
Flax fiber	Epoxy		Vacuum infusion	[112]
Flax fiber	Vinyl ester	Basalt fiber	Vacuum assisted resin infusion	[113]
Flax fiber	Epoxy resin	Glass fiber	Compression molding	[114]
Flax fiber	Vinyl ester	Glass fiber	Resin infusion	[109]
Flax fiber	Barium sulphate	Woven aloevera	Compression molding	[115]
		Ramie fiber (RF)		
Ramie fiber	Polylactic acid	Poly(ε-caprolactone)	Compression molding	[116]
Ramie fiber	PVA	Glass fiber	Compression molding	[117]
Ramie fiber	Vinyl ester	Jute fiber	Hand lay-up	[97]
Ramie woven	Epoxy		Hand lay-up	[118]
Ramie cloth	Unsaturated polyester resin		Resin casting	[119]

Table 2. *Cont.*

Natural Fiber	Matrix	Hybrids	Process	Ref.
Abaca/banana fiber (ABF)				
Abaca/banana fiber	Polypropylene		Mixer-injection, mixer compression, and direct compression moldings	[120]
Abaca fiber	Cement	Silica		[121]
Enzyme modified abaca fiber	Polypropylene		Injection molding	[122]
Abaca fiber	Polyethylene	Banana fiber	Rotational molding	[123]
Banana fiber	Low density polyethylene		Compression molding	[124]
Abaca fiber	Polystyrene		Compression molding	[125]
Plain weave abaca fiber	Polyester resin		Hand lay-up	[126]
Banana fiber	Polyvinyl alcohol resin		Hand lay-up	[127]
Sisal fiber (SF)				
Sisal fiber	Phenolic resin	Aramid fiber	Compression molding	[128]
Sisal fiber	Bioepoxy	Hemp fiber	Hand lay-up	[104]
Sisal fiber	Polyester	Bamboo fiber	Hand lay-up	[91]
Sisal fiber	PLA	Banana fiber	Injection molding	[129]
Sisal fiber	Unsaturated polyester	Carbon fibers	Hand lay-up	[130]
Sisal fiber	Waste carbon	Glass fiber	Single extrusion and press consolidation	[131]
Sisal fiber	Epoxy	Jute fiber	Hand lay-up	[132]

3.1. Sugar Palm Fiber Reinforced Hybrid Composites

Arenga Pinnata (also known as sugar palm) is a versatile palm species with wide applications in foods and beverages [133], timber commodities [134], biofibers [135–141], biopolymers [142,143] and biocomposites [144–153]. Sugar palm fibers are recognized for their great durability, as well as their resistance to seawater. Sugar palm fibers have been used to produce ropes for ship cordages that have confirmed the good performance in saltwater [154]. Via the hand-lay-up technique, Misri et al. [28] manufactured a small boat using innovative material, a hybrid of sugar palm fiber and fiberglass-reinforced unsaturated polyester. The mechanical properties of the hybrid boat were investigated via the tensile and impact tests and were found the increased impact strength of 2.471 kJ/m^2 and tensile modulus of 1840.6 MPa. Sanyang et al. [155] reported that the sugar palm fiber demonstrated a lower density than the commercial E-glass fiber of 1.22–1.26 kg/m^3 and 2.55 kg/m^3, respectively. This consequently resulted in the weight reduction of the manufactured boat by 50%. Recently, sugar palm fiber has been investigated as a hybrid reinforcement [154,156–159]. Certain precautions must be considered in the development of these novel natural fiber composites in terms of applicability. For instance, critical assessment and characterization of these composites for practical use in more comprehensive applications. Figure 3 presents the schematic diagram of layout segmentation and reinforcement layout of sugar palm/glass fiber designed by Nurazzi et al. [139] The results revealed improvements in thermal stability, char residue, as well as decomposition temperature as the glass fibers and sugar palm ratios, were raised to 50/50 for both 30 wt.% and 40 wt.% of fiber loadings.

Afzaluddin et al. [160] investigated the influence of the different treatments with 2% silane (TSSP), 6% alkaline (TNSP), and a combination of 6% alkaline–2% silane (TNSSP) on the thermal and physical characteristics of sugar palm/glass/thermoplastic polyurethane hybrid composites. The findings showed that the combined alkaline–silane-treated hybrid composites (TNSSP) displayed the minimum water absorption, thickness swelling, and density as with other hybrid composites. Besides this, good thermal stability was observed in the treated sugar palm fiber-based composites compared to the untreated ones. It is suggested that treated sugar palm/glass/thermoplastic polyurethane hybrid composites can fit automotive component applications. The results of this research are aligned with other studies conducted on the treated sugar palm fiber-reinforced polymer hybrid composites [33,35,36,161].

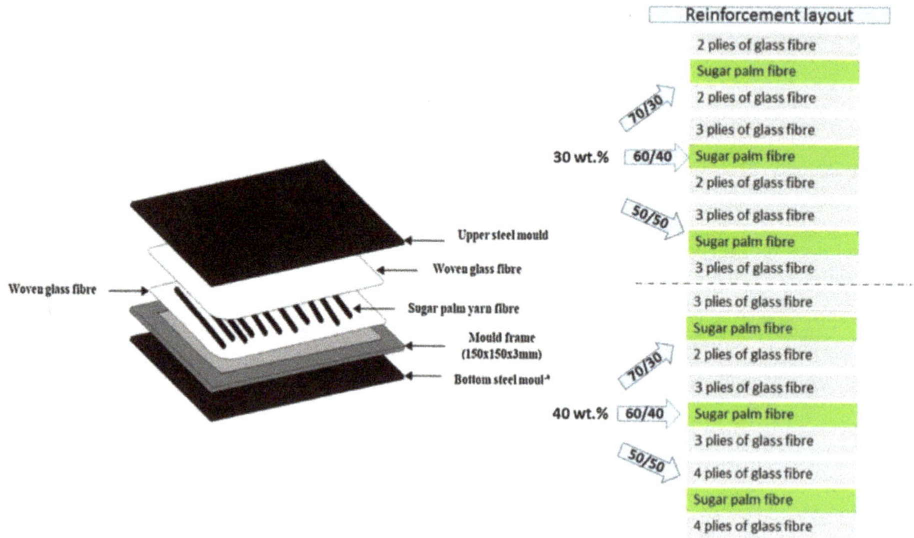

Figure 3. Schematic diagram of layup segmentation and reinforcement layout [139]. Extracted with permission from Elsevier.

3.2. Kenaf Fiber Reinforced Hybrid Composites

Kenaf (*Hibiscus cannabinus* L.) is among the most common natural fibers used as polymer matrix composite (PMC) reinforcement. It is an annual herbaceous plant that can be cultivated in a variety of climates and grows to more than 3 m in 3 months, even in temperate climates [162]. Davoodi et al. [48] replaced an automobile bumper beam with a hybrid kenaf/glass-reinforced epoxy composite to reduce environmental impact while maintaining the requisite strength.

The development of kenaf-glass (KG) fiber reinforced unsaturated polyester (UPE) hybrid composite was performed by Atiqah et al. [49] via the process of sheet molding compound for structural applications. The ratio of 70:30 (by volume) of UPE and KG fibers in a mat form is used using untreated and treated kenaf fiber. During the mercerisation process, the kenaf fiber was alkaline treated for 3 h using a 6% sodium hydroxide (NaOH) diluted solution. Figure 4 shows the sequence of kenaf and glass fibers and matrix in between a mild steel mold for the fabrication of a hybrid composite. The result demonstrated that the highest tensile, flexural and impact strengths were attained from the treated kenaf containing 15/15 v/v KG fibers reinforced UPE hybrid composite. Besides this, the main fracture mode of composites observed under the scanning electron microscopy fractography was fiber debonding, cracking, and pull-out. Better interfacial bonding between the matrix was found in the kenaf treated 15/15 v/v KG reinforced hybrid composite than with other combinations. The hybridization of natural fibers, particularly synthetic and kenaf fibers, is an excellent method to improve the mechanical characteristics of the fabricated hybrid composite, as reported in many works [51–59,163].

The fabrication of kenaf fiber reinforced polypropylene (PP) sheets into a sheet form have been successfully carried out via thermoforming, where the optimum process is the compression molding that employs a layered sifting of a micro-fine PP powder and chopped kenaf fibers [164]. 30 and 40 wt.% fiber contents provide sufficient reinforcement which improves the PP matrix's strength. The strength of the molded kenaf–PP composites was evidenced to possess better flexural and tensile strengths compared to the strength of other molded natural fiber composites, e.g., coir, kenaf, and sisal reinforced thermoplastics. The economic advantage of using kenaf composites over E-glass and other natural fibers is the possibility to analyze the elastic modulus data. The fabricated kenaf maleated PP composites exhibited a greater modulus/cost and an advanced specific modulus compared to coir, sisal, and E-glass. Therefore, they deliver a choice for substituting the currently used materials with a lower-cost alternative with a higher strength that is also environmentally friendly.

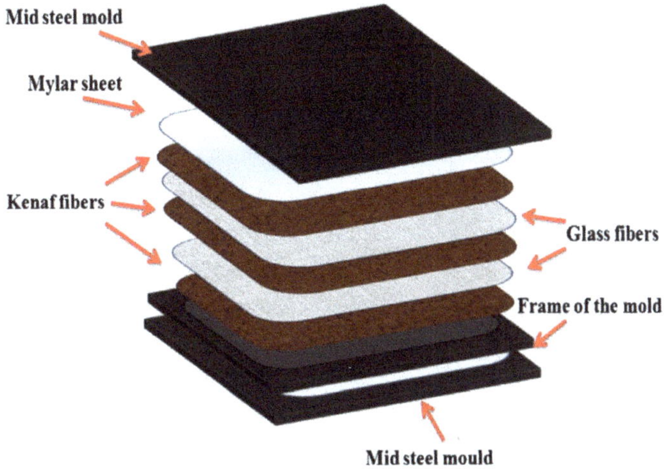

Figure 4. The sequence of kenaf and glass fibers and matrix in between mold for hybridization conducted by Atiqah et al. [48]. Extracted with permission from Elsevier.

The wood flour/kenaf fiber and PP hybrid composites were set to evaluate the hybrid outcome on the properties of the composites [67]. The findings demonstrated that non-hybrid composites (wood flour and kenaf fiber) revealed the lowest moduli compared with the hybrid composites; in addition, moduli of the hybrid composites strictly adhered with the relationship between the fiber reinforcement to wood filler. It was more likely to estimate the elastic modulus of composites using the hybrid mixtures equation rather than with the Halpin–Tsai equation. The influence of natural rubber toughening with polyester resin as the matrix on kenaf fibers were also studied by Bonnia et al. [165].

3.3. Oil Palm Fiber Reinforced Hybrid Composites (OPRPC)

Oil palm, *Elaeis guineensis* consist of two *Arecaceae* or palm family species. Oil palm empty fruit bunch fibers are among potential reinforcement fibers for polymer composites [166,167]. Agarwal et al. [167] examined the stress relaxation behavior in phenol-formaldehyde resin reinforced with oil palm empty fruit bunch fibers and hybrid composites composed of oil palm fibers and glass fibers. The examination of the influences of fiber treatment, loading, strain level, and physical aging on the stress relaxation behavior and the calculation of the rate of relaxation at different time intervals were performed to describe the progressive alterations in the relaxation mechanisms [168].

Suriani et al. [69] introduced the oil palm empty fruit bunch (OPEFB) fiber and $Mg(OH)_2$ into epoxy resin to obtain a hybrid composite, as shown in Figure 5. Four specimens were considered; (1) specimen A (blank, 0% fiber), (2) specimen B (20% fiber), (3) specimen C (35% fiber), and (4) specimen D (50% fiber). The used reinforcing and fire retarding additives were the PET yarn and magnesium hydroxide, respectively. The burning test result exhibited better flammability in specimen B, with the lowest average burning rate of 11.47 mm/min. Specimen A demonstrated the highest tensile strength of 10.79 N/mm^2. An SEM morphological test revealed rising surface defects by the rupture that resulted in the decline of the composites' tensile properties. The authors summarized that the tensile properties and flammability of OPEFB fiber-reinforced fire-retardant epoxy composites weakened with the increments in the fiber volume content at the optimum

20% loading of 11.47 mm/min and 4.29 kPa, respectively. Another study conducted by Farah et al. [70] for the characterization of hybrid epoxy composites containing oil palm empty fruit bunch/woven kenaf fabric reinforcement demonstrated that the increased oil palm fiber content leads to an increase in the impact strength of the hybrid composite. It is described by the other circumstance in which randomly oriented empty fiber bunches (EFB) has a moderate interfacial interaction with epoxy that is vital to attaining a higher impact strength. An investigation into the impact of oil extraction, compounding processes and fiber loading [76], as well as matrix alteration on the mechanical characteristics of oil palm empty fruit bunch filled PP composites was also conducted [71]. Moreover, oil palm empty fruit bunch fiber/PP composites and oil palm-originated cellulose/PP composites were compared [169].

The effect of chemical alteration of the composites containing oil palm/phenol formaldehyde was studied by comparing polyester and epoxy matrices. In addition, the dielectric relaxation and the fiber orientation effect on the dynamic electrical properties of palm tree fiber-reinforced polyester composites were studied [69,170–172].

Figure 5. Extraction of OPEBF, and fabrication of OPEBF/polyester yarn, magnesium hydroxide reinforced epoxy resin hybrid composite. Extracted from [69] with permission.

3.4. Pineapple Leaf Fiber Reinforced Hybrid Composites (PARPC)

Pineapple—*Ananas comosus*—is a tropical plant native to Brazil, with long leaves containing fibers that have a high cellulose content. They are cheap and easily available. In addition, pineapple leaves possess the possibility to be used as a reinforcing agent in polymers. At present, pineapple leaf fibers are the by-products of pineapple farming, making these inexpensive fibers accessible for industrial use, especially for the reinforcement of polycarbonate to manufacture composites [173,174]. The composite fabricated from silane-treated pineapple leaf fibers revealed the most excellent impact and tensile strengths. Thermogravimetric analysis data demonstrated that the composites' thermal stability was poorer than neat polycarbonate resin, which also declined with the rising content of pineapple leaf fiber. The Transient Plane Source (TPS) technique was employed to study the thermal conductivity and diffusivity of phenol-formaldehyde composites reinforced with pineapple leaf fibers [175]. The composites' effective thermal diffusivity and conductivity were found to decrease compared to pure phenol-formaldehyde due to the increment in the fiber loading fraction.

Various efforts to improve pineapple leaf fiber's quality have been carried out via several surface alterations, e.g., alkali treatment, dewaxing, cyanoethylation, and grafting acrylonitrile onto dewaxed fibers [176]. The mechanical characteristics were optimum at 30 wt.% fiber loading. From all surface modifications, 10% acrylonitrile grafted fiber-reinforced polyester composite exhibited a maximum tensile strength of 48.36 MPa. However, cyano-ethylated fiber composites demonstrated a better impact and flexural strengths of 27% and 41% more, respectively, compared to unmodified composite. The effect fiber content and surface treatment were also studied using natural rubber and PP as the matrices [177,178].

Hashim et al. [77] conducted a study using a vacuum infusion technique on the influence of stacking sequence and ply orientation on the mechanical characteristics of pineapple leaf fiber (PALF)/Carbon hybrid laminate composites. The tensile and flexural tests' findings displayed that the laminate with inner carbon plies and ply orientation [0°, 90°] resulted in the maximum tensile strength as well as modulus of 187.67 MPa and 5.23 GPa, respectively. Fracture properties of the composite laminates were investigated using scanning electron microscopy and it was discovered that the failure was started at the weakest fiber layer. This phenomenon might be due to the failure modes, including delamination, debonding, matrix crack, fiber breaking, and fiber pull-out [179–186].

In a work conducted by Sathees Kumar et al. [82], the effects of fiber loading on the mechanical characteristics of reinforced polyester reinforced with sisal and pineapple leaf (PALF) fibers using an injection molding technique were studied, as shown in Figure 6. Figure 6 showed that equal weight % share of PALF and sisal enhanced the overall mechanical attributes, e.g., ductile strength (207 MPa), bending strength (90.3 MPa), impact (29 J/m^2), and hardness (83.7). The mechanical test results revealed a regular trend of an increase in flexural, tensile, impact, and hardness with the addition of PALF fibers, and this was supported by various works [80,81,187,188]. Besides that, they concluded that this type of composite material could be valuable for multiple industries, including automotive and construction fields.

Figure 6. The fabrication process of pineapple leaf (PALF) and sisal fiber reinforced polyester composites using injection molding technique. Extracted from ref. [82] with permission.

3.5. *Bamboo Fiber Reinforced Hybrid Composites (BRPC)*

Bamboo (*Bambusa* Shreb.) is a perennial plant that is able to reach a height of 40 m in monsoon climates. Figure 7 displays the morphological structure of the bamboo fiber [189]. Bamboo is used in carpentry, construction, plaiting, and weaving. Curtains are made from bamboo fiber and absorb various wavelengths of ultraviolet radiation, resulting in less harmful radiation to the human body.

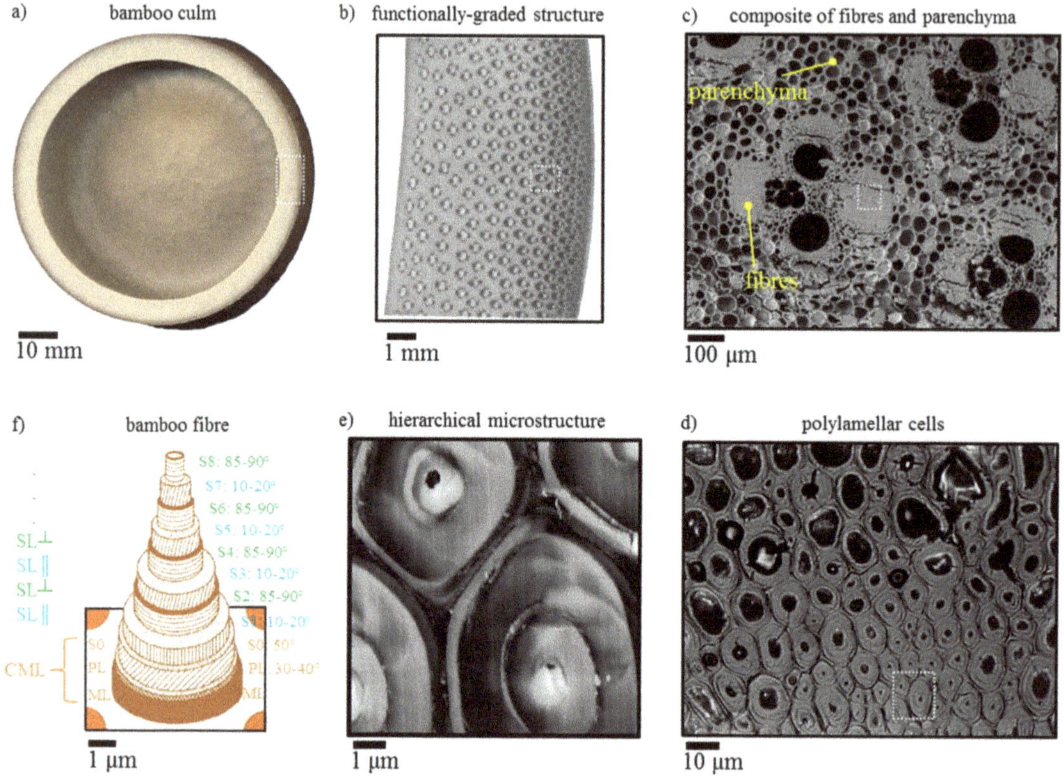

Figure 7. (**a**) Bamboo culm, (**b**) bamboo culm cross-section, (**c**) vascular bundle, (**d**) polylamellar cells, (**e**) microstructure fibers, and (**f**) bamboo's model of polylamellae structure. Extracted with permission from Ref. [189].

Osorio et al. [83] developed a novel mechanical extraction process of long bamboo fibers (*Guadua angustifolia*) for use as a reinforcing agent in structural composites. The effectiveness of the new reinforcement was evaluated by fabricating the composites containing unidirectional bamboo fiber/epoxy (BFC) with alkali-treated and untreated fibers. Two orientations of fiber (transverse and longitudinal) were employed in the flexural tests. When untreated fibers were utilized, the composite's longitudinal flexural strength was greater, whereas treatment increased the longitudinal flexural stiffness. For untreated bamboo in epoxy, the transverse strength rose with the decreasing alkali concentrations, while its three-point bending strength was already extremely high at approximately 33 MPa. They concluded that bamboo fiber offers a natural and renewable alternative to glass fiber and is helpful as traditional natural fiber reinforcement in a variety of applications where glass fiber and conventional natural fibers are already in use.

When preparing bamboo fiber-reinforced composites, characteristics of material and method affect the produced bamboo hybrid composite's quality [190]. A novel composite material fabricated from a right reinforcement material and the matrix combination is able to fulfill a specific application's requirements [191]. The benefits offered by composite materials include their excellent strength, their lightweight, and their moldability. In contrast, polymeric fiber composites have a high raw material cost. Numerous methods of fabrication have been developed to manufacture bamboo-reinforced plastics as well as hybrid composites, e.g., cold and hot presses, and injection molding. These procedures have been used on various bamboo-reinforced polymeric materials to make hybrid composites [84–87,192].

Bamboo fiber, aliphatic polyester, and polyolefin blends are particularly appealing. Blending bamboo fiber with polypropylene (PP) and polylactic acid (PLA) will lead to enhanced chemical, mechanical and thermal characteristics. The materials that result may be turned into products with more convenience and at a lower cost. The development of novel composites using a polypropylene (PP)/polylactic acid (PLA) matrix and filler bamboo fiber (BF) results in modifications in the raw thermoplastic's processability, morphology, and rheological characteristics [193]. Maleic anhydride grafted polypropylene (MAH-g-PP) was used at the filler–matrix interface to increase PP, PLA, and BF interface strength and to improve PLA dispersion and composite toughness. The addition of MAH-g-PP to composites resulted in positive morphological and rheological alterations, which were linked to enhanced PLA dispersion and increased bamboo fiber–matrix interactions.

Glass and bamboo fibers were used to create hybrid composites made of isophthalate polyester and vinyl ester resin. The optimized glass fiber reinforced composites were submitted to dynamical mechanical analysis to evaluate the dynamic characteristics as a function of temperature and frequency with 25, 50, and 75% of glass fibers substituted by bamboo fibers. The storage modulus E' was spotted to drop as the wt.% of bamboo fibers increased. The loss modulus was observed to reduce with loading; however, the damping property increased significantly. Fiber–matrix bonding was visible in scanning electron micrographs of composite flexural fracture surfaces.

3.6. *Jute Fiber Reinforced Hybrid Composites (JRPC)*

Jute is obtained from *Corchorus* genus plants that have about 100 species. At present, jute dominates the highest production volume among bast fibers and is globally available as one of the cheapest natural fibers. Jute is best grown in India, Bangladesh, and China. Figure 8 shows that jute plants are being cultivated for fiber production. Sarkar and Ray [94] studied the alkali-treated jute fiber reinforced with vinyl ester resin using the compression molding technique, as shown in Figure 9. The mechanical, dynamic, thermal, and impact fatigue behavior were compared with the untreated jute fiber–vinyl ester composites. Better fiber dispersion resulted from an extended alkali treatment that eliminated hemicelluloses, hence improving the crystallinity. All properties of mechanical, dynamic, thermal and impact were excellent due to the longer treatment period, concentration, and conditions during the alkali treatment [94].

Jute fiber reinforced hybrid composites have a number of advantages, e.g., a low specific gravity, increased tensile and compressive strength and modulus, and improved fatigue strength [194]. In work conducted by Prasath et al. [102], polyester-based polymer composites were developed by a compression molding technique with different stacking sequences of basalt and jute fabrics into the general-purpose polyester matrix. The result showed that a combination of pure basalt fiber maintained higher values during flexural and tensile tests. However, in the impact test, basalt fiber was somewhat lesser than jute fiber-reinforced composites [102].

Figure 8. Jute plant. Reproduced with permission from Ramesh et al. [195].

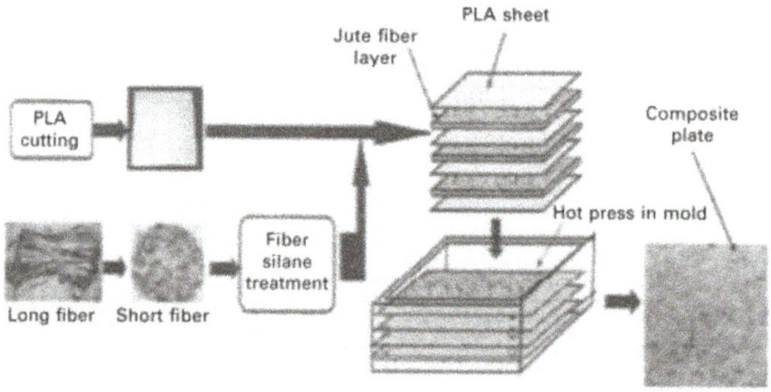

Figure 9. Fabrication procedure of jute fiber reinforced polymer composites [196]. Reproduced with permission.

Ramana and Ramprasad [95] conducted a study on a new hybrid composite developed from jute and carbon fiber reinforced epoxy composite and discussed its superiority or inferiority compared to jute-epoxy and carbon-epoxy composites so that the extent of the utility of the newly developed composite could be established. The hand layup technique was utilized for the composite preparation, and the total fiber content considered was 45%. The newly developed composites, for instance, jute and carbon-epoxy hybrid composites, can replace carbon-epoxy composites without much loss of tensile and flexural strengths as well as a flexural modulus and with improved ductility and impact strength [95].

Mohanty et al. [197] studied the surface modification influence on the biodegradability and mechanical properties of jute/Biopol and jute/PA (Poly Amide) composites. More than 50%, 30%, and 90% in tensile, bending, and impact strengths were found and compared to the values obtained for pure Biopol sheets. In addition, greater than 50% weight loss was observed after 150 days of compost burial of the jute/Biopol composites. The hybridization effects on tensile characteristics of jute–cotton woven fabric reinforced polyester composites were investigated as functions of fiber orientation, content, and texture of roving. Tensile characteristics along the alignment direction of jute roving (transverse to cotton roving alignment) rose continuously with fiber content until 50% before showing a tendency to decline. The composites' tensile strength value at 50% fiber content parallel to the jute roving was approximately 220% greater than pure polyester resin [101].

The evaluations conducted on jute fiber reinforced PP composites include a matrix modification effect, gamma radiation influence, interfacial adhesion effect on creep and dynamic mechanical behaviors, silane coupling agent influence, and natural rubber effect [198–201]. The jute/plastic composites properties were studied, comprising crystallinity, thermal stability, transesterification, modification, durability, weathering, eco-design of automotive components, fiber orientation on frictional and wear behaviors, and alkylation [198,202–207]. Jute fiber reinforced composites used polyester resin matrix, and examinations were carried out on the water absorption and dielectric behavior relationship [208], properties of elasticity, fracture criteria and notched strength [96], characterization of impact damage [209], thermal behavior and weathering [210], and silane treatment effect [211].

3.7. Hemp Fiber Reinforced Hybrid Composites (HRPC)

Hemp is another renowned bast fiber crop, an annual plant in the Cannabis family that cultivates in temperate climates. As a European Union subsidy for non-food agriculture, many current initiatives are progressing for its development in Europe. PP composites with hemp fibers were functionalized by the reactions of melt grafting using glycidyl methacrylate (GMA) and were manufactured via batch mixing [212]. The fibers and PP matrix modifications and various compatibilizer additions were conducted to enhance the interactions of the fiber–matrix. In comparison with the unaltered composite, chemical bonding between the fiber and the polymer (PP/Hemp) resulted in improved fiber distribution in the PP matrix as well as higher interfacial adhesion in the modified composite. Matrix and fiber modifications highly influenced the phase behavior and thermal stability of the composites. The alterations in the crystallization behavior and spherulitic morphology of PP in the composites were analyzed due to the hemp fibers' nucleating effect. Additionally, with increasing modified hemp content, a significant rise in the PP isothermal crystallization rate (120–138 °C) was observed. All composites demonstrated a higher tensile modulus (about 2.9 GPa) and lower elongation at break when compared to plain PP. Still, compatibilization with modified PP (10 phr) boosted the stiffness of the composites due to better fiber–matrix interfacial adhesion.

Ramesh et al. [103] fabricated hybrid composites using carbon, alkaline-treated, and untreated hemp fibers and investigated their properties. The hybrid composites possessed maximum tensile, flexural, impact, and shear strengths of 61.4 MPa, 122.4 MPa, 4.2 J/mm^2, and 25.5 MPa, respectively. In addition, from the composites' mechanical properties, the alkaline-treated composites exhibited better performance [103]. Thiagamani [104] fabricated hybrid bio-composites using the green epoxy matrix, reinforced with sisal (S) and hemp (H) fiber mats via the cost-effective hand lay-up method and hot press employing different stacking sequences, as presented in Figure 10. As the stacking sequence was changed, the tensile strength varied slightly, where the intercalated arrangement (HSHS) hybrid composite demonstrated a maximum tensile modulus compared with the other hybrid counterparts. Hybrid composites (SHHS and HSSH) possessed a compressive strength that was 40% more than the other layering configurations, and the HHHH sample had the maximum ILSS of 4.08 MPa [104].

Li and co-workers [213–215] investigated the effects of chelators, white rot fungi, and enzyme treatments towards hemp fiber separation from bundles and enhanced the hemp fibers' interfacial interaction with the PP matrix. The findings indicated that treated fiber composites had a greater interfacial shear strength than untreated fiber composites, a conclusion that was corroborated by a large body of literature [92,174,216–219]. This demonstrates that the white rot fungal treatment increased the interfacial attachment of hemp fiber to PP. Composites made of chelator concentrate treated hemp fibers exhibited the maximum tensile strength, measuring 42 MPa, a 19% improvement above composites made of untreated hemp fiber. Additionally, hemp fiber reinforced PP composites showed fascinating recyclability [220]. The findings demonstrated that despite the high number of reprocessing cycles, the mechanical properties of hemp fiber/PP composites were well maintained. Newtonian viscosity reduced as the number of cycles increased, indicating a decline in chain scissions and molecular weight caused by reprocessing. Another possible explanation for the decrease in viscosity was the shortening of the fibers during reprocessing.

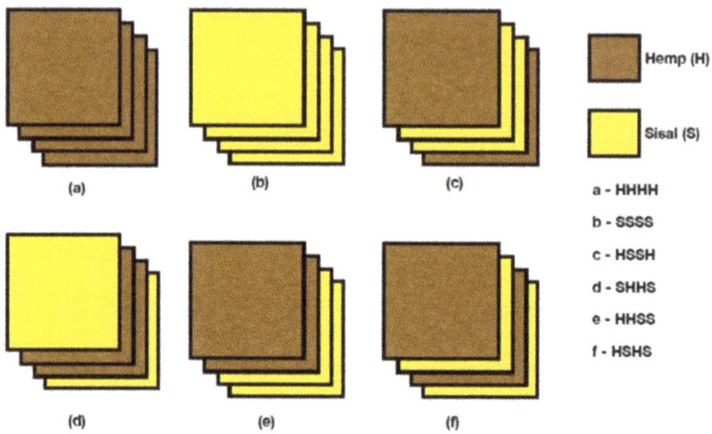

Figure 10. Different layering arrangements of sisal and hemp fiber mats [104]. Reproduced with permission.

In addition, several investigations on the hemp-based composites have also been conducted in terms of their effect on the falling weight impact properties [221], composites' properties and performances for curved pipes [222], impact load performance of resin transfer molded composites [223], composites' micromechanics [224], the influence of soybean oil and nano clay hybrid blends [105], as well as the practicality of untreated hemp as the biocomposites' fiber source [225]. Kunanopparat et al. [226] investigated the viability of using wheat gluten as a hemp fiber-reinforced composite matrix, focusing on the effect of thermal treatment and plasticization on mechanical properties.

3.8. Flax Fiber Reinforced Hybrid Composites (FRPC)

Flax is among the world's oldest fiber crops, containing bast fiber that is cultivated in temperate regions. Flax bast fiber is often utilized for applications in the higher value-added textile industries. Recently, flax has been broadly used in composites. The dynamic and static mechanical characteristics of nonwoven-based flax fiber reinforced PP composites were investigated while taking into account the effect of zein coupling agent, a zein protein [227]. It was discovered that composites containing zein protein as a coupling agent have improved mechanical properties. The composites' storage modulus increased with the addition of a zein coupling agent due to the increased interfacial adhesion. The diameter and position of flax fibers in the stems are used to evaluate their tensile mechanical properties. The substantial dispersion of these attributes is a result of the fiber's longitudinal axis size variation. The increased mechanical qualities of the fibers originating from the stem's center are related to their cell walls' chemical composition. The mechanical characteristics of unidirectional flax fiber/epoxy matrix composites were investigated in terms of their fiber content. The composites' properties were poorer than predicted from the characteristics of a single fiber.

Chaudhary et al. [100] developed and characterized the composites made from natural fibers (hemp/epoxy, jute/epoxy, flax/epoxy) and their hybrid composites (hemp/flax/epoxy, jute/hemp/epoxy, and jute/hemp/flax/epoxy). Among hemp/epoxy, jute/epoxy, and flax/epoxy, a higher hardness (98 Shore-D) and tensile strength (46.2 MPa) was shown by flax/epoxy composite. In contrast, better impact and flexural strengths were exhibited by jute/epoxy (7.68 kJ/m^2) and hemp/epoxy (85.59 MPa) composites, respectively. In general, hybrid composites exhibited better mechanical performance. For example, jute/hemp/flax/epoxy hybrid composite demonstrated the highest tensile modulus strength and sn impact strength of 1.88 GPa, 58.59 MPa, and 10.19, kJ/m^2, respectively. In contrast, the flexural strength of jute/hemp/epoxy hybrid composite was maximum,

86.6 MPa [100]. A similar trend has been shown by Fiore et al. [228], fabricating jute-basalt reinforced hybrid composites via the hand-lay-up method, as presented in Figure 11, for structural applications.

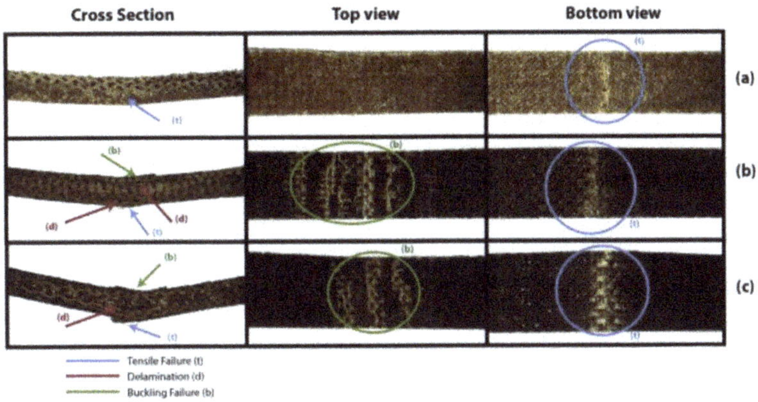

Figure 11. Jute-basalt reinforced hybrid composites including (**a**) jute fibre, (**b**) intercalate and (**c**) sandwich laminates [228]. Reproduced with permission, Elsevier.

Paturel and Dhakal [109] studied the moisture absorption influence on flax and flax/glass hybrid laminates to investigate their low-velocity impact behavior. Three different composite laminates, (1) flax fiber reinforced vinyl ester, (2) flax fiber hybridized glass fiber, and (3) glass fiber reinforced vinyl ester, were manufactured via the resin infusion method. Moisture immersion tests were conducted by immersing various specimens in seawater baths at room temperature and 70 °C at various periods of time. The low velocity falling weight impact test was conducted at a 25 J incident energy level, and the impact damage behavior was analyzed using scanning electron microscopy (SEM) and X-ray microcomputed tomography (micro CT) under both aging circumstances. With glass fiber hybridization, the percentage of moisture taken in by flax vinyl ester specimens was lowered. The maximum weight growth percentages for flax fiber, flax/glass hybrid, and glass fiber reinforced composites immersed in water at room temperature for 696 h were 3.97%, 1.93%, and 0.431%, respectively. When compared to a flax/vinyl ester composite without hybridization, the hybrid composite demonstrated increased load and energy, demonstrating that the hybrid system is a viable technique for improving the structural performance of natural fiber composites. At room temperature, the composites' moisture absorption behavior was found to follow Fickian behavior [109].

Numerous studies on the composites of flax fiber/polypropylene have been conducted. However, these researches concentrated on various variables, natural fiber thermoplastic mat (NMT) and glass fiber thermoplastic mat (GMT) comparison [229], the effect of glass fiber hybridization and fiber/matrix modification [230], the influence of fiber treatment on crystallization and thermal properties [110], surface treatment influence on the interface by thermoplastic starch, glycerol triacetate, boiled flax yarn, and -methacryl oxypropyl trimethoxy-silane [231], matrices comparison (PP and PLA) on the properties of composites [111], material and processing parameters effects [232], and processing methods influence [233]. Buttlar [234] reported the viability of flax fiber composite applications in the bus and automotive industries.

The bio-technical fiber modification effects are ascribed with: (i) toughness and fracture behavior, (ii) alkaline fiber treatment influence on unidirectional composites, and (iii) processing parameter influence on the successive flax fiber's decortication steps (retting, scotching, and hackling) towards the flax fiber reinforced epoxy composites [112,235–237]. Thermal degradation and fire resistance of flax fiber composites reinforced with polyester resin were studied, as well as the influence of chemical treatments on surface properties and adhesion, and also the influence of chemical treatments on the water absorption and mechanical characteristics [238,239]. Three soybean oil-based resins, methacrylic anhydride modified soybean oil, methacrylated soybean oil, and acetic anhydride modified soybean oil, were also used as matrices for the flax fiber-reinforced biocomposites.

3.9. Ramie Fiber Reinforced Hybrid Composites (RRPC)

Ramie is a plant from the Urticaceae (*Boehmeria* spp.) family that comprises approximately 100 species. The exploitation of ramie is for use as textile fiber with two limiting factors: production regions as well as a need for more considerable pre-treatment than other commercial bast fibers [240–243]. Ramie fiber/sugar palm fiber reinforced epoxy hybrid composites were manufactured using a combination of melt mixing and injection molding techniques as shown in Figure 12 [45]. Numerous ramie fiber/PP composites were manufactured by changing the fiber length, content, and pretreatment technique. Increments in fiber length and content were associated with significant increases in tensile, flexural, and compression strengths. Nonetheless, they negatively affected the elongation behavior and impact strength of composites. The preparation of thermoplastic biodegradable composites containing ramie fibers and a PLA/PCL matrix was carried out via in situ polymerization [116]. The influences of fiber content and length on the impact and tensile strengths of this biodegradable composite reinforced with natural fibers were studied along with the effect of a silane coupling agent towards improving interfacial adhesion. Tensile and impact strengths were found to be highest with the use of a silane coupling agent, ramie fiber length of 5–6 mm, and 45 wt.% fiber content.

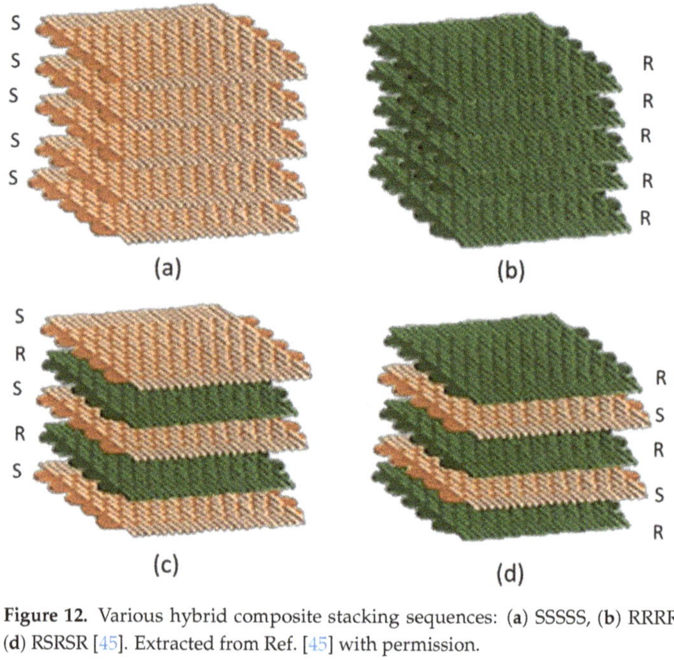

Figure 12. Various hybrid composite stacking sequences: (**a**) SSSSS, (**b**) RRRRR, (**c**) SRSRS, and (**d**) RSRSR [45]. Extracted from Ref. [45] with permission.

When compared to other natural fibers, the use of ramie fibers as reinforcement in hybrid composites is favored due to their superior mechanical qualities. Romanzini et al. [117] investigated the changes in chemical composition and thermal stability of ramie fibers post washing with distilled water. Apart from this, research on glass and washed ramie fiber composites was carried out, with an emphasis on the effects of using different fiber lengths (25, 35, 45, and 55 mm) and the fiber compositions, while the fiber loading was set at 21 vol.%. They reported that composites could be potentially produced from washed ramie fibers. The composite containing fiber length of 45 mm exhibited higher flexural strength despite the insignificant difference observed in lower volume fractions of glass fiber of 0:100 and 25:75. Better impact and flexural properties were also obtained from the increased glass fiber's relative volume fraction up to a limit of 75% [117].

The major problem with employing natural fibers is that they are incompatible with a polymer matrix, which reduces the mechanical performance [244–247]. Tezara et al. [97] investigated the influence of stacking sequences, alkali treatment, and orientations of fiber on the mechanical characteristics of hybrid jute (J) and ramie (R) reinforced vinyl ester (VE) composites. First, woven fibers were made using three- and four-layer stacking sequences with a 0° orientation. A higher tensile strength value of 298.90 MPa was observed from the RJJR stacking sequence fabricated from different fiber orientations, e.g., 0°, 30°, 45°, and 90°. This was done to study the influence of fiber orientation on the flexural and tensile characteristics. 0° fiber orientation possessed significantly flexural and tensile strengths compared with other orientations of 28.90 MPa and 66.81 MPa, correspondingly. Enhancement of mechanical properties was also conducted via 5 wt.% and 10 wt.% alkali treatments, resulting in a maximum flexural strength (34.50%) increment in 0° RJJR with 5 wt.% compared with the untreated RJJR. They concluded that the fiber orientation and a lower alkali treatment concentration (5 wt.%) combination had significantly improved the mechanical characteristics of fiber hybrid composites.

Hand lay-up method employing epoxy as a matrix is used to manufacture bulletproof panels, where the prototype is more lightweight and economical compared to the conventional ones made of steel-based materials, Kevlar/aramid composite, and ceramic plates used in military antiballistic equipment [118]. The findings from bullet testing revealed the panels' ability to resist high-impact projectile (level II) penetration and resulted in minimal fractures. However, level IV ballistic testing demonstrated the failure of all prototype panels to resist the projectile's high-impact velocity. From the tests, ramie fiber has enough breaking strength and toughness to pass level II bullet testing. Among the matrices used to reinforce ramie fiber are included polyester [119], epoxy–bioresin [248], soy protein [249], epoxy [250] and PP [251].

3.10. Abaca/Banana Fiber Reinforced Hybrid Composites (ARPC)

The banana plant produces abaca/banana fiber, the strongest commercially available cellulose fiber, which is strong and seawater-resistant. Abaca is a native of the Philippines, where it is currently grown, as well as in Ecuador, and was then the most chosen rope fiber in marine applications.

Bledzki et al. [120] studied the mechanical characteristics of abaca fiber reinforced PP composites with varying fiber lengths (5, 25, and 40 mm) and compounding procedures (mixer-injection, mixer compression, and direct compression moldings). When the length of the fibers was increased from 5 to 40 mm, the tensile and flexural characteristics were increased slightly, but not significantly. The mixer-injection molding technique outperformed the other two compounding procedures in mechanical performance (tensile strength was roughly 90% greater). The comparison of the composites of abaca fiber PP with the composites of jute and flax fiber PP revealed that the best falling weight impact properties and notched Charpy (Figure 13) were possessed by abaca fiber composites. Figure 14 shows the higher odor concentration of abaca fiber composites compared to flax and jute fiber composites.

The effects of fiber loading, frequency, and temperature on the polarity of banana fiber reinforced polyester composites were studied in a dynamic mechanical analysis [252]. The composites' storage modulus at 40% fiber loading was the greatest, showing that the inclusion of abaca fiber in the polyester matrix resulted in reinforcing effects at higher temperatures. Enhanced fiber and matrix interactions were confirmed by the increased dynamic modulus and low damping values. Abaca fibers were reinforced with the matrices of cement [121], polyurethane [253], aliphatic polyester resin [254], PP [122,255], urea-formaldehyde [256], PE [123,124], polyester [125,126], and polyvinyl alcohol [127] to evaluate the properties of the produced composites.

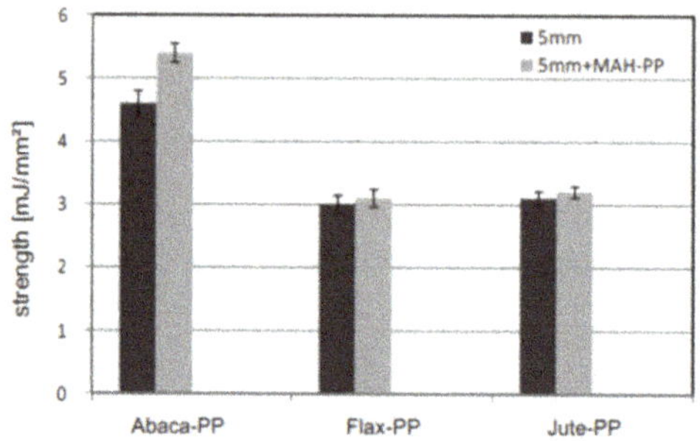

Figure 13. Notch Charpy strength of abaca/jute/flax fiber–PP composites comparison with and without MAH–PP. Reprinted with permission, Elsevier.

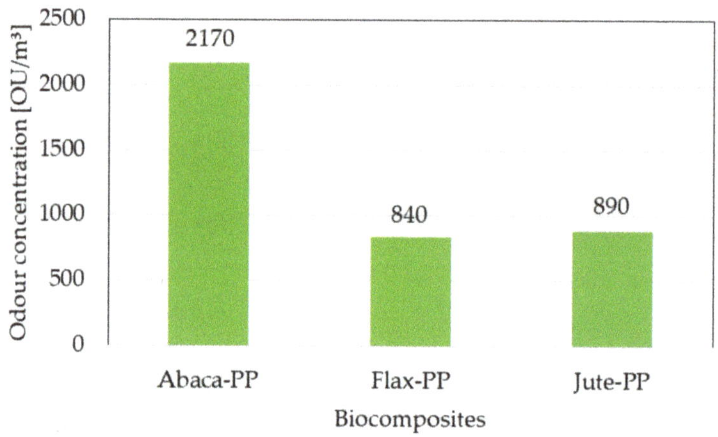

Figure 14. Comparison of odor emission concentration abaca/jute/flax fiber–PP composites. Reprinted with permission, Elsevier.

3.11. *Sisal Fiber Reinforced Hybrid Composites (SRPC)*

Sisal is a type of *agave* (*Agave sisalana*) mostly grown in Brazil and East Africa. Between 1998 and 2010, global demand for sisal fibers and products was predicted to fall by 2.3% each year. Synthetic replacements and harvesting systems adoption that use less or no twine continued to undercut the conventional market for fibers. Sisal fiber will be used to make a wide range of non-structural and structural industrial goods using various polymer matrices.

The composites' mechanical properties are heavily impacted by the bonding between the fiber and matrix, as reported by Senthilkumar et al. [257] and Ilyas et al. [258]. Good interfacial bonding induces transfer of the applied stress by the reinforced polymer composites to fibers. The hydrophilicity and hydrophobicity of the fibers and resin, respectively, usually result in poor bonding of the plant-based fibers that could be overcome via mechanical interlocking, chemical, inter-diffusion and electrostatic bondings, chemical pretreatment, as well as coupling agent [259]. Compression molding (CM), resin transfer molding (RTM), and injection molding are among the common techniques of natural fiber composite fabrication [260–262]. These methods differ from each other in terms of processing temperature, pressure, and speed. Sreekumar [263] studied the mechanical properties of the fabricated sisal fiber polyester composites via resin transfer and compression moldings. The RTM composites demonstrated a higher Young's modulus, tensile and tensile flexural strengths, and flexural modulus. CM composites, on the other hand, possessed more water absorption and voids due to the weaker adhesion of fiber-matrix compared to RTM composites.

Getu et al. [91] reported that composite materials possessed a low density with a high strength to weight ratio, stiffness to weight, strength ratios, and fatigue strength to weight ratio than conventional engineering materials, allowing them to be used in wide structural constructions applications. Lightweight natural fibers produce lightweight composite materials that in automotive applications improve fuel economy by minimizing harmful emissions. As shown in Figure 15, Getu et al. [91] prepared and characterized the performance of sisal and bamboo reinforced polyester hybrid composite (BSFRHC) with various sisal and unidirectional (UD) bamboo fiber orientations. Following that, BSFRHC was developed with a total fiber volume percentage of 20% via hand lay-up method using 3:1 bamboo to sisal fibers ratio prior to compressive, tensile, flexural and impact tests. It was concluded that varying fiber orientation resulted in variation in tensile strength; a higher tensile strength was found in the composite of bamboo/sisal fiber with 0° fiber orientation. The 0° fiber orientation composite demonstrated a higher compressive strength than the 90° fiber orientation composite and the bidirectional (0°/90°) fiber orientation composite. Higher tensile and flexural strengths were observed in the unidirectional 90° fiber orientation, whereas almost similar tensile strengths were obtained from the unidirectional 90° and bidirectional (0°/90°), and bidirectional (0°/90°) possesses higher flexural strength compared to unidirectional 90° fiber orientation. ANSYS Software was used to carry out the impact analysis of BSFRHC based vehicle internal door panel and the potential for the applications of interior automotive parts was revealed from the composites of sisal and bamboo fibers in unidirectional 0°.

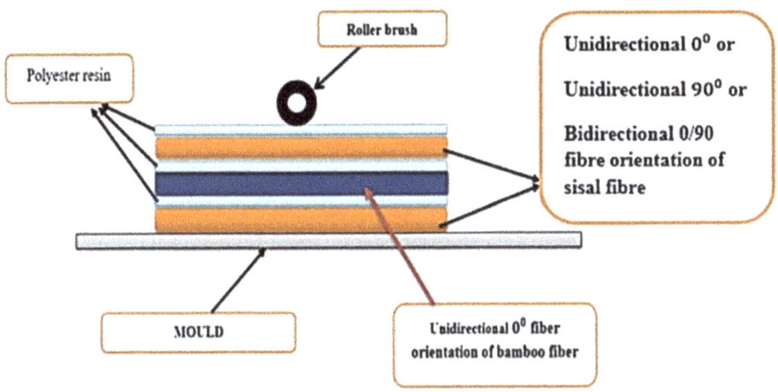

Figure 15. Fabrication of composite with varied orientations of sisal fiber [91]. Reprinted with permission, Elsevier.

Asaithambi et al. [129] conducted a study on the effect of Benzoyl Peroxide (BP) fiber surface treatment towards the mechanical characteristics of banana/sisal fiber (BSF) reinforced PLA composites [129]. BSF underwent BP treatment for the purpose of improving fiber and matrix adhesion. Twin-screw extrusion of BSF (30 wt.%) reinforced PLA (70 wt.%) hybrid composites was performed, followed by injection molding. The findings revealed that treated BSF possessed better bonding and wettability, resulting in the PLA matrix's restricted motion. When comparing the composites of BSF-reinforced PLA with untreated BSF reinforced PLA and virgin PLA, the mechanical characteristics, e.g., flexural and tensile moduli, were improved.

Noorunnisa Khanam et al. [130] investigated the fluctuation of mechanical characteristics, e.g., flexural and tensile properties of the hybrid composites comprising randomly oriented unsaturated polyester-based sisal/carbon fibers varying fiber weight ratios by the hand lay-up approach. These hybrid composites were tested for chemical resistance to different solvents, acids, and alkalis. The influence of treating sisal fibers with NaOH on the tensile, flexural, and chemical resistance characteristics of these sisal/carbon hybrid composites was also investigated. The flexural and tensile characteristics of the hybrid composites were improved with the rising loading of carbon fiber, where the tensile and flexural characteristics of these hybrid composites were found to be superior to those of the matrix. Alkali treatment resulted in significant improvements in the tensile and flexural characteristics of the sisal/carbon hybrid composites. All compounds, excluding carbon tetrachloride, were resistant to these untreated and alkali-treated hybrid composites in chemical resistance tests.

Incorporation of zinc borate and magnesium hydroxide into sisal/PP composites as flame retardants was performed to improve the composites' thermal stability as well as to reduce the composites' burning rate [264]. The same study reported no synergistic effect from incorporating magnesium hydroxide and zinc borate into sisal/PP composites. Furthermore, at high shear rates, the sisal/PP composites showed substantial changes in shear viscosity, showing that the flame retardants utilized in this investigation did not affect the composites' processability. The sisal/PP composites that had flame retardants added to them had tensile and flexural properties comparable to those of the sisal/PP composites without flame retardants.

Environmental impacts of degradation behavior [265], coupling agent influence on abrasive wear qualities, and the ageing effect [266] on mechanical characteristics have all been examined with sisal/PP composites. All plant fiber composites were developed by Zhang et al. [267] by transforming wood flour using a proper benzylation procedure and compounding of both discontinuous and continuous sisal fibers to create composites from renewable resources. The developed sisal/plasticized wood flour composites were found to be fully biodegradable from the degradation tests. The process of decomposition was accelerated by taking into account both lignin and cellulose in the composites. When it comes to practical applications, composites' hydrophobicity and flame resistance are vital; therefore, molecular modification and/or integration of inorganic additives are appropriate approaches as long as the composite's biodegradability is not compromised.

Many studies were performed on the composites of sisal fiber reinforced polyester concerning their characteristics of moisture absorption [268], as well as treatment of fiber with admicellar [269]. A few investigations were conducted on the composites of sisal fiber-reinforced phenolic resin, e.g., chemical alteration of such with lignins [270], hydroxyl-terminated polybutadiene rubber modification [271], cure cycles effect [272], employing glyoxal from natural resources [272], and alkali treatment effect [273]. Nevertheless, epoxy resin was employed as a matrix for sisal fiber-reinforced composites, and the effects of fiber orientation on electrical characteristics [274] and reinforcing degree [275] were investigated. A different matrix (cement) was also used in the sisal fiber-reinforced composites to study their cracking micro-mechanisms [276] and the influence of accelerated carbonation on cementitious roofing [277].

Towo et al. [278] prepared composites using treated sisal fibers with epoxy and polyester resin matrices. Dynamic thermal analysis and fatigue evaluation tests were conducted on the produced composites and revealed better mechanical characteristics in alkali-treated fiber bundle composites than untreated fiber bundle composites. The polyester resin matrices were most affected by alkali treatment, where improvements in the composites' fatigue lives were found for the alkali-treated sisal fiber bundles. The superiority of alkali-treated fiber composites was analyzed and was associated with low cycle fatigue. Epoxy matrix composites possessed a longer fatigue life than polyester matrix composites. The chemical treatment had significantly and positively impacted the fatigue life of polyester matrix composites; however, it demonstrated a lesser effect on epoxy matrix composites. Studies on sisal fibers were also conducted with other matrices, e.g., rubber [279], phenol-formaldehyde [256], cellulose acetate [280], bio polyurethane [281], and polyethylene [282] in terms of their morphological, mechanical, cure, and chemical properties.

4. Mechanical Properties of NF Reinforcement Hybrid Composites

Researchers have been focusing their research interests on composites of natural fibers, e.g., biocomposites made of natural or synthetic resins reinforced by natural fibers. Natural fibers have numerous advantages, including their low density, which results in comparatively lightweight composites having excellent specific properties [244,283,284]. Additionally, these fibers offer significant cost savings and ease of processing, as well as being a highly renewable resource, thereby reducing reliance on domestic and foreign petroleum oil. Researchers have reviewed recent advances in natural fiber (e.g., flax, hemp, jute, kenaf, straw, bamboo, and coir) applications in composites [8,285,286]. Nilza et al. [287] designed and manufactured composites from three Jamaican natural cellulosic fiber: sugar cane bagasse, banana trunk, and coconut husk coir. The prepared samples were tested for carbon and ash contents, moisture content, water absorption, elemental and chemical analyses, and tensile strength.

4.1. Tensile Properties

Natural fiber-reinforced composites possess similar mechanical characteristics to synthetic fibers, as reported by Van De Velde and Kiekens [288] for hemp, flax, sisal, and jute fibers, in terms of strength and modulus compared to glass fiber. Srinivasan et al. [289] researched the ultimate tensile strength of the composites of glass fiber and banana/flax reinforced polymer (GFRP). A higher ultimate tensile strength (39 N/mm^2) was observed in the flax banana-GFRP hybrid composite compared to the banana-GFRP and flax-GFRP composites with 30 N/mm^2 and 32 N/mm^2, respectively. Paul et al. revealed the mechanical characteristics of the composites of kenaf reinforced polypropylene showing improvements in ultimate tensile stress and tensile modulus with a rising fraction of fiber weight [23]. Table 3 displays the tensile properties comparison of different natural fibers with synthetic fibers.

Table 3. Tensile properties of natural and inorganic fibers.

Fibers	Density (kg/m^3)	Diameter (μm)	Tensile Strength (MPa)	Tensile Modulus (GPa)	% Elongation	Ref.
Sugar Palm	1290	99–311	190.29	3.69	19.6	[157]
Jute	1460	-	393–800	10–30	1.5–1.8	[290]
Sisal	1450	50–300	227–400	9–20	2–14	[290]
Kenaf	1400	81	250	4.3	-	[290]
Flax	1500	-	345–1500	27.6–80	1.2–2.3	[291]
Hemp	1480	-	550–900	70	1.6	[291]
Banana	1350	80–250	529–759	8.20	1–3.5	[292]
Coir	1150	100–460	108–252	4–6	15–40	[292]
Bamboo	910	-	503	35.91	1.4	[292]
Cotton	1600	-	287–597	5.5–12.6	3–10	[293]
E-glass	2550	<17	3400	73	3.4	[294]
S-glass	2500	-	4580	85	4.6	[294]
Carbon (Std. PAN-based)	1400	-	4000	230–240	1.4–1.8	[294]

4.2. Flexural Properties

The potential of composite materials' use in structural applications is determined via a few parameters. Major strength is the flexural properties that include flexural strength, modulus, and load as well as deflection at the break. Flexural strength is related to the fiber content/fiber length, as reported in a few studies. Satyanarayana et al. [293] demonstrated that improvements in toughness and ductility of bamboo-mesh reinforced cement composites as well as significant enhancements in the tensile, flexural, and impact strengths. Banana and glass fibers were fabricated at different fiber lengths and loadings in the phenol-formaldehyde composites, and the mechanical properties were compared. From the composites of flexural property analysis, the optimum length of fiber needed for banana and glass fibers was different from the phenol-formaldehyde resole matrix reinforcement [295]. Aziz et al. observed the influence of random and long kenaf and hemp fibers alkalization and alignment on the formed composite fabricated via a combination of the fibers with polyester resin hot-pressed [296]. Long and alkalized fiber composites exhibited higher flexural strength and modulus compared with the as-received fiber composites. The decline in the flexural properties was due to water absorption that damaged and degraded fiber-matrix interfacial bonding; however, the maximum strain was simultaneously increased [297–299].

Shibata and team reported that the densified structure of kenaf fibers contributes to their composites' higher flexural strength compared to the porous bagasse fibers [300]. Other researchers studied the effect of hybridizing the composites of jute/glass-reinforced epoxy on their mechanical properties. The E-glass fabric layers added to the composites' outer layers revealed improvements in the properties of bending, tensile, and impact of the jute-reinforced composites [98]. A summary of specific moduli of natural and glass fibers is presented in Figure 16.

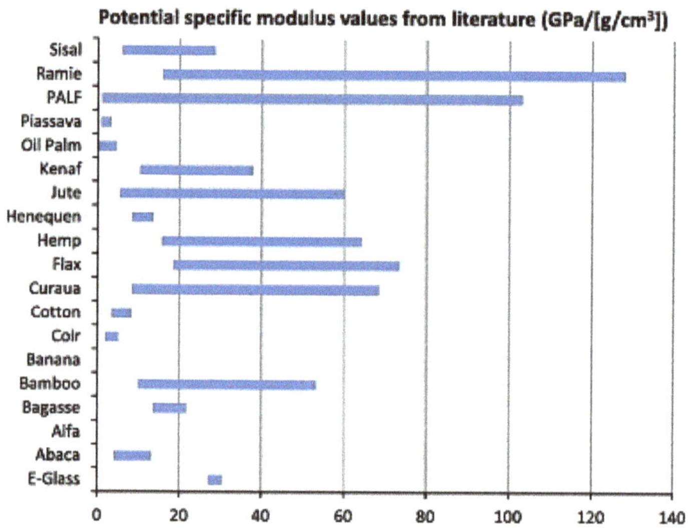

Figure 16. Comparison of values and ranges of potential specific modulus between natural and glass fibers [299]. Extracted with permission from Elsevier.

4.3. Impact Properties

Pothan et al. [301] examined the composites prepared from short banana fiber reinforced polyester, with the aim of studying the influence of fiber lengths and content on the composites' impact strength. A 40 mm fiber length yielded the highest impact strength, while 40% incorporation of untreated fibers resulted in a 34% improvement of impact strength. Another study on the impact behavior of 35% jute/vinyl ester composites reinforced with alkali-treated and untreated fibers revealed hemicellulose removal, improving the crystallinity and, consequently, better fiber distribution [302]. Sanjay et al. [303] compared different compositions of laminates to investigate the impact behavior of the composites' banana/E-glass fabrics reinforced polyester hybrid and found 6 J impact strength in the hybrid laminate, which was the highest value attained.

4.4. Hardness Properties

Zampaloni et al. [304] discussed the excellent potential of the current materials substitute by the Kenaf–maleated polypropylene composites that demonstrated more efficient modulus/cost as well as better specific strength and modulus at a cheaper cost compared to the materials reinforced with E-glass, coir, and sisal (Figure 17). The hardness of various laminates fabricated from hybrid composites of banana/E-glass fabrics reinforced polyester utilizing different stacking sequences was measured by Sanjay et al. [303] Laminate L1, or the composite of pure glass fiber, possessed the hardness of 26.72 HV, while laminate L2 (composites of pure banana fiber) exhibited the poorest hardness of 12.36 HV.

Figure 17. Kenaf–PP/MAPP composite hybrid composites tested for hardness properties [304]. Extracted with permission.

5. Current Application on NF Reinforcement Hybrid Composites

Due to its low manufacturing cost, max strength ratio, and simple manufacturing process, hybrid natural fiber composites have already been widely extensively utilized in numerous textile and engineering applications. Furthermore, natural fiber composites demonstrated a good combination of mechanical qualities for aerospace and automotive applications, including its enhanced impact strength, tensile, bending and compressive behavior, as well as improved fatigue properties. Bio-based hybrid composites are a rapidly increasing product in the industrial sectors as a means of reducing environmental effects in today's society.

5.1. Automotive

The automotive industry demands composite materials in order to comply with new regulations and to remain competitive. At present, plant fibers are used in the exterior and interior components in semi- or non-structural applications, fulfilling the performance standards, e.g., elongation, ultimate breaking force, impact strength, flexural properties, flammability, fogging characteristics, acoustic absorption, odor, dimensional stability, aptness for processing dwell time and temperature, crash resistance, and water absorption. A few renowned automakers, e.g., Volkswagen-Audi, Daimler-Chrysler, and Opel-GM, have already begun incorporating natural fiber composites into their passenger car parts, including rear parcel shelf, door trim panels, and seat squabs [305,306]. Table 4 shows the present use of natural fiber in the automobile sector by the big automobile companies.

The prospects for a lightweight design from plant fiber composites are demonstrated, for instance, in Mercedes E-Class's panels and even external underbody panels and Volkswagen's door structures (phenol-formaldehyde/flax composite) [307]. Given the unique qualities of natural fiber composites, an approximately 15% weight reduction in components is feasible compared to glass fiber reinforced composites.

Table 4. Lists of natural fibers reinforced hybrid composites currently being used in the automotive industry. Extracted from Ref. [308] with permission.

Automobile Model	Natural Fiber Utilized	Applications
Audi A2	Flax, sisal fibers and polyurethane	Door trim panels
BMW 7 Series	Sisal fiber	Door trim panels
Chevrolet Impala	Flax fiber and polypropylene	Rear shelf compartment
Ford Focus and Fiesta	Kenaf fiber and sun	Interior door panels
Honda Pilot	Wood fiber	Floor area parts
Mercedes Benz A, C, E and S-class	Flax, hemp, sisal, cotton, abaca and jute fibers	Underbody panels, seat back rests, engine and transmission cover and rear panel shelves
Toyota Prius and Raum	Corn biopolymer, starch and kenaf fiber	Instrument panels, sun visors, ceiling surface skins and spare tire cover

5.2. Aerospace

During the early stages of the aircraft industry's development, aircraft structures were invented using wire, wood (natural composite), and fabric compositions. Aluminum alloys have been the dominating material in the aerospace industry since the 1930s. The newest components of civil aircraft are made of natural fibers that are also used as a substructure in conjunction with fibers containing composites and other synthetic fibers, e.g., glass, carbon, and Kevlar. The design of V22 Osprey tiltrotor's wings is extremely rigid and risky, and is most likely to be constructed using fiber composites with low-density materials. In defense aircraft, a fascinating advancement called "stealth" has emerged, which requires the designer to achieve the smallest possible radar cross-section (RCS) by reducing the potentials of early detection via defending radar sets. Constant radius changes are required to create the airframe's essential compound curvatures, which are much easier to build using composites compared to metal and radar-absorbent material (RAM).

5.3. Oil and Gas

Hybrid natural fiber-reinforced composites of natural fiber-reinforced composites have been found to have less critical environmental impacts than glass fiber reinforced composites in some applications [309]. Natural fibers have been used with glass fiber for underground pipes; this application is faster and offers adequate strength. However, certain issues, e.g., water absorption and strength, have yet to be studied [310].

5.4. Maritime

According to Moreau et al. [311], fiber-reinforced plastic (FRP) structures in boat construction uses primarily thermosetting resins (e.g., vinyl ester, epoxy, polyester, etc.) Only lately have thermoplastic resins (polypropylene, polyamide, PBT, PET, etc.) started being applied in fittings or boat-building. In recent times, the resin structure has evolved in 2 forms: both low styrene content and emission are commercially available despite the evolution of bio-based resins; however, conventional resins are still relevant in the nautical area [174].

The environmental benefits of adhesives and bio-based resins are found in their elimination of toxins in common, their emphasis on human health and the environment, their reduction of hazardous as well as toxic materials and waste, recycling capabilities, and the decline of polluting air emissions. Natural fibers are also gaining popularity in the composites sector. However, their application in structural components is limited due to their generally poor physical characteristics, whereas at present, they are applied in filling functions. Glass fibers provide for 89% of the fiber's capacity used in composites on a global scale. In contrast, natural fibers account for just 10%. Simultaneously, several R&D studies of natural fiber uses are underway, resulting in more industrial uses in the foreseeable future.

Glass fibers were accessible during World War II, shortly after polyester resins were created, as a result of the accidental discovery of a manufacturing method involving blown air on a molten glass stream. Soon after, glass-reinforced plastic became popular, and in the early 1950s, GRP boats became accessible. Fiberglass boats are significantly more appealing due to their strength, including their high vibration damping ability, lightweight, low construction costs, high impact resistance, ease of fabrication, maintenance, and repair.

Fiberglass boat construction has evolved over the years to include a variety of means with the sole objective of an improvement in boat construction skills and techniques to fulfill the aims of fiberglass boat building, e.g., to produce lightweight products, corrosion and impact resistance, vibration damping, low cost, and ease of construction. Now, hybrid natural fiber (NF) composites are broadly utilized in a variety of technical applications that comprise marine applications, particularly in boat construction.

5.5. Textiles

Pineapple leaves (PAL) have been utilized as threads and fabrics in a variety of nations for centuries. Excellent pina fibers are derived from pineapple leaves in the Philippines and are used to create textiles for table linens, dresses, mats, bags, and other apparel items. Applications of PAL in the textile industry are established in Indonesia, while in Malaysia, the efforts in PAL employment in Malaysia are still in their infancy. The PAL dresses, however, are recognized to be expensive, which is understandable because of the tedious procedures involved.

6. Estimated Costing for NF as Reinforcement in Hybrid Composites

Excellent specific properties and cheap natural fiber composites are the primary factors of their attractiveness for wide applications, as reported by Sanjay et al. [303] and Ho et al. [292] The price lists of NF, synthetic fibers, and matrices in the US and Malaysia currencies, USD and MYR, respectively, are tabulated in Table 5. Table 6 tabulated the estimated cost for hybrid NF/synthetic fibers and matrices for numerous applications.

Table 5. Estimated cost of natural, synthetic fiber and matrices in US Dollar (USD) and Malaysia Ringgit (MYR). Extracted from Ref. [292] with permission.

No.	Types of Natural Fiber	Cost (Money per Tons)	
		USD	**MYR**
1.	Bamboo	500	2092.80
2.	Banana	890	3725.18
3.	Flax	3150	13,184.64
4.	Hemp	1550	6487.68
5.	Jute	950	3976.32
6.	Kenaf	400	1674.24
7.	Pineapple	455	1904.45
8.	Sisal	650	2720.64
9.	Sugar palm	4000	16,742.40
No.	**Types of Synthetic Fiber**	**Cost (Money per Tons)**	
		USD	**MYR**
1.	Carbon	12,500	52,320.00
2.	Kevlar (aramid)	20,000	83,712.00
3.	Fiber Glass	980	4101.89
4.	Glass	1500	6278.40
No.	**Types of Matrices**	**Cost (Money per Tons)**	
		USD	**MYR**
1.	Epoxy	2650	11,091.84
2.	Polyester	550	2302.08
3.	Vinyl ester	1550	6487.68
4.	Polyurethane	2750	11,510.40

Currency United States Dollar (USD) to Malaysian Ringgit (MYR) on date 8 June 2021.

Table 6. Estimated costing for NF as reinforcement in hybrid composites (in US Dollar/USD and Malaysia Ringgit/MYR) [303].

No.	Reinforced Composite			Estimated Cost (Money per Tons)	
	Natural Fiber	Synthetic Fiber	Matrices	USD	MYR
1.	Bamboo	Carbon	Epoxy	15,650	65,504.64
2.	Bamboo	Carbon	Polyester	15,700	65,713.92
3.	Bamboo	Carbon	Vinyl ester	16,700	69,899.52
4.	Bamboo	Carbon	Polyurethane	17,900	74,922.24
5.	Bamboo	Kevlar	Epoxy	25,300	105,895.68
6.	Bamboo	Kevlar	Polyester	21,050	88,106.88
7.	Bamboo	Kevlar	Vinyl ester	22,100	92,501.76
8.	Bamboo	Kevlar	Polyurethane	23,250	97,315.20
9.	Bamboo	Fiber Glass	Epoxy	4130	17,286.53
10.	Bamboo	Fiber Glass	Polyester	2030	8511.59
11.	Bamboo	Fiber Glass	Vinyl ester	3030	12,704.49
12.	Bamboo	Fiber Glass	Polyurethane	4230	17,735.97
13.	Bamboo	Glass	Epoxy	4650	19,496.99
14.	Bamboo	Glass	Polyester	2550	10,691.90
15.	Bamboo	Glass	Vinyl ester	3550	14,884.80
16.	Bamboo	Glass	Polyurethane	4750	19,916.27
17.	Banana	Carbon	Epoxy	16,040	67,254.12
18.	Banana	Carbon	Polyester	13,940	58,449.03
19.	Banana	Carbon	Vinyl ester	14,940	62,641.93
20.	Banana	Carbon	Polyurethane	16,140	67,673.41
21.	Banana	Kevlar	Epoxy	23,540	98,700.87
22.	Banana	Kevlar	Polyester	21,440	89,895.78
23.	Banana	Kevlar	Vinyl ester	22,440	94,088.68
24.	Banana	Kevlar	Polyurethane	23,640	99,120.16
25.	Banana	Fiber Glass	Epoxy	4520	18,951.91
26.	Banana	Fiber Glass	Polyester	2420	10,146.82
27.	Banana	Fiber Glass	Vinyl ester	3420	14,339.72
28.	Banana	Fiber Glass	Polyurethane	4620	19,371.20
29.	Banana	Glass	Epoxy	5040	21,132.22
30.	Banana	Glass	Polyester	2940	12,327.13
31.	Banana	Glass	Vinyl ester	3940	16,520.03
32.	Banana	Glass	Polyurethane	5140	21,551.51
33.	Flax	Carbon	Epoxy	18,300	76,730.07
34.	Flax	Carbon	Polyester	16,200	67,924.98
35.	Flax	Carbon	Vinyl ester	17,200	72,117.88
36.	Flax	Carbon	Polyurethane	18,400	77,149.36
37.	Flax	Kevlar	Epoxy	25,800	108,176.82
38.	Flax	Kevlar	Polyester	23,700	99,371.73
39.	Flax	Kevlar	Vinyl ester	24,700	103,564.63
40.	Flax	Kevlar	Polyurethane	25,900	108,596.11
41.	Flax	Fiber Glass	Epoxy	6780	28,427.86
42.	Flax	Fiber Glass	Polyester	4680	19,622.77
43.	Flax	Fiber Glass	Vinyl ester	5680	23,815.67
44.	Flax	Fiber Glass	Polyurethane	6880	28,847.15
45.	Flax	Glass	Epoxy	7300	30,608.17
46.	Flax	Glass	Polyester	5200	21,803.08
47.	Flax	Glass	Vinyl ester	6200	25,995.98
48.	Flax	Glass	Polyurethane	7400	31,027.46
49.	Hemp	Carbon	Epoxy	16,700	70,021.43
50.	Hemp	Carbon	Polyester	14,600	61,216.34
51.	Hemp	Carbon	Vinyl ester	15,600	65,409.24
52.	Hemp	Carbon	Polyurethane	16,800	70,440.72
53.	Hemp	Kevlar	Epoxy	24,200	101,468.18

Table 6. *Cont.*

No.	Reinforced Composite			Estimated Cost (Money per Tons)	
	Natural Fiber	Synthetic Fiber	Matrices	USD	MYR
54.	Hemp	Kevlar	Polyester	24,750	107,967.18
55.	Hemp	Kevlar	Vinyl ester	26,300	110,273.27
56.	Hemp	Kevlar	Polyurethane	29,050	121,803.74
57.	Hemp	Fiber Glass	Epoxy	5180	21,719.22
58.	Hemp	Fiber Glass	Polyester	3080	12,914.13
59.	Hemp	Fiber Glass	Vinyl ester	4080	17,107.03
60.	Hemp	Fiber Glass	Polyurethane	5280	22,138.51
61.	Hemp	Glass	Epoxy	5700	23,899.53
62.	Hemp	Glass	Polyester	3600	15,094.44
63.	Hemp	Glass	Vinyl ester	4600	19,287.34
64.	Hemp	Glass	Polyurethane	5800	24,318.82
65.	Jute	Carbon	Epoxy	16,100	67,505.69
66.	Jute	Carbon	Polyester	14,000	58,700.60
67.	Jute	Carbon	Vinyl ester	15,000	62,893.50
68.	Jute	Carbon	Polyurethane	16,200	67,924.98
69.	Jute	Kevlar	Epoxy	23,600	98,952.44
70.	Jute	Kevlar	Polyester	21,500	90,147.35
71.	Jute	Kevlar	Vinyl ester	22,500	94,340.25
72.	Jute	Kevlar	Polyurethane	23,700	99,371.73
73.	Jute	Fiber Glass	Epoxy	4580	19,203.48
74.	Jute	Fiber Glass	Polyester	2480	10,398.39
75.	Jute	Fiber Glass	Vinyl ester	3480	14,591.29
76.	Jute	Fiber Glass	Polyurethane	4680	19,622.77
77.	Jute	Glass	Epoxy	5100	21,383.79
78.	Jute	Glass	Polyester	3000	12,578.70
79.	Jute	Glass	Vinyl ester	4000	16,771.60
80.	Jute	Glass	Polyurethane	5200	21,803.08
81.	Kenaf	Carbon	Epoxy	15,550	65,199.60
82.	Kenaf	Carbon	Polyester	13,450	56,394.50
83.	Kenaf	Carbon	Vinyl ester	14,450	60,587.40
84.	Kenaf	Carbon	Polyurethane	15,650	65,618.88
85.	Kenaf	Kevlar	Epoxy	23,050	96,646.35
86.	Kenaf	Kevlar	Polyester	20,950	87,841.25
87.	Kenaf	Kevlar	Vinyl ester	21,950	92,034.15
88.	Kenaf	Kevlar	Polyurethane	23,150	97,065.63
89.	Kenaf	Fiber Glass	Epoxy	4030	16,897.39
90.	Kenaf	Fiber Glass	Polyester	1930	8092.30
91.	Kenaf	Fiber Glass	Vinyl ester	2930	12,285.20
92.	Kenaf	Fiber Glass	Polyurethane	4130	17,316.68
93.	Kenaf	Glass	Epoxy	4550	19,077.69
94.	Kenaf	Glass	Polyester	2450	10,272.60
95.	Kenaf	Glass	Vinyl ester	3450	14,465.50
96.	Kenaf	Glass	Polyurethane	4650	19,496.99
97.	Pineapple	Carbon	Epoxy	15,605	65,430.20
98.	Pineapple	Carbon	Polyester	13,505	56,625.11
99.	Pineapple	Carbon	Vinyl ester	14,505	60,818.01
100.	Pineapple	Carbon	Polyurethane	15,705	65,849.49
101.	Pineapple	Kevlar	Epoxy	23,105	96,876.95
102.	Pineapple	Kevlar	Polyester	21,005	88,071.86
103.	Pineapple	Kevlar	Vinyl ester	22,005	92,264.76
104.	Pineapple	Kevlar	Polyurethane	23,205	97,296.24
105.	Pineapple	Fiber Glass	Epoxy	4085	17,128.00
106.	Pineapple	Fiber Glass	Polyester	1985	8322.91
107.	Pineapple	Fiber Glass	Vinyl ester	2985	12,515.81
108.	Pineapple	Fiber Glass	Polyurethane	4185	17,547.29

Table 6. *Cont.*

No.	Reinforced Composite			Estimated Cost (Money per Tons)	
	Natural Fiber	Synthetic Fiber	Matrices	USD	MYR
109.	Pineapple	Glass	Epoxy	4605	19,308.30
110.	Pineapple	Glass	Polyester	2505	10,503.21
111.	Pineapple	Glass	Vinyl ester	3505	14,696.11
112.	Pineapple	Glass	Polyurethane	4705	19,727.59
113.	Sisal	Carbon	Epoxy	15,800	66,247.82
114.	Sisal	Carbon	Polyester	13,700	57,442.73
115.	Sisal	Carbon	Vinyl ester	14,700	61,635.63
116.	Sisal	Carbon	Polyurethane	15,900	66,667.11
117.	Sisal	Kevlar	Epoxy	23,300	97,694.57
118.	Sisal	Kevlar	Polyester	21,200	88,889.48
119.	Sisal	Kevlar	Vinyl ester	22,200	93,082.38
120.	Sisal	Kevlar	Polyurethane	23,400	98,113.86
121.	Sisal	Fiber Glass	Epoxy	4280	17,945.61
122.	Sisal	Fiber Glass	Polyester	2180	9140.52
123.	Sisal	Fiber Glass	Vinyl ester	3180	13,333.42
124.	Sisal	Fiber Glass	Polyurethane	4380	18,364.90
125.	Sisal	Glass	Epoxy	4800	20,125.92
126.	Sisal	Glass	Polyester	2700	11,320.83
127.	Sisal	Glass	Vinyl ester	3700	15,513.73
128.	Sisal	Glass	Polyurethane	4900	20,545.21
129.	Sugar palm	Carbon	Epoxy	19,150	80,294.04
130.	Sugar palm	Carbon	Polyester	17,050	71,488.94
131.	Sugar palm	Carbon	Vinyl ester	18,050	75,681.85
132.	Sugar palm	Carbon	Polyurethane	19,250	80,713.32
133.	Sugar palm	Kevlar	Epoxy	26,650	111,740.78
134.	Sugar palm	Kevlar	Polyester	24,550	102,935.69
135.	Sugar palm	Kevlar	Vinyl ester	25,550	107,128.60
136.	Sugar palm	Kevlar	Polyurethane	26,750	112,160.07
137.	Sugar palm	Fiber Glass	Epoxy	7630	31,991.83
138.	Sugar palm	Fiber Glass	Polyester	5530	23,186.74
139.	Sugar palm	Fiber Glass	Vinyl ester	6530	27,379.64
140.	Sugar palm	Fiber Glass	Polyurethane	7730	32,411.12
141.	Sugar palm	Glass	Epoxy	8150	34,172.14
142.	Sugar palm	Glass	Polyester	6050	25,367.04
143.	Sugar palm	Glass	Vinyl ester	7050	29,559.94
144.	Sugar palm	Glass	Polyurethane	8250	34,591.42

Currency United States Dollar (USD) to Malaysian Ringgit (MYR) on date 8 June 2021.

7. Conclusions and Future Directions

Currently, the production of many bio-based plastics has been demonstrated at the demo and pilot scale, and some had been hugely commercialized. Some of the products are partly bio-based (i.e., polyamides, polyols bio-based polyethylene (PE), polypropylene (PP), or polyethylene terephthalate (PET)), and some of them are entirely new plastics (i.e., thermoplastic starch (TPS), polyhydroxybutyrate (PHB), polyhydroxyalkanoates (PHAs), or poly (lactic acid) (PLA). Partly bio-based plastics often require petrochemical monomers that cannot be resembled by bio-based substitutions, at least not at a reasonable price. Today's bio-based plastics are sophisticated materials that could technically be a substitute for around 90% of the plastics we use today.

Natural fiber reinforced hybrid composites are superior to petroleum-based composites because they have a higher strength-to-weight ratio, a low manufacturing cost due to their facile processes, and are environmentally beneficial. As a result, natural fiber composites have numerous advantages in commercial, industrial and engineering applications. Natural fibers, on the other hand, have lower strength than synthetic composites, but when combined with synthetic or biosynthetic composites, they offer high strength with a lower environmental impact. This study outlines the properties of natural fiber, composite hybridization, estimated costing, and natural fiber applications in various industries. This gives a comprehensive idea of how natural fibers are processed and commercialized. The characteristics of natural fibers such as sisal, jute, abaca, sugar palm, kenaf and hemp, were studied. In addition, the numerous applications of the hybrid composites in various sectors were described.

Author Contributions: Writing—original draft preparation, M.J.S., R.A.I., M.M.H.; conceptualization, M.J.S., R.A.I.; methodology, M.J.S., R.A.I.; writing—review and editing, M.J.S., R.A.I., M.M.H., A.K., M.T.H.S., S.M.S., C.M.R., F.N.W., F.Z., M.A.A., F.S.M.R., S.S.; supervision, M.J.S., R.A.I.; project administration, M.J.S., R.A.I.; Funding—M.Y.M.Z. All authors have read and agreed to the published version of the manuscript.

Funding: This study was funded by Fundamental Research Grant Scheme (FRGS) Vot 59624/Correlation of Manufacturing Defects, Interfacial Adhesion, Physical and Mechanical Properties of Plant Fiber Reinforced Hybrid Composite Material Towards Compatibility Behaviors. Besides this, the authors would like express gratitude for the financial support received from Universiti Teknologi Malaysia, project CRG 30.3: Retardant coating using graphene/bamboo aerogel mixtures on SAR robotics system, grant number PY/2020/03495—R.J130000.7351.4B534. The research has been carried out under program Research Excellence Consortium (JPT (BPKI) 1000/016/018/25 (57)) provided by Ministry of Higher Education Malaysia (MOHE). The authors also would like to thank Universiti Teknologi Malaysia (UTM) for work and facilities support. The article processing fee of this manuscript is funded by Research Management Centre, Universiti Putra Malaysia.

Institutional Review Board Statement: Not applicable.

Informed Consent Statement: Not applicable.

Acknowledgments: The authors would like to thank all the technicians and support staffs from Maritime Technology Lab, Universiti Malaysia Terengganu and Biocomposite Lab, INTROP, Universiti Putra Malaysia for their great contribution and help. This study was funded by Fundamental Research Grant Scheme (FRGS) Vot 59624/Correlation of Manufacturing Defects, Interfacial Adhesion, Physical and Mechanical Properties of Plant Fiber Reinforced Hybrid Composite Material Towards Compatibility Behaviors. Besides this, the authors would like express gratitude for the financial support received from Universiti Teknologi Malaysia, project CRG 30.3: Retardant coating using graphene/bamboo aerogel mixtures on SAR robotics system, grant number PY/2020/03495—R.J130000.7351.4B534. The research has been carried out under program Research Excellence Consortium (JPT (BPKI) 1000/016/018/25 (57)) provided by Ministry of Higher Education Malaysia (MOHE). The author also would like to thank Universiti Teknologi Malaysia (UTM) for work and facilities support.

Conflicts of Interest: The authors declare no conflict of interest.

References

1. Haris, N.I.N.; Ilyas, R.A.; Hassan, M.Z.; Sapuan, S.M.; Afdzaluddin, A.; Jamaludin, K.R.; Zaki, S.A.; Ramlie, F. Dynamic Mechanical Properties and Thermal Properties of Longitudinal Basalt/Woven Glass Fiber Reinforced Unsaturated Polyester Hybrid Composites. *Polymers* **2021**, *13*, 3343. [CrossRef]
2. Sapuan, S.M.; Aulia, H.S.; Ilyas, R.A.; Atiqah, A.; Dele-Afolabi, T.T.; Nurazzi, M.N.; Supian, A.B.M.; Atikah, M.S.N. Mechanical properties of longitudinal basalt/woven-glass-fiber-reinforced unsaturated polyester-resin hybrid composites. *Polymers* **2020**, *12*, 2211. [CrossRef] [PubMed]
3. Sanjay, M.R.; Arpitha, G.R.; Naik, L.L.; Gopalakrishna, K.; Yogesha, B. Applications of Natural Fibers and Its Composites: An Overview. *Nat. Resour.* **2016**, *7*, 108–114. [CrossRef]
4. Sanjay, M.R.; Arpitha, G.R.; Senthamaraikannan, P.; Kathiresan, M.; Saibalaji, M.A.; Yogesha, B. The Hybrid Effect of Jute/Kenaf/E-Glass Woven Fabric Epoxy Composites for Medium Load Applications: Impact, Inter-Laminar Strength, and Failure Surface Characterization. *J. Nat. Fibers* **2019**, *16*, 600–612. [CrossRef]
5. Madhu, P.; Sanjay, M.R.; Senthamaraikannan, P.; Pradeep, S.; Saravanakumar, S.S.; Yogesha, B. A review on synthesis and characterization of commercially available natural fibers: Part II. *J. Nat. Fibers* **2019**, *16*, 25–36. [CrossRef]
6. Ghelli, D.; Minak, G. Low velocity impact and compression after impact tests on thin carbon/epoxy laminates. *Compos. Part. B Eng.* **2011**, *42*, 2067–2079. [CrossRef]

7. Boria, S.; Pavlovic, A.; Fragassa, C.; Santulli, C. Modeling of falling weight impact behavior of hybrid basalt/flax vinylester composites. *Procedia Eng.* **2016**, *167*, 223–230. [CrossRef]
8. Harussani, M.M.; Sapuan, S.M.; Rashid, U.; Khalina, A. Development and Characterization of Polypropylene Waste from Personal Protective Equipment (PPE)-Derived Char-Filled Sugar Palm Starch Biocomposite Briquettes. *Polymers* **2021**, *13*, 1707. [CrossRef]
9. Chan, J.X.; Wong, J.F.; Petrů, M.; Hassan, A.; Nirmal, U.; Othman, N.; Ilyas, R.A. Effect of Nanofillers on Tribological Properties of Polymer Nanocomposites: A Review on Recent Development. *Polymers* **2021**, *13*, 2867. [CrossRef]
10. Nurazzi, N.M.; Sabaruddin, F.A.; Harussani, M.M.; Kamarudin, S.H.; Rayung, M.; Asyraf, M.R.M.; Aisyah, H.A.; Norrrahim, M.N.F.; Ilyas, R.A.; Abdullah, N.; et al. Mechanical Performance and Applications of CNTs Reinforced Polymer Composites—A Review. *Nanomaterials* **2021**, *11*, 2186. [CrossRef]
11. Poodts, E.; Minak, G.; Mazzocchetti, L.; Giorgini, L. Fabrication, process simulation and testing of a thick CFRP component using the RTM process. *Compos. Part. B Eng.* **2014**, *56*, 673–680. [CrossRef]
12. Malkapuram, R.; Kumar, V.; Singh Negi, Y. Recent development in natural fiber reinforced polypropylene composites. *J. Reinf. Plast. Compos.* **2009**, *28*, 1169–1189. [CrossRef]
13. Harussani, M.M.; Sapuan, S.M.; Rashid, U.; Khalina, A.; Ilyas, R.A. Pyrolysis of polypropylene plastic waste into carbonaceous char: Priority of plastic waste management amidst COVID-19 pandemic. *Sci. Total Environ.* **2021**, 149911. [CrossRef]
14. De Paola, S.; Fragassa, C.; Minak, G.; Pavlovic, A. Green Composites: A review of state of art. *Carbon N.Y.* **2013**, *1*, 230–240.
15. Hyseni, A.; De Paola, S.; Minak, G.; Fragassa, C. Mechanical characterization of ecocomposites. In Proceedings of the 30th Danubia Adria Symposium on Advanced Mechanics, Damir Semenski, Zagreb, Croatia, 25 September 2013; pp. 25–28.
16. Alotaibi, M.D.; Alshammari, B.A.; Saba, N.; Alothman, O.Y.; Sanjay, M.R.; Almutairi, Z.; Jawaid, M. Characterization of natural fiber obtained from different parts of date palm tree (*Phoenix dactylifera* L.). *Int. J. Biol. Macromol.* **2019**, *135*, 69–76. [CrossRef]
17. Vinod, A.; Vijay, R.; Singaravelu, D.L.; Sanjay, M.R.; Siengchin, S.; Yagnaraj, Y.; Khan, S. Extraction and characterization of natural fiber from stem of cardiospermum halicababum. *J. Nat. Fibers* **2021**, *18*, 898–908. [CrossRef]
18. Kadier, A.; Ilyas, R.A.; Huzaifah, M.R.M.; Harihastuti, N.; Sapuan, S.M.; Harussani, M.M.; Azlin, M.N.M.; Yuliasni, R.; Ibrahim, R.; Atikah, M.S.N.; et al. Use of Industrial Wastes as Sustainable Nutrient Sources for Bacterial Cellulose (BC) Production: Mechanism, Advances, and Future Perspectives. *Polymers* **2021**, *13*, 3365. [CrossRef]
19. Faruk, O.; Bledzki, A.K.; Fink, H.-P.; Sain, M. Biocomposites reinforced with natural fibers: 2000. *Prog. Polym. Sci.* **2012**, *37*, 1552–1596. [CrossRef]
20. Garkhail, S.K.; Heijenrath, R.W.H.; Peijs, T. Mechanical properties of natural-fibre-mat-reinforced thermoplastics based on flax fibres and polypropylene. *Appl. Compos. Mater.* **2000**, *7*, 351–372. [CrossRef]
21. Lee, C.H.; Khalina, A.; Nurazzi, N.M.; Norli, A.; Harussani, M.M.; Rafiqah, S.; Aisyah, H.A.; Ramli, N. The Challenges and Future Perspective of Woven Kenaf Reinforcement in Thermoset Polymer Composites in Malaysia: A Review. *Polymers* **2021**, *13*, 1390. [CrossRef]
22. Wambua, P.; Ivens, J.; Verpoest, I. Natural fibres: Can they replace glass in fibre reinforced plastics? *Compos. Sci. Technol.* **2003**, *63*, 1259–1264. [CrossRef]
23. Rao, K.M.M.; Rao, K.M. Extraction and tensile properties of natural fibers: Vakka, date and bamboo. *Compos. Struct.* **2007**, *77*, 288–295. [CrossRef]
24. Jha, K.; Tyagi, Y.K.; Kumar, R.; Sharma, S.; Roslim, M.; Huzaifah, M.; Li, C.; Ilyas, R.A.; Dwivedi, S.P.; Saxena, A.; et al. Assessment of Dimensional Stability, Biodegradability, and Fracture Energy of Bio-Composites Reinforced with Novel Pine Cone. *Polymers* **2021**, *13*, 3260. [CrossRef]
25. Kumari, N.; Bangar, S.P.; Petrů, M.; Ilyas, R.A.; Singh, A.; Kumar, P. Development and Characterization of Fenugreek Protein-Based Edible Film. *Foods* **2021**, *10*, 1976. [CrossRef]
26. Tarique, J.; Sapuan, S.M.; Khalina, A.; Sherwani, S.F.K.; Yusuf, J.; Ilyas, R.A. Recent developments in sustainable arrowroot (Maranta arundinacea Linn) starch biopolymers, fibres, biopolymer composites and their potential industrial applications: A review. *J. Mater. Res. Technol.* **2021**, *13*, 1191–1219. [CrossRef]
27. Misri, S.; Leman, Z.; Sapuan, S.M.; Ishak, M.R. Mechanical properties and fabrication of small boat using woven glass/sugar palm fibres reinforced unsaturated polyester hybrid composite. *IOP Conf. Ser. Mater. Sci. Eng.* **2010**, *11*, 012015. [CrossRef]
28. Kamaruddin, Z.H.; Jumaidin, R.; Selamat, M.Z.; Ilyas, R.A. Characteristics and Properties of Lemongrass (Cymbopogan Citratus): A Comprehensive Review. *J. Nat. Fibers* **2021**, *18*, 1–18. [CrossRef]
29. Dashtizadeh, Z.; Abdan, K.; Jawaid, M.; Khan, M.A.; Behmanesh, M.; Dashtizadeh, M.; Cardona, F.; Ishak, M. Mechanical and thermal properties of natural fibre based hybrid composites: A review. *Pertanika J. Sci. Technol.* **2017**, *25*, 1103–1122.
30. Nurazzi, N.M.; Asyraf, M.R.M.; Rayung, M.; Norrrahim, M.N.F.; Shazleen, S.S.; Rani, M.S.A.; Shafi, A.R.; Aisyah, H.A.; Radzi, M.H.M.; Sabaruddin, F.A.; et al. Thermogravimetric Analysis Properties of Cellulosic Natural Fiber Polymer Composites: A Review on Influence of Chemical Treatments. *Polymers* **2021**, *13*, 2710. [CrossRef]
31. Nurazzi, N.M.; Asyraf, M.R.M.; Fatimah Athiyah, S.; Shazleen, S.S.; Rafiqah, S.A.; Harussani, M.M.; Kamarudin, S.H.; Razman, M.R.; Rahmah, M.; Zainudin, E.S.; et al. A Review on Mechanical Performance of Hybrid Natural Fiber Polymer Composites for Structural Applications. *Polymers* **2021**, *13*, 2170. [CrossRef]
32. Sapuan, S.M.; Lok, H.Y.; Ishak, M.R.; Misri, S. Mechanical properties of hybrid glass/sugar palm fibre reinforced unsaturated polyester composites. *Chin. J. Polym. Sci.* **2013**, *31*, 1394–1403. [CrossRef]

33. Afzaluddin, A.; Jawaid, M.; Salit, M.S.; Ishak, M.R. Physical and mechanical properties of sugar palm/glass fiber reinforced thermoplastic polyurethane hybrid composites. *J. Mater. Res. Technol.* **2019**, *8*, 950–959. [CrossRef]
34. Baihaqi, N.M.Z.N.; Khalina, A.; Nurazzi, N.M.; Aisyah, H.A.; Sapuan, S.M.; Ilyas, R.A. Effect of fiber content and their hybridization on bending and torsional strength of hybrid epoxy composites reinforced with carbon and sugar palm fibers. *Polimery* **2021**, *66*, 36–43. [CrossRef]
35. Safri, S.N.A.; Sultan, M.T.H.; Shah, A.U.M. Characterization of benzoyl treated sugar palm/glass fibre hybrid composites. *J. Mater. Res. Technol.* **2020**, *9*, 11563–11573. [CrossRef]
36. Safri, S.N.A.; Sultan, M.T.H.; Jawaid, M.; Abdul Majid, M.S. Analysis of dynamic mechanical, low-velocity impact and compression after impact behaviour of benzoyl treated sugar palm/glass/epoxy composites. *Compos. Struct.* **2019**, *226*, 111308. [CrossRef]
37. Jumaidin, R.; Sapuan, S.M.; Jawaid, M.; Ishak, M.R.; Sahari, J. Thermal, mechanical, and physical properties of seaweed/sugar palm fibre reinforced thermoplastic sugar palm Starch/Agar hybrid composites. *Int. J. Biol. Macromol.* **2017**, *97*, 606–615. [CrossRef]
38. Radzi, A.M.; Sapuan, S.M.; Jawaid, M.; Mansor, M.R. Water absorption, thickness swelling and thermal properties of roselle/sugar palm fibre reinforced thermoplastic polyurethane hybrid composites. *J. Mater. Res. Technol.* **2019**. [CrossRef]
39. Ibrahim, M.I.J.; Sapuan, S.M.; Zainudin, E.S.; Zuhri, M.Y.M. Preparation and characterization of cornhusk/sugar palm fiber reinforced Cornstarch-based hybrid composites. *J. Mater. Res. Technol.* **2020**, *9*, 200–211. [CrossRef]
40. Mohd Nurazzi, N.; Khalina, A.; Sapuan, S.M.; Rahmah, M. Development of sugar palm yarn/glass fibre reinforced unsaturated polyester hybrid composites. *Mater. Res. Express* **2018**, *5*, 045308. [CrossRef]
41. Edhirej, A.; Sapuan, S.M.; Jawaid, M.; Zahari, N.I. Cassava/sugar palm fiber reinforced cassava starch hybrid composites: Physical, thermal and structural properties. *Int. J. Biol. Macromol.* **2017**, *101*, 75–83. [CrossRef]
42. Bachtiar, D.; Siregar, J.P.; bin Sulaiman, A.S.; bin Mat Rejab, M.R. Tensile Properties of Hybrid Sugar Palm/Kenaf Fibre Reinforced Polypropylene Composites. *Appl. Mech. Mater.* **2014**, *695*, 155–158.
43. Ibrahim, M.I.J.; Sapuan, S.M.; Zainudin, E.S.; Zuhri, M.Y.M.; Edhirej, A. Processing and Characterization of Cornstalk/Sugar Palm Fiber Reinforced Cornstarch Biopolymer Hybrid Composites. In *Advanced Processing, Properties, and Applications of Starch and Other Bio-Based Polymers*; Elsevier: Amsterdam, The Netherlands, 2020; pp. 35–46.
44. Siregar, J.P.; Zalinawati, M.; Cionita, T.; Rejab, M.R.M.; Mawarnie, I.; Jaafar, J.; Hamdan, M.H.M. Mechanical properties of hybrid sugar palm/ramie fibre reinforced epoxy composites. *Mater. Today Proc.* **2021**, *46*, 1729–1734. [CrossRef]
45. Razali, N.; Sapuan, S.M.; Razali, N. Mechanical Properties and Morphological Analysis of Roselle/Sugar Palm Fiber Reinforced Vinyl Ester Hybrid Composites. In *Natural Fibre Reinforced Vinyl Ester and Vinyl Polymer Composites*; Elsevier: Amsterdam, The Netherlands, 2018; pp. 169–180.
46. Mukhtar, I.; Leman, Z.; Zainudin, E.S.; Ishak, M.R. Hybrid and Nonhybrid Laminate Composites of Sugar Palm and Glass Fibre-Reinforced Polypropylene: Effect of Alkali and Sodium Bicarbonate Treatments. *Int. J. Polym. Sci.* **2019**, *2019*, 12. [CrossRef]
47. Davoodi, M.M.; Sapuan, S.M.; Ahmad, D.; Ali, A.; Khalina, A.; Jonoobi, M. Mechanical properties of hybrid kenaf/glass reinforced epoxy composite for passenger car bumper beam. *Mater. Des.* **2010**, *31*, 4927–4932. [CrossRef]
48. Atiqah, A.; Maleque, M.A.; Jawaid, M.; Iqbal, M. Development of kenaf-glass reinforced unsaturated polyester hybrid composite for structural applications. *Compos. Part. B Eng.* **2014**, *56*. [CrossRef]
49. Jaafar, C.N.A.; Zainol, I.; Rizal, M.A.M. Preparation and characterisation of epoxy/silica/kenaf composite using hand lay-up method. In Proceedings of the 27th Scientific Conference of the Microscopy Society Malaysia (27th SCMSM 2018), Melaka, Malaysia, 3 December 2018; pp. 2–6.
50. Aiza Jaafar, C.N.; Zainol, I.; Ishak, N.S.; Ilyas, R.A.; Sapuan, S.M. Effects of the Liquid Natural Rubber (LNR) on Mechanical Properties and Microstructure of Epoxy/Silica/Kenaf Hybrid Composite for Potential Automotive Applications. *J. Mater. Res. Technol.* **2021**, *12*, 1026–1038. [CrossRef]
51. Chee, S.S.; Jawaid, M.; Alothman, O.Y.; Fouad, H. Effects of Nanoclay on Mechanical and Dynamic Mechanical Properties of Bamboo/Kenaf Reinforced Epoxy Hybrid Composites. *Polymers* **2021**, *13*, 395. [CrossRef]
52. Hammami, H.; Jawaid, M.; Kallel, A. Effects of oil palm and montmorillonite nanofillers on stiffness and interfacial adhesion of kenaf/epoxy hybrid nanocomposites. *Polym. Compos.* **2021**, *42*, 2948–2957. [CrossRef]
53. Mirzaei, J.; Fereidoon, A.; Ghasemi-Ghalebahman, A. Experimental study on mechanical properties of polypropylene nanocomposites reinforced with a hybrid graphene/PP-g-MA/kenaf fiber by response surface methodology. *J. Elastomers Plast.* **2021**, *11*, 009524432110153. [CrossRef]
54. Sabaruddin, F.A.; Paridah, M.T.; Sapuan, S.M.; Ilyas, R.A.; Lee, S.H.; Abdan, K.; Mazlan, N.; Roseley, A.S.M.; Abdul Khalil, H.P.S. The effects of unbleached and bleached nanocellulose on the thermal and flammability of polypropylene-reinforced kenaf core hybrid polymer bionanocomposites. *Polymers* **2020**, *13*, 116. [CrossRef]
55. Supian, A.B.M.; Sapuan, S.M.; Zuhri, M.Y.M.; Zainudin, E.S.; Ya, H.H.; Hisham, H.N. Effect of winding orientation on energy absorption and failure modes of filament wound kenaf/glass fibre reinforced epoxy hybrid composite tubes under intermediate-velocity impact (IVI) load. *J. Mater. Res. Technol.* **2021**, *10*, 1–14. [CrossRef]
56. Suriani, M.J.; Zainudin, H.A.; Ilyas, R.A.; Petrů, M.; Sapuan, S.M.; Ruzaidi, C.M.; Mustapha, R. Kenaf Fiber/Pet Yarn Reinforced Epoxy Hybrid Polymer Composites: Morphological, Tensile, and Flammability Properties. *Polymers* **2021**, *13*, 1532. [CrossRef] [PubMed]

57. Tamrakar, S.; Kiziltas, A.; Mielewski, D.; Zander, R. Characterization of kenaf and glass fiber reinforced hybrid composites for underbody shield applications. *Compos. Part. B Eng.* **2021**, *216*, 108805. [CrossRef]
58. Yahaya, R.; Sapuan, S.M.; Jawaid, M.; Leman, Z.; Zainudin, E.S. Mechanical performance of woven kenaf-Kevlar hybrid composites. *J. Reinf. Plast. Compos.* **2014**, *33*, 2242–2254. [CrossRef]
59. Alavudeen, A.; Rajini, N.; Karthikeyan, S.; Thiruchitrambalam, M.; Venkateshwaren, N. Mechanical properties of banana/kenaf fiber-reinforced hybrid polyester composites: Effect of woven fabric and random orientation. *Mater. Des.* **2015**, *66*, 246–257. [CrossRef]
60. Nampoothiri, E.N.; Bensam Raj, J.; Thanigaivelan, R.; Karuppasamy, R. Experimental investigation on mechanical and biodegradation properties of indian almond–kenaf fiber-reinforced hybrid composites for construction applications. *J. Nat. Fibers* **2020**, *17*, 1–11. [CrossRef]
61. Prabhu, L.; Krishnaraj, V.; Gokulkumar, S.; Sathish, S.; Sanjay, M.R.; Siengchin, S. Mechanical, chemical and sound absorption properties of glass/kenaf/waste tea leaf fiber-reinforced hybrid epoxy composites. *J. Ind. Text.* **2020**, 1528083720957392. [CrossRef]
62. Hanan, F.; Jawaid, M.; Md Tahir, P. Mechanical performance of oil palm/kenaf fiber-reinforced epoxy-based bilayer hybrid composites. *J. Nat. Fibers* **2020**, *17*, 155–167. [CrossRef]
63. Dhar Malingam, S.; Ng, L.F.; Chan, K.H.; Subramaniam, K.; Selamat, M.Z.; Zakaria, K.A. The static and dynamic mechanical properties of kenaf/glass fibre reinforced hybrid composites. *Mater. Res. Express* **2018**, *5*, 95304. [CrossRef]
64. Sosiati, H.; Shofie, Y.A.; Nugroho, A.W. Tensile properties of Kenaf/E-glass reinforced hybrid polypropylene (PP) composites with different fiber loading. Ph.D. Thesis, Kyushu University, Fukuoka, Japan, 2018.
65. Ismail, A.S.; Jawaid, M.; Naveen, J. Void content, tensile, vibration and acoustic properties of kenaf/bamboo fiber reinforced epoxy hybrid composites. *Materials* **2019**, *12*, 2094. [CrossRef]
66. Mirbagheri, J.; Tajvidi, M.; Hermanson, J.C.; Ghasemi, I. Tensile Properties of Wood Flour/Kenaf Fiber Polypropylene Hybrid Composites. *J. Appl. Polym. Sci.* **2007**, *105*, 3054–3059. [CrossRef]
67. Haryati, A.; Razali, N.; Petrů, M.; Taha, M.; Muhammad, N.; Ilyas, R.A. Effect of Chemically Treated Kenaf Fibre on Mechanical and Thermal Properties of PLA Composites Prepared through Fused Deposition Modeling (FDM). *Polymers* **2021**, *13*, 3299. [CrossRef]
68. Suriani, M.J.; Radzi, F.S.M.; Ilyas, R.A.; Petrů, M.; Sapuan, S.M.; Ruzaidi, C.M. Flammability, Tensile, and Morphological Properties of Oil Palm Empty Fruit Bunches Fiber/Pet Yarn-Reinforced Epoxy Fire Retardant Hybrid Polymer Composites. *Polymers* **2021**, *13*, 1282. [CrossRef]
69. Hanan, F.; Jawaid, M.; Paridah, M.T.; Naveen, J. Characterization of Hybrid Oil Palm Empty Fruit Bunch/Woven Kenaf Fabric-Reinforced Epoxy Composites. *Polymers* **2020**, *12*, 2052. [CrossRef]
70. Suradi, S.S.; Yunus, R.M.; Beg, M.D.H.; Rivai, M.; Yusof, Z.A.M. Oil palm bio-fiber reinforced thermoplastic composites-effects of matrix modification on mechanical and thermal properties. *J. Appl. Sci.* **2010**, *10*, 3271–3276. [CrossRef]
71. Ramlee, N.A.; Jawaid, M.; Zainudin, E.S.; Yamani, S.A.K. Tensile, physical and morphological properties of oil palm empty fruit bunch/sugarcane bagasse fibre reinforced phenolic hybrid composites. *J. Mater. Res. Technol.* **2019**, *8*, 3466–3474. [CrossRef]
72. Muhammad Amir, S.M.; Hameed Sultan, M.T.; Md Shah, A.U.; Jawaid, M.; Safri, S.N.A.; Mohd, S.; Mohd Salleh, K.A. Low Velocity Impact and Compression after Impact Properties on Gamma Irradiated Kevlar/Oil Palm Empty Fruit Bunch Hybrid Composites. *Coatings* **2020**, *10*, 646. [CrossRef]
73. Islam, M.; Gupta, A.; Rivai, M.; Beg, M. Characterization of microwave-treated oil palm empty fruit bunch/glass fibre/polypropylene composites. *J. Thermoplast. Compos. Mater.* **2017**, *30*, 986–1002. [CrossRef]
74. Khalil, H.P.S.A.; Hanida, S.; Kang, C.W.; Fuaad, N.A.N. Agro-hybrid composite: The effects on mechanical and physical properties of oil palm fiber (EFB)/glass hybrid reinforced polyester composites. *J. Reinf. Plast. Compos.* **2007**, *26*, 203–218. [CrossRef]
75. Rozman, H.D.; Tay, G.S.; Kumar, R.N.; Abusamah, A.; Ismail, H.; Mohd, I.Z.A. The effect of oil extraction of the oil palm empty fruit bunch on the mechanical properties of polypropylene–oil palm empty fruit bunch–glass fibre hybrid composites. *Polym. Plast. Technol. Eng.* **2001**, *40*, 103–115. [CrossRef]
76. Hashim, M.K.R.; Abdul Majid, M.S.; Jamir, M.R.M.; Kasim, F.H.; Sultan, M.T.H. The Effect of Stacking Sequence and Ply Orientation on the Mechanical Properties of Pineapple Leaf Fibre (PALF)/Carbon Hybrid Laminate Composites. *Polymers* **2021**, *13*, 455. [CrossRef]
77. Siakeng, R.; Jawaid, M.; Asim, M.; Saba, N.; Sanjay, M.R.; Siengchin, S.; Fouad, H. Alkali treated coir/pineapple leaf fibres reinforced PLA hybrid composites: Evaluation of mechanical, morphological, thermal and physical properties. *XPRESS Polym. Lett.* **2020**, *14*, 717–730. [CrossRef]
78. Zin, M.H.; Abdan, K.; Mazlan, N.; Zainudin, E.S.; Liew, K.E.; Norizan, M.N. Automated spray up process for Pineapple Leaf Fibre hybrid biocomposites. *Compos. Part. B Eng.* **2019**, *177*, 107306. [CrossRef]
79. Venkata Deepthi, P.; Sita Rama Raju, K.; Indra Reddy, M. Dynamic mechanical analysis of banana, pineapple leaf and glass fibre reinforced hybrid polyester composites. *Mater. Today Proc.* **2019**, *18*, 2114–2117. [CrossRef]
80. Asim, M.; Paridah, M.T.; Saba, N.; Jawaid, M.; Alothman, O.Y.; Nasir, M.; Almutairi, Z. Thermal, physical properties and flammability of silane treated kenaf/pineapple leaf fibres phenolic hybrid composites. *Compos. Struct.* **2018**, *202*, 1330–1338. [CrossRef]
81. Sathees Kumar, S.; Muthalagu, R.; Nithin Chakravarthy, C. Effects of fiber loading on mechanical characterization of pineapple leaf and sisal fibers reinforced polyester composites for various applications. *Mater. Today Proc.* **2021**, *44*, 546–553. [CrossRef]

82. Osorio, L.; Trujillo, E.; Van Vuure, A.W.; Verpoest, I. Morphological aspects and mechanical properties of single bamboo fibers and flexural characterization of bamboo/epoxy composites. *J. Reinf. Plast. Compos.* **2011**, *30*, 396–408. [CrossRef]
83. Nayak, S.K.; Mohanty, S.; Samal, S.K. Influence of short bamboo/glass fiber on the thermal, dynamic mechanical and rheological properties of polypropylene hybrid composites. *Mater. Sci. Eng. A* **2009**, *523*, 32–38. [CrossRef]
84. Samal, S.K.; Mohanty, S.; Nayak, S.K. Polypropylene bamboo/glass fiber hybrid composites: Fabrication and analysis of mechanical, morphological, thermal, and dynamic mechanical behavior. *J. Reinf. Plast. Compos.* **2009**, *28*, 2729–2747. [CrossRef]
85. Thwe, M.M.; Liao, K. Environmental effects on bamboo-glass/polypropylene hybrid composites. *J. Mater. Sci.* **2003**, *38*, 363–376. [CrossRef]
86. Thwe, M.M.; Liao, K. Durability of bamboo-glass fiber reinforced polymer matrix hybrid composites. *Compos. Sci. Technol.* **2003**, *63*, 375–387. [CrossRef]
87. Biswas, S.; Satapathy, A.; Patnaik, A. Effect of ceramic Fillers On Mechanical Properties of Bamboo Fiber Reinforced Epoxy Composites: A Comparative Study. In *Proceedings of the Advanced Materials Research*; Trans Tech Publications Ltd.: Freienbach, Switzerland, 2010; Volume 123, pp. 1031–1034.
88. Krishnan, G.S.; Velmurugan, P. Investigation on the characteristics of bamboo/jute reinforced hybrid epoxy polymer composites. *Mater. Res. Express* **2019**, *6*, 105346.
89. Shukla, R.; Srivastava, A.K. FTIR analysis of Bamboo and Flax (BFBF) Mat Reinforced Epoxy Hybrid Composite. *Int. J. Sci. Res. Sci. Eng. Technol.* **2019**, *6*, 119–124.
90. Getu, D.; Nallamothu, R.B.; Masresha, M.; Nallamothu, S.K.; Nallamothu, A.K. Production and characterization of bamboo and sisal fiber reinforced hybrid composite for interior automotive body application. *Mater. Today Proc.* **2021**, *38*, 2853–2860. [CrossRef]
91. Aruchamy, K.; Pavayee Subramani, S.; Palaniappan, S.K.; Sethuraman, B.; Velu Kaliyannan, G. Study on mechanical characteristics of woven cotton/bamboo hybrid reinforced composite laminates. *J. Mater. Res. Technol.* **2020**, *9*, 718–726. [CrossRef]
92. Olaniran, O.; Uwaifo, O.; Bamidele, E.; Olaniran, B. An investigation of the mechanical properties of organic silica, bamboo leaf ash and rice husk reinforced aluminium hybrid composite. *Mater. Sci. Eng. Int. J.* **2019**, *3*, 129–134. [CrossRef]
93. Sarkar, B.K.; Ray, D. Effect of the defect concentration on the impact fatigue endurance of untreated and alkali treated jute–vinylester composites under normal and liquid nitrogen atmosphere. *Compos. Sci. Technol.* **2004**, *64*, 2213–2219. [CrossRef]
94. Ramana, M.V.; Ramprasad, S. Experimental investigation on jute/carbon fibre reinforced epoxy based hybrid composites. *Mater. Today Proc.* **2017**, *4*, 8654–8664. [CrossRef]
95. Ahmed, K.S.; Vijayarangan, S.; Naidu, A.C.B. Elastic properties, notched strength and fracture criterion in untreated woven jute–glass fabric reinforced polyester hybrid composites. *Mater. Des.* **2007**, *28*, 2287–2294. [CrossRef]
96. Tezara, C.; Zalinawati, M.; Siregar, J.P.; Jaafar, J.; Hamdan, M.H.M.; Oumer, A.N.; Chuah, K.H. Effect of Stacking Sequences, Fabric Orientations, and Chemical Treatment on the Mechanical Properties of Hybrid Woven Jute–Ramie Composites. *Int. J. Precis. Eng. Manuf. Technol.* **2021**, *8*, 1–13. [CrossRef]
97. Pandita, S.D.; Yuan, X.; Manan, M.A.; Lau, C.H.; Subramanian, A.S.; Wei, J. Evaluation of jute/glass hybrid composite sandwich: Water resistance, impact properties and life cycle assessment. *J. Reinf. Plast. Compos.* **2014**, *33*, 14–25. [CrossRef]
98. Akil, H.M.; Santulli, C.; Sarasini, F.; Tirillò, J.; Valente, T. Environmental effects on the mechanical behaviour of pultruded jute/glass fibre-reinforced polyester hybrid composites. *Compos. Sci. Technol.* **2014**, *94*, 62–70. [CrossRef]
99. Chaudhary, V.; Bajpai, P.K.; Maheshwari, S. Studies on mechanical and morphological characterization of developed jute/hemp/flax reinforced hybrid composites for structural applications. *J. Nat. Fibers* **2018**, *15*, 80–97. [CrossRef]
100. Joseph, K.; De Carvalho, L.H. Jute/cotton woven Fabric Reinforced Polyester Composites: Effect of Hybridization. In *Proceedings of the ISNaPol/2000: Third International Symposium on Natural Polymers and Composites and the Workshop on Progress in Production and Processing of Cellulosic Fibres and Natural Polymers*; Embrapa Instrumentação Agropecuária: Sao Carlos, Brazil, 2000; pp. 454–459.
101. Prasath, K.A.; Krishnan, B.R.; Arun, C.K. Mechanical properties of woven fabric Basalt/jute fibre reinforced polymer hybrid composites. *Int. J. Mech. Eng.* **2013**, *2*, 279–290.
102. Ramesh, M.; Deepa, C.; Arpitha, G.R.; Gopinath, V. Effect of hybridization on properties of hemp-carbon fibre-reinforced hybrid polymer composites using experimental and finite element analysis. *World J. Eng.* **2019**, *16*, 248–259. [CrossRef]
103. Thiagamani, S.M.K.; Krishnasamy, S.; Muthukumar, C.; Tengsuthiwat, J.; Nagarajan, R.; Siengchin, S.; Ismail, S.O. Investigation into mechanical, absorption and swelling behaviour of hemp/sisal fibre reinforced bioepoxy hybrid composites: Effects of stacking sequences. *Int. J. Biol. Macromol.* **2019**, *140*, 637–646. [CrossRef]
104. Haq, M.; Burgueño, R.; Mohanty, A.K.; Misra, M. Hybrid bio-based composites from blends of unsaturated polyester and soybean oil reinforced with nanoclay and natural fibers. *Compos. Sci. Technol.* **2008**, *68*, 3344–3351. [CrossRef]
105. Pappu, A.; Pickering, K.L.; Thakur, V.K. Manufacturing and characterization of sustainable hybrid composites using sisal and hemp fibres as reinforcement of poly(lactic acid) via injection moulding. *Ind. Crop. Prod.* **2019**, *137*, 260–269. [CrossRef]
106. Sarasini, F.; Tirillò, J.; Sergi, C.; Seghini, M.C.; Cozzarini, L.; Graupner, N. Effect of basalt fibre hybridisation and sizing removal on mechanical and thermal properties of hemp fibre reinforced HDPE composites. *Compos. Struct.* **2018**, *188*, 394–406. [CrossRef]
107. Ahmad, M.A.A.; Abdul Majid, M.S.; Ridzuan, M.J.M.; Mazlee, M.N.; Gibson, A.G. Dynamic mechanical analysis and effects of moisture on mechanical properties of interwoven hemp/polyethylene terephthalate (PET) hybrid composites. *Constr. Build. Mater.* **2018**, *179*, 265–276. [CrossRef]

108. Paturel, A.; Dhakal, H.N. Influence of water absorption on the low velocity falling weight impact damage behaviour of flax/glass reinforced vinyl ester hybrid composites. *Molecules* **2020**, *25*, 278. [CrossRef]
109. Arbelaiz, A.; Fernandez, B.; Ramos, J.A.; Mondragon, I. Thermal and crystallization studies of short flax fibre reinforced polypropylene matrix composites: Effect of treatments. *Thermochim. Acta* **2006**, *440*, 111–121. [CrossRef]
110. Di Landro, L.; Lorenzi, W. Static and dynamic properties of thermoplastic matrix/natural fiber composites. *J. Biobased Mater. Bioenergy* **2009**, *3*, 238–244. [CrossRef]
111. Liu, Q.; Hughes, M. The fracture behaviour and toughness of woven flax fibre reinforced epoxy composites. *Compos. Part. A Appl. Sci. Manuf.* **2008**, *39*, 1644–1652. [CrossRef]
112. Almansour, F.A.; Dhakal, H.N.; Zhang, Z.Y. Investigation into Mode II interlaminar fracture toughness characteristics of flax/basalt reinforced vinyl ester hybrid composites. *Compos. Sci. Technol.* **2018**, *154*, 117–127. [CrossRef]
113. Cheour, K.; Assarar, M.; Scida, D.; Ayad, R.; Gong, X.-L. Long-term immersion in water of flax-glass fibre hybrid composites: Effect of stacking sequence on the mechanical and damping properties. *Fibers Polym.* **2020**, *21*, 162–169. [CrossRef]
114. Arulmurugan, M.; Selvakumar, A.S.; Prabu, K.; Rajamurugan, G. Effect of barium sulphate on mechanical, DMA and thermal behaviour of woven aloevera/flax hybrid composites. *Bull. Mater. Sci.* **2020**, *43*, 58. [CrossRef]
115. Xu, H.; Wang, L.; Teng, C.; Yu, M. Biodegradable composites: Ramie fibre reinforced PLLA-PCL composite prepared by in situ polymerization process. *Polym. Bull.* **2008**, *61*, 663–670. [CrossRef]
116. Romanzini, D.; Ornaghi Junior, H.L.; Amico, S.C.; Zattera, A.J. Preparation and characterization of ramie-glass fiber reinforced polymer matrix hybrid composites. *Mater. Res.* **2012**, *15*, 415–420. [CrossRef]
117. Marsyahyo, E.; Jamasri; Rochardjo, H.S.B.; Soekrisno. Preliminary investigation on bulletproof panels made from ramie fiber reinforced composites for NIJ level II, IIA, and IV. *J. Ind. Text.* **2009**, *39*, 13–26. [CrossRef]
118. Wen, L.E.I.; LEI, W.; Chao, R.E.N. Effect of volume fraction of ramie cloth on physical and mechanical properties of ramie cloth/UP resin composite. *Trans. Nonferrous Met. Soc. China* **2006**, *16*, s474–s477.
119. Bledzki, A.K.; Mamun, A.A.; Faruk, O. Abaca fibre reinforced PP composites and comparison with jute and flax fibre PP composites. *Express Polym. Lett.* **2007**. [CrossRef]
120. Savastano, H., Jr.; Santos, S.F.; Radonjic, M.; Soboyejo, W.O. Fracture and fatigue of natural fiber-reinforced cementitious composites. *Cem. Concr. Compos.* **2009**, *31*, 232–243. [CrossRef]
121. Bledzki, A.K.; Mamun, A.A.; Jaszkiewicz, A.; Erdmann, K. Polypropylene composites with enzyme modified abaca fibre. *Compos. Sci. Technol.* **2010**, *70*, 854–860. [CrossRef]
122. Ortega, Z.; Monzón, M.D.; Benítez, A.N.; Kearns, M.; McCourt, M.; Hornsby, P.R. Banana and abaca fiber-reinforced plastic composites obtained by rotational molding process. *Mater. Manuf. Process.* **2013**, *28*, 879–883. [CrossRef]
123. Prasad, N.; Agarwal, V.K.; Sinha, S. Banana fiber reinforced low-density polyethylene composites: Effect of chemical treatment and compatibilizer addition. *Iran. Polym. J.* **2016**, *25*, 229–241. [CrossRef]
124. Agung, E.H.; Sapuan, S.M.; Hamdan, M.M.; Zaman, H.; Mustofa, U. Optimization of the mechanical properties of abaca fibre-reinforced high impact polystyrene (HIPS) composites using box-behnken design of experiments. *Polym. Polym. Compos.* **2011**, *19*, 697–710. [CrossRef]
125. Iqbal, M.; Aminanda, Y.; Firsa, T.; Ali, M. Bending Strength of Polyester Composites Reinforced with Stitched Random Orientation And Plain Weave Abaca Fiber. In *Proceedings of the IOP Conference Series: Materials science and Engineering*; IOP Publishing: Bristol, UK, 2020; Volume 739, p. 12035.
126. Rana, R.S.; Rana, S.; Nigrawal, A. Preparation and mechanical properties evaluation of polyvinyl alcohol and banana fibres composite. *Mater. Today Proc.* **2020**, *26*, 3145–3147. [CrossRef]
127. Zhong, L.X.; Fu, S.Y.; Zhou, X.S.; Zhan, H.Y. Effect of surface microfibrillation of sisal fibre on the mechanical properties of sisal/aramid fibre hybrid composites. *Compos. Part. A Appl. Sci. Manuf.* **2011**, *42*, 244–252. [CrossRef]
128. Asaithambi, B.; Ganesan, G.; Ananda Kumar, S. Bio-composites: Development and mechanical characterization of banana/sisal fibre reinforced poly lactic acid (PLA) hybrid composites. *Fibers Polym.* **2014**, *15*, 847–854. [CrossRef]
129. Noorunnisa Khanam, P.; Abdul Khalil, H.P.S.; Jawaid, M.; Ramachandra Reddy, G.; Surya Narayana, C.; Venkata Naidu, S. Sisal/Carbon Fibre Reinforced Hybrid Composites: Tensile, Flexural and Chemical Resistance Properties. *J. Polym. Environ.* **2010**, *18*, 727–733. [CrossRef]
130. Aslan, M.; Tufan, M.; Küçükömeroğlu, T. Tribological and mechanical performance of sisal-filled waste carbon and glass fibre hybrid composites. *Compos. Part. B Eng.* **2018**, *140*, 241–249. [CrossRef]
131. Gupta, M.K. Effect of frequencies on dynamic mechanical properties of hybrid jute/sisal fibre reinforced epoxy composite. *Adv. Mater. Process. Technol.* **2017**, *3*, 651–664. [CrossRef]
132. Sapuan, S.M.; Ishak, M.R.; Leman, Z.; Huzaifah, M.R.M.; Ilyas, R.A.; Ammar, I.M.; Ishak, M.H.A.; Yana, I. *Pokok Enau: Potensi Dan Pembangunan Produk*, 1st ed.; Sapuan, S.M., Ed.; Penerbit Universiti Putra Malaysia: Serdang, Selangor, 2017; ISBN 9789673447473.
133. Ishak, M.R.; Sapuan, S.M.; Leman, Z.; Rahman, M.Z.A.A.; Anwar, U.M.K.K.; Siregar, J.P. Sugar palm (Arenga pinnata): Its fibres, polymers and composites. *Carbohydr. Polym.* **2013**, *91*, 699–710. [CrossRef]
134. Ilyas, R.A.; Sapuan, S.M.; Ishak, M.R.; Zainudin, E.S. Sugar palm nanofibrillated cellulose (Arenga pinnata (Wurmb.) Merr): Effect of cycles on their yield, physic-chemical, morphological and thermal behavior. *Int. J. Biol. Macromol.* **2019**, *123*, 379–388. [CrossRef]

135. Ilyas, R.A.; Sapuan, S.M.; Ishak, M.R. Isolation and characterization of nanocrystalline cellulose from sugar palm fibres (Arenga Pinnata). *Carbohydr. Polym.* **2018**, *181*, 1038–1051. [CrossRef]
136. Ilyas, R.A.; Sapuan, S.M.; Ibrahim, R.; Abral, H.; Ishak, M.R.; Zainudin, E.S.; Asrofi, M.; Atikah, M.S.N.; Huzaifah, M.R.M.; Radzi, A.M.; et al. Sugar palm (Arenga pinnata (Wurmb.) Merr) cellulosic fibre hierarchy: A comprehensive approach from macro to nano scale. *J. Mater. Res. Technol.* **2019**, *8*, 2753–2766. [CrossRef]
137. Norizan, M.N.; Abdan, K.; Ilyas, R.A.; Biofibers, S.P. Effect of fiber orientation and fiber loading on the mechanical and thermal properties of sugar palm yarn fiber reinforced unsaturated polyester resin composites. *Polimery* **2020**, *65*, 34–43. [CrossRef]
138. Nurazzi, N.M.; Khalina, A.; Sapuan, S.M.; Ilyas, R.A.; Rafiqah, S.A.; Hanafee, Z.M. Thermal properties of treated sugar palm yarn/glass fiber reinforced unsaturated polyester hybrid composites. *J. Mater. Res. Technol.* **2020**, *9*, 1606–1618. [CrossRef]
139. Ilyas, R.A.; Sapuan, S.M.; Ishak, M.R.; Zainudin, E.S. Effect of delignification on the physical, thermal, chemical, and structural properties of sugar palm fibre. *BioResources* **2017**, *12*, 8734–8754. [CrossRef]
140. Ilyas, R.A.; Sapuan, S.M.; Atikah, M.S.N.; Asyraf, M.R.M.; Rafiqah, S.A.; Aisyah, H.A.; Nurazzi, N.M.; Norrrahim, M.N.F. Effect of hydrolysis time on the morphological, physical, chemical, and thermal behavior of sugar palm nanocrystalline cellulose (Arenga pinnata (Wurmb.) Merr). *Text. Res. J.* **2021**, *91*, 152–167. [CrossRef]
141. Sahari, J.; Sapuan, S.M.; Zainudin, E.S.; Maleque, M.A. Physico-chemical and thermal properties of starch derived from sugar palm tree (Arenga pinnata). *Asian J. Chem.* **2014**, *26*, 955–959. [CrossRef]
142. Sanyang, M.L.; Sapuan, S.M.; Jawaid, M.; Ishak, M.R.; Sahari, J. Effect of plasticizer type and concentration on physical properties of biodegradable films based on sugar palm (arenga pinnata) starch for food packaging. *J. Food Sci. Technol.* **2016**, *53*, 326–336. [CrossRef] [PubMed]
143. Ilyas, R.A.; Sapuan, S.M.; Ishak, M.R.; Zainudin, E.S. Sugar palm nanocrystalline cellulose reinforced sugar palm starch composite: Degradation and water-barrier properties. *IOP Conf. Ser. Mater. Sci. Eng.* **2018**, *368*, 012006. [CrossRef]
144. Ilyas, R.A.; Sapuan, S.M.; Ishak, M.R.; Zainudin, E.S. Development and characterization of sugar palm nanocrystalline cellulose reinforced sugar palm starch bionanocomposites. *Carbohydr. Polym.* **2018**, *202*, 186–202. [CrossRef]
145. Ilyas, R.A.; Sapuan, S.M.; Ishak, M.R.; Zainudin, E.S. Water transport properties of bio-nanocomposites reinforced by sugar palm (Arenga Pinnata) nanofibrillated cellulose. *J. Adv. Res. Fluid Mech. Therm. Sci. J.* **2018**, *51*, 234–246.
146. Nazrin, A.; Sapuan, S.M.; Zuhri, M.Y.M.; Tawakkal, I.S.M.A.; Ilyas, R.A. Water barrier and mechanical properties of sugar palm crystalline nanocellulose reinforced thermoplastic sugar palm starch (TPS)/poly(lactic acid) (PLA) blend bionanocomposites. *Nanotechnol. Rev.* **2021**, *10*, 431–442. [CrossRef]
147. Ilyas, R.A.; Sapuan, S.M.; Atiqah, A.; Ibrahim, R.; Abral, H.; Ishak, M.R.; Zainudin, E.S.; Nurazzi, N.M.; Atikah, M.S.N.; Ansari, M.N.M.; et al. Sugar palm (Arenga pinnata [Wurmb.] Merr) starch films containing sugar palm nanofibrillated cellulose as reinforcement: Water barrier properties. *Polym. Compos.* **2020**, *41*, 459–467. [CrossRef]
148. Hazrol, M.D.; Sapuan, S.M.; Ilyas, R.A.; Othman, M.L.; Sherwani, S.F.K. Electrical properties of sugar palm nanocrystalline cellulose reinforced sugar palm starch nanocomposites. *Polimery* **2020**, *65*, 363–370. [CrossRef]
149. Atikah, M.S.N.; Ilyas, R.A.; Sapuan, S.M.; Ishak, M.R.; Zainudin, E.S.; Ibrahim, R.; Atiqah, A.; Ansari, M.N.M.; Jumaidin, R. Degradation and physical properties of sugar palm starch/sugar palm nanofibrillated cellulose bionanocomposite. *Polimery* **2019**, *64*, 680–689. [CrossRef]
150. Ilyas, R.A.; Sapuan, S.M.; Ibrahim, R.; Abral, H.; Ishak, M.R.; Zainudin, E.S.; Atiqah, A.; Atikah, M.S.N.; Syafri, E.; Asrofi, M.; et al. Thermal, Biodegradability and Water Barrier Properties of Bio-Nanocomposites Based on Plasticised Sugar Palm Starch and Nanofibrillated Celluloses from Sugar Palm Fibres. *J. Biobased Mater. Bioenergy* **2020**, *14*, 234–248. [CrossRef]
151. Ilyas, R.A.; Sapuan, S.M.; Ibrahim, R.; Abral, H.; Ishak, M.R.; Zainudin, E.S.; Atikah, M.S.N.; Mohd Nurazzi, N.; Atiqah, A.; Ansari, M.N.M.; et al. Effect of sugar palm nanofibrillated cellulose concentrations on morphological, mechanical and physical properties of biodegradable films based on agro-waste sugar palm (Arenga pinnata (Wurmb.) Merr) starch. *J. Mater. Res. Technol.* **2019**, *8*, 4819–4830. [CrossRef]
152. Suriani, M.J.; Sapuan, S.M.; Ruzaidi, C.M.; Nair, D.S.; Ilyas, R.A. Flammability, morphological and mechanical properties of sugar palm fiber/polyester yarn-reinforced epoxy hybrid biocomposites with magnesium hydroxide flame retardant filler. *Text. Res. J.* **2021**. [CrossRef]
153. Ishak, M.R.; Leman, Z.; Sapuan, S.M.; Salleh, M.Y.; Misri, S. Effect of Sea Water Treatment on the Impact and Flexural Strength of Sugar Palm Fibre Reinforced Epoxy Composites. *Int. J. Mech. Mater. Eng. (IJMME)* **2009**, *4*, 316–320.
154. Sanyang, M.L.; Sapuan, S.M.; Jawaid, M.; Ishak, M.R.; Sahari, J. Recent developments in sugar palm (Arenga pinnata) based biocomposites and their potential industrial applications: A review. *Renew. Sustain. Energy Rev.* **2016**, *54*, 533–549. [CrossRef]
155. Leman, Z.; Sastra, H.Y.; Sapuan, S.M.; Hamdan, M.M.H.M.; Maleque, M.A. Study on impact properties of Arenga pinnata fibre reinforced epoxy composites. *J. Appl. Technol.* **2005**, *3*, 14–19.
156. Bachtiar, D.; Sapuan, S.M.; Hamdan, M.M. Flexural properties of alkaline treated sugar palm fibre reinforced epoxy composites. *Int. J. Automot. Mech. Eng.* **2010**, *1*, 79–90. [CrossRef]
157. Ishak, M.R.; Sapuan, S.M.; Leman, Z.; Rahman, M.Z.A.; Anwar, U.M.K. Characterization of sugar palm (Arenga pinnata) fibres. *J. Therm. Anal. Calorim.* **2012**, *109*, 981–989. [CrossRef]
158. Nurazzi, N.M.; Harussani, M.M.; Zulaikha, N.D.S.; Norhana, A.H.; Syakir, M.I.; Norli, A. Composites based on conductive polymer with carbon nanotubes in DMMP gas sensors—An overview. *Polimery* **2021**, *66*, 85–97. [CrossRef]

159. Rozilah, A.; Jaafar, C.N.A.; Sapuan, S.M.; Zainol, I.; Ilyas, R.A. The Effects of Silver Nanoparticles Compositions on the Mechanical, Physiochemical, Antibacterial, and Morphology Properties of Sugar Palm Starch Biocomposites for Antibacterial Coating. *Polymers* **2020**, *12*, 2605. [CrossRef]
160. Atiqah, A.; Jawaid, M.; Sapuan, S.M.; Ishak, M.R.; Alothman, O.Y. Thermal Properties of Sugar Palm/Glass Fiber Reinforced Thermoplastic Polyurethane hybrid composites. *Compos. Struct.* **2018**. [CrossRef]
161. Atiqah, A.; Jawaid, M.; Sapuan, S.M.; Ishak, M.R.; Ansari, M.N.M.; Ilyas, R.A. Physical and thermal properties of treated sugar palm/glass fibre reinforced thermoplastic polyurethane hybrid composites. *J. Mater. Res. Technol.* **2019**, *8*, 3726–3732. [CrossRef]
162. Nishino, T.; Hirao, K.; Kotera, M.; Nakamae, K.; Inagaki, H. Kenaf reinforced biodegradable composite. *Compos. Sci. Technol.* **2003**, *63*, 1281–1286. [CrossRef]
163. Nor, A.F.M.; Hassan, M.Z.; Rasid, Z.A.; Aziz, S.A.; Sarip, S.; Md Daud, M.Y. Optimization on Tensile Properties of Kenaf/Multi-walled CNT Hybrid Composites with Box-Behnken Design. *Appl. Compos. Mater.* **2021**, *28*, 607–632. [CrossRef]
164. Du, Y.; Zhang, J.; Yu, J.; Lacy, T.E., Jr.; Xue, Y.; Toghiani, H.; Horstemeyer, M.F.; Pittman, C.U., Jr. Kenaf bast fiber bundle–reinforced unsaturated polyester composites. IV: Effects of fiber loadings and aspect ratios on composite tensile properties. *For. Prod. J.* **2010**, *60*, 582–591. [CrossRef]
165. Bonnia, N.N.; Ahmad, S.H.; Zainol, I.; Mamun, A.A.; Beg, M.D.H.; Bledzki, A.K. Mechanical properties and environmental stress cracking resistance of rubber toughened polyester/kenaf composite. *Express Polym. Lett* **2010**, *4*, 55–61. [CrossRef]
166. Sreekala, M.S.; Kumaran, M.G.; Thomas, S. Stress relaxation behaviour in oil palm fibres. *Mater. Lett.* **2001**, *50*, 263–273. [CrossRef]
167. Agarwal, R.; Saxena, N.S.; Sharma, K.B.; Thomas, S.; Sreekala, M.S. Effect of different treatments on the thermal behavior of reinforced phenol–formaldehyde polymer composites. *J. Appl. Polym. Sci.* **2000**, *78*, 603–608. [CrossRef]
168. Norrrahim, M.N.F.; Huzaifah, M.R.M.; Farid, M.A.A.; Shazleen, S.S.; Misenan, M.S.M.; Yasim-Anuar, T.A.T.; Naveen, J.; Nurazzi, N.M.; Rani, M.S.A.; Hakimi, M.I.; et al. Greener Pretreatment Approaches for the Valorisation of Natural Fibre Biomass into Bioproducts. *Polymers* **2021**, *13*, 2971. [CrossRef]
169. Khalid, M.; Ratnam, C.T.; Chuah, T.G.; Ali, S.; Choong, T.S.Y. Comparative study of polypropylene composites reinforced with oil palm empty fruit bunch fiber and oil palm derived cellulose. *Mater. Des.* **2008**, *29*, 173–178. [CrossRef]
170. Khalil, H.P.S.A.; Issam, A.M.; Shakri, M.T.A.; Suriani, R.; Awang, A.Y. Conventional agro-composites from chemically modified fibres. *Ind. Crop. Prod.* **2007**, *26*, 315–323. [CrossRef]
171. Kaddami, H.; Dufresne, A.; Khelifi, B.; Bendahou, A.; Taourirte, M.; Raihane, M.; Issartel, N.; Sautereau, H.; Gerard, J.-F.; Sami, N. Short palm tree fibers–Thermoset matrices composites. *Compos. Part. A Appl. Sci. Manuf.* **2006**, *37*, 1413–1422. [CrossRef]
172. Amor, I.B.; Rekik, H.; Kaddami, H.; Raihane, M.; Arous, M.; Kallel, A. Studies of dielectric relaxation in natural fiber–polymer composites. *J. Electrostat.* **2009**, *67*, 717–722. [CrossRef]
173. Threepopnatkul, P.; Kaerkitcha, N.; Athipongarporn, N. Polycarbonate with pineapple leaf fiber to produce functional composites. In *Proceedings of the Advanced Materials Research*; Trans Tech Publisher: Bristol, UK, 2008; Volume 47, pp. 674–677.
174. Norizan, M.N.; Moklis, M.H.; Alias, A.H.; Rushdan, A.I.; Norrrahim, M.N.F.; Abdan, K.; Abdullah, N. Treatments of Natural Fibre as Reinforcement in Polymer Composites-Short Review. *Funct. Compos. Struct.* **2021**, *3*, 024002. [CrossRef]
175. Mangal, R.; Saxena, N.S.; Sreekala, M.S.; Thomas, S.; Singh, K. Thermal properties of pineapple leaf fiber reinforced composites. *Mater. Sci. Eng. A* **2003**, *339*, 281–285. [CrossRef]
176. Mishra, S.; Misra, M.; Tripathy, S.S.; Nayak, S.K.; Mohanty, A.K. Potentiality of pineapple leaf fibre as reinforcement in PALF-polyester composite: Surface modification and mechanical performance. *J. Reinf. Plast. Compos.* **2001**, *20*, 321–334. [CrossRef]
177. Shih, Y. Mechanical and thermal properties of waste water bamboo husk fiber reinforced epoxy composites. *Mater. Sci. Eng. A* **2007**, *446*, 289–295. [CrossRef]
178. Lopattananon, N.; Panawarangkul, K.; Sahakaro, K.; Ellis, B. Performance of pineapple leaf fiber–natural rubber composites: The effect of fiber surface treatments. *J. Appl. Polym. Sci.* **2006**, *102*, 1974–1984. [CrossRef]
179. Saidane, E.H.; Scida, D.; Pac, M.-J.; Ayad, R. Mode-I interlaminar fracture toughness of flax, glass and hybrid flax-glass fibre woven composites: Failure mechanism evaluation using acoustic emission analysis. *Polym. Test.* **2019**, *75*, 246–253. [CrossRef]
180. Arun, P.V.R.; Rajadurai, A. Inter laminar shear strength behavior of acid, base and silane treated E-glass fibre epoxy resin composites on drilling process. *Def. Technol.* **2017**, *13*, 40–46. [CrossRef]
181. Suriani, M.J.; Rapi, H.Z.; Ilyas, R.A.; Petrů, M.; Sapuan, S.M. Delamination and Manufacturing Defects in Natural Fiber-Reinforced Hybrid Composite: A Review. *Polymers* **2021**, *13*, 1323. [CrossRef]
182. Riccio, A. Delamination in the context of composite structural design. *Delamination Behav. Compos. A Vol. Woodhead Publ. Ser. Compos. Sci. Eng.* **2008**, 28–64. [CrossRef]
183. Chermoshentseva, A.S.; Pokrovskiy, A.M.; Bokhoeva, L.A. The behavior of delaminations in composite materials—Experimental results. *IOP Conf. Ser. Mater. Sci. Eng.* **2016**, *116*. [CrossRef]
184. Khan, S.U.; Kim, J.K. Impact and delamination failure of multiscale carbon nanotube-fiber reinforced polymer composites: A review. *Int. J. Aeronaut. Sp. Sci.* **2011**, *12*, 115–133. [CrossRef]
185. Kim, J.; Technology, M.S.-C.S. *Undefined Impact and Delamination Failure of Woven-Fabric Composites*; Elsevier: Amsterdam, The Netherlands, 2020.
186. Dhanawade, A.; Kumar, S. Experimental study of delamination and kerf geometry of carbon epoxy composite machined by abrasive water jet. *J. Compos. Mater.* **2017**, *51*, 3373–3390. [CrossRef]

187. Senthilkumar, K.; Saba, N.; Chandrasekar, M.; Jawaid, M.; Rajini, N.; Alothman, O.Y.; Siengchin, S. Evaluation of mechanical and free vibration properties of the pineapple leaf fibre reinforced polyester composites. *Constr. Build. Mater.* **2019**, *195*, 423–431. [CrossRef]
188. Feng, N.L.; Malingam, S.D.; R'azali, N.; Subramonian, S. Alkali and Silane Treatments towards Exemplary Mechanical Properties of Kenaf and Pineapple Leaf Fibre-reinforced Composites. *J. Bionic Eng.* **2020**, *17*, 380–392. [CrossRef]
189. Shah, D.U.; Konnerth, J.; Ramage, M.H.; Gusenbauer, C. Mapping thermal conductivity across bamboo cell walls with scanning thermal microscopy. *Sci. Rep.* **2019**, *9*, 1–8. [CrossRef] [PubMed]
190. Zakikhani, P.; Zahari, R.; Sultan, M.T.H.; Majid, D.L. Extraction and preparation of bamboo fibre-reinforced composites. *Mater. Des.* **2014**, *63*, 820–828. [CrossRef]
191. Azammi, A.M.N.; Ilyas, R.A.; Sapuan, S.M.; Ibrahim, R.; Atikah, M.S.N.; Asrofi, M.; Atiqah, A. Characterization Studies Of Biopolymeric Matrix And Cellulose Fibres Based Composites Related To Functionalized Fibre-Matrix Interface. In *Interfaces in Particle and Fibre Reinforced Composites*; Elsevier: London, UK, 2020; pp. 29–93, ISBN 9780081026656.
192. Thwe, M.M.; Liao, K. Effects of environmental aging on the mechanical properties of bamboo–glass fiber reinforced polymer matrix hybrid composites. *Compos. Part. A Appl. Sci. Manuf.* **2002**, *33*, 43–52. [CrossRef]
193. Ying-Chen, Z.; Hong-Yan, W.; Yi-Ping, Q. Morphology and properties of hybrid composites based on polypropylene/polylactic acid blend and bamboo fiber. *Bioresour. Technol.* **2010**, *101*, 7944–7950. [CrossRef]
194. Thyavihalli Girijappa, Y.G.; Mavinkere Rangappa, S.; Parameswaranpillai, J.; Siengchin, S. Natural Fibers as Sustainable and Renewable Resource for Development of Eco-Friendly Composites: A Comprehensive Review. *Front. Mater.* **2019**, *6*, 1–14. [CrossRef]
195. Ramesh, M.; Palanikumar, K.; Reddy, K.H. Plant fibre based bio-composites: Sustainable and renewable green materials. *Renew. Sustain. Energy Rev.* **2017**, *79*, 558–584. [CrossRef]
196. Lim, J.K. Stress corrosion cracking (SCC) in polymer composites. In *Stress Corrosion Cracking*; Elsevier: Amsterdam, The Netherlands, 2011; pp. 485–536.
197. Mohanty, A.K.; Khan, M.A.; Hinrichsen, G. Surface modification of jute and its influence on performance of biodegradable jute-fabric/Biopol composites. *Compos. Sci. Technol.* **2000**, *60*, 1115–1124. [CrossRef]
198. Zaman, H.U.; Khan, R.A.; Haque, M.E.; Khan, M.A.; Khan, A.; Huq, T.; Noor, N.; Rahman, M.; Mustafizur Rahman, K.; Huq, D. Preparation and mechanical characterization of jute reinforced polypropylene/natural rubber composite. *J. Reinf. Plast. Compos.* **2010**, *29*, 3064–3065. [CrossRef]
199. Acha, B.A.; Reboredo, M.M.; Marcovich, N.E. Creep and dynamic mechanical behavior of PP–jute composites: Effect of the interfacial adhesion. *Compos. Part. A Appl. Sci. Manuf.* **2007**, *38*, 1507–1516. [CrossRef]
200. Wang, X.; Cui, Y.; Xu, Q.; Xie, B.; Li, W. Effects of alkali and silane treatment on the mechanical properties of jute-fiber-reinforced recycled polypropylene composites. *J. Vinyl Addit. Technol.* **2010**, *16*, 183–188. [CrossRef]
201. Hong, C.K.; Hwang, I.; Kim, N.; Park, D.H.; Hwang, B.S.; Nah, C. Mechanical properties of silanized jute–polypropylene composites. *J. Ind. Eng. Chem.* **2008**, *14*, 71–76. [CrossRef]
202. Sampath, A.; Martin, G.C. Enhancement of Natural Fiber-Epoxy Interaction Using Bi-Functional Surface Modifiers (963). In *Proceedings of the Technical Papers of the Annual Technical Conference*; Society of Plastics Engineers (SPE) Incorporated: Danbury, CT, USA, 2000; Volume 2, pp. 2274–2278.
203. Dwivedi, U.K.; Chand, N. Influence of fibre orientation on friction and sliding wear behaviour of jute fibre reinforced polyester composite. *Appl. Compos. Mater.* **2009**, *16*, 93–100. [CrossRef]
204. Singh, B.; Gupta, M.; Verma, A. The durability of jute fibre-reinforced phenolic composites. *Compos. Sci. Technol.* **2000**, *60*, 581–589. [CrossRef]
205. Alves, C.; Ferrão, P.M.C.; Silva, A.J.; Reis, L.G.; Freitas, M.; Rodrigues, L.B.; Alves, D.E. Ecodesign of automotive components making use of natural jute fiber composites. *J. Clean. Prod.* **2010**, *18*, 313–327. [CrossRef]
206. Mir, A.; Zitoune, R.; Collombet, F.; Bezzazi, B. Study of mechanical and thermomechanical properties of jute/epoxy composite laminate. *J. Reinf. Plast. Compos.* **2010**, *29*, 1669–1680. [CrossRef]
207. Sarikanat, M. The influence of oligomeric siloxane concentration on the mechanical behaviors of alkalized jute/modified epoxy composites. *J. Reinf. Plast. Compos.* **2010**, *29*, 807–817. [CrossRef]
208. Fraga, A.N.; Frullloni, E.; De la Osa, O.; Kenny, J.M.; Vázquez, A. Relationship between water absorption and dielectric behaviour of natural fibre composite materials. *Polym. Test.* **2006**, *25*, 181–187. [CrossRef]
209. Santulli, C. Post-impact damage characterisation on natural fibre reinforced composites using acoustic emission. *NDT E Int.* **2001**, *34*, 531–536. [CrossRef]
210. Dash, B.N.; Rana, A.K.; Mishra, H.K.; Nayak, S.K.; Tripathy, S.S. Novel low-cost jute–polyester composites. III. Weathering and thermal behavior. *J. Appl. Polym. Sci.* **2000**, *78*, 1671–1679. [CrossRef]
211. Sever, K.; Sarikanat, M.; Seki, Y.; Erkan, G.; Erdoğan, Ü.H. The mechanical properties of γ-methacryloxypropyltrimethoxy silane-treated jute/polyester composites. *J. Compos. Mater.* **2010**, *44*, 1913–1924. [CrossRef]
212. Pracella, M.; Chionna, D.; Anguillesi, I.; Kulinski, Z.; Piorkowska, E. Functionalization, compatibilization and properties of polypropylene composites with hemp fibres. *Compos. Sci. Technol.* **2006**, *66*, 2218–2230. [CrossRef]
213. Li, Y.; Pickering, K.L. The effect of chelator and white rot fungi treatments on long hemp fibre-reinforced composites. *Compos. Sci. Technol.* **2009**, *69*, 1265–1270. [CrossRef]

214. Li, Y.; Pickering, K.L. Hemp fibre reinforced composites using chelator and enzyme treatments. *Compos. Sci. Technol.* **2008**, *68*, 3293–3298. [CrossRef]
215. Li, Y.; Pickering, K.L.; Farrell, R.L. Determination of interfacial shear strength of white rot fungi treated hemp fibre reinforced polypropylene. *Compos. Sci. Technol.* **2009**, *69*, 1165–1171. [CrossRef]
216. Wang, C.; Zuo, Q.; Lin, T.; Anuar, N.I.S.; Mohd Salleh, K.; Gan, S.; Yousfani, S.H.S.; Zuo, H.; Zakaria, S. Predicting thermal conductivity and mechanical property of bamboo fibers/polypropylene nonwovens reinforced composites based on regression analysis. *Int. Commun. Heat Mass Transf.* **2020**, *118*, 104895. [CrossRef]
217. Amin, K.F.; Sharif, A.; Hoque, M.E. Bonding Mechanism and Interface Enhancement of Bamboo Fiber Reinforced Composites. In *Bamboo Fiber Composites*; Springer: Singapore, 2021; pp. 215–233.
218. Lobo, A.; Haseebuddin, M.R.; Harsha, S.; Acharya, K.G.; Balaji, G.; Pal, B. Mechanical behavior of disposed fiberglass filled bamboo mat reinforced polyester composite. *Mater. Today Proc.* **2021**. [CrossRef]
219. de Araujo Alves Lima, R.; Kawasaki Cavalcanti, D.; de Souza e Silva Neto, J.; Meneses da Costa, H.; Banea, M.D. Effect of surface treatments on interfacial properties of natural intralaminar hybrid composites. *Polym. Compos.* **2020**, *41*, 314–325. [CrossRef]
220. Bourmaud, A.; Baley, C. Rigidity analysis of polypropylene/vegetal fibre composites after recycling. *Polym. Degrad. Stab.* **2009**, *94*, 297–305. [CrossRef]
221. Santulli, C.; Caruso, A.P. Effect of fibre architecture on the falling weight impact properties of hemp/epoxy composites. *J. Biobased Mater. Bioenergy* **2009**, *3*, 291–297. [CrossRef]
222. Cicala, G.; Cristaldi, G.; Recca, G.; Ziegmann, G.; El-Sabbagh, A.; Dickert, M. Properties and performances of various hybrid glass/natural fibre composites for curved pipes. *Mater. Des.* **2009**, *30*, 2538–2542. [CrossRef]
223. Scarponi, C.; Pizzinelli, C.S.; Sánchez-Sáez, S.; Barbero, E. Impact load behaviour of resin transfer moulding (RTM) hemp fibre composite laminates. *J. Biobased Mater. Bioenergy* **2009**, *3*, 298–310. [CrossRef]
224. Eichhorn, S.J.; Young, R.J. Composite micromechanics of hemp fibres and epoxy resin microdroplets. *Compos. Sci. Technol.* **2004**, *64*, 767–772. [CrossRef]
225. Hepworth, D.G.; Hobson, R.N.; Bruce, D.M.; Farrent, J.W. The use of unretted hemp fibre in composite manufacture. *Compos. Part. A Appl. Sci. Manuf.* **2000**, *31*, 1279–1283. [CrossRef]
226. Kunanopparat, T.; Menut, P.; Morel, M.-H.; Guilbert, S. Reinforcement of plasticized wheat gluten with natural fibers: From mechanical improvement to deplasticizing effect. *Compos. Part. A Appl. Sci. Manuf.* **2008**, *39*, 777–785. [CrossRef]
227. John, M.J.; Anandjiwala, R.D. Chemical modification of flax reinforced polypropylene composites. *Compos. Part. A Appl. Sci. Manuf.* **2009**, *40*, 442–448. [CrossRef]
228. Fiore, V.; Scalici, T.; Sarasini, F.; Tirilló, J.; Calabrese, L. Salt-fog spray aging of jute-basalt reinforced hybrid structures: Flexural and low velocity impact response. *Compos. Part. B Eng.* **2017**, *116*, 99–112. [CrossRef]
229. Oksman, K. Mechanical properties of natural fibre mat reinforced thermoplastic. *Appl. Compos. Mater.* **2000**, *7*, 403–414. [CrossRef]
230. Arbelaiz, A.; Fernández, B.; Cantero, G.; Llano-Ponte, R.; Valea, A.; Mondragon, I. Mechanical properties of flax fibre/polypropylene composites. Influence of fibre/matrix modification and glass fibre hybridization. *Compos. Part. A Appl. Sci. Manuf.* **2005**, *36*, 1637–1644. [CrossRef]
231. Duhovic, M.; Horbach, S.; Bhattacharyya, D. Improving the interface strength in flax fibre poly(lactic) acid composites. *J. Biobased Mater. Bioenergy* **2009**, *3*, 188–198. [CrossRef]
232. Van de Velde, K.; Kiekens, P. Effect of material and process parameters on the mechanical properties of unidirectional and multidirectional flax/polypropylene composites. *Compos. Struct.* **2003**, *62*, 443–448. [CrossRef]
233. Bos, H.L.; Müssig, J.; van den Oever, M.J.A. Mechanical properties of short-flax-fibre reinforced compounds. *Compos. Part. A Appl. Sci. Manuf.* **2006**, *37*, 1591–1604. [CrossRef]
234. Buttlar, H.B. Natural fibre reinforced construction materials for SMC applications. In *Proceedings of the Conference RIKO-2005*; Niedersachsen, German, 2005; pp. 1–24.
235. Stuart, T.; Liu, Q.; Hughes, M.; McCall, R.D.; Sharma, H.S.S.; Norton, A. Structural biocomposites from flax—Part I: Effect of bio-technical fibre modification on composite properties. *Compos. Part. A Appl. Sci. Manuf.* **2006**, *37*, 393–404. [CrossRef]
236. Van de Weyenberg, I.; Chi Truong, T.; Vangrimde, B.; Verpoest, I. Improving the properties of UD flax fibre reinforced composites by applying an alkaline fibre treatment. *Compos. Part. A Appl. Sci. Manuf.* **2006**, *37*, 1368–1376. [CrossRef]
237. Van de Weyenberg, I.; Ivens, J.; De Coster, A.; Kino, B.; Baetens, E.; Verpoest, I. Influence of processing and chemical treatment of flax fibres on their composites. *Compos. Sci. Technol.* **2003**, *63*, 1241–1246. [CrossRef]
238. Manfredi, L.B.; Rodríguez, E.S.; Wladyka-Przybylak, M.; Vázquez, A. Thermal degradation and fire resistance of unsaturated polyester, modified acrylic resins and their composites with natural fibres. *Polym. Degrad. Stab.* **2006**, *91*, 255–261. [CrossRef]
239. Baley, C.; Busnel, F.; Grohens, Y.; Sire, O. Influence of chemical treatments on surface properties and adhesion of flax fibre–polyester resin. *Compos. Part. A Appl. Sci. Manuf.* **2006**, *37*, 1626–1637. [CrossRef]
240. Yu, T.; Ren, J.; Li, S.; Yuan, H.; Li, Y. Effect of fiber surface-treatments on the properties of poly(lactic acid)/ramie composites. *Compos. Part. A Appl. Sci. Manuf.* **2010**, *41*, 499–505. [CrossRef]
241. Nam, S.; Netravali, A.N. Green composites. I. physical properties of ramie fibers for environment-friendly green composites. *Fibers Polym.* **2006**, *7*, 372–379. [CrossRef]
242. Ishikawa, A.; Okano, T.; Sugiyama, J. Fine structure and tensile properties of ramie fibres in the crystalline form of cellulose I., II, IIII and IVI. *Polymers* **1997**, *38*, 463–468. [CrossRef]

243. Goda, K.; Sreekala, M.S.; Gomes, A.; Kaji, T.; Ohgi, J. Improvement of plant based natural fibers for toughening green composites—Effect of load application during mercerization of ramie fibers. *Compos. Part. A Appl. Sci. Manuf.* **2006**, *37*, 2213–2220. [CrossRef]
244. Ilyas, R.A.; Sapuan, S.M. Biopolymers and Biocomposites: Chemistry and Technology. *Curr. Anal. Chem.* **2020**, *16*, 500–503. [CrossRef]
245. Nurazzi, N.M.; Asyraf, M.R.M.; Khalina, A.; Abdullah, N.; Aisyah, H.A.; Rafiqah, S.A.; Sabaruddin, F.A.; Kamarudin, S.H.; Norrrahim, M.N.F.; Ilyas, R.A.; et al. A Review on Natural Fiber Reinforced Polymer Composite for Bullet Proof and Ballistic Applications. *Polymers* **2021**, *13*, 646. [CrossRef]
246. Ilyas, R.A.; Sapuan, S.M. The Preparation Methods and Processing of Natural Fibre Bio-polymer Composites. *Curr. Org. Synth.* **2020**, *16*, 1068–1070. [CrossRef]
247. Aisyah, H.A.; Paridah, M.T.; Sapuan, S.M.; Ilyas, R.A.; Khalina, A.; Nurazzi, N.M.; Lee, S.H.; Lee, C.H. A comprehensive review on advanced sustainable woven natural fibre polymer composites. *Polymers* **2021**, *13*, 471. [CrossRef]
248. Müssig, J. Cotton fibre-reinforced thermosets versus ramie composites: A comparative study using petrochemical-and agro-based resins. *J. Polym. Environ.* **2008**, *16*, 94–102. [CrossRef]
249. Lodha, P.; Netravali, A.N. Characterization of stearic acid modified soy protein isolate resin and ramie fiber reinforced 'green'composites. *Compos. Sci. Technol.* **2005**, *65*, 1211–1225. [CrossRef]
250. Kishi, H.; Fujita, A. Wood-based epoxy resins and the ramie fiber reinforced composites. *Environ. Eng. Manag. J.* **2008**, *7*, 517–523. [CrossRef]
251. Long, C.-G.; He, L.-P.; Zhong, Z.-H.; Chen, S.-G. Studies on the polypropylene composites reinforced by ramier fiber and K2Ti6O13 whisker. *Res. Lett. Mater. Sci.* **2007**, *2007*. [CrossRef]
252. Pothan, L.A.; Thomas, S.; Groeninckx, G. The role of fibre/matrix interactions on the dynamic mechanical properties of chemically modified banana fibre/polyester composites. *Compos. Part. A Appl. Sci. Manuf.* **2006**, *37*, 1260–1269. [CrossRef]
253. El-Meligy, M.G.; Mohamed, S.H.; Mahani, R.M. Study mechanical, swelling and dielectric properties of prehydrolysed banana fiber–Waste polyurethane foam composites. *Carbohydr. Polym.* **2010**, *80*, 366–372. [CrossRef]
254. Teramoto, N.; Urata, K.; Ozawa, K.; Shibata, M. Biodegradation of aliphatic polyester composites reinforced by abaca fiber. *Polym. Degrad. Stab.* **2004**. [CrossRef]
255. Harussani, M.M.; Sapuan, S.M.; Khalina, A.; Rashid, U.; Tarique, J. Slow pyrolysis of disinfected COVID-19 non-woven polypropylene (PP) waste. In *Proceedings of the International Symposium on Applied Sciences and Engineering ISASE2021*; Office of International Affairs, Atatürk University: Erzurum, Turkey, 2021; pp. 310–312.
256. Zhong, J.B.; Lv, J.; Wei, C. Mechanical properties of sisal fibre reinforced urea formaldehyde resin composites. *Express Polym. Lett.* **2007**, *1*, 681–687. [CrossRef]
257. Senthilkumar, K.; Saba, N.; Rajini, N.; Chandrasekar, M.; Jawaid, M.; Siengchin, S.; Alotman, O.Y. Mechanical properties evaluation of sisal fibre reinforced polymer composites: A review. *Constr. Build. Mater.* **2018**, *174*, 713–729. [CrossRef]
258. Ilyas, R.A.; Sapuan, S.M.; Norrrahim, M.N.F.; Yasim-Anuar, T.A.T.; Kadier, A.; Kalil, M.S.; Atikah, M.S.N.; Ibrahim, R.; Asrofi, M.; Abral, H.; et al. Nanocellulose/starch biopolymer nanocomposites: Processing, manufacturing, and applications. In *Advanced Processing, Properties, and Application of Strach and Other Bio-Based Polymer*; Al-Oqla, F.M., Ed.; Elsevier Inc.: Amsterdam, The Netherlands, 2021.
259. Azammi, A.M.N.; Ilyas, R.A.; Sapuan, S.M.; Ibrahim, R.; Atikah, M.S.N.; Asrofi, M.; Atiqah, A. Characterization studies of biopolymeric matrix and cellulose fibres based composites related to functionalized fibre-matrix interface. In *Interfaces in Particle and Fibre Reinforced Composites- From Macro to Nano Scales*; Woodhead Publishing: London, UK, 2019; pp. 1–68, ISBN 9780081026656.
260. Faruk, O.; Bledzki, A.K.; Fink, H.-P.; Sain, M. Progress Report on Natural Fiber Reinforced Composites. *Macromol. Mater. Eng.* **2014**, *299*, 9–26. [CrossRef]
261. Ilyas, R.A.; Sapuan, S.M.; Asyraf, M.R.M.; Atikah, M.S.N.; Ibrahim, R.; Dele-Afolabi, T.T.; Hazrol, M.D. Introduction to Biofiller-Reinforced Degradable Polymer Composites. In *Biofiller-Reinforced Biodegradable Polymer Composites*; Jumaidin, R., Sapuan, S.M., Ismail, H., Eds.; CRC Press: Boca Raton, FL, USA, 2020; pp. 1–23.
262. Jumaidin, R.; Saidi, Z.A.S.; Ilyas, R.A.; Ahmad, M.N.; Wahid, M.K.; Yaakob, M.Y.; Maidin, N.A.; Rahman, M.H.A.; Osman, M.H. Characteristics of Cogon Grass Fibre Reinforced Thermoplastic Cassava Starch Biocomposite: Water Absorption and Physical Properties. *J. Adv. Res. Fluid Mech. Therm. Sci.* **2019**, *62*, 43–52.
263. Sreekumar. *Matrices for Natural-Fibre Reinforced Composites*; Woodhead Publishing Limited: Brimingham, UK, 2008; ISBN 9781845692674.
264. Suppakarn, N.; Jarukumjorn, K. Mechanical properties and flammability of sisal/PP composites: Effect of flame retardant type and content. *Compos. Part. B Eng.* **2009**, *40*, 613–618. [CrossRef]
265. Joseph, P.V.; Rabello, M.S.; Mattoso, L.H.C.; Joseph, K.; Thomas, S. Environmental effects on the degradation behaviour of sisal fibre reinforced polypropylene composites. *Compos. Sci. Technol.* **2002**, *62*, 1357–1372. [CrossRef]
266. Dwivedi, U.K.; Chand, N. Influence of MA-g-PP on abrasive wear behaviour of chopped sisal fibre reinforced polypropylene composites. *J. Mater. Process. Technol.* **2009**, *209*, 5371–5375. [CrossRef]
267. Zhang, M.; Rong, M.; Lu, X. Fully biodegradable natural fiber composites from renewable resources: All-plant fiber composites. *Compos. Sci. Technol.* **2005**, *65*, 2514–2525. [CrossRef]

268. Athijayamani, A.; Thiruchitrambalam, M.; Natarajan, U.; Pazhanivel, B. Effect of moisture absorption on the mechanical properties of randomly oriented natural fibers/polyester hybrid composite. *Mater. Sci. Eng. A* **2009**, *517*, 344–353. [CrossRef]
269. Sangthong, S.; Pongprayoon, T.; Yanumet, N. Mechanical property improvement of unsaturated polyester composite reinforced with admicellar-treated sisal fibers. *Compos. Part. A Appl. Sci. Manuf.* **2009**, *40*, 687–694. [CrossRef]
270. Megiatto, J.D., Jr.; Silva, C.G.; Rosa, D.S.; Frollini, E. Sisal chemically modified with lignins: Correlation between fibers and phenolic composites properties. *Polym. Degrad. Stab.* **2008**, *93*, 1109–1121. [CrossRef]
271. Megiatto, J.D., Jr.; Ramires, E.C.; Frollini, E. Phenolic matrices and sisal fibers modified with hydroxy terminated polybutadiene rubber: Impact strength, water absorption, and morphological aspects of thermosets and composites. *Ind. Crop. Prod.* **2010**, *31*, 178–184. [CrossRef]
272. Ramires, E.C.; Megiatto, J.D., Jr.; Gardrat, C.; Castellan, A.; Frollini, E. Biobased composites from glyoxal–phenolic resins and sisal fibers. *Bioresour. Technol.* **2010**, *101*, 1998–2006. [CrossRef]
273. Athijayamani, A.; Thiruchitrambalam, M.; Natarajan, U.; Pazhanivel, B. Influence of alkali-treated fibers on the mechanical properties and machinability of roselle and sisal fiber hybrid polyester composite. *Polym. Compos.* **2010**, *31*, 723–731. [CrossRef]
274. Chand, N.; Jain, D. Effect of sisal fibre orientation on electrical properties of sisal fibre reinforced epoxy composites. *Compos. Part. A Appl. Sci. Manuf.* **2005**, *36*, 594–602. [CrossRef]
275. Meddahi, A.; Ait Tahar, K.; Bibi, M. Studies of sisal fiber-containing composites. *J. Nat. Fibers* **2008**, *5*, 36–46. [CrossRef]
276. de Andrade Silva, F.; Mobasher, B.; Toledo Filho, R.D. Cracking mechanisms in durable sisal fiber reinforced cement composites. *Cem. Concr. Compos.* **2009**, *31*, 721–730. [CrossRef]
277. Tonoli, G.H.D.; Santos, S.F.; Joaquim, A.P.; Savastano, H., Jr. Effect of accelerated carbonation on cementitious roofing tiles reinforced with lignocellulosic fibre. *Constr. Build. Mater.* **2010**, *24*, 193–201. [CrossRef]
278. Towo, A.N.; Ansell, M.P. Fatigue of sisal fibre reinforced composites: Constant-life diagrams and hysteresis loop capture. *Compos. Sci. Technol.* **2008**, *68*, 915–924. [CrossRef]
279. Wongsorat, W.; Suppakarn, N.; Jarukumjorn, K. Mechanical Properties, Morphological Properties, and Cure Characteristics of Sisal Fiber/Natural Rubber Composites: Effects of Fiber and Compatibilizer Content. *Adv. Mater. Res.* **2010**, *123*, 1171–1174.
280. de Paula, M.P.; Lacerda, T.M.; Frollini, E. Sisal cellulose acetates obtained from heterogeneous reactions. *Express Polym. Lett* **2008**, *2*, 423–428. [CrossRef]
281. Bakare, I.O.; Okieimen, F.E.; Pavithran, C.; Khalil, H.P.S.A.; Brahmakumar, M. Mechanical and thermal properties of sisal fiber-reinforced rubber seed oil-based polyurethane composites. *Mater. Des.* **2010**, *31*, 4274–4280. [CrossRef]
282. Favaro, S.L.; Ganzerli, T.A.; de Carvalho Neto, A.G.V.; da Silva, O.; Radovanovic, E. Chemical, morphological and mechanical analysis of sisal fiber-reinforced recycled high-density polyethylene composites. *Express Polym. Lett.* **2010**, *4*, 465–473. [CrossRef]
283. Diyana, Z.N.; Jumaidin, R.; Selamat, M.Z.; Ghazali, I.; Julmohammad, N.; Huda, N.; Ilyas, R.A. Physical Properties of Thermoplastic Starch Derived from Natural Resources and Its Blends: A Review. *Polymers* **2021**, *13*, 1396. [CrossRef]
284. Supian, A.B.M.; Sapuan, S.M.; Jawaid, M.; Zuhri, M.Y.M.; Ilyas, R.A.; Syamsir, A. Crashworthiness Response of Filament Wound Kenaf/Glass Fibre-reinforced Epoxy Composite Tubes with Influence of Stacking Sequence under Intermediate-velocity Impact Load. *Fibers Polym.* **2021**, 1–12. [CrossRef]
285. Kandachar, P.; Brouwer, R. Applications of bio-composites in industrial products. *MRS Online Proc. Libr.* **2001**, *702*, 1–12. [CrossRef]
286. Eichhorn, S.J.; Baillie, C.A.; Zafeiropoulos, N.; Mwaikambo, L.Y.; Ansell, M.P.; Dufresne, A.; Entwistle, K.M.; Herrera-Franco, P.J.; Escamilla, G.C.; Groom, L. Current international research into cellulosic fibres and composites. *J. Mater. Sci.* **2001**, *36*, 2107–2131. [CrossRef]
287. Jústiz-Smith, N.G.; Virgo, G.J.; Buchanan, V.E. Potential of Jamaican banana, coconut coir and bagasse fibres as composite materials. *Mater. Charact.* **2008**, *59*, 1273–1278. [CrossRef]
288. Van De Velde, K.; Kiekens, P. Thermal degradation of flax: The determination of kinetic parameters with thermogravimetric analysis. *J. Appl. Polym. Sci.* **2002**, *83*, 2634–2643. [CrossRef]
289. Srinivasan, V.; Rajendra Boopathy, S.; Sangeetha, D.; Vijaya Ramnath, B. Evaluation of mechanical and thermal properties of banana–flax based natural fibre composite. *Mater. Des.* **2014**, *60*, 620–627. [CrossRef]
290. Olusegun, D.S.; Stephen, A.; Adekanye, T.A. Assessing mechanical properties of natural fibre reinforced composites for engineering applications. *J. Miner. Mater. Charact. Eng.* **2012**, *11*, 780–784.
291. Yan, L.; Chouw, N.; Jayaraman, K. Flax fibre and its composites—A review. *Compos. Part. B Eng.* **2014**, *56*, 296–317. [CrossRef]
292. Ho, M.; Wang, H.; Lee, J.-H.; Ho, C.; Lau, K.; Leng, J.; Hui, D. Critical factors on manufacturing processes of natural fibre composites. *Compos. Part. B Eng.* **2012**, *43*, 3549–3562. [CrossRef]
293. Satyanarayana, K.G.; Sukumaran, K.; Mukherjee, P.S.; Pavithran, C.; Pillai, S.G.K. Natural fibre-polymer composites. *Cem. Concr. Compos.* **1990**, *12*, 117–136. [CrossRef]
294. Brischetto, S. A comparative study of composite structures reinforced with carbon, glass or natural fibers. *Multidiscip. Model. Mater. Struct.* **2017**, *17*, 165–187. [CrossRef]
295. Joseph, S.; Sreekala, M.S.; Oommen, Z.; Koshy, P.; Thomas, S. A comparison of the mechanical properties of phenol formaldehyde composites reinforced with banana fibres and glass fibres. *Compos. Sci. Technol.* **2002**, *62*, 1857–1868. [CrossRef]
296. Aziz, S.H.; Ansell, M.P. The effect of alkalization and fibre alignment on the mechanical and thermal properties of kenaf and hemp bast fibre composites: Part 1—Polyester resin matrix. *Compos. Sci. Technol.* **2004**, *64*, 1219–1230. [CrossRef]

297. Alamri, H.; Low, I.M. Effect of water absorption on the mechanical properties of n-SiC filled recycled cellulose fibre reinforced epoxy eco-nanocomposites. *Polym. Test.* **2012**, *31*, 810–818. [CrossRef]
298. Alamri, H.; Low, I.M. Mechanical properties and water absorption behaviour of recycled cellulose fibre reinforced epoxy composites. *Polym. Test.* **2012**, *31*, 620–628. [CrossRef]
299. Dittenber, D.B.; Gangarao, H.V.S. Critical review of recent publications on use of natural composites in infrastructure. *Compos. Part. A Appl. Sci. Manuf.* **2012**, *43*, 1419–1429. [CrossRef]
300. Shibata, S.; Cao, Y.; Fukumoto, I. Press forming of short natural fiber-reinforced biodegradable resin: Effects of fiber volume and length on flexural properties. *Polym. Test.* **2005**, *24*, 1005–1011. [CrossRef]
301. Pothan, L.A.; Thomas, S.; Neelakantan, N.R. Short banana fiber reinforced polyester composites: Mechanical, failure and aging characteristics. *J. Reinf. Plast. Compos.* **1997**, *16*, 744–765. [CrossRef]
302. Ray, D.; Sarkar, B.K.; Bose, N.R. Impact fatigue behaviour of vinylester resin matrix composites reinforced with alkali treated jute fibres. *Compos. Part. A Appl. Sci. Manuf.* **2002**, *33*, 233–241. [CrossRef]
303. Sanjay, M.R.; Arpitha, G.R.; Laxmana Naik, L.; Gopalakrishna, K.; Yogesh, B. Studies on mechanical properties of banana/e-glass fabrics reinforced polyester hybrid composites. *J. Mater. Environ. Sci.* **2016**, *7*, 3179–3192.
304. Zampaloni, M.; Pourboghrat, F.; Yankovich, S.A.; Rodgers, B.N.; Moore, J.; Drzal, L.T.; Mohanty, A.K.; Misra, M. Kenaf natural fiber reinforced polypropylene composites: A discussion on manufacturing problems and solutions. *Compos. Part. A Appl. Sci. Manuf.* **2007**, *38*, 1569–1580. [CrossRef]
305. Magurno, A. Vegetable fibres in automotive interior components. *Die Angew. Makromol. Chem.* **1999**, *272*, 99–107. [CrossRef]
306. Schuh, T.G. *Renewable Materials for Automotive Applications*; Daimler-Chrysler, A., Ed.; CRC Press: Boca Raton, FL, USA, 1999.
307. Wahab, M.; Sapuan, S.M.; Harussani, M.M.; Zuhri, M.Y.M.; Saleh, A.A. Conceptual Design of Glass/Renewable Natural Fibre-Reinforced Polymer Hybrid Composite Motorcycle Side Cover. *J. Renew. Mater.* **2021**, *9*, 1973–1989. [CrossRef]
308. Prabhu, L.; Krishnaraj, V.; Sathish, S.; Gokulkumar, S.; Karthi, N.; Rajeshkumar, L.; Balaji, D.; Vigneshkumar, N.; Elango, K.S. A review on natural fiber reinforced hybrid composites: Chemical treatments, manufacturing methods and potential applications. *Mater. Today Proc.* **2021**, *45*, 8080–8085. [CrossRef]
309. Joshi, S.V.; Drzal, L.T.; Mohanty, A.K.; Arora, S. Are natural fiber composites environmentally superior to glass fiber reinforced composites? *Compos. Part. A Appl. Sci. Manuf.* **2004**, *35*, 371–376. [CrossRef]
310. Yu, H.N.; Kim, S.S.; Hwang, I.U. Application of natural fiber reinforced composites to trenchless rehabilitation of underground pipes. *Compos. Struct.* **2008**, *86*, 285–290. [CrossRef]
311. Moreau, R. Nautical Activities: What Impact on the Environment? A Life Cycle Approach for "Clear Blue" Boating-Commissioned by the European Confederation of Nautical Industries-ECNI. 2009. Available online: https://www.europeanboatingindustry.eu/images/Documents/For_publications/Nautical-activities_what-impact-on-the-environment.pdf (accessed on 8 October 2021).

polymers

MDPI

Article

Quasi-Static Compression Properties of Bamboo and PVC Tube Reinforced Polymer Foam Structures

J. J. N. Amelia [1], M. Y. M. Zuhri [1,2,*], Z. Leman [1], N. I. Zahari [1], A. As'arry [3] and R. A. Ilyas [4,5]

[1] Advanced Engineering Materials and Composites Research Centre, Department of Mechanical and Manufacturing Engineering, Faculty of Engineering, Universiti Putra Malaysia, Serdang 43400, Malaysia; ameliajuria96@gmail.com (J.J.N.A.); zleman@upm.edu.my (Z.L.); rubie@upm.edu.my (N.I.Z.)

[2] Laboratory of Biocomposite Technology, Institute of Tropical Forestry and Forest Product (INTROP), Universiti Putra Malaysia, Serdang 43400, Malaysia

[3] Department of Mechanical and Manufacturing Engineering, Faculty of Engineering, Universiti Putra Malaysia, Serdang 43400, Malaysia; zizan@upm.edu.my

[4] School of Chemical and Energy Engineering, Faculty of Engineering, Universiti Teknologi Malaysia, UTM, Johor Bahru 81310, Malaysia; ahmadilyas@utm.my

[5] Centre for Advanced Composite Materials (CACM), Universiti Teknologi Malaysia, UTM, Johor Bahru 81310, Malaysia

* Correspondence: zuhri@upm.edu.my

Abstract: In recent years, there has been a growing interest for composite materials due to the superior capability to absorb energy and lightweight factor. These properties are compatible to be utilized in the development for transportation system as it can reduce the fuel consumption and also minimize the effect of crash to the passenger. Therefore, the aim for this project is to study the compression strength and energy absorbing capability for Polyvinyl chloride (PVC) and bamboo tubes reinforced with foam. Several parameters are being considered, these being the effect of single and multiple tube reinforced foam structure, foam density, diameter of the tube as well as effect of different crosshead speed. The results showed that increasing the relative foam density will led to an increase in the compression strength and specific energy absorption (SEA) values. Furthermore, a significant increase of compression strength can be seen when several tubes are introduced into the foam while SEA remained almost the same. Finally, the influence of crosshead below 20 mm/min did not vary significantly for both compression strength and SEA.

Keywords: compression strength; foam; specific energy absorption; tube

Citation: Amelia, J.J.N.; Zuhri, M.Y.M.; Leman, Z.; Zahari, N.I.; As'arry, A.; Ilyas, R.A. Quasi-Static Compression Properties of Bamboo and PVC Tube Reinforced Polymer Foam Structures. *Polymers* **2021**, *13*, 3603. https://doi.org/10.3390/polym13203603

Academic Editor: Emanoil Linul

Received: 30 September 2021
Accepted: 13 October 2021
Published: 19 October 2021

1. Introduction

An energy absorbent structure can be a structure that exchanges half or all of the mechanical energy into another sort of energy. The energy recover will either be reversible similarly in the case of elastic strain energy in solids or it will be irreversible like plastic dissemination of energy related with lasting distortion of the strong structure. Energy absorbing structure design and analysis differ significantly from standard structural design and analysis. Energy absorbing structures must withstand high impact loads due to the complicated deformation and failure by significant geometry changes, strain hardening effects, strain-rate effects, and interactions between distinct deformation modes including bending and stretching [1]. There are several types of sandwich cores that have been studied in recent years. The determination for the core material and its design are crucial and might be vary depending on the application. In addition, structural elements are inserted into the core structures such as foams and honeycombs to modify mechanical properties of the sandwich structures such as core compression strength, buckling instability and in-plane shear [2]. In typical structural applications, the thickness of the face sheet rarely exceeds several millimetres, whereas the thickness of core can be over 50 mm [3]. In a study by Colloca et al. [4] they have reported that the compressive modulus of the foam (Divinycell

PVC) increases as the relative density of the foam increases, but the densification strain decreases. Also, a comparison of absorbed energy shows that the energy absorption during compression process rise up to 600% more than under tension due to greater strain value.

To develop crashworthy structures, tubular structures have been commonly used in the design of tube reinforced foam structure. The most frequent and oldest type of foldable energy absorber is thin tubes. When a thin-walled steel tube is subjected to an axial compressive load, it can generate either a concertina (axisymmetric buckle) or a diamond (non-axisymmetric buckle). A concertina deformation mode is most efficient energy absorbing mode [5]. Khan et al. [6] explained that failure mode for thin wall cylindrical composite tube can be divided into three modes; (1) mushrooming failure, (2) brittle fracture of the composite tube and (3) increasing folding and hinging. In addition, the energy absorbing capacity is highest at the first mode, medium at third mode and the least at second mode. Zuhri et al. [7] examined the energy absorbing properties of bamboo-based structure by conducting quasi-static and dynamic tests. According to the data, the energy absorbing capability of individual bamboo tubes increases when the diameter-to-thickness ratio (D/t) decreases. Also, the small diameter of bamboo tubes showed more noticeable crushing than the larger size. A study from Zhou et al. [8] suggested that by embedding tubes in a foam panel, it will influence the failure process within the composite tubes by significantly improving their ability to absorb energy. However, the SEA values of the hybrid tube reinforced foams were found to be insensitive to variations in foam density. It is important to note that tube-based foams have a larger energy-absorbing capacity than many comparable core systems, emphasising its potential for usage in extreme crushing situations. In a study conducted by Alhawamdeh et al. [9] shows that the failure mode of the axially loaded hollow pultruded fibre-reinforced polymer (PFRP) profiles varied depending on their cross-sectional shape. The hollow box profiles are dominated by local buckling of the walls, whereas the hollow circular profiles dominated by compressive and shear failure at the profiles ends. The results stated that the hollow circular PFRP profiles, the failure mode is the same for all length-to-width (L/D) ratios.

This paper introduces a new combination structure for the purpose to enhance an energy absorption of the current foam structure. Attention is focused on understanding the energy absorbing characteristics of single and multiple tubes reinforced foam structure under quasi-static loading conditions.

2. Materials and Methods

2.1. Material

Two types of crosslinked PVC foams with different densities are used in this study, which are Divinycell F50 and Divinycell HP80 foams with a density of 50 kg/m^3 and 80 kg/m^3, respectively. The foams are supplied by DiabGroup and comes in the shape of flat panel, which are color-coded to differentiate the foam type and both are having thickness of 20 mm. The selected tubes are Polyvinyl chloride (PVC) and eco-friendly bamboo tube. The commercial PVC tubes used are round conduit where it is mainly used for construction, electrical, etc. Two different sizes of tubing are selected, where the outer diameters (D_o) are 20 mm and 25 mm. Then, it is cut into 20 mm length to ensure it having the same height as the foam. The length of the specimen is similar to work conducted by Alia et al. [10] and Cinar [2]. On the other hand, bamboo tubes used are readily available in the laboratory and it is light brown in colour. This bamboo is originally used to create a beautiful garden edging and is chosen due to their highly versatile natural resource that is easily to sustained and eco-friendly. The selection of D_o for the bamboo tube is based on diameter size that is close to 25 mm, due to it is not uniform in nature. Prior to testing, the tubes are cut into 20 mm length (similar reason as for PVC tube) using the circular saw from the internode parts of the bamboo. It is known that the diameter of the node part is slightly larger than the internode part and can affect the structural performance, as investigated by Molari et al. [11]. However, in this study, the node is not considered to ensure the insertion tube having an interference fit to the foam. The physical properties of the tested tubes are shown in Table 1.

Table 1. Physical properties of PVC and Bamboo tubes.

Tube	Outer Diameter, D_o (mm)	Inner Diameter, D_i (mm)	Thickness (mm)	D_i/t Ratio
PVC 20	20	17	1.5	11.3
PVC 25	25	21	2.0	10.5
Bamboo	25	18	3.5	5.1

2.2. Fabrication and Mechanical Testing

Initially, foam thickness of 20 mm is cut into block of square with a dimension of 50 × 50 mm. Then, a hole is formed using circular drill at the centre of the foam to allow an identical size of tube to be inserted. Single tube-reinforced foam samples are designed by embedding a 20 mm long PVC and bamboo tubes into the Divinycell F50 and HP80 foam. In addition, for multiple tubes, the size of the square block is double than the single tube and using the same method as for single tube. Figure 1 shows the illustration and sample of the structure (the unit is in mm).

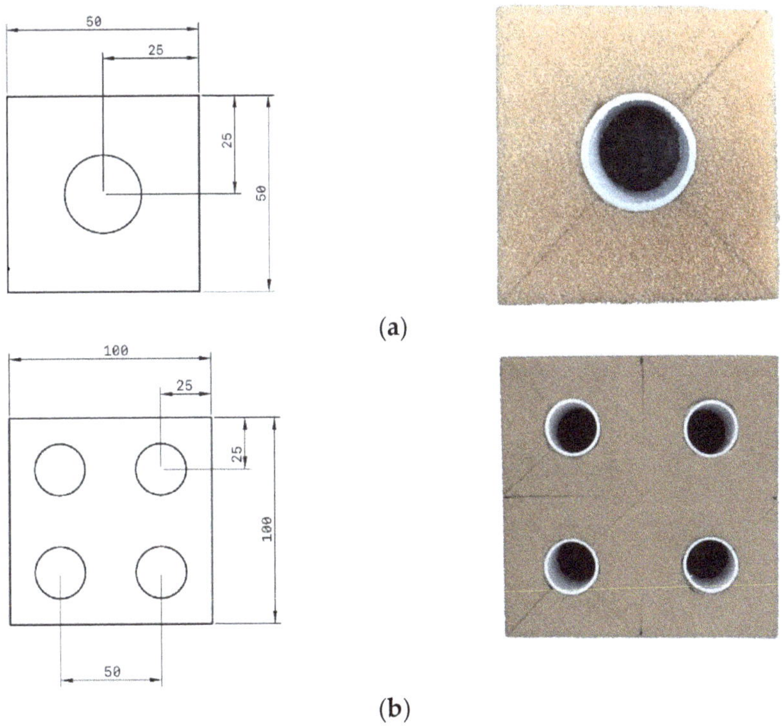

Figure 1. Dimension and positioning for (**a**) single tube and (**b**) multiple tube reinforced foam structure.

Later, a series of axial quasi-static compression testing is performed using Universal Testing Machine Instron model 3382 with a load cell of 100 kN as shown in Figure 2. Each of the specimens are tested at a constant crosshead speed of 2 mm/min. For this purpose, compression test is carried out in accordance to ASTM C365 [12]. For each of the test configurations, three specimens are replicated.

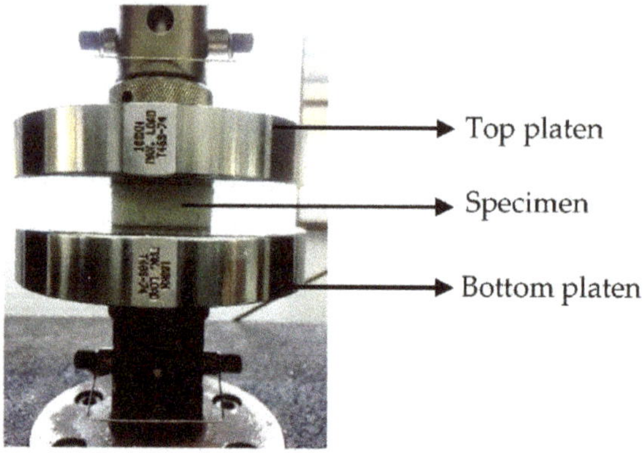

Figure 2. Quasi-static compression testing of specimen.

The specimen is axially crushed between the parallel steel flat platens by placing it on the bottom platen. The crosshead is then lowered until the specimen come into touched with the surface of the top platen. The quasi-static tests are continued until it reaches a compaction point where the curve begins to rise up steeply after completing the sustained crushing [13]. For analysis purposes, the mechanism of failures is monitored, and the deformation process images are taken throughout the tests. The load-displacement raw data is used to determine the compression strength and specific energy absorption characteristics of the structures. Finally, the multiple tubes reinforced foam structures also being tested using different crosshead speed of 5 and 20 mm/min. Prior to testing the specimens, each specimen is labelled with a code for easily identification. Table 2 shows the code used for single and multiple tube reinforced foam structures.

Table 2. Code used for the tube reinforced foam structures.

Type of Specimen		Code	
Foam	**Tube**	**Single**	**Multiple**
F50	Bamboo Ø 25 mm	FBS	FBM
F50	PVC Ø 20 mm	FSS	FSM
F50	PVC Ø 25 mm	FLS	FLM
HP80	Bamboo Ø 25 mm	HBS	HBM
HP80	PVC Ø 20 mm	HSS	HSM
HP80	PVC Ø 25 mm	HLS	HLM

3. Results and Discussion

3.1. Compressive Behaviour of Foam Material

A typical stress-strain curves following quasi-static test on the F50 and HP80 foam is shown in Figure 3. There are three phases during the compression process. Initially, in the elastic regime, the material response is roughly linearly occurred up to approximately 4% strain and near to the yield point where the elastic response ends. This is following the Hooke's law, which stated that the strain is proportionate to the applied stress. Next, at the beginning of the crushing regime, a constant plateau stress is forming after the first substantial deviation from the linear regime. Finally, the densification regime begins where the force increases drastically with little deformation. This illustrates on how the foam materials have unique properties such as the ability to deform extensively while sustaining modest amounts of stress before reaching the densification regime [1,14].

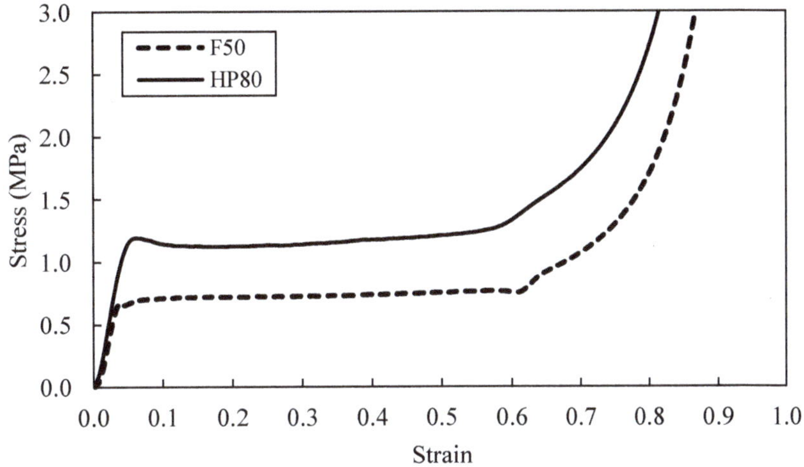

Figure 3. Stress-strain curves of F50 and HP80 foam.

Finally, the plateau region gradually ends as stiffening occurs when the cell walls collapse and started to interact with the neighbouring cell walls of the foam. This continuous interaction condition results in a rapidly increasing strain where the densification take place at 70 to 80% strain. It is also can be seen that an increase in density will also increase the compressive strength and the SEA. For example, an increase in density from 50 to 80 kg/m^3 resulted in 71% increase in the compressive strength. The values obtained from the experimental testing is summarised in Table 3.

Table 3. Compression properties of foam.

Foam	Density (kg/m^3)	Compressive Load (kN)	Compressive Strength (MPa)	Compressive Modulus (MPa)	SEA (kJ/kg)
F50	50	1.76	0.70	26.45	7.89
HP80	80	2.99	1.20	29.66	9.46

3.2. Compression Behaviour of Single Tube

In Figure 4, the load-displacement curves present the behaviour of PVC and bamboo tubes. As for PVC tubes, identical traces of compressive force applied is observed, where the force rose until it reached a point before the tube wall buckled and formed a fold due to interpenetration collapsed. The development of the first fold in the PVC tube occurred at the same time that the initial peak force is obtained. Following this, the two sections of formation plastic hinges at cell wall junctions are compacted together while folding process continued to occur. Compaction of the PVC tube would typically cause to increase the load as the curve displayed a second peak. However, folding process will weaken the PVC tubes which causes the load to drop after second peak until the tube is fully compacted as shown in Figure 5a,b. Khan et al. [6] suggested that compaction and delamination can be happened at the same time, where they balanced each other out which yielding in a relatively sustained crushed until reached the densification point.

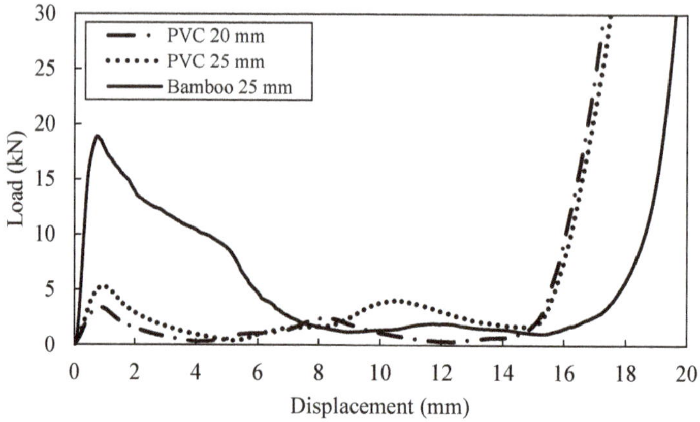

Figure 4. Load-displacement curves for three types of tube.

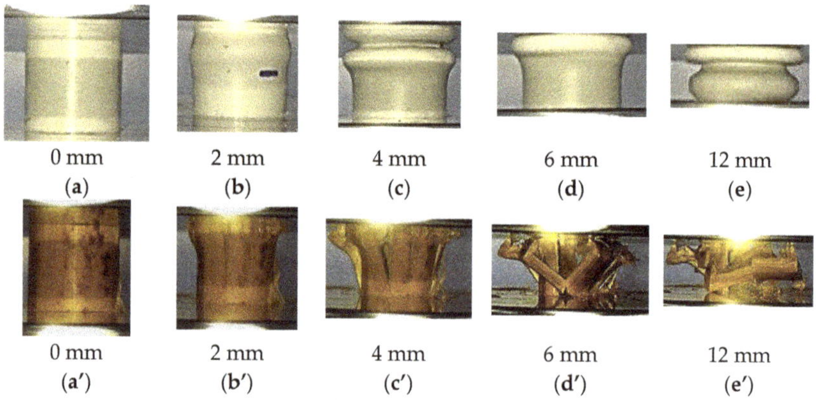

Figure 5. The compression modes at several displacement values of (**a**)–(**e**) PVC tube and (**a′**)–(**e′**) Bamboo tube.

Similarly for the individual bamboo tube, during the crushing process of the bamboo tube, the tube wall did not fail as in PVC tube, this is due to the bamboo has higher stiffness compared to PVC. The longitudinal fractures of bamboo developed when the load exerted on the tube resulting in tube wall separated into several part as shown in Figure 5a′–e′ or Figure 6c. Thus, this event further clarifies why the second peak did not occur for bamboo tube as in the case with the PVC tube.

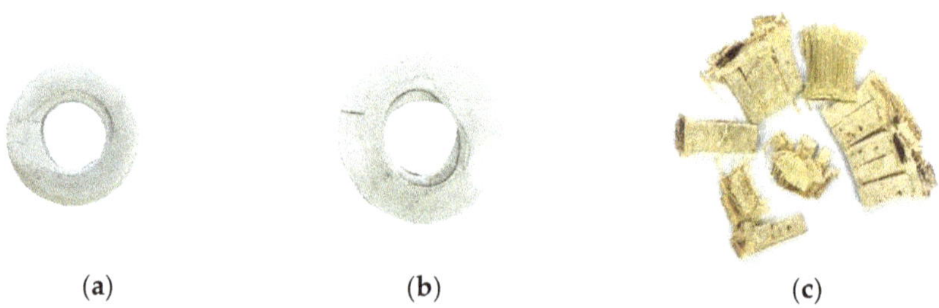

Figure 6. Shape of tube after deformation: (**a**) PVC Ø 20 (**b**) PVC Ø 25 and (**c**) Bamboo.

In comparison, the individual 25 mm diameter of PVC tube (D/t = 12.5) offer greater compression strength compared to the 20 mm of PVC tube (D/t = 13.4) counterpart. It is noted that the SEA value for 20 mm PVC tube is 7.74 kJ/kg which is less than 25 mm PVC tube, that is 9.58 kJ/kg. With the increasing of D/t ratio, lower value in compressive strength and the specific energy absorption have been recorded during the testing process. In addition, the SEA value for bamboo tube (D/t = 4.82) is approximately 21.95 kJ/kg, which is greater than both PVC tubes due to the lower diameter-to-thickness ratio as suggested by [7]. In terms of compressive strength and modulus, the bamboo tube dominates the response of axial compression loading as shown in Figure 4. The failure deformation of the PVC and bamboo tubes are photographed in Figures 5 and 6.

3.3. Single Tube Reinforced Foam Structures

A series of tests are undertaken to characterize the energy-absorbing behaviour of the single-tube reinforced foam structure. Figure 7 shows a comparison of typical compressive load-displacement curves following tests on F50, FBS, FSS and FLS. The curves for all tube reinforced structures exhibited similar traits, which is linear response before the fracture occurred to the structure at approximately 1 mm. During this phenomenon, the stiffness for the tube reinforced foam structure will reduce. Non-linear response appears in the curve after the peak load is attained until the compression is in a stable mode with an approximately constant force, before declining throughout the last phase of the test. An examination of the curves for FBS structure indicates that the load increases linearly up to roughly 26.6 kN. This is followed by FLS and FSS which have the peak load at 6.9 kN and 4.5 kN respectively. The benchmark sample, which is F50, provide the lowest maximum load when it started to fracture compared to the other three types of tube reinforced foam structure.

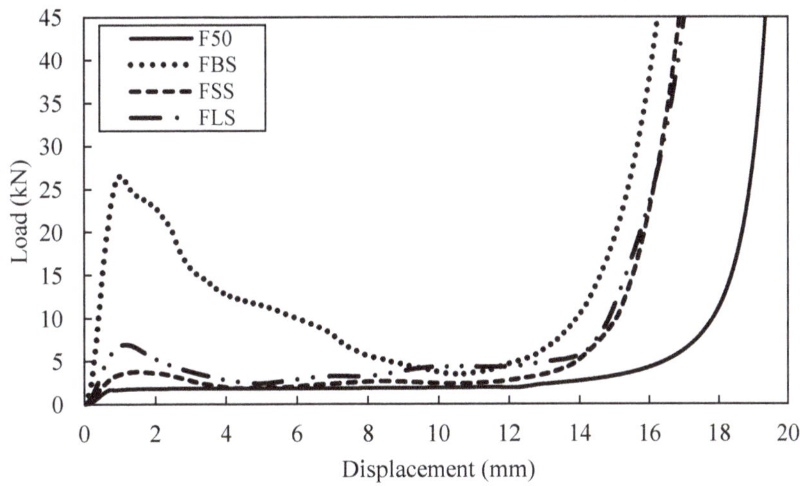

Figure 7. Load-displacement curve for F50 reinforced with different type of single tubes.

A typical load-displacement curve for tube reinforced with HP80 foam is presented in Figure 8. The density of the foam used is 80 kg/m^3 which is denser than the F50 foam. Here, the HBS structure load clearly rise up to 27 kN at 1.3 mm and drop gradually before the densification point at approximately 12.5 mm. Conversely, the curve for HSS and HLS exhibit similar traits under compression load. The initial fracture of HSS structure takes place when the load increases up to 6.3 kN while the HLS structure starts to fail at 7.3 kN which is slightly higher than HSS structure. The curve for three tube reinforced foam structures obviously showed that the strength of the reinforced structure is better than the benchmark sample, i.e., HP80. The increase of density foam in the tube reinforced foam structure leading to slightly changes in the strength of the structure. The increment of strength value for HBS and HLS structures are not more than 6% when compared to FBS and FLS structures. However, for the HSS structure, an increment of approximately 40% in strength is recorded.

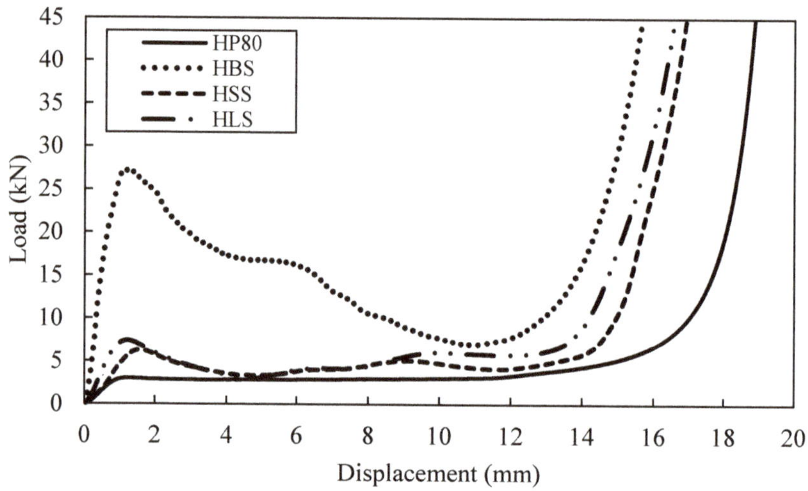

Figure 8. Load-displacement curve for HP80 reinforced with different type of single tubes.

Based on the energy absorption under the load-displacement curves for both F50 and HP80 foams, the bamboo reinforced foam structure offers an energy-absorbing capability greater than both PVC reinforced foam structures. The SEA values for FBS, FLS and FSS structures are found to be 17.9, 10.0 and 8.2 kJ/kg, respectively. In contrast, the SEA value for HBS, HLS and HSS are 24.5, 10.9 and 10.1 kJ/kg, respectively. Previous study by Umer et al. [15] has found that the SEA values is around 20 kJ/kg for single bamboo tube (D/t = 4.6) reinforced foam structure which is almost similar to the value for FBS and HBS. However, it is interesting to note that the SEA increases with the increasing density foam, similar to study conducted by Alia [1].

The deformation of the single tube reinforced foam structure after completing the compression process is shown in Figure 9. All the tubes were compacted and locked up at densification region when the tube walls have fully collapsed, as suggested by Rajput et al. [14]. Here, the lateral movement of tube reinforced structure is limited by increasing the foam density from 50 to 80 kg/m^3. In Figure 9b, there is only a small amount of bamboo tube splitting is visible on the outside diameter. A closer look reveals that numerous of these fractures have penetrated to the neighbouring foam, similar to [15]. For PVC reinforced foam structures, the buckling and folding failure also being restricted with the increasing of foam density. Thus, the foam has been successful in constraining the lateral movement and failure mechanism.

3.4. Multi Tube Reinforced Foam Structures

The following part of this study focused into the effect of multiple tube reinforced foam structures on its energy-absorbing response. Here, four tubes are used and embedded into the foam similar to the single tube.

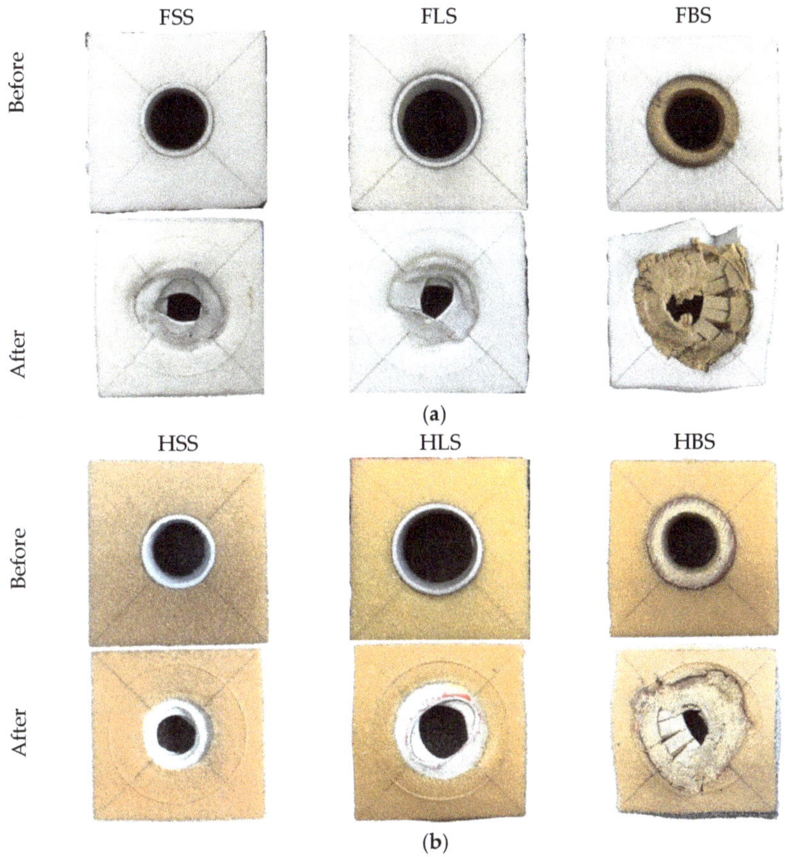

Figure 9. The top view of single tube reinforced with (**a**) F50, and (**b**) HP80 foam structures.

In this study, the multiple tube reinforced with HP80 foam offer better energy absorption than F50 foam due to the higher density in HP80 as shown in Figures 10 and 11. It is observed that the energy absorption for FBM (F50 foam) and HLM (HP80 foam) are the most efficient. Both structures offer greater area of load-displacement curves that is related to SEA value. An increase value of SEA is recorded from 17.9 kJ/kg to 20.9 kJ/kg for FBS and FBM structures (from single to multiple). Relatively, for HLS and HLM structures, the SEA value increase from 10.9 kJ/kg to 11.5 kJ/kg. However, the SEA values from single to multiple tube reinforced foam structure did not vary significantly due to the influence of the initial mass of the structure as it is inversely proportional to the SEA value, as proposed by Alantali et al. [16]. Invariably, the bamboo structures always dominated in any cases. However, the limitation for the testing machine used in this work is not capable of achieving compression loads exceeding 100 kN. This explained why the HBM specimen is not carry on being tested and included in Figure 11 as it has exceeded the machine ability.

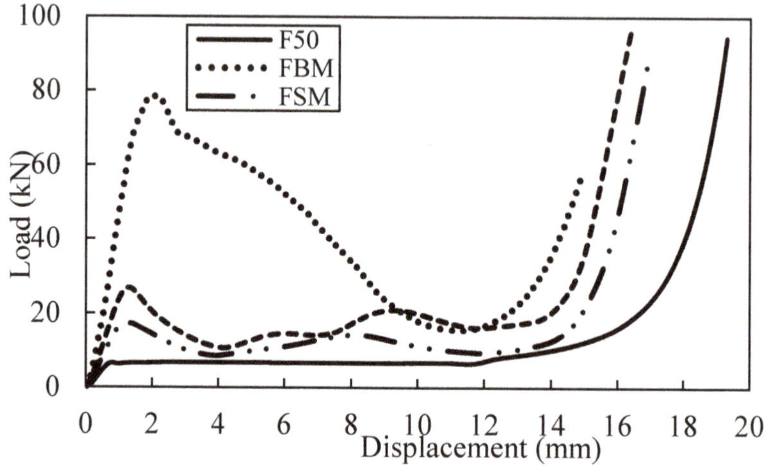

Figure 10. Load-displacement curve for multiple tube reinforced with F50 foam structure.

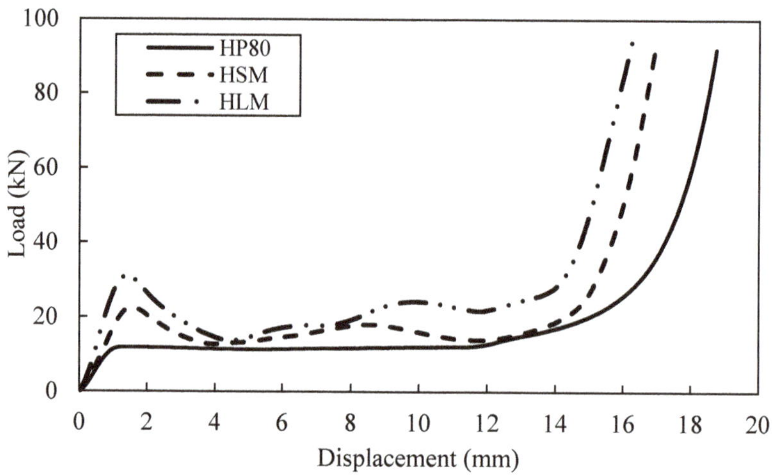

Figure 11. Load-displacement curve for multiple tube reinforced with HP80 foam structure.

Furthermore, the SEA value and strength found to be increased when there is an increasing in tube diameter as well as the foam density. This shows that the local limitation imposed by the F50 and HP80 foam has a significant impact on the failure processes in tubes. In Figure 13, the deformation failure of multiple tube reinforced foam structures after testing is presented. The top section of the PVC tube shift slightly to one side when interpenetration occurred due to the transverse shear while the fracture of fibre occurred in the bamboo tube causes the tube to fractured into radial cracks and longitudinal splitting [7,17]. The failure of the bamboo tube can be seen clearly from the top view. As comparison to the Figure 9a, the failure of the single bamboo tube reinforced foam structure (FBS) is similar to the multiple bamboo tube reinforced foam structure (FBM) as shown in Figure 13a. The failed bamboo tube pushed the foam and leads to a crack in the foam structure. This means that the neighbouring tube properties can be affected as well. Figure 13 compares the single and multiple tubes reinforced foam structures.

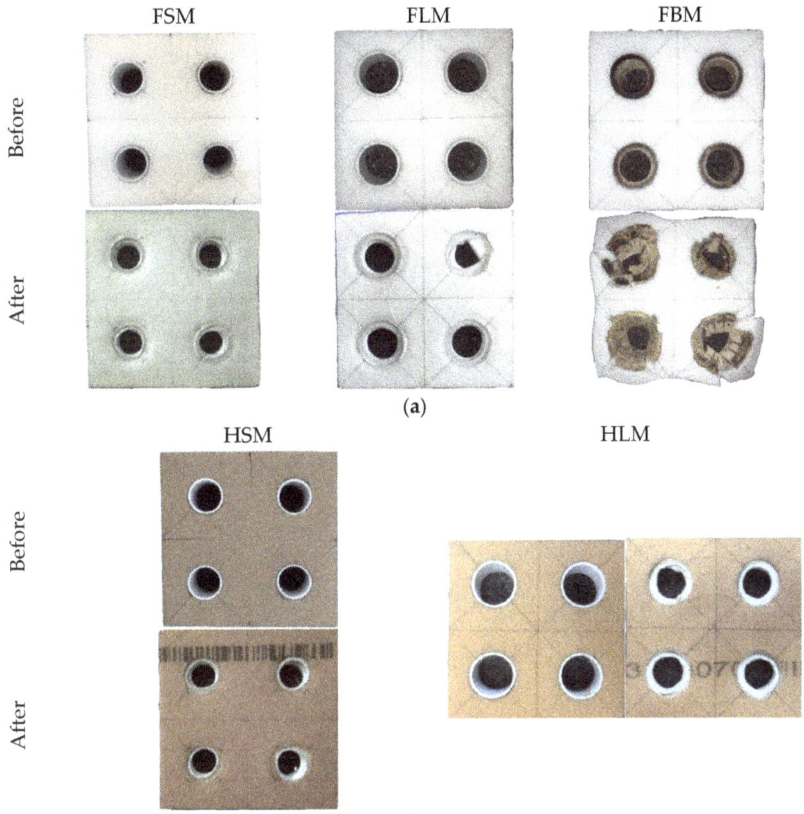

(b)

Figure 12. The top view of multiple tube structure with (**a**) F50 and (**b**) HP80 foam strucctures.

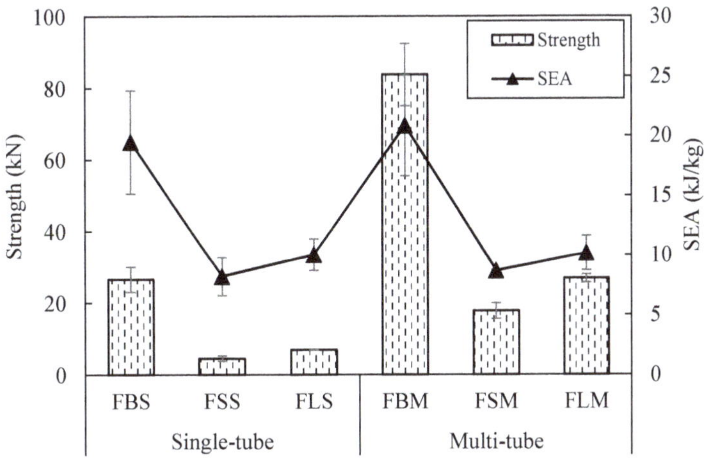

(a)

Figure 13. *Cont.*

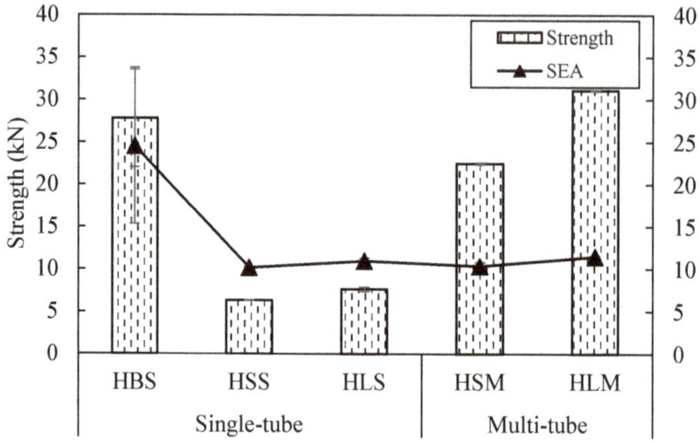

(**b**)

Figure 13. Compressive strength and SEA for single and multiple tube embedded in (**a**) F50 and (**b**) HP80 foams.

3.5. *Influence of Crosshead Speed on Multiple Tube Reinforced Foam Structures*

Figure 14 presents the results for HLM structure under different crosshead speed displacement. At the initial stage, the peak force of the crosshead speed at 20 mm/min is only 3.3 kN higher than the 2 mm/min. This is maybe due to the crosshead speed used is not much in difference. However, the crosshead speed of 20 mm/min shows the highest peak force in comparison to the other, followed by 5 mm/min and lastly 2 mm/min. Here, as the crosshead speed is increasing, a higher value of compressive strength is obtained which is similar to the study by [18]. Additionally, by increasing the strain-rate will resulting a slight increase in the SEA capability of the reinforced foams, possibly due to the rate-sensitivity of the failure processes occurring in the reinforced structure.

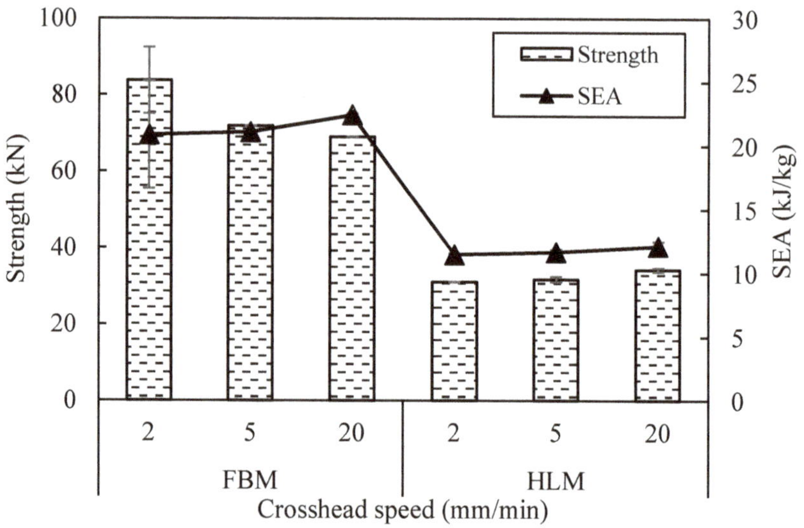

Figure 14. Compressive strength and SEA for FBM and HLM structure with three different compression crosshead speed.

4. Conclusions

The conclusions of this study can be drawn as follow:

1. Single bamboo tube reinforced foam structure offers highest value in specific energy absorption and compression strength. By increasing the foam density, it will increase the specific energy absorption and compressive strength value. A larger size of diameter of tube can enhance the specific energy absorption and compressive strength.
2. The compressive strength increases significantly when several tubes are introduced inside the foam. In contrast, the specific energy absorption does not change significantly for multiple tube reinforced foam structure. The multiple bamboo tube reinforced F50 foam structure offers better value of compression properties. In addition, the multiple PVC tube reinforced foam structure with a diameter of 25 mm gives the highest energy absorbing capability in HP80 foam.
3. The specific energy absorption does not vary significantly under the different crosshead speed below 20 mm/min. However, an increase in crosshead speed during testing can leads a slightly higher value of specific energy absorption. Therefore, the structures are believed to have a strain-rate effect when a higher speed is applied.

Author Contributions: Project administration, M.Y.M.Z. and J.J.N.A.; Visualization, J.J.N.A.; Writing—original draft, J.J.N.A. and M.Y.M.Z.; Writing—review & editing, Z.L., N.I.Z., A.A. and R.A.I. All authors have read and agreed to the published version of the manuscript.

Funding: This work is supported by Universiti Putra Malaysia through Putra Grant (GP-IPS/2018/9663200).

Acknowledgments: Gratitude also expressed to Muhammad Wildan Ilyas Mohamed Ghazali for his assistance in conducting this experimental work.

Conflicts of Interest: The authors declare no conflict of interest.

References

1. Alia, R.A. The Energy-Absorbing Characteristics of Novel Tube-Reinforced Sandwich Structures. Ph.D. Thesis, University of Liverpool, Liverpool, UK, 2015.
2. Cinar, K. Evaluation of sandwich panels with composite tube-reinforced foam core under bending and flatwise compression. *J. Sandw. Struct. Mater.* **2020**, *22*, 480–493. [CrossRef]
3. Birman, V.; Kardomateas, G.A. Review of current trends in research and applications of sandwich structures. *Compos. Part B Eng.* **2018**, *142*, 221–240. [CrossRef]
4. Colloca, M.; Dorogokupets, G.; Gupta, N.; Porfiri, M. Mechanical properties and failure mechanisms of closed-cell PVC foams. *Int. J. Crashworthiness* **2012**, *17*, 327–336. [CrossRef]
5. Shinde, R.B.; Mali, K.D. An Overview on Impact Behaviour and Energy Absorption of Collapsible Metallic and Non-Metallic Energy Absorbers used in Automotive Applications. *IOP Conf. Ser. Mater. Sci. Eng.* **2018**, *346*, 2018. [CrossRef]
6. Khan, R.A.; Mahdi, E.; Cabibihan, J.J. Effect of fibre orientation on the quasi-static axial crushing behaviour of glass fibre reinforced polyvinyl chloride composite tubes. *Materials* **2021**, *14*, 2235. [CrossRef] [PubMed]
7. Zuhri, M.Y.M.; Liao, Y.; Wang, Q.Y.; Guan, Z.W. The energy absorbing properties of bamboo-based structures. *J. Sandw. Struct. Mater.* **2019**, *21*, 1032–1054. [CrossRef]
8. Zhou, J.; Guan, Z.; Cantwell, W.J. The energy-absorbing behaviour of composite tube-reinforced foams. *Compos. Part B Eng.* **2018**, *139*, 227–237. [CrossRef]
9. Alhawamdeh, M.; Alajarmeh, O.; Aravinthan, T.; Shelley, T.; Schubel, P.; Kemp, M.; Zeng, X. Modelling hollow pultruded FRP profiles under axial compression: Local buckling and progressive failure. *Compos. Struct.* **2021**, *262*, 113650. [CrossRef]
10. Alia, R.A.; Cantwell, W.J.; Langdon, G.S.; Yuen, S.C.K.; Nurick, G.N. The energy-absorbing characteristics of composite tube-reinforced foam structures. *Compos. Part B Eng.* **2014**, *61*, 127–135. [CrossRef]
11. Molari, L.; Mentrasti, L.; Fabiani, M. Mechanical characterization of five species of Italian bamboo. *Structures* **2020**, *24*, 59–72. [CrossRef]
12. ASTM. Standard Test Method for Flatwise Compressive Properties of Sandwich Cores 1. *Am. Soc. Test. Mater.* **2003**, 2–4. [CrossRef]
13. Li, Q.M.; Magkiriadis, I.; Harrigan, J.J. Compressive strain at the onset of densification of cellular solids. *J. Cell. Plast.* **2006**, *42*, 371–392. [CrossRef]
14. Rajput, M.S.; Burman, M.; Köll, J.; Hallström, S. Compression of structural foam materials—Experimental and numerical assessment of test procedure and specimen size effects. *J. Sandw. Struct. Mater.* **2019**, *21*, 260–288. [CrossRef]

15. Umer, R.; Balawi, S.; Raja, P.; Cantwell, W.J. The energy-absorbing characteristics of polymer foams reinforced with bamboo tubes. *J. Sandw. Struct. Mater.* **2014**, *16*, 108–122. [CrossRef]
16. Alantali, A.; Alia, R.A.; Umer, R.; Cantwell, W.J. Energy absorption in aluminium honeycomb cores reinforced with carbon fibre reinforced plastic tubes. *J. Sandw. Struct. Mater.* **2019**, *21*, 2801–2815. [CrossRef]
17. Wei, Y.; Zhou, M.; Zhao, K.; Zhao, K.; Li, G. Stress–strain relationship model of glulam bamboo under axial loading. *Adv. Compos. Lett.* **2020**, *29*, 1–11. [CrossRef]
18. Xiang, C.; Xiao, Z.; Ding, H.; Wang, Z. Compressive Properties and Energy Absorption Characteristics of Extruded Mg-Al-Ca-Mn Alloy at Various High Strain Rates. *Materials* **2020**, *14*, 87. [CrossRef] [PubMed]

polymers

MDPI

Article

Assessment of Dimensional Stability, Biodegradability, and Fracture Energy of Bio-Composites Reinforced with Novel Pine Cone

Kanishka Jha [1,*], Yogesh K. Tyagi [2], Rajeev Kumar [1], Shubham Sharma [3,*], Muhammad Roslim Muhammad Huzaifah [4,*], Changhe Li [5], Rushdan Ahmad Ilyas [6,7], Shashi Prakash Dwivedi [8], Ambuj Saxena [8] and Alokesh Pramanik [9]

1 School of Mechanical Engineering, Lovely Professional University, Phagwara 144411, India; rajeev.14584@lpu.co.in
2 Department of Mechanical Engineering, DIT University, Dehradun 248009, India; yogesh_tyagi30@yahoo.co.in
3 Department of Mechanical Engineering, IK Gujral Punjab Technical University, Main Campus-Kapurthala, Ibban 144603, India
4 Department of Crop Science, Faculty of Agricultural Science and Forestry, Universiti Putra Malaysia Bintulu Campus, Bintulu 97000, Malaysia
5 School of Mechanical and Automotive Engineering, Qingdao University of Technology, Qingdao 266520, China; sy_lichanghe@163.com
6 School of Chemical and Energy Engineering, Faculty of Engineering, Universiti Teknologi Malaysia, Johor Bahru 81310, Malaysia; ahmadilyas@utm.my
7 Centre for Advanced Composite Materials, Universiti Teknologi Malaysia, Johor Bahru 81310, Malaysia
8 Department of Mechanical Engineering, G.L. Bajaj Institute of Technology and Management, Greater Noida 201306, India; spdglb@gmail.com (S.P.D.); ambuj.saxena1@gmail.com (A.S.)
9 School of Civil and Mechanical Engineering, Curtin University, Perth 6102, Australia; Alokesh.Pramanik@curtin.edu.au
* Correspondence: kanishka.21537@lpu.co.in (K.J.); shubham543sharma@gmail.com or shubhamsharmacsirclri@gmail.com (S.S.); muhammadhuzaifah@upm.edu.my (M.R.M.H.)

Citation: Jha, K.; Tyagi, Y.K.; Kumar, R.; Sharma, S.; Huzaifah, M.R.M.; Li, C.; Ilyas, R.A.; Dwivedi, S.P.; Saxena, A.; Pramanik, A. Assessment of Dimensional Stability, Biodegradability, and Fracture Energy of Bio-Composites Reinforced with Novel Pine Cone. *Polymers* **2021**, *13*, 3260. https://doi.org/10.3390/polym13193260

Academic Editors: Emin Bayraktar, S. M. Sapuan and R. A. Ilyas

Received: 14 August 2021
Accepted: 16 September 2021
Published: 24 September 2021

Abstract: In this investigation, biodegradable composites were fabricated with polycaprolactone (PCL) matrix reinforced with pine cone powder (15%, 30%, and 45% by weight) and compatibilized with graphite powder (0%, 5%, 10%, and 15% by weight) in polycaprolactone matrix by compression molding technique. The samples were prepared as per ASTM standard and tested for dimensional stability, biodegradability, and fracture energy with scanning electron micrographs. Water-absorption and thickness-swelling were performed to examine the dimensional stability and tests were performed at 23 °C and 50% humidity. Results revealed that the composites with 15 wt % of pine cone powder (PCP) have shown higher dimensional stability as compared to other composites. Bio-composites containing 15–45 wt % of PCP with low graphite content have shown higher disintegration rate than neat PCL. Fracture energy for crack initiation in bio-composites was increased by 68% with 30% PCP. Scanning electron microscopy (SEM) of the composites have shown evenly-distributed PCP particles throughout PCL-matrix at significantly high-degrees or quantities of reinforcing.

Keywords: polycaprolactone; pine cone powder; graphite; dimensional stability; biodegradability; fracture energy; SEM

1. Introduction

From the very beginning, fire outbreak and rate of spread of forest fire were dominantly affected by the various forest waste and *Pines roxburghii* (pine cone) flower in northern part of India is one of these main sources [1,2]. The devastating effect of forest fires not only effects the vegetation but also devastates the entire ecosystem in that geographic region. The burgeoning awareness regarding forest fuel in spreading fire has lead to the development of numerous solutions [3]. One such kernels of an idea is to use dried pine cones to develop composite materials for non-structural applications. These days, forest waste is either parched, or even better high-end uses are found [4]. In past decade, polymer composites with natural fibre have been characterized by many researchers to investigate their potential use in structural and non-structural applications. Among various natural fibres, pine cones' contribution to forest fuel for spreading wild fires in northern India has been reported in numerous studies [5–7]. Currently, the demand for biodegradable composites is growing tremendously and they have found specific applications in automobile and packaging industry [8,9]. Recently, polymer composite with natural fibre has been characterized by many researchers to investigate their potential use in structural and non-structural applications. Among various natural fibres, pine cones contribution to forest fuel for spreading fire in northern India is reported in numerous literatures. Current scenario demand of NFRP's is growing tremendously and finds their specific applications in automobile and packaging industry [10–13]. Materials for biodegradable matrices are also an important aspect in selecting biodegradable composites. Polycaprolactone belongs to the aliphatic polyester family and considered a competitive candidate among other biodegradable polymers [14–18]. It has also been reported that, due to mechanical incompatibility between the two blended media, polycaprolactone tensile strength decreases when blended with starch. In another study, PCL and calcium sulphate (CS) whisker composites were fabricated with different whisker weight percentages and the authors reported that lower weight fractions of reinforcement resulted throughout excellent massive enhancement in (21%) flexibility and (22%) toughness, while the thermal characteristics were unlikely to be affected by the existence of CS-whiskers. They also found numerous applications as construction material [19,20]. In another study by Jha et al., they have utilized the pine cone powder in biodegradable PCL matrix. They have found that pine cone powder at higher loading showed poor performance due to poor interfacial bonding resulted from agglomeration of the microparticles [21–23]. Samy Yousef et al., made an attempt to analyse the mechanical and thermal properties of non-metallic components of recycled woven fibreglass and epoxy resin from waste printed circuit boards. The unmodified samples (without holes) had the most stress with 92 Mpa and strain more by 4.7% and sample with hole had reduction of 41% and 1.55% in stress and strain respectively, in thermal properties melting temperature was around 146–175 c for plane the temperature was 165.12 c and crystalline degree decreased by 17%. Due to the presence of notches, the strength of recovered fibreglass declined by 48% [24].

In the present research study, polycaprolactone (PCL) was utilized as a continuous phase and pine cone particles (PCP) with 0–45% weight fraction and graphite powder 0–15% weight fraction were employed as the discontinuous phase. Extracted pine cone was reinforced with PCL by altering weight-fraction and improved the biodegradation characteristics. It was discovered that elongation and toughness characteristics were first increased and then decreased on raising the pine cone weight fraction. The foremost objectives of this investigation are to observe water absorptivity, biodegradability, and fracture energy. Further mechanical properties were enhanced by modifying the continuous phase through graphite addition.

2. Experimentation

Pine cones were collected from lower Himalayas of Northern India. The density of extracted fibres was calculated by ASTM D792-91 and reported as 0.168 gm/cm^3 [25]. The extraction of pine cone fibre from collected pine flower was depicted in Figure 1. Extracted pine fibres were first treated with an alkali solution to wash out the unwanted biological extracts such as cellulose, hemi-cellulose, and lignin (Figure 2). Treated fibres (Figure 1c) were then ground down to a 200-micron particle size before being incorporated in polycaprolactone, which was purchased from Sigma Aldrich Inc., Anekal Taluk, Bangalore, India and specimens were fabricated with varying weight fractions of fibres and designations as illustrated in Table 1. Pine cones in their ground form were used as reinforcement.

Required specimen sheets were prepared by compression molding (100 T) at 150 °C, and the thickness was maintained at 3.2 ± 0.4 mm for characterization. Graphite was used as a matrix modifier, which enhances the matrix and fibre interaction. Graphite powder was procured from Loba Chemicals Pvt Ltd., Colaba, Mumbai, India with molecular weight of 12.01 g/mol and density of 1.8 g/cm^3. Graphite was added as a percentage of matrix addition.

Figure 1. Pine cone flowers in their natural state (**1**,**2**) and extraction of pine cone fibres (**a**–**d**).

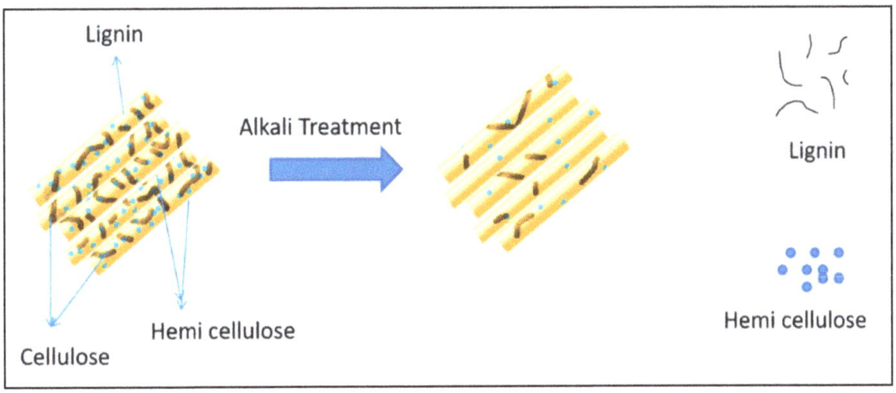

Figure 2. Schematic representation of alkali treatment of secondary wall of fibre.

Table 1. Composite designations.

In %	CD-1	CD-2	CD-3	CD-4	CD-5	CD-6	CD-7
Fibre	0	15	30	45	45	45	45
Matrix	100	85	70	55	52.25	49.5	46.75
Graphite	0	0	0	0	2.75	5.5	8.25

Weak interfacial adhesion between pine cone and PCL was reported from previous studies [7]. To compatiblize the present combination of fibre and resin, graphite powder was used. Water absorption was performed with the specimens as per ASTM D570.

Following standards were used for calculations

Diffusion-coefficient:

$$D = \pi\left(m^2 l^2 / 16 W_\infty^2\right) \quad (1)$$

Sorption-coefficient:

$$S = W_\infty / W_t \quad (2)$$

Permeability-coefficient:

$$P = D \times S \quad (3)$$

where m is gradient of the linearity-portion of the sorption-curvature and l is the initial thickness of the sample.

W_∞ and W_t are molar-percentages of water-uptake at infinite-duration and at time t.

Water-absorption value was evaluated as per the formula:

$$\frac{W_t - W_0}{W_t} \times 100\% \quad (4)$$

Thickness-swelling was determined as per the formula:

$$\frac{T_t - T_0}{T_t} \times 100\% \quad (5)$$

where, T_0 and T_t is the specimen thickness without absorption and at time t.

Disintegration tests of developed specimens were conducted in composting condition according to ISO 20200 standard-procedure, by using commercialized composting (Figure 3) with sawdust, rabbit-food, starches, oils, and urea [26].

Figure 3. Compost box as per standards.

Tested samples were cut as per the standard (15 × 15 × 0.2 mm^3 Figure 4) and buried at 10 cm depth in perforated boxes and incubated at 25 °C, represented in Figure 5. Periodical addition of water and proper proportion of compost guaranteed the aerobic conditions. After disintegration experiments (0, 10, 20, 30, 40, 50, 60, 70, and 75 days), samples have been expelled from composting and subsequently cleaned with filtered water to eradicate remaining residues of compost and also to prevent additional microorganism attacks.

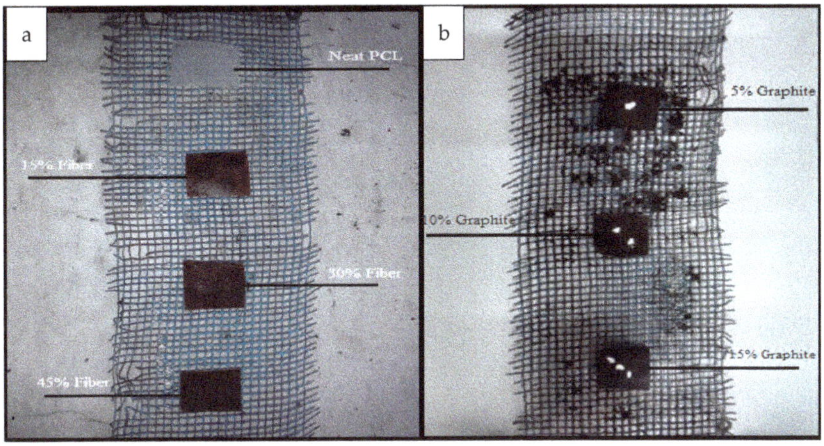

Figure 4. Mounting of bio-specimen on wire gauge (**a**) for unmodified matrix, (**b**) Modified matrix with graphite content.

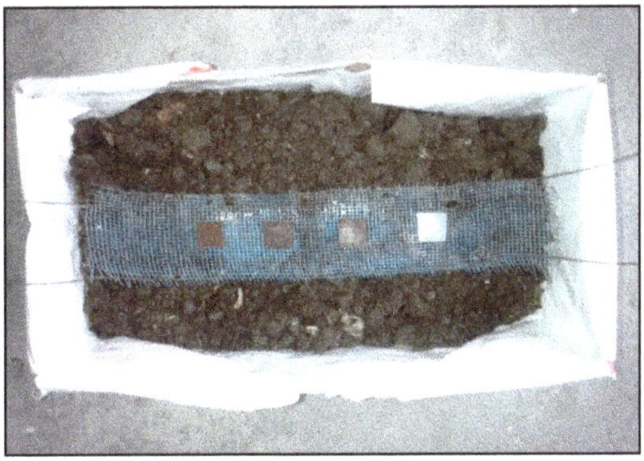

Figure 5. Compost box with samples placed inside.

The samples were dehydrated over 24 h at 23 °C and 50 percent relative-humidity prior to analysis.

The disintegrability values for every buried specimen have been determined by employing the accompanying formula:

$$\text{Disintegrability } (\%) = \frac{W_0 - W_t}{W_0} \times 100\% \tag{6}$$

The percent volume-fraction of void-spaces in composites was estimated employing the underlying correlation:

$$V_v = \left(\frac{\rho_{th} - \rho_{ac}}{\rho_{th}}\right) \times 100 \tag{7}$$

3. Results and Discussion

3.1. Water Absorptivity

Water resistance tests were performed at 23 °C and 50% humidity. Figures 6 and 7 illustrated the variation of water-absorption (WA) and thickness-swelling (TS) respectively with time for different wt % of PCP in PCL matrix. The percentage values of WA of prepared composites were revealed to be raised [27] with an increase in weight percentage of PCP. Higher absorption percentages were observed for higher wt % of PC as compared to neat PCL. However, water absorption values for 15 wt % loading of PC were found to stabilize at 2% for the entire period of observation. TS tests have revealed that the composites with 15 wt % of PCP have shown higher dimensional stability as compared to other composites. The diffusion coefficient is a material property that describes how solvent molecules migrate through solids, whereas the sorption coefficient is correlated to a saturation of water absorbed by composites. Higher values of sorption coefficient mean that a composite gets saturated in less time, whereas lower sorption coefficient values indicate a longer period until saturation. The cumulative influence of the diffusion coefficient and the sorption coefficient is given by that of the permeability coefficient. Fick's law has been utilized to elucidate the diffusing characteristics of composites [28]. Table 2 showed the values of water absorption parameters for different composite designations. It was evident from the results that neat PCL had maximum sorption coefficient as compared to the samples with PCP content, which displays that pine cone enhanced the hydrophobicity of the composite. Results of absorption tests also revealed that, among the composite fractions, a 30% weight fraction produces better results for sorption, diffusion, and permeability coefficients which indicates that this specimen is best suited for practical applications and this trend was also supported by void volume results presented earlier.

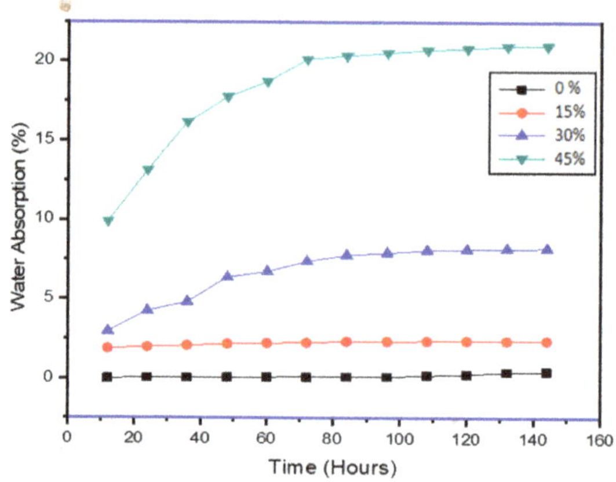

Figure 6. Water Absorption properties of PCL–PCP composites.

Where, 0% is the neat matrix with no fibre; 15% is the fibre wt %; 30% is the fibre wt %; and 45% is the fibre wt % and the rest is the matrix with additives.

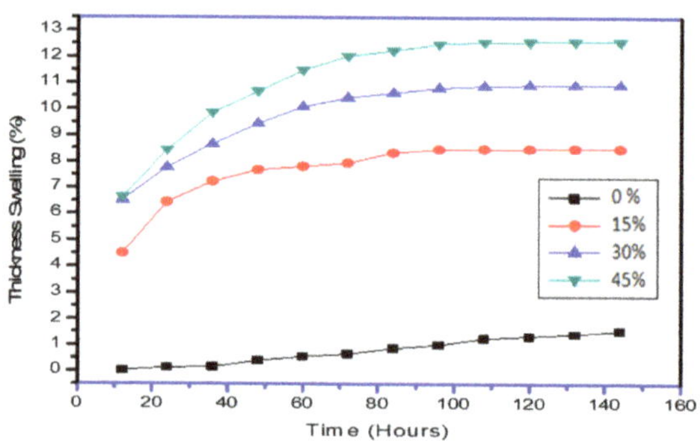

Figure 7. Thickness swelling properties of PCL-PCP composites.

Where, 0% is the neat-matrix with no fibre; 15% is the fibre wt %; 30% is the fibre wt %; and 45% is the fibre wt % and rest is the matrix with additives.

Table 2. Diffusion, sorption, and permeability study of composites.

Samples	Percentages of Water-Uptake at Infinite Time (W_∞)	Sorption Coefficient (S)	Diffusion Coefficient (D) (mm^2/s)	Permeability Coefficient (P) (mm^2/s)
0%	0.43	26.88	6.44×10^{-8}	1.73×10^{-6}
15%	2.36	1.25	2.97×10^{-5}	3.72×10^{-5}
30%	8.21	2.77	6.05×10^{-6}	1.68×10^{-5}
45%	20.99	2.12	1.04×10^{-5}	2.20×10^{-5}

Figure 6 also gives a relation between curve behavior and hydrophilic character of the developed composites. P45 with maximum fibre wt % showed the most water absorption as compared to the samples with low fibre content (P0, P15, P30) which concludes that exposed fibre increased the hydrophilic character in the composite. The region of the curves for all specimens above square root of time from 25 to 60 h show that the substantial rise in the water absorption is due to hydrophilicity of the pine fibre and also due to cellulosic content present in the fibre which causes swelling of the fibre. The hydroxyl group present in the material structure reacts with the hydrogen bond of water molecules and results in high water absorption [29,30]. From the time of 90 h on the X-axis, it can be seen that nearly every specimen reached the saturation point of water uptake and therefore the curves start to flatten, following the Fickian diffusion. Further incorporation of graphite in P45 specimens was also tested for water absorption and thickness swelling, results revealed that graphite micro particles get settled in void, shown in SEM images, thus reducing the water uptake of the specimens as the graphite content increased, as illustrated in Figure 7.

For graphite samples, the water absorption rate becomes constant after or around 90 h. Whereas, for the P45G0 sample, it was around 84 h. Hybridisation of PCL–PCP composites with graphite has somewhat declines the moisture-uptake performance of pine cone fibre composite. It was also evaluated from the Figure 8 that at 5%, 10%, and 15% of graphite the water uptake percentage decreases by 67%, 51.5%, and 61% respectively. This behaviour of graphite was depicting that presence of graphite reduces the hydrophobicity of the pine cone and fill the void-spaces existing in the vicinity of the fibres.

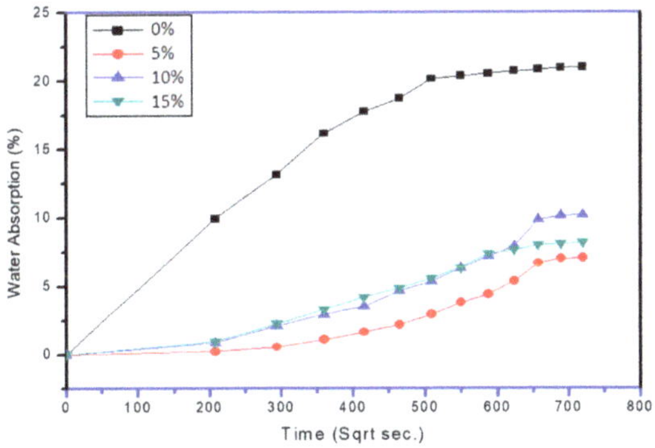

Figure 8. Water absorption properties of graphite-loaded composites.

Where, 0% is the neat-matrix with no fibre; 5% is the fibre wt %; 10% is the fibre wt %; and 15% is the fibre wt %, and rest is the matrix with additives.

TS results also shows the same behaviour of graphite loading and dimensional stability alters by a large amount for 15% loading. All the samples with graphite loading show dimensional stability after or around 90 h of water absorption (Figure 9). Table 3 shows the values of water-absorption parameters for diverse composite samples loaded with graphite. It was evident from the results that samples without graphite content have maximum sorption coefficient as compared to the samples with graphite content, which displays that graphite enhanced the hydrophobicity of the composites.

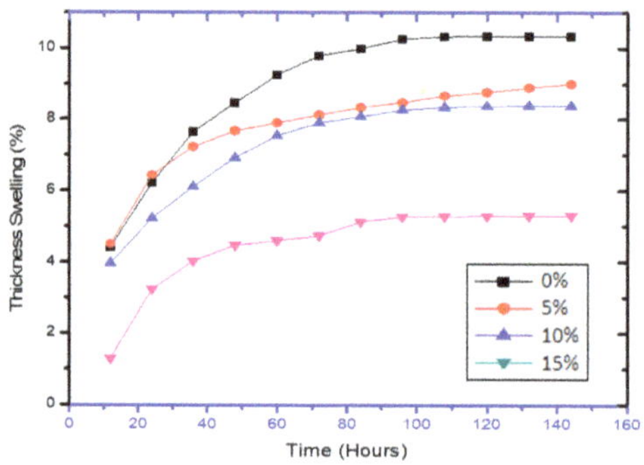

Figure 9. Thickness swelling properties of graphite-loaded composites.

Where, 0% is the neat-matrix with no fibre; 5% is the fibre wt %; 10% is the fibre wt %; and 15% is the fibre wt %, and rest is the matrix with additives.

Table 3. Diffusion, sorption, and permeability study of graphite-loaded composites.

Samples	% of Water-Uptake at Infinite-Time (W_∞)	Diffusion Coefficient (D) (mm^2/s)	Sorption Coefficient (S)	Permeability Coefficient (P) (mm^2/s)
CD-4	0.43	6.44064×10^{-8}	26.875	3.71538×10^{-5}
CD-5	7.04	4.34×10^{-8}	32.74419	1.42×10^{-6}
CD-6	10.2	3.1×10^{-7}	12.24049	3.8×10^{-6}
CD-7	8.17	6.09×10^{-7}	8.742643	5.32×10^{-6}

3.2. Biodegradability

Use of ligno-cellulosic material as a natural fibre reinforcement improves the microbial-attack and bio-degradation by endorsing bio-fouling. Biodegradation rate generally depends on the interfacial adhesion of fibre-matrix interactions and hydrophilicity of the polymeric matrix [31,32]. The disintegration study was taken for 75 days, when PCL/PCP biocomposites were 90% disintegrated, according to the ISO 20200 (ISO 20200:2006), for a biodegradable material. Biocomposites comprising 15–45 wt % of PCP presented massive disintegration-rate that neat PCL as showed in Figure 9. Test specimens displayed substantial change in their disintegration rate after 20 days of the burial with a nominal roughing and hole formation. Initial PCL degradation was due to ester cleavage and diffusion of oligomeric species causing bulk weight loss [33]. Further biodegradation process was due to the water absorption by PCL matrix. Slow initial degradation was resulted due to hydrophobic nature of PCL. In this sense, addition of PCP accelerated the rate of water-absorption and facilitate the transfer of water to PCL matrix, and higher PCP content enhances the biodegradation of PCL matrix [34]. In later stages, breakdown of cellulosic chains contributes to the higher weight reduction suffered by the biocomposites.

Lignocellulosic natural fibre reinforcement modifies the microbe-based attack and enhances the biodegradation by initiating bio-fouling. The rate of biodegradation greatly depends on the interfacial adhesion of fibre–matrix interactions and hydrophilicity of the polymeric matrix [35]. In the present work developed biodegradable polymer composite was compatibilized with graphite at different weight fraction. Disintegration study of the developed composites were taken for 75 days, when PCL-G-PCP bio-composites were

90% disintegrated, according to the ISO 20200 [26], for a biodegradable material. Bio-composites containing low graphite content showed higher disintegration rate as showed in Figure 10. PCL-G-PCP specimens showed considerable change in their disintegration rate after 30 days of the burial with a nominal surface roughing and holes formations. Initial matrix degradation was due to cleavage of ester bonds and diffusion of oligomeric species causing considerable weight loss [31]. Further biodegradation process was due to the water absorption by PCL matrix. Slow initial degradation was resulted due to hydrophobic nature of PCL approximately for first 10 days. In this sense, addition of PCP accelerated the rate of water absorption and facilitate the transfer of water to PCL matrix, and higher PCP content enhances the biodegradation of PCL matrix [36]. In later stages (encircled in figure and in last 5–10 days), breakdown of cellulosic chains contributes to the higher weight reduction suffered by the bio-composites.

Figure 10. Bio-degradation test with 0, 15, 30, and 45 wt % of PCP.

3.3. Fracture Energy

Fracture energy of the developed bio-composite was found to be optimized at 30% weight fraction of pine cone particle as shown in Figure 11. Initially the fracture energy was decreased as introducing fibres at 15% weight fraction in the matrix [37]. After increasing the weight fraction to 30%, the energy required for crack initiation was increased by 68%. Then it further decreases as weight fraction was increased to 45%, due to the increase in void content. This trend of the tear results was supported by water absorption results (permeability coefficient), in which a 30% weight fraction specimen of bio-composite shows minimum water permeability which was due to the lower void content in the vicinity of the particles clearly observed in SEM results shown in Figure 12. The presence of voids hampers the stress transfer from matrix to fibre phase resulting in higher fracture energy at 30% than at 45% weight fraction [38–40].

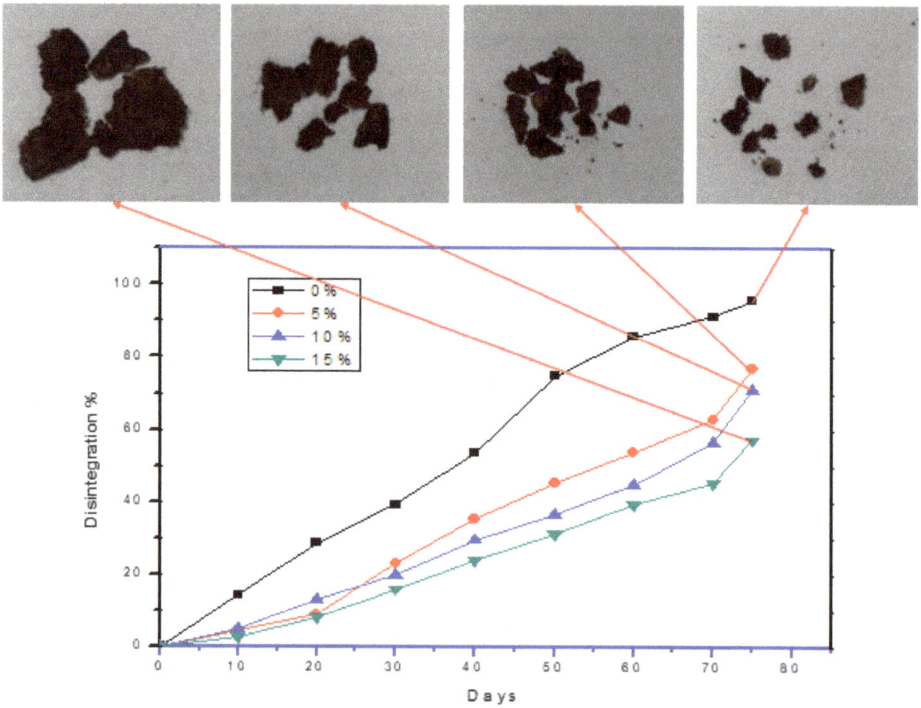

Figure 11. Bio-degradation test with 0, 5, 10, and 15 wt % of graphite with PCP.

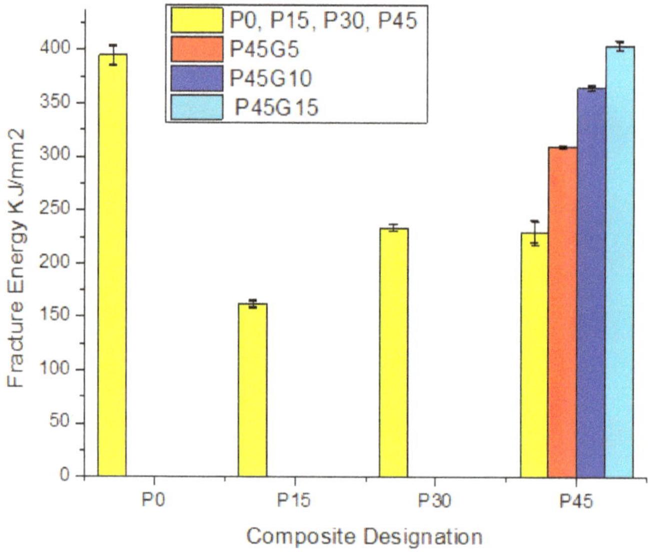

Figure 12. Fracture energy dependency on composite designation.

The experimentally measured and theoretical densities of developed composites were depicted in Table 4. The difference calculated between experimental and theoretical densities of the developed composites gives an idea of the voids in the fabricated composites which adversely affect the properties significantly [36]. The difference in surface tension of matrix and fibre is one of the reasons for void creation in mechanical stirring. Increasing the content, increases the void content. Another reason for increased void content is agglomeration of particles at higher weight fractions.

Table 4. Density and void content (%) of developed composites. Actual density by ASTM C693.

Samples	Theoretical Density (g/cm^3)	Actual Density (g/cm^3)	Void's Volume (%)
CD-1	1.145	1.141	0.3
CD-2	0.9707	0.92	5.2
CD-3	0.9360	0.83	1.13
CD-4	0.7983	0.805	6.8

3.4. Morphology

Micrographs for various weight fractions (15%, 30%, and 45%) of PCP in PCL have been shown in Figure 13 and for different wt % (5%, 10%, and 15%) of graphite loading have been shown in Figure 14. Composites with 15 weight fraction reinforcement shows better interfacial adhesion with the PCP particles as compared to 30 and 45 weight fractions [31]. A closer observation at higher magnifications of graphite loaded samples shows that all the graphite granules were well-connected and the pores which are clearly visible in unmodified samples were filled with graphite filler (Figure 13). Uniform blending of graphite at higher graphite loading percentages can be easily seen by micrographs. Micrographs had shown sites of voids for water accumulation which further hamper adhesion and water resistance properties [41–46].

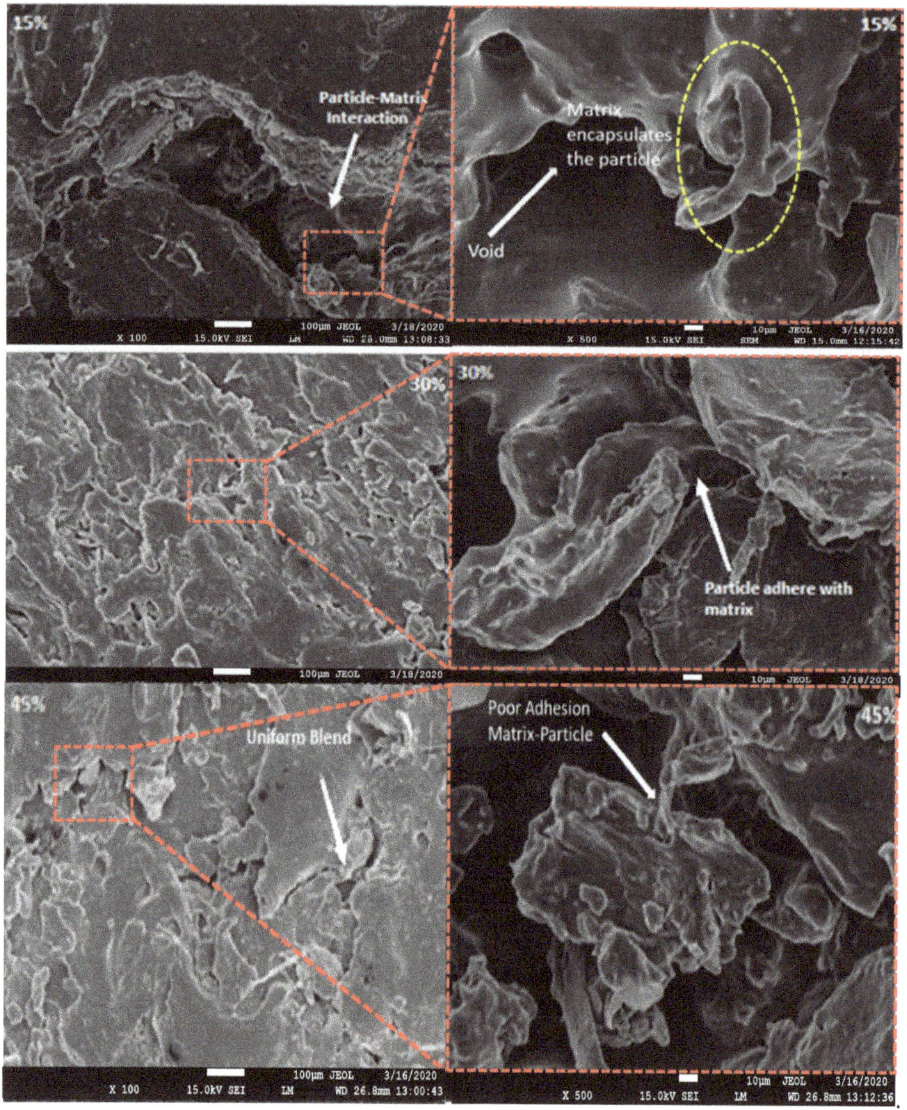

Figure 13. Microscopic images of 15, 30, and 45 wt % of PCP composite and neat PCL at 500× and 1000×.

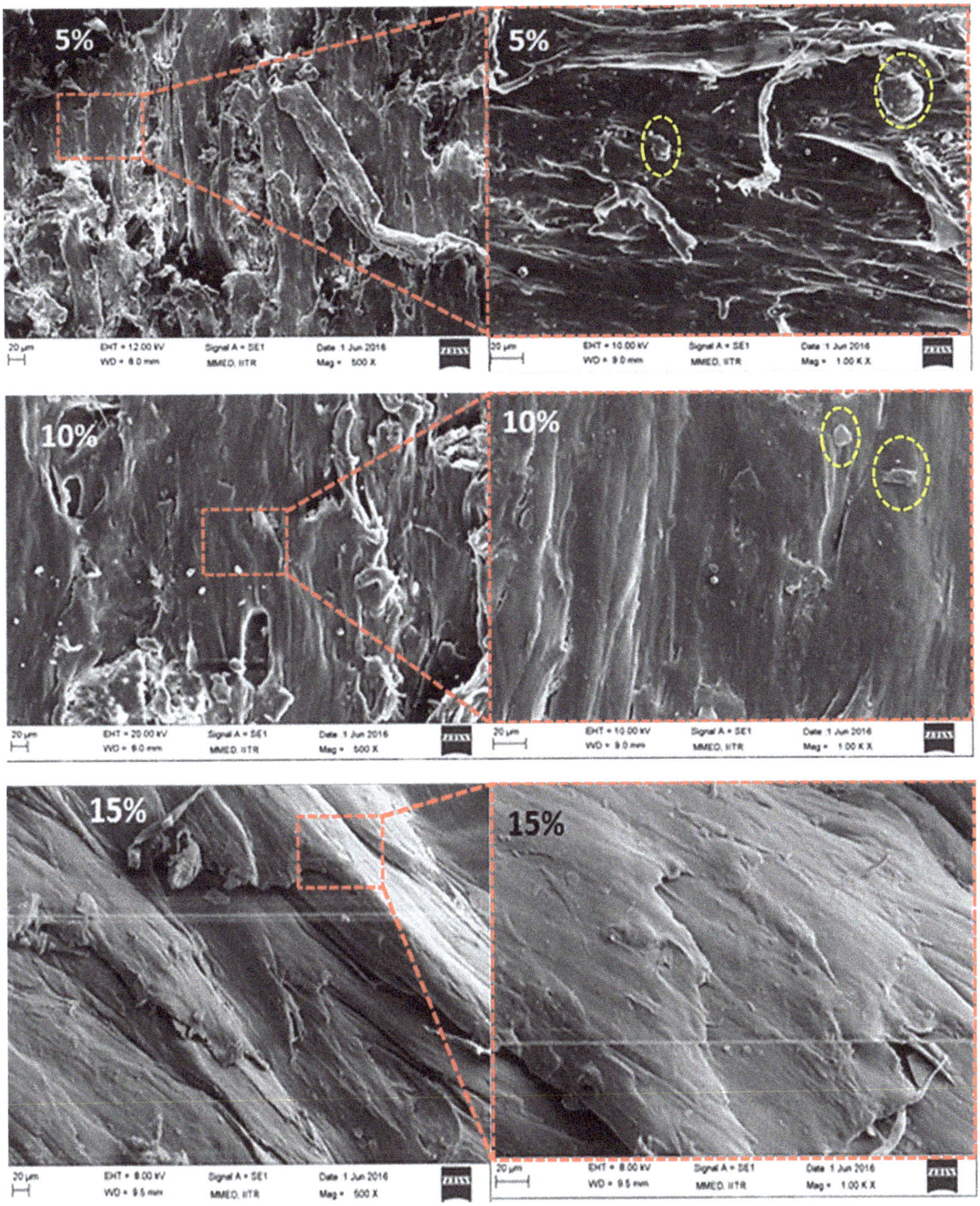

Figure 14. Microscopic images of 5, 10, and 15 wt % of Graphite loaded composite at 500× and 1000×.

4. Conclusions

Novel bio-composites were developed based on polycaprolactone (PCL) and plant-based residue pine cone particles based on graphite compatibilization in polycaprolactone (PCL-G) and PCP.

i. Compatibilising with graphite reduces the effect of the hydrophobic nature of pine cone and improves the interfacial adhesion at a molecular level, as well as diminishing the voids in the agglomerated fibres, thus imparting graphite composites with lower tendency for water absorption.

ii. Bio-composites reinforced with 15–45 wt % of PCP showed higher bio-disintegration than neat PCL.

iii. Bio-composites containing low graphite content also showed higher disintegration rate.

iv. Fracture energy was found to have negative slope with increasing fibres from 0–45 weight fraction in the matrix. After increasing the weight fraction to 30%, the energy required for crack initiation was increased by 68%. Then it further diminishes as weight fraction of fibre was increased to 45%, due to the increase in void content.

v. Microscopy of the composite fractured surfaces depicts the uniform dispersion of PCP particle embedded in PCL matrix at higher fraction of reinforcement. Pine cone particles (PCP) at 15 weight fractions in PCL matrix was observed.

This study presented a novel approach to utilize pine forest fuel as an alternative for synthetic reinforcements in a polymer matrix. Tensile, flexural, water absorption and morphology for the developed material were analysed and reported. Experimental values depict the behavior of reinforcement over the evaluated properties, and it was found that all specimens have achieved at-par performances for utilization as non-structural panels.

Author Contributions: Conceptualization, K.J., Y.K.T., R.K. and S.S.; Methodology, K.J., Y.K.T., R.K. and S.S.; Formal analysis, K.J., Y.K.T., R.K. and S.S.; Investigation, K.J. and S.S.; Resources, K.J., Y.K.T., R.K., S.S., M.R.M.H., C.L. and R.A.I.; Writing—original draft preparation, K.J., Y.K.T., R.K. and S.S.; Writing—review and editing, S.S., M.R.M.H., C.L., R.A.I., S.P.D., A.S. and A.P.; Supervision, K.J., Y.K.T., R.K. and S.S.; Funding acquisition, S.S., M.R.M.H. and R.A.I. All authors have read and agreed to the published version of the manuscript.

Funding: The article processing fee of this manuscript is funded by Research Management Centre, Univesiti Putra Malaysia.

Institutional Review Board Statement: Not applicable.

Informed Consent Statement: Not applicable.

Data Availability Statement: The data presented in this study are available on request from the corresponding author.

Conflicts of Interest: The authors declare no conflict of interest.

References

1. Mohanty, A.K.; Misra, M.; Drzal, L.T. Sustainable bio-composites from renewable resources: Opportunity and challenges in the green material world. *J. Polym. Environ.* **2002**, *10*, 19–26. [CrossRef]
2. Bachtiar, D.; Sapuan, S.M.; Hamdan, M.M. The effect of alkaline treatment on tensile properties of sugar palm fiber reinforced epoxy composites. *Mater. Des.* **2008**, *29*, 1285–1290. [CrossRef]
3. Bledzki, A.K.; Gassan, J. Composites Reinforced with Cellulose based Fabric. *Polym. Sci.* **1999**, *24*, 221–274.
4. Sharifah, H.A.; Martin, P.A. The effect of alkalization and fiber alignment on the mechanical and thermal properties of kenaf and hemp Bast fiber composites: Part 1—Polyester resin matrix. *Compos. Sci. Technol.* **2004**, *64*, 1219–1230.
5. Shaik, D.S.; Kant, Y.; Mitra, D.; Singh, A.; Chandola, H.C.; Sateesh, M.; Chauhan, P. Impact of biomass burning on regional aerosol optical properties: A case study over northern India. *J. Environ. Manag.* **2019**, *244*, 328–343. [CrossRef] [PubMed]
6. Saranya, K.R.L.; Reddy, C.S.; Rao, P.P. Estimating carbon emissions from forest fires over a decade in Similipal Biosphere Reserve, India. *Remote Sens. Appl. Soc. Environ.* **2016**, *4*, 61–67. [CrossRef]
7. Schmerbeck, J.; Kohli, A.; Seeland, K. Ecosystem services and forest fires in India-context and policy implications from a case study in Andhra Pradesh. *For. Policy Econ.* **2015**, *50*, 337–346. [CrossRef]
8. Hung, S.J.; Edelman, P.G. An overview of biodegradable polymers and biodegradation of polymers. In *Degradable Polymers: Principles and Application*; Scott, G., Gilead, D., Eds.; Springer: Berlin/Heidelberg, Germany, 1995; pp. 8–24.
9. Wu, C.S. Physical Properties and Biodegradability of Maleated Polycaprolactone / Starch Composite. *Polym. Degrad. Stab.* **2003**, *80*, 127–134. [CrossRef]
10. Sarikaya, E.; Çallioğlu, H.; Demirel, H. Production of epoxy composites reinforced by different natural fibers and their mechanical properties. *Compos. Part B Eng.* **2019**, *167*, 461–466. [CrossRef]

11. Flynn, J.; Amiri, A.; Ulven, C. Hybridized carbon and flax fiber composites for tailored performance. *Mater. Des.* **2016**, *102*, 21–29. [CrossRef]
12. Jha, K.; Tyagi, Y.K.; Yadav, A.S. Mechanical and thermal behaviour of biodegradable composites based on polycaprolactone with pine cone particle. *Sādhanā* **2018**, *43*, 135. [CrossRef]
13. Echeverria, C.; Pahlevani, F.; Gaikwad, V.; Sahajwalla, V. The effect of microstructure, filler load and surface adhesion of marine bio-fillers, in the performance of hybrid wood-polypropylene particulate bio-composite. *J. Clean. Prod.* **2017**, *154*, 284–294. [CrossRef]
14. Wu, K.J.; Wu, C.S.; Chang, J.S. Biodegradability and Mechanical Properties of Polycaprolactone Composites Encapsulating Phosphate-Solubilizing Bacterium Bacillus Sp. PG01. *Process. Biochem.* **2007**, *42*, 669–675. [CrossRef]
15. Naira, L.S.; Laurencin, C.T. Biodegradable Polymers as Biomaterials. *Prog. Polym. Sci.* **2007**, *32*, 762–798. [CrossRef]
16. Jha, K.; Kataria, R.; Verma, J.; Pradhan, S. Potential biodegradable matrices and fiber treatment for green composites: A review. *AIMS Mater. Sci.* **2019**, *6*, 119–138. [CrossRef]
17. Kolybaba, M.; Tabil, L.G.; Panigrahi, S.; Crerar, W.J.; Powell, T.; Wang, B. Biodegradable polymers: Past, present and future. In *ASABE/CSBE North Central Intersectional Meeting*; American Society of Agricultural and Biological Engineers: St. Joseph, MI, USA, 2003; pp. 3–4.
18. Liu, J.Y.; Reni, L.; Wei, Q.; Wu, J.L.; Liu, S.; Wang, Y.J.; Li, Y. Fabrication and Characterization of Polycaprolactone/Calcium Sulfate Whisker Composites. *eXPRESS Polym. Lett.* **2011**, *5*, 742–752. [CrossRef]
19. Hajiha, H.; Sain, M.; Mei, L.H. Modification and characterization of Hemp and Sisal fibers. *J. Nat. Fibers* **2014**, *11*, 144–168. [CrossRef]
20. Riedel, U. Natural fiber-reinforced biopolymers as construction materials—New discoveries. In Proceedings of the 2nd International Wood and Natural Fiber Composites Symposium, Kassel, Germany, 28–29 June 1999; pp. 28–29.
21. Jha, K.; Kumar, R.; Verma, K.; Chaudhary, B.; Tyagi, Y.K.; Singh, S. Application of modified TOPSIS technique in deciding optimal combination for bio-degradable composite. *Vacuum* **2018**, *157*, 259–267. [CrossRef]
22. Singh, H.; Singh, T. Effect of fillers of various sizes on mechanical characterization of natural fiber polymer hybrid composites: A review. *Mater. Today Proc.* **2019**, *18*, 5345–5350. [CrossRef]
23. Pappu, A.; Thakur, V.K.; Patidar, R.; Asolekar, S.R.; Saxena, M. Recycling marble wastes and Jarosite wastes into sustainable hybrid composite materials and validation through Response Surface Methodology. *J. Clean. Prod.* **2019**, *240*, 118249. [CrossRef]
24. Yousef, S.; Tatariants, M.; Bendikiene, R.; Denafas, G. Mechanical and thermal characterizations of non-metallic components recycled from waste printed circuit boards. *J. Clean. Prod.* **2017**, *167*, 271–280. [CrossRef]
25. Mohan, P.K.; Kumar, A.; Mohite, P.M. Development of In-house Unidirectional Carbon/epoxy Prepregs and its Characterization for Aerospace Applications. *Procedia Struct. Integr.* **2019**, *14*, 176–183. [CrossRef]
26. ISO. *ISO 20200:2004. Plastics—Determination of the Degree of Disintegration of Plastic Materials Under Simulated Composting Conditions in a Laboratory-Scale Test*; ISO: Geneva, Switzerland, 2004.
27. Huang, G.; Sun, H. Effect of water absorption on the mechanical properties of glass/polyester composites. *Mater. Des.* **2007**, *28*, 1647–1650. [CrossRef]
28. Shahinur, S.; Hasan, M. Jute/Coir/Banana Fiber Reinforced Bio-Composites: Critical Review of Design, Fabrication, Properties and Applications. In *Encyclopedia of Renewable and Sustainable Materials*; Elsevier Ltd.: Oxford, UK, 2019; pp. 751–756. [CrossRef]
29. Bismarck, A.; Aranberri-Askargorta, I.; Springer, J. Surface characterization of flax, hemp and cellulose fibers; surface properties and the water uptake behavior. *Polym. Compos.* **2002**, *23*, 872–894. [CrossRef]
30. Espert, A.; Vilaplana, F.; Karlsson, S. Comparison of water absorption in natural cellulosic fibers from wood and one-year crops in polypropylene composites and its influence on their mechanical properties. *Compos. Part. A Appl. Sci. Manuf.* **2004**, *35*, 1267–1276. [CrossRef]
31. Chouzouri, G.; Xanthos, M. Degradation of aliphatic polyesters in the presence of inorganic fillers. *J. Plast. Film. Sheeting* **2007**, *23*, 19–36. [CrossRef]
32. Dahiya, J.B.; Rana, S. Thermal degradation and morphological studies on cotton cellulose modified with various aryl phosphorodichloridites. *Polym. Int.* **2004**, *53*, 995–1002. [CrossRef]
33. Wahit, M.U.; Akos, N.I.; Laftah, W.A. Influence of Natural Fibers on the Mechanical Properties and Biodegradation of Poly (lactic acid) and Poly(e-caprolactone) Composites: A Review. *Polym. Compos.* **2012**, *33*, 1045–1053. [CrossRef]
34. Honma, T.; Zhao, L.; Asakawa, N.; Inoue, Y. Poly (ε-caprolactone)/chitin and Poly (e-caprolactone)/chitosan Blend Films with Compositional Gradients: Fabrication and their Biodegradability. *Macromol. Biol. Sci.* **2006**, *6*, 241–249. [CrossRef]
35. Ludueña, L.; Vázquez, A.; Alvarez, V. Effect of lignocellulosic filler type and content on the behavior of polycaprolactone based eco-composites for packaging applications. *Carbohydr. Polym.* **2012**, *87*, 411–421. [CrossRef]
36. Wu, C.S. A Comparison of the Structure, Thermal Properties, And Biodegradability of Polycaprolactone/Chitosan and Acrylic Acid Grafted Polycaprolactone/Chitosan. *Polymer* **2006**, *46*, 147–155. [CrossRef]
37. Ilyas, R.A.; Sapuan, S.M.; Asyraf, M.R.M.; Dayana, D.A.Z.N.; Amelia, J.J.N.; Rani, M.S.A.; Norrrahim, M.N.F.; Nurazzi, N.M.; Aisyah, H.A.; Sharma, S.; et al. Polymer Composites Filled with Metal. Derivatives: A Review of Flame Retardants. *Polymers* **2021**, *13*, 1701. [CrossRef]

38. Chohan, J.S.; Mittal, N.; Kumar, R.; Singh, S.; Sharma, S.; Dwivedi, S.P.; Saxena, A.; Chattopadhyaya, S.; Ilyas, R.A.; Le, C.H.; et al. Optimization of FFF Process Parameters by Naked Mole-Rat Algorithms with Enhanced Exploration and Exploitation Capabilities. *Polymers* **2021**, *13*, 1702. [CrossRef]
39. Chohan, J.S.; Mittal, N.; Kumar, R.; Singh, S.; Sharma, S.; Singh, J.; Rao, K.V.; Mia, M.; Pimenov, D.Y.; Dwivedi, S.P. Mechanical Strength Enhancement of 3D Printed Acrylonitrile Butadiene Styrene Polymer Components Using Neural Network Optimization Algorithm. *Polymers* **2020**, *12*, 2250. [CrossRef] [PubMed]
40. Singh, Y.; Singh, J.; Sharma, S. Multi-objective Optimization of Kerf-taper and Surface-roughness Quality Characteristics for Cutting-operation On Coir and Carbon Fibre Reinforced Epoxy Hybrid Polymeric Composites During CO_2-Pulsed Laser-cutting Using RSM. *Lasers Manuf. Mater. Process.* **2021**, *8*, 157–182. [CrossRef]
41. Sharma, S.; Singh, J.; Kumar, H.; Sharma, A.; Aggarwal, V.; Gill, A.; Jayarambabu, N.; Kailasa, S.; Rao, K.V. Utilization of rapid prototyping technology for the fabrication of an orthopedic shoe inserts for foot pain reprieve using thermo-softening viscoelastic polymers: A novel experimental approach. *Meas. Control* **2020**, *53*, 519–530. [CrossRef]
42. Singh, Y.; Singh, J.; Sharma, S.; Sharma, A.; Chohan, J.S. Process Parameter Optimization in Laser Cutting of Coir Fiber Reinforced Epoxy Composite—A Review. *Mater. Today Proc.* **2021**, *47*, 4738–4744. [CrossRef]
43. Chohan, J.S.; Kumar, R.; Singh, T.B.; Singh, S.; Sharma, S.; Singh, J.; Mia, M.; Pimenov, D.Y.; Chattopadhyaya, S.; Dwivedi, S.P.; et al. Taguchi S/N and TOPSIS Based Optimization of Fused Deposition Modelling and Vapor Finishing Process for Manufacturing of ABS Plastic Parts. *Materials* **2020**, *13*, 5176. [CrossRef]
44. Singh, Y.; Singh, J.; Sharma, S.; Nguyen, D.N. Fabrication and characterization of coir/carbon-fiber reinforced epoxy-based hybrid composite for helmet shells and sports-good applications: Influence of fiber surface modifications on the mechanical, thermal and morphological properties. *J. Mater. Res. Technol.* **2020**, *9*, 15593–15603. [CrossRef]
45. Sharma, S.; Sudhakara, P.; Singh, J.; Ilyas, R.A.; Asyraf, M.R.M.; Razman, M.R. Critical Review of Biodegradable and Bioactive Polymer Composites for Bone Tissue Engineering and Drug Delivery Applications. *Polymers* **2021**, *13*, 2623. [CrossRef]
46. Sharma, S.; Sudhakara, P.; Omran, A.A.B.; Singh, J.; Ilyas, R.A. Recent Trends and Developments in Conducting Polymer Nanocomposites for Multifunctional Applications. *Polymers* **2021**, *13*, 2898. [CrossRef] [PubMed]

polymers

MDPI

Article

Comparative Analysis of Erosive Wear Behaviour of Epoxy, Polyester and Vinyl Esters Based Thermosetting Polymer Composites for Human Prosthetic Applications Using Taguchi Design

Jeetendra Mohan Khare [1], Sanjeev Dahiya [1], Brijesh Gangil [2], Lalit Ranakoti [3], Shubham Sharma [4,*], Muhammad Roslim Muhammad Huzaifah [5,*], Rushdan Ahmad Ilyas [6,7], Shashi Prakash Dwivedi [8], Somnath Chattopadhyaya [9], Huseyin Cagan Kilinc [10] and Changhe Li [11]

1 School of Engineering & Technology, University of Technology, Rajasthan, Jaipur 303903, India; jmkhare2010@gmail.com (J.M.K.); dahiyasanjeev764@gmail.com (S.D.)
2 Department of Mechanical Engineering, H.N.B. Garhwal University, Garhwal, Srinagar 246174, India; brijeshgangil7@gmail.com
3 Department of Mechanical Engineering, National Institute of Technology (NIT Uttarakhand), Srinagar 246174, India; lalitranakoti9000@gmail.com
4 Department of Mechanical Engineering, IK Gujral Punjab Technical University, Main Campus, Kapurthala 144603, India
5 Department of Crop Science, Faculty of Agricultural Science and Forestry, Universiti Putra Malaysia Bintulu Campus, Bintulu 97000, Malaysia
6 School of Chemical and Energy Engineering, Faculty of Engineering, Universiti Teknologi Malaysia, Johor Bahru 81310, Malaysia; ahmadilyas@utm.my
7 Centre for Advanced Composite Materials, Universiti Teknologi Malaysia, Johor Bahru 81310, Malaysia
8 G.L. Bajaj Institute of Technology & Management, Greater Noida 201310, India; spdglb@gmail.com
9 Department of Mechanical Engineering, Indian Institute of Technology (ISM) Dhanbad, Dhanbad 826004, India; somnathchattopadhyaya@iitism.ac.in
10 Civil Engineering Department, Istanbul Esenyurt University, Istanbul 34510, Turkey; huseyincagankilinc@esenyurt.edu.tr
11 School of Mechanical and Automotive Engineering, Qingdao University of Technology, Qingdao 266520, China; sy_lichanghe@163.com
* Correspondence: shubham543sharma@gmail.com or shubhamsharmacsirclri@gmail.com (S.S.); muhammadhuzaifah@upm.edu.my (M.R.M.H.)

Citation: Khare, J.M.; Dahiya, S.; Gangil, B.; Ranakoti, L.; Sharma, S.; Huzaifah, M.R.M.; Ilyas, R.A.; Dwivedi, S.P.; Chattopadhyaya, S.; Kilinc, H.C.; et al. Comparative Analysis of Erosive Wear Behaviour of Epoxy, Polyester and Vinyl Esters Based Thermosetting Polymer Composites for Human Prosthetic Applications Using Taguchi Design. *Polymers* **2021**, *13*, 3607. https://doi.org/10.3390/polym13203607

Academic Editors: Emin Bayraktar, S. M. Sapuan and R. A. Ilyas

Received: 25 September 2021
Accepted: 15 October 2021
Published: 19 October 2021

Abstract: In polymer composites, synthetic fibers are primarily used as a chief reinforcing material, with a wide range of applications, and are therefore essential to study. In the present work, we carried out the erosive wear of natural and synthetic fiber-based polymer composites. Glass fiber with jute and *Grewia optiva* fiber was reinforced in three different polymer resins: epoxy, vinyl ester and polyester. The hand lay-up method was used for the fabrication of composites. L_{16} orthogonal array of Taguchi method used to identify the most significant parameters (impact velocity, fiber content, and impingement angle) in the analysis of erosive wear. ANOVA analysis revealed that the most influential parameter was in the erosive wear analysis was impact velocity followed by fiber content and impingement angle. It was also observed that polyester-based composites exhibited the highest erosive wear followed by vinyl ester-based composites, and epoxy-based composites showed the lowest erosive wear. From the present study, it may be attributed that the low hardness of the polyester resulting in low resistance against the impact of erodent particles. The SEM analysis furthermore illustrates the mechanism took place during the wear examination of all three types of composites at highest fiber loading. A thorough assessment uncovers brittle fractures in certain regions, implying that a marginal amount of impact forces was also acting on the fabricated samples. The developed fiber-reinforced polymer sandwich composite materials possess excellent biocompatibility, desirable promising properties for prosthetic, orthopaedic, and bone-fracture implant uses.

Keywords: natural and synthetic fibers; thermosetting polymers; L_{16} orthogonal array; Taguchi method; erosion mechanism; SEM analysis; prosthetic applications

1. Introduction

Fiber-reinforced polymer composite materials are the modern materials applied in various applications such as automobile interiors, construction articles, transportation materials, packaging, and household ware [1–3]. The prerequisite necessary for all the aforementioned applications is good strength, stiffness, durability, flexibility, etc., which a composite does exhibits as reported by several literatures in the past decades [4–6]. Synthetic fiber-reinforced polymer composites have several advantages such as extremely high strength comparable to metals but have several demerits such as high carbon release in the environment, non-disposable leading to soil degradation, high cost, etc. [7]. On the other hand, natural fiber reinforced polymer composites exhibit quite remarkable mechanical properties but far lower than synthetic counterparts [8–10]. To optimize parameters such as strength, cost, etc., and minimize the hazardous environmental effects, hybridization of natural and synthetic fiber has been performed. For instance, the hybrid composite was prepared by incorporating Cocos nucifera and Lufa cylindrical fiber at a weight fraction of 30% in the mix of MEKP and cobalt naphthenate and reported that mechanical properties improved by 31%. It was also reported that changing the fiber ratio in the composition led to alteration of property from ductile to brittle [11]. The addition of cellulosic fiber (banana, abaca, jute, and hemp, wood) in glass fiber reinforced polymer hybrid composite yields higher mechanical strength than composite containing only glass fiber [12–14]. Analysis of the hybrid composite of synthetic-synthetic fiber [15] and plant-animal fiber [16,17] have also been carried out and reported improved results regarding mechanical properties. Kevlar-kenaf fiber reinforced polymer hybrid composite was fabricated to investigate mechanical properties [18,19] and reported that significant enhancement can be obtained in the impact properties at 40 wt% of fiber reinforcement. Automobile door panels and several automotive parts are under consideration to be manufactured from the hybrid composite of hemp and kenaf fiber [20].

Glass fiber is one of the strongest known fiber primarily used as a chief reinforcing material in the polymer composite with a wide range of applications. However, due to its high cost, several natural fibers were reinforced to make the composite economic. In addition, glass fiber is highly brittle and, therefore, mostly reinforced with ductile natural fiber to reduce the brittleness of hybrid polymer composite [21]. In recent times, various natural fibers have been reinforced with glass fiber to enhance the mechanical properties and wear resistance of glass fiber polymer composite. For instance, reinforcement of fibers such as bamboo, sisal, and wood in glass fiber composite enhanced tensile, flexural, and impact properties of hybrid composite [22–24]. Silica nanoparticles filled glass fiber reinforced epoxy composites yield five times higher fatigue strength than virgin glass fiber composite [25]. Fracture strength, interlaminar shear strength, flexural strength, and impact strength were improved by incorporating flax, basalt, and jute fiber in glass fiber reinforced polymer composite [26–29]. It has been observed that the weightage of natural fiber was kept below the weightage of glass fiber in the hybrid composite to obtain optimum mechanical strength [30–32]. The mechanical and tribological properties of a hybrid composite comprising more than two fibers, primarily two natural and one glass fiber, were also investigated. In this regard, glass fiber/sisal fiber/chitosan reinforced polymer sandwich were fabricated for orthopaedic fracture application and reported impressive wear resistance and modulus [33]. Hybridization of 10% jute and 10% tea leaf fiber in glass yielded mechanical strength of such value that can even replace virgin glass fiber composite [34]. Reinforcing jute, sisal, kenaf, or combining the two in glass fiber reinforced polymer composites also improves the mechanical properties. [35,36].

Material erosion caused by hard particles is one of several types of material degradation categorized as wear. Polymers composites also work under different working conditions, requiring an analysis of their wear activities before they are located in a real environment. Several studies have been reported in the past for the investigation of erosive wear of polymer composite containing different types of fiber and fillers [37]. For instance, glass fiber reinforced polymer composite filled with micro silica and zinc oxide was fabricated via vacuum-assisted method and investigated for erosive wear behavior at a different impingent angle ranging from 20° to 90° [38]. It was reported that silica fumes enhanced the composite's erosive wear resistance, while zinc oxide promoted erosive wear.

Moreover, increasing the impingement angle and size of erodent particles increased the erosive wear. The incorporation of marble dust in glass fiber reinforced polymer composite reduces erosive wear of the fabricated composite, as marble dust increases the hardness and stiffness of the composite surface [39]. Diversifying the process parameters and input variables such as matrix type, filling material, and manufacturing method can reduce the erosive wear of the manufactured composites. Calcium carbonate ($CaCO_3$), barium sulphate ($BaSO_4$), and tungsten carbide (WC) filled glass fiber reinforced PA/ABS composite prepared by injection molding exhibited relatively lower erosive wear at higher impingement angle, i.e., 90° or 75° [40–42]. However, the incorporation of $CaCO_3$ and $BaSO_4$ promoted ductile and semi ductile erosive wear while WC promoted brittle wear of the composite. Treatment of fiber by chemical agents plays a crucial role in enhancing the resistance of composite against erosive wear. Benzoylated treated areca sheath fiber reinforced polyvinyl chloride exhibited lower erosive wear than untreated areca sheath fiber-reinforced composite [43]. In addition, fiber treatment encourages good bonding between fiber and matrix resulted in low erosive wear efficiency. The erosive wear of carbon and glass fibre reinforced composites demonstrated that they could attain excellent erosive wear resistance without using any ceramic filler; however, this was not the case with natural fibre reinforced composites. [44]. It is interesting to observe from the literature that synthetic fiber reinforced composites exhibited lower erosive wear than natural fiber-reinforced polymer composite. Jute fiber has excellent strength, good UV protection, low thermal conduction and attractive anti-static properties which qualifies it a good reinforcing material. In addition, low cost *Grewia optiva* fiber, low density, easily available in the Himalayan region contains high amount of pectin (jelling and thickening agent) which is advantageous in making good bonding with polymeric chain and can be useful in a way of making cheaper and lighter prosthesis with good mechanical and tribological properties. As discussed in the literature, research on the erosive wear of composites combining both natural and synthetic fibers has been limited. As a result, the current research focuses on the hybridization of synthetic and natural fibres and compares the erosive wear of composites comprising three different resins: epoxy, polyester, and vinyl ester, using the Taguchi methodology as shown in the Figure 1.

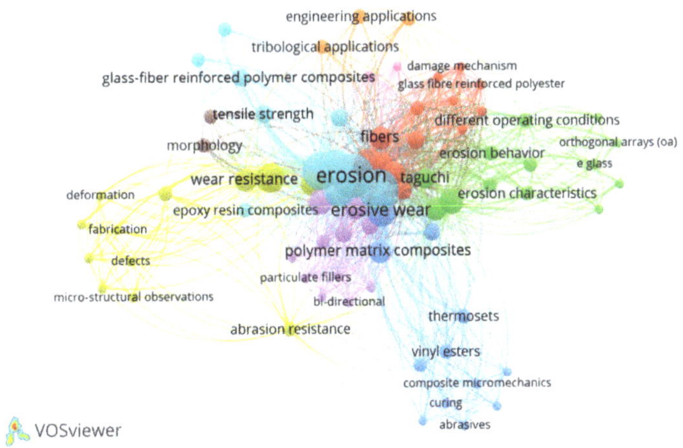

Figure 1. Bibliometrics visualisation overview of technology-based progressions on the erosion-behaviour of composites obtained from the various thermoset-resins (epoxy, polyester, and vinyl-ester) for a multitude of scenarios.

2. Experimentation

2.1. Materials

Bisphenol resin and epichlorohydrin were purchased from HEXION Specialty Chemicals Pvt. Ltd. Karnataka and mixed in the ratio of 5:1 for the preparation of epoxy (Epikote Resin 828), having good chemical resistance, internal adhesion, and appropriate wetting pigment. Esterification of epoxy with unsaturated mono carboxylic acid purchased from Amtech Ester Pvt. Ltd. Delhi was performed to prepare vinyl ester and dissolve the reactant in the solution of the solution styrene to provide stability to the prepared vinyl ester. Dibasic organic acids with polyhydric alcohols were purchased from Yes composites India Ltd. New Delhi and mixed in the appropriate ratio for the preparation of polyester.

Strand of chopped glass fiber having good strength and high insulating properties were procured from Yes composite Ltd. Natural fibers, i.e., jute and *Grewia optiva* were purchased in the form of bi-directional mat locally from the Uttarakhand Bamboo Board (India). These fibers were treated from NaOH solution with 8% concentration and then washed in distilled to remove dirt and dust present on the surface of fibers. Fibers used in the present investigation are shown in Figure 2.

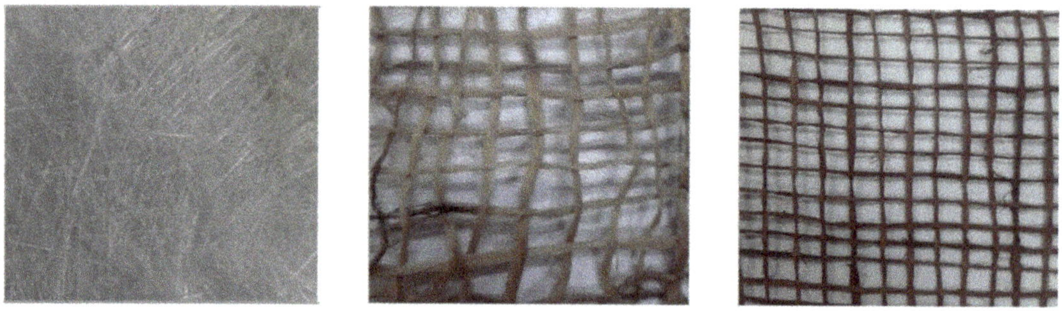

(a) (b) (c)

Figure 2. Showing (**a**) Glass fiber, (**b**) Jute fiber and (**c**) *Grewia optiva* fiber.

2.2. Methods

The composite samples were made using the hand lay-up approach, as shown in Figure 3. Glass plates measuring 500 × 300 × 4 mm^3 were utilized as molding plates for composite manufacturing. Double-sided tape was used on all sides of the molding plate to achieve the desired thickness and secure the side bidding of the fabricated composite. Silicon spray was used over the molding plates to avoid the sticking of the sample with the plates. Firstly the resin was poured over the molding plates and evenly dispersed with the help of a steel roller, after which the natural and synthetic fibre mat of known percentage were placed one by one over the resin. Subsequently, the remaining resin was spread evenly over the mat with the help of a roller. Finally, a 15 kg load was held above the sample and left to cure. The composite sample was taken out of the mould and cut to the appropriate dimension for erosive wear characterization after 24 h of curing, as shown in Figure 4. Samples fabricated by varying fiber weightage are illustrated in Table 1.

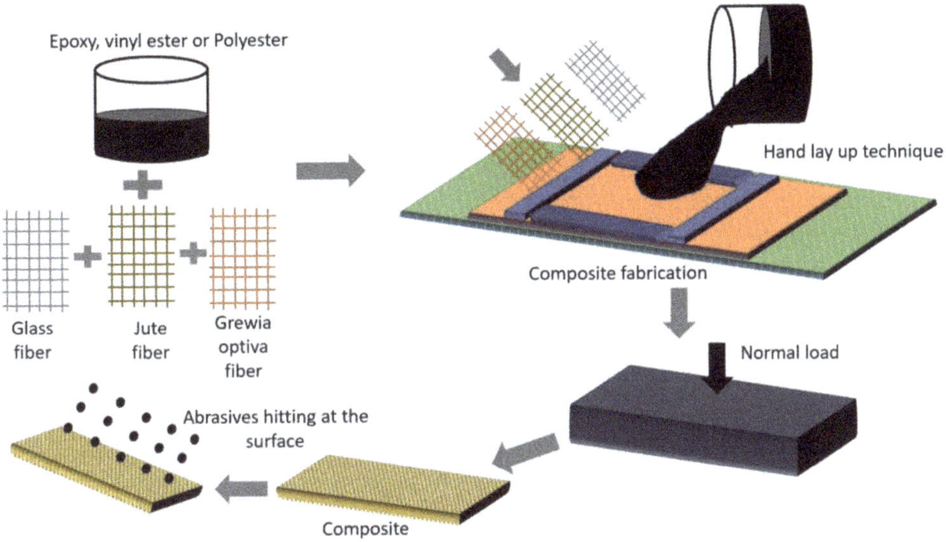

Figure 3. Process of fabrication of composite.

Figure 4. (**a**) Epoxy based, (**b**) vinyl ester based and (**c**) poly ester-based sample for erosive wear test.

Table 1. Composition of fabricated composites.

S. No.	Designation	Compositions
1	E0	Epoxy + Glass fiber (5 wt%)
2	E1	Epoxy + glass fiber (5 wt%) + jute (2.5 wt%) + *Grewia optiva* (2.5 wt%)
3	E2	Epoxy + glass fiber (5 wt%) + jute (5 wt%) + *Grewia optiva* (5 wt%)
4	E3	Epoxy + glass fiber (5 wt%) + jute (7.5 wt%) + *Grewia optiva* (7.5 wt%)
5	V0	Vinyl ester + glass fiber (5 wt%)
6	V1	Vinyl ester + glass fiber (5 wt%) + jute (2.5 wt%) + *Grewia optiva* (2.5 wt%)
7	V2	Vinyl ester + glass fiber (5 wt%) + jute (5 wt%) + *Grewia optiva* (5 wt%)
8	V3	Vinyl ester + glass fiber (5 wt%) + jute (7.5 wt%) + *Grewia optiva* (7.5 wt%)
9	P0	Polyester + glass fiber (5 wt%)
10	P1	Polyester + glass fiber (5 wt%) + jute (2.5 wt%) + *Grewia optiva* (2.5 wt%)
11	P2	Polyester + glass fiber (5 wt%) + jute (5 wt%) + *Grewia optiva* (5 wt%)
12	P3	Polyester + glass fiber (5 wt%) + jute (7.5 wt%) + *Grewia optiva* (7.5 wt%)

2.3. Erosive Wear Analysis

The analysis of erosive wear for the fabricated samples was carried out as per ASTM-G 76 standard of size 30 × 30 × 5 mm^3 on air jet erosion tester supplied by DUCOM, India as shown in Figure 5. The erodent particles used in the experiment were silica of size varying from 100 to 250 μm. Silica particles were forced to impinge at the surface of the sample through a tungsten carbide nozzle for 15 min at different experimental conditions. After completing the test, the surface of the samples was cleaned by using acetone, and an electronic weighing machine measured its weight.

Figure 5. Air jet erosion test machine set up.

2.4. Taguchi Experiment Design

Various control factors influence the erosive wear, such as the size of the erosive particle, velocity of impact, angle of impingement, filler content, concentration, etc. In this present investigation, erosive wear is assessed by evaluating three control factors, each having four levels as tabulated in Table 2.

Table 2. Control factors and respective levels.

Control Factors	Levels				
	I	II	III	IV	Units
Impact Velocity	30	40	50	60	m/s
Impingement angle	45	60	75	90	degree
Fiber content					
(For composites E_1, E_2, E_3)	0	05	10	15	wt%
(For composites V_1, V_2, V_3)	0	05	10	15	
(For composites P_1, P_2, P_3)	0	05	10	15	

Considering Table 2, if all the experiments are to be performed, it will be 81 numbers for three control factors and four levels. It will become cumbersome and laborious. Moreover, a lot of time and energy will be required, which makes it a costly deal. Alternatively, the Taguchi approach can be applied, which uses an orthogonal array to break down the 81 numbers in just a handful of experiments offering enough control factors as provided by 81 experiments. Here, L_{16} orthogonal array has been constructed as shown in Table 3 for the investigation in the Minitab to analyse erosive wear. Moreover, signal to noise (S/N) ratio was analysed using lower the better characteristics as per equation 1to examine the erosive wear of the composite samples.

Table 3. Design-matrix array arrangement.

S. No.	Impact Velocity (m/s)	Natural Fiber (wt%)	Impingent Angle (Degree)
1	30	0	45
2	30	5	60
3	30	10	75
4	30	15	90
5	40	0	60
6	40	5	45
7	40	10	90
8	40	15	75
9	50	0	75
10	50	5	90
11	50	10	45
12	50	15	60
13	60	0	90
14	60	5	75
15	60	10	60
16	60	15	45

Lower-the-better characteristic:

$$\text{S/N ratio} = -10log\frac{1}{n}\left(\sum_{i=1}^{n} y_i\right) \tag{1}$$

Here, y is erosive wear and n is number of experiments.

Literature-Based on Erosive Wear Analysis of Polymer Composites Using Taguchi Approach

The erosive wear analysis of fiber/filler reinforced polymer composites has been carried out recently at different impact velocities, filler content, impingement angle, erodent size, etc., as shown in Table 4. Several types of orthogonal array have been used to examine the effect of different control factors on the erosive wear of composite. Table 4 suggests that erosive wear gets significantly affected by altering the controlling factors. Initially, orthogonal arrays were used only for composite containing epoxy or polyester based composites. But limited or none of the studies were done on vinyl ester-based composites.

Furthermore, the effect of three fibers in a composite with different resins has rarely been investigated.

Table 4. List of work carried out for erosive wear using Taguchi experiment.

S. No.	Composition	Optimization Technique (Taguchi)	Control Factor with Corresponding Level					Highlights of Work	References
1	*Grewia optiva*–glass fiber–dolomite filler–epoxy	L_{16} orthogonal array	Impact velocity:	10	20	30	40	Highest influence on erosive wear was shown by impact velocity followed by dolomite content and erodent size. Lowest wear was obtained at impingent angle of 30°	[45]
			Dolomite content:	0	5	10	15		
			Impingement angle	30°	45°	60°	90°		
			Erodent size:	100	150	200	250		
2	Jute–SiC–epoxy	L_9 orthogonal array	Impact velocity	32	44	58		Erosive wear increased with the increase in impact velocity, fiber content and impingement angle. Erodent size showed least effect on erosive wear.	[46]
			Impingement angle	30	60	90			
			Erodent size:	200	300	400			
			Fiber content	20	30	40			
3	Al_2O_3-glass fiber-polyester	L_{27} orthogonal array	Impact Velocity	43	54	65		At low impact velocity, the composite responded in a semi ductile manner while at high velocity, the composite responded in a ductile manner	[47]
			Filler content	0	10	20			
			Impingement angle	30	60	90			
			Stand-off distance	65	75	85			
			Erodent size	250	350	450			
4	AlN-glass fiber-epoxy	L_9 orthogonal array	Impact Velocity	33	47	57		Most influential parameter in the analysis of erosive was impact velocity followed by temperature and filler content respectively.	[48]
			Filler content	5	10	15			
			Impingement angle	30	60	90			
			Temperature	50	75	100			
5	Bagasse fiber-epoxy	L_{27} orthogonal array	Impact Velocity	30	50	70		Around 80% influence of fiber weightage was observed in the erosive wear followed by 14% impingement angle and 4% impact velocity. Maximum erosive wear was observed at high fiber weightage of impingement angle of 60°	[49]
			Filler content	10	20	30			
			Impingement angle	30	60	90			
			Stand-off distance	65	75	85			
			Erodent size	250	350	450			
6	Needle punched Polyester fiber mat-epoxy	L_{27} orthogonal array	Impact Velocity	43	54	65		Erosive wear increased with the increase in impingement angle till 60° but as the impingement angle increased beyond 60°, the erosive wear decreased. Composite exhibited semi ductile erosive wear.	[50]
			Filler content	10	20	30			
			Impingement angle	30	60	90			
			Stand-off distance	65	75	85			
			Erodent size	250	350	450			

Table 4. *Cont.*

S. No.	Composition	Optimization Technique (Taguchi)	Control Factor with Corresponding Level					Highlights of Work	References
7	Palm leaf fiber-epoxy-palm leaf powder	L_{16} orthogonal array	Impact Velocity	40	50	60	70	Composite with 15% palm leaf fiber at 60° impingement angle and impact velocity of 80 m/s showed the highest wear erosion resistance.	[51]
			Filler content	0	5	10	15		
			Impingement angle	45	60	75	90		
			Erodent size	40	60	80	100		
8	E glass fiber-SiC-epoxy	L_{27} orthogonal array	Impact Velocity		32	45	58	A significant reduction in erosive wear was observed by the addition of SiC in glass fiber composite. Maximum wear erosion has occurred at 60°. Composite transform in brittle structure with the incorporation of SiC.	[52]
			SiC content		0	10	20		
			Impingement angle		30	60	90		
			Stand-off distance		120	180	240		
			Erodent size		300	500	800		
9	Himalayan agave fiber-polyester	L_{16} orthogonal array	Sliding Velocity	1.5	2.5	3.5	4.5	Composite with fiber of 5 mm length exhibited the highest erosive wear resistance. Longer fiber (7 mm) reinforced composite exhibited greater erosive wear due to fiber fracture and surface damage. Optimum parameters efficient erosive wear resistance were reported as sliding velocity: 15 m/s, normal load: 20 N, fiber length: 5 mm, and sliding distance: 1500 m.	[53]
			Fiber Length	0	3	5	7		
			Sliding Distance	1000	2000	3000	4000		
			Normal Load	10	15	20	25		
10	Glass fiber-fly ash-polyester	L_{27} orthogonal array	Impact Velocity		32	45	58	The highest erosive wear of composite occurred at 60° of impingement angle and showed semi ductile behavior. Fly ash content has the highest influence on erosive wear in terms of influencing factors, followed by impingement angle and erodent size. Impact velocity has minimum impact on erosive wear, as reported in the study.	[54]
			Fly ash Content		0	10	20		
			Impingement angle		45	60	90		
			Stand-off distance		120	180	240		
			Erodent size		300	500	800		

Note: Impact velocity: (m/s), erodent size: (mm), impingement angle: (degree), stand-off distance: (mm), normal load: (N), sliding distance: (m), fiber/filler content: (weight percentage), fiber length: (mm).

3. Results and Discussions

3.1. Mechanical Properties

The properties shown in Table 5 are the average of 3 readings taken for each sample which has taken from previous research [55]. When the tensile strength of composites was compared, it was discovered that epoxy-based composites had the maximum tensile strength, 72 MPa, at 15 wt% loading of jute and *Grewia optiva* fibre. However, in the case of flexural strength, the vinyl ester based composites outperformed both epoxy and polyester-based composites with the highest value of 48 MPa at 5 wt% loading of both jute and *Grewia optiva* fiber. The impact and hardness values were found to be higher for epoxy-based composites. The properties depicted in the Table 5 shows that the fiber reinforcement in epoxy is more advantageous than reinforcement in vinyl ester and polyester due to higher overall enhancement of mechanical properties. The reduction of the mechanical property in viny ester and polyester composites can be related to the photochemical degradation, plasticizing effect, and weak interfacial adhesion, weakening the interface between the matrix and fillers. On the other hand, epoxy-based composites accomplish efficient mechanical interlocking between the fibre and epoxy, resulting in good stress transfer from the epoxy to the fiber [56,57].

Table 5. Mechanical properties of composites.

Sample	Mechanical Properties			
	Tensile Strength (MPa)	Flexural Strength (MPa)	Impact Strength (J)	Hardness (HRL)
E0	38	22	1.2	52
E1	54	38	1.4	68
E2	68	42	1.68	76
E3	72	36	2.1	57
V0	32	24	1.1	38
V1	47	42	1.2	48
V2	62	48	1.5	54
V3	69	38	1.9	42
P0	28	18	0.9	44
P1	42	28	1.05	56
P2	56	35	1.26	64
P3	60	31	1.71	51

3.2. *Taguchi Analysis of Erosive Wear*

On the prepared composites, three control factors and four levels were investigated using the L_{16} orthogonal array. The analysis of several combinations of control factors was performed using Minitab 15. Erosive wear rates of different resin-based composite and their corresponding S/N ratio have been presented in Table 6. Furthermore, the effect of control factor on the erosive wear with the respective ranking is tabulated in Table 6. Observations revealed that the control factor which influenced erosive the most was found to be impact velocity. The effect of fiber content on erosive wear was low as compared to impact velocity but quite considerable. Impingement angle has the least effect on erosive wear among all the three control factors. Further, a graph of control factors at different levels is shown in Figure 4. For epoxy-based composites, it can be concluded from Table 6 that erosive wear increases with the increase in impact velocity and is found to be minimum at an Impact velocity of 30 m/s, 15 wt% fiber reinforcement and impingement angle of 90° whereas maximum erosive wear was obtained at 60 m/s 0 wt% fiber reinforcement and impingement angle of 90°. This indicates that the addition of fiber in epoxy does not necessarily influence the erosive wear but also the impingement angle. When the erodent particles hit the composite surface, the epoxy material is first contacted and then the reinforcement after the erosion. The Rockwell hardness of epoxy is relatively high, with an outstanding value of 80 number, which is capable of bearing the impact of erodent particle at such high velocity. The respective S/N ratio at 3rd run and 13th run erosive wear is also the evidence for minimum and maximum erosive wear.

Table 6. Erosive wear and corresponding S/N ratio of the composites.

S. No.	Erosive Wear of Epoxy (mg/kg)	S/N Ratios	Erosive Wear of Vinyl Ester (mg/kg)	S/N Ratios	Erosive Wear of Polyester (mg/kg)	S/N Ratios
1	298.96	−49.5123	304.67	−49.6766	309.88	−49.8239
2	218.47	−46.7878	221.08	−46.8910	232.10	−47.3135
3	219.06	−46.8113	236.11	−47.4623	228.07	−47.1614
4	209.65	−46.4299	305.44	−49.6985	253.45	−48.0778
5	338.75	−50.5976	445.21	−52.9713	318.85	−50.0717
6	260.37	−48.3118	278.021	−48.8816	297.92	−49.4820
7	309.44	−49.8115	211.67	−46.5132	247.81	−47.8824
8	311.23	−49.8616	309.37	−49.8096	384.53	−51.6986
9	461.64	−53.2861	361.09	−51.1523	506.43	−54.0904
10	416.86	−52.3998	398.56	−52.0099	434.67	−52.7632
11	306.99	−49.7425	380.88	−51.6158	371.09	−51.3896
12	386.25	−51.7374	432.01	−52.7099	489.56	−53.7961
13	527.65	−54.4469	550.84	−54.8205	588.19	−55.3904
14	520.97	−54.3363	554.41	−54.8766	566.71	−55.0672
15	364.06	−51.2235	381.56	−51.6313	408.34	−52.2204
16	411.85	−52.2948	415.51	−52.3716	421.01	−52.4858

It can also be observed from Table 6 that the erosive wear of vinyl ester composites among the 16 runs was found to be minimum at 7th run maximum at 14th run, which is at an impact velocity of 40 m/s, fiber weightage of 10 wt%, 90° impingement angle and impact velocity of 60 m/s, fiber weightage of 0 wt%, impingement angle of 75°. The magnitude of erosive wear was approximately similar to epoxy-based composites but 0.7% higher in magnitude. Interestingly, the most increased erosive wear occurs at 75° impingement angle and 0 wt%. This shows that fiber inclusion in the vinyl ester composites reduces the erosive wear, and impingement angles from 75° to 90° showed the same erosive wear. Moreover, the upsurge in impact velocity increases the erosive wear of the vinyl ester composite.

Erosive wear for polyester-based composites was found to be higher as compared to both epoxy and vinyl ester-based composites. It was observed that minimum and maximum erosive wear were obtained at an impact velocity of 30 m/s, fiber reinforcement of 0 wt%, 75° impingement angle (3rd run) and impact velocity of 60 m/s, fiber reinforcement of 0 wt% and impingement angle of 90° (13th run), respectively. The reason of higher wear may be attributed to the low hardness of polyester resulting in low resistance against the impact of erodent particles [58].

The analysis of results as shown in Figure 6 concludes that the combination at an impact velocity of 30 m/s, fiber content of 10 wt% and impingement angles of 45° yields the lowest erosive wear in epoxy-based composites. For the vinyl ester composites, the lowest wear rate was obtained for the combination at impact velocity of 30 m/s and fiber content of 10 wt% and impingement angles of 45°. Additionally, in the case of polyester-based composite, the lowest erosive wear was obtained at 30 m/s, fiber content of 10 wt% and impingement angles of 45°. Furthermore, as described in Table 7, the order of effectiveness of control factors is impact velocity (1st), fiber content (2nd), and impingement angle (3rd).

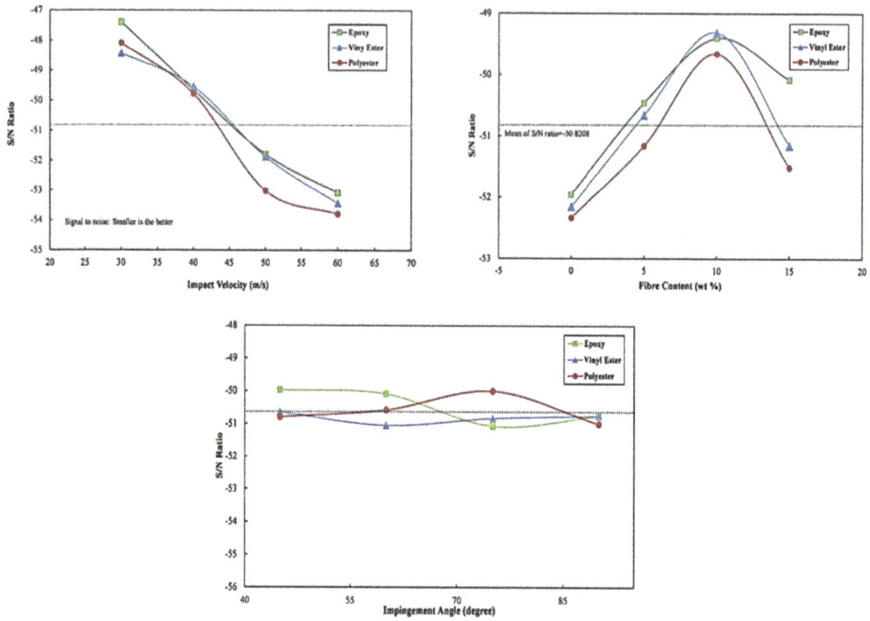

Figure 6. Mean of S/N ratio for epoxy (ep), vinyl ester (ve) and polyester (pe) based composites.

Table 7. Response table for composites.

Levels	Epoxy			Vinyl Ester			Polyester		
	Impact Velocity (m/s)	Fiber Content (wt%)	Impingement Angle (Degree)	Impact Velocity (m/s)	Fiber Content (wt%)	Impingement Angle (Degree)	Impact Velocity (m/s)	Fiber Content (wt%)	Impingement Angle (Degree)
1	−47.39	−51.96	−49.97	−48.43	−52.16	−50.64	−48.09	−52.34	−50.8
2	−49.65	−50.46	−50.09	−49.54	−50.66	−51.05	−49.78	−51.16	−50.58
3	−51.79	−49.40	−51.07	−51.87	−49.31	−50.83	−53.01	−49.66	−50
4	−53.08	−50.08	−50.77	−53.43	−51.15	−50.76	−53.79	−51.51	−51
Delta	5.69	2.56	1.11	4.99	2.85	0.41	5.7	2.68	1.21
Rank	1	2	3	1	2	3	1	2	3

Analysis of Variance

Analysis of variance (ANOVA) was used for the analysis of the statistical significance of control factors at 95% confidence level. The p-value of control factors or epoxy-based composites is shown in Table 8 which shows the degree of significance of the control factor on erosive wear. It was observed that the value of p for impact velocity is 0.001 which is lower than assumed p-value, i.e., 0.05 and is the most significant factor for the analysis of erosive wear. The next significant factor was fiber content with the p-value of 0.046. Impingement angle was observed to be least significant factor as per the ANOVA Table 8. The contribution of all the factors for epoxy-based composites with the highest contribution of impact velocity has a value of 76.35%, followed by fiber content with the contribution factor of 14.36% and the least contribution of impingement angle with 3.49% is shown in Figure 7.

Table 8. Analysis of variance for SNRA2, using adjusted SS for tests epoxy-based composites.

Source	DF	Seq SS	Adj SS	Adj MS	F-Value	*p*-Value
Impact Velocity (m/s)	3	74.915	74.915	24.972	26.43	0.001
Fiber content (wt%)	3	14.098	14.098	4.699	4.97	0.046
Impingement angle (Degree)	3	3.43	3.43	1.143	1.21	0.384
Error	6	5.67	5.67	0.945		
Total	15	98.113				

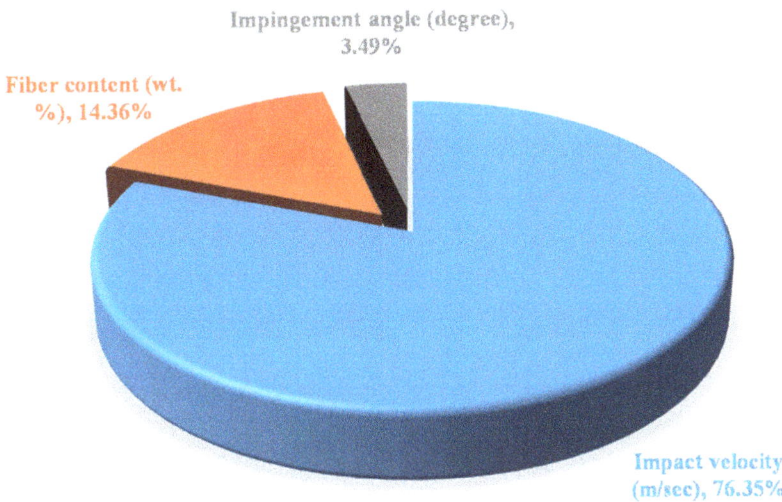

Figure 7. Contribution chart of factors for epoxy-based composites.

The *p*-value of control factors or vinyl ester-based composites is shown in Table 9. It was observed that the value of *p* for impact velocity is 0.0031, which is lower than the assumed *p*-value, i.e., 0.05, and is the most significant factor for the analysis of erosive wear. The next significant factor was fiber content with the *p*-value of 0.275. Impingement angle was observed to be the least significant factor as per the ANOVA Table 9. The contribution of all the factors for vinyl ester-based composites with the highest contribution of impact velocity has a value of 61.82%, followed by fiber content with the contribution factor of 17.15% and the least contribution of impingement angle with 0.37% is shown in Figure 8.

Table 9. Analysis of Variance for SNRA2, using adjusted SS for tests vinylester based composites.

Source	DF	Seq SS	Adj SS	Adj MS	F-Value	*p*-Value
Impactvelocity (m/s)	3	60.893	60.893	20.298	5.97	0.031
Fiber content (wt%)	3	16.829	16.829	5.61	1.65	0.275
Impingement angle (Degree)	3	0.362	0.362	0.121	0.04	0.99
Error	6	20.405	20.405	3.401		
Total	15	98.489				

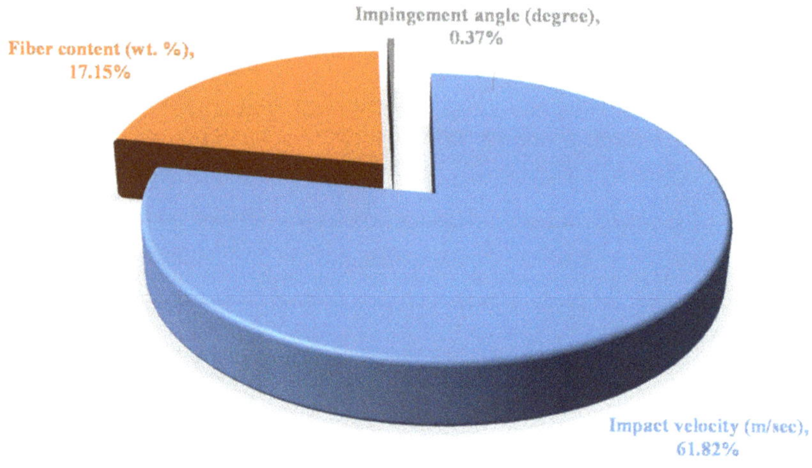

Figure 8. Contribution chart of factors for vinyl ester-based composites.

The p-value of control factors or polyester-based composites is shown in Table 10. It was observed that the value of p for impact velocity is 0.001, which is lower than the assumed p-value, i.e., 0.05, and is the most significant factor for the analysis of erosive wear. The next significant factor was fiber content with a p-value of 0.043. Impingement angle was observed to be least significant factor as per the ANOVA Table 10. The contribution of all the factors for epoxy-based composites with the highest contribution of impact velocity has a value of 77.75%, followed by fiber content with the contribution factor of 13.53% and least contribution of impingement angle with 3.44% is shown in Figure 9.

Table 10. Analysis of variance for SNRA2, using adjusted SS for tests polyester.

Source	DF	Seq SS	Adj SS	Adj MS	F-Value	p-Value
Impact Velocity (m/s)	3	86.549	86.549	28.85	29.58	0.001
Fiber content (wt%)	3	15.069	15.069	5.023	5.15	0.043
Impingement angle (Degree)	3	3.835	3.835	1.278	1.31	0.355
Error	6	5.853	5.853	0.975		
Total	15	111.305				

3.3. Morphological Analysis

The Figure 10 illustrates the mechanism took place during the wear examination of all three types of composites at highest fiber loading. Epoxy based composites as shown in Figure 10a exhibited fiber pull out leading to exposure of fibers with the wear surface [56–62]. The patches of ploughing at macro level have also been observed which may be considered as the major factor in the wear of the composite. However, the overall wear of epoxy-based composites reduced by the interaction of fiber with the mating surface took place by fiber pull out. It is to be noted that addition of jute and grewia fibers in the glass fiber-epoxy composites increases the erosive wear till 10 wt% of loading. However, on increasing the natural fiber weightage beyond 10% loading, the erosive wear behaviour reduces significantly. Large wear debris was spotted in vinyl ester-based composites (Figure 10b). These wear debris formed due to the detachment of sub polymeric material from base material due to low van der wall forces. Here, the exposure of fiber is negligible which somehow can be linked to the comparative higher wear than epoxy-based composites. In case of polyester composites (Figure 10c). The bonding between the natural fiber and ester group of vinyl ester and polyester matrix is comparatively low as compared to epoxy-natural fiber bonding which triggered more fiber detachment. Interestingly, at higher natural fiber loading (15 wt% jute and grewia), the improvement in the erosive wear was not significant as obtained in the study. Apparently, the erosive wear examination revealed that main wear mechanism responsible for material removal is groove formation and micro-ploughing [63–65]. A close examination also discloses brittle fractures at some parts, which shows that a small amount of impact forces was also acting on the samples.

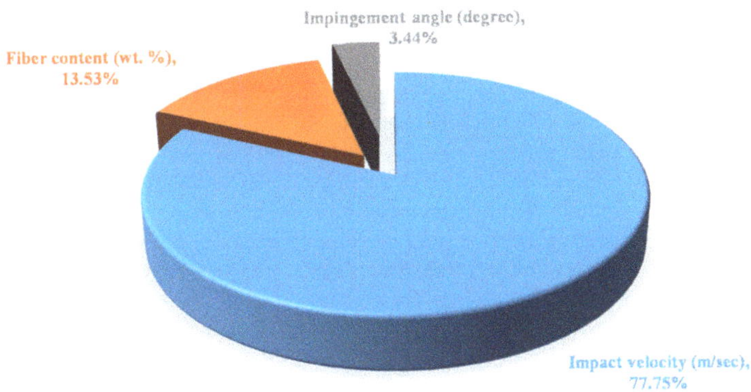

Figure 9. Contribution chart of factors for polyester based composites.

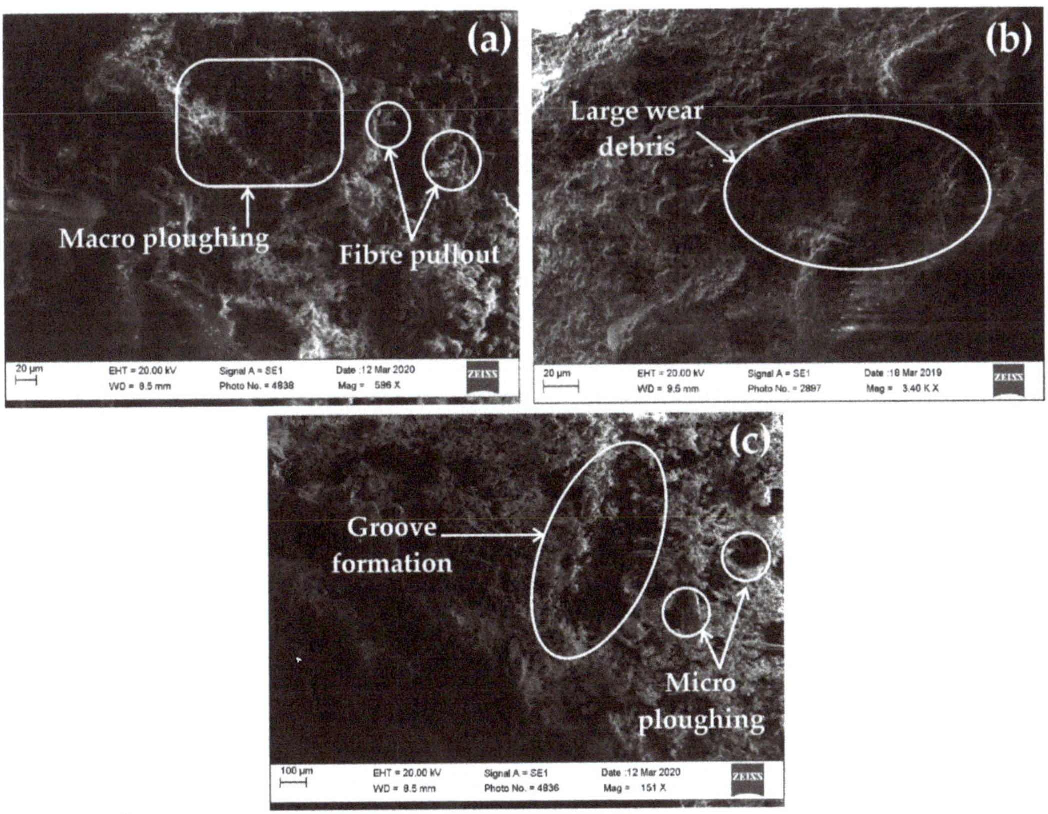

Figure 10. SEM images at highest fiber loading (**a**) epoxy based composite, (**b**) vinyl ester and (**c**) polyester based composite.

The scientometric analysis emphasizing the highlights especially on several of the most notable biocomposite material findings available in the literary works with a prominence on the biocompatibility, and material characteristics of biocomposites for artificial-limp/prosthetic applications as displayed in the Figure 11. Novel bio-based composites have mostly been evolved in response with a burgeoning consumption for eco-friendly sustainable materials and the willingness to minimise the expenditure with conventional fibres reinforced fossil-fuel derived composites. Investigators had already focused primarily on biocomposites, which are constituted of naturally or synthetic resins derived from the natural fibre-reinforcements. Natural fibres have significant upsides as they are a light-density material which tends to produce comparatively light-weight composites with slightly elevated unique characteristics [63–70]. Such filaments now provide substantial savings, efficient use of resources, and processability, and seems to be a profoundly renewable energy source, aiding to curtail dependence on international as well as household petroleum products. In the current context, self-sustaining environmentally sound approaches to traditional materials, explicitly glass-fibres, are now being deemed to be used in the lamination of artificial-limbs or prostheses connectors.

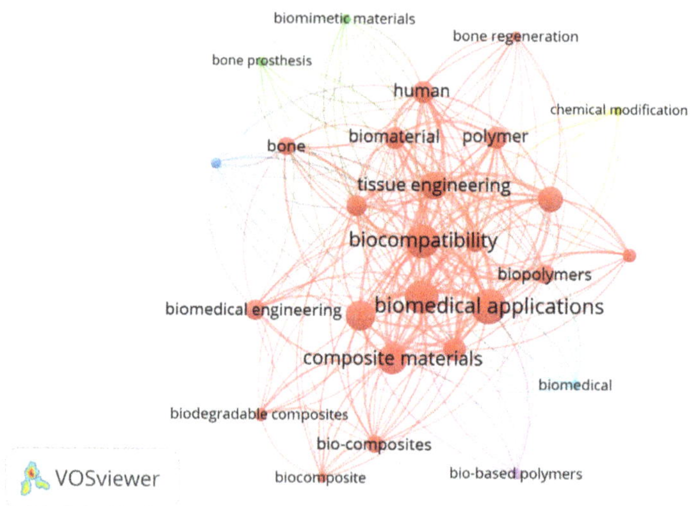

Figure 11. Bibliometric assessment on the applications of natural fiber/synthetic fiber reinforced thermosetting polymeric composites for biomedical and human prosthetic applications.

Utilizing polymer's higher-thermal conversion power-density, delivery methods that rely on these compounds were being used in photo-thermal therapeutic-treatment [71]. The investigators had summarized the findings, emphasized that, whilst area is still very much in beginning phases, conjugated-polymeric/poly-electrolyte interfacial-interactions have vast potential for healthcare applications [71].

Throughout this perspective, one of the most important aspects of sustainable-development is being utilised the organic-matter (biomass), and its compounds as a predecessor of carbon-materials [72]. A concise summarization of current developments in the synthesizing approach of self-sustaining carbon-compounds and their promising implications has been investigated. This report discusses fundamental observations and crucial recommendations for the eventual development of green carbon-materials and their burgeoning usage in catalytic and healthcare [72].

Metabolism fingerprinting of biological fluids record a wide range of disorders, and urinary-detecting, in particularly, provides ideal non-invasiveness towards upcoming diagnoses [73]. Owing with a restricted bio-markers and higher sampling intricacy, existing urinary identification provides major shortcomings and necessitates use of sophisticated materials to extract biomolecular data. Polymer@Ag generated urinary meta-bolic finger-printing (UMFs) by LDI-MS within seconds employing approximately One Litre of urinary without enriching or purifying [73].

Selective target, trans-membrane distribution, transport, and stimulation responsiveness could all be integrated within peptide-based theranostic nanostructures [74]. Throughout the article, the researchers had discussed generalized principles for synthesising peptide-based therapeutic, and diagnostic nano-materials, with a focus on performance, design, and numerous bio-medical possibilities, and they have illustrated their significant development over the last five years [74].

The developed microchip identifies tiny metabolite-molecules in human plasma rapidly, sensitively, and preferentially without enriching or purifying [75]. On-chip plasma fingerprinting enable further distinction among women having ovarian/colorectal cancer & control-subjects, and also therapeutic assessment for significant medical surveillance. The research explored the use of laser-de-sorption or ionisation mass-spectroscopy in huge therapeutic medicinal towards vitro-testing [75].

The co-workers have proposed synthesized Palladium–Gold alloys using mass-spectroscopy-based metabolite fingerprint, and assessment throughout the diagnostic and radiation treatment of medullo-blastoma [76]. Deep learning has been employed to identify medullo-blastoma individuals, whereas radiation therapy has been observed and an initial array of plasma metabolites-biomarkers was discovered exhibiting progressive alterations [76].

4. Conclusions

Natural fiber/synthetic fiber reinforced polymeric composites have been successfully fabricated with three different polymer resins by hand lay-up process for the analysis of erosive wear behavior. Taguchi technique L_{16} orthogonal array was used to optimize the number of experiments. The obtained results from the investigation revealed that impact velocity is the most significant control factor in the analysis of erosive wear. The second most considerable control was fiber content, followed by impingement angle for all the composites. Polyester-based composites exhibited the highest erosive wear among all the composites, followed by vinyl ester. Epoxy-based composites showed the least erosive wear among all. The contribution charts found that impact velocity has the highest contribution in polyester-based composites, fiber content in vinyl ester composites, and impingement angle in epoxy composites. The morphological analysis exemplified the overarching wear-performance of epoxy-based composites reduced by the interaction of fibers with the contact mating-surface took place by fiber pull-out. Large wear debris was spotted in vinyl ester-based composites as revealed by the SEM analysis. These wear debris formed due to the detachment of sub polymeric material from base material due to low van der wall forces. In case of polyester composites, the examination revealed that main wear mechanism responsible for material removal is groove formation and micro-ploughing. The developed fiber-reinforced polymer sandwich composites are showing significant properties for orthopaedic, bone-fracture fixation applications. The present work can also be investigated for various other polymers such as PE, PC, PLA and PHB for further new possibilities of the composite for better performance. Fillers such as dolomite and marble dust powder can also be added in the composite to further enhance the erosive behaviour of the present composite.

Author Contributions: Conceptualization, J.M.K., S.D., B.G., L.R. and S.S.; Methodology, J.M.K., S.D., B.G., L.R. and S.S.; Formal Analysis, J.M.K., S.D., B.G., L.R. and S.S.; Investigation, J.M.K., S.D., B.G., L.R. and S.S.; Writing—Original Draft Preparation, J.M.K., S.D., B.G., L.R. and S.S.; Writing—Review and editing, S.S., M.R.M.H., R.A.I., S.P.D., S.C., H.C.K. and C.L.; Supervision, J.M.K., S.D., B.G., L.R. and S.S.; Software, J.M.K. and S.S.; Funding acquisition, S.S., M.R.M.H. and R.A.I. All authors have read and agreed to the published version of the manuscript.

Funding: The article processing fee of this manuscript is funded by Research Management Centre, Univesiti Putra Malaysia.

Institutional Review Board Statement: Not applicable.

Informed Consent Statement: Not applicable.

Data Availability Statement: The data presented in this study are available on request from the corresponding author.

Conflicts of Interest: The authors declare no conflict of interest.

References

1. Asim, M.; Jawaid, M.; Saba, N.; Nasir, M.; Sultan, M.T.H. Processing of hybrid polymer composites—A review. In *Hybrid Polymer Composite Materials*; Elsevier: Amsterdam, The Netherlands, 2017; pp. 1–22.
2. Kumar, S.; Mer, K.K.S.; Gangil, B.; Patel, V.K. Synergistic effect of hybrid Himalayan Nettle/Bauhinia-vahlii fibers on physico-mechanical and sliding wear properties of epoxy composites. *Def. Technol.* **2020**, *16*, 762–776. [CrossRef]
3. Andrady, A.L.; Pandey, K.K.; Heikkilä, A.M. Interactive effects of solar UV radiation and climate change on material damage. *Photochem. Photobiol. Sci.* **2019**, *18*, 804–825. [CrossRef]
4. Baig, Z.; Mamat, O.; Mustapha, M. Recent progress on the dispersion and the strengthening effect of carbon nanotubes and graphene-reinforced metal nanocomposites: A review. *Crit. Rev. Solid State Mater. Sci.* **2018**, *43*, 1–46. [CrossRef]
5. Lalit, R.; Mayank, P.; Ankur, K. Natural Fibers and Biopolymers Characterization: A Future Potential Composite Material. *Stroj. Cas. J. Mech. Eng.* **2018**, *68*, 33–50. [CrossRef]
6. Karami, P.; Khasraghi, S.S.; Hashemi, M.; Rabiei, S.; Shojaei, A. Polymer/nanodiamond composites—A comprehensive review from synthesis and fabrication to properties and applications. *Adv. Colloid Interface Sci.* **2019**, *269*, 122–151. [CrossRef]
7. Mudenur, C.; Mondal, K.; Singh, U.; Katiyar, V. Production of Polyhydroxyalkanoates and Its Potential Applications. In *Advances in Sustainable Polymers*; Springer: Singapore, 2019; pp. 131–164.
8. Rajak, D.; Pagar, D.; Menezes, P.; Linul, E. Fiber-Reinforced Polymer Composites: Manufacturing, Properties, and Applications. *Polymers* **2019**, *11*, 1667. [CrossRef] [PubMed]
9. Gangil, B.; Ranakoti, L.; Verma, S.; Singh, T.; Kumar, S. Natural and Synthetic Fibers for Hybrid Composites. In *Hybrid Fiber Composites: Materials, Manufacturing, Process Engineering*; John Wiley & Sons: New York, NY, USA, 2020; pp. 1–15.
10. Kureemun, U.; Ravandi, M.; Tran, L.Q.N.; Teo, W.S.; Tay, T.E.; Lee, H.P. Effects of hybridization and hybrid fibre dispersion on the mechanical properties of woven flax-carbon epoxy at low carbon fibre volume fractions. *Compos. Eng.* **2018**, *134*, 28–38. [CrossRef]
11. Nagarajaganesh, B.; Rekha, B. Intrinsic cellulosic fiber architecture and their effect on the mechanical properties of hybrid composites. *Arch. Civ. Mech. Eng.* **2020**, *20*, 125. [CrossRef]
12. Ramnath, B.V.; Kokan, S.J.; Raja, R.N.; Sathyanarayanan, R.; Elanchezhian, C.; Prasad, A.R.; Manickavasagam, V.M. Evaluation of mechanical properties of abaca-jute-glass fibre reinforced epoxy composite. *Mater. Des.* **2013**, *51*, 357–366. [CrossRef]
13. Bhoopathi, R.; Ramesh, M.; Deepa, C. Fabrication and Property Evaluation of Banana-Hemp-Glass Fiber Reinforced Composites. *Procedia Eng.* **2014**, *97*, 2032–2041. [CrossRef]
14. Gairola, S.P.; Tyagi, Y.K.; Gangil, B.; Sharma, A. Fabrication and mechanical property evaluation of non-woven banana fibre epoxy-based polymer composite. *Mater. Today Proc.* **2021**, *44*, 3990–3996. [CrossRef]
15. Burks, B.; Armentrout, D.; Kumosa, M. Characterization of the fatigue properties of a hybrid composite utilized in high voltage electric transmission. *Compos. Appl. Sci. Manuf.* **2011**, *42*, 1138–1147. [CrossRef]
16. Singh, N.P.; Aggarwal, L.; Gupta, V.K. Tensile Behavior of Sisal/Hemp Reinforced High Density Polyethylene Hybrid Composite. *Mater. Today Proc.* **2015**, *2*, 3140–3148. [CrossRef]
17. Ranakoti, L.; Rakesh, P.K. Physio-mechanical characterization of tasar silk waste/jute fiber hybrid composite. *Compos. Commun.* **2020**, *22*, 100526. [CrossRef]
18. Yahaya, R.; Sapuan, S.M.; Jawaid, M.; Leman, Z.; Zainudin, E.S. Mechanical performance of woven kenaf-Kevlar hybrid composites. *J. Reinf. Plast. Compos.* **2014**, *33*, 2242–2254. [CrossRef]
19. Yahaya, R.; Sapuan, S.M.; Jawaid, M.; Leman, Z.; Zainudin, E.S. Effect of layering sequence and chemical treatment on the mechanical properties of woven kenaf-aramid hybrid laminated composites. *Mater. Des.* **2015**, *67*, 173–179. [CrossRef]
20. Jawaid, M.; Khalil, H.P.S.A. Cellulosic/synthetic fibre reinforced polymer hybrid composites: A review. *Carbohydr. Polym.* **2011**, *86*, 1–18. [CrossRef]
21. Sathishkumar, T.P.; Satheeshkumar, S.; Naveen, J. Glass fiber-reinforced polymer composites—A review. *J. Reinf. Plast. Compos.* **2014**, *33*, 1258–1275. [CrossRef]
22. Raghavendra Rao, H.; Varada Rajulu, A.; Ramachandra Reddy, G.; Hemachandra Reddy, K. Flexural and compressive properties of bamboo and glass fiber-reinforced epoxy hybrid composites. *J. Reinf. Plast. Compos.* **2010**, *29*, 1446–1450. [CrossRef]
23. Naidu, V.N.P.; Reddy, G.R.; Kumar, M.A.; Reddy, M.M.; Khanam, P.N.; Naidu, S.V. Compressive & impact properties of sisal/glass fiber reinforced hybrid composites. *Int. J. Fibre Text. Res.* **2011**, *1*, 11–14.
24. Almaadeed, M.A.; Kahraman, R.; Khanam, P.N.; Madi, N. Date palm wood flour/glass fibre reinforced hybrid composites of recycled polypropylene: Mechanical and thermal properties. *Mater. Des.* **2012**, *42*, 289–294. [CrossRef]

25. Manjunatha, C.M.; Bojja, R.; Jagannathan, N.; Kinloch, A.J.; Taylor, A.C. Enhanced fatigue behavior of a glass fiber reinforced hybrid particles modified epoxy nanocomposite under WISPERX spectrum load sequence. *Int. J. Fatigue* **2013**, *54*, 25–31. [CrossRef]
26. Zhang, Y.; Li, Y.; Ma, H.; Yu, T. Tensile and interfacial properties of unidirectional flax/glass fiber reinforced hybrid composites. *Compos. Sci. Technol.* **2013**, *88*, 172–177. [CrossRef]
27. Amuthakkannan, P.; Manikandan, V.; Uthayakumar, M. Mechanical properties of basalt and glass fiber reinforced polymer hybrid composites. *J. Adv. Microsc. Res.* **2014**, *9*, 44–49. [CrossRef]
28. Vinayagamoorthy, R.; Rajeswari, N. Mechanical performance studies on Vetiveria zizanioides/jute/glass fiber-reinforced hybrid polymeric composites. *J. Reinf. Plast. Compos.* **2014**, *33*, 81–92. [CrossRef]
29. Sanjay, M.R.; Yogesha, B. Studies on Mechanical Properties of Jute/E-Glass Fiber Reinforced Epoxy Hybrid Composites. *J. Miner. Mater. Charact. Eng.* **2016**, *4*, 15–25. [CrossRef]
30. Haque, M.M.; Hasan, M. Influence of fiber surface treatment on physico-mechanical properties of betel nut and glass fiber reinforced hybrid polyethylene composites. *Adv. Mater. Process. Technol.* **2018**, *4*, 511–525. [CrossRef]
31. Sosiati, H.; Shofie, Y.A.; Nugroho, A.W. Tensile properties of Kenaf/E-glass reinforced hybrid polypropylene (PP) composites with different fiber loading. *Evergr. Jt. J. Nov. Carbon Resour. Sci. Green Asia Strategy* **2018**, *5*, 1–5. [CrossRef]
32. Mittal, M.; Chaudhary, R. Development of PALF/Glass and COIR/Glass fiber reinforced hybrid epoxy composites. *J. Mater. Sci. Surf. Eng.* **2018**, *6*, 851–861.
33. Arumugam, S.; Kandasamy, J.; Shah, A.U.M.; Sultan, M.T.H.; Safri, S.N.A.; Majid, M.S.A.; Basri, A.A.; Mustapha, F. Investigations on the Mechanical Properties of Glass Fiber/Sisal Fiber/Chitosan Reinforced Hybrid Polymer Sandwich Composite Scaffolds for Bone Fracture Fixation Applications. *Polymers* **2020**, *12*, 1501. [CrossRef]
34. Prabhu, L.; Krishnaraj, V.; Sathish, S.; Gokulkumar, S.; Karthi, N. Study of mechanical and morphological properties of jute-tea leaf fiber reinforced hybrid composites: Effect of glass fiber hybridization. *Mater. Today Proc.* **2020**, *27*, 2372–2375. [CrossRef]
35. Ramesh, M.; Palanikumar, K.; Reddy, K.H. Evaluation of Mechanical and Interfacial Properties of Sisal/Jute/Glass Hybrid Fiber Reinforced Polymer Composites. *Trans. Indian Inst. Met.* **2016**, *69*, 1851–1859. [CrossRef]
36. Prabhu, L.; Krishnaraj, V.; Gokulkumar, S.; Sathish, S.; Sanjay, M.R.; Siengchin, S. Mechanical, chemical and sound absorption properties of glass/kenaf/waste tea leaf fiber-reinforced hybrid epoxy composites. *J. Ind. Text.* **2020**, 1528083720957392. [CrossRef]
37. Vigneshwaran, S.; Uthayakumar, M.; Arumugaprabu, V. A review on erosion studies of fiber-reinforced polymer composites. *J. Reinf. Plast. Compos.* **2017**, *36*, 1019–1027. [CrossRef]
38. Öztürk, B.; Gedikli, H.; Kılıçarslan, Y.S. Erosive wear characteristics of E-glass fiber reinforced silica fume and zinc oxide-filled epoxy resin composites. *Polym. Compos.* **2020**, *41*, 326–337. [CrossRef]
39. Choudhary, M.; Singh, T.; Dwivedi, M.; Patnaik, A. Waste marble dust-filled glass fiber-reinforced polymer composite Part I: Physical, thermomechanical, and erosive wear properties. *Polym. Compos.* **2019**, *40*, 4113–4124. [CrossRef]
40. Karsli, N.G.; Yilmaz, T.; Aytac, A.; Ozkoc, G. Investigation of erosive wear behavior and physical properties of SGF and/or calcite reinforced ABS/PA6 composites. *Compos. Eng.* **2013**, *44*, 385–393. [CrossRef]
41. Choudhary, M.; Sharma, A.; Agarwal, P.; Singh, T.; Patnaik, T.; Patnaik, A. Experimental and numerical investigation of mechanical and erosion behavior of barium sulphate filled glass fiber reinforced polymer composites. *Polym. Compos.* **2021**, *42*, 753–773. [CrossRef]
42. Mohan, N.; Mahesha, C.R.; Rajaprakash, B.M. Erosive Wear Behaviour of WC Filled Glass Epoxy Composites. *Procedia Eng.* **2013**, *68*, 694–702. [CrossRef]
43. Nayak, S.; Mohanty, J. Erosion wear behavior of benzoyl chloride modified areca sheath fiber reinforced polymer composites. *Compos. Commun.* **2020**, *18*, 19–25. [CrossRef]
44. Qian, D.; Bao, L.; Takatera, M.; Kemmochi, K.; Yamanaka, A. Fiber-reinforced polymer composite materials with high specific strength and excellent solid particle erosion resistance. *Wear* **2010**, *268*, 637–642. [CrossRef]
45. Verma, S.K.; Gupta, A.; Singh, T.; Gangil, B.; Jánosi, E.; Fekete, G. Influence of dolomite on mechanical, physical and erosive wear properties of natural-synthetic fiber reinforced epoxy composites. *Mater. Res. Express* **2019**, *6*, 125704. [CrossRef]
46. Jha, A.K.; Mantry, S.; Satapathy, A.; Patnaik, A. Erosive Wear Performance Analysis of Jute-Epoxy-SiC Hybrid Composites. *J. Compos. Mater.* **2010**, *44*, 1623–1641.
47. Kaundal, R.; Patnaik, A.; Satapathy, A. Mechanical characterizations and development of erosive wear model for Al_2O_3-filled short glass fiber-reinforced polymer composites. *Proc. Inst. Mech. Eng. J. Mater. Des. Appl.* **2018**, *232*, 893–908. [CrossRef]
48. Panda, P.; Mantry, S.; Mohapatra, S.; Singh, S.K.; Satapathy, A. A study on erosive wear analysis of glass fiber-epoxy-AlN hybrid composites. *J. Compos. Mater.* **2014**, *48*, 107–118. [CrossRef]

49. Singh, T.; Tejyan, S.; Patnaik, A.; Singh, V.; Zsoldos, I.; Fekete, G. Fabrication of waste bagasse fiber-reinforced epoxy composites: Study of physical, mechanical, and erosion properties. *Polym. Compos.* **2019**, *40*, 3777–3786. [CrossRef]
50. Tejyan, S.; Singh, T.; Patnaik, A.; Fekete, G.; Gangil, B. Physico-mechanical and erosive wear analysis of polyester fibre-based nonwoven fabric-reinforced polymer composites. *J. Ind. Text.* **2019**, *49*, 447–464. [CrossRef]
51. Kar, J.; Rout, A.K.; Sutar, A.K. Physical, Mechanical, and Erosion Characterization of Palm Leaf Stalk Fiber Reinforced Epoxy Composites Filled with Palm Leaf Stem Stalk (PLSS) Powder. *BioResources* **2018**, *13*, 7212–7231. [CrossRef]
52. Patnaik, A.; Satapathy, A.; Mahapatra, S.S.; Dash, R.R. Implementation of Taguchi Design for Erosion of Fiber-Reinforced Polyester Composite Systems with SiC Filler. *J. Reinf. Plast. Compos.* **2008**, *27*, 1093–1111. [CrossRef]
53. Kumar, S.; Prasad, L.; Kumar, S.; Patel, V.K. Physico-mechanical and Taguchi-designed sliding wear properties of Himalayan agave fiber reinforced polyester composite. *J. Mater. Res. Technol.* **2019**, *8*, 3662–3671. [CrossRef]
54. Patnaik, A.; Satapathy, A.; Mahapatra, S.S.; Dash, R.R. Erosive Wear Assessment of Glass Reinforced Polyester-Flyash Composites Using Taguchi Method. *Int. Polym. Process.* **2008**, *23*, 192–199. [CrossRef]
55. Khare, J.M.; Dahiya, S.; Gangil, B.; Ranakoti, L. Influence of different resins on Physico-Mechanical properties of hybrid fiber reinforced polymer composites used in human prosthetics. *Mater. Today Proc.* **2021**, *38*, 345–349. [CrossRef]
56. Islam, M.S.; Pickering, K.L.; Foreman, N.J. The Effect of Accelerated Weathering on the Mechanical Properties of Alkali Treated Hemp Fibre/Epoxy Composites. *J. Adhes. Sci. Technol.* **2011**, *25*, 1947–1959. [CrossRef]
57. Sethi, S.; Ray, B.C. Environmental effects on fibre reinforced polymeric composites: Evolving reasons and remarks on interfacial strength and stability. *Adv. Colloid Interface Sci.* **2015**, *217*, 43–67. [CrossRef]
58. Rout, A.; Satapathy, A.; Mantry, S.; Sahoo, A.; Mohanty, T. Erosion Wear Performance Analysis of Polyester-GF-Granite Hybrid Composites using the Taguchi Method. *Procedia Eng.* **2012**, *38*, 1863–1882. [CrossRef]
59. Chohan, J.S.; Mittal, N.; Kumar, R.; Singh, S.; Sharma, S.; Dwivedi, S.P.; Saxena, A.; Chattopadhyaya, S.; Ilyas, R.A.; Le, C.H.; et al. Optimization of FFF Process Parameters by Naked Mole-Rat Algorithms with Enhanced Exploration and Exploitation Capabilities. *Polymers* **2021**, *13*, 1702. [CrossRef]
60. Ilyas, R.A.; Sapuan, S.M.; Asyraf, M.R.M.; Dayana, D.A.Z.N.; Amelia, J.J.N.; Rani, M.S.A.; Norrrahim, M.N.F.; Nurazzi, N.M.; Aisyah, H.A.; Sharma, S.; et al. Polymer Composites Filled with Metal Derivatives: A Review of Flame Retardants. *Polymers* **2021**, *13*, 1701. [CrossRef]
61. Chohan, J.S.; Mittal, N.; Kumar, R.; Singh, S.; Sharma, S.; Singh, J.; Rao, K.V.; Mia, M.; Pimenov, D.Y.; Dwivedi, S.P. Mechanical Strength Enhancement of 3D Printed Acrylonitrile Butadiene Styrene Polymer Components Using Neural Network Optimization Algorithm. *Polymers* **2020**, *12*, 2250. [CrossRef]
62. Singh, Y.; Singh, J.; Sharma, S.; Aggarwal, V.; Pruncu, C.I. Multi-objective Optimization of Kerf-taper and Surface-roughness Quality Characteristics for Cutting-operation On Coir and Carbon Fibre Reinforced Epoxy Hybrid Polymeric Composites During CO_2-Pulsed Laser-cutting Using RSM. *Lasers Manuf. Mater. Process* **2021**, *8*, 157–182. [CrossRef]
63. Sharma, S.; Singh, J.; Kumar, H.; Sharma, A.; Aggarwal, V.; Gill, A.; Jayarambabu, N.; Kailasa, S.; Rao, K. Utilization of rapid prototyping technology for the fabrication of an orthopedic shoe inserts for foot pain reprieve using thermo-softening viscoelastic polymers: A novel experimental approach. *Meas. Control* **2020**, *53*, 519–530. [CrossRef]
64. Singh, Y.; Singh, J.; Sharma, S.; Sharma, A.; Chohan, J.S. Process Parameter Optimization in Laser Cutting of Coir Fiber Reinforced Epoxy Composite—A Review. *Mater. Today Proc.* **2021**, in press. [CrossRef]
65. Chohan, J.S.; Kumar, R.; Singh, T.B.; Singh, S.; Sharma, S.; Singh, J.; Mia, M.; Pimenov, D.Y.; Chattopadhyaya, S.; Dwivedi, S.P.; et al. Taguchi S/N and TOPSIS Based Optimization of Fused Deposition Modelling and Vapor Finishing Process for Manufacturing of ABS Plastic Parts. *Materials* **2020**, *13*, 5176. [CrossRef] [PubMed]
66. Prabhakaran, S.; Vijayan, K.; Sharma, S.; Mouleeswaran, S.K.; Ramasamy, J.K.; Redoune, Z. Experimental study on thermal and morphological analysis of Green composite sandwich made of Flax and agglomerated cork. *J. Therm. Anal. Calorim.* **2020**, *139*, 3003–3012. [CrossRef]
67. Sharma, S.; Sudhakara, P.; Singh, J.; Ilyas, R.A.; Asyraf, M.R.M.; Razman, M.R. Critical Review of Biodegradable and Bioactive Polymer Composites for Bone Tissue Engineering and Drug Delivery Applications. *Polymers* **2021**, *13*, 2623. [CrossRef] [PubMed]
68. Sharma, S.; Sudhakara, P.; Omran, A.A.B.; Singh, J.; Ilyas, R.A. Recent Trends and Developments in Conducting Polymer Nanocomposites for Multifunctional Applications. *Polymers* **2021**, *13*, 2898. [CrossRef] [PubMed]
69. Jha, K.; Tyagi, Y.K.; Kumar, R.; Sharma, S.; Huzaifah, M.R.M.; Li, C.; Ilyas, R.A.; Dwivedi, S.P.; Saxena, A.; Pramanik, A. Assessment of Dimensional Stability, Biodegradability, and Fracture Energy of Bio-Composites Reinforced with Novel Pine Cone. *Polymers* **2021**, *13*, 3260. [CrossRef]
70. Kadier, B.; Ilyas, R.A.; Huzaifah, M.R.M.; Harihastuti, N.; Sapuan, S.M.; Harussani, N.M.; Azlin, M.N.M.; Yuliasni, R.; Ibrahim, R.; Atikah, M.S.N.; et al. Use of Industrial Wastes as Sustainable Nutrient Sources for Bacterial Cellulose (BC) Production: Mechanism, Advances, and Future Perspectives. *Polymers* **2021**, *19*, 3365. [CrossRef] [PubMed]
71. Jephcott, L.; Eslami, M.; Travaglini, L.; Lauto, A.; Mawad, D. A conjugated polymer-liposome complex: A contiguous water-stable, electronic, and optical interface. *View* **2020**, *2*, 20200081. [CrossRef]
72. Lan, G.; Yang, J.; Ye, R.-P.; Boyjoo, Y.; Liang, J.; Liu, X.; Li, Y.; Liu, J.; Qian, K. Sustainable Carbon Materials toward Emerging Applications. *Small Methods* **2021**, *5*, 2001250. [CrossRef]

73. Yang, J.; Wang, R.; Huang, L.; Zhang, M.; Niu, J.; Bao, C.; Shen, N.; Dai, M.; Guo, Q.; Wang, Q.; et al. Urine metabolic fingerprints encode subtypes of kidney diseases. *Angew. Chem. Int. Ed.* **2020**, *59*, 1703. [CrossRef]
74. Rong, L.; Lei, Q.; Zhang, X.-Z. Recent advances on peptide-based theranostic nanomaterials. *View* **2020**, *1*, 20200050. [CrossRef]
75. Shu, W.; Wang, Y.; Liu, C.; Li, R.; Pei, C.; Lou, W.; Lin, S.; Di, W.; Wan, J. Construction of a Plasmonic Chip for Metabolic Analysis in Cervical Cancer Screening and Evaluation. *Small Methods* **2020**, *4*, 1900469. [CrossRef]
76. Cao, J.; Shi, X.; Gurav, D.D.; Huang, L.; Su, H.; Li, K.; Niu, J.; Zhang, M.; Wang, Q.; Jiang, M.; et al. Medulloblastoma: Metabolic Fingerprinting on Synthetic Alloys for Medulloblastoma Diagnosis and Radiotherapy Evaluation. *Adv. Mater.* **2020**, *32*, 2070178. [CrossRef]

polymers MDPI

Article

Dynamic Mechanical Properties and Thermal Properties of Longitudinal Basalt/Woven Glass Fiber Reinforced Unsaturated Polyester Hybrid Composites

Nur Izzah Nabilah Haris [1], R. A. Ilyas [2,3], Mohamad Zaki Hassan [4,*], S. M. Sapuan [5,6], Atiqah Afdzaluddin [7], Khairur Rijal Jamaludin [4], Sheikh Ahmad Zaki [8] and Faizir Ramlie [4]

1 Institute of Advanced Technology, Universiti Putra Malaysia, Serdang 43400, Malaysia; nurizzahnabilah.haris@gmail.com
2 School of Chemical and Energy Engineering, Faculty of Engineering, Universiti Teknologi Malaysia, Johor Bahru 81310, Malaysia; ahmadilyas@utm.my
3 Centre for Advanced Composite Materials (CACM), Universiti Teknologi Malaysia, Johor Bahru 81310, Malaysia
4 Razak Faculty of Technology and Informatics, Universiti Teknologi Malaysia, Jalan Sultan Yahya Petra, Kuala Lumpur 54100, Malaysia; khairur.kl@utm.my (K.R.J.); faizir.kl@utm.my (F.R.)
5 Laboratory of Biocomposite Technology, Institute of Tropical Forestry and Forest Products (INTROP), Universiti Putra Malaysia, Serdang 43400, Malaysia; sapuan@upm.edu.my
6 Advanced Engineering Materials and Composites Research Centre (AEMC), Department of Mechanical and Manufacturing Engineering, Faculty of Engineering, Universiti Putra Malaysia, Serdang 43400, Malaysia
7 Institute of Microengineering and Nanoelectronics, Universiti Kebangsaan Malaysia, Bangi 43600, Malaysia; a.atiqah@ukm.edu.my
8 Malaysia-Japan International Institute of Technology, Universiti Teknologi Malaysia, Jalan Sultan Yahya Petra, Kuala Lumpur 54100, Malaysia; sheikh.kl@utm.my
* Correspondence: mzaki.kl@utm.my

Citation: Haris, N.I.N.; Ilyas, R.A.; Hassan, M.Z.; Sapuan, S.M.; Afdzaluddin, A.; Jamaludin, K.R.; Zaki, S.A.; Ramlie, F. Dynamic Mechanical Properties and Thermal Properties of Longitudinal Basalt/Woven Glass Fiber Reinforced Unsaturated Polyester Hybrid Composites. *Polymers* **2021**, *13*, 3343. https://doi.org/10.3390/polym13193343

Academic Editor: Vincenzo Fiore

Received: 8 September 2021
Accepted: 22 September 2021
Published: 29 September 2021

Abstract: This study investigates the mechanical, thermal, and chemical properties of basalt/woven glass fiber reinforced polymer (BGRP) hybrid polyester composites. The Fourier transform infrared spectroscopy (FTIR) was used to explore the chemical aspect, whereas the dynamic mechanical analysis (DMA) and thermomechanical analysis (TMA) were performed to determine the mechanical and thermal properties. The dynamic mechanical properties were evaluated in terms of the storage modulus, loss modulus, and damping factor. The FTIR results showed that incorporating single and hybrid fibers in the matrix did not change the chemical properties. The DMA findings revealed that the B7.5/G22.5 composite with 7.5 wt% of basalt fiber (B) and 22.5 wt% of glass fiber (G) exhibited the highest elastic and viscous properties, as it exhibited the higher storage modulus (8.04×10^9 MPa) and loss modulus (1.32×10^9 MPa) compared to the other samples. All the reinforced composites had better damping behavior than the neat matrix, but no further enhancement was obtained upon hybridization. The analysis also revealed that the B22.5/G7.5 composite with 22.5 wt% of basalt fiber and 7.5 wt% of glass fiber had the highest T_g at 70.80 °C, and increased by 15 °C compared to the neat matrix. TMA data suggested that the reinforced composites had relatively low dimensional stabilities than the neat matrix, particularly between 50 to 80 °C. Overall, the hybridization of basalt and glass fibers in unsaturated polyester formed composites with higher mechanical and thermal properties than single reinforced composites.

Keywords: hybrid composite; glass fiber; basalt fiber; DMA; TMA

1. Introduction

Fiber-reinforced polymer (FRP) composites are fabricated by blending reinforcing fibers with a polymer matrix [1]. In nearly all cases, combining a matrix and reinforcing fiber results in composites with higher mechanical properties compared to the single components [2,3]. Optimization of the fabrication parameters are often used to achieve an optimum level of overall composite quality [4–6]. However, under certain circumstances, the properties of single FRP composites might fail to meet the desired level of characteristics. Accordingly, scientists began to incorporate more than one type of reinforcing fibers in a single matrix, forming hybrid FRP composites [7–9]. The synergistic effects of hybridization between more than one fiber could overcome the drawbacks of single FRP composites. This technique is quickly gaining in popularity due to its ability to give the flexibility to alter the composite behavior, which is not possible with single FRB composites. It is essential to note that if the hybrid arrangement is not carefully planned, the resulting hybrid composite may have a lesser strength than its parts.

Glass fiber is a fine synthetic fiber made up of glass that typically comprises more than 50% of silica along with various mineral oxides such as calcium, iron, and aluminum oxides. It is extensively used as thermal, acoustical, and electrical insulation on its own [10]. Moreover, it is utilized as a reinforcing material for polymer composites owing to its lightweight characteristic, strength, versatility, and robustness [11]. Glass fiber reinforced composites are good in terms of performance and cost. Due to its reasonable price and unique properties, it is one of the most prominent reinforcing agents in the current and future segments [12]. Basalt fiber is a mineral-sourced fiber that originates from volcanic magma. The basalt fiber is manufactured by crushing, melting, and extruding molten basalt rocks through small nozzles, producing a fibrous form of basalt. It offers high thermal and chemical resistance, good adhesion to polymer, and high elastic modulus [13].

Recently, several experiments have been conducted to investigate the mechanical characteristics of the basalt/glass fiber reinforced polymer (BGRP), as shown in Table 1.

Basalt fiber is one of the materials considered as reinforcement with polymer composite. The incorporation of this fiber within the polymer composite, has resulted in high operating temperature range, excellent heat, good strength, good chemical resistance, low water absorption, and good sound insulation properties [14–19]. Barczewski et al. [20] evaluated the effect of hybridizing mineral basalt (B) powder with glass fiber in the polypropylene (PP) matrix using injection molding. It was discovered that increasing the amount of B filler resulted in decreasing the mechanical properties of the composites. A similar finding was also reported by Mahesh Babu et al. [21], whereby an increase in B filler caused a lower impact strength. Furthermore, the influence of the stacking sequence of hybrid BGRP laminates in terms of mechanical properties was also observed by El-Wazery [22]. It was revealed that there was an enhancement in the mechanical properties of basalt/carbon-reinforced polymer (BCRP) composites with stacking sequences (Carbon/Basalt/Carbon/Basalt/Carbon) as compared to the other stacking sequences (Glass/Basalt/Glass/Basalt/Glass). Moreover, the layering arrangement such as (Glass/Glass/Basalt/Glass/Basalt/Basalt) offered an outstanding tensile performance than the other laminate configurations. However, the hybrid composite still portrayed a lower performance as compared to the plain BFRP and GFRP composites [23]. Most of the studies reported that lamination between the basalt and glass fibers could offer a lower mechanical property as compared with the plain basalt or glass composite laminates [24–26]. Bozkurt et al. [24] evaluated the influence of the woven basalt fiber with an E-glass fabric reinforced with epoxy resin using vacuum-assisted resin transfer molding. By increasing the basalt fiber content in all the hybrid composite samples, considerable deteriorations in tensile strength and modulus were observed. In addition, Abd El-Baky et al. [25] mentioned that the tensile characteristics of BGRP composites were inferior compared to GFRP laminates. The authors advised that increasing the G-fiber content enhances the mechanical behavior. Likewise, Patel et al. [24] had concluded that the hybrid BGRP laminate showed the highest tensile strength. However, it demonstrated a lower impact strength as compared to the neat glass laminate. Interestingly, Jain et al. [27] developed hybrid BGRP composites with better flexural strength and lower tensile strength than the BRFP laminate. It runs contrary to the observations reported by Abd El-Baky et al. [25]. Moreover, Ozbek et al. [26] reported a lower value of impact strength of the BGRP sample. However, the result was comparably low in comparison to the levels described in the literature. In contrast, Sapuan et al. [28] stated that the findings obtained in terms of the mechanical characteristics surpass the previous works in this field. The BGRP samples significantly offered high tensile properties as well as flexural and impact strength values at a composition of 22.5 wt% of G and 7.55 wt% of B fabrics.

Table 1. Maximum mechanical properties of hybrid BGRP composites from previous studies.

No.	Resin	Hybrid	Fabrication Method	Tensile Strength (MPa)	Flexural Strength (MPa)	Impact Strength (kJ/m^2)	Remark	Ref.
1	PP	Filler	Extrusion	31	-	-	Increasing basalt content led to a decrease in tensile strength.	[20]
2	Epoxy	Filler	Hand layup	273	497	426	The addition of a large amount of basalt reduced the impact strength.	[21]
3	Polyester	Laminate	Hand layup	78	175	-	Mechanical properties of BGRP composite are lower than BCRP laminates.	[22]
4	Epoxy	Laminate	Vacuum begging	448	-	-	The tensile strength of BGRP composite was lower than GFRP laminates.	[24]
5	Polyester	Laminate	Hand layup	293	302	192	The stacking sequence of BGRP improved mechanical properties.	[23]
6	Epoxy	Laminate	Vacuum begging	225	195	212	Increased glass fiber increased the tensile strength (BGRP < GFRP composite).	[25]
7	Epoxy	Laminate	Vacuum begging	-	-	4	Hybrid BGRP had a lower impact strength than the BFRP composite.	[26]
8	Polyester	Laminate	Hand layup	270	-	946	Hybrid BGRP offered the highest value of mechanical properties compared to GFRP and BFRP laminates.	[28]
9	Polyester	Laminate	Hand layup	246	-	204	Hybrid BGRP laminate had the highest tensile strength.	[29]

Despite the fact that numerous studies have attempted to fabricate BGRP hybrid composites, only a few or no extensive investigation was reported on the mechanical, thermal, and chemical properties of BGRP hybrid unsaturated polyester (UP) composite. On that account, this study was performed to fabricate hybrid unsaturated polyester composites at various compositions of reinforcing fibers using the hand layup technique. The mechanical and thermal properties of fabricated composites were analyzed using DMA and TMA.

2. Materials and Methods

2.1. Materials

The hybrid composites comprise roving basalt fiber, woven E-glass fiber, unsaturated polyester resin matrix, and methyl ethyl ketone peroxide (MEKP) catalyst. The roving basalt fiber was purchased from Basaltex NV (Wevelgem, Belgium), whereas the woven E-glass fiber was provided by Innovative Pultrusion Sdn. Bhd. (Negeri Sembilan, Malaysia). The chemical constituents and physical properties of the fibers are listed in Tables 2 and 3, respectively.

Table 2. Chemical constituents of basalt and E-glass fibers.

Chemical Components	Composition (wt%)	
	Basalt	E-Glass
SiO_2 (silica)	57.5	55
Al_2O_3 (alumina)	16.9	15
Fe_2O_3 (ferric oxide)	9.5	0.3
MgO	3.7	3
Na_2O	2.5	0.8
TiO_2	1.1	-
K_2O	0.8	0.2
B_2O_3	-	7
F	-	0.3

Table 3. Physical properties of basalt and E-glass fibers.

Properties	Basalt	E-Glass
Density	2.67	2.55–2.58
Modulus (GPa)	85–89	78–80
Strength (MPa)	2900–3100	2000–2500
Moisture (%)	0.008	0.1

2.2. Fabrication of B/G/UP Hybrid Composites

Initially, the basalt fiber was cut into 300 mm pieces, while the woven glass fiber was cut into 300 × 300 mm squares to fit into a 300 × 300 × 5 mm steel mold. Individual layers of fiber were manually laid in the mold and compressed at a compression load of 40 tons. Figure 1 shows the method of laminating basalt fiber with glass fiber plies.

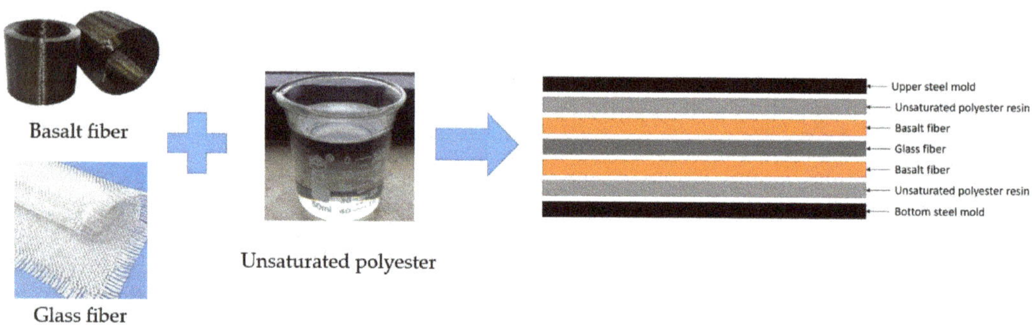

Figure 1. Typical lamination process of basalt fiber and glass fiber plies.

The unsaturated polyester resin and the catalyst were combined with compressed fiber, yielding sandwich-structured hybrid composites. The compositions of fabricated composites are depicted in Table 4.

Table 4. Basalt/glass fiber composition in hybrid composites.

No.	Composites	Sample Designation	Composition (wt%)		
			Matrix	Basalt	Glass
1	Five layers BF	BF	70	30	0
2	Four layers BF + two layers woven GF	B22.5/G7.5	70	22.5	7.5
3	Three layers woven GF + Three layers of BF	B15/G15	70	15	15
4	Two layers woven BF + three layers of GF	B7.5/G22.5	70	7.5	22.5
5	Five layers of woven GF	GF	70	0	30
6	UPE resin	UP	100	0	0

The numbers in the sample designation indicate the composition of basalt and glass fibers. One neat unsaturated polyester resin and two non-hybrid reinforced composites were also fabricated. The matrix to fiber ratios for all the hybrid and single reinforced composites were maintained at 7:3. An example calculation of hybrid composite (sample B15/G15) is tabulated in Table 5.

Table 5. Example calculation of hybrid composite (sample B15/G15).

Weight Percentage (wt%)	
Glass Fiber (wt%)	= 15 wt%
Basalt Fiber (wt%)	= 15 wt%
Unsaturated Polyester Resin (wt%)	= 70 wt%
MEKP (wt%) to UP resin	= 0.5% to UP resin
Density	
Glass Fiber	= 2.2 kg/L
Basalt Fiber	= 2.65 kg/L
Unsaturated Polyester Resin	= 1.87 kg/L
MEKP	= 1.152 kg/L
Volume	
Steel Mould	= 300 × 300 × 5 mm^3 = 0.45 L
Glass Fiber	= 15% × 0.45 L = 0.0675 L
Basalt Fiber	= 15% × 0.45 L = 0.0675 L
Unsaturated Polyester Resin	= 70% × 0.45 L = 0.315 L
MEKP	= 0.5% × 0.315 L = 0.001575 L
Weight	
Glass Fiber	= 0.00675 L × 2.2 kg/L = 0.01485 kg = 148.5 g
Basalt Fiber	= 0.00675 L × 2.65 kg/L = 0.01788 kg = 178.8 g
Unsaturated Polyester Resin	= 0.315 L × 1.87 kg/L = 0.589 kg = 589 g
MEKP	= 0.001575 L × 1.152 kg/L = 0.0018144 kg = 1.8144 g

2.3. Chemical Structure of Composites

The FTIR spectra of B/GF/UP composites were obtained using an IR spectrometer (Thermo Nicolet Corporation, Nicolet 6700 AEM, Madison, WI, USA) in the range of 4000 to 400 cm^{-1} at a 4 cm^{-1} resolution. The composite samples were mixed with KBr, compacted into pellets, and analyzed afterwards.

2.4. Dynamic Mechanical Analysis

A dynamic mechanical analyzer (TA Instruments, Q800 DMA, New Castle, DE, USA) was used to determine the storage modulus, loss modulus, and damping factor of B/GF/UP composites as a function of temperature (−100 to 150 °C). The frequency and heating rate were fixed at 1 Hz and 10 °C·min^{-1}, respectively.

2.5. Thermomechanical Analysis

The dimensional changes of B/GF/UP composites as a function of temperature were measured using a thermomechanical analyzer (TA Instruments, TMA Q400, New Castle, DE, USA) per ASTM E831 [30]. The analysis was performed under a constant flow of nitrogen gas within a temperature range of −100 to 150 °C at a 5 °C·min^{-1} heating rate.

3. Results and Discussion

3.1. Chemical Structure of Composites

The FTIR spectra for the neat polyester and basalt/woven glass fiber reinforced composites from 4000 to 400 cm^{-1} wavenumbers are shown in Figure 2. Based on the UP spectrum, some low intensity peaks appeared at high-range wavenumbers (>2000 cm^{-1}) at 3530 cm^{-1} (O–H stretching), 2920 cm^{-1}, and 2850 cm^{-1} (C–H stretching) [31]. At lower wavenumbers, numerous intense bands were observed at 1715 cm^{-1} (ester C=O stretching), 1600 cm^{-1} (ester C=O stretching), 1462 cm^{-1}, 1450 cm^{-1}, (C–H bending), 1250 cm^{-1} (aromatic ester C–O stretching), 1066 cm^{-1} (C–O stretching), and 703 cm^{-1} (C–H bending) [31–33]. All the identified bands are attributed to the chemical moieties in the unsaturated polyester chemical structure. Similar neat UP spectra were also reported by Arrieta et al., Chukwu et al., and Koto and Soegijono [31–33].

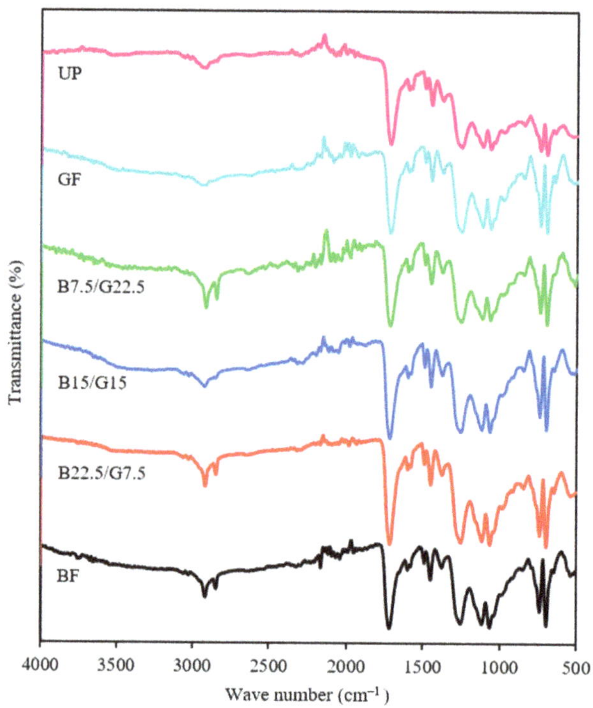

Figure 2. FTIR spectra of BF, B22.5/G7.5, B15/G15, B7.5/G22.5, GF, and UP composites.

Upon incorporating fiber reinforcements, slight changes in the composites' chemical characteristics were observed. For the single glass fiber reinforced composite (GF), the intensity of all the absorbance peaks appeared to be lower than the neat UP, but no new peak was identified. This observation suggests that the addition of glass fiber did not modify the chemical properties of the composite. As no functional groups are present on the glass fiber's surface, no chemical modification can occur unless the glass fiber is chemically modified or grafted prior to incorporation in the polymer matrix [34]. Regarding the single basalt fiber reinforced composite (BF), the only apparent difference in the spectrum is the intensified peak at 2920 and 2850 cm^{-1} (C–H stretching). It was previously reported that the basalt fiber would exhibit a single absorbance peak at 1000 to 1200 cm^{-1} due to the Si-O-Si bond vibration and a minor peak at 2800 to 2900 cm^{-1} due to C–H stretching, which explains the intensified peak for the C–H stretching bands [35,36].

With regards to the hybrid fiber reinforced composites FTIR spectra (B22.5/G7.5, B15/G15, and B7.5/G22.5), no apparent differences between the single and hybrid composites were observed. All the characteristic bands that were observed in the spectra also appeared in the neat UP matrix. On this basis, it can be inferred that the hybridization of basalt and glass fibers in the matrix does not alter the chemical properties of the composites. Several studies have proposed to chemically treat the basalt fiber to improve the chemical bonding and fiber/matrix adhesion. In particular, Seghini et al. [36] conducted plasma polymerization on basalt fiber to enhance the adhesion with an epoxy matrix and found some chemical changes on the fiber surface via FTIR.

3.2. Dynamic Mechanical Properties

3.2.1. Storage Modulus

The storage modulus (E′) is a measure of the stored energy in the elastic structure of a material that corresponds to the material's elastic response. This variable is particularly beneficial to evaluate the stiffness and elastic behavior of composites. The variation of the storage modulus with respect to the temperature for the fabricated composites are illustrated in Figure 3.

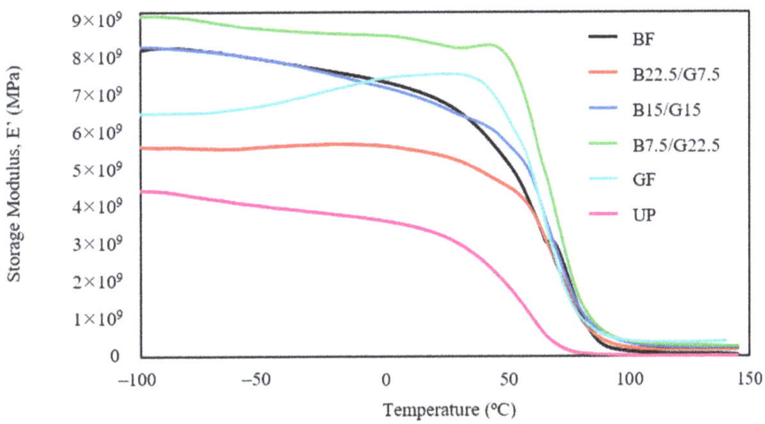

Figure 3. Storage modulus (E′) of BF, B22.5/G7.5, B15/G15, B7.5/G22.5, GF, and UP composites.

As evident from the storage modulus plot, the single reinforced composites (BF and GF) had higher storage modulus values compared to the neat UP. In particular, the GF composite showed a more substantial increment of storage modulus, from 3.70 × 10^9 MPa (UP) to 7.37 × 10^9 MPa at a near transition temperature range within 0 to −50 °C. Upon hybridization of both fibers, only the B7.5/G22.5 hybrid composite had a higher storage modulus (8.04 × 10^9 MPa) than BF (6.40 × 10^9 MPa) and GF (7.37 × 10^9 MPa). It appeared that the hybridization composition in B22.5/G7.5 and B15/G15 had a negative effect on the composites' elasticity as revealed by the reduction of storage modulus as compared to BF and GF. The high storage modulus of B7.5/G22.5 (high proportion of glass fiber) indicates that the glass fiber primarily contributes to the enhanced elastic property of the hybrid composite. This statement is supported by the higher storage modulus GF than the BF composite, as shown in Figure 2. It can be deduced that a high proportion of glass fiber and a small amount of basalt in the hybrid composite promotes an efficient stress transfer between the reinforcing fibers and the unsaturated polyester composite, creating a positive hybrid effect on the elasticity [37,38]. A similar observation was also reported by Barczewski et al. for the glass fiber/basalt powder polypropylene hybrid composite, whereby a hybrid composite with higher glass fiber content portrayed the highest storage modulus [20]. The presence of basalt powder in the composite caused a gradual reduction of elasticity. The current finding also accords with the elastic modulus reported in their work. At a temperature range beyond 90 °C, the storage modulus values of all the composites dropped to a minimum and remained plateau, which portrays that the matrix has reached a rubbery state [39].

3.2.2. Loss Modulus

Contrary to the storage modulus, the loss modulus (E") is helpful to evaluate the viscous attribute of a polymeric material [40]. The quantification of the parameter is made based on the energy lost, that is dissipated as heat, when subjected to a load cycle [41]. The loss modulus plots of single and hybrid reinforced unsaturated polyester composites within −100 to 150 °C are depicted in Figure 4.

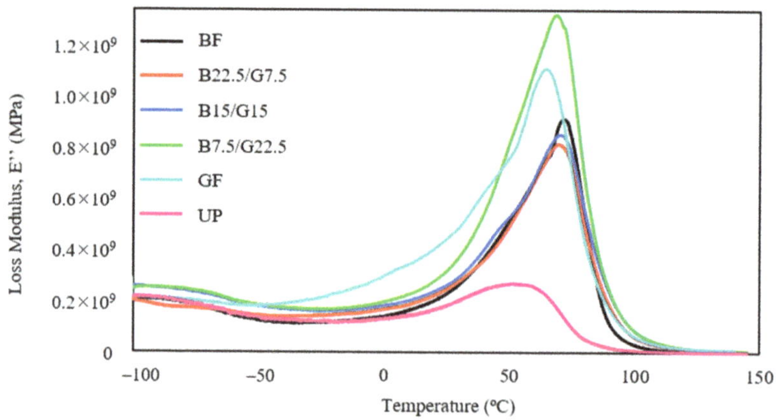

Figure 4. Loss modulus (E") of BF, B22.5/G7.5, B15/G15, B7.5/G22.5, GF, and UP composites.

Apparently, all the composites exhibited similar loss modulus curve trends. A gradual increase at the low-temperature range (<50 °C), a linear surge at the transition region (50–70 °C), and eventually a steep drop of loss modulus at higher temperatures. The highest point of the loss modulus curve can be used to compute the glass transition temperature (T_g). The neat UP showed the lowest loss modulus peak at 2.74×10^8 MPa, implying that the polyester molecular chains are more mobile without the fiber reinforcement [42]. All the reinforced composites exhibited remarkably high loss modulus particularly B7.5/G22.5 and GF composites with 1.32×10^9 MPa and 1.10×10^9 MPa peak modulus values. The higher peak values suggest that both composites, especially B7.5/G22.5, have a good fiber-matrix bond strength compared to the rest. Both composites comprised higher loadings of glass fiber than the other composites, indicating that the glass fiber reinforcement is capable of improving the viscous property. The hybridization of basalt at 7.5 wt% and glass fiber at 22.5 wt% seemed to have a notable enhancement effect on the loss modulus. The enhancement of glass fiber reinforced composites' viscous response for the sisal/woven glass fiber reinforced polyester composite with higher glass fiber loading was also reported by Gupta and Deep [43]. The B22.5/G7.5 and B15/G15 hybrid composites showed marginally lower loss modulus peaks than the BF composite. Although exhibiting no viscous response improvement, the addition of basalt fiber at higher composition in the matrix had improved the composites' thermal stability, as evidenced by the peak shifting to a higher temperature range. This thermal stability enhancement is ascribable to the polymer chains segmental motion restriction caused by the reinforcing fiber [44]. The T_g derived from the loss modulus curve will be described shortly.

3.2.3. Tan δ

The tan δ or damping factor is the ratio of loss modulus to storage modulus that represents a material's impact resistance and its elastic and viscous phase weightage [39,45]. A high tan δ value indicates that the material is more viscous and has high energy dissipation potential, whereas a low tan δ is associated with a highly elastic material capable of storing energy within its structure [46]. Figure 5 displays the tan δ plots of the fabricated single and hybrid reinforced unsaturated polyester composites within −100 to 150 °C. The peak tan δ values along with the T_g derived from the tan δ and loss modulus curves are summarized in Table 6.

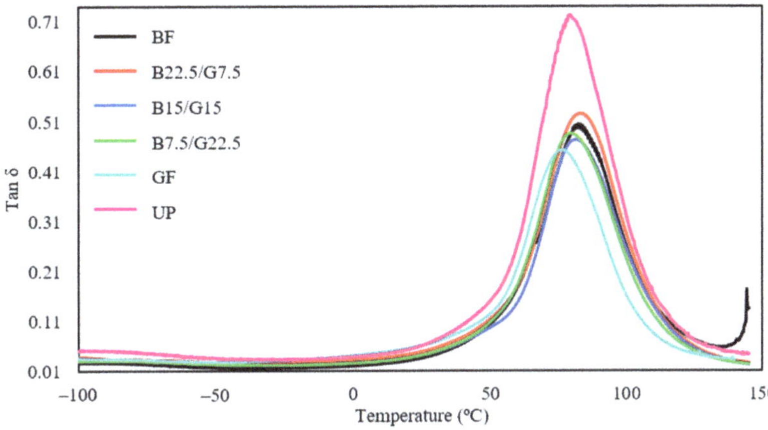

Figure 5. Tan δ of BF, B22.5/G7.5, B15/G15, B7.5/G22.5, GF, and UP composites.

Table 6. Peak height and glass transition temperatures of hybrid basalt/glass reinforced unsaturated polyester resin composites.

Sample ID	Peak Height of Tan δ	Temperature (°C)	
		T_g from Tan δ	T_g from Loss Modulus
BF	0.49	81.37	71.48
B22.5/G7.5	0.52	84.47	70.80
B15/G15	0.46	83.36	69.85
B7.5/G22.5	0.48	80.24	69.70
GF	0.44	76.98	66.17
UP	0.71	80.24	55.72

As evident in the plot, UP showed the highest tan δ peak, indicating a more viscous characteristic with high polymeric chain mobility. Single and hybrid fiber reinforcements in the matrix have substantially affected the damping property of the composites, making them more elastic than the neat matrix as observable by the lowered peaks of tan δ. Among all the reinforced composites, the GF composite showed the most significant tan δ reduction relative to the UP composite by 38% from 0.71 to 0.44. The BF composite portrayed a slightly lower tan δ peak reduction, which was 32% compared to the neat matrix. The incorporation of reinforcing fibers has led to restricted polymeric chain mobility due to strong adhesion at the fiber-matrix interface, and thus lowers the flexibility of the composite [46]. All the hybrid composites, namely B22.5/G7.5, B15/G15, and B7.5/G22.5 exhibited roughly similar reductions of tan δ. Based on this observation, it is inferred that the hybrid reinforcement has a negligible effect on the damping property of the composite.

With respect to the T_g derived from the tan δ plot, it appeared that the incorporation of single basalt fiber had marginally improved the thermal stability as evidenced by the increase of T_g from 80.24 to 81.37 °C. However, the inclusion of single glass fiber reinforcement had reduced the T_g from 80.24 to 76.98 °C. Higher T_g values were obtained for hybrid composites with higher loadings of basalt fiber relative to glass fiber. This finding suggests that basalt fiber is more prominent than glass fiber when it comes to enhancing the thermal stability of the composites. The higher thermal stability indicates that the composite has better molecular chain crosslinking in its structure [47]. A similar trend was found based on the T_g values derived from loss modulus peaks. The hybridization of both fibers had resulted in composites with intermediate T_g values within 70.80 and 69.70 °C, indicating a moderate improvement of interface bonding between the matrix and fiber [48]. The inclusion of basalt and glass fibers had decreased the mobility of the polymer chain, yielding higher T_g for single and hybrid composites than the UP specimen. Amuthakkannan and Manikandan reported a similar T_g of basalt fiber reinforced UP composite at approximately 68 °C [49]. The T_g of glass fiber reinforced UP composite from this study also accords with those reported by Subrata et al. at 71.35 °C [50]. Comparing the T_g values derived from both methods, the trend exhibited by the data of loss modulus is more realistic, whereby all the reinforced fiber showed improved thermal stability. In particular, single reinforced composites showed the highest and lowest increment, whereas the hybrid composites showed an intermediate improvement of thermal stability. The validity and more realistic data of T_g obtained from the loss modulus was also reported by numerous researchers that dealt with hybrid reinforced composites [42,43].

3.3. Thermomechanical Properties

TMA is a straightforward and convenient tool to evaluate the thermal properties of a polymeric material. The instrument measures and records the changes of a sample's dimension as it is subjected to heat or load, revealing information on the structure, composition, and potential application of the material [51]. The TMA result of the fabricated single and hybrid basalt/woven glass fiber reinforced unsaturated polyester composites is presented in Figure 6.

Based on the plot, it is evident that the UP composite exhibited the lowest extent of expansion and shrinkage among all the examined composites. This finding suggests that the neat matrix has better dimensional stability when subjected to temperature variation [52]. The reinforced composites portrayed a more notable expansion and shrinkage upon temperature changes, implying that fiber reinforcement at all fiber loadings caused an undesired effect on the dimensional properties. The detrimental effect is more obvious between 50 to 80 °C than the other temperatures, whereby all the reinforced composites exhibited steep increments of dimension. BF, B15/G15, B7.5/G22.5, and GF showed a similar behavior of dimensional change as evidenced by the overlapping curves in the TMA plot. As the temperature gradually increased, the dimension increased correspondingly, implying the occurrence of faster and unrestrained molecular movements and lack of fiber/matrix interaction [53]. On the other hand, B22.5/G7.5 had a slightly different response, whereby a sudden shrinkage occurred from 60 to 67 °C. The shrinkage indicates the penetration of reinforcing fibers into the polymer matrix [54]. Asim et al. [54] reported a similar behavior of dimensional change for the pineapple leaf/kenaf fiber reinforced phenolic composite. Overall, it can be deduced that the inclusion of basalt and glass fibers in the matrix negatively impacts the composites' dimensional stability, particularly at higher temperatures. Despite that fact, the dimensional changes beyond 80 °C occurred gradually and steadily.

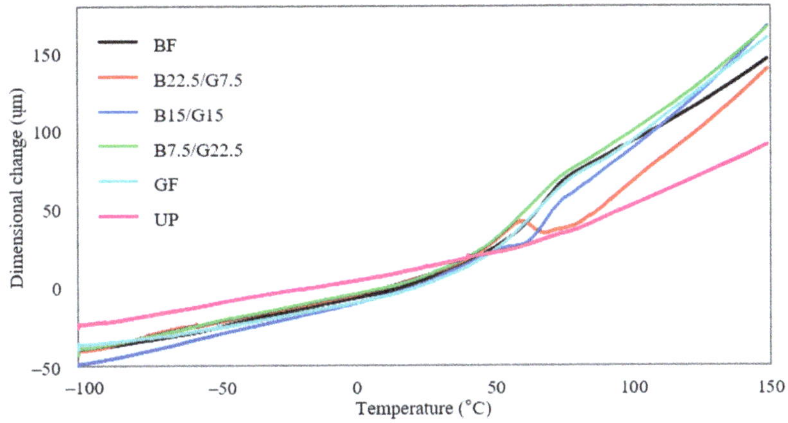

Figure 6. Thermomechanical analysis plots of BF, B22.5/G7.5, B15/G15, B7.5/G22.5, GF, and UP composites.

4. Conclusions

This study was undertaken to investigate the mechanical, thermal, and chemical properties of hybrid BGRP composites. The FTIR data revealed that the incorporation of single and hybrid basalt and glass fibers in the matrix do not eloquently alter the composites' chemical properties as no additional peaks were identified in the spectra. In future studies, chemical modification and grafting on the fiber prior to composite fabrication can be performed to promote chemical interactions between the components.

Moreover, the dynamic mechanical and thermal properties of hybrid BGRP composites were investigated using DMA. Among all the hybrid composites, only the B7.5/G22.5 composite portrayed a positive hybrid effect on the elasticity behavior, evidenced by a significantly high storage modulus value. It was deduced that an efficient stress transfer between the reinforcing fibers and the unsaturated polyester is achievable with a high proportion of glass fiber and a small amount of basalt fiber in the composite. The B7.5/G22.5 composite is also good in terms of viscous property, as revealed by the highest loss modulus value compared to the other hybrid BGRP composites. A higher glass fiber ratio yields remarkably better hybrid composite viscosity. Concerning the damping property, it appeared that all the hybrid BGRP composites showed no substantial improvement compared to the single reinforced GF composite. Despite having lower tan δ reductions than the single reinforced GF composite, the damping behavior of all the hybrid composites are still better than the neat matrix. The T_g values derived from the tan δ and loss modulus plots have revealed that the hybridization has marginally improved the thermal stability. The T_g had improved by up to 15 °C with glass and basalt fibers hybridization in a single matrix.

Lastly, the dimensional stability of the composites were evaluated using the thermomechanical analysis. Based on the result, it was found that the inclusion of basalt and glass fibers in the matrix negatively impacts the composites' dimensional stability, particularly at higher temperatures. Therefore, the hybrid BGRP composite is deemed suitable for lower temperature applications.

Overall, the findings suggest that the hybridization of basalt and glass fibers in the unsaturated polyester formed composites with better mechanical and thermal properties than the single reinforced composites and a neat matrix.

Author Contributions: Funding acquisition, F.R.; resources, S.M.S. and K.R.J.; supervision, A.A.; visualization, S.A.Z.; writing—original draft, N.I.N.H.; writing—review and editing, R.A.I. and M.Z.H. All authors have read and agreed to the published version of the manuscript.

Funding: This study was funded by Universiti Teknologi Malaysia through the 'Geran Universiti Penyelidik' (GUP) UTMFR Scheme Q.K.130000.2656.21H13 and the 'Collaborative Research Grant' Project Q.K.130000.2456.08G38, as well as the Ministry of Higher Education (MOHE) through the Fundamental Research Grant Scheme FRGS/1/2019/TK08/UTM/02/4. In addition, the authors would like to express their gratitude for the financial support received from Universiti Teknologi Malaysia, project CRG 30.3: Retardant coating using graphene/bamboo aerogel mixtures on SAR robotics system, grant number PY/2020/03495—R.J130000.7351.4B534. The research has been carried out under the program Research Excellent Consortium (JPT)(BKI)1000/016/018/25(57) provided by the Ministry of Higher Education Malaysia (MOHE).

Institutional Review Board Statement: Not applicable.

Informed Consent Statement: Not applicable.

Data Availability Statement: The data presented in this study are available on request from the corresponding author.

Acknowledgments: The authors are thankful to Universiti Teknologi Malaysia for the research facilities provided.

Conflicts of Interest: The authors declare no conflict of interest.

References

1. Ballo, A.; Närhi, T. 3-Biocompatibility of fiber-reinforced composites for dental applications. In *Biocompatibility of Dental Biomaterials*; Woodhead Publishing Series in Biomaterials; Shelton, R., Ed.; Woodhead Publishing: Sawston, UK, 2017; pp. 23–39. ISBN 9780081008843.
2. Wang, G.; Yu, D.; Kelkar, A.D.; Zhang, L. Electrospun nanofiber: Emerging reinforcing filler in polymer matrix composite materials. *Prog. Polym. Sci.* **2017**, *75*, 73–107. [CrossRef]
3. Suhot, M.A.; Hassan, M.Z.; Aziz, S.A.A.; Md Daud, M.Y. Recent Progress of Rice Husk Reinforced Polymer Composites: A Review. *Polymers* **2021**, *13*, 2391. [CrossRef] [PubMed]
4. Hassan, M.Z.; Sapuan, S.M.; Roslan, S.A.; Aziz, S.A.; Sarip, S. Optimization of tensile behavior of banana pseudo-stem (Musa acuminate) fiber reinforced epoxy composites using response surface methodology. *J. Mater. Res. Technol.* **2019**, *8*, 3517–3528. [CrossRef]
5. Hassan, M.Z.; Roslan, S.A.; Sapuan, S.M.; Rasid, Z.A.; Mohd Nor, A.F.; Md Daud, M.Y.; Dolah, R.; Mohamed Yusoff, M.Z. Mercerization Optimization of Bamboo (Bambusa vulgaris) Fiber-Reinforced Epoxy Composite Structures Using a Box–Behnken Design. *Polymers* **2020**, *12*, 1367. [CrossRef]
6. Nor, A.F.M.; Hassan, M.Z.; Rasid, Z.A.; Aziz, S.A.; Sarip, S.; Md Daud, M.Y. Optimization on Tensile Properties of Kenaf/Multi-walled CNT Hybrid Composites with Box-Behnken Design. *Appl. Compos. Mater.* **2021**, *28*, 607–632. [CrossRef]
7. Prasad, V.; Suresh, D.; Joseph, M.A.; Sekar, K.; Ali, M. Development of Flax Fibre Reinforced Epoxy Composite with Nano Tio2 Addition Into Matrix To Enhance Mechanical Properties. *Mater. Today Proc.* **2018**, *5*, 11569–11575. [CrossRef]
8. Adesina, O.T.; Jamiru, T.; Sadiku, E.R.; Ogunbiyi, O.F.; Beneke, L.W. Mechanical evaluation of hybrid natural fibre–reinforced polymeric composites for automotive bumper beam: A review. *Int. J. Adv. Manuf. Technol.* **2019**, *103*, 1781–1797. [CrossRef]
9. Norhasnan, N.H.; Hassan, M.Z.; Nor, A.F.; Zaki, S.A.; Dolah, R.; Jamaludin, K.R.; Aziz, S.A. Physicomechanical Properties of Rice Husk/Coco Peat Reinforced Acrylonitrile Butadiene Styrene Blend Composites. *Polymers* **2021**, *13*, 1171. [CrossRef]
10. Shioya, M.; Kikutani, T. Chapter 7—Synthetic textile fibres: Non-polymer fibres. In *Textiles and Fashion*; Woodhead Publishing Series in Textiles; Sinclair, R., Ed.; Woodhead Publishing: Sawston, UK, 2015; pp. 139–155. ISBN 9781845699314.
11. Sanjay, M.R.; Yogesha, B. Studies on Natural/Glass Fiber Reinforced Polymer Hybrid Composites: An Evolution. *Mater. Today Proc.* **2017**, *4*, 2739–2747. [CrossRef]
12. Chawla, K.K. Glass fibers. In *Reference Module in Materials Science and Materials Engineering*; Elsevier: Amsterdam, The Netherlands, 2016; ISBN 9780128035818.
13. De Araújo, M. 1—Natural and man-made fibres: Physical and mechanical properties. In *Fibrous and Composite Materials for Civil Engineering Applications*; Woodhead Publishing Series in Textiles; Fangueiro, R., Ed.; Woodhead Publishing: Sawston, UK, 2011; pp. 3–28. ISBN 9781845695583.
14. Fan, Y.; Guo, J.; Wang, X.; Xia, Y.; Han, P.; Shangguan, L.; Zhang, M. Comparative Failure Study of Different Bonded Basalt Fiber-Reinforced Polymer (BFRP)-AL Joints in a Humid and Hot Environment. *Polymers* **2021**, *13*, 2593. [CrossRef]
15. Ogrodowska, K.; Łuszcz, K.; Garbacz, A. Nanomodification, Hybridization and Temperature Impact on Shear Strength of Basalt Fiber-Reinforced Polymer Bars. *Polymers* **2021**, *13*, 2585. [CrossRef]
16. Ricciardi, M.R.; Papa, I.; Coppola, G.; Lopresto, V.; Sansone, L.; Antonucci, V. Effect of Plasma Treatment on the Impact Behavior of Epoxy/Basalt Fiber-Reinforced Composites: A Preliminary Study. *Polymers* **2021**, *13*, 1293. [CrossRef]
17. Le, C.H.; Louda, P.; Buczkowska, K.E.; Dufkova, I. Investigation on Flexural Behavior of Geopolymer-Based Carbon Textile/Basalt Fiber Hybrid Composite. *Polymers* **2021**, *13*, 751. [CrossRef] [PubMed]

18. Motoc, D.L.; Ferri, J.M.; Ferrandiz-Bou, S.; Garcia-Garcia, D.; Balart, R. Dynamic–Mechanical and Decomposition Properties of Flax/Basalt Hybrid Laminates Based on an Epoxidized Linseed Oil Polymer. *Polymers* **2021**, *13*, 479. [CrossRef]
19. Glaskova-Kuzmina, T.; Zotti, A.; Borriello, A.; Zarrelli, M.; Aniskevich, A. Basalt Fibre Composite with Carbon Nanomodified Epoxy Matrix under Hydrothermal Ageing. *Polymers* **2021**, *13*, 532. [CrossRef] [PubMed]
20. Barczewski, M.; Matykiewicz, D.; Mysiukiewicz, O.; Maciejewski, P. Evaluation of polypropylene hybrid composites containing glass fiber and basalt powder. *J. Polym. Eng.* **2018**, *38*, 281–289. [CrossRef]
21. Babu, S.M.; Rao, M.V. Effect of basalt powder on mechanical properties and dynamic mechanical thermal analysis of hybrid epoxy composites reinforced with glass fiber. *J. Chin. Adv. Mater. Soc.* **2018**, *6*, 311–328. [CrossRef]
22. El-Wazery, M.S. Mechanical Characterization of Glass-Basalt-Carbon/Polyester Hybrid Composites. *Int. J. Eng.* **2018**, *31*, 1139–1145. [CrossRef]
23. Patel, N.I.; Patel, K.; Gohil, P.; Chaudhry, V. Investigations on Mechanical Strength of Hybrid Basalt/Glass Polyester Composites. *Int. J. Appl. Eng. Res.* **2018**, *16*, 4083–4088.
24. Bozkurt, Ö.Y.; Erkliğ, A.; Bozkurt, Y.T. Influence of basalt fiber hybridization on the vibration-damping properties of glass fiber reinforced epoxy laminates. *Mater. Res. Express* **2018**, *6*, 15301. [CrossRef]
25. El-Baky, M.A.A.; Attia, M.A.; Abdelhaleem, M.M.; Hassan, M.A. Flax/basalt/E-glass Fibers Reinforced Epoxy Composites with Enhanced Mechanical Properties. *J. Nat. Fibers* **2020**, 1–15. [CrossRef]
26. Özbek, Ö.; Bozkurt, Ö.Y.; Erkliğ, A. Effect of glass fiber hybridization on low velocity impact behaviors of basalt fiber reinforced composites laminates. *Int. J. Mater. Eng. Technol.* **2020**, *3*, 21–29.
27. Jain, A.; Singh, B.; Sharma, K.; Shrivastava, Y. Fabrication, Testing and Machining of Hybrid Basalt-Glass Fiber Reinforced Plastic composite. *Indian J. Pure Appl. Phys.* **2021**, *59*, 258–262.
28. Sapuan, S.M.; Aulia, H.S.; Ilyas, R.A.; Atiqah, A.; Dele-Afolabi, T.T.; Nurazzi, M.N.; Supian, A.B.M.; Atikah, M.S.N. Mechanical Properties of Longitudinal Basalt/Woven-Glass-Fiber-reinforced Unsaturated Polyester-Resin Hybrid Composites. *Polymers* **2020**, *12*, 2211. [CrossRef]
29. Patel, N.; Patel, K.; Chaudhary, V.; Gohil, P. Investigations on drilling of hybrid basalt/glass polyester composites. *Aust. J. Mech. Eng.* **2020**, 1–10. [CrossRef]
30. *ASTM E831-19 Standard Test Method for Linear Thermal Expansion of Solid Materials by Thermomechanical Analysis*; Techstreet: Pennsylvania, PA, USA, 2019.
31. Arrieta, J.S.; Richaud, E.; Fayolle, B.; Nizeyimana, F. Thermal oxidation of vinyl ester and unsaturated polyester resins. *Polym. Degrad. Stab.* **2016**, *129*, 142–155. [CrossRef]
32. Koto, N.; Soegijono, B. Effect of Rice Husk Ash Filler of Resistance Against of High-Speed Projectile Impact on Polyester-Fiberglass Double Panel Composites. *J. Phys. Conf. Ser.* **2019**, *1191*, 12058. [CrossRef]
33. Chukwu, M.; Madueke, C.; Ekebafe, L. Effects of Snail Shell as Filler on the Mechanical Properties of Terephthalic Unsaturated Polyester Resin. *Niger. Res. J. Chem. Sci.* **2019**, *6*, 21–32.
34. Prakash, V.A.; Rajadurai, A. Inter laminar shear strength behavior of acid, base and silane treated E-glass fibre epoxy resin composites on drilling process. *Def. Technol.* **2017**, *13*, 40–46. [CrossRef]
35. Matykiewicz, D.; Barczewski, M. On the impact of flax fibers as an internal layer on the properties of basalt-epoxy composites modified with silanized basalt powder. *Compos. Commun.* **2020**, *20*, 100360. [CrossRef]
36. Seghini, M.C.; Touchard, F.; Sarasini, F.; Cech, V.; Chocinski-Arnault, L.; Mellier, D.; Tirillò, J.; Bracciale, M.P.; Zvonek, M. Engineering the interfacial adhesion in basalt/epoxy composites by plasma polymerization. *Compos. Part A Appl. Sci. Manuf.* **2019**, *122*, 67–76. [CrossRef]
37. Alothman, O.Y.; Jawaid, M.; Senthilkumar, K.; Chandrasekar, M.; Alshammari, B.A.; Fouad, H.; Hashem, M.; Siengchin, S. Thermal characterization of date palm/epoxy composites with fillers from different parts of the tree. *J. Mater. Res. Technol.* **2020**, *9*, 15537–15546. [CrossRef]
38. Atiqah, A.; Jawaid, M.; Sapuan, S.M.; Ishak, M.R.; Alothman, O.Y. Thermal properties of sugar palm/glass fiber reinforced thermoplastic polyurethane hybrid composites. *Compos. Struct.* **2018**, *202*, 954–958. [CrossRef]
39. Manoharan, S.; Suresha, B.; Ramadoss, G.; Bharath, B. Effect of Short Fiber Reinforcement on Mechanical Properties of Hybrid Phenolic Composites. *J. Mater.* **2014**, *2014*, 478549. [CrossRef]
40. Nurazzi, N.M.; Khalina, A.; Sapuan, S.M.; Ilyas, R.A.; Rafiqah, S.A.; Hanafee, Z.M. Thermal properties of treated sugar palm yarn/glass fiber reinforced unsaturated polyester hybrid composites. *J. Mater. Res. Technol.* **2020**, *9*, 1606–1618. [CrossRef]
41. Boparai, K.S.; Singh, R. Thermoplastic composites for fused deposition modeling filament: Challenges and applications. In *Reference Module in Materials Science and Materials Engineering*; Elsevier: Amsterdam, The Netherlands, 2018; ISBN 9780128035818.
42. Vinu Kumar, S.M.; Senthil Kumar, K.L.; Siddhi Jailani, H.; Rajamurugan, G. Mechanical, DMA and Sound Acoustic behaviour of Flax woven fabric reinforced Epoxy composites. *Mater. Res. Express* **2020**, *7*, 085302. [CrossRef]
43. Gupta, M.; Deep, V. Effect of Stacking Sequence on Flexural and Dynamic Mechanical Properties of Hybrid Sisal/Glass Polyester Composite. *Am. J. Polym. Sci. Eng.* **2017**, *5*, 53–6253.
44. Negawo, T.A.; Polat, Y.; Buyuknalcaci, F.N.; Kilic, A.; Saba, N.; Jawaid, M. Mechanical, morphological, structural and dynamic mechanical properties of alkali treated Ensete stem fibers reinforced unsaturated polyester composites. *Compos. Struct.* **2019**, *207*, 589–597. [CrossRef]

45. Arulmurugan, M.; Prabu, K.; Rajamurugan, G.; Selvakumar, A.S. Impact of $BaSO_4$ filler on woven Aloevera/Hemp hybrid composite: Dynamic mechanical analysis. *Mater. Res. Express* **2019**, *6*, 45309. [CrossRef]
46. Atiqah, A.; Ansari, M.N.M.; Jawaid, M.; Hamdan, A. Viscoelastic Properties of Kenaf/Basalt Reinforced Epoxy Hybrid Composites through Vacuum Infusion Methods. *Test Eng. Manag.* **2020**, *83*, 1226–1231.
47. Zhang, D.; He, M.; He, W.; Zhou, Y.; Qin, S.; Yu, J. Influence of Thermo-Oxidative Ageing on the Thermal and Dynamical Mechanical Properties of Long Glass Fibre-Reinforced Poly(Butylene Terephthalate) Composites Filled with DOPO. *Materials* **2017**, *10*, 500. [CrossRef]
48. Norizan, M.N.; Abdan, K.; Ilyas, R.A.; Zin, M.; Muthukumar, C.; Rafiqah, S.; Aisyah, H. Effect of fiber orientation and fiber loading on the mechanical and thermal properties of sugar palm yarn fiber reinforced unsaturated polyester resin composites. *Polim. Warsaw* **2019**, *64*, 34–43. [CrossRef]
49. Amuthakkannan, P.; Manikandan, V. Free vibration and dynamic mechanical properties of basalt fiber reinforced polymer composites. *Indian J. Eng. Mater. Sci.* **2018**, *25*, 265–270.
50. Das, S.C.; Paul, D.; Grammatikos, S.A.; Siddiquee, M.A.B.; Papatzani, S.; Koralli, P.; Islam, J.M.M.; Khan, M.A.; Shauddin, S.M.; Khan, R.A.; et al. Effect of stacking sequence on the performance of hybrid natural/synthetic fiber reinforced polymer composite laminates. *Compos. Struct.* **2021**, *276*, 114525. [CrossRef]
51. James, J. Chapter 7—Thermomechanical analysis and its applications. In *Thermal and Rheological Measurement Techniques for Nanomaterials Characterization*; Micro and Nano Technologies; Thomas, S., Thomas, R., Zachariah, A.K., Mishra, R.K., Eds.; Elsevier: Amsterdam, The Netherlands, 2017; pp. 159–171. ISBN 9780323461399.
52. Espinach, F.X.; Boufi, S.; Delgado-Aguilar, M.; Julián, F.; Mutjé, P.; Méndez, J.A. Composites from poly(lactic acid) and bleached chemical fibres: Thermal properties. *Compos. Part B Eng.* **2018**, *134*, 169–176. [CrossRef]
53. Vinod, A.; Yashas Gowda, T.G.; Vijay, R.; Sanjay, M.R.; Gupta, M.K.; Jamil, M.; Kushvaha, V.; Siengchin, S. Novel Muntingia Calabura bark fiber reinforced green-epoxy composite: A sustainable and green material for cleaner production. *J. Clean. Prod.* **2021**, *294*, 126337. [CrossRef]
54. Asim, M.; Jawaid, M.; Paridah, M.T.; Saba, N.; Nasir, M.; Shahroze, R.M. Dynamic and thermo-mechanical properties of hybridized kenaf/PALF reinforced phenolic composites. *Polym. Compos.* **2019**, *40*, 3814–3822. [CrossRef]

polymers

MDPI

Article

Thermal Stability and Dynamic Mechanical Analysis of Benzoylation Treated Sugar Palm/Kenaf Fiber Reinforced Polypropylene Hybrid Composites

S. Mohd Izwan [1], S.M. Sapuan [1,2,*], M.Y.M. Zuhri [1,2] and A.R. Mohamed [3]

1 Centre of Advanced Engineering Materials and Composites Research, Department of Mechanical and Manufacturing Engineering, Universiti Putra Malaysia, UPM, Serdang 43400, Malaysia; mr_iez88@yahoo.com (S.M.I.); zuhri@upm.edu.my (M.Y.M.Z.)
2 Laboratory of Bio Composite Technology, Institute of Tropical Forestry and Forest Products, Universiti Putra Malaysia, UPM, Serdang 43400, Malaysia
3 Department of Manufacturing and Material Engineering, Kulliyyah of Engineering, International Islamic University Malaysia, Kuala Lumpur 50728, Malaysia; mrahman@iium.edu.my
* Correspondence: sapuan@upm.edu.my

Abstract: This research was performed to evaluate the mechanical and thermal properties of sugar palm fiber (SPF)- and kenaf fiber (KF)-reinforced polypropylene (PP) composites. Sugar palm/kenaf was successfully treated by benzoylation treatment. The hybridized bio-composites (PP/SPF/KF) were fabricated with overall 10 weight percentage (wt%) relatively with three different fibers ratios between sugar palm-treated and kenaf-treated (7:3, 5:5, 3:7) and vice versa. The investigations of thermal stability were then carried out by using diffraction scanning calorimetry (DSC) and thermogravimetry analysis (TGA). The result of a flammability test showed that the treated hybrid composite (PP/SPF/KF) was the specimen that exhibited the best flammability properties, having the lowest average burning rate of 28 mm/min. The stiffness storage modulus (E′), loss modulus (E″), and damping factor (Tan δ) were examined by using dynamic mechanical analysis (DMA). The hybrid composite with the best ratio (PP/SPF/KF), T-SP5K5, showed a loss modulus (E″) of 86.2 MPa and a damping factor of 0.058. In addition, thermomechanical analysis (TMA) of the studies of the dimension coefficient (μm) against temperature were successfully recorded, with T-SP5K5 achieving the highest dimensional coefficient of 30.11 μm at 105 °C.

Keywords: biocomposites; kenaf; sugar palm; thermal; dynamic mechanical analysis; benzoylation

Citation: Mohd Izwan, S.; Sapuan, S.M.; Zuhri, M.Y.M.; Mohamed, A.R. Thermal Stability and Dynamic Mechanical Analysis of Benzoylation Treated Sugar Palm/Kenaf Fiber Reinforced Polypropylene Hybrid Composites. *Polymers* **2021**, *13*, 2961. https://doi.org/10.3390/polym13172961

Academic Editor: Markus Gahleitner

Received: 30 July 2021
Accepted: 27 August 2021
Published: 31 August 2021

1. Introduction

For decades, synthetic fibers have been the leading commodity in the composites industry. However, synthetic fibers possess many disadvantages, as they catch fire easily very hydrophobic and non-biodegradable. Since synthetic fibers have many shortcomings, researchers have had growing interest in producing polymers that incorporate natural fibers. Natural fibers are becoming more common as a viable option due to the harmful environmental and health consequences of synthetic fibers. Concerns about the environment and the rising greenhouse effect and increasing interest in the use of sustainable materials has motivated researchers to investigate biocomposite materials. In today's manufacturing environment, natural fiber composites are playing a prominent role in many vital applications, such as in fuselages and propellers in the aerospace industry, racing car bodies, wings of wind turbines, bicycle frames, and automobile interiors, seat cushions, and door panels, etc. [1–3]. The great interest in natural fiber composites is due to their high performance, biodegradability, non-abrasive light weight, and low cost [4,5]. Moreover, the widespread adoption of natural fibers and biopolymers as green materials is being motivated by the rapid depletion of petroleum supplies, as well as by a growing recognition of global environmental issues associated with the use of traditional plastics. [6–8]. The successful application of biopolymers and the promise of alternative pathways with a reduced carbon footprint arising from the use of green materials bodes well for the future design and development of ever more sophisticated green materials [9].

Natural fibers and biopolymers have attracted scientists and industry because of their environmentally beneficial and long-lasting properties. Natural fibers such as sugar palm fiber, corn husk fiber [10], jute, and wheat arrowroot, as well as cassava bagasse, are used as reinforcement materials in polymer composites for a variety of reasons, including their ability to be reusable and their low cost, and because they are environmentally sustainable and have good strength and stiffness properties [11]. For material applications, a broad variety of naturally occurring biopolymers extracted from renewable materials are available. Some bacteria and plants (chitin, starch, and cellulose) are currently used in commercial products, whereas others are underutilized [12]. Starch has been explored as a possible alternative to traditional plastic packaging. These starch (or cellulose) biopolymers include animal-based (chitin) polymers and microbial (exopolysaccharides and polyhydroxyalkanoate) polymers [13] that are chemically synthesized from agro-based resource monomer (poly lactic acid) as well as chemically synthesized from conventionally synthesized monomers. Despite their current use, starch biopolymers have been characterized as having weak mechanical properties and a low water barrier resistance [14,15]. These drawbacks have significantly hindered a wider range of their application, especially in packaging [16,17]. Much research has been performed in an attempt to improve the mechanical properties of starch biopolymers without affecting their biodegradation properties. Research has found that reinforcing starch biopolymer with natural fiber is one way to strengthen both its mechanical and thermal properties.

Among the many different types of natural resources, kenaf plants have been extensively exploited over the past few years [18]. One of the reasons for this growing interest is that natural fibers such as kenaf have a higher specific strength as glass fiber and a similar specific modulus strength [19]. Kenaf, which is from the Hibiscus cannabinus family, was selected due to its low cost, low density, good toughness, recyclability, good sound absorption performance, acceptable strength properties, and biodegradability [20,21]. Furthermore, kenaf fiber was selected because it has already been utilized in the automotive industry and because it has a good surface and produces a lightweight material with high mechanical properties and thus does not have to prove itself as a reliable product. Since kenaf brings a lot of promising qualities, a study on the hybridization of two types of natural fiber-reinforced thermoplastic composites was an alternative option for a novel breakthrough. The hybridization of composite fibers refers to the merger of two or more reinforcing materials, such as filler, to enhance the overall properties of a material [22].

Meanwhile, sugar palm (also known as Arenga pinnata) is a tropical tree that belongs to the Palmae family. Apart from the production of its sugar and starch extract [23], this tree was also known for the fiber from its trunk and from its fronds: sugar palm fiber (SPF). Traditionally, sugar palm fiber was used for various domestic materials [24]. This is due to the excellent characteristics of sugar palm fiber that improve tensile strength and reduce the wettability degree of a composite surface. Owing to that, incorporation of treated fibers in a polymer matrix promised a good thermal resistance and reduced thermal degradation [25]. As the research has grown, sugar palm fiber has shown many significant advantages to be considered: it is abundant and widespread, and it shows a promising ability to enrich physical and mechanical strength, thermal stability, and density, as well as showing excellent biodegradability [26].

The hybridization of sugar palm fiber and kenaf fiber as a filler was not a recent finding for composite reinforcement. Polymer composites with reinforced fibers usually consist of more than one type of particle fiber compounded together with a polymer as part of their matrix system [27]. The properties of a hybrid composite are influenced by the fiber content, length, and orientation. The selection of the fiber constituent for hybrid composites affects the hybridization and the requirement of the material being fabricated.

Several studies have shown that hybridization of natural and synthetic fiber can improve mechanical and thermal properties. A study by Devi et al. [28] showed that the dynamic mechanical properties—including the storage modulus, loss modulus and tan δ—of pineapple/glass hybrid-reinforced unsaturated polyester resin composites were enhanced when more content of glass fiber was added to the composites. A previous study examined the effects on a composite's behavior of combining rattan nanoparticles into polypropylene with filler contents ranging from 2% to 20% [29]. The study found that 5% was the most optimal filler content for achieving better mechanical properties of the composite. Furthermore, mechanical performance decreased when the filler content was increased from 5% to 20%. Another research study observed hybrid composites with different composition that were prepared with different amounts of fibers (i.e., 10%, 20% and 30% by weight percent), in which the ratios between sugar palm and kenaf fiber were 30:70, 50:50, and 70:30. The study found that the tensile strength of composites tended to decrease when the content of loading fibers increased [30]. Therefore, the current study focused only on the implementation of 10% of kenaf/sugar palm as the composite filler content in order to offer better bonding between the fibers and polypropylene matrix.

The selection of compatible fibers and fiber properties, therefore, contributes a critical aspect in designing a better hybrid composite. The previous research has thoroughly examined the effects of benzoylation treatment of SPF with different parameters. A previous study found that kenaf and sugar palm fibers were compatible combinations for hybrid composites due to the outcomes of high tensile strength and toughness of the kenaf/sugar palm composites [31–33]. As reported, the benzoylation of fibers improves fiber–matrix adhesion, thus improving thermal stability, increasing composite strength, and decreasing water absorption [34–36]. Benzoyl chloride was used in this research for SPF and KF treatment. This benzoyl chloride helps to decrease the hydrophilic nature of SPF and KF and improves the interaction with the resin matrix [37].

Hence, in this paper, the hybridized polypropylene composite with kenaf/sugar palm fiber was further examined for its thermal stability using dynamic thermal analysis. The preparation and characterization of thermosetting and thermoplastic composite materials reinforced with kenaf and sugar palm fibers with and without treatment using benzoylation methods were conducted. As presumed, treatment using alkaline on the surface of the fiber changed the surface wettability, altering the mechanical and physical properties of the natural fibers. Moreover, the benzoylation of kenaf and sugar palm fibers treatment also successfully confirmed an incremental increase in tensile strength. Thus, the main objective for this paper was an investigation of thermosetting composites based on their thermal stability, thermal degradation, flammability, and modulus stress by using instrumentation such as diffraction calorimetry (DSC), thermogravimetry analysis (TGA), dynamic mechanical analysis (DMA), thermomechanical analysis (TMA), and flammability analysis.

2. Material and Methods

2.1. Materials

Polypropylene pellet and benzoyl chloride were supplied by Mecha Solve Engineering (M) Sdn Bhd. Sugar palm fiber (SPF) was purchased from Jempol, Negeri Sembilan, Malaysia, and kenaf fiber (KF) was obtained from Lembaga Kenaf and Tembakau Negara (LKTN) Kelantan, Malaysia. Kenaf (*Hibiscus cannabinus*) and the sugar palm tree (*Arenga pinnata*) were used. In addition, raw kenaf palm and kenaf fibers were washed with deionized water and rinsed. They were then pulverized, cleaned, and dried at 70 °C in an oven. All other chemicals and solvents that were used in this work were at 98% purity. Polypropylene pellet crystals with 0.946 g/cm^3 density were used. The pellets that were used were whitish gray, ovular, and 5mm long and 3mm diameter.

2.2. Alkalization and Benzoylation of Kenaf and Sugar Palm Fiber (KF/SPF)

The clean dried kenaf and sugar palm fibers were then soaked. An amount of this fiber was soaked in 18% concentration of NaOH solution as a pre-treatment for 30 min. After that, the partially treated kenaf and sugar palm fibers were filtered and rinsed with ionized water and dried in an oven at 70 °C [34,35]. The treated fibers were then immersed in 10% concentration NaOH solution agitated well with 50 mL benzoyl chloride for 15 min. The treated KF and SPF was then soaked with ethanol for 1 h and rinsed with tap water in order to remove unreacted benzoyl chloride and excess dirt. Treated KF and SPF were then dried at 60 °C for 24 h [36]. Alkalization of the samples was performed to remove impurities and benzoylation was performed to enhance the melting point of the samples [37].

2.3. *Compounding of Kenaf and Sugar Palm Fibers (KF/SPF) and Preparation of Particle Composite*

The treated and untreated kenaf and sugar palm fibers were ground into short-form fibers with an approximate length of 0.1–0.5 mm by using a pulverizing machine (Pulveriseet P-19). In order to obtain uniformly cut fiber particle sizes, the fibers were sieved by using a 40 mesh electronic sieve (Endecotts). Finally, ground KF and SPF fibers were dried at 60 °C for 12 h to avoid contamination. Next, a melt extruder was used to compound the treated and untreated ground KF and SPF with polypropylene as their polymer matrix by using a Brabender plastograph (Model 815651, Brabender GmbH & Co. KG, Duisburg, Germany). An amount of 20 g of mixture was prepared for each cycle of extrusion, and the compositions of the hybrid composites of SPF/KF/PP are presented in Table 1. The initial 'U' and 'T' indicate untreated and treated hybrid composites, respectively.

Table 1. Compositions of sugar palm fiber, kenaf fiber, and polypropylene hybrid composite (SPF/KF/PP).

Hybrid Composites	SPF (g)	KF (g)	PP (g)	Total Weight (g)
U-SP3K7	0.6	1.4	18	20
U-SP5K5	1	1	18	20
U-SP7K3	1.4	0.6	18	20
T-SP3K7	0.6	1.4	18	20
T-SP5K5	1	1	18	20
T-SP7K3	1.4	0.6	18	20

The mixing temperature was set at 180 °C, while the rotor speed of the rotating screw was set at 50 rpm. Polypropylene was discharged in the chamber and melted for 3 min before the compounding took place. The KF and SPF particle fibers and the polymer were extruded over approximately 10 min of holistic mixing. Thermoset composites of SPF/KF/PP were crushed into granular size, followed by the use of a hot mold pressing. Customized samples of hybrid composites were then pre-heated at 180 °C for 5 min and pressed at 190 °C for 7 min by using a hot press machine. After that, the composite samples were cold pressed at 25 °C for 5 min and chopped into plain composite sheets sized 150 × 150 × 3 mm before being cut into a standard shape for TGA, DMA, TMA, DSC, and flammability test, as illustrated in Figure 1.

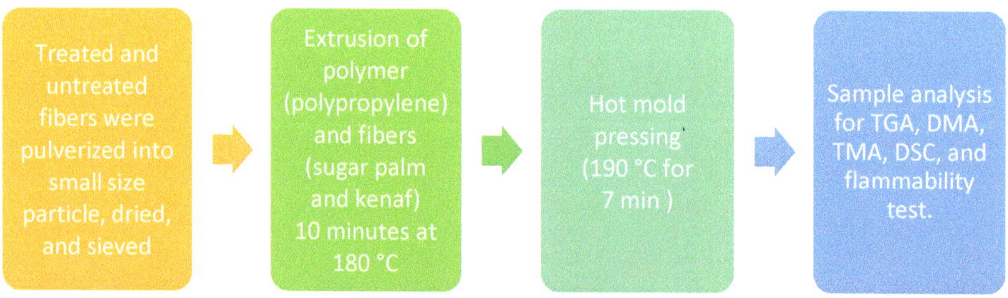

Figure 1. Flowchart of different steps for polypropylene with sugar palm and kenaf fibers (PP/KF/SPF) composite.

2.4. Thermal Instrumentations

2.4.1. Thermogravimetric Analysis (TGA and DTG)

Thermogravimetric analysis (TGA) was performed to examine the structural properties and thermal stability of the hybrid composite materials. The analysis was carried out to calculate the degradation curve of the SPF/KF/PP composites towards their degradation temperature (°C). A Mettler Toledo machine (TGA/SDTA 851e, USA) was used, and all composites (SPF/KF/PP) were observed between 30 to 600 °C at a heating rate of 20 °C/min. Nitrogen gas flow was recorded at 50 mL/min. The weight of the samples varied from between 6 and 20 mg.

2.4.2. Differential Scanning Calorimetry Analysis (DSC)

Differential scanning calorimetry (DSC) analysis of the samples was carried out with a PerkinElmer (USA) Diamond thermogravimetric (TG)/DSC analyzer. The work was carried out with 20 milligram of the composite fibers sample filled in aluminum pans under a dynamic nitrogen atmosphere in a temperature range of 25–600 °C and a heating rate of 5 °C/min. The percentage of crystallinity X (%) was calculated as Equation (1):

$$\frac{\Delta H}{\Delta H_{100}} = X^{C} \quad (1)$$

where ΔH is the heat of crystallization of the sample analyzed (J/g), and ΔH_{100} is a reference value that represents the heat of crystallization for a 100% crystalline polymer. For PP, ΔH_{100} is taken as 209 J/g.

2.4.3. Dynamic Mechanical Analysis (DMA)

A dynamic mechanical analyzer (TA Instrument, Q800, USA) was used for the evaluation of the storage modulus, loss modulus, and mechanical damping factor (tan δ). The storage modulus (E′), loss modulus (E″), and loss factor (tan δ) of the composite specimen were evaluated as a function of temperature (−100 °C to 100 °C) using TA 2980 software (TA Universal Analysis, USA). A dynamic mechanical analyzer was equipped with a dual cantilever bending fixture at the frequency of 1 Hz with the heating constant rate at 10 °C/min. Three-point bending mode was examined. The heating rate used was at 5 °C/min under an amplitude frequency of 1 Hz. Liquid nitrogen was used as the cooling agent, and the temperature range was from −150 °C to 150 °C. The amplitude was set at 7–10 mm, depending on the thickness of the samples. The samples had a thickness of 4–5 mm, width of 9–10 mm, and length of 50–60 mm.

2.4.4. Thermal Mechanical Analysis (TMA)

The coefficient of thermal expansion (CTE) was measured by heating the specimen from −50 °C to 100 °C at a rate of 5 °C/min under a nitrogen atmosphere with a flow rate of 100 mL/min. The probe was applied with a 0.05 N loading to measure the strain in the specimens and their temperature. The coefficient of thermal expansion was estimated from the linear slope of the strain–temperature curve using a thermomechanical analyzer (TMA Q 400, TA Instruments, New Castle, DE, USA). The specimen dimensions were 7 mm × 7 mm × 1.8 mm.

2.4.5. Flammability Analysis

A flammability test of pure PP and SPF/KF/PP composites was carried out for all samples via a horizontal burning test according to ASTM D635. Three specimens from each composite ratio, with dimensions of 125 mm × 13 mm × 3 mm, were prepared, and two lines at 25 and 100 mm from one end of the sample were drawn as the reference marks as shown in Figure 2a [38]. Then, the burning time from the first reference mark to the second reference mark (25 mm from the end and 100 mm from the end, respectively) was recorded as shown in Figure 2b. The linear rate of the burning samples was calculated using Equation (2).

$$V = \frac{L}{t} 60 \tag{2}$$

where V is the linear burning rate (mm/min), L is the burnt length (mm), and t is the time (minutes) [35].

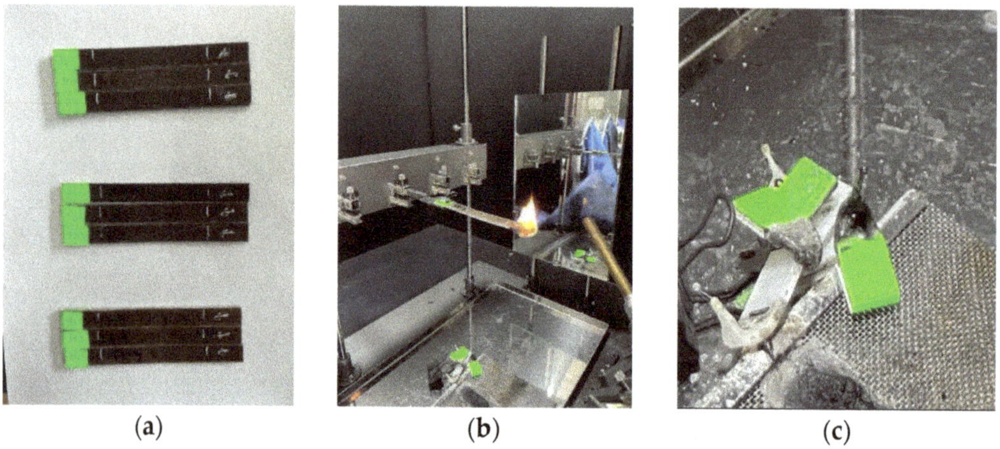

Figure 2. (**a**) Composite hybrid (SP/K/PP); (**b**) Composite hybrid (SP/K/PP) performance in flammability test; (**c**) Sample debris after test 3.

3. Results and Discussions

3.1. Thermogravimetric Analysis (TGA and DTG)

Thermogravimetry is primarily influenced by an accurate heating rate and conditions. Thermogravimetry offers a quantifiable analysis of the amount of moisture and volatile compounds present in fibers, the weight loss, and the thermal breakdown. It also assists in determining the degradation mechanism. Thermogravimetric analysis (TGA) and derivative thermogravimetric (DTG) curves of hybrid composites of sugar palm, kenaf, and polypropylene (SPF/KF/PP) were employed to investigate the thermal stability and decomposition of the polymeric system. As shown in Table 2, weight loss (%) at T_{max} and percentage residual (%) were investigated with different composition weight ratios of hybrid samples.

Table 2. Onset temperature, maximum temperature, weight loss, and maximum temperature and residual at 800 °C, analyzed in TGA analysis.

Samples	T_{on} (°C)	T_{max} (°C)	Weight Loss at T_{max} (wt%)	Char at 800 °C (wt%)
PP	-	439	99.46	0.53
U-SP3K7	274	421.8	93.37	2.02
U-SP5K5	276.1	442.9	91.22	2.357
U-SP7K3	298.6	437.3	90.84	1.86
T-SP3K7	294.2	425.8	92.65	2.1
T-SP5K5	285.8	443.13	85.02	5.22
T-SP7K3	279.4	442.7	83.05	3.196

Pure PP [39], treated, and untreated fibers hybrid composite SPF/KF/PP with varied ratios were compared, as illustrated in Figure 3. The thermogram in Figure 3a shows that at the first quartile degradation, there was a slight weight loss for all SPF/KF/PP hybrid composites ratios. Owing to that, the reduction in percentage weight was due to the release of moisture content in the SPF/KF/PP hybrid composites [40]. In addition, because of the differences in the chemical topology of the fiber components—mostly lignin, hemicellulose, and cellulose—kenaf and sugar palm fibers presumed to decomposes at significant temperatures. In this study, it was confirmed that the thermal degradation of PP kenaf/SP composites had a multi-stepped degradation. The initial transition between 30 and 150 °C indicates the water loss of SPF/KF/PP hybrid composites [41].

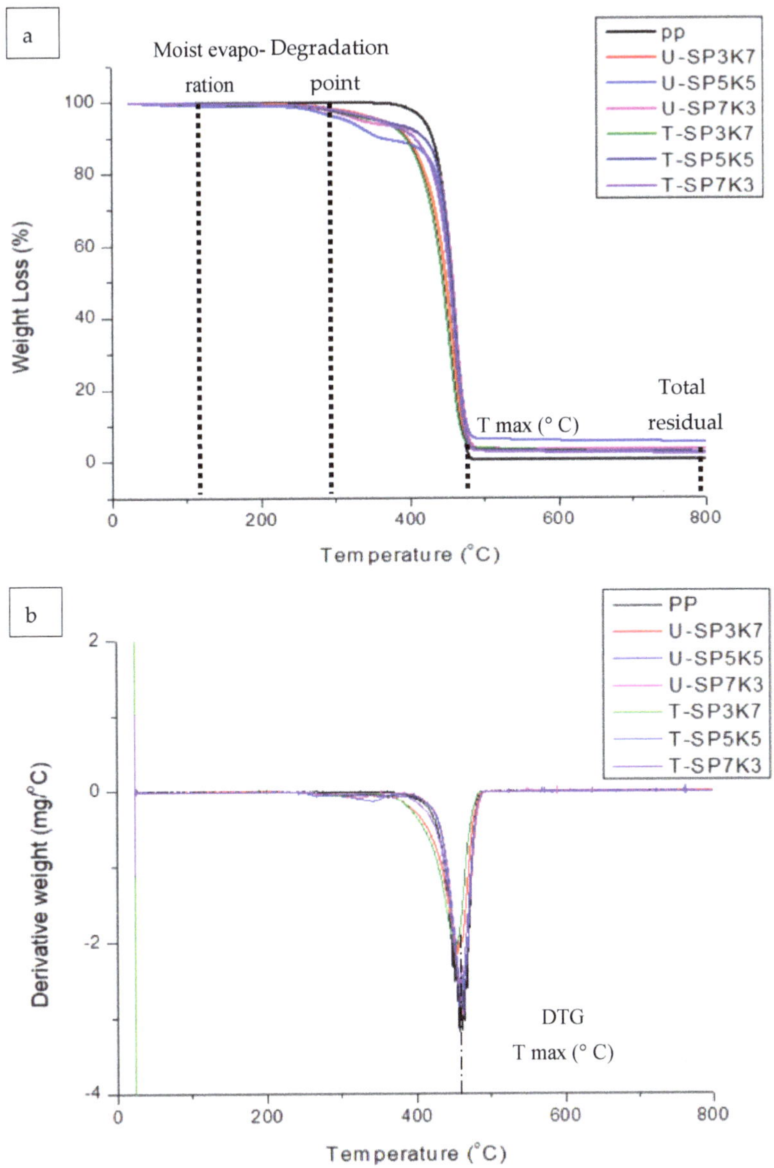

Figure 3. TGA composite hybrid SPF/KF/PP with varied composition: (**a**) Weight loss (%) analysis of SPF/KF/PP; (**b**) Derivative thermogravimetric (DTG) of SPF/KF/PP.

Different natural fibers exhibit different decomposition profiles, as shown in Figure 2. The curves for pure PP are also presented for comparison. For all SPF/KF/PP hybrid composite formulations, the TG curves (Figure 3a) indicate that composites containing kenaf fibers and sugar palm exhibited very similar weight loss trends until the temperature range of 400–450 °C, where the second quartile of degradation took place. During the second quartile, the weight loss can be seen at max, which is when composite tends to decompose at a higher rate. This is explained by the fact that other elements inside the fiber, such as lignin, cellulosic material, and detangled of hydrogen bond of polypropylene, are decomposing [41].

As shown in Table 2, pure polypropylene showed the highest weight loss at T_{max}(wt%), which can be seen at 99.46%, with only 0.53% total residual char at 800 °C. In addition, the untreated hybrid composite with 5 percent of sugar palm and kenaf fiber (U-SP5K5) showed an increment of total residual after 800 °C. However, the highest total residual of SPF/KF/PP hybrid composite T-SP5K5 was recorded at a total weight residue of 15.8% at T_{max} and 5.22% char at 800 °C. First, as shown in TGA analysis, the addition of fibers increased thermal stability by lowering the total weight loss at T_{max}, which can be seen by the weight loss (%) trends at T_{max} for all compositions of the SPF/KF/PP hybrid composite. Furthermore, the benzoylation treatment gave an additional retardancy to the thermal stability [42].

Figure 3b shows the derivative thermogram (DTG) analysis. The DTG curves of the composites reveal that their degradation process occurred in three stages. Figure 3b shows zero degradation at first and the derivative weight decreasing at around 350–400 °C degradation. It starts with the decomposition of the PP, followed by that of the U-SPK and T-SPK. U-SP5K5 showed the least derivative weight loss for untreated hybrid composite. Meanwhile, the best degradation behavior of the SPF and KF hybrid composite was for T-SP5K5. The peak of degradation for all samples showed that organic elements start first to deteriorate. This is explained by the presence of moisture in the fiber and its loss at the first quartile, 100–200 °C. U-SPFKF responds to the decomposition process earlier than the T-SPFKF composites because the benzoylation treatment increased the thermal stability of SPF/KF/PP. The rearrangement of the hydroxyl group after benzoylation treatment inside the cellulosic moieties of the fibers presumes to increase the heat retardancy of the hybrid composites and to slow the thermal degradation [43].

3.2. *Differential Scanning Calorimetry Analysis*

The DSC curves of the hybrid composites SPF/KF/PP are presented in Figure 4. Information on the DSC analysis is listed in Table 3 and discussion of the analysis follows. As observed from the graph in Figure 4, all samples shared comparable values around 95–125 °C, which was due to the loss of moisture from all composites samples. Polypropylene did not show any transition glass temperature, as it is in a crystalline state. In comparison with the treated and untreated composite SPF/KF/PP, the glass transition temperature of all hybrid composite compositions showed a slight peak of transition glass (T_g) temperature. The transition glass temperature was shown to be the highest for the T-SP5K5 hybrid composite at 121.43 °C and a melting point at 161.43 °C, compared with the other composition ratios. According to Phiri, Khoathane, and Sadiku [44], the melting point (T_m) of polymer PP occurs at around 146.43 °C and increases gradually after incorporation of kenaf and sugar palm fibers. Hybrid composites U-SP7K3, U-SP5K5, and U-SP3K7 showed T_m at 149.43, 155.63, and 148.33 °C, respectively. Table 3 also shows the results of treated sugar palm and kenaf fibers with benzoyl treatment, which possessed a better melting point and transition glass temperature.

A significant trend was shown in the results of treated kenaf and sugar palm filler with T-SP7K3, T-SP5K5, and T-SP3K7, exceeding 122.53, 127.63, 125.43 °C, respectively, for their transition glass temperature (T_g). T-SP5K5 achieved the highest values of T_g and T_m at 127.6 and 165.63 °C, which might correspond to the additional increased interaction between matrix and filler and might have led to a restriction in the polymer chain of the composites [45]. A more noticeable effect on the thermal properties of the hybrid composites can be observed through the enthalpy peaks of the DSC curves. All peaks occurred at a T_m approximately in the same temperature range but at different enthalpy intensity, which took place at around 140–185 °C [46]. These results were in good agreement with the above discussion, where the effect of the benzoyl group on the surface of KF and SPF after benzoylation treatment increased the composites' thermal endurance compared with the untreated fibers. In addition, as we have proved in the research, the existence of a T_g peak shows that the hybrid composite SPF/KF/PP in all composition is amorphous [47].

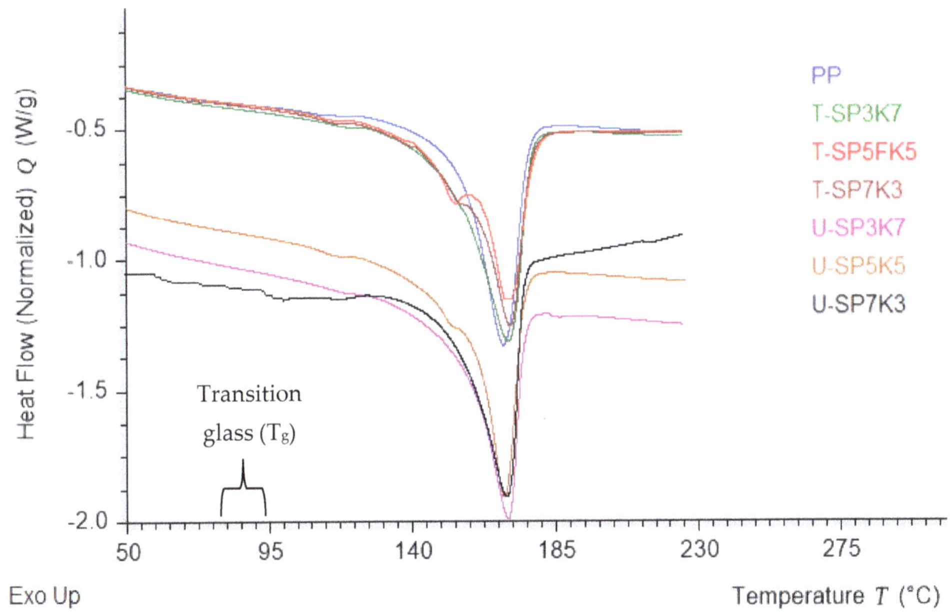

Figure 4. DSC curve of SPF/KF/PP hybrid composite with different composition ratios.

Table 3. Transition glass temperature, melting point temperature, enthalpy, and degree crystallinity of SPF/KF/PP hybrid composite.

Sample	Transition Glass Temperature (T_g)	Melting Point Temperature (T_m)	Enthalpy ΔH (J/g)	Degree Crystallinity (%)
PP	-	146.43	126.11	60.00
U-SP7K3	115.23	149.43	97.460	46.63
U-SP5K5	118.23	155.63	120.95	57.87
U-SP3K7	119.13	148.33	99.160	47.44
T-SP7K3	122.53	160.53	111.23	53.22
T-SP5K5	127.63	165.63	116.78	55.88
T-SP3K7	125.43	161.43	105.89	50.66

The ΔHc (crystallization enthalpy) values of the PP were obtained at 126.7 J/g. The result demonstrates that the ΔHc of composites decreased with the absence of both treated and untreated sugar palm and kenaf fibers. This trend is in agreement with the results of Huda et al., where lower melting temperature and crystallization enthalpies of the composites was observed to decrease with the addition of recycled newspaper cellulose fibers and talc, compared with neat PP [48]. Table 3 also demonstrates that the degree of crystallinity of SP/K/PP composites was lower compared with neat PP, which was below

60%. A significant trend was shown in the results of treated kenaf and sugar palm filler with U-SP7K3, U-SP5K5, and U-SP3K7, which had overall greater crystallinity degree at 46.63%, 57.87%, and 47.44%, compared with treated composites T-SP7K3, T-SP5K5, and T-SP3K7 with values of 53.22%, 52.88%, and 50.66%, respectively. Furthermore, the addition of fibers content in PP resulted in a decrease in the degree of crystallinity of the polymer, which happened due to both treated and untreated fillers kenaf and sugar palm fibers obstructing the mobilization of the PP macromolecular chain and preventing the macromolecular segment from obtaining an ordered alignment of the crystal lattice. Cellulose is also meant to hinder the formation of crystallinity in polymer. Thus, the crystallinity of composites was decreased [49].

3.3. Flammability Analysis (FA)

One of the characterizations of plastics resin is that they can easily flare up when exposed to sufficient heat in the presence of oxygen. Because of the rate of burning for plastics, considerable work has been directed to the study and minimization of the flammability of these materials, such as by the addition of flame retardant chemicals to prevent or minimize the combustion of these materials. A test was done to classify and measure the burning characteristics of the plastics resin. The burning rates of PP and PP composites measured by a horizontal burning test are shown in a bar chart in Figure 5.

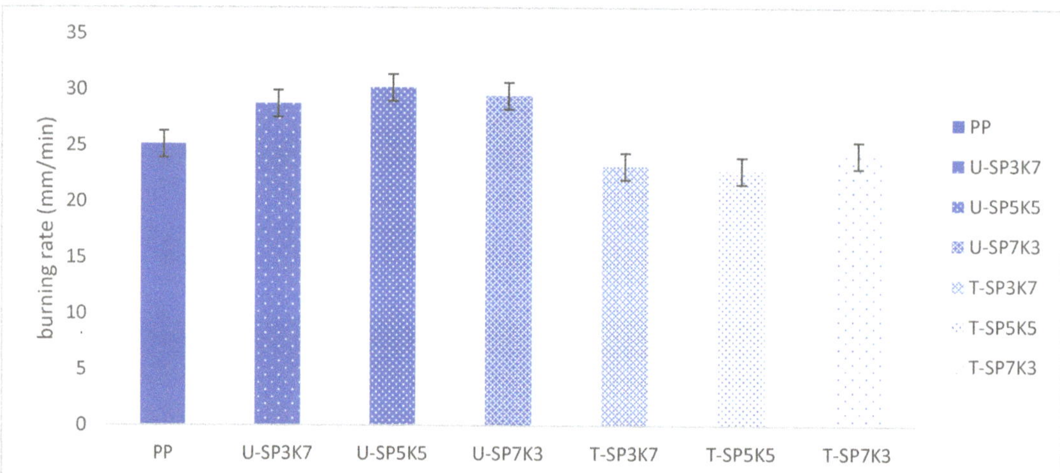

Figure 5. The effect of different fiber loadings on the burning rate of (SP/K/SP) hybrid composites.

Overall, neat PP and untreated and treated kenaf and sugar palm composites U-(SPF/KF) /T-(SPF/KF) showed a burning rate (mm/min) for neat polypropylene of 25.12 (mm) min^{-1}. Most polymer resins, including PP, are extremely flammable. During the burning process, the untreated kenaf and sugar palm hybrid composite formed a non-protective oil layer on the surface of the matrix, serving as an oxygen conductor and permitting heat to penetrate the matrix [50,51]. Therefore, the quantity of decomposed volatiles that escaped the interior polymer matrix was increased, resulting in a shorter burning time and thus increasing the linear burning rate.

With the incorporation of untreated kenaf and sugar palm fibers, a higher burning rate was recorded. Examination shows that composite hybrid samples U-SP3K7, U-SP5K5, and U-SP7K3 achieved 28.72, 30.16, and 29.43 (mm) min^{-1} burning rates, respectively. In these cases, natural fibers are expected to act as combustion sources for the composites. As we know, fibers have high sensitivity towards flame; thus, the incorporation of fibers indeed increased the flammability rate. The high lignin content of kenaf, as compared with some other natural fibers such as flax, hemp, and sisal [48], promotes high heat of combustion and initiates ignition by reducing the thermal stability, which promotes ignition. Generally, kenaf and sugar palm fibers, like other natural fibers, consist of 60–80% cellulose, 5–20% lignin (pectin), and up to 20% moisture [52,53].

However, with the incorporation of benzoyl treatment of palm and kenaf fibers. The flammability rate was reduced consistently. The treated hybrid composites T-SP3K7, T-SP5K5, and T-SP7K3 showed 23.16, 22.76, and 24.16 (mm) min^{-1} burning rate, respectively, which indicates that PP composite incorporating benzoyl-treated fibers has improved flame retardancy properties, compared with the untreated sample of hybrid composites [54]. In this experiment, it can be deduced that the incorporation of untreated fibers increases the burning rate of the fibers and lowers the burning rate of treated fibers.

3.4. Dynamic Mechanical Analysis (DMA)

Typically, DMA is conducted to assess differences in the stiffness, damping, and T_g of polymeric composites during curing [55]. A DMA exhibits the outcomes on storage modulus (E′), which is related to the Young's modulus of the composite. The storage modulus, or E′, is exploited by material researchers to identify the stiffness of a composite. In general, the E′ describes the ability of a material/composite to store energy for the upcoming application [56]. A viscous response of a material/composite is referred to as loss modulus (E″) or dynamic loss modulus [57,58]. E″ establishes output data on the tendency of material/composites to release the applied energy, and it is frequently linked with the term internal friction. E″ is sensitive to distinct types of relaxation processes, morphology, transitions, molecular motions, and other heterogeneities of the material structure. DMA aids material engineers and researchers in estimating the amount of polymer chains immobilized by the filler surface [57]. Figure 6 shows the storage modulus E′(Pa) for pure PP and for treated and untreated U(SPF/KF) /T(SPF/KF). From Table 4, the highest E′ at 20, 40, and 60 °C was recorded at 1360, 1002, and 741 MPa, respectively, which belonged to hybrid composite sample T-SP7K3. In comparison with the hybrid sample with untreated loading fiber U-SP3K7, the lowest E′ was shown at 20, 40, and 60 °C with 1200, 879, and 622 MPa, respectively. The storage modulus trend proposed a reduction in storage modulus proportional to the increase of temperature for all ratios sampled. There were no significant differences in storage modulus (E′) for any of the compositions of hybrid composite (SPF/KF/PP). It is evident that incorporation of kenaf and sugar palm fibers results in an increase in the storage modulus of the biocomposite which reveals the effective stress transfer from the fiber to the matrix at the interface.

Table 4. Hybrid composite sample (SP/K/PP) with storage modulus (MPa).

	Sample with Storage Modulus (MPa)						
Temperature	**PP**	**U-SP3K7**	**U-SP5K5**	**U-SP7K3**	**T-SP3K7**	**T-SP5K5**	**T-SP7K3**
20 °C	1600	1290	1300	1200	1100	1360	1312
40 °C	1200	954	958	879	813	991	968
60 °C	989	701	695	622	601	711	709

The loss modulus E″ (Pa) was examined, which confirmed that alkalization and benzoylation on kenaf fiber aids in increasing the surface area of the fiber via the fibrillation effect, as the process splits the single-fiber bundle into small ones. At higher temperature, due to loss in stiffness of both the fiber and the matrix, the loss modulus drops. It is worth noticing that composites reinforced with benzoyl chloride and NaOH-treated fibers had a lower reduction in the value of E″ when temperature was increased compared both with composites reinforced with untreated fibers and with neat residual, as shown in Figure 7a.

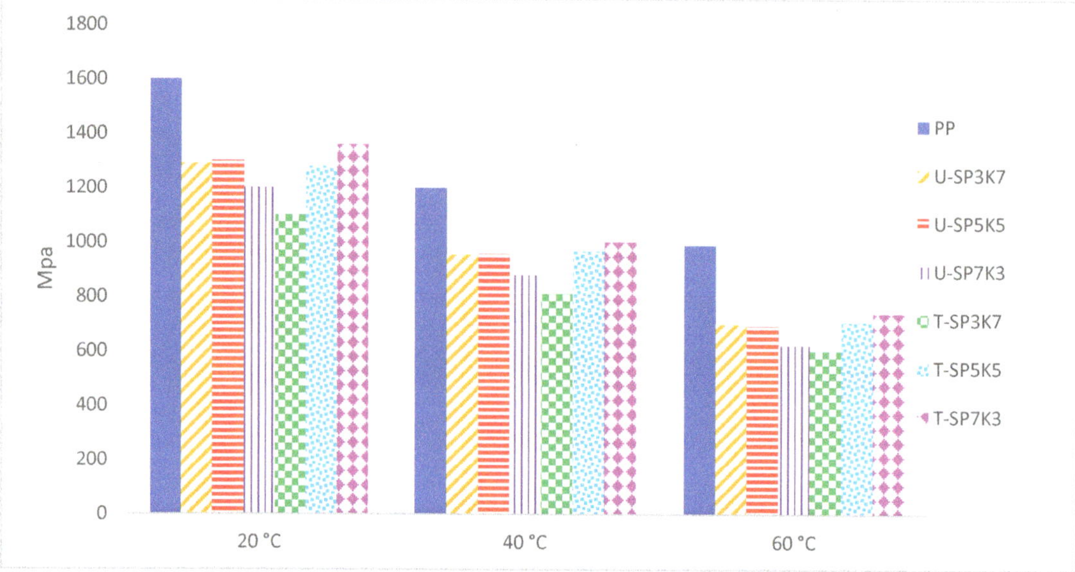

Figure 6. Temperature dependence of the storage moduli (E′) MPa of varied compositions SPF/KF/PP hybrid composite at 20, 40 and 60 °C.

In particular, the loss modulus was tabulated as shown in Table 5. The loss modulus at peak (E″) was determined at peak 13 °C. The loss modulus (E″) of polypropylene (PP) was notably higher than those of the untreated and treated composites SPF/KF/PP at 105.3 MPa. Untreated hybrid composites U-SP7K3, U-SP5K5, and U-SP3K7 showed varied E″ at 81.5, 80.7, and 79.8 MPa, respectively. On the other hand, treated hybrid composites T-SP7K3, T-SP5K5, and T-SP3K7 showed 71.3, 86.2, and 91.2 MPa, respectively. The higher loss modulus for the two composition ratios of treated hybrid composites T-SP5K5 and T-SP3K7 indicates that benzoylation treatment affected their mechanical properties, especially the loss modulus E″ of SPF/KF/PP. Hence, this treatment escalated the effective area for the mechanical interlocking between the two phases of composites, which are fibers (kenaf and sugar palm) and polymer, and subsequently led to increased interfacial loading, which contributed to improved dynamic mechanical properties. Figure 7a shows that T-SP7K3 presented the highest loss modulus at peak, with 91.2 MPa.

Table 5. Loss modulus at peak (E″) (MPa) and damping at peak (Tan δ).

Sample	Loss Modulus at Peak (E″) (MPa)	Damping at Peak (Tan δ) (Pa)
PP	105.3 ± 2.16	0.0617 ± 0.012
U-SP7K3	81.5 ± 1.34	0.0564 ± 0.034
U-SP5K5	80.7 ± 1.14	0.0585 ± 0.041
U-SP3K7	79.8 ± 1.27	0.0572 ± 0.032
T-SP7K3	71.3 ± 1.62	0.0513 ± 0.023
T-SP5K5	86.2 ± 1.06	0.0531 ± 0.041
T-SP3K7	85.2 ± 1.11	0.0529 ± 0.022

In addition, Figure 7b depicts the tan δ curves of the neat PP with all composition ratios of SPF/KF/PP composites. It was observed that incorporation of kenaf and sugar palm fibers led to a pronounced decrease in the maximum value of tan δ. Neat PP showed a damping value at 0.060 Pa. For the incorporation of untreated fibers, U-SP7K3, U-SP5K5, and U-SP3K7 showed damping values of 0.056, 0.058, and 0.057 Pa, respectively.

In comparison with the treated hybrid composite SPF/KF/PP, T-SP7K3, T-SP5K5, and T-SP3K7 achieved damping factors at 0.051, 0.053, and 0.052, respectively.

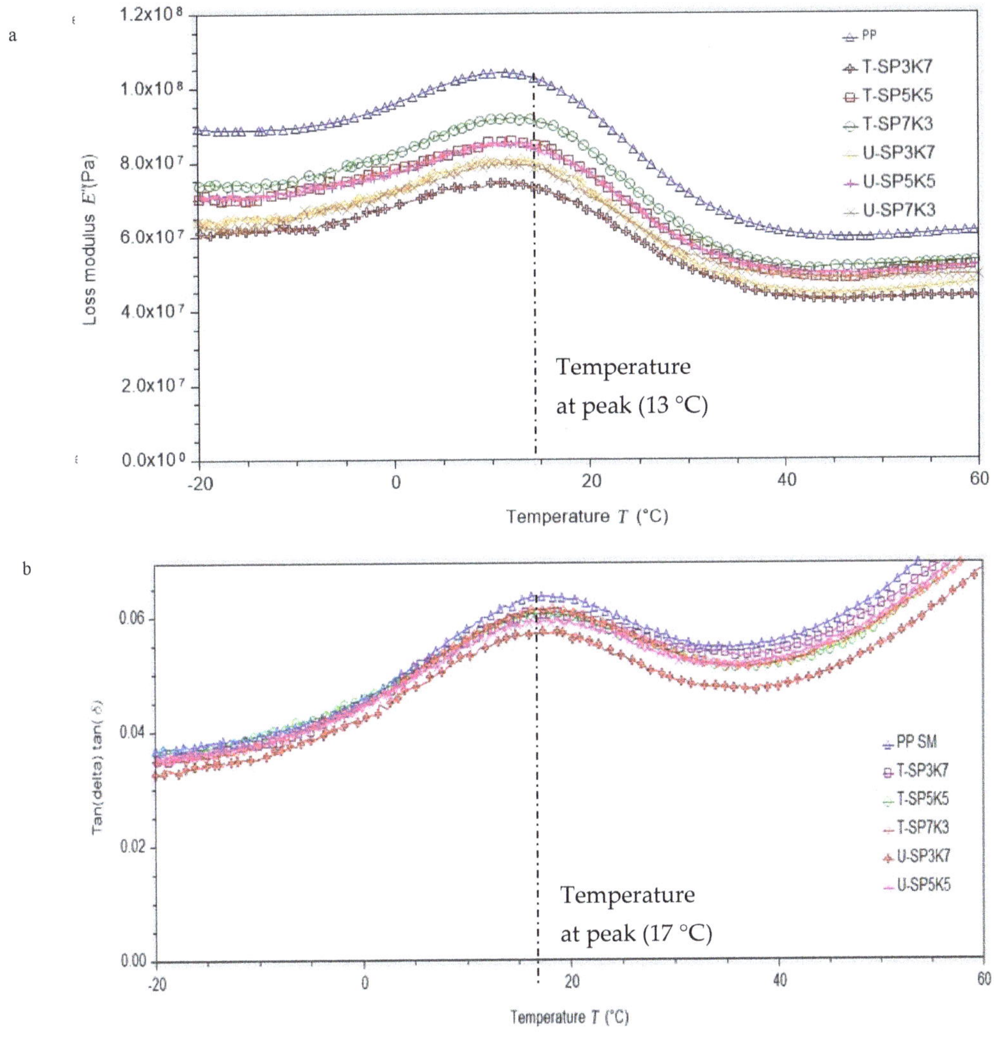

Figure 7. (**a**) Loss modulus at peak E" (Pa) and proportional to the temperature of varied composition of SP/K/SP hybrid composites; (**b**) Loss factor (damping) tan δ at peak in relationship with the temperature of varied compositions of SP/K/SP hybrid composite.

As observed in Figure 7b, the fibers contribution to the damping were low as compared with those of the neat PP matrix. This suggests that the combined attenuation of sugar palm and kenaf fiber reinforced composites would be primarily caused by the molecular motion of PP and the strong interaction between the fibers surface and the matrix interface. Moreover, the removal of the lignin in mercerized fibers led to a change in the extent of hydrogen-bonding, affecting the tan δ in the hybrid composites. Additionally, the width of the tan δ peak was increased in all the biocomposites; this phenomenon can be attributed to molecular relaxations taking place in the composite, which were not present in the matrix.

It shows that the presence of the treated sugar palm and kenaf fibers dramatically reduced tan δ, thus indicating the presence of good adhesion, resulting in low damping. Conversely, the damping of U-SPFKF composites was found to be higher than that of neat PP resin due to weak adhesion between the hydrophilic untreated fibers and the hydrophobic polymer used as the matrix. These results also confirm the good effect of the mercerization performed on the fiber/matrix compatibility, resulting in improved stress transfer and good interfacial adhesion.

3.5. Thermomechanical Analysis (TMA)

Thermomechanical analysis of the pure PP as well as their hybrid composites with treated and untreated sugar palm/kenaf composites were examined at different fiber ratios. They were carried out to explore the dimensional changes or the coefficient of thermal expansion (CTE) in both regions. Due to the stretching and quenching of the composite during fabrication, internal stress in composites was created. In the testing, when an external load was applied to sample in the axial direction with temperature, the porosity of sample started to collapse or 'shrink', and the sample showed three phases of deformities.

The deformities began with the positive strain due to elastic creep. Near or at the T_g of the polymer, in between 45 and 105 °C, the creep strain was recovered, followed by shrinking. As demonstrated in Table 6, thermal expansion coefficient (CTE) after 45 °C shows pure PP with 10.23. In addition, untreated reinforced composite U-SP7K3, U-SP5K5, and U-SP3K7 were shown CTE at 1.13, 3.21, and 2.14 respectively. In addition, the treated reinforced-composite, T-SP7K3, T-SP5K5, and T-SP3K7 were shown CTE at 1.23–7.32 and 6.14. Overall, the small amount of CTE in both treated and untreated composite portrayed low CTE value as the transition state and moisture evaporation which hindered an extreme expansion. On the other hand after 105° C demonstrate U-SP7K3, U-SP5K5, and U-SP3K7 with 11.31, 24.93 and 24.74 value of CTE respectively. However, T-SP7K3, T-SP5K5, and T-SP3K7 shown a huge expansion at 12.74, 30.11, and 18.23 respectively. Due to the melting of the polymer composite, (SPF/KF/PP), CTE were shown higher at 105°

Table 6. Thermal expansion after 45 °C and 105 °C.

Samples	Thermal Expansion (CTE) after 45 °C	Thermal Expansion (CTE) after 105 °C
PP	10.21 ± 0.12	31.23 ± 1.82
U-SP7K3	1.13 ± 0.12	11.31 ± 0.49
U-SP5K5	3.21 ± 0.14	24.93 ± 0.74
U-SP3K7	2.14 ± 0.41	24.74 ± 0.45
T-SP7K3	1.23 ± 0.53	12.74 ± 0.61
T-SP5K5	7.32± 0.81	30.11 ± 0.43
T-SP3K7	6.14 ± 0.73	18.23 ± 0.72

Highly cross-linked polymers and the large number of stretched tie chains contributed to the high modulus of elasticity and reversible (recoverable) deformation [59,60]. The negative strain showed pore shrinkage, termed $T_{deformation}$ at $T_{deformation}$, and strain induced necking began and propagated along the drawing direction until the sample ruptured. From the TMA graphs, it may be noted that the pure PP and untreated and treated duo fiber composites showed different patterns of TMA curves. Figure 8 demonstrates that the glass transition temperature could not be detected for the neat PP, and the curves showed a steep drop for the untreated hybrid composites, which is associated with the low cross-linking in these materials compared with the treated SPF/KF/PP [61]. SPF/KF/PP composite turned out to have highly mobile materials in the rubbery stage. Benzoyl-treated sugar palm- and kenaf-reinforced composites revealed better interfacial bonding between the PP matrix, causing better surface adhesion and cross-linking in composites. Compared with the untreated fibers, the hybrid composite demonstrated a lower dimensional μm change due to the poor compatibility between untreated fibers and the polymer matrix.

The curve of the T-SP7K3 hybrid composite showed rigidity in the rubbery region, which is an indication of the high degree of cross-linking of fiber [62]. The obtained results also demonstrate better mechanical properties of the T-SP7K3 hybrid composite.

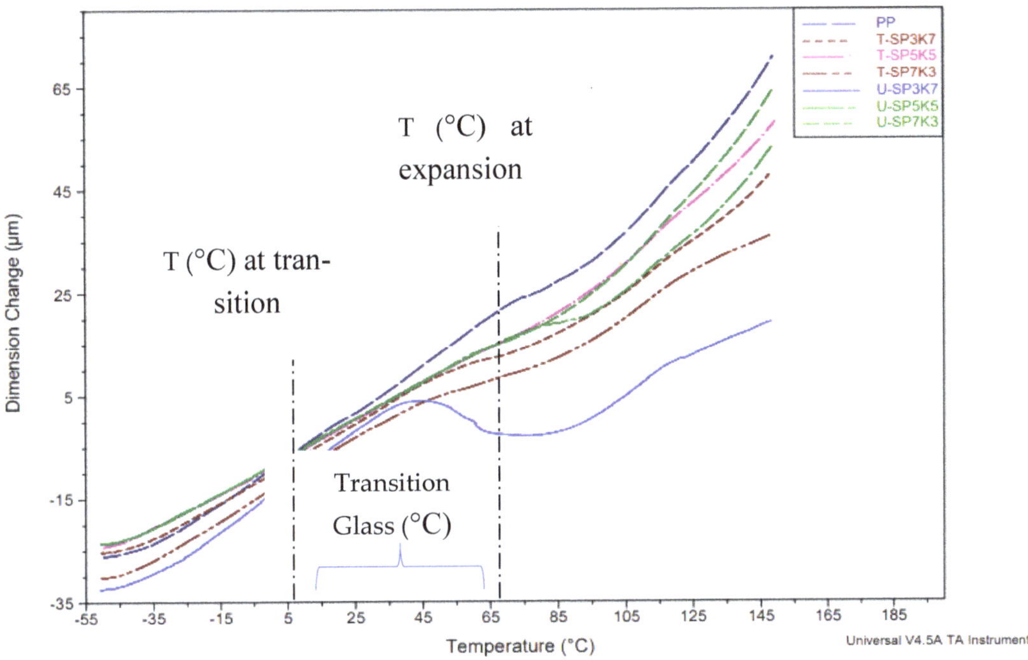

Figure 8. Thermogram of the dimension change in varied compositions of SP/K/SP hybrid composites.

4. Conclusions

This research investigated the thermal stability and the dynamic and thermomechanical properties of treated and untreated sugar palm- and kenaf fibers-reinforced polypropylene composites and found that composition T-SPF/KF/PP showed better thermal stability in comparison with all untreated SPF/KF/PP hybrid composite ratios. After incorporation with treated kenaf and sugar palm fiber, thermal properties of the hybrid composites were improved. The lowest weight loss wt% at T_{max} was for hybrid composite T-SP5K5, with 85.02% total residue char at 800 °C (wt%) recorded at 5.22%. The benzoylation treatment towards the fibers gave a good interfacial bonding, where the polymer acted as a barrier to prevent the degradation of the natural fibers. These results were in good agreement with the above discussion, where the effect of the benzoyl group on the surface of KF and SPF after benzoylation treatment increased the thermal stability of the composites, compared with the untreated composites. A flammability test it showed a reduction in the burning rate (mm/min) with the incorporation of treated fibers in the SPF/KF/PP hybrid composite. T-SP5K5 was determined to achieve the lowest burning rate at 22.53 (mm/min). In addition, as proved by the DSC curve, the existence of a T_g peak showed that the hybrid composite SPF/KF/PP in all compositions was amorphous, except for neat polypropylene, which showed no transition glass peak. The DSC curve also confirmed that the highest transition glass and melting point was for T-SP5K5, with a T_g at 127.63 and a T_m at 165.63.

Storage modulus analysis showed that hybrid composite T-SP5K5 at 20, 40, and 60 °C with 1360, 991, and 711 MPa (SPF/KF/PP) with benzoylation treatment, respectively, showed the highest storage modulus (E′). Loss modulus (E″) and damping tan δ at peak rating showed that the incorporation of fibers into a polymer restricts the mobility of the polymer chains, leading to lower flexibility, which ultimately decreases the damping characteristics. T-SP5K5 showed an E″ and tan δ of 86.2 MPa and 0.0531 Pa, respectively. Furthermore, T-SP5K5 showed coefficients of thermal expansion (CTE) values after 45 and 105 °C of 7.323 and 30.11, respectively.

Author Contributions: Conceptualization, S.M.I. and S.M.S.; methodology, S.M.I.; software, S.M.I. and M.Y.M.Z.; validation, S.M.S. and A.R.M.; formal analysis, S.M.I. and M.Y.M.Z.; investigation, S.M.I.; resources, S.M.I. and S.M.S.; data curation, S.M.I.; writing—original draft preparation, S.M.I.; writing—review and editing, S.M.I. and M.Y.M.Z.; visualization, S.M.I.; supervision, S.M.S., M.Y.M.Z. and A.R.M.; project administration, S.M.S.; funding acquisition, S.M.S. All authors have read and agreed to the published version of the manuscript.

Funding: The authors would like to thank the Universiti Putra Malaysia for the financial support provided through the Universiti Putra Malaysia Grant Putra Berimpak (vote number 9679800) and through Grant Inisiatif Putra Siswazah (vote number 9663200), as well as thanking the Ministry of Higher Education (MOHE) for providing the award to the principal author in this project.

Institutional Review Board Statement: Not applicable.

Informed Consent Statement: Not applicable.

Data Availability Statement: The data presented in this study are available on request from the corresponding author.

Acknowledgments: The authors would like to thank the Universiti Putra Malaysia for the financial support provided through the Universiti Putra Malaysia Grant Putra Berimpak (vote number 9679800) and through Grant Inisiatif Putra Siswazah (vote number 9663200), as well as thanking the Ministry of Higher Education (MOHE) for providing the award to the principal author in this project.

Conflicts of Interest: The authors declare no conflict of interest.

References

1. Sreenivas, H.T.; Krishnamurthy, N.; Murali, M.S.; Arpitha, G.R. Influence of stacking sequence and orientation of the fabric on mechanical properties of twill Kenaf/Kevlar reinforced unsaturated polyester hybrid composites. *J. Ind. Text.* **2021**, *29*, 179–189.
2. Aisyah, H.A.; Paridah, M.T.; Sapuan, S.M.; Ilyas, R.A.; Khalina, A.; Nurazzi, N.M.; Lee, S.H.; Lee, C.H. A comprehensive review on advanced sustainable woven natural fibre polymer composites. *Polymers* **2021**, *13*, 471. [CrossRef] [PubMed]
3. Baihaqi, N.N.; Khalina, A.; Nurazzi, N.M.; Aisyah, H.A.; Sapuan, S.M.; Ilyas, R.A. Effect of fiber content and their hybridization on bending and torsional strength of hybrid epoxy composites reinforced with carbon and sugar palm fibers. *Polimery* **2021**, *66*, 36–43. [CrossRef]
4. Singh, K.; Das, D.; Nayak, R.K.; Khandai, S.; Kumar, R.; Routara, B.C. Effect of silanizion on mechanical and tribological properties of kenaf-carbon and kenaf-glass hybrid polymer composites. *Mater. Today Proc.* **2020**, *26*, 2094–2098. [CrossRef]
5. Alsubari, S.; Zuhri, M.Y.M.; Sapuan, S.M.; Ishak, M.R.; Ilyas, R.A.; Asyraf, M.R.M. Potential of natural fiber reinforced polymer composites in sandwich structures: A review on its mechanical properties. *Polymers* **2021**, *13*, 423. [CrossRef] [PubMed]
6. Sabaruddin, F.A.; Paridah, M.T.; Sapuan, S.M.; Ilyas, R.A.; Lee, S.H.; Abdan, K.; Mazlan, N.; Roseley, A.S.M.; Abdul Khalil, H.P.S. The Effects of Unbleached and Bleached Nanocellulose on the Thermal and Flammability of Polypropylene-Reinforced Kenaf Core Hybrid Polymer Bionanocomposites. *Polymers* **2021**, *13*, 116. [CrossRef] [PubMed]
7. Nazrin, A.; Sapuan, S.M.; Zuhri, M.Y.M.; Ilyas, R.A.; Syafiq, R.; Sherwani, S.F.K. Nanocellulose reinforced thermoplastic starch (TPS), polylactic acid (PLA), and polybutylene succinate (PBS) for food packaging applications. *Front. Chem.* **2020**, *8*, 213. [CrossRef]
8. Rashid, M.; Chetehouna, K.; Cablé, A.; Gascoin, N. Analysing Flammability Characteristics of Green Biocomposites: An Overview. *Fire Technol.* **2021**, *57*, 31–67. [CrossRef]
9. Sukumaran, N.P.; Gopi, S. Overview of biopolymers: Resources, demands, sustainability, and life cycle assessment modeling and simulation. In *Biopolymers and Their Industrial Applications*; Elsevier: Amsterdam, The Netherlands, 2021; pp. 1–19.
10. Edhirej, R.A.; Aisyah, H.A.; Ibrahim, R.; Atikah, M.S.N.; Salwa, H.N.; Aung, M.M.; SaifulAzry, S.O.A.; Megashah, L.N.; Ainun, Z.M.A. Renewable Sources for Packaging Materials. In *Bio-Based Packaging: Material, Environmental and Economic Aspects*; John Wiley & Sons: Hoboken, NJ, USA, 2021; pp. 353–370. [CrossRef]
11. Edhirej, A.; Sapuan, S.M.; Jawaid, M.; Zahari, N.I. Preparation and characterization of cassava bagasse reinforced thermoplastic cassava starch. *Fibers Polym.* **2017**, *18*, 162–171. [CrossRef]
12. Thomas, S.K.; Parameswaranpillai, J.; Krishnasamy, S.; Begam, P.M.S.; Nandi, D.; Siengchin, S.; George, J.J.; Hameed, N.; Salim, N.V.; Sienkiewicz, N. A comprehensive review on cellulose, chitin, and starch as fillers in natural rubber biocomposites. *Carbohydr. Polym. Technol. Appl.* **2021**, *2*, 100095.

13. Adeyeye, O.A.; Sadiku, E.R.; Reddy, A.B.; Ndamase, A.S.; Makgatho, G.; Sellamuthu, P.S.; Perumal, A.B. The use of biopolymers in food packaging. In *Green Biopolymers and Their Nanocomposites*; Springer: Singapore, 2019; pp. 137–158.
14. Ilyas, R.A.; Sapuan, S.M.; Ibrahim, R.; Abral, H.; Ishak, M.R.; Zainudin, E.S.; Atikah, M.S.N. Effect of sugar palm nanofibrillated cellulose concentrations on morphological, mechanical and physical properties of biodegradable films based on agro-waste sugar palm (*Arenga pinnata* (Wurmb.) *Merr*) starch. *J. Mater. Res. Technol.* **2019**, *8*, 4819–4830. [CrossRef]
15. Verma, A.; Gaur, A.; Singh, V.K. Mechanical properties and microstructure of starch and sisal fiber biocomposite modified with epoxy resin. *Mater. Perform. Charact.* **2017**, *6*, 500–520. [CrossRef]
16. Amin, U.; Khan, M.U.; Majeed, Y.; Rebezov, M.; Khayrullin, M.; Bobkova, E.; Shariati, M.A.; Chung, I.M.; Thiruvengadam, M. Potentials of polysaccharides, lipids and proteins in biodegradable food packaging applications. *Int. J. Biol. Macromol.* **2021**, *183*, 2148–2198. [CrossRef]
17. Villadiego, K.M.; Tapia, M.J.A.; Useche, J.; Macías, D.E. Thermoplastic Starch (TPS)/Polylactic Acid (PLA) Blending Methodologies: A Review. *J. Polym. Environ.* **2021**, *323*, 1–17.
18. Akil, H.; Omar, M.F.; Mazuki, A.A.M.; Safiee, S.Z.A.M.; Ishak, Z.A.M.; Bakar, A.A. Kenaf fiber reinforced composites: A review. *Mater. Des.* **2011**, *32*, 4107–4121. [CrossRef]
19. Bledzki, A.K.; Gassan, J. Composites reinforced with cellulose based fibres. *Prog. Polym. Sci.* **1999**, *24*, 221–274. [CrossRef]
20. Saad, M.J.; Kamal, I. Kenaf core particleboard and its sound absorbing properties. *J. Sci. Technol.* **2012**, *4*, 23–34.
21. Asdrubali, F. Survey on the acoustical properties of new sustainable materials for noise control. In Proceedings of the Euronoise, Tampere, Finland, 30 May–1 June 2006; Volume 30, pp. 1–10.
22. Fu, S.-Y.; Xu, G.; Mai, Y. On the elastic modulus of hybrid particle/short-fiber/polymer composites. *Compos. Part. B Eng.* **2002**, *33*, 291–299. [CrossRef]
23. Sahari, J.; Sapuan, S.M.; Zainudin, E.S.; Maleque, M.A. Thermo-mechanical behaviors of thermoplastic starch derived from sugar palm tree (*Arenga pinnata*). *Carbohydr. Polym.* **2013**, *92*, 1711–1716. [CrossRef]
24. Lumingkewas, R.H.; Setyadi, R.; Yanita, R.; Akbar, S.; Yuwono, A.H. Tensile behavior of composite concrete reinforced sugar palm fiber. In *Key Engineering Materials*; Trans Tech Publications Ltd.: Stafa-Zurich, Switzerland, 2018; Volume 777, pp. 471–475.
25. Ishak, M.R.; Sapuan, S.M.; Leman, Z.; Rahman, M.Z.A.; Anwar, U.M.K.; Siregar, J.P. Sugar palm (*Arenga pinnata*): Its fibres, polymers and composites. *Carbohydr. Polym.* **2013**, *91*, 699–710. [CrossRef]
26. Syafiq, R.; Sapuan, S.M.; Zuhri, M.R.M. Antimicrobial activity, physical, mechanical and barrier properties of sugar palm based nanocellulose/starch biocomposite films incorporated with cinnamon essential oil. *J. Mater. Res. Technol.* **2021**, *11*, 144–157. [CrossRef]
27. Sathishkumar, T.P.; Naveen, J.; Satheeshkumar, S. Hybrid fiber reinforced polymer composites–a review. *J. Reinf. Plast. Compos.* **2014**, *33*, 454–471. [CrossRef]
28. Devi, L.U.; Bhagawan, S.S.; Thomas, S. Dynamic mechanical analysis of pineapple leaf/glass hybrid fiber reinforced polyester composites. *Polym. Compos.* **2010**, *31*, 956–965. [CrossRef]
29. Nikmatin, S.; Syafiuddin, A.; Hong Kueh, A.B.; Maddu, A. Physical, thermal, and mechanical properties of polypropylene composites filled with rattan nanoparticles. *J. Appl. Res. Technol.* **2017**, *15*, 386–395. [CrossRef]
30. Bachtiar, D.; Siregar, J.P.; bin Sulaiman, A.S.; bin Mat Rejab, M.R. Tensile Properties of Hybrid Sugar Palm/Kenaf Fibre Reinforced Polypropylene Composites. *Appl. Mech. Mater.* **2015**, *695*, 155–158. [CrossRef]
31. Izwan, S.M.; Sapuan, S.M.; Zuhri, M.Y.M.; Mohamed, A.R. Effects of benzoyl treatment on NaOH treated sugar palm fiber: Tensile, thermal, and morphological properties. *J. Mater. Res. Technol.* **2020**, *9*, 5805–5814. [CrossRef]
32. Clyne, T.W.; Hull, D. *An Introduction to Composite Materials*; University of Cambridge UK: Cambridge, UK, 2019.
33. Safri, S.N.A.; Sultan, M.T.H.; Saba, N.; Jawaid, M. Effect of benzoyl treatment on flexural and compressive properties of sugar palm/glass fibres/epoxy hybrid composites. *Polym. Test.* **2018**, *71*, 362–369. [CrossRef]
34. Kushwaha, P.K.; Kumar, R. Influence of chemical treatments on the mechanical and water absorption properties of bamboo fiber composites. *J. Reinf. Plast. Compos.* **2011**, *30*, 73–85. [CrossRef]
35. Mittal, V.; Saini, R.; Sinha, S. Natural fiber-mediated epoxy composites–A review. *Compos. Part. B Eng.* **2016**, *99*, 425–435. [CrossRef]
36. Singha, A.S.; Rana, A.K. A study on benzoylation and graft copolymerization of lignocellulosic cannabis indica fiber. *J. Polym. Environ.* **2012**, *20*, 361–371. [CrossRef]
37. Asim, M.; Paridah, M.T.; Chandrasekar, M.; Shahroze, R.M.; Jawaid, M.; Nasir, M.; Siakeng, R. Thermal stability of natural fibers and their polymer composites. *Iran. Polym. J.* **2020**, *29*, 625–648. [CrossRef]
38. Pilarska, A.; Bula, K.; Myszka, K.; Rozmanowski, T.; Szwarc-Rzepka, K.; Pilarski, K.; Chrzanowski, Ł.; Czaczyk, K.; Jesionowski, T. Functional polypropylene composites filled with ultra-fine magnesium hydroxide. *Open Chem.* **2015**, *13*, 161. [CrossRef]
39. Abbas-Abadi, M.S. The effect of process and structural parameters on the stability, thermo-mechanical and thermal degradation of polymers with hydrocarbon skeleton containing PE, PP, PS, PVC, NR, PBR and SBR. *J. Therm. Anal. Calorim.* **2021**, *143*, 2867–2882. [CrossRef]
40. Nazrin, A.; Sapuan, S.M.; Zuhri, M.Y.M. Mechanical, Physical and Thermal Properties of Sugar Palm Nanocellulose Reinforced Thermoplastic Starch (TPS)/Poly (Lactic Acid) (PLA) Blend Bionanocomposites. *Polymers* **2020**, *12*, 2216. [CrossRef]

41. Saba, N.; Safwan, A.; Sanyang, M.L.; Mohammad, F.; Pervaiz, M.; Jawaid, M.; Alothman, O.Y.; Sain, M. Thermal and dynamic mechanical properties of cellulose nanofibers reinforced epoxy composites. *Int. J. Biol. Macromol.* **2017**, *102*, 822–828. [CrossRef] [PubMed]
42. Khan, T.; Sultan, M.T.H.; Jawaid, M.; Safri, S.N.A.; Shah, A.U.M.; Majid, M.S.A.; Zulkepli, N.N.; Jaya, H. The Effects of Stacking Sequence on Dynamic Mechanical Properties and Thermal Degradation of Kenaf/Jute Hybrid Composites. *J. Renew. Mater.* **2021**, *9*, 73–84. [CrossRef]
43. Veerasimman, A.; Shanmugam, V.; Rajendran, S.; Johnson, D.J.; Subbiah, A.; Koilpichai, J.; Marimuthu, U. Thermal Properties of Natural Fiber Sisal Based Hybrid Composites–A Brief Review. *J. Nat. Fibers* **2021**, 1–11. [CrossRef]
44. Phiri, G.; Khoathane, M.C.; Sadiku, E.R. Effect of fibre loading on mechanical and thermal properties of sisal and kenaf fibre-reinforced injection moulded composites. *J. Reinf. Plast. Compos.* **2014**, *33*, 283–293. [CrossRef]
45. Gargol, M.; Klepka, T.; Klapiszewski, Ł.; Podkościelna, B. Synthesis and Thermo-Mechanical Study of Epoxy Resin-Based Composites with Waste Fibers of Hemp as an Eco-Friendly Filler. *Polymers* **2021**, *13*, 503. [CrossRef] [PubMed]
46. Ghori, S.W.; Rao, G.S. Mechanical and thermal properties of date palm/kenaf fiber-reinforced epoxy hybrid composites. *Polym. Compos.* **2021**, *42*, 2217–2224. [CrossRef]
47. Ismail, N.F.; Mohd Radzuan, N.A.; Sulong, A.B.; Muhamad, N.; Che Haron, C.H. The Effect of Alkali Treatment on Physical, Mechanical and Thermal Properties of Kenaf Fiber and Polymer Epoxy Composites. *Polymers* **2021**, *13*, 2005. [CrossRef]
48. Kathirselvam, M.; Kumaravel, A.; Arthanarieswaran, V.P.; Saravanakumar, S.S. Characterization of Cellulose Fibers in Thespesia Populnea Barks: Influence of Alkali Treatment. *Carbohydr. Polym.* **2019**, *217*, 178–189. [CrossRef]
49. AlMaadeed, M.A.; Kahraman, R.; Khanam, P.N.; Madi, N. Date Palm Wood Flour/Glass Fibre Reinforced Hybrid Composites of Recycled Polypropylene: Mechanical and Thermal Properties. *Mater. Des.* **2012**, *42*, 289–294. [CrossRef]
50. Zhang, Z.X.; Zhang, J.; Lu, B.-X.; Xin, Z.X.; Kang, C.K.; Kim, J.K. Effect of flame retardants on mechanical properties, flammability and foamability of PP/wood-fiber composites. *Compos. Part. B Eng.* **2012**, *43*, 150–158. [CrossRef]
51. Nishimura, N.; Izumi, A.; Kuroda, K. Structural characterization of kenaf lignin: Differences among kenaf varieties. *Ind. Crop. Prod.* **2002**, *15*, 115–122. [CrossRef]
52. Li, B.; He, J. Investigation of mechanical property, flame retardancy and thermal degradation of LLDPE-wood-fibre composites. *Polym. Degrad. Stab.* **2004**, *83*, 241–246. [CrossRef]
53. Swolfs, Y.; Gorbatikh, L.; Verpoest, I. Fibre Hybridisation in Polymer Composites: A Review Part A Applied Science and Manufacturing. *Composites* **2014**, *67*, 181–200. [CrossRef]
54. Suriani, M.J.; Sapuan, S.M.; Ruzaidi, C.M.; Nair, D.S.; Ilyas, R.A. Flammability, morphological and mechanical properties of sugar palm fiber/polyester yarn-reinforced epoxy hybrid biocomposites with magnesium hydroxide flame retardant filler. *Text. Res. J.* **2021**. [CrossRef]
55. Anuar, H.; Zuraida, A. Thermal properties of injection moulded polylactic acid–kenaf fibre biocomposite. *Malays. Polym. J.* **2011**, *6*, 51–57.
56. Majhi, S.K.; Sanjay, N.K.; Mohanty, S.; Unnikrishnan, L. Mechanical and fracture behavior of banana fiber reinforced Polylactic acid biocomposites. *Int. J. Plast. Technol.* **2010**, *14*, 57–75. [CrossRef]
57. Jawaid, M.; Khalil, H.P.S.A.; Hassan, A.; Dungani, R.; Hadiyane, A. Effect of jute fibre loading on tensile and dynamic mechanical properties of oil palm epoxy composites. *Compos. Part. B Eng.* **2013**, *45*, 619–624. [CrossRef]
58. Behazin, E.; Misra, M.; Mohanty, A.K. Sustainable biocomposites from pyrolyzed grass and toughened polypropylene: Structure-property relationships. *ACS Omega* **2017**, *2*, 2191–2199. [CrossRef]
59. Asyraf, M.R.M.; Rafidah, M.; Azrina, A.; Razman, M.R. Dynamic mechanical behaviour of kenaf cellulosic fibre biocomposites: A comprehensive review on chemical treatments. *Cellulose* **2021**, *28*, 2675–2695. [CrossRef]
60. Lee, S.-Y.; Park, S.Y.; Song, H.S. Lamellar crystalline structure of hard elastic HDPE films and its influence on microporous membrane formation. *Polymer* **2006**, *47*, 3540–3547. [CrossRef]
61. Ju, J.W.; Lee, H.-K. A micromechanical damage model for effective elastoplastic behavior of partially debonded ductile matrix composites. *Int. J. Solids Struct.* **2001**, *38*, 6307–6332. [CrossRef]
62. Saba, N.; Jawaid, M. A review on thermomechanical properties of polymers and fibers reinforced polymer composites. *J. Ind. Eng. Chem.* **2018**, *67*, 1–11. [CrossRef]

polymers

MDPI

Review

Recent Trends and Developments in Conducting Polymer Nanocomposites for Multifunctional Applications

Shubham Sharma [1,2,*], P. Sudhakara [1,*], Abdoulhdi A. Borhana Omran [3,4,*], Jujhar Singh [5] and R. A. Ilyas [6,7]

1 Regional Centre for Extension and Development, CSIR-Central Leather Research Institute, Leather Complex, Kapurthala Road, Jalandhar 144021, Punjab, India
2 PhD Research Scholar, IK Gujral Punjab Technical University, Jalandhar-Kapurthala, Highway, VPO, Ibban 144603, Punjab, India
3 Department of Mechanical Engineering, College of Engineering, Universiti Tenaga Nasional, Jalan Ikram-Uniten, Kajang 43000, Selangor, Malaysia
4 Department of Mechanical Engineering, College of Engineering Science & Technology, Sebha University, Sabha 00218, Libya
5 Department of Mechanical Engineering, IK Gujral Punjab Technical University, Jalandhar-Kapurthala, Highway, VPO, Ibban 144603, Punjab, India; jujharsingh2085@gmail.com
6 School of Chemical and Energy Engineering, Faculty of Engineering, Universiti Teknologi Malaysia, Johor Bahru 81310, Johor, Malaysia; ahmadilyas@utm.my
7 Centre for Advanced Composite Materials, Universiti Teknologi Malaysia, Johor Bahru 81310, Johor, Malaysia
* Correspondence: shubham543sharma@gmail.com or shubhamsharmacsirclri@gmail.com (S.S.); sudhakarp@clri.res.in (P.S.); amhmad@uniten.edu.my (A.A.B.O.); Tel.: +91-7009239473 (S.S.)

Abstract: Electrically-conducting polymers (CPs) were first developed as a revolutionary class of organic compounds that possess optical and electrical properties comparable to that of metals as well as inorganic semiconductors and display the commendable properties correlated with traditional polymers, like the ease of manufacture along with resilience in processing. Polymer nanocomposites are designed and manufactured to ensure excellent promising properties for anti-static (electrically conducting), anti-corrosion, actuators, sensors, shape memory alloys, biomedical, flexible electronics, solar cells, fuel cells, supercapacitors, LEDs, and adhesive applications with desired-appealing and cost-effective, functional surface coatings. The distinctive properties of nanocomposite materials involve significantly improved mechanical characteristics, barrier-properties, weight-reduction, and increased, long-lasting performance in terms of heat, wear, and scratch-resistant. Constraint in availability of power due to continuous depletion in the reservoirs of fossil fuels has affected the performance and functioning of electronic and energy storage appliances. For such reasons, efforts to modify the performance of such appliances are under way through blending design engineering with organic electronics. Unlike conventional inorganic semiconductors, organic electronic materials are developed from conducting polymers (CPs), dyes and charge transfer complexes. However, the conductive polymers are perhaps more bio-compatible rather than conventional metals or semi-conductive materials. Such characteristics make it more fascinating for bio-engineering investigators to conduct research on polymers possessing antistatic properties for various applications. An extensive overview of different techniques of synthesis and the applications of polymer bio-nanocomposites in various fields of sensors, actuators, shape memory polymers, flexible electronics, optical limiting, electrical properties (batteries, solar cells, fuel cells, supercapacitors, LEDs), corrosion-protection and biomedical application are well-summarized from the findings all across the world in more than 150 references, exclusively from the past four years. This paper also presents recent advancements in composites of rare-earth oxides based on conducting polymer composites. Across a variety of biological and medical applications, the fact that numerous tissues were receptive to electric fields and stimuli made CPs more enticing.

Keywords: biomedical; conducting polymers; corrosion; doped; electronics; shape memory polymers; sensors; actuators; optical limiting

Citation: Sharma, S.; Sudhakara, P.; Omran, A.A.B.; Singh, J.; Ilyas, R.A. Recent Trends and Developments in Conducting Polymer Nanocomposites for Multifunctional Applications. *Polymers* **2021**, *13*, 2898. https://doi.org/10.3390/polym13172898

Academic Editors: Emin Bayraktar

Received: 26 July 2021
Accepted: 10 August 2021
Published: 28 August 2021

1. Introduction

Approximately three decades ago, intrinsically conducting polymers were discovered and this discovery drew the attention of researchers because of countless applications of these polymers in the scientific field. These are also called synthetic metals as their electrical conductivity is very high, similar to those of metals. Examples of various conducting polymers (CPs) are polyacetylene, poly furan, polypyrrole, and polythiophene, which are insulators in their neutral state, as illustrated in Figure 1. The insulating behavior of polymers can be converted into conducting by carrying out doping of different salts by chemical and electrochemical redox reactions.

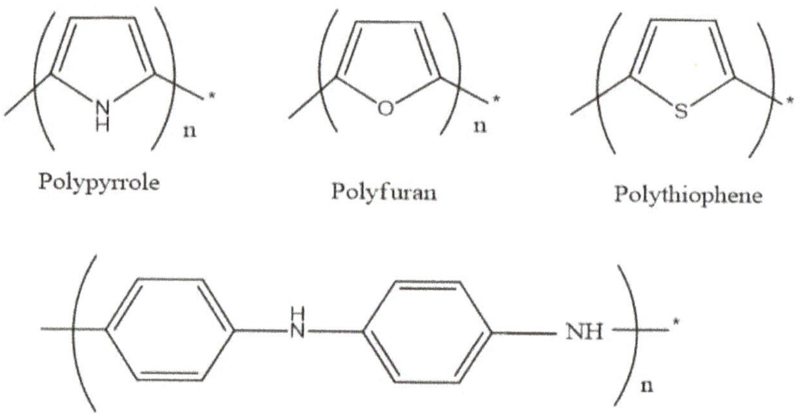

Figure 1. Intrinsically conducting polymers. Reproduced with permission from [1–19].

The highly conducting polysulfur nitride $[SN]_x$ was discovered by Walatka et al. in 1973 [1]. MacDiarmid, Shirakawa, and Heeger enhanced the semiconducting behavior of organic polyacetylene in late 1970, which was synthesized by the chemical polymerization method. Their work on doping of polyacetylene with halogen derivatives was noticed and published in the chemical communication journal in 1977. These three scientists were conferred with the Nobel Prize in Chemistry in 2000 for discovering conducting polymers (CPs). After discovering conducting polyacetylene, scientists became interested in making other conducting polymers like polythiophene, polyaniline, polypyrrole, and polyfuran. Contrary to metals, these polymers can be processed at low temperatures, but their main problem is their stability. The conducting nature of these polymers is intrinsic due to their structure rather than by adding any conducting materials.

Fillers also play a pivotal role towards modification in the semiconducting and electrochemical performance of CPs. Since past decades, a wide range of chalcogenides derived from transition metals has been employed as fillers for modification in semiconducting [2] and electrochemical performance of CPs [3,20–49], as illustrated in Table 1. The product derived through the blending of such fillers with CPs is defined as PNCs. The electrochemical supercapacitance of such PNCs is well documented [50–78], whereas limited research has been made on the implication of rare earth oxides (REOs) as fillers for semiconducting and electrochemical applications, as demonstrated in Table 1 [79–110].

Table 1. Applications of various rare earth oxides based conducting polymer nanocomposites.

Polymer	REO	Applications	References
Poly(ethylene oxide) (PEO)	La_2O_3	Semiconductor and Solid Polymer Electrolyte (SPE)	[4]
Polyaniline (PANI)	La0.67Sr0.33MnO$_3$	Sensor	[110]
PANI	Sm_2O_3, La_2O_3	Thermally stable material	[111]
PANI	CeO_2	Thermally stable material	[112]
PANI	CeO_2	Semiconductor and supercapacitor	[112]
PANI	La-Nd	Electromagnetic Interference	[11]
PANI	Ce-TiO_2	Sensor	[113]
PANI	CeO_2, Dy_2O_3	Thermally stable material	[114]
PANI	Terbium(iii)	Light Emitting Diode	[10]
PANI	WO_3	Sensing	[115]
PANI	Nd_2O_3:Al_2O_3	Dielectric constant	[116]
Polycarbazole	-	Semiconductor	[117]
Polyindole (PIN)	TiO_2	Semiconductor	[118]
PIN	Y_2O_3	Dielectric constant	[119]
Polypyrrole (PPY)	CeO_2	Semiconductor	[120]
PPY	CeO_2	Sensor	[121]
PPY	Nb_2O_5	Semiconductor	[122]
PPY	Y_2O_3	Semiconductor	[2]
PPY	Sm_2O_3	Supercapacitor	[6]
PPY	Y_2O_3	Batteries, sensors and actuators	[123]
PPY	La^{3+}, Sm^{3+}, Tb^{3+}, Eu^{3+}	Supercapacitor	[124]
PPY	RuO_2	Supercapacitor	[9]
PPY	Eu_2O_3	Supercapacitor	[125]
PPY	Y_2O_3	Dielectric constant	[123]
Polyvinyl Alcohol (PVA)	Ho^{3+}, Gd^{3+}	Optical display	[126]
PVA/PPY	-	Dielectric	[127]
Polyvinylidene fluoride (PVDF)	La_2O_3	Thermally stable material	[128]

The electrical and electrochemical characteristics of REOs vary in a size dependent manner. Furthermore, PNCs derived through blending REOs with CPs offer a wide spectrum of applications such as solid polymer electrolytes [4], semiconductors [2], windows in dye-sensitized cells [5], electrochemical behavior, and charge storage [6–8]. Common CPs involved in synthesis of REOs based PNCs are PPY [9], PTh [10], PANI [11], and PIN [12].

The electrical conductivity of electrodes is routinely recorded at variable temperatures without taking the cognizance variations in their microstructure. The electronic and electrochemical significance of CPs has been well documented since decades [11,13–15]. CPs commonly utilized for charging a storage battery and semi-conducting applications are PPY, PIN, PCbz, PAc, PANI, and PTh [16], as illustrated in Figure 2a–f.

In 1987, Heeger and his coworkers used polythiophene for making diodes for electronic devices applications and then developed high efficiency polymer-based LEDs. These polymer LEDs have been used to make emission displays, which were used in cell phones in 2003 [17].

The various applications of conducting polymers can be increased by doping with other functional materials to form polymer composites [18]. These are used in different fields like physics, chemistry, electronics, and biomedical science [19].

Conducting polymers containing metal particles possess interesting properties of scientific and practical interests [20]. During the past few decades, researchers are paying more attention in conducting polymer composites to develop some new properties that were not observed in individual materials [21,22]. Researchers are more interested in developing the three-dimensional structure of conducting polymers, hybrid and nano hybrid materials of conducting polymers. The hybrid and nanohybrid conducting polymers are synthesized by adding metal, metal oxides, graphene, graphene oxide in conducting polymers. These new materials improve functionality in different areas like sensors, electronic devices, and biomedical application. The graphene nano hybrid of these polymers is used as an electrode in synthesis of capacitors. These nanohybrid materials increase stability, flexibility, and capacitances of capacitors [23]. Such polymers can be deposited either chemically or electrochemically on the metal. The different properties of polymers, like thermal stability, mechanical properties, conductivity, and corrosion protection properties on steel and aluminium, can be improved by doping. The doped conducting polymers have more capability for corrosion protection than undoped polymers because they give a suitable environment for corrosion protection on metal surfaces by restricting movements of corrosive agents or forming a uniform passive layer of doped polymers on metal surfaces [24–26]. Figure 3 reveals the bio-informatic visualization, which exhibits current progressions on the polymeric nanocomposites in a broader spectrum of anti-static, anti-corrosion, actuators, sensors, shape memory alloys, biomedical application, flexible electronics, solar cells, fuel cells, supercapacitors, LEDs, and adhesive domains using the Vosviewer Scientometric analysis.

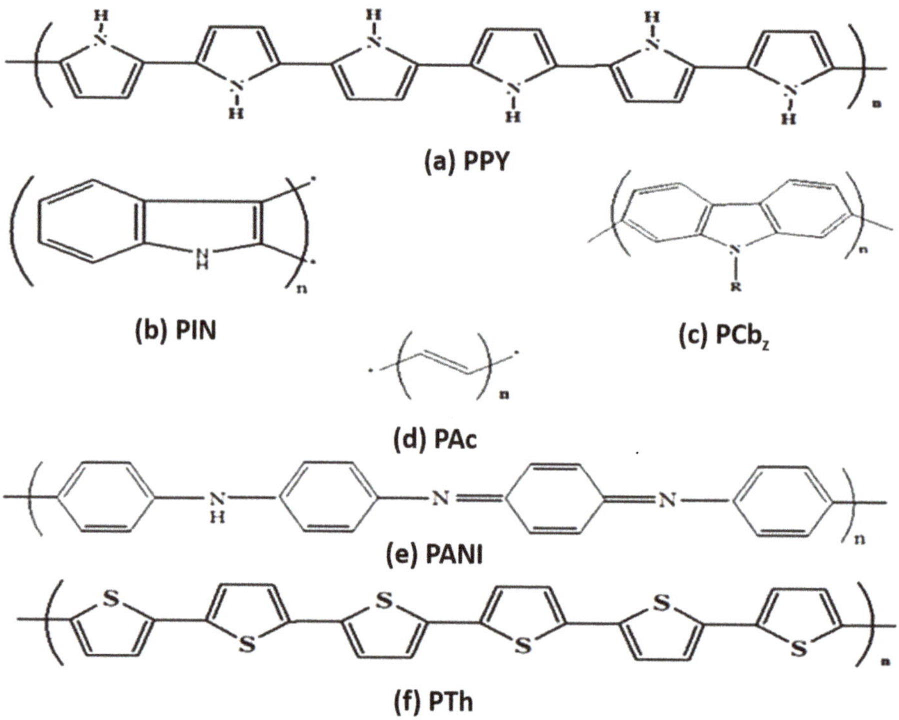

Figure 2. (**a**). PPY; (**b**). PIN; (**c**). PCbz; (**d**). Pac; (**e**). PANI; and (**f**). PTh-based CPs are used for semiconducting and electrochemical applications. Reproduced with permission from [9–16].

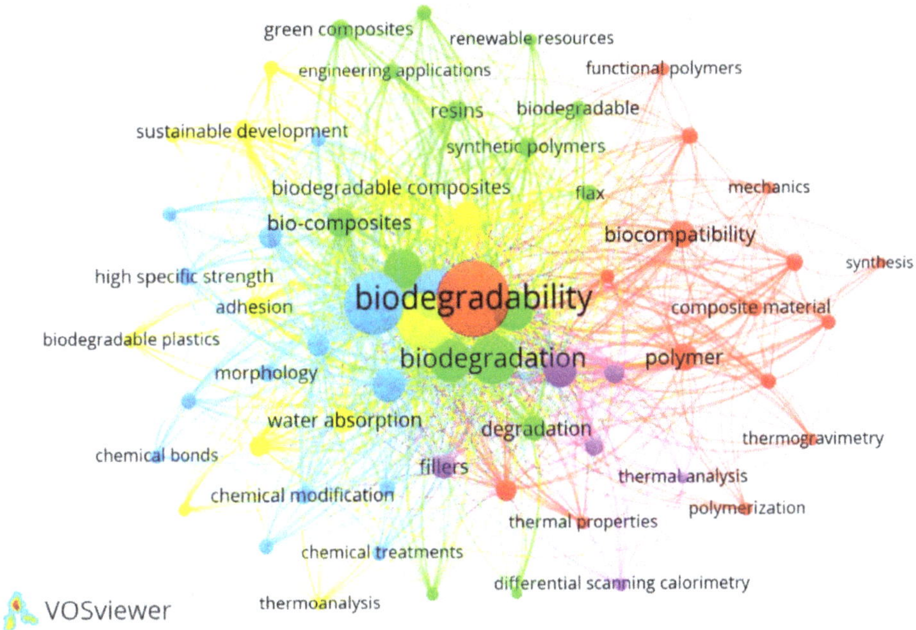

Figure 3. Systematic mapping summary of scientific advancements on polymer bio-nanocomposites for Multifunctional applications in anti-static, anti-corrosion, actuators, sensors, shape memory alloys, biomedical application, flexible electronics, solar cells, fuel cells, supercapacitors, LEDs, and adhesive domains.

2. Synthesis of Conducting Polymers

In the available literature, different ways to produce inductively coupled plasma (ICP) have been demonstrated. The polymerization process forms a solution containing the monomer is either a chemical or electrochemical process.

2.1. Chemical Polymerization

In this polymerization, monomers can be polymerized by various oxidizing agents like ammonium per sulphate, hydrogen per oxides, etc. [27]. The chemical polymerization of aniline is shown in Figure 4. This type of polymerization occurs by any of the methods: Addition polymerization and step growth polymerization. An oxidant is used to polymerize the monomer, and anions are doped as a counter part of the oxidative CP. This method to produce ICPs is widely used in industry.

Polyaniline and polypyrrole were synthesized on various substrates such as Pt, Au, Fe, Al, stainless steel, carbon fibers, brass, and zinc [28–30]. Isomers of poly-toluidine have been synthesized using the chemical oxidation method at 0 °C using potassium dichromate as oxidant and hydrochloric acid as dopant [31]. Polyaniline composites doped with TiO_2 were also synthesized by this method [32]. Polypyrrole doped with various dopants like Lithium per Chlorate ($LiClO_4$), para-Toluene Sulfonate (p-TS), and Naphthalene Sulfonic acid (NSA) was synthesized by chemical polymerization [33]. Polyaniline doped with tungstate was also chemically synthesized and characterized by various techniques [34].

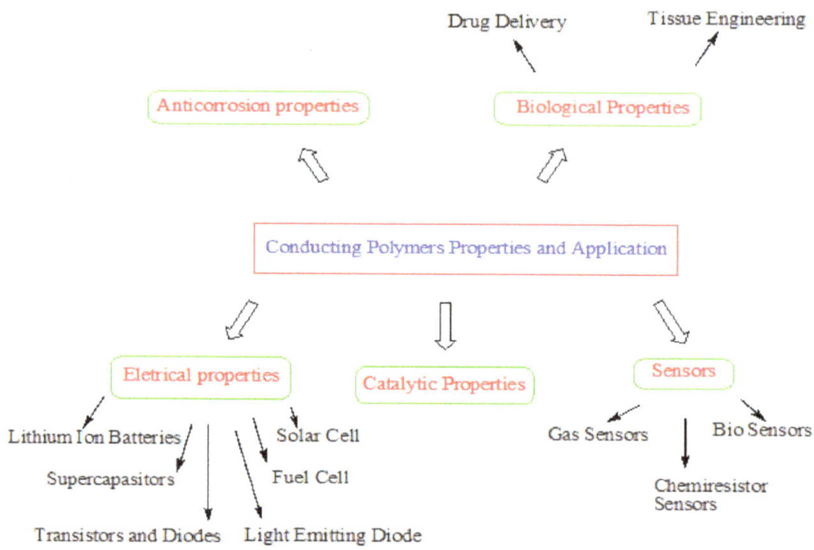

Figure 4. Reaction showing Chemical Polymerization of aniline. Reproduced with permission from [27–30].

The composite films of polypyrrole and polyvinylidine fluoride were formed by the chemical oxidation method and ammonium per sulphate used as an oxidant [35]. The nanocomposites of polypyrrole with copper sulfide were synthesized and characterized by various techniques [36].

Polymerization of furan by acidic catalysts has been reported by various researchers [37,38]. Armour et al. observed the electrical conductivity of polyfuran, which was synthesized chemically utilizing trichloroacetic acid [39]. Polyfuran was synthesized using pyridinium chlorochromate (PCC) as oxidizing agent [40].

Pyrrole was polymerized by chemical an oxidation method in the presence of $Fe_2(SO_4)_3$ and surfactant. The surfactant and oxidizing agent increased the conductivity and yield of polypyrrole [41]. Polypyrrole doped with tungstate or vanadate was synthesized by the chemical polymerization method and characterized by various techniques [42].

2.2. Electrochemical Synthesis

Electrochemical methods also synthesize the conducting polymers. It is very simple and a better technique for the preparation of conducting polymers because, in this technique, polymerization and doping levels could be controlled [43]. In this technique, three electrodes, working, counter, and reference electrodes, are required.

The physical properties of CPs coating are affected by the nature and size of counter ions used. The properties of conducting polymers like thermal and mechanical can be improved by the incorporating sulfonated aromatic ions [44]. The coating of poly (N-methyl pyrrole) doped with TiO_2 deposited on steel substrates by this method was studied [45]. PPy/TiO_2 nanocomposites were synthesized, and these composites are used for paint application [46]. The electro-polymerization of polyaniline, polypyrrole, and their composites was carried out on stainless steel and aluminium using the cyclic voltammetry technique [47,48]. Oxalic acid and tungstate doped polypyrrole films were potentiostatically electro-polymerized on the surface of aluminum alloy 1100 [49]. Polyaniline composites doped with tungstate and molybdate were synthesized by the electrochemical method [50,51]. Polypyrrole composites doped with zinc phosphate were deposited on AISI 1010 steel [52]. The copper doped polypyrrole was synthesized on steel by an electrochemical method for corrosion protection [53]. Table 2 exhibits the comparison of chemical and electrochemical polymerization synthesis methods with their respective merits and demerits.

Table 2. Comparison of Chemical and Electrochemical Polymerization.

Chemical Polymerization	Electrochemical Polymerization
Yield of the product is large in amount	Yield is less, and synthesis of the thin film is possible
Synthesis is difficult	Synthesis is quite easy
They do not offer control of polymerization and doping level	In this method, polymerization and doping levels can be controlled
Doping and polymerization do not occur simultaneously	Doping and polymerization occur simultaneously
Polymer is easily collected and packed	Difficult to remove the film from the electrode surface

The semiconducting and electrochemical performance of PNCs is evaluated as to their working electrode (WE) fabricated through coating composition of carbonaceous material with CPs and a polymeric binder. Common carbonaceous materials employed for the fabrication of WE are graphite and its tubular nanostructures. CPs used are either synthesized or commercially procured. The common conduction mechanism in rare earth oxides (REOs) (polar) has been explained through polaron theory, as revealed in Figure 5.

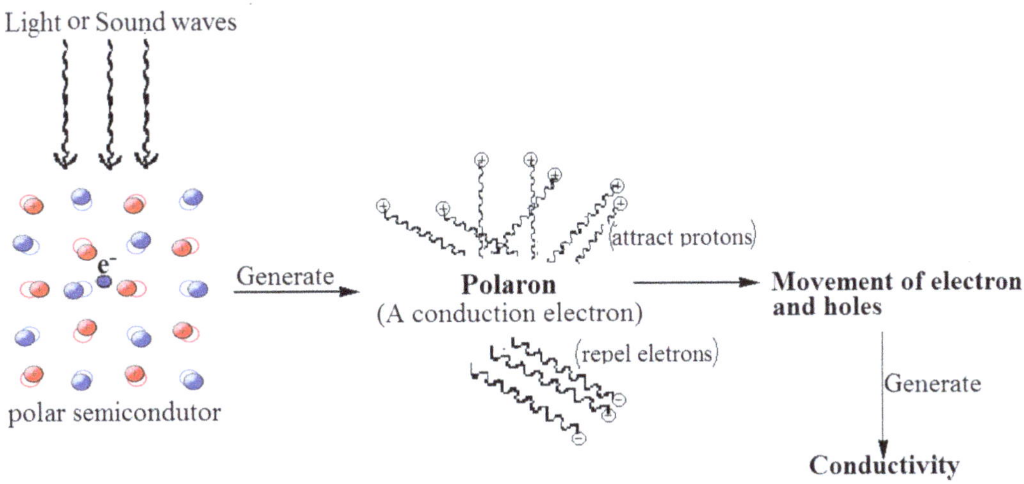

Figure 5. Conduction mechanism through polaron theory.

Interaction involving conformation degree of freedom through semi-conducting polymerics could localize an electric charge, and also have a substantial impact on carrier mobility, energy, and mass-transfer. Polarons could arise whenever charged particles cause aberrations, distortions, and abnormalities in the surrounding media, like locally stretching vibratory patterns or dielectric polarization. Such deformations, subsequently, come into contact with the charge-particles in an appealing manner, striving to localise it. First, the development of vibrating polarons in conducting polymers using a firmly binding model paradigm for charge hops across adjoining rings, which is linked to ring deformities. The coupling constants employing theoretical calculations, and molecular dynamics, for ring aberrations, and the couplings to the charged particles and carrier mobility. One such mechanism produces mainly wide, loosely knit/linked polarons on individual chain-rings. Furthermore, the polarons stabilized via dielectric polarization, which have been semi-classically characterized by a charged polarizable continua, are then interacting with the charge-carrier wave-function. Di-electrically stabilised polarons, in contrary to vibrating polarons, are narrower, more firmly coupled, and persistent, and stabilized in two-dimensional crystallographic layers.

An excitation activating energy level indicates the presence of a relaxed mechanism (a conduction activity), which can be characterized as polaron-hopping among adjacent locations in within the crystalline lattice structure. Through use of lattice oscillating, it is necessary to mobilise the confined localized electron. Among these instances, electrons are assumed to travel through hopping movement stimulated by lattice crystalline oscillating, i.e., a conducting method is supposed to be multiphonon-assisted hopping of tiny polarons across localized regions.

Doping with acidic functionalities reduces the band gap that enhances the conductivity of PPY [54]. Semiconducting components and electrodes for charge storage devices are routinely developed through either of chemical or physical vapor deposition over semiconducting wafers. Alternatively, PPY electrodes are produced through electrode position over metals or by hand laying of a composition of PPY over metallic substrates in presence of graphite, dopants and polymeric binders [11,13,14]. Common polymeric binders employed for the development of WE are polyvinylidene fluoride (PVDF), polytetrafluoroethylene (PTFE), and sulphonated polysulphone (SPS), as shown in Figure 6.

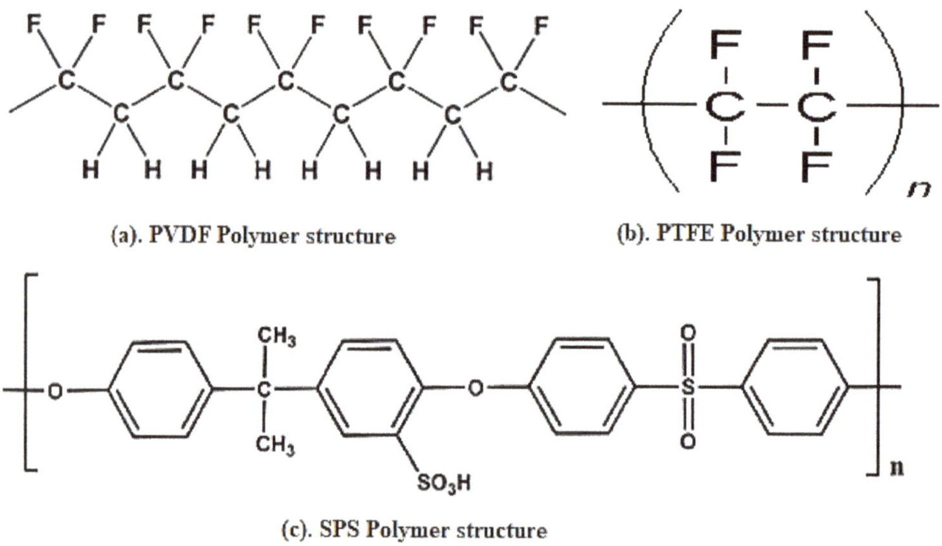

Figure 6. Binders, (**a**). PVDF-polymeric structure; (**b**). PTFE-polymeric structure; and (**c**). SPS-polymeric structures, used for electrode fabrication. Reproduced with permission from [11,13,14].

3. Properties and Multifunctional Applications of Conducting Polymers

Conducting polymers have wide applications in various fields such as supercapacitors, electrochromic devices, biosensors, and electrocatalysts [55–57]. Conducting polymer nanocomposites of inorganic oxides have various applications in the field of chemistry and physics due to their electro-optical properties [58]. Various applications and properties of conducting polymers are shown in Figure 7. Several properties of conducting polymers like processability, conductivity, permeability, and mechanical properties are increased by dopant anions [59–61].

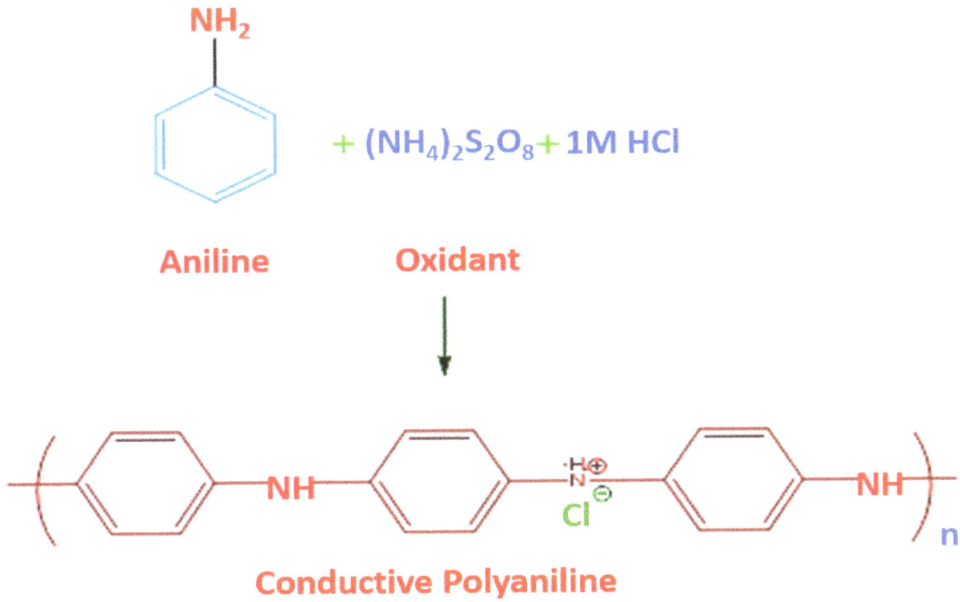

Figure 7. Numerous applications and properties of conducting polymers. Reproduced with permission from [62,63].

3.1. Electrical Properties

Conducting polymers have various applications in electronic devices like batteries, solar cells, fuel cells, and supercapacitors due to the highly conducting nature of polymers. Various applications of CPs are described below.

3.1.1. Lithium-Ion Batteries

Conducting polymers have been used in batteries. Several polymers like polypyrrole, polyaniline, and polyacetylene are used as electrodes in batteries. PPy composites doped with $MnCo_2O_4$ are used as anode in lithium-ion batteries. These composites have good stability, a high-performance rate, and are light weight [62]. These batteries are used in electrical vehicles, mobile phone, and tablets.

Since lithium-ion batteries have such a complicated intricate architecture framework, all electrode elements, comprising active materials, binding additives, conductive agents, electrolytes, and membranes, must be designed in an inventive manner. One such review demonstrated how physicomechanical deterioration might have been whittled down by replacing the conventional PVDF/C electrodes matrices with CP-based bindings as elastic and adherent conducting materials. This is one way whereby research towards passive components and composite materials extends the electrode surface, active material, absorber layer area through shifting prerequisites and demands for active materials, like lower volumetric changes, to passive components, which could handle those demands, like flexibility, adhesion-conducting [62].

Multi-functionalized CP-based binding adhesives have demonstrated intriguing characteristics like suitability with solvent-treatment, appropriate electrochemical-ionic, carrier-mobility properties, significant adherence to active components, and probable capacities contributions. Through formation of composites with nonpolar hydrophile polymerics, altering CP side-branches with hydrophilic groups, and manufacturing a hydrogel-derived three-dimensional electrode structure of conducting polymeric active components through in-situ-polymerization, investigators have already also addressed the inherent and profoundly poor processing of pristine CPs. The previous two methods are appealing due to their ease of synthesis and scaling reproducibility at an affordable price. Another technique inherited the benefits of hydrogel materials, like adjustable nanostructures and centralized hierarchical three-dimensional electrode structures comprising interconnecting ionic and electronic conductions. Furthermore, consequentially, the above-mentioned designing notion for CP-based binding adhesives, specifically hydrogel-based CP composites, might discover usage in energy storage devices, rechargeable-batteries, super-capacitors, and flexible/elastic equipments [62].

3.1.2. Solar Cells

Conducting polymers have been used in solar cells. These are used as an electro catalyst in solar cells. PPy aluminium oxide composites were used as electro catalyst for solar cells. Dye sensitized and photovoltaic solar cells based on conducting polymers are used in the place of silicon solar cells because these have high energy conversion efficiency and a lower cost than silicon-based solar cells. These are also used as energy transfer mediators in solar cells [63].

Conducting polymers have found widespread application in solar-sectors like dye sensitized solar cells (D.S.S.C.), perovskite-structured solar cells (P.S.C.), and organic/polymer-photovoltaics (O.P.V.).

Polymers could be utilised in D.S.S.C. as not only an elastic adjustable surface, but rather as pore-forming and films-forming agents of photo-anode-films; likewise, owing to their rising catalytic-properties, conductive polymerics as well as comparative composite materials are being employed to manufacture Pt-free counter-electrode materials; and in PSC, polymers could indeed be utilized to promote nucleating polymerization, crystallization, and control crystal-growth. Because of its remarkable energy-gap and charge carrier mobility, polymerics could also be employed as hole-transporting materials. Developing innovative polymeric hole-transporting materials with higher charge carrier mobility and optimal binding energy-level design layout is problematic [63].

Conducting polymerics are extensively employed as active absorber layered-structure in O.P.V.s to affect photo-harvesting efficiencies and equipment operating performance parameters. The technique of obtaining escalating efficiency O.P.V.s seems to be the emergence of various polymeric donors/drivers with a narrower band-gap as well as optimal energy barrier frameworks [63].

3.1.3. Fuel Cells

In the past few decades, fuel cells had various advantages for applications in electric vehicles, such as automobiles [64]. Polymer fuel cells are of two types: low temperature and high temperature fuel cells. The membrane of high temperature fuel cells is made with poly (benzimidazoles). Direct methanol fuel cells (DMFCs) have been used in the field of energy applications because they have fuel portability and high energy conversion efficiency [65]. Conducting polymers with 1D-nanostructures are used as electro catalyst supports in cells [66].

Even though conducting polymeric materials have distinctive characteristics such as rising electron-carrier charge mobility, conductivity, and electrochemical performance, owing to the existence of the organic conjugated polymeric backbone, excellent electron de-localization from of the CPs to the hybrid material, and higher surface-area, such characteristics assist in making conducting polymeric-based nano-hybrids (C.P.N.H.s) an appealing material for analyzing their physical and chemical characteristics but also continuing to expand their application areas (P.E.M.F.C.s) to various sorts of polymeric electrolytic-membrane fuel cells. Till date, relatively few polymer multimetallic effectual electro-catalyst-supported CPs have been explored, and transitions material-based C.P.N.H.s have also still not been disclosed. These steep prices have had the ability to strengthen the catalysis performances and functionality and efficiency of fuel cell energy storage equipment [65,66].

3.1.4. Light Emitting Diodes (LEDs)

Conducting polymers like Poly(p-phenylenevinylenes) (PPVs), Poly(dialkylfluorenes) (PFs), Polythiophenes (PThs), and their derivatives exhibit potential for polymer light-emitting diodes (PLED) applications [67,68]. By introducing bulky phenyl side groups in the polymer, the performance of PLED could be improved [69,70].

3.1.5. Supercapacitors

These are energy storage devices and are used in solar arrays, hybrid electric vehicles. They have intermediate specific energy between batteries and capacitors. They have a high charging and discharging capability. The supercapacitors based on conducting polymers have high charge storage efficiency so they can store a large amount of energy [71]. Conducting polymers have been used as active electrode materials for supercapacitors due to high conductivity, flexibility, stability, and low cost [72–74]. The hybrid type supercapacitor is shown in Figure 8.

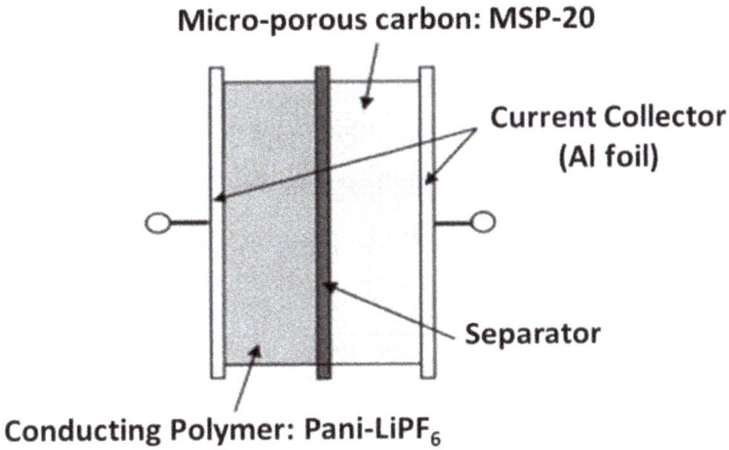

Figure 8. Hybrid type supercapacitor. Reproduced with permission from [75].

3.2. Anticorrosion Properties

These days, conducting polymers and their composites have been used as corrosion protecting agents on metal surfaces [76,77]. Corrosion protecting behavior of these composites is due to the capacity of inhibit the movement of corrosion causing agents on surface of metals [78–80]. It was studied that the polyaniline coating on the steel surface protects from corrosion by forming the passive film [81]. Polyaniline epoxy blended coating on steel as a corrosion inhibitor has been studied [82]. Polyaniline/polypyrrole and polyaniline-polypyrrole phosphotungstate composites were used as corrosion protecting agents on the mild steel surface. Composite films give better corrosion protection than bare polyaniline and polypyrrole [83]. Polyaniline doped with TiO_2 nanoparticles (PTC) were used as corrosion protectors, and they were more effective than undoped polyaniline [32].

Oxalate, as well as tungstate doped-PPy, used as a corrosion protector on aluminum has been observed [49]. It was observed that the PANI-MoO_4^{2-} coating acts as a better corrosion inhibitor as compared to pure PANI coating [51]. Polyaniline and its composites films possess corrosion inhibition properties [84,85]. It was studied that zinc phosphate-doped PPy gives better corrosion protection than undoped PPy [52,53]. Poly-6-amino-m-cresol doped with copper nanocomposites give corrosion protection of mild steel. These composites give better performance than bare polymers [86]. The corrosion behavior of 7075 aluminum, copper modified Al, polypyrrole modified Al, and copper/polypyrrole modified Al samples were noticed.

3.3. Catalytic Properties

Conducting polymers have been used as electrocatalysts and photocatalysts in biosensors, cells, and energy-related devices because of high conductivity and electroactive properties of conducting polymers. The high conductivity of conducting polymers increases the efficiency of charge transfer between electrode and electrolyte, which improves catalytic activity. These are used as a catalyst for enzymes in electrochemical sensors. The nanocomposites of polymers were used as photocatalysts [87]. The nanocomposites of polypyrrole-titanium dioxide showed more photocatalytic activities in the degradation of Rhodamine B than pure TiO_2 [88]. Fe_3O_4/Pd@PPy composites showed superior catalytic activity and better stability in successive cycling tests [89].

3.4. Sensors

In the recent past, polymer nanocomposites have emerged as the most promising materials for cost effective sensors with excellent sensitivity and selectivity with fast and reliable sensing techniques. The unique electrical, thermal, and optical properties of graphene when combined with the light weight, good processability, and excellent mechanical properties of polymers, offer a new class of materials capable of fulfilling the stringent requirements for a wide variety of sensors. Graphene–polymer nanocomposites have attracted enormous interest, mainly due to the fact that exceptionally high performing composites can be prepared by the use of extremely small quantities of the filler due to their nano-level dispersion in the polymer matrix. Basically, 'Sensors' detect the changes in any of its physical properties and convert them into a measurable signal. Based on the changes in their optical, electrical, chemical, electrochemical, and mechanical properties, graphene–polymer nanocomposites can be used as biosensors, chemical sensors, gas/vapor sensors, strain sensors, etc.

Conducting polymers have wide applications in sensors like gas sensors, bio sensors, optical sensors, strain sensors, and chemiresistor sensors.

3.4.1. Gas Sensors

Conducting polymers are used in gas sensors. These are employed in forming the active layer in sensors due to the conductive and flexible nature of conducting polymers. Gas sensors have a wide range of applications in different areas, like industrial production, food processing, environmental monitoring, and health care, etc. [90–92]. Conducting polymers and doped conducting polymers with different metal salts have been used in gas sensors. Polypyrrole film with various dopants has been used in gas sensors [93,94].

Biochemical sensors are made of a responsive component towards a specified analyte (molecular-identification) and transduction processor, which converts analyte-concentrations into measurable input-physical electrical signals like voltage, absorption, current, or mass. Gas sensor devices based on CPs were being categorized by IUPAC dependent on input-signaling waveform transduction. Sensors appertaining to the chemically modulating of electrochemical, optoelectronic, or physicomechanical transmission and propagation processes of CPs would be thoroughly addressed through consideration with gas-sensing applications.

Around ambient temperature, the contact interface among the conducting polymer as well as the gas analyte is still quite significant. As a result, sensors predicated on conducting polymerics could provide spectacular signal output responses, whereas those dependent on synthetic transition metal-oxide have almost zero responsiveness at room temperature. Consequentially, conducting polymeric sensing devices require less energy and also have a simplified device architecture.

3.4.2. Bio Sensors

Conducting polymers are very useful for the expansion of biosensors because these are good materials for the immobilization of biomolecules. Conducting polymers and their composites are used in fabrication of different biosensors and also improve speed and sensitivity of biosensors. Delocalization of electrons in conducting polymers is very fast, which is good for efficient biosensors [95]. The conducting polymers provide suitable environment for immobilization of enzymes and biomolecules on the electrode surface. The enzymes and biomolecules may be amalgamated in conducting polymer films during electrochemical deposition on electrodes. The amalgamation of enzymes in conducting polymers gives proximity between active site of the enzyme and conducting surface of the electrode, making it suitable for biosensor construction. Glucose oxidase can be successfully amalgamated in polypyrrole films for glucose detection [96]. The biosensors based on conducting polymers were discovered to treat penicillin and detect innumerable chromosomal disorders. A highly sensitive and rapid flow injection system for urea analysis was fabricated using the composite film of polypyrrole and a polyion complex [97]. Glucose biosensors are used to estimate glucose by the arrest of glucose oxidase enzymes with conducting polymers. The DNA biosensors based on conducting polymers have been investigated for diagnosing and treating of various diseases like chromosomal disorder by repairing, degradation, or multiplication. Biosensors have a great role in environmental monitoring by controlling various hazardous chemicals like formaldehyde and hydrogen peroxide, which cause pollution in the environment [98].

3.4.3. Chemiresistor Sensors

The conducting polymers have also been used in chemiresistor sensors due to their conductivity. The conducting polymers play an important role in sensors because they can act as both electron donors and electron acceptors when interacting with the gaseous form. The conductivity of conducting polymers increases when it acts as an electron donor to gas and decreases when it acts as an electron acceptor to gas [99]. Pt, Au, and Pt-Ni IDAs pre-patterned over alumina, quartz, glass, acrylic strip, silicon chips, Si_3N_4/Si are normally used as chemiresistor sensors [100].

3.4.4. Strain Sensors

Strain sensors measure a local deformation due to an applied strain. Strain sensors mainly find application in damage detection, structural health monitoring, and structural and fatigue studies of materials. The outstanding electrical properties of graphene have made it the most promising material for developing strain sensors. In general, graphene-based strain sensors can work mainly based on three mechanisms:

a. based on structure deformation of graphene,
b. based on over connected graphene sheets and finally,
c. based on tunneling effect of neighboring graphene sheets.

The structural deformation of graphene due to an applied strain can induce attractive changes to its electronic band structure and electrical properties. There have been a lot of theoretical calculations to find out the effect of different types of strain on graphene. Experimentally, strain sensors based on the structural deformation of graphene have already been demonstrated by many researchers. CVD-grown graphene, transferred on flexible polydimethylsiloxane (PDMS) substrate, showed a gauge factor of ~151 [101]. When the strain was applied, initially, the resistance was decreased, which was due to the relaxation of pre-existing wrinkles on the graphene sheet.

Further increase in strain resulted in increase in resistance due to distortion of the hexagonal honeycomb crystal structure of graphene. Even though these sensors have high sensitivity, the large strains can cause unrecoverable structural deformation, limiting their practical application. Researchers made efforts to overcome this limitation by forming rippled graphene layers on pre-strained PDMS [102]. The graphene layer was deposited on a pre-strained PDMS, and when the strain of the PDMS was released, the graphene layer above the PDMS formed a rippled structure. The resistance of this sensor was found to linearly decrease with increasing strain. The resistance changed from 5.9 kΩ to 3.6 kΩ when the strain was increased from 0% to 20% till the graphene layer was completely flat.

3.5. Actuators

These features make them the most promising candidates for many applications, including artificial muscles and robotics. Graphene polymer composite actuators are considered one of the most versatile actuator systems.

The electrically triggered polymer actuators can follow two different mechanisms: either utilizing Maxwell's stress, which is generated as a result of the electrostatic attraction between the two electrodes, or by pure electrostrictive effect [103]. However, there can be another class of polymer actuators in which the large mechanical deformation is caused by ionic charge movement at lower voltages like in ionic polymer based metallic composites [104]. During the last few years, there have been several reports to develop electromechanical actuators with excellent actuator performance by combining the unique properties of graphene with different polymer systems. Unlike the conventional polymer based electromechanical actuators, graphene–polymer composite actuators follow different mechanisms in different systems depending on the materials and the design of the actuators. This makes them the most versatile actuator materials for various practical applications.

In a study by Chen et al. using poly(methylmethacrylate) functionalized graphene–polyurethane (MG-PU) composite actuators, it was found that the introduction of functionalized graphene into the PU matrix significantly improved the electric field induced strain behavior when compared to pure PU films. The electrical field-induced straintronics increased from 17.6 per cent for pure PU to around 32.8 per cent for 1.5 wt. per cent, MG-PU composite film and was nearly two-timing that of pure film as presented in Figure 9 [105].

Graphene can undergo contraction on heating due to its negative temperature coefficient of thermal expansion. This unique property was utilized by Zhu et al. for designing a bimorph actuator using a graphene–epoxy hybrid system [106]. The graphene-on-organic film actuator was developed as a cantilever in which graphene acted both as the conducting layer and heating layer. Upon applying electric power, the graphene was directly heated, and the epoxy was warmed up by diffused heating. Due to the mismatch in thermal expansion of graphene and epoxy, the cantilever exhibited a deflection towards the graphene coated side. The device exhibited high actuation behavior at very low power. For instance, the cantilever tip deflected 1 μm with an input voltage as low as 1 V within 0.02 s and returned back to its original position within 0.1 s. They have also reported that the flapping and bending motion of the actuators can be controlled by changing the frequency and duration of applied voltage. They have demonstrated the working of this graphene on organic film actuator in the form of a dragon fly wing, as illustrated in Figure 10.

A similar bimorph actuator based on a bilayer of graphene and polydiacetylene (PDA) was reported by Liang et al. [107]. The actuator generated a large actuation motion under a low electric current in response to both dc and ac signals. For example, for a bimorph of size 10 mm length by 2.7 mm width, at a very low dc current of 20 mA, a displacement as large as 1.8 mm and curvature of 0.37 cm^{-1} were obtained. Similarly, at applied DC of just 0.29 A/mm^2, actuation stress as high as 160 MPa/g was obtained. Under the ac signal, this actuator displayed reversible swing behavior.

Recently, the electroactive performance of graphene-loaded cellulose composite actuators was reported by Sen et al. [108]. The films of microcrystalline cellulose (MCC) loaded with graphene nanoplatelets were prepared by the solvent casting method. An ionic liquid, 1-butyl-3-methylimidazolium chloride (BMCl), was used as the solvent. The composite films were converted into actuator strips by forming electrodes using gold leaf. The incorporation of graphene enhanced the conductivity and mechanical properties of the polymer composite. The actuator performance was measured at 3 to 7 V. The study reported that the graphene-based polymer composite loading reduced the actuation speed but operating at higher excitation voltages. Moreover, an increase of 267% in the maximum displacement at an excitation voltage of 3 V was achieved with the addition of graphene, as revealed from the Figure 11a–d. This clearly indicates that the loading of graphene nanoplatelets has led to improved electro-active actuator efficiency of the polymer composites.

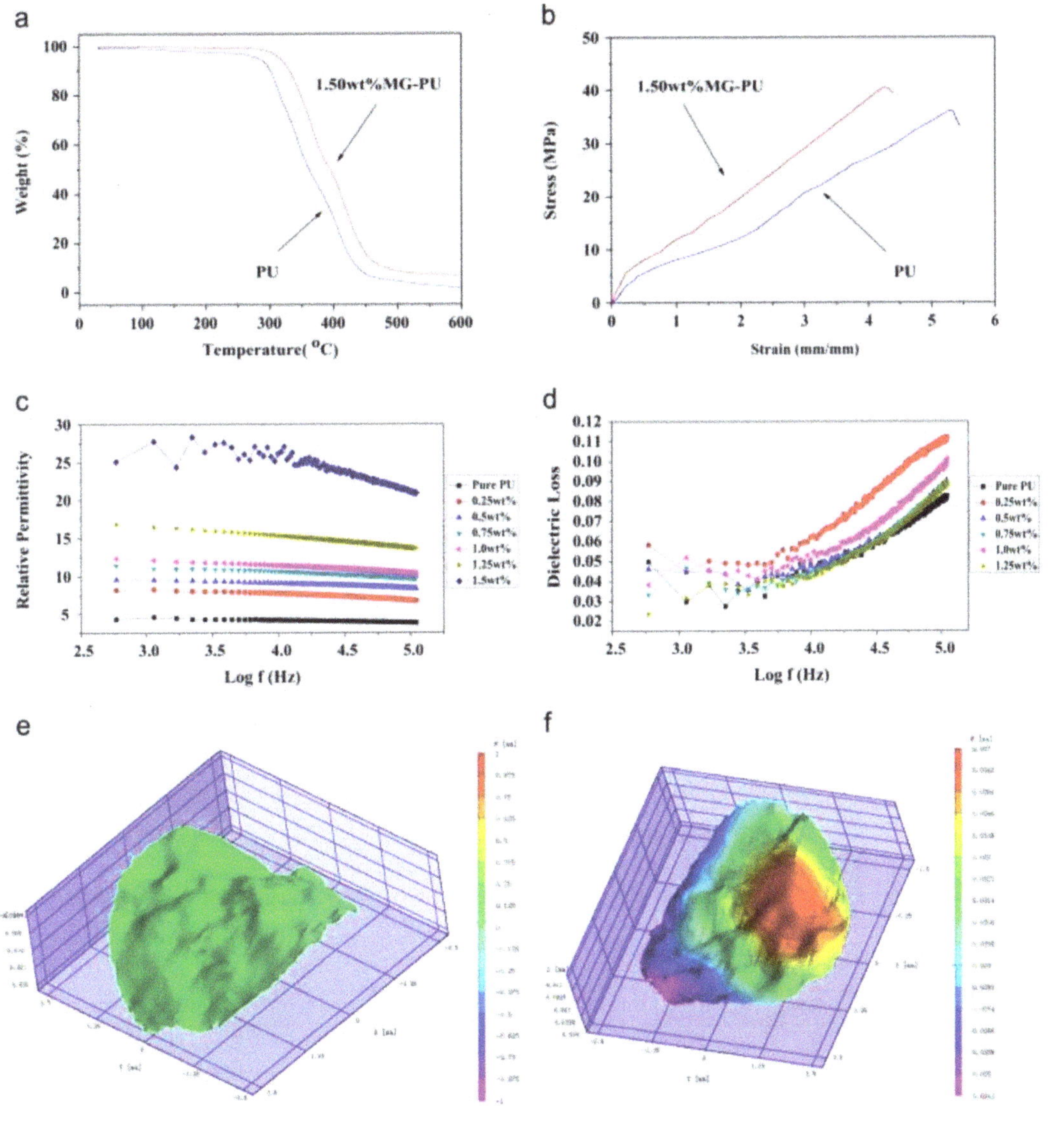

Figure 9. DTA curve (**a**). Load-deflection curves; (**b**) of Polyurethane and 1.5 wt. percent poly(methylmethacrylate) functionalized graphene–polyurethane, dielectric-constant; (**c**) and dielectric-loss; (**d**) of poly(methylmethacrylate) functionalized graphene–polyurethane composites, electrical field induced strain-nephograms; ((**e**) Voltage-off, (**f**) Voltage-on) of 1.50 wt. percent poly(methylmethacrylate) functionalized graphene–polyurethane composite film. Reproduced with permission from [105].

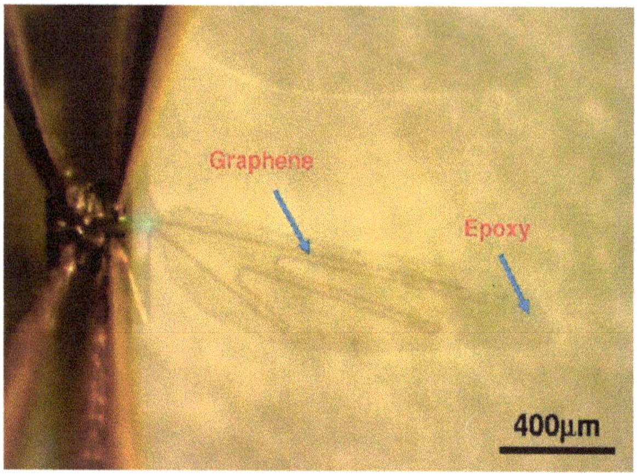

Figure 10. A Graphene on organic film in the form of a dragonfly wing. Reproduced with permission from [106].

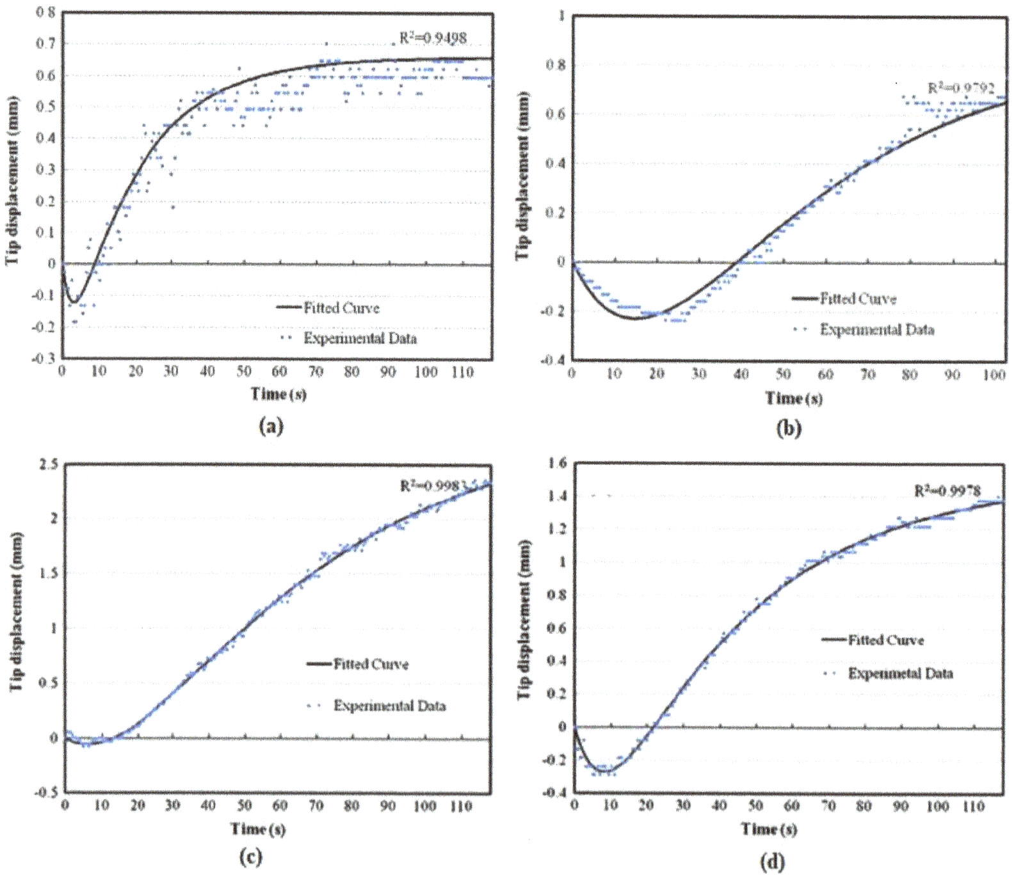

Figure 11. (**a**). Tip-displacement of the graphene-based polymer composite actuator in excitation-voltage of 3 V DC; (**b**). At 0.1 wt. percent graphene loading; (**c**). At 0.25 percent graphene loading and (**d**). At 0.5 percent graphene loading. Reproduced with permission from [108].

Polymer nanocomposite based optical actuator is one of the fast-developing fields in contemporary research.

In a graphene–polymer composite actuator, the homogeneous dispersion of graphene is one of the important aspects to be ensured for good optical actuation behavior. The homogeneity of graphene dispersion was achieved by many researchers using functionalization techniques [109,110]. The IR triggered actuation in graphene–polydimethylsiloxane (PDMS) systems showed less photomechanical stress and strain. For example, 2 wt.% graphene nanoplatelet-PDMS system exhibited a change in stress of only less than 40 kPa [111]. Even with single layer graphene (1 wt.%), the photomechanical stress of the PDMS system was <50 kPa [112]. However, graphene-PDMS systems exhibited excellent reversibility.

Seema et al. reported a Thermally reduced graphene oxide (TRGO)-PDMS system with higher photomechanical stress of 133.44 kPa [113]. Even though the strain obtained was only 7.17%, the actuation behavior was 100% reversible.

Similarly, graphene/styrene-isoprene-styrene (SIS) copolymer composite was studied as an optical actuator [114]. The maximum photomechanical stress and strain obtained were 28.34 kPa and 3.1%, respectively. However, the on-off cycle of the actuator exhibited a marching behavior due to creep deformation, which became less prominent after a few on-off cycles.

Graphene–polymer composite actuators which can be driven by solvents, pH, chemicals, etc., can find a wide range of applications particularly in the biomedical field. A solvent driven actuator was developed by Deng et al. by pattering few-layer graphene (FLG) on an epoxy-based photoresist polymer (SU-8) [115]. The FLG/SU-8 bilayer can fold when immersed in water due to de-solvation of SU-8 or can unfold when immersed in acetone due to re-solvation of SU-8. Even after solvation and de-solvation, the graphene layer is held intact. This also puts up an opportunity to integrate graphene-based sensors on solvent driven self-folding polymer actuators.

3.6. Flexible Electronics

In recent years, there has been intensive research on flexible and stretchable electronics especially in the field of wearable electronic devices, stretchable circuits, flexible batteries, membrane keyboards, biomedical sensors, and artificial tissues. One of the critical parameters to realize flexible electronics is to retain the high conductivity under mechanical deformations.

Wong et al. have developed a process to fabricate highly conductive and flexible graphene aerogel/PDMS composites. [116] A spontaneous reduction process prepared graphene aerogel, and due to its porous nature of this aerogel the incorporation of PDMS into the graphene framework was easier. Graphene aerogel/PDMS composite showed a high conductivity of 95 S/m at small filler loading of 0.8 wt%, as exhibited in Figure 12. The flexibility of the composites was studied by electromechanical tests and achieved good retention in conductivity (of about 80%) under various bending conditions.

3.7. Shape Memory Polymers

Incorporating conducting fillers into the shape memory polymer (SMP) is one of the most widely adopted methods to develop smarter shape memory materials. Many researchers have already succeeded in improving the shape memory effect by incorporating conducting fillers like graphene into the polymer matrix. In addition to imparting mechanical strength due to the high thermal conductivity of graphene, it can uniformly heat the SMP composites, which results in faster response and better shape recoverability.

A composite with triple stage shape memory performance was demonstrated by combining chemically modified graphene oxide with an interpenetrating polymer network of polyurethane with two different molecular weights [117].

The shape memory effect (SME) of graphene oxide (GO) incorporated shape memory polyurethane (SMPU) nanofibers were studied by thermal cyclic tests [118]. The SMPU/GO nanofibrous mats exhibited a better SME and lowered thermal shrinkage when compared with pristine SMPU nanofibrous mats. With a GO loading of 4.0 wt.%, a very low thermal shrinkage ratio of 4.7 ± 0.3%, a very high average fixation ratio, and a recovery ratio of 92.1% and 96.5% were obtained, exhibited from the Figures 13 and 14, respectively [118]. The GO cross-linked SMPU molecular chains are not free to shrink and, hence, can be stabilized much more quickly for reaching the final structure, which results in high fixation and recovery ratio.

In a similar study, the correlation between the temperature dependent SME, and the cross-linking density in GO/polyurethane nano composites was established by Ponnamma et al. [119]. As the filler concentration was varied, the crosslinking density was also varied affecting the shape memory properties.

Graphene–polymer composites have emerged as one of the most promising materials that can revolutionize the field of electronics and optoelectronics.

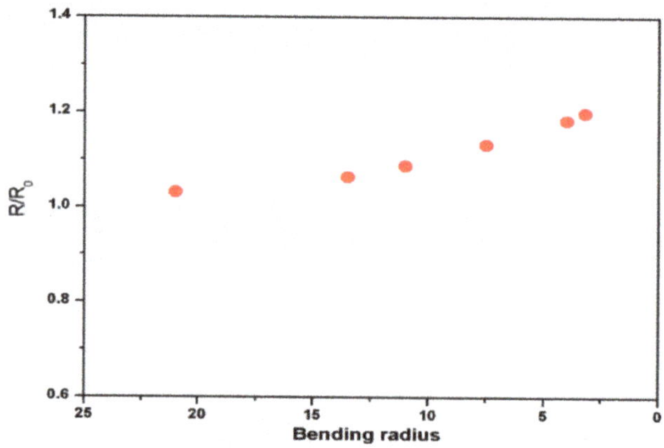

Figure 12. Variations in resistance as a function of bending-radius for the graphene aerogel/PDMS composites. Reproduced with permission from [116].

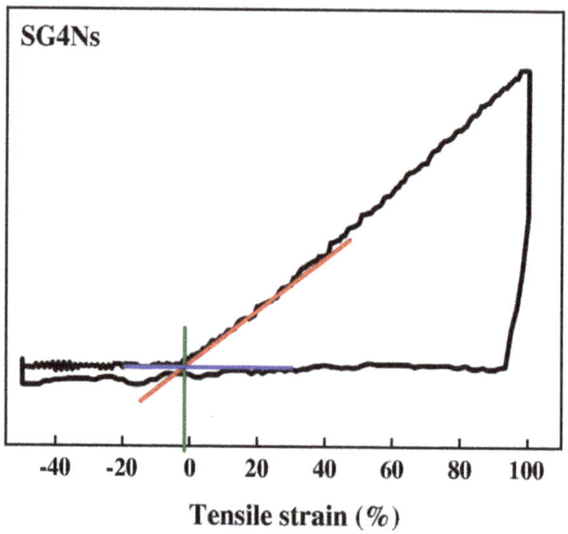

Figure 13. Thermal-shrinkage during physical-testing of pristine SMPU/GO nanofibrous mats. Reproduced with permission from [118].

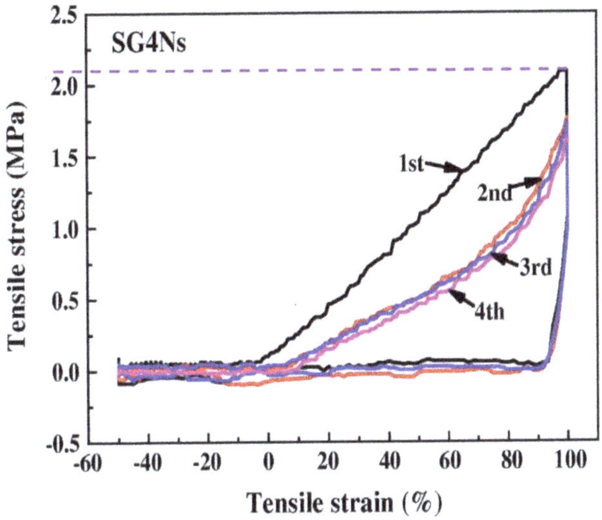

Figure 14. Average fixation-ratio of SMPU/GO nanofibrous mats during cyclic-tensile test resembling shape-memory effect. Reproduced with permission from [118].

3.8. Optical Limiting Applications

Optical limiting is a mechanism by which certain materials, which are transparent to light at low intensities, restrict the transmission of light above a threshold input intensity. Graphene, GO, and RGO exhibited high nonlinear optical absorption, which makes them suitable as good optical limiters. Most of the studies were based on dispersions in liquid. However, for practical device applications, these need to be used as solid materials, and the dispersion of graphene or GO or RGO in the solid matrix is very important. There have been a few studies on the nonlinear optical properties of RGO in various solid matrices including glass and polymers. The nonlinear optical properties of covalently functionalized GO in silica gel glasses were studied by Tao et al. [120]. They have observed that the nonlinear optical response of functionalized GO was better in silica gel glass than in deionized water.

Similarly, RGO with porphyrin incorporated in polymers was studied as optical limiters by a few researchers [121]. There have been a few reports on the optical limiting properties of graphene–polymer composites also. However, in order to avoid the dispersibility issues of graphene, functionalization was required in most of the cases [122,123].

3.9. Biomedical Applications

Conducting polymers have innumerable applications in the field of medical science like drug delivery, biomedical implants, tissue engineering, and diabetic monitoring.

3.9.1. Drug Delivery

During last few years conducting polymers have been used in drug delivery due to biomedical compatibility. These are good for drug release applications [124]. The choice of drug delivery method depends upon the types of drug and types of treatment required. The routes of drug delivery are peroral, gastrointestinal, rectal, ocular, intravaginal, transdermal, vascular injection, nasal, and pulmonary [125]. The different types of drugs like anionic, cationic or neutral can be injected into the polymer backbone [126].

The surface area of conducting polymer films can be increased by titanium and carbon nanotubes to store drugs [127,128]. Many therapeutic drugs, such as 2-ethylhexyl phosphate, dopamine, naproxen, heparin, and dexamethasone, have been bound and successfully released from these polymers [129–144].

3.9.2. Tissue Engineering

Conducting polymers also find applications in tissue engineering due to their stimulus-responsive property. The composites of these polymers act as substrates that promote cell growth, adhesion, and proliferation at the polymer-tissue interface [130,131,145–158].

3.9.3. Diabetic Monitoring

Conducting polymers and their nanocomposites have advantages in the diagnosis and treatment of diabetes. These are used for the manufacturing of devices that are needed in the diabetes treatment of human beings. The pros of employing conducting polymers in diabetic-treatment are that their physical and chemical characteristics are being modified by doping with numerous chemical agents whenever required. The glucose biosensor is used in the treatment of diabetes. Conducting polymers are applied in closed loop delivery devices, which are needed for diabetic patients [132].

Thus, conducting polymers containing metal particles possess interesting properties of scientific and practical interests [132–148]. Researchers are more interested in developing the three-dimensional structure of conducting polymers, hybrid, and nanohybrid materials of conducting polymers. The hybrid and nanohybrid conducting polymers are synthesized by adding metal, metal oxides, graphene, and graphene oxide to conducting polymers. These new materials improve functionality in different areas like sensors, electronic devices, and biomedical application. The graphene nanohybrid of these polymers is used as an electrode in synthesis of capacitors. These nanohybrid materials increase stability, flexibility, and capacity of capacitors [149–161]. The different properties of polymers like thermal stability, mechanical properties, conductivity, corrosion protection properties on steel and aluminium can be escalated to a firm extent by doping.

4. Novel Polymer Nanocomposite Materials for Multifunctional Engineering Applications

Li et al. outline the synthetic–polymeric composite biomedical-coatings comprising of inorganic elements, but also present their design methodologies and manufacturing processes. For developing composites coatings, synthetic as well as polymeric elements can indeed be blended prior to applying coating methods, yet they can be subsequently coated and sprayed onto surfaces. Generally, the operations really are not challenging [162].

For most scenarios, polymerics serve as hosting matrixes while inorganic particles act as dispersing filling materials, emphasizing the existing concern that elements of composite coatings generally exhibit their merits and take measures in such a discrete way instead of in a spontaneous synergistic manner. Additionally, study on intermolecular-level interfacial contact and interactions among synthetic and polymeric elements is inadequate [162].

How else to optimize the proportion and focus on improving functionality and suitability of the elements, as well as how to strengthen the stimulatory synergic effect to develop novel bio-functions, might just be the upcoming challenging issue for polymeric composite coatings, which also relies heavily on a comprehensive understanding of the interactions among polymeric materials as well as inorganic particles. It is indeed worthwhile to examine alternative prospects for synthetic–polymeric composite coatings, including their realistic medical applications [162].

Rikhari et al. mixed graphene–oxide (GO) with a conducting poly-pyrrole (PPy) polymeric and then deposited it on titanium material, employing a traditional electropolymerization process. Such hybrid coatings exhibited excellent resistance to corrosion, and bio-compatibility, rendering them attractive for orthopaedic implants, bone-regeneration, and biomedical applications [163].

Sulfonic acids were being utilized as doping agents in the electro-chemical synthesis of poly-pyrrole (PPy) coatings on carbon-steel as reported by the Nautiyal et al. [164]. The influence of acidity doping (p-toluene sulfonic-acid (p-T.S.A.), sulfuric-acid (S.A.), camphor-sulfonic-acid (C.S.A.)) upon carbon-steel surface modification passivation was analyzed employing linear-potentio-dynamics and contrasted to a coating's microstructure as well as corrosive shielding effectiveness as showed in Figure 15. The sorts of doping agents utilized had a substantial influence on the coating's potential to shield the surface of the metal against chloride-ion attacks. The corrosive behavior of a polymeric backbone is primarily determined by the volume, structure, magnitude, size, positioning, and orientation of the doping agent. Furthermore, both p-T.S.A. and sodium-do-decyl-benzene-sulfonate (S.D.B.S.) have included additional benzene-rings, which stacked around each other to produce lamellar-membranous sheet-like barriers to chloride-ions, rendering them effective doping agents for PPy coatings for reducing corrosion to a considerable extent. Additionally, adhesive ability was strengthened via directly introducing long-chain carboxylic-acid into to the monomer's solvent. Besides that, PPy coatings coated with SDBS indicated remarkable bio-cidal characteristics against Staphylococci. PPy coatings on carbon-steels with double anticorrosion and superior bio-cidal abilities have such a promising possibility to be used throughout the industries for anticorrosion and anti-microbial applications [164].

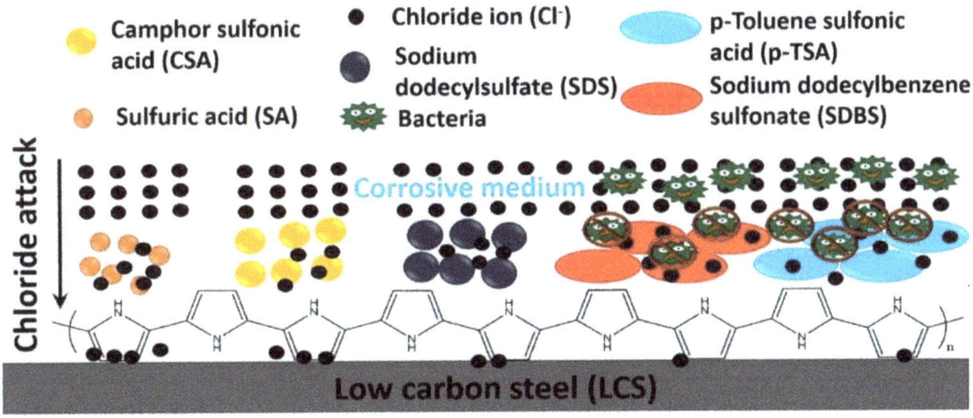

Figure 15. Poly-pyrrole (PPy) polymeric-coatings on carbon-steel for bio-medical applications. Reproduced with permission from [164].

The phenomena of static-charge are unforeseeable, especially when an airplane is traveling at higher latitudes, inflicting the accumulation of static charge-carriers to surpass a specified threshold, resulting in the breakdown of its components and subsystems, involving devastating explosive blasts and radio-transmission malfunction loss, as revealed by Yadav et al. [165]. The accumulating of static charge-carriers on aeroplanes is caused by the interaction among the aircraft's outermost exterior surface-layer as well as the external atmospheric characteristics, which would include surrounding air, snow, hailstorm, dirt, and volcanism eruption-ash, in addition to its tribo-electric charging. Because of the lightweight materials, and similar physicomechanical characteristics, innovative polymeric-based composites or nanocomposites have become prevalent structural constituents for aviation sectors throughout the last few decades. However, such polymeric composites do not provide low-resistance trajectory for electric charge transport, thus, attempting to make them susceptible to influence of thunderstorms, lightning-attacks, and p-static interference. For this perspective, nano-filler formulations are critical for developing conducting polymeric composite structures using nonconductive polymeric matrices. With the emergence of carbon-derived polymeric nanocomposites, certain challenges pertaining to non-conductive polymeric matrices have indeed been satisfactorily resolved, and the composites have evolved into an avantgarde genre [165].

Owing to its remarkable electrical properties, redox characteristic, but also functionality at ambient temperatures, novel nanocomposites appertaining to graphene and conducting polymerics like poly-aniline (P.A.N.i.), poly-pyrrole (PPy), poly (3,4 ethyl-dioxy-thiophene) (P.E.D.O.T.), poly-thiophene (PTh), as well as their compound-derivatives have notably become extensively utilized as active materials throughout gas-sensing applications as enumerated by the Zamiri et al. [166] Whenever these two materials have been blended, they excelled, pure graphene as well as conducting polymerics, in terms of sensors-based characteristics. This could be attributable to the nanocomposites' high specific surface area, and indeed the combined synergic effects of graphene and conductive polymerics. A kind of graphene, as well as conductive polymeric nanocomposites processing techniques, including in situ polymerization, electro-polymerization, solution-blending, selfassemblage, and many others, have been revealed, and utilization of such nanocomposites as sensing materials has also been shown to enhance the effectiveness of gas-sensing.

A kind of graphene, as well as conductive polymeric nano-composites processing techniques, including in situ polymerization, electro-polymerization, solution-blending, self-assemblage, and many others, have been revealed, and utilisation of such nanocomposites as sensing materials has also been shown to enhance the effectiveness of gas-sensing [166].

Findings unveiled that the reduced graphene oxide–P.A.N.I. blends packed on a flexible-based poly-ethylene-terephthalates thin film demonstrated the maximum response of 344.12 to 100 ppm ammonia, the gas sensor activity of P.E.D.O.T./reduced graphene oxide nanocomposites processed through in situ-polymerization method unveiled outstanding sensing effectiveness to nitrogen-dioxide, and reduced graphene oxide/PPy nanocomposite demonstrated the maximum sensitivity of 102 percent under 50 ppm as displayed in Figure 16 [166].

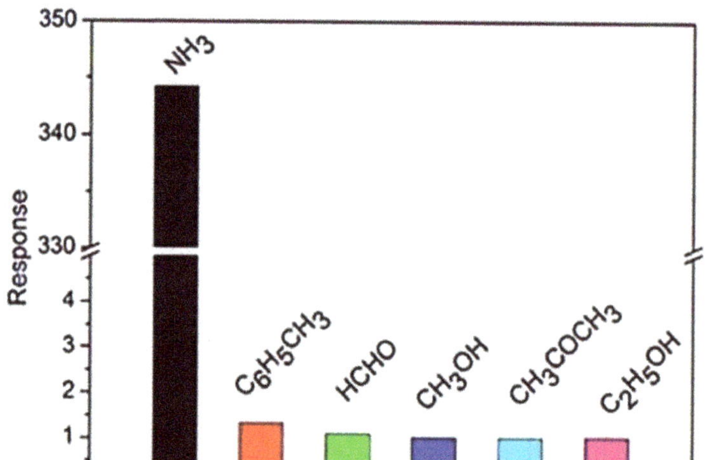

Figure 16. Sensitivity response of reduced graphene-oxide–PANI hybridized thin films towards various gases at 100 ppm. Reproduced with permission from [166].

Conducting-polymerics with ultra-high conductivity, and electro-chemical characteristics have attracted the attention of investigators as catalysis accelerators for polymeric-electrolytic-membrane-based fuel-cells (P.E.M.F.C.'s) including microbial fuel-cells (M.F.C.'s) as discussed by Ghosh et al. [167]. Furthermore, metallic or metallic-oxide catalytic-accelerators can be immobilized on the surface of a virgin polymer, or a biocompatible-polymer to produce conductive polymeric-based nano-hybrids (C.P.N.H.'s) having excellent catalyzed activity and durability. Transition metal-oxides, which provide greater superficial surface area and porosity and permeable nanostructures, exhibited distinctive synergic activity with conducting polymerics. As a result, C.P.N.H.'s could be employed to produce a stabilized, environment-friendly bio- or electrocatalyst, exhibiting enhanced catalytic performance as well as an increased electron charge particle transfer rate. Palladium/poly-pyrrole (PPy) C.P.N.H.'s possesses 7.5 and 78 times more mass-activities of commercialized palladium/carbon, and bulk palladium/PPy used as anode materials for ethanol-oxidation, respectively. The electro-catalytic activities of Palladium-rich multi-metallic alloy-compositions placed on PPy nanofibrous was roughly 5.5 times more than the mono-metallic counter-parts. Likewise, binary and ternary platinum-rich electro-catalysts displayed higher catalysis performance for the methyl-alcohol-oxidation, as well as the catalysis behavior of Pt-24Pd-26Au-50/PPy considerably enhanced up to 12.45 A/mg platinum, which itself is around fifteen times superior than those of commercialized platinum/carbon (0.85 A/mg platinum) [167].

Outcomes also reported that such a novel class of C.P.s-based bio-materials can still be employed in cellular investigations, like tissue-engineered, regenerative-medicine, bioengineering, bio-medical equipment, scaffoldings, and so on. It might be immensely favorable to the progress of such a domain [168].

5. Drawbacks of Conducting-Polymers

Conducting-polymerics (C.P.'s) with multifunctional-applications have had some constraints owing to their toxic-carcinogenic character and disparities in in vivo and in vitro investigations. In general, the crucial challenges for C.P.'s include physical characteristics, cytocompatibility, bioactivity, and bio-compatibility.

Though C.P.'s have tremendous promising opportunities in multi-functional applications, they have numerous downsides due to the initial bursting drug-delivery release-rate, and the hydrophobic characteristics of the polymeric, thus, hindering their usage. Nonetheless, drug-releasing mechanisms seem to be of enormous interest to scientists as well as providing possibilities in treating cancer, and in minimal invasive-methods for myriads of neuro, and cardio-vascular treatments.

Another key drawback of existing C.P.'s are as follows: the cyclic-stabilisation is worse than that of carbon materials, and the energy-capacity, power-density, specific-capacitive, and specific-power are significantly low while comparing to transition metal-oxides.

6. Concluding Remarks and Future-Outlook

Nowadays, great advancements in conducting polymers and rare earth oxides have been achieved in electrochemical and conductors by modifying the surface of the working electrode. This review paper gives sound information about the chemical and electrochemical synthesis methods and applications of conducting polymers in different fields, like electronic devices, sensors, protection of corrosion, shape memory polymers, actuators, flexible electronics, optical limiting, drug delivery, and tissue engineering. The nanocomposites and nanohybrid materials of conducting polymers improve the useful properties of polymers in different fields. An improved thermal and cyclic stability with a low internal resistance of the composites was observed with the application as dielectric, antistatic properties, semiconductor, and energy storage devices. Biopolymeric nanocomposites for storing energy, power-generation, energy production, conservation of energy, and corrosion inhibitors purposes have indeed been presented in this article. ICP nanocomposites have piqued the curiosity of researchers due to their high electrical properties, being relatively inexpensive, the ease of manufacture, and their biodegradability. Graphene and certain other nano-fillers are proliferating and gaining momentum as appealing possibilities for nanocomposites. Progressions in the fabrication, evaluation, testing, characterization, and computation modeling of nanostructures have offered ample and endless possibilities for tailoring the engineering characteristics of PNC frameworks.

Corrosive inhibitors and anodic shielding are the two foremost significant methods for reducing the metal corrosion rates. Composite-conducting polymerics (C.C.P.'s) were shown to offer excellent corrosion-resistant characteristics compared to conductive polymeric-coating materials. This seems to be attributable to the nanoadditive's increased area of surface for dopant-releasing and the development of a barriers impact versus diffusion transport. It is predicted that in upcoming investigations, a plethora of reinforcing particulates will be at the forefront of interest, with a heavy emphasis on the application of C.C.P.'s on certain metal surfaces and in diverse domains. Furthermore, since corrosive resistance by C.P.'s is predominantly focused on anodic shielding, the stabilization of the passive oxide layer underneath the polymeric coating and the inhibitors of aggressive anion-ions from penetrating the polymeric membrane must, therefore, be thoroughly studied.

The toxic effect of electro-active poly(aniline)-based oligomers has indeed been thoroughly examined, and it is a restrictive concern for bio-medical applications. Throughout this domain, suggestions for novel electro-active oligomers predicated on better stabilized, and bio-compatible electro-active monomeric-subunits and/or oligomers could be of interest. A further approach would be to utilize a relatively minimal, aniline-based oligomer concentration, or even to explore newer possibilities to escalate bio-compatibility.

Electro-active macro-monomers appear as a promising strategy for obtaining grafting co-polymers with conductance, and bio-degradability. Substantial research is required, nevertheless, before these bio-materials can still be employed in cellular investigations, like tissue-engineered, regenerative-medicine, bioengineering, bio-medical equipment, scaffoldings, and so on. It might be immensely favorable to the progress of such a domain.

Some other crucial trend is that it would be worthwhile not only to synthesize novel materials, but also to delve deeply into their characterization. Significant investigations have recently concentrated on the bio-interface of materials with cellular, either by investigating nano-sized scaling characteristics or interacting with bio-molecules, in order to properly comprehend what could be better for cellular interface interaction. Subsequently, the insights could contribute to developing intelligent bio-materials inside the upcoming years.

C.P.'s are generally infusible, and unsolvable, rendering treatment of such materials challenging, particularly when mixed with the other polymeric matrices for anti-static applications. Even so, with advancement of manufacturing methods, the afore-mentioned concern is now being overcome, and C.P.'s are showing promise in conductive materials in anti-static applications, owing to their convenience of synthesizing, their lower-density, their light-color, and the manageable electrical characteristics, between whereby the water-based conductive coatings have been remarkably alluring, deserving additional endeavour for useful anti-static devices.

Nevertheless, numerous impediments have yet to be overcome. The first one is the paucity of elevated nano-fillers, as well as the exorbitant operating costs and complexity in scaling-up. The next difficulty is correlated with the handling and treatment of PNCs. Controlling a dispersal scattering and orientations of the nano-fillers is paramount for optimising the functionality efficacy of PNCs. The final barrier is a dearth of comprehensive insight and prediction capabilities of the recurring viable fundamental processing structural–characteristic interactions necessary to completely maximize the commercialization prospects of such innovative biomaterials. The generated models frequently failed to anticipate scientific findings. Furthermore, substantial extensive research attempts to generate breakthrough innovative biopolymer nano-composites employing cleaner sustainable, economical, environmentally-sound, and alternative renewable sources of energy are desired from the standpoint perspective of attaining sustainable development.

Author Contributions: Conceptualization, S.S.; methodology, S.S.; formal analysis, S.S., P.S., J.S.; investigation, S.S., P.S., J.S.; resources, S.S., P.S., J.S., R.A.I.; writing—original draft preparation, S.S.; writing—review and editing, S.S., P.S., A.A.B.O., J.S., R.A.I.; supervision, S.S., P.S., J.S., R.A.I.; project administration, S.S., P.S., J.S.; funding acquisition, S.S., A.A.B.O., R.A.I. All authors have read and agreed to the published version of the manuscript.

Funding: The APC was covered by Universiti Tenaga Nasional (UNITEN), Malaysia, through BOLD refresh publication fund 2021(J510050002-BOLDRefresh2025-Centre of Excellence).

Institutional Review Board Statement: Not applicable.

Informed Consent Statement: Not applicable.

Data Availability Statement: No data were used to support this study.

Acknowledgments: The APC was covered by Universiti Tenaga Nasional (UNITEN), Malaysia through BOLD refresh publication fund 2021(J510050002-BOLDRefresh2025-Centre of Excellence). The author Shubham Sharma wish to acknowledge the Department of RIC, IKGPTU, Kapurthala, Punjab, India for providing opportunity to conduct this research task. The author P. Sudhakara gratefully acknowledges the support from Science and Engineering Research Board (SERB-YSS/2015/001294)), New Delhi, India.

Conflicts of Interest: The authors declare no conflict of interest.

References

1. Walatka, V.V.; Labes Perlstein, J.H. Polysulfur nitride—A one-dimensional chain with a metallic ground state. *Phys. Rev. Lett.* **1973**, *31*, 1139–1142. [CrossRef]
2. Cheng, Q.; Pavlinek, V.; Li, C.; Lengalova, A.; He, Y.; Sáha, P. Synthesis and structural properties of polypyrrole/nano-Y_2O_3 conducting composite. *Appl. Surf. Sci.* **2006**, *253*, 1736–1740. [CrossRef]
3. Wang, X.; Wang, T.; Liu, D.; Guo, J.; Liu, P. Synthesis and electrochemical performance of CeO_2/PPy nanocomposites: Interfacial effect. *Ind. Eng. Chem. Res.* **2016**, *55*, 866–874. [CrossRef]
4. Kumar, P.R.; Venkateswarlu, M.; Satyanarayana, N. Synthesis and ac conductivity studies of PEO+ $LiClO_{4+}$ La_2O_{3+} MoO_3 nanocomposite polymer solid electrolyte. *NSTI-Nanotech 2012* **2011**, *1*, 546–549.
5. Zamborini, F.P.; Alphenaar, B.W.; Kharel, P.L. Enhancing the Photovoltaic Performance of Dye-Sensitized Solar Cells with Rare-Earth Metal Oxide Nanoparticles. *Meet. Abstr.* **2017**, *15*, 879.
6. Liu, P.; Wang, Y.; Wang, X.; Yang, C.; Yi, Y. Polypyrrole-coated samarium oxide nanobelts: Fabrication, characterization, and application in supercapacitors. *J. Nanopart. Res.* **2012**, *14*, 1232. [CrossRef]
7. Mo, Z.; He, J.; Wang, J.; Feng, C.; Guo, R. Preparation and characterization of PPy/NanoGs/Fe_3O_4 conductive and magnetic nanocomposites. *J. Exp. Nanosci.* **2013**, *8*, 113–120. [CrossRef]
8. Shiri, H.M.; Ehsani, A. A novel and facile route for the electrosynthesis of Ho_2O_3 nanoparticles and its nanocomposite with p-type conductive polymer: Characterisation and electrochemical performance. *Bull. Chem. Soc. Jpn.* **2016**, *89*, 1201–1206. [CrossRef]
9. Zang, J.; Bao, S.J.; Li, C.M.; Bian, H.; Cui, X.; Bao, Q.; Sun, C.Q.; Guo, J.; Lian, K. Well-aligned cone-shaped nanostructure of polypyrrole/RuO_2 and its electrochemical supercapacitor. *J. Phys. Chem. C* **2008**, *112*, 14843–14847. [CrossRef]
10. Rafiqi, F.A.; Majid, K. Synthesis, characterization, luminescence properties and thermal studies of polyaniline and polythiophene composites with rare earth terbium (III) complex. *Synth. Met.* **2015**, *202*, 147–156. [CrossRef]
11. Wang, Y.; Wang, J.; Zhang, X.F.; Liu, Y.Q. Synthesis. Characterization and Properties of PANI (La-Nd Doped $BaFe_{12}O_{19}$) Composites. *Key Eng. Mater.* **2017**, *27*, 327–334. [CrossRef]
12. Maji, P.; Choudhary, R.B.; Majhi, M. Effect of Y_2O_3 on polyindole for high frequency capacitor application. *AIP Conf. Proc.* **2017**, *1832*, 070003.
13. Meng, Q.; Cai, K.; Chen, Y.; Chen, L. Research progress on conducting polymer-based supercapacitor electrode materials. *Nano Energy* **2017**, *36*, 268–285. [CrossRef]
14. Le, T.H.; Kim, Y.; Yoon, H. Electrical and electrochemical properties of conducting polymers. *Polymers* **2017**, *9*, 150. [CrossRef] [PubMed]
15. Naveen, M.H.; Gurudatt, N.G.; Shim, Y.B. Applications of conducting polymer composites to electrochemical sensors: A review. *Appl. Mater. Today* **2017**, *9*, 419–433. [CrossRef]
16. Gurunathan, K.; Murugan, A.V.; Marimuthu, R.; Mulik, U.P.; Amalnerkar, D. Pelectrochemically synthesized conducting polymeric materials for applications towards technology in electronics, optoelectronics and energy storage devices. *Mater. Chem. Phys.* **1999**, *61*, 73–191. [CrossRef]
17. Shirakawa, H.; Louis, E.G.; MacDiarmid, A.G.; Chiang, C.K.; Heeger, A.J. Synthesis of electrically conducting organic polymers: Halogen derivatives of polyacetylene $(CH)_x$. *J. Chem. Soc. Chem. Commun.* **1977**, *16*, 578–580. [CrossRef]
18. Holze, R.; Inzelt, G. Conducting polymers. *J. Appl. Electrochem.* **2009**, *39*, 953–954. [CrossRef]
19. Jang, J. Conducting polymer nanomaterials and their application. *Emiss. Mater. Nanomater.* **2006**, *199*, 189–260.
20. Brezoi, D.V.; Ion, R.M. Phase evolution induced by polypyrrole in iron oxide polypyrrole nanocomposite. *Sens. Actuators B Chem.* **2005**, *109*, 171–175. [CrossRef]
21. Babazadeh, M.; Gohari, F.R.; Olad, A. Characterization and physical properties Investigation of conducting polypyrrole/TiO_2 nanocomposites prepared through one step in situ polymerization method. *J. Appl. Polym. Sci.* **2012**, *123*, 1922–1927. [CrossRef]
22. Hosseini, M.G.; Bagheri, R.; Najjar, R. Electropolymerization of polypyrrole and polypyrrole ZnO nanocomposites on mild steel and its corrosion protection performance. *J. Appl. Polym. Sci.* **2011**, *121*, 3159–3166. [CrossRef]
23. Nguyen, D.N.; Yoon, H. Recent Advances in Nanostructured Conducting polymers: From Synthesis to Practical Applications. *Polymers* **2016**, *8*, 118. [CrossRef] [PubMed]
24. Meneguzzi, A.; Pham, M.C.; Lacroix, J.C.; Ferreir, C.A. Electroactive polyaromatic amine films for iron protection in sulfate medium. *J. Electrochem. Soc.* **2001**, *148*, B121. [CrossRef]

25. Lenz, D.M.; Delamar, M.; Ferreira, C.A. Application of polypyrrole/TiO_2 composite Films as corrosion protection of mild steel. *J. Electroanal. Chem.* **2003**, *54*, 35–44. [CrossRef]
26. Lenz, D.M.; Ferreira, C.A.; Delamar, M. Distribution analysis of TiO_2 and commercial zinc phosphate in polypyrrole matrix by XPS. *Synth. Met.* **2002**, *126*, 179–182. [CrossRef]
27. Machida, S.; Miyata, S.; Techagumpuch, A. Chemical synthesis of high electrically conductive polypyrrole. *Synth. Met.* **1989**, *31*, 311–318. [CrossRef]
28. Çolak, N.; Ozyilmaz, A. Polyaniline Coating on Iron–Synthesis and Characterization. *Polym. Plast. Technol. Eng.* **2005**, *44*, 1547–1558. [CrossRef]
29. Camalet, J.L.; Lacroix, J.C.; Aeiyach, S.; Chane-Ching, K.; Lacaze, P.C. Electrode-posoition of protective polyaniline films on mild steel. *J. Electroanal. Chem.* **1996**, *416*, 179–182. [CrossRef]
30. Lacroix, J.C.; Camalet, J.L.; Aeiyach, S.; Chane-Ching, K.L.; Petitjean, J.; Chauveau, E.; Lacaze, P.C. Aniline electro-polymerization on mild steel and zinc in a two step process. *J. Electroanal. Chem.* **2000**, *481*, 76–81. [CrossRef]
31. Madhusudhana, G.; Santhi, R.J. Synthesis, characterization and corrosion behavior of isomers of conducting poly—Toluidine on mild steel in acid medium. *Int. J. Sci. Res.* **2015**, *4*, 1645–1650.
32. Sathiyanarayanan, S.; Azim, S.S.; Venkatachari, G. Preparation of polyaniline TiO_2 composite and its comparative corrosion protection performance with polyaniline. *Synth. Met.* **2007**, *157*, 205–213. [CrossRef]
33. Chitte, H.K.; Bhat, N.V.; Gore, A.V.; Shind, G.N. Synthesis of polypyrrole using ammonium peroxydisulfate (APS) as oxidant together with some dopants for use in gas sensors. *Mater. Sci. Appl.* **2011**, *2*, S1491–S1498.
34. Kamaraj, K.; Karpakam, V.; Sathiyanarayanan, S.; Syazim, S.; Venkatachari, G. Synthesis of tungstate doped polyaniline and its usefulness in corrosion protective coating. *Electrochim. Acta* **2011**, *56*, 9262–9268. [CrossRef]
35. Taunk, M.; Kapil, A.; Chand, S. Synthesis and electrical characterization self-conducting polypyrrole-poly(vinylidene fluoride) composite films. *Macromol. J.* **2008**, *2*, 74–79. [CrossRef]
36. Ramesan, M.T. Synthesis, characterization and conductivity studies of polypyrrole copper sulfide nanocomposites. *J. Appl. Polym. Sci.* **2013**, *128*, 1540–1546. [CrossRef]
37. Livingston, H.K.; Senkus, R.; Hsieh, J.T.T.; Kresta, J. The polymerization of furan on Surfaces. *Macromol. Chem. Phys.* **1972**, *161*, 101–111. [CrossRef]
38. Granzow, A.; Wenedenburg, J.; Henglein, A. Die γ-radiolyse des furans and Thiophens. *Z. Nat. B* **1964**, *19*, 1015–1017. [CrossRef]
39. Armour, M.; Davies, A.G.; Upadhyay, J.; Wassermann, A. Coloured electrically conducting polymers from furan, pyrrole and thiophene. *J. Polym. Sci.* **1967**, *5*, 1527–1538. [CrossRef]
40. McConnell, R.; Godwin, W.R.; Baker, S.E.; Powell, K.; Baskett, A. Morara Polyfuran and copolymers a chemical synthesis. *Int. J. Polym. Mater. Polym. Biomater.* **2004**, *53*, 697–708. [CrossRef]
41. Kudoh, Y. Properties of polypyrrole prepared by chemical polymerization using aqueous solution containing $Fe_2(SO_4)_3$ and anionic surfactant. *Synth. Met.* **1996**, *79*, 17–22. [CrossRef]
42. Jadhav, N.; Jensen, M.B.; Gelling, V. Tungstate and vanadate doped polypyrrole Aluminum flake composite coatings for the corrosion protection of aluminum 2024-T3. *J. Coat. Technol. Res.* **2015**, *12*, 259–276. [CrossRef]
43. Bahrami, A.; Talib, Z.A.; Shahriari, E.; Yunus, W.M.A.; Kasim, A.; Behzad, K. Characterization of electrosynthesized conjugated polymer carbon nanotube composite optical nonlinearity and electrical property. *Int. J. Mol. Sci.* **2012**, *13*, 918–928. [CrossRef] [PubMed]
44. Hallik, A.; Alumaa, A.; Kurig, H.; Janes, A.; Lust, E.; Tamm, J. On the porosity of polypyrrole films. *Synth. Met.* **2007**, *157*, 1085–1090. [CrossRef]
45. Mahmoudian, M.R.; Basirun, W.J.; Alias, Y. Synthesis and characterization of Poly (N-methypyrrole/TiO_2 composites on steel. *Appl. Surf. Sci.* **2011**, *257*, 3702–3708. [CrossRef]
46. Ferreira, C.A.; Domenech, S.C.; Lacaze, P.C. Synthesis and characterization of polypyrrole/TiO_2 composites on mild steel. *J. Appl. Electrochem.* **2001**, *31*, 49–56. [CrossRef]
47. Subathira, A.; Meyyappan, R.M. Anticorrosion behavior of polyaniline/polypyrrole composite coating on stainless steel. *Int. J. Chem. Sci.* **2011**, *9*, 493–502.
48. Akundy, G.S.; Rajagopalan, R.; Iroh, J.O. Electrochemical deposition of polyaniline polypyrrole composite coatings on aluminum. *J. Appl. Polym. Sci.* **2002**, *83*, 1970–1977. [CrossRef]
49. Castagno, K.R.L.; Azambuja, D.S.; Dalmoro, V. Polypyrrole electropolymerized on aluminum alloy 1100 doped with oxalate and tungstate anions. *J. Appl. Electrochem.* **2009**, *39*, 93–100. [CrossRef]
50. Sabouri, M.; Shahrabi, T.; Hosseini, M.G. Influence of tungstate ion dopants in corrosion Protection behavior of polyaniline coating on mild steel. *Mater. Corros.* **2008**, *59*, 814–818. [CrossRef]
51. Karpakam, V.; Kamaraj, K.; Sathiyanarayanan, S.; Venkatachari, G.; Ramu, S. Electro synthesis of polyaniline molybdate coating on steel and its corrosion protection performance. *Electrochim. Acta* **2011**, *56*, 2165–2173. [CrossRef]
52. Lenz, D.M.; Delamar, M.; Ferreira, C.A. Improvement of the anticorrosion properties of polypyrrole by zinc phosphate pigment incorporation. *Prog. Org. Coat.* **2007**, *58*, 64–69. [CrossRef]
53. Aravindan, N.; Sangaranarayanan, M.V. Influence of solvent composition on the Anticorrosion performance of copper polypyrrole (Cu–PPy) coated 304 stainless steel steel. *Prog. Org. Coat.* **2016**, *95*, 38–45. [CrossRef]

54. Shaktawat, V.; Jain, N.; Dixit, M.; Saxena, N.S.; Sharma, K.; Sharma, T.P. Temperature dependence of conductivity of polypyrrole doped with sulphuric acid. *Indian. J. Pure Appl. Phys.* **2008**, *46*, 427–430.
55. Shankar, K.; Mor, G.K.; Paulose, M.; Varghese, O.K.; Grimes, C.A. Effect of device geometry on the performance of TiO_2 nanotube array-organic semiconductor double heterojunction solar cells. *J. Non-Cryst. Solids* **2008**, *354*, 2767–2771. [CrossRef]
56. Abidian, M.R.; Martin, D.C. Experimental and theoretical characterization of Implantable neural microelectrodes modified with conducting polymers nanotubes. *Biomaterials* **2008**, *28*, 1273–1283. [CrossRef]
57. Zeng, T.W.; Lo, H.H.; Lin, Y.Y.; Chen, C.W.; Su, W.F. Hybrid poly(3-hexylthio Phene)/titanium dioxide nanorods material for solar cell applications. *Sol. Energy Mater. Sol. Cell* **2009**, *93*, 952–957. [CrossRef]
58. Rupali, G.; Amitabha, D. Conducting polymer nanocomposites: A Brief Overview. *Chem. Mater.* **2000**, *12*, 608–622.
59. Cao, Y.; Qiu, J.; Smith, P. Effect of solvent and co-solvents on the processability of polyaniline-spectroscopic and diffraction studies. *Synth. Met.* **1995**, *69*, 187–190. [CrossRef]
60. Reghu, M.; Yoon, C.O.; Yang, C.Y.; Moses, D.; Heger, A.J. Superlocalization of the electronic wave functions in conductive polymer blends at concentrations near the percolation threshold. *Macromolecules* **1993**, *26*, 7245–7249. [CrossRef]
61. Olinga, T.E.; Fraysse, J.; Travers, J.P.; Dufresne, A.; Pron, A. Highly conducting and solution-processable polyaniline obtained via protonation with a new sulfonic acid containing plasticizing functional groups. *Macromolecules* **2000**, *33*, 2107–2113. [CrossRef]
62. Zhan, L.; Chen, H.; Fang, J.; Wang, S.; Ding, L.X.; Li, Z.; Ashman, P.J.; Wang, H. Coaxial Co_3O_4 polypyrrole core-shell nanowire arrays for high performance lithium ion. *Electrochim. Acta* **2016**, *209*, 192–200. [CrossRef]
63. McGehee, D.G.; Topinka, M.A. Conducting Polymers Applications for Electronic Devices and Sensors. *Nat. Mater.* **2006**, *5*, 675–684. [CrossRef] [PubMed]
64. Costamagna, P.; Srinivasan, S. Quantum jumps in the PEMFC science and techno-logy from the 1960s to the year 2000 Part II. Engineering, technology development and application aspects. *J. Power Sources* **2001**, *102*, 253–269. [CrossRef]
65. Quartarone, E.; Angioni, S.; Mustarelli, P. Polymer and Composite Membranes for Proton-Conducting High-Temperature Fuel Cells. *Materials* **2017**, *10*, 687. [CrossRef] [PubMed]
66. Rajesh, B.; Thampi, K.R.; Bonard, J.M.; Mathieu, J.H.; Xanthopoulos, S.N.; Viswanathan, S. Conducting polymeric nanotubules as high-performance methanol oxidation catalyst catalyst Support. *Chem. Commun.* **2003**, *16*, 2022–2023. [CrossRef] [PubMed]
67. Becker, H.; Spreitzer, H.; Kreuder, W.; Kluge, E.; Schenk, H.; Parker, I.; Cao, Y. Soluble PPVs with enhanced performance A mechanistic approach. *Adv. Mater.* **2000**, *12*, 42–48. [CrossRef]
68. Lee, S.H.; Jang, B.B.; Tsutsui, T. Sterically hindered fluorenyl substituted poly (p-phenylenevinylenes) for light-emitting diodes. *Macromolecules* **2002**, *35*, 1356–1364. [CrossRef]
69. Jin, Y.; Kang, J.H.; Song, S.; Park, S.H.; Moon, J.; Woo, H.Y.; Lee, K.H.; Suh, H.S. Poly-(phenylenevinylene) derivatives containing a new electron-withdrawing phenyl group for LEDs. *Bull. Korean Chem. Soc.* **2008**, *29*, 139–147.
70. Sokolik, I.; Yang, Z.; Karasz, F.E.; Morton, D.C. Blue-light electroluminescence from phenylene vinylene-based copolymers. *J. Appl. Phys.* **1993**, *74*, 3584–3586. [CrossRef]
71. Pasquier, A.D.; Laforgue, A.; Simon, P.; Amatucci, G.G.; Fauvarque, J.F. A non-aqueous Asymmetric Hybrid $Li_4Ti_5O_{12}$/Poly (fluorophenylthiophene) Energy Storage Device. *J. Electrochem. Soc.* **2002**, *149*, A302–A306. [CrossRef]
72. Sarangapani, S.; Tilak, B.V.; Chen, C.P. Materials for electrochemical capacitors. *J. Electrochem. Soc.* **1996**, *143*, 3791–3799. [CrossRef]
73. Faggioli, E.; Rena, P.; Danel, V.; Andrieu, X.; Mallant, R.; Kahlen, H. Supercapacitors for the energy management of electric vehicles. *J. Power Sources* **1999**, *84*, 261–269. [CrossRef]
74. Ates, M.; Sarac, A.S. Electrochemical impedance spectroscopic study of poly-thiophene on carbon materials. *Polym. Plast. Technol. Eng.* **2011**, *50*, 1130–1148. [CrossRef]
75. Ryu, K.S.; Lee, Y.; Has, K.S.; Park, Y.J.; Kang, M.G.; Park, N.G.; Chang, S.H. Electrochemial supercapacitor based on polyaniline doped with lithium salt and active carbon electrodes. *Solid State Ion.* **2004**, *175*, 765–769. [CrossRef]
76. Garcia, M.L.A.; Smit, M.A. Study of electrodeposited polypyrrole coatings for the corrosion protection of stainless-steel bipolar plates for the PEM fuel cell. *J. Power Sources* **2006**, *158*, 397–402. [CrossRef]
77. Ren, Y.J.; Chen, J.; Zeng, C.L. Corrosion protection of type 304 stainless steel bipolar plates of proton-exchange membrane fuel cells by doped polyaniline coating. *J. Power Sources* **2010**, *195*, 1914–1919. [CrossRef]
78. Kilmartin, P.A.; Trier, L.; Wright, G.A. Corrosion inhibition of polyaniline and poly (o-methoxyaniline) on stainless steels. *Synth. Met.* **2002**, *131*, 99–109. [CrossRef]
79. Ocon, P.; Ibanez, A.; Fatas, E. Electrochemical and mechanical properties of poly-pyrrole coatings on steel. *Electrochim. Acta* **2004**, *49*, 3693–3699.
80. Tuken, T.; Ozyilmaz, A.T.; Yazic, B.; Erbil, M. Electrochemical synthesis of polyaniline on mild steel in acetonitrile-$LiClO_4$ and corrosion performance. *Appl. Surf. Sci.* **2004**, *236*, 292–305. [CrossRef]
81. Santros, J.R.; Mattoso, L.H.C.; Mothed, A.J. Investigation of corrosion protection of steel by polyaniline films. *Electrochim. Acta* **1998**, *43*, 309–313. [CrossRef]
82. Talo, A.; Forsen, O.; Yeassani, S. Corrosion protective polyaniline epoxy blend coating on mild steel. *Synth. Met.* **1999**, *102*, 1394–1396. [CrossRef]
83. Xu, J.; Zhang, Y.; Zhang, D.; Tang, Y.; Cang, H. Electrosynthesis of PANI/PPy coating doped by phosphotungstate on mild steel and their corrosion resistances. *Prog. Org. Coat.* **2015**, *88*, 84–91. [CrossRef]

84. Sazou, D.; Georgolios, C. Formation of conducting polyaniline coatings on iron surfaces by electropolymerization of aniline in aqueous solutions. *J. Electroanal. Chem.* **1997**, *429*, 81–93. [CrossRef]
85. Sazou, D. Electrodeposition of ring-substituted polyanilines on Fe surfaces from aqueous oxalic acid solutions and corrosion protection of Fe. *Synth. Met.* **2001**, *118*, 133–147. [CrossRef]
86. Keles, H.; Solmaz, R.; Ozcan, M.; Kardas, G.; Dehri, I. Copper modified poly-6-amino m-cresol (poly-AmC/Cu) coating for mild steel protection. *Surf. Coat. Technol.* **2009**, *203*, 1469–1473. [CrossRef]
87. Zhou, Q.; Shi, G. Conducting Polymer-Based Catalysts. *J. Am. Chem. Soc.* **2016**, *138*, 2868–2876. [CrossRef]
88. Gao, F.; Hou, X.; Wang, A.; Chu, G.; Wu, W.; Chen, J.; Zou, H. Preparation of Polyyrrole/TiO_2 nanocomposites with enhanced photocatalytic performance. *Particuology* **2016**, *26*, 73–78. [CrossRef]
89. Zhang, H.; Liu, Y.; Wu, J.; Xin, B. One step preparation of Fe_3O_4/Pd@ polypyrrole composites with enhanced catalytic activity and stability. *J. Colloid Interface Sci.* **2016**, *476*, 214–221. [CrossRef]
90. Hopkins, A.R.; Lewis, N.S. Detection and classification characteristics of arrays of carbon black/organic polymer composite chemiresistive vapor detectors for the nerve agent stimulants dimethyl methyl phosphonate and di isopropyl methyl phosponate. *Anal. Chem.* **2001**, *73*, 884–892. [CrossRef]
91. Doleman, B.J.; Lewis, N.S. Comparison of odor detection thresholds and odor discriminablities of a conducting polymer composite electronic nose versus mammalian olfaction. *Sens. Actuators B* **2001**, *72*, 41–50. [CrossRef]
92. Jin, G.; Norrish, J.; Too, C.; Wallace, G. Polypyrrole filament sensors for gases and vapors. *Curr. Appl. Phys.* **2004**, *4*, 366–369. [CrossRef]
93. Fang, Q.; Chetwynd, D.G.; Gardner, J.W. Conducting polymer films by UV photo Processing. *Sens. Actuators A Phys.* **2002**, *99*, 74–77. [CrossRef]
94. Mashat, L.A.; Tran, H.D.; Wlodarski, W.; Kaner, R.B.; Zadeh, K.K. Conductometric hydrogen gas sensors based on polypyrrole nanofibers. *Sens. J.* **2008**, *4*, 365–370. [CrossRef]
95. Nambiar, S.; Yeow, J. Conductive polymer-based sensors for biomedical applications. *Biosens. Bioelectron.* **2011**, *26*, 1825–1832. [CrossRef]
96. Umana, M.; Waller, J. Protein modified electrodes: The glucose oxidase/polypyrrole system. *Anal. Chem.* **1986**, *58*, 2979–2983. [CrossRef]
97. Osaka, T.; Komaba, S.; Fujino, Y.; Matsuda, T.; Satoh, I. High sensitivity flow injection analysis of urea using composite electropolymerized polypyrrole-polyion complex film. *J. Electrochem. Soc.* **1999**, *146*, 615–619. [CrossRef]
98. Marco, M.P.; Barcelo, D. Environment applications of analytical biosensors. *Meas. Sci. Technol.* **1996**, *7*, 1547–1572. [CrossRef]
99. Zheng, W. Effect of organic vapors on the molecular conformation of non-doped polyaniline. *Synth. Met.* **1997**, *84*, 63–64. [CrossRef]
100. Selvakumar, S.; Somanathan, N.; Reddy, K.A. Chemi resistor sensors based on conducting polymers for hypergolic propellants and acidic vapors of rocket exhaust plumes—A review. *Propellants Explos. Pyrotech.* **2013**, *38*, 176–189. [CrossRef]
101. Fu, X.W.; Liao, Z.M.; Zhou, J.X.; Zhou, Y.B.; Wu, H.C.; Zhang, R.; Yu, D. Strain dependent resistance in chemical vapor deposition grown graphene. *Appl. Phys. Lett.* **2011**, *99*, 213107. [CrossRef]
102. Wang, Y.; Yang, R.; Shi, Z.; Zhang, L.; Shi, D.; Wang, E.; Zhang, G. Super-elastic graphene ripples for flexible strain sensors. *ACS Nano* **2011**, *5*, 3645–3650. [CrossRef] [PubMed]
103. Diaconu, I.; Dorohoi, D.O.; Topoliceanu, F. Electrostriction of a polyurethane elastomer-based polyester. *IEEE Sens. J.* **2006**, *6*, 876–880. [CrossRef]
104. Shahinpoor, M. Ionic polymer–conductor composites as biomimetic sensors, robotic actuators and artificial muscles—A review. *Electrochim. Acta* **2003**, *48*, 2343–2353. [CrossRef]
105. Chen, T.; Qiu, J.; Zhu, K.; He, X.; Kang, X.; Dong, E.L. Poly (methyl methacrylate)-functionalized graphene/polyurethane dielectric elastomer composites with superior electric field induced strain. *Mater. Lett.* **2014**, *128*, 19–22. [CrossRef]
106. Zhu, S.E.; Shabani, R.; Rho, J.; Kim, Y.; Hong, B.H.; Ahn, J.H.; Cho, H.J. Graphene-based bimorph microactuators. *Nano Lett.* **2011**, *11*, 977–981. [CrossRef]
107. Liang, J.; Huang, L.; Li, N.; Huang, Y.; Wu, Y.; Fang, S.; Baughman, R. Electromechanical actuator with controllable motion, fast response rate, and high-frequency resonance based on graphene and polydiacetylene. *ACS Nano* **2012**, *6*, 4508–4519. [CrossRef] [PubMed]
108. Sen, I.; Seki, Y.; Sarikanat, M.; Cetin, L.; Gurses, B.O.; Ozdemir, O.; Yilmaz, O.C.; Sever, K.; Akar, E.; Mermer, O. Electroactive behavior of graphene nanoplatelets loaded cellulose composite actuators. *Compos. Part B Eng.* **2015**, *69*, 369–377. [CrossRef]
109. Jung, J.H.; Jeon, J.H.; Sridhar, V.; Oh, I.K. Electro-active graphene–Nafion actuators. *Carbon* **2011**, *49*, 1279–1289. [CrossRef]
110. Surana, K.; Singh, P.K.; Bhattacharya, B.; Verma, C.S.; Mehra, R.M. Synthesis of graphene oxide coated Nafion membrane for actuator application. *Ceram. Int.* **2015**, *41*, 5093–5099. [CrossRef]
111. Loomis, J.; King, B.; Burkhead, T.; Xu, P.; Bessler, N.; Terentjev, E.; Panchapakesan, B. Graphene-nanoplatelet-based photomechanical actuators. *Nanotechnology* **2012**, *23*, 045501. [CrossRef]
112. Loomis, J.; King, B.; Panchapakesan, B. Layer dependent mechanical responses of graphene composites to near-infrared light. *Appl. Phys. Lett.* **2012**, *100*, 073108. [CrossRef]
113. Ansari, S.; Rahima, C.; Muralidharan, M.N. Photomechanical Characteristics of Thermally Reduced Graphene Oxide–Polydimethylsiloxane Nanocomposites. *Polym. Plast. Technol. Eng.* **2013**, *52*, 1604–1610. [CrossRef]

114. Ansari, S.; Muralidharan, M.N.; Ushus, D. Graphene/poly (styrene–b–isoprene–b–styrene) nanocomposite optical actuators. *J. Appl. Polym. Sci.* **2013**, *130*, 3902–3908. [CrossRef]
115. Deng, T.; Yoon, C.; Jin, Q.; Li, M.; Liu, Z.; Gracias, D.H. Selffolding graphene-polymer bilayers. *Appl. Phys. Lett.* **2015**, *106*, 203108. [CrossRef]
116. Song, B.; Wu, Z.; Zhu, Y.; Moon, K.S.; Wong, C.P. Three-dimensional graphene-based composite for flexible electronic applications. In Proceedings of the 2015 IEEE 65th Electronic Components and Technology Conference (ECTC), San Diego, CA, USA, 26–29 May 2015; pp. 1803–1807.
117. Kim, Y.J.; Park, H.C.; Kim, B.K. Triple shape-memory effect bysilanized polyurethane/silane-functionalized graphene oxide nanocomposites bilayer. *High Perform. Polym.* **2015**, *27*, 886–897. [CrossRef]
118. Tan, L.; Gan, L.; Hu, J.; Zhu, Y.; Han, J. Functional shape memory composite nanofibers with graphene oxide filler. *Compos. Part A Appl. Sci. Manuf.* **2015**, *76*, 115–123. [CrossRef]
119. Ponnamma, D.; Sadasivuni, K.K.; Strankowski, M.; Moldenaers, P.; Thomas, S.; Grohens, Y. Interrelated shape memory and Payne effect in polyurethane/graphene oxide nanocomposites. *RSC Adv.* **2013**, *3*, 16068–16079. [CrossRef]
120. Tao, L.; Zhou, B.; Bai, G.; Wang, Y.; Yu, S.F.; Lau, S.P.; Xu, D. Fabrication of covalently functionalized graphene oxide incorporated solid-state hybrid silica gel glasses and their improved nonlinear optical response. *J. Phys. Chem. C* **2013**, *117*, 23108–23116. [CrossRef]
121. Du, Y.; Dong, N.; Zhang, M.; Zhu, K.; Na, R.; Zhang, S.; Wang, J. Covalent functionalization of graphene oxide with porphyrin and porphyrin incorporated polymers for optical limiting. *Phys. Chem. Chem. Phys.* **2017**, *19*, 2252–2260. [CrossRef]
122. Gan, Y.; Feng, M.; Zhan, H. Enhanced optical limiting effects of graphene materials in polyimide. *Appl. Phys. Lett.* **2014**, *104*, 171105. [CrossRef]
123. Husaini, S.; Slagle, J.E.; Murray, J.M.; Guha, S.; Gonzalez, L.P.; Bedford, R.G. Broadband saturable absorption and optical limiting in graphene-polymer composites. *Appl. Phys. Lett.* **2013**, *102*, 191112. [CrossRef]
124. Elie, A.G. Electroconductive hydrogels: Synthesis, characterization and biomedical applications. *Biomaterials* **2010**, *31*, 2701–2716. [CrossRef]
125. Owens, D.R.; Zinman, B.; Bolli, G. Alternative routes of insulin delivery. *Diabet. Med.* **2003**, *20*, 886–898. [CrossRef] [PubMed]
126. Thompson, B.C.; Moulton, S.E.; Ding, J.; Richardson, R.; Cameron, A.; Leary, S.O.; Wallace, G.G.; Clark, G.M. Optimizing the incorporation and release of a neurotrophic factor using conducting polypyrrole. *J. Control Release* **2006**, *116*, 285–294. [CrossRef] [PubMed]
127. Luo, X.; Matranga, C.; Tan, S.; Alba, N.; Cui, X.T. Carbon nanotube nano-reservoir for controlled release of anti-inflammatory dexamethasone. *Biomaterials* **2011**, *32*, 6316–6323. [CrossRef]
128. Herrasti, P.; Kulak, A.N.; Bavykin, D.V.; Ponce, C.; Walsh, F.C. Electrodeposition of polypyrrole titanate nanotube composites coatings and their corrosion resistance. *Electrochim. Acta* **2011**, *56*, 1323–1328. [CrossRef]
129. Massoumi, B.; Entezami, A.A. Electrochemically stimulated 2-ethyl hexyl phosphate (EHP) release through redox switching of conducting polypyrrole film and polypyrrole/poly (N-methylpyrrole) self-doped polyaniline bilayers. *Polym. Int. J.* **2002**, *51*, 555–560. [CrossRef]
130. Miller, L.L.; Zhou, X.U. Poly (N-methyl pyrrolylium) poly (styrene sulfonate) a conductive electrically switchable cation exchanger that cathodically binds and anodically releases dopamine. *Macromolecules* **1987**, *20*, 1594–1597. [CrossRef]
131. Bendrea, A.D.; Cianga, L.; Cianga, I. Progress in the field of conducting polymers for tissue engineering applications. *J. Biomater. Appl.* **2011**, *26*, 3–84. [CrossRef]
132. Zhao, Y.; Cao, L.; Li, L.; Cheng, W.; Xu, L.; Ping, X.; Pan, L.; Shi, Y. Conducting Polymers and Their Applications in Diabetes Management. *Sensors* **2016**, *16*, 1787. [CrossRef]
133. Gupta, K.; Jana, P.; Meikap, A.; Nath, T. Synthesis of La0.67Sr0.33MnO$_3$ and polyaniline nanocomposite with its electrical and magneto-transport properties. *J. Appl. Phys.* **2010**, *107*, 073704. [CrossRef]
134. Huang, Z.; Wang, S.; Li, H.; Zhang, S.; Tan, Z.-C. Thermal stability of several polyaniline/rare earth oxide composites: Part IV. Polyaniline/La_2O_3 and polyaniline/Sm_2O_3 composites. *J. Therm. Anal. Calorim.* **2014**, *115*, 259–266. [CrossRef]
135. Wang, S.; Huang, Z.; Wang, J.; Li, Y.; Tan, Z. Thermal stability of several polyaniline/rare earth oxide composites (I): Polyaniline/CeO_2 composites. *J. Therm. Anal. Calorim.* **2012**, *107*, 1199–1203. [CrossRef]
136. Thayyil, S.; Al-Maghrabi, M.; Bahuleyan, B.; Ramesan, M.T. Synthesis, characterization, thermal properties, conductivity and sensor application study of polyaniline/cerium-doped titanium dioxide nanocomposites. *J. Mater. Sci.* **2018**, *53*, 591–603. [CrossRef]
137. Zhang, S.; Wang, S.; Huang, Z.; Li, Y.; Tan, Z. A kinetic analysis of thermal decomposition of polyaniline and its composites with rare earth oxides. *J. Therm. Anal. Calorim.* **2014**, *119*, 1853–1860. [CrossRef]
138. Parvatikar, N.; Jain, S.; Khasim, S.; Revansiddappa, M.; Bhoraskar, S.; Prasad, M. Electrical and humidity sensing properties of polyaniline/WO_3 composites. *Sens. Actuators B Chem.* **2006**, *114*, 599–603. [CrossRef]
139. Khasim, S.; Ansari, J.; Parveen, A.; Al-Hartomy, O.; Khattar, Z.; Badi, N.; Roy, A. Synthesis, characterization, dielectric and rectification properties of PANI/Nd_2O_3:Al_2O_3 nanocomposites. *Polym. Adv. Technol.* **2016**, *27*, 1064–1071. [CrossRef]
140. Hoshino, K.; Yazawa, N.; Tanaka, Y.; Chiba, T.; Izumizawa, T.; Kubo, M. Polycarbazole nanocomposites with conducting metal oxides for transparent electrode applications. *ACS Appl. Mater. Interfaces* **2010**, *22*, 413–424. [CrossRef]

141. Sever, E.; Unal, H. Colloidal properties of surface functionalized nanocube-TiO_2/poly(3-octylthiophene) core/shell conducting nanocomposite. *Appl. Surf. Sci.* **2015**, *355*, 1028–1036. [CrossRef]
142. Erdönmez, S.; Karabul, Y.; Kilic, M.; Özdemir, Z.; Esmer, K. Structural characterization and dielectric parameters of Polyindole/WO_3 nanocomposites. *Polym. Compos.* **2021**, *42*, 1347–1355. [CrossRef]
143. Galembeck, A.; Alves, O. Chemical polymerization of pyrrole on CeO_2 films. *Synth. Met.* **1997**, *84*, 151–152. [CrossRef]
144. Karimi, A.; Husain, S.W.; Hosseini, M.; Azar, P.A.; Ganjali, M.R. Rapid and sensitive detection of hydrogen peroxide in milk by Enzyme-free electrochemiluminescence sensor based on a polypyrrole-cerium oxide nanocomposite. *Sens. Actuators B Chem.* **2018**, *271*, 90–96. [CrossRef]
145. Seemaa, S.; Ambika Prasad, M.V.N. Synthesis, Characterisation and Conductivity Studies of Polypyrrole-Nb_2O_5 Composites. *Int. J. Ethics Eng. Manag.* **2014**, *1*, 270–272.
146. Vishnuvardhan, T.K.; Kulkarni, V.R.; Basavaraja, C.; Raghavendra, S.C. Synthesis, characterization and A.C. conductivity of polypyrrole/Y_2O_3 composites. *Bull. Mater. Sci.* **2006**, *29*, 77–83. [CrossRef]
147. Sun, W.; Mo, Z. Polypyrrole coated graphene nanoplatelets and the effect of rare earth ions with nanocomposites. *J. Polym. Res.* **2014**, *21*, 516. [CrossRef]
148. Majumder, M.; Choudhary, R.; Thakur, A.; Karbhal, I. Impact of rare-earth metal oxide (Eu_2O_3) on the electrochemical properties of a polypyrrole/CuO polymeric composite for supercapacitor applications. *RSC Adv.* **2017**, *7*, 20037–20048. [CrossRef]
149. Reddy, M.O.; Chandra Babu, B. Structural, Optical, Electrical, and Magnetic Properties of PVA: Gd^{3+} and PVA: Ho^{3+} Polymer Films. *Indian J. Mater. Sci.* **2015**, *2015*, 927364. [CrossRef]
150. Mohanapriya, M.K.; Deshmukh, K.; Ahamed, M.B.; Chidambaram, K.; Khadheer Pasha, S.K. Influence of cerium oxide (CeO_2) nanoparticles on the structural, Morphological, mechanical and dielectric properties of PVA/PPy blend Nanocomposites. *Mater. Today Proc.* **2016**, *3*, 1864–1873. [CrossRef]
151. Song, J.; Lu, C.; Xu, D.; Ni, Y.; Liu, Y.; Xu, Z.; Liu, J. The effect of lanthanum oxide (La_2O_3) on the structure and crystallization of poly(vinylidene fluoride). *Polym. Int.* **2010**, *59*, 954–960. [CrossRef]
152. Ilyas, R.A.; Sapuan, S.M.; Asyraf, M.R.M.; Dayana, D.A.Z.N.; Amelia, J.J.N.; Rani, M.S.A.; Norrrahim, M.N.F.; Nurazzi, N.M.; Aisyah, H.A.; Sharma, S.; et al. Polymer Composites Filled with Metal Derivatives: A Review of Flame Retardants. *Polymers* **2021**, *13*, 1701. [CrossRef] [PubMed]
153. Chohan, J.S.; Mittal, N.; Kumar, R.; Singh, S.; Sharma, S.; Dwivedi, S.P.; Saxena, A.; Chattopadhyaya, S.; Ilyas, R.A.; Le, C.H.; et al. Optimization of FFF Process Parameters by Naked Mole-Rat Algorithms with Enhanced Exploration and Exploitation Capabilities. *Polymers* **2021**, *13*, 1702. [CrossRef]
154. Chohan, J.S.; Mittal, N.; Kumar, R.; Singh, S.; Sharma, S.; Singh, J.; Rao, K.V.; Mia, M.; Pimenov, D.Y.; Dwivedi, S.P. Mechanical Strength Enhancement of 3D Printed Acrylonitrile Butadiene Styrene Polymer Components Using Neural Network Optimization Algorithm. *Polymers* **2020**, *12*, 2250. [CrossRef]
155. Singh, Y.; Singh, J.; Sharma, S. Multi-objective Optimization of Kerf-taper and Surface-roughness Quality Characteristics for Cutting-operation On Coir and Carbon Fibre Reinforced Epoxy Hybrid Polymeric Composites During CO2-Pulsed Laser-cutting Using RSM. *Lasers Manuf. Mater. Process.* **2021**, *8*, 157–182. [CrossRef]
156. Sharma, S.; Singh, J.; Kumar, H.; Sharma, A.; Aggarwal, V.; Gill, A.; Jayarambabu, N.; Kailasa, S.; Rao, K.V. Utilization of rapid prototyping technology for the fabrication of an orthopedic shoe inserts for foot pain reprieve using thermo-softening viscoelastic polymers: A novel experimental approach. *Meas. Control* **2020**, *53*, 519–530. [CrossRef]
157. Singh, Y.; Singh, J.; Sharma, S.; Sharma, A.; Chohan, J. Process Parameter Optimization in Laser Cutting of Coir Fiber Reinforced Epoxy Composite—A Review. *Mater. Today Proc.* **2021**. [CrossRef]
158. Chohan, J.S.; Kumar, R.; Singh, T.B.; Singh, S.; Sharma, S.; Singh, J.; Mia, M.; Pimenov, D.Y.; Chattopadhyaya, S.; Dwivedi, S.P.; et al. Taguchi S/N and TOPSIS Based Optimization of Fused Deposition Modelling and Vapor Finishing Process for Manufacturing of ABS Plastic Parts. *Materials* **2020**, *13*, 5176. [CrossRef] [PubMed]
159. Omran, A.A.B.; Mohammed, A.A.B.A.; Sapuan, S.M.; Ilyas, R.A.; Asyraf, M.R.M.; Rahimian Koloor, S.S.; Petrů, M. Micro- and Nanocellulose in Polymer Composite Materials: A Review. *Polymers* **2021**, *13*, 231. [CrossRef]
160. Prabhakaran, S.; Vijayan, K.; Sharma, S.; Mouleeswaran, S.K.; Ramasamy, J.K.; Redoune, Z. Experimental study on thermal and morphological analysis of Green composite sandwich made of Flax and agglomerated cork. *J. Therm. Anal. Calorim.* **2020**, *139*, 3003–3012. [CrossRef]
161. Singh, Y.; Singh, J.; Sharma, S.; Lam, T.D.; Nguyen, D.N. Fabrication and characterization of coir/carbon-fiber reinforced epoxy-based hybrid composite for helmet shells and sports-good applications: Influence of fiber surface modifications on the mechanical, thermal and morphological properties. *J. Mater. Res. Technol.* **2020**, *9*, 15593–15603. [CrossRef]
162. Li, H.-Y.; Huang, D.-N.; Ji, J. Inorganic-polymer composite coatings for biomedical devices. *Smart Mater. Med.* **2021**, *2*, 1–14. [CrossRef]
163. Rikhari, B.; Mani, S.P.; Rajendran, N. Polypyrrole/graphene oxide composite coating on Ti implants: A promising material for biomedical applications. *J. Mater. Sci.* **2020**, *55*, 5211–5229. [CrossRef]
164. Nautiyal, A.; Qiao, M.; Cook, J.; Zhang, X.; Huang, T. High Performance Polypyrrole Coating for Corrosion Protection and Biocidal Applications. *Appl. Surf. Sci.* **2018**, *427*, 922–930. [CrossRef]
165. Yadav, R.; Tirumali, M.; Wang, X.; Naebe, M.; Kandasubramanian, B. Polymer composite for antistatic application in aerospace. *Def. Technol.* **2020**, *16*, 107–118. [CrossRef]

166. Zamiri, G.; Haseeb, A.S.M.A. Recent Trends and Developments in Graphene/Conducting Polymer Nanocomposites Chemiresistive Sensors. *Materials* **2020**, *13*, 3311. [CrossRef]
167. Ghosh, S.; Das, S.; Mosquera, M.E.G. Conducting Polymer-Based Nanohybrids for Fuel Cell Application. *Polymers* **2020**, *12*, 2993. [CrossRef]
168. Sharma, S.; Sudhakara, P.; Singh, J.; Ilyas, R.A.; Asyraf, M.R.M.; Razman, M.R. Critical Review of Biodegradable and Bioactive Polymer Composites for Bone Tissue Engineering and Drug Delivery Applications. *Polymers* **2021**, *13*, 2623. [CrossRef]

polymers MDPI

Review

Polymer Composites Filled with Metal Derivatives: A Review of Flame Retardants

R. A. Ilyas [1,2,*], S. M. Sapuan [3,4], M. R. M. Asyraf [5], D. A. Z. N. Dayana [4], J. J. N. Amelia [4], M. S. A. Rani [6,7], Mohd Nor Faiz Norrahim [8], N. M. Nurazzi [9], H. A. Aisyah [3,*], Shubham Sharma [10], M. R. Ishak [5], M. Rafidah [11] and M. R. Razman [12,*]

1 School of Chemical and Energy Engineering, Faculty of Engineering, Universiti Teknologi Malaysia (UTM), Johor Bahru 81310, Johor, Malaysia
2 Centre for Advanced Composite Materials (CACM), Universiti Teknologi Malaysia (UTM), Johor Bahru 81310, Johor, Malaysia
3 Laboratory of Biocomposite Technology, Institute of Tropical Forestry and Forest Products (INTROP), Universiti Putra Malaysia (UPM), Serdang 43400, Selangor, Malaysia; sapuan@upm.edu.my
4 Advanced Engineering Materials and Composites (AEMC), Department of Mechanical and Manufacturing Engineering, Faculty of Engineering, Universiti Putra Malaysia (UPM), Serdang 43400, Selangor, Malaysia; nuruldayanaa64@gmail.com (D.A.Z.N.D.); ameliajuria96@gmail.com (J.J.N.A.)
5 Department of Aerospace Engineering, Faculty of Engineering, Universiti Putra Malaysia (UPM), Serdang 43400, Selangor, Malaysia; asyrafriz96@gmail.com (M.R.M.A.); mohdridzwan@upm.edu.my (M.R.I.)
6 Solar Energy Research Institute (SERI), Universiti Kebangsaan Malaysia (UKM), Bangi 43600, Selangor, Malaysia; saifulasmal@gmail.com
7 Centre for Tropicalisation, National Defence University of Malaysia, Kem Sungai Besi, Kuala Lumpur 57000, Malaysia
8 Research Center for Chemical Defence, Universiti Pertahanan Nasional Malaysia (UPNM), Kem Perdana Sungai Besi, Kuala Lumpur 57000, Malaysia; faiznorrrahim@gmail.com
9 Centre for Defence Foundation Studies, Universiti Pertahanan Nasional Malaysia (UPNM), Kem Perdana Sungai Besi, Kuala Lumpur 57000, Malaysia; mohd.nurazzi@gmail.com
10 Department of Mechanical Engineering, Main Campus, IK Gujral Punjab Technical University, Kapurthala 144603, India; shubham543sharma@gmail.com or shubhamsharmacsirclri@gmail.com
11 Department of Civil Engineering, Faculty of Engineering, Universiti Putra Malaysia (UPM), Serdang 43400, Selangor, Malaysia; rafidahmazlan16@gmail.com
12 Research Centre for Sustainability Science and Governance (SGK), Institute for Environment and Development (LESTARI), Universiti Kebangsaan Malaysia (UKM), Bangi 43600, Selangor, Malaysia
* Correspondence: ahmadilyas@utm.my (R.A.I.); a.humaira.aisyah@gmail.com (H.A.A.); mrizal@ukm.edu.my (M.R.R.)

Citation: Ilyas, R.A.; Sapuan, S.M.; Asyraf, M.R.M.; Dayana, D.A.Z.N.; Amelia, J.J.N.; Rani, M.S.A.; Norrahim, M.N.F.; Nurazzi, N.M.; Aisyah, H.A.; Sharma, S.; et al. Polymer Composites Filled with Metal Derivatives: A Review of Flame Retardants. *Polymers* **2021**, *13*, 1701. https://doi.org/10.3390/polym13111701

Academic Editor: Paolo Ferruti

Received: 29 April 2021
Accepted: 19 May 2021
Published: 23 May 2021

Abstract: Polymer composites filled with metal derivatives have been widely used in recent years, particularly as flame retardants, due to their superior characteristics, including high thermal behavior, low environmental degradation, and good fire resistance. The hybridization of metal and polymer composites produces various favorable properties, making them ideal materials for various advanced applications. The fire resistance performance of polymer composites can be enhanced by increasing the combustion capability of composite materials through the inclusion of metallic fireproof materials to protect the composites. The final properties of the metal-filled thermoplastic composites depend on several factors, including pore shape and distribution and morphology of metal particles. For example, fire safety equipment uses polyester thermoplastic and antimony sources with halogenated additives. The use of metals as additives in composites has captured the attention of researchers worldwide due to safety concern in consideration of people's life and public properties. This review establishes the state-of-art flame resistance properties of metals/polymer composites for numerous industrial applications.

Keywords: flame retardant; polymer composites; metal; metal components; characterization; combustion mechanism

1. Introduction

Polymer composites are globally recognized due to their thermal insulation properties. To improve their thermal and heat resistant performance further, certain metallic materials are added to polymers, such as copper [1], nickel [2], magnesium [3–5], and zinc [6]. The inclusion of metal component in polymer-based composites has produced new promising materials with high potential in various engineering sectors. Metal-filled composites offer numerous advantages, such as heat conduction, static electricity discharge, conversion of mechanical signal into electrical signal, and electromagnetic interference shielding signal [7]. Metal-filled polymer composites have been widely used in recent years, particularly as flame retardants, due to their superior characteristics. Figure 1 shows the trending of research conducted on flame retardant metal filled polymer composite. It can be seen that the trending of this research is increasing by 2000% over 20 years. These conductive polymer composites potentially combine significant advantageous characteristics of plastics and metals, offering less cost with high production rate [8], design flexibility [9], noncorrosive [10], and lightweight properties [11]. Processing methods, including the use of an internal mixer and extrusion and injection molding, can be adopted to fabricate these compounds [12–14].

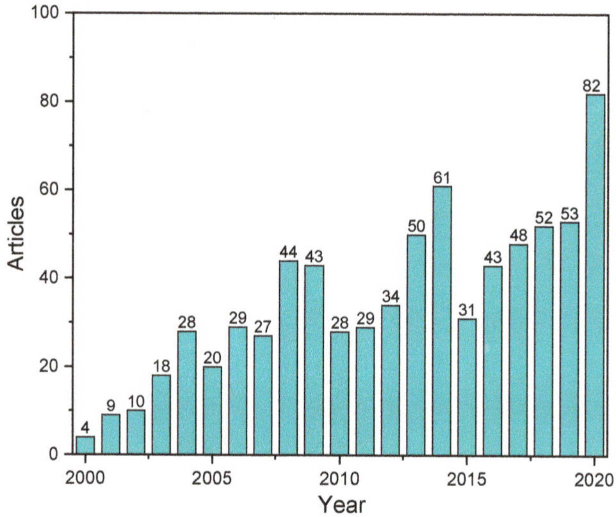

Figure 1. The number of publications on flame retardant metal polymer composite in the last two decades indicating the increasing interest from (Scopus, May 2021, Search: "Metal" and "Flame" and "Polymer").

Fiber reinforced composites can be classified into four groups according to their matrices: metal matrix composites (MMCs), ceramic matrix composites (CMCs), carbon/carbon composites (C/C), and polymer matrix composites (PMCs) or polymeric composites [15]. The four forms of polymer composite materials are widely used in vehicles, aircraft, spacecraft, boats [16], civil engineering [17], packing [18], cross arms in transmission towers [19], and portable fire extinguisher [20]. The use of polymer composites is increasing rapidly due to their excellent mechanical properties, such as creep [21–23], flexural [24], chemical resistance [18], and corrosion resistance [25]. Polymer materials are formed from hydrocarbon chains, which burn easily under intense heat; they can also burst into flames or emit smoke when exposed to light [26]. Numerous incidents have occurred previously in aircraft; at present, however, a remarkable increase in the fire tolerance of polymer composite materials can be observed during collisions [27]. To increase environmental sustainability, engineers and scientists are currently seeking to replace nonbiodegradable fibers (e.g., glass and carbon–aramids) with biodegradable fibers (e.g., corn [28],water hyacinth [29], coir [30], ginger [31,32], cotton [33,34], kenaf [11,35–39], sugarcane [40–43], flax [44], ramie [45], hemp [46], kapok [47], sisal [48], wood [17], oil palm [3,49], banana [50], and sugar palm [4,51–60]. However, polymer composites reinforced with natural fibers frequently heat up efficiently [61] and exhibit high thermal conductivity [62].

In this review, the term "flame retardant" (FR) is applied to various industrial and consumer products. The requirements of FRs that are relevant to the product's quality and longevity must be fulfilled. FRs provide fire protection by restricting the flow of oxygen to the flames. Expanded polyurethane foams and their compound foams are frequently used in fire suppressants, increasing metal combustion temperature, and minimizing flame diffusion [63]. Meanwhile, other important features, such as mechanical and thermal performance and environmental friendliness, e.g., not posing hazards to humans and the environment and capable of being recycled and reused, must be maintained. The usage of halogenated materials to prepare FRs is an efficient process. However, a gradual decrease in the acceptability of these products has been noted due to the increasing emphasis on environmental and health issues involving FRs.

Numerous metal particles, such as aluminum, copper, zinc, stainless steel, silver, gold, and nickel, are used in different polymer matrices [64]. Metal-filled polymer composites are reported to have increased electrical and thermal conductivity. Table 1 lists the metallic fillers used in FR applications. In some polymer composites, the metal and natural fibers such as kenaf [65], flax [66], and jute [67] have been added together with the polymeric resin to enhance the structural and thermal stability. Krishnasamy et al. [65] reported that the addition of aluminum and copper in jute epoxy hybrid composite resulted in the excellent thermal stability, as well as improved in their mechanical strength such as tensile and flexural performance. According to El-sabbagh et al. [68], by adding some amount of magnesium hydroxide ($Mg(OH)_2$)—about 20–30 wt% to the flax reinforced polypropylene composite improved the onset of decomposition temperature and LOI values. The composite comprising 50 wt% flax and 30 wt% flame retardant, in particular, achieved a 27% LOI score and a V-2 grade from the UL-94 test, with a long burning period and no dripping. By releasing a large amount of water, $Mg(OH)_2$ have efficient flame retardant efficiency by diluting the amount of fuel required to support combustion. They also reported that the addition of fibers and $Mg(OH)_2$ increased tensile stiffness in this study.

Table 1. Various metallic fillers use for fire retardant applications [69].

Flame Retardant Chemical Nature	Example of Flame Retardants	Working Mechanism
Metal oxides and hydroxide	Magnesium hydroxide, Aluminum hydroxide, alumina trihydrate, calcium carbonate	Heat sink
Boron based	Boric acid, borax, Zink borate, boron phosphate	By forming the insulating layer
Halogen based	TCPA, TBPA, Polybrominated diphenyl ethers, Polybrominated biphenyl	Gas-phase
Phosphorus based	THPC	Condense phase
Synergistic	P/N, Halogen/Antimony tri-oxide, P/halogen	The presence of other compounds would increase the slowness of the flame emitted by the major compound.
Intumescent	Acid donor (ex-phosphoric acid, ammonium polyphosphate), carbonizing agent (ex-pentaerythritol), bowling agent (ex-melamine, urea)	Both in the gas and condensed phase

Thus, the importance of flame retardants of metal polymers composite is discussed in this review. The impact of various addition of metal component, such as zinc, copper, aluminum, and nickel, on the flammability and fire retardancy of polymer composites are examined, with an emphasis on natural fiber reinforced polymer composites. The method of combustion and the commonly employed flammability measurement methods for polymer-based composites are also discussed. Finally, this review aims to provide state-of-the-art views of the fire resistance performance of various metal filler–polymer composites.

2. Flame Retardancy of Polymer Reinforced Composites

Using a mixture of recyclable waste polymer and PP fiber-reinforced materials has been demonstrated to be the most suitable method for the economical production of an entirely recyclable fire safety engineering design [70]. Figure 2 shows the metal particle distributions in polymer composites. In the case of natural fiber-reinforced polymer composites, their susceptibility to heat and flame retardant is one of the limitations. This is because of the presence of cellulose in plant fibers and hydrocarbon-based polymers causes the composites to be highly flammable [71]. Understanding the thermal decomposition and flammability of bio-based fibers, polymers, and their composites is crucial. Furthermore, appropriate flame retardant treatments and addition of metal component in these composites have been shown to significantly improve their thermal stability and fire resistance. According to Girisha et al. [72], adding amount of cellulosic fiber raises flammability of sisal/coir fiber reinforced epoxy composites since natural fiber embraces combustion. It is a weak flame retardant owing to the formation of a surface layer during the pyrolysis of the cellulosic material, which has a low fire resistance and act as a fire supporter, preventing heat from being transmitted to unpyrolized products.

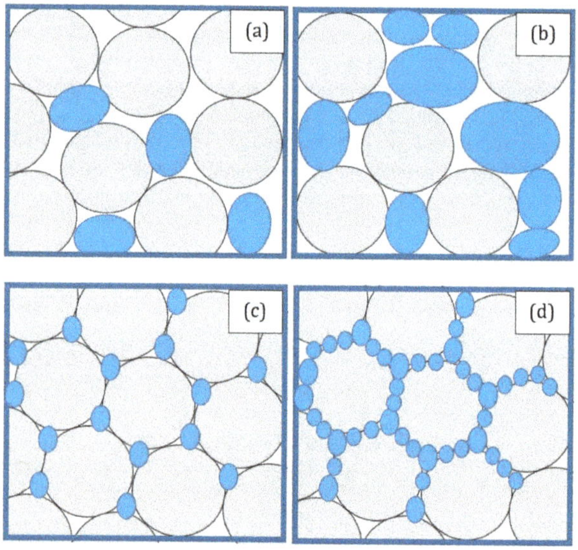

Figure 2. Illustration of random (**a**,**b**) and segregated (**c**,**d**) distributions of metal particles in polymer composites.

To date, several studies have investigated the flame retardancy of polymer composites filled with metal derivatives. Bar et al. [69] examined the flame retardation of polymer compounds. They determined that a particular fire retardation method and the effect of polymer composites were enhanced by FR with different composite properties. Figure 3 illustrates the FR synthesis process in composites via melt condensation reaction. Methodological approaches for improving the fireproofing effect are based on firefighting chemicals. The insertion of FR compounds or micro/nano FR fillers into a polymer backbone can increase the polymer matrix's flame slope. To achieve a highly FR composite material, the polymer frame matrix must exhibit more than 15% fireproofing filtration; this condition compromises its mechanical properties. In the current analysis, halogen-based FRs can increase the flame retardation effect of the formulation at lower concentrations relative to metal hydroxides. However, halogen is toxic to the atmosphere, and thus, its use has been banned. The size of the nano type filer would be very harmful to human health.

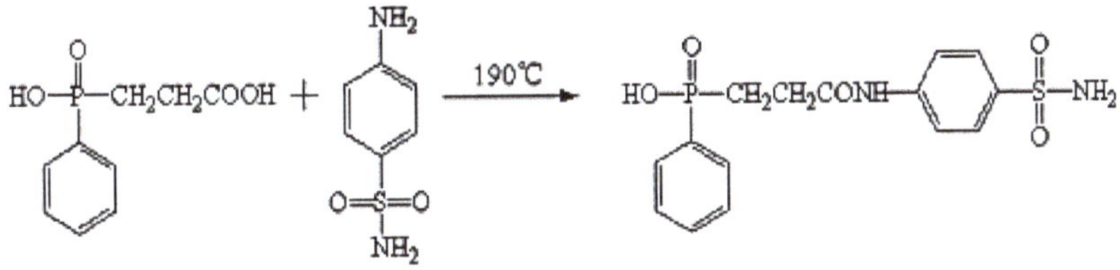

Figure 3. Chemical reaction of flame retardant synthesis of composites. Reproduced with copyright permission from Zhao et al. [73], Elsevier.

Zadeh et al. [74] researched the recovery of polymerized FR mixtures enhanced with palm fiber. Their findings demonstrated that magnesium hydroxide ($Mg(OH)_2$) is used on artificial composites; the material was fireproofed. This research was conducted using a conical calorimeter to measure fire resistance and analyze limiting oxygen index (LOI). The exhaust gas emission rate, overall heat release rate of the composite material comprising the refractory filler were tested in accordance with oxygen consumption theory. The most important development was that recycled materials and palm oil waste can produce composites that are affordable and environmentally safe. Although palm oil retains the mechanical properties of ternary composite mixtures, FRs exhibit heat resistance. This study showed that fire protection decreases mechanical efficiency. However, palm fiber increases the total strength of the construction material, helping achieve physical and mechanical properties. Composite products with 10% fibers and 1% binder (a combination of polyvinyl anhydride and maleic acid) exhibit mechanical strength and thermal tolerance.

Recently, the literature that presents the flame retardancy findings of metal composites has emerged. Yuan et al. [75] experimented on melamine (MA)-modified graphene oxide (GO). They found that slowing the combustion of PP altered MA-modified GO via heavy Δ–Δ interactions and hydrogen and electrostatic bonds. In their thesis, GO used the modified Hummers process to oxidize graphite powder. In particular, 0.6 g of GO and 3 g of MA were combined to create FGO. Interestingly, the findings of the transmission electron microscopy (TEM) and scanning electron microscopy (SEM) demonstrated that FGO nanosheets are evenly scattered in polymer matrices with embedded and flaky microstructures. FGO/PP nanocomposites exhibited better thermal stability and flame tolerance relative to their GO counterparts.

Fiber metal laminates (FMLs) are often used in the manufacture of hybrid natural fiber/metal polymer composites because of their strong electrical and thermal conductivity. FMLs are lightweight construction structures made up of thin metal contrasting with thin composite plies of metal as exterior surfaces (0.3–0.5 mm in thickness). Rather than improved thermal properties, various studies on hybridization of natural fiber with FML has been reported to enhanced their mechanical properties. Because of the inclusion of aluminum layers in the composite structure, tensile, flexural, and impact properties in sisal fiber reinforced aluminum laminates is significantly improved [76]. The tensile strength and dimensional stability of kenaf woven fabric reinforced with polypropylene also improved when utilizing FML aluminum, as reported by Ishak et al. [77].

3. Combustion Mechanism and Flame Retardancy of Composites

Considering the continuous increase in plastic waste and environmental degradation, biodegradables that use renewable products as a substitute for traditional petroleum plastics are becoming increasingly common. In recent years, halogen-free FRs have elicited considerable attention because halogen fuels used for combustion produce large and fast volumes of fires. Some types of fire prevention equipment include thermoplastic polyester and commonly use halogenation agents, special bromine, polymers, and antimony sources.

Halogen-free systems pose problems, some of which are new to municipal authorities; halogen-free phosphorus compounds, surface treatments, and reaction processes have been used in polyethylene terephthalate (PET) textiles for many years [78]. Figure 4 depicts the combustion cycle and potential flame retardancy approach.

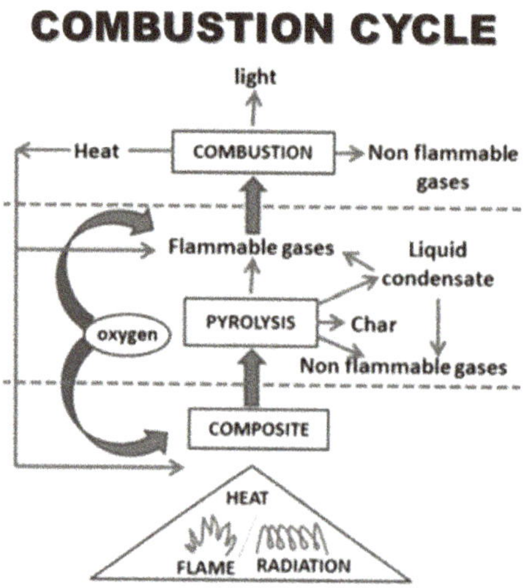

Figure 4. Flame retardancy approach and combustion cycle. Reproduced with copyright permission from Bar et al. [73], Springer.

3.1. Combustion Mechanism

Several researchers have also investigated the burning of metals in the shapes of chains, rods, and ribbons. Figure 5 illustrates the combustion mechanism of polymers. A motion picture technique is used to calculate burning times in different atmospheres. The addition of small amounts of water vapor exhibits a remarkably significant effect. Effort has been exerted to quantify burning times on the basis of the fact that transport processes are considerably slower than chemical reactions, and thus, the pace of the burning phase can be controlled. Burn instability processes in functional combustion systems are highly complicated due to coupled correlations and intrinsic nonlinearities correlated with the involved phenomena. Consequently, most instability processes cannot be modelled or represented using conventional analytical techniques unless several simplification processes are introduced to solve the issue.

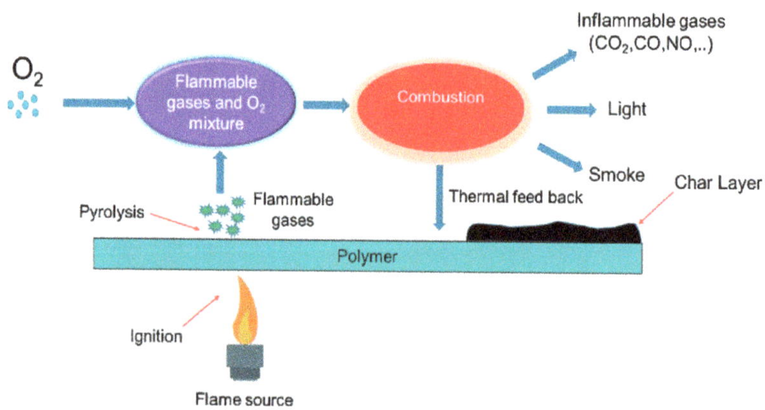

Figure 5. Schematic diagram for the combustion process cycle of polymers.

3.2. Flame Retardant Techniques

Advanced fire protection equipment may include three components: (1) acid source, including ammonium polyphosphate (APP); (2) Fourier transform infrared (FTIR) output (metals may increase the release of ammonia and carbon dioxide); and (3) fuel gas. Figure 6 illustrates flame retardancy techniques. The finding is consistent with microscale combustion calorimetry (MCC) and cone calorimetry (CONE) products. That is, if oxygen amounts can be diluted more easily, then ammonia and carbon dioxide emissions will increase. The reduced gas emissions can result in a decline in a material's heat release. The preceding experimental results and previous experiments have been developed for potential fire safety methods involving metals (iron, magnesium, aluminum, and zinc) in paraffin/intumescent FR (IFR) systems. Applications may produce polyphosphoric acid at high temperatures, and polyphosphoric acid may react with the pentaerythritol OH group. The FRs filler would emit ammonia gas to suppress the oxygen. Simultaneously, polyphosphoric acid can interact with metal oxide (metal +), in which this structure may be extended to the stability of polyphosphoric acid, and increasing the molecular weight of polyphosphoric acid can increase the viscosity of the FR layer, making the protective layer more efficient in shielding the polymer matrix [79].

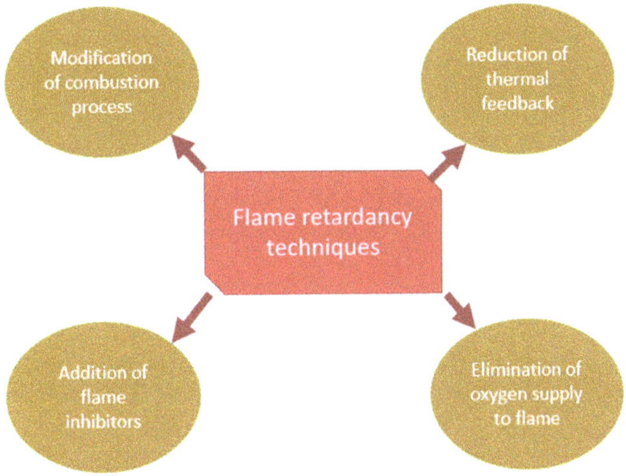

Figure 6. Methods of flame retardancy.

One of the disadvantages of polymer composites is their high flammability, which prevents their use in a variety of areas. As a result, improving their flame retardant properties is important, and a lot of effort has gone into it. The most effective approach for modifying the flame retardant properties is to incorporate FRs during the compounding phase. Recent findings show that the development in flame-retardant additives has been developed rapidly and new trends were discovered. The use of Ammonium polyphosphate (APP) in order to boost the flame retardant properties has found to increase the fire properties in polymer-based composites [80,81]. In addition, a combination of APP with other flame retardants, such as expandable graphite, SiO_2, or $CaCO_3$ [82–84], recorded to increase the effectiveness of fire retardancy. Furthermore, there are a variety of inorganic additives [85], organic flame retardants, nano-fillers [86], and anionic nano-clays [87] that were reported can improve the flame retardancy.

4. Characterization of Composites after Flame Retardant Treatment

The fire efficiency of the reinforcement components is enhanced by treating with FR chemicals. However, the fireproofing performance of the polymer matrix can be increased by using micro/nano FR fillers or by adding FR composites to the polymer backbone.

This process resolves the degradation of micro-compounds and nanofillers used in newly developed polymers and polymer replacements, along with their effects on composite features, such as automatic and thermal effects. However, the production of all FR material composites is the same during the development stage; for further innovations, researchers must focus on the production of healthy and environmentally sustainable FRs that can increase the firing efficiency of composite products at minimal concentration levels [69]. Table 2 summarizes metal particle functions as FR additives in various composites.

Table 2. Reported type for metal components of flame retardant on different types of composites.

Metal Components	Composites	Effect of Reinforcement	Reference
Metal hydroxide	Ethylene-vinyl acetate (EVA)	Form new layer that acts as insulation to flame	[88]
Silicon-containing, metal hydrate and oxide	Polypropylene (PP)	Decreasing the flow rate of the burning surface	[89]
Metal oxides	Thermoplastic polyurethane (TPU)	Low flammability and smoke emission	[90]
Zinc borate and magnesium hydroxide	Sawdust/rice husk filled polypropylene	A marginal reduction in mechanical properties and reduce flammability	[91]
Magnesium hydroxide ($Mg(OH)_2$ and zinc borate (Zb)	Fiber/polypropylene	Improved thermal stability and flame retardancy	[92]
Magnesium hydroxide	Ethylene-vinyl acetate (EVA)	Better water resistance, flame retardancy, and higher pyrolysis temperature	[93]
Salicylaldoxime and chelated copper(II)salicylaldehyde	Polyethylene (PE)	Provide good flame retardant behavior	[94]
Metal chelates	Polyvinyl alcohol (PVA)	Promotes thermal stability and improve flame-retardant	[95]
Aluminum and magnesium hydroxides	Rubbers and ethylene-vinyl acetate (EVA)	No corrosive or potentially toxic substances occur and reducing the smoke level	[96]
Zinc phosphonate	Glass-fiber reinforced poly(butylene terephthalate)	No improvement on fire behavior satisfactorily	[97]
Manganese (IV) oxide (MnO_2), zinc oxide (ZnO), and nickel(III) oxide (Ni_2O_3)	Polypropylene (PP)	Enhance the charring and corresponds well to the gas release with increasing temperature	[98]
Magnesium hydroxide and alumina trihydrate	Low-density polyethylene (LDPE) and ethylene-vinyl acetate (EVA)	Superior thermal stability and reduction of gases produced during burning	[99]
Nanometer titanium dioxide (nano-TiO_2), aluminum oxide (Al_2O_3), and magnesium aluminate spinel ($MgAl_2O_4$)	Ammonium polyphosphate-pentaerythritol-melamine (APP-PER-MEL)	Enhance fire-resistant and anti-aging properties of the APP-PER-MEL coating	[100]
Ionic liquid-based metal-organic hybrid (PMAIL)	Epoxy resin (EP)	Total smoke production was reduced	[101]
Aluminum phosphonate (AlPi), antimony oxide, and nanometric iron oxide	Poly(3-hydroxy-butyrate-co-3-hydroxyvalerate) /poly(butylene adipate-co-terephthalate) (PHBV/PBAT)	Great pyrolysis and the fire retardancy	[102]
Aluminum trihydrate	Ethylene-vinyl acetate (EVA) and montmorillonites (MMT)	Improvement of thermal stability and flame retardancy	[103]
Iron, magnesium, aluminum, and zinc	Paraffin	Increase the char yield and decrease volatilization for the combustible gases	[79]
Metal hydroxides and antimony trioxide	Thermoplastics	Improvements in thermal stability and pigmentation properties	[104]
Metal-based organic (MBO)	Polyvinylchloride (PVC)	Improved resistance to ignition, flame spread, and smoke generation	[105]
Aluminum trihydrate	Ethylene-vinyl acetate (EVA)	Reduction in heat release rates	[106]
Silicon-containing materials and metal oxides	Aliphatic and aromatic phosphonates	Good smoke suppressant effects	[107]
Zinc borate (ZnB)	Polyamides, polyesters, polyolefin, and boron Compounds	Lower heat release and lower total heat evolved	[108]

Table 2. *Cont.*

Metal Components	Composites	Effect of Reinforcement	Reference
Metal Phosphonates and Aluminum Oxide Hydroxide	Polyamide, Polyesters, and phosphorous	Improved flame-retardant and mechanical or electrical performance	[109]
Aluminum hydroxide ($Al(OH)_3$)	Cycloaliphatic polyamine, epoxy resins	Small burned area and better tensile strength properties	[110]
Metal chelates, chromium acetylacetonate, and zinc acetylacetonate	Polypropylene and poly(4,4-diamino diphenyl methane-O-bicyclicpentaerythritol phosphate-phosphate)	A denser char layer was established on the composite	[111]
Metal hydroxides	Silicon	Improve the thermal protective layer build on the polymer's surface	[78]
Alumina trihydrate, montmorillonite (MMT)	Ethylene-vinyl acetate and nanocomposite	Improve the fiber-matrix adhesion	[112]
Zinc borate, and magnesium hydroxide ($Mg(OH)_2$)	Polypropylene and ammonium polyphosphate	Thermal stability and fire retardancy were improved	[92]
Titanium dioxide	Ammonium polyphosphate-pentaerythritol-melamine (APP-PER-MEL)	Anti-aging properties of the flame-retardant coating were improved	[100]
Melamine poly (zinc phosphate) (MPZnP)	Epoxy resin (EP) and polyphosphate	Earlier decomposition and slightly changed evolved gas	[113]
Metallic oxide and Metal hydroxide	Graphene foam	Better flame retardant and compressible structure	[114]
Manganese and metal salts	Ammonium polyphosphate and cellulose	Enhancing flame retardant efficiency	[115]
Zinc hydroxyl stannate and alumina trihydrate	Ethylene-vinyl acetate, polyurethane, styrene-butadiene rubber, silicone rubber, and polychloroprene rubber.	Improvement of fire resistance and better mechanical and thermal properties of the elastomer	[116]
Ammonium bromide, manganese(II), iron(II), calcium, zinc oxalate, and metal oxalates	Polyamide and cotton	Reduction of combustion rate for cotton	[117]
Nickel-metal hydride, nickel-cadmium (Ni-Cd), and metal oxide	Graphites	Excellent ability for flame-retardance, cell performance, and wettability improvement	[118]
Copper metal complex	Polyurethane	Superior flame retardant and antimicrobial properties	[119]
Diphenyl phosphates and calcium hypophosphite,	Polycarbonates and polyurethanes	Good thermal stability and low volatility	[120]
Metal hydroxides, metal hydrate, and alumina trihydrate	Ethylene-vinyl acetate and octadecylamine	Improvement of tensile and flame-resistance properties	[121]
Cupric and zinc ions	Polyethylenimine and ramie fabric	Improved thermal stability and reduced flammability	[122]
Zinc Borate and metal hydroxide	Polyethylene terephthalate, woven and organophosphorus	Decrease smoke release but no flammability improvement	[123]

Interests in the flame retardancy of metal composites have been renewed recently. Kusakli et al. [110] improved the flame retardancy and mechanical properties of epoxy composites by using FR with red mud (RM) waste to demonstrate the FR properties of these polymer composites and to prove that FR systems are safe to use because of their high chemical and thermal resistance. The effects of ammonium tetrafluoroborate (ATFB), RM waste and aluminum hydroxide ($Al(OH)_3$) on the composites' mechanical and flame-resistant characteristics were investigated. RM waste was ground and sieved into particles measuring less than 63 μm to prepare the ER-based composite materials. Subsequently, different amounts of ATFB, RM waste, and $Al(OH)_3$ were mixed with the ER matrix at 2000 rpm via mechanical stirring and ultrasonication for 1.5 h at 60 °C to achieve strong dispersion. The combustion test demonstrated that the RM–ATFB–$Al(OH)_3$ mixture can be efficient if halogen-free FR is used in coating and construction areas for materials based on epoxy. Al_2O_3, which is formed by $Al(OH)_3$ decomposition reaction, prevents heat and oxygen from being transferred between the material and the environment, and thus, additional oxygen is required to ignite the sample. This previous study indicated that the burned area of the composite was only a small proportion of the total. For this composite, the experimental and estimated LOI values were 26 and 29, respectively. Burning studies were conducted to test the flammability of hydroxide and boron retardants.

The extensive literature signifies that numerous studies have examined the flame retardancy of metal composites. Song et al. [111] investigated the effect of metal chelates on flame retardancy of polypropylene (PP)/PDBPP. This study demonstrated the synthesis of the new oligomeric phosphorus-nitrogen containing intumescent flame retardant, poly(4,4-diamino diphenyl methane-O-bicydicpentaerythritol phosphate–phosphate) (PDBPP). Moreover, this study assessed whether the presence of metal chelates can enhance the flame retardancy of PP/PDBPP systems. Two metal chelates (zinc and chromium acetylacetone) are commercially available for the purpose of analysis. They were used as synergic agents without additional purification and other starting materials and solvents. The LOI value of PP/PDBPP (80/20) increased to 25, indicating a substantial improvement in PP flame retardancy in the presence of PDBPP. As demonstrated via Raman spectroscopy, infrared spectroscopy, and electron scanning microscopy, metal chelates (a decomposition product of PDBPP) may react with polyphosphoric acid as a cross-related network. A more compact layer, which produced PP/PDBPP with enhanced thermal and FR performance, was formed via salt bridges. This result showed that highly valuable metal chelates may improve the delays of FRs. Chang et al. [121] studied the flame retardancy and thermal stability of ethylene-vinyl acetate (EVA) copolymer nanocomposites when reinforced with alumina trihydrate (ATH) and montmorillonite (MMT). Organoclay (OMMT) was prepared by adding 20 g of MMT to 92 meq100 g^{-1} to 1000 mL of deionized (DI) water, with cationic exchange capacity. The mixture was agitated for 6 h and labelled as Solution A. Then, 4.96 g of octadecyl amine was dissolved in 50 mL of DI water, stirred for 3 h and called Solution B. Solutions A and B were mixed and heated for 3 h at 80 °C. OMMT was stored after 24 h of filtration, washing, and vacuum drying. The best FR quality (40/60%) of the total cable wire included a small amount of MMT. This study indicated that superior tensile strength was achieved at 3 wt% MMT. Furthermore, EVA's flame retardancy is free from halogen, with 3% OMMT and 47% ATH achieving optimum deformation and flame resistance (LOI = 28). The tensile and fire inhibition characteristics of the nanocomposites were improved significantly.

Researchers have attempted to evaluate the effect of the flame retardancy of metal composites. Suppakarn and Jarukumjorn [124] examined the mechanical and thermal properties of sisal/PP composites and determined the effects of FR type and content. The objective of this research was to add FRs $Mg(OH)_2$ and zinc borate (ZnB) to enhance the flame resistance of the morphological and mechanical features of sisal/PP composites. The ratio of $Mg(OH)_2$ to ZnB was different in each sisal/PP composite location, while the overall content was maintained frequently at 30 wt%. Maleic anhydride grafted PP (MAPP) was also used as a compatibilizer to enhance adhesion between PP/sisal and PP/FRs. The flammability of PP and PP composites was investigated using ASTM D635 (standard test method for rate of burning and/or extent and time of burning of plastics in a horizontal position). The specimen was held horizontally, and a flame was applied to one end of the sample. Marking time was recorded from the first mark, 25 mm from the end of the mark to the second mark, and 100 mm from the end of the mark. Three specimens were tested for each composite. The composites were then measured for burning speed. The burning rate of the 30 ZnB composite was close to that of the clean PP. Meanwhile, the burning rate was immediately below that of pure PP for the 15 Mg/15 ZnB composite stage. Consequently, $Mg(OH)_2$ more efficiently decreased PP composite's burning intensity than ZnB addition. This study demonstrated comparable tensile and flexural properties with the addition of $Mg(OH)_2$ and ZnB without FRs for the sisal/PP composites. The addition of $Mg(OH)_2$ and ZnB enhanced the flame retardancy of sisal/PP composites without losing their mechanical properties.

The past 30 years have witnessed increasingly rapid advances in the flame retardancy of metal composites. Davies et al. [115] conducted a study on the sensitization of the heat treatment of APP by selected metal ions and their potential to improved cotton fabric flame retardancy. The effect of adding a series of metal salts on the thermal behavior of APP as a means of sensitizing FR behavior was determined. The addition of metal salts apparently improved the FR efficiency of APP as part of the pentaerythritol FR method in a PP matrix. Dry APP (MCM) mixtures were prepared from various dry mixtures of 2% w/w of each metal salt. Ferric sulphate APP mixtures with salt amounts ranging from 1% to 5% w/w were also prepared. Interestingly, the sodium and magnesium salts produce the highest increases with $\Delta LOI_{(salt)} \geq 1.8$. Salts, such as manganese and zinc sulphates, the largest of which existed at the same DTG transition temperatures, exhibited lower DLOI (salt) values of $0.9 \times 10^{1.1}$. This study demonstrated that some metal ions, particularly Mn^{2+} and Zn^{2+}, are absent when facilitating the thermal degradation of APP, improving the performance of flame retardation in the polymer at lower temperatures. The metal ion-doped APP did not only exhibit higher sensitization to cellulose decomposition in the presence of cellulose, but it also improved flame retardancy by limiting the oxygen index of cotton fabric.

Studies on composite materials have demonstrated the importance of the flame retardancy of metal composites. Beyer [103] investigated the fire-resistant property of EVA nanocomposites and advancements in the combination of nanofillers with ATH. FR nanocomposites were found to be formulated with modified layered silicates by melt blending ethylene-vinyl acetate (EVA) copolymers (MMT). Thermogravimetry (TGA) was conducted in various atmospheres, such as nitrogen and air. A major improvement in the thermal stability of the nanocomposites based on silicate was demonstrated. Moreover, a cone calorimeter was used to examine the fire properties of materials. The observation from the results showed a reduction in the cone calorimeter's heat release peak, indicating that the char formation of the nanocomposites was enhanced and was responsible for the improved flame retardation. The thermal properties of EVA were reportedly improved. Moreover, EVA nanocomposites combined with metal hydroxides, such as ATH, presented the possibility of FRs as new compounds with reduced total filler contents.

The problem of metal-filled polymer composites flame retardancy has received considerable attention. Yen et al. [88] conducted research on the synergistic FR effects of metal hydroxides and nano-clay on EVA composites. The results of the observation indicated that LOI value was significantly improved when 1–2% weight nano-clay was replaced with aluminum hydroxide or $Mg(OH)_2$ in the EVA blend, while maintaining the V–0 rating. The CONE test data showed that the peak heat release rate decrease was approximately 28% to 47%. Smoke density data registered a decrease of approximately 16–25%. TGA data also showed that the thermal stability and char residue of the EVA samples were improved by nano-clay. The metal oxide layer on the burning surface was also suggested to be reinforced by creating a silicate layer. Lujan-Acosta et al. [99] studied the synergistic effects of organo-modified MMT and metal hydroxides, namely, $Mg(OH)_2$ and ATH, as FRs in low-density polyethylene (LDPE)/EVA nanocomposites integrated with amino alcohol. Grafted polyethylene was found to be compatible with LDPE/EVA/clay/FR nanocomposites (PEgDMAE). The structural characterization of nanocomposites was performed via X-ray diffraction (XRD) analysis and scanning transmission electron microscopy (STEM). In addition, horizontal burning and CONE tests for UL-94 and LOI were conducted to analyze the FR properties of nanocomposites. Thermal degradation output was also tested via FTIR coupled with TGA (TG-FTIR). The XRD analysis showed a change in the d001 plane to the lower-angle characteristic of the clay peak, indicating an intercalated–exfoliated microstructure. In the polymer matrix, which was expressed in FR properties, a significant dispersion of FRs by $Mg(OH)_2$ and ATH was observed in the STEM images. Lujan-Acosta et al. [99] reported that the TG-FTIR result showed excellent thermal stability of the nanocomposites, and a major reduction was observed in the gases emitted during combustion. Therefore, the

FR mechanism of LDPE/PEgDMAE/EVA/clay/$Mg(OH)_2$ nanocomposite was proposed on the basis of the findings of thermal degradation and thermal stability.

Jeencham et al. [92] examined the effect of FRs on the mechanical and thermal properties of sisal fiber/PP composites. The FR performance of APP, $Mg(OH)_2$, ZnB, and sisal fiber/PP composite combination was presented. The experiment was performed via vertical and horizontal burning tests. Moreover, MAPP was used as an integration enhancer for the PP/fiber and PP/FR systems. The result indicated that the addition of FRs to the composites decelerated the burning rate of the PP composite. Among several types of FRs, APP showed that the most powerful FR improvement was achieved by the PP composites during the vertical and horizontal burning tests. Jeencham et al. [92] reported that the flame retardancy and thermal stability of PP composites were enhanced without weakening their mechanical properties.

Li et al. [125] investigated the varying effects of flame retardancy and aluminum phosphonate (AlPi) mechanisms on poly (p-phenylene oxide) (PPO), thermoplastic polyurethane (TPU), and PP. The influence of AlPi on the flame retardancy of the three polymers (PPO, TPU, and PP) was determined. Experiments using LOI, SEM, vertical burning test (UL-94), CONE, and TGA were conducted. The results showed that the addition of AlPi substantially increased the LOI values of PPO and PP, but had nearly no effect on the LOI of TPU. In addition, although PP increased, the peak heat release rates of PPO and TPU decreased. A dense char layer was developed by the PPO composite, demonstrating the best flame retardancy. Meanwhile, a thinner char layer was developed by the TPU composite. During combustion, however, the PP composite did not form any char layer. The addition of AlPi effectively reduced the TPU matrix's melt dripping and improved flame retardation. AlPi/PP composites acted as a fire resistor, decreasing the productive combustion heat of the volatiles, and increasing the amount of released carbon monoxide. Sharma and Saxena [105] studied the FR smoke suppressant protection provided by polyvinyl chloride (PVC). The metal-based organic (MBO) complexes were synthesized to be used as FR smoke suppressants in PVC formulations. FR smoke suppressant ingredients with 325–400 mesh size were mixed with a 2–5% thickener solution and appropriate amounts of wetting, anti-settling and anti-foaming agents. Vinyl acetate and vinyl versatate copolymer emulsions (binders) were modified by reacting with a polymeric plasticizer and dihydroxydimethylol ethylene urea. The observation results showed that the smoke suppression output achieved outstanding results when either of the two MBOs was used. Moreover, LOI increased, particularly when the PVC samples were plasticized using a phosphate plasticizer. The coated cables did not exhibit any surface flame spread when exposed. Moreover, the generation of smoke was extremely poor for the coated cables. The coatings were highly efficient in minimizing the burning actions of power cables, significantly improving circuit failure time.

The effect of FR ZnB or boric acid mixed with $Mg(OH)_2$ was observed in Sain et al. [91] on the FR and mechanical properties of natural fiber/PP composites with $Mg(OH)_2$. The experiment was conducted using the horizontal burning rate test. The specimen was held horizontally, and a flame ignited by gas was added to flare up the end of the specimen. In addition, LOI analysis was performed by placing the sample vertically in a glass chamber wherein nitrogen and oxygen flow was controlled. The observation results indicated that 25% of $Mg(OH)_2$ can significantly minimize the filled composite's flammability to approximately 50% without FR. The partial substitution of 5% $Mg(OH)_2$ with ZnB or boric acid exhibited a retarding effect on the flame retardancy properties of $Mg(OH)_2$. $Mg(OH)_2$ can affect the flammability of natural fiber-filled PP composites by reducing the capability to ignite the composites. Even when $Mg(OH)_2$ was used with ZnB and boric acid, no synergetic effect was observed. Finally, a small reduction in the mechanical properties of the composites was observed with the combination of FRs.

Braun et al. [97] investigated the fire retardancy mechanisms of metal phosphonates and metal phosphonates combined with MA cyanide (MC) in glass-fiber-reinforced poly (1,4-butylene terephthalate) (PBT/GF). The result showed the pyrolysis and fire activity of PBT/GF with two distinct metal phosphonates as fire retardants with and without MC. An experiment was performed via TGA and TGA coupled with infrared spectroscopy. The analysis data were collected from flammability tests, CONE tests, and SEM/energy-dispersive X-ray spectroscopy (EDS) and X-ray fluorescence (XRF) spectroscopy. Dosages of approximately 13% to 20% of halogen-free FR aluminum phosphonate or aluminum phosphonate with MC in PBT/GF were able to meet the requirements for electrical engineering and computer applications (UL 94 1⁄4 V–0; LOI > 42%). Meanwhile, the average 16% for zinc phosphonate with MC did not satisfactorily increase fire behavior (UL 94 $\frac{1}{4}$ HB; LOI $\frac{1}{4}$ 27–28%). The AlPi content indicated that the residue remained mechanically intact in the examined specimen and covered the polymeric materials from pyrolysis. This phenomenon created superior flame retardancy in the AlPi materials and met the application test criteria. Gallo et al. [102] studied the synergistic effects between nanometric metal oxides and phosphonate. They determined that for petroleum-based plastics, the FR synergy between phosphorus-based additives and metal oxides was used and applied to bio-based materials. The pyrolysis and flame retardancy properties of AlPi, along with antimony oxide and nanometric iron oxide, on a blend of poly (3-hydroxybutyrate-co-3-hydroxyvalerate)/poly(butyleneadipate-co-terephthalate) (PHBV/PBAT) were analyzed. Crystallinity changed, and the reaction between the polymer and additives may influence biodegradation because biodegradation occurred first in the unstructured polymer region. Moreover, AlPi decomposed separately from PHBV/Ecoflex, which was primarily in gas phase as phosphonic acid. In the solid phase, AlPi was partly retained as inorganic phosphate. However, the addition of metal oxide did not considerably affect the thermal and combustion activities of PHBV/Ecoflex. Only the synchronous inclusion of AlPi and metal oxide with a global filler content of 10 wt% contributed to the good effect on flame retardancy, increasing the value for UL 94 rating and inducing additional char formation. Fire retardation improvement was due to the increase in char production and the preferred changes in the classification of UL 94. Moreover, the nanofiller and phosphorus components worked together in the FR mechanism, with the primary mechanism behaving like FR in gas state.

Xiao et al. [101] examined the effect of ionic liquid-based metal–organic hybrid on the thermal degradation, fire retardancy, and smoke suppression properties of ER composites. An anion exchange occurred between phosphomolybdic acid and phosphonate-based ionic liquid. A new multifunctional ionic liquid-based metal–organic mixture (PMAIL) was developed and applied to ER as an effective FR. PMAIL-based ER composite was prepared. Firstly, 1.24 g of PMAIL was dispersed into 15 mL of ethanol. Then, 100 g bisphenol A diglycidyl ether ER was added with magnetic stirring. Secondly, the mixture was stirred at 100 °C for 1 h to eliminate ethanol. Lastly, the mixture was cast into preheated molds and cured at 100 °C for 2 h. The carbonized yield of ER-PMAIL1 (1 wt% addition) composite at 700 °C was dramatically increased by 108% from 12% and 25% for ER. Meanwhile, ER-PMAIL6 (6 wt% addition) composite could reach V–0 rating in the UL-94 vertical burning experiment. The total smoke output and peak heat release rate of ER-PMAIL6 decreased by 15.4% and 31%, respectively, compared with ER. The carbonized yield of ER-PMAIL6 was improved by nearly 160% from 9% to 25% compared with ER for the CONE test, indicating strong mechanical properties and intumescent carbonized layer for superior flame retardance. Suriani et al. [3] investigated the horizontal burning rate by using $Mg(OH)_2$ to determine its capability as FR composite. Different percentages of oil palm empty fruit bunch fiber (OPEFB) were added, with PET yarn and $Mg(OH)_2$ as controls. The burning test showed that the specimen with 20% OPEFB exhibited better flammability properties, with the lowest average burning rate (11.47 mm/min). Figure 7 depicts the sample of specimens after the horizontal burning test. A conclusion was drawn that the flammability and tensile properties of OPEFB fiber-reinforced epoxy composites were reduced when fiber volume contents were increased at an optimal loading of 20%, with values of 11.47 mm/min and 4.29 MPa, respectively.

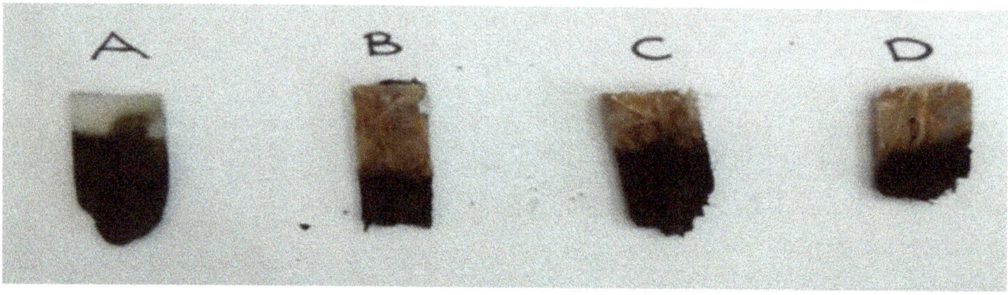

Figure 7. Sample of OPEFB/ PET yarn/magnesium hydroxide reinforced epoxy hybrid composites after the horizontal burning test. (**A**): 0% OPEFB; (**B**): 20% OPEFB; (**C**): 20% OPEFB; 35% OPEFB; and (**D**): 50% OPEFB. Reproduced with copyright permission from Suriani et al. [3], Polymers MDPI.

5. The Economic Analysis of Metal-Filled Polymer Composites

In the last few decades, there has been a surge of concern in the effect of metal materials on composite manufacturing particularly in thermal stability component. Using metal as a reinforcing medium has the ability to improve recycling rates while still locating high-value uses for polymer composites. For the case of polymer composites filled with metal derivatives, by eliminating the exploration, processing, and shipping, it will dramatically minimize environmental impacts. The latest study by Bulei et al. [126] found that recycle process of aluminum alloys for metal matrix composite has economic and technological advantages. It is also mention in this paper that it is not cost-effective to use fresh raw materials to produce high-value-added goods, where they can be obtained from scrap by competitive non-ferrous metal recycling technologies. In terms of waste utilization, the strategies adopted are achievable from an economic standpoint. As a result, the cost-benefit ratio satisfies the economic feasibility criterion. In addition, in a study by Dong et al. [127] concerning environmental effects modelling and the economic effects of composite recycling for fiber reinforced polymers, which motivated both environmental and economic factors in creating recycling routes for increasing quantity of fiber reinforced polymer scrap generated. Both glass and carbon fiber reinforced polymers recycling methods were compared to low-value end-of-life options, with pyrolysis appearing to be an appealing solution to recycling carbon fiber reinforced polymers that meets both environmental and economic benefits. However, in terms of cost, using metal as a reinforcement must be considered. Newest study on the evaluation and techno-economic analysis of the metal alloys tubes of multi-effect distillation (MED) for seawater desalination process, using titanium tubes enhanced with polymer (polyethylene (PE)-expanded graphite (EG)) composite were conducted by Tahir et al. [128]. In the study, they discovered that MED built on polymer composite tubes preferred economic and carbon pollution indicators, with the ability to reduce the cost of the MED evaporator by 40% less than the cost of the titanium evaporator.

6. Drawbacks and Challenges

According to a review of polymer composites filled with metal derivatives, it is clear that the metal filled polymer composites are favorable materials in terms of thermal and fire retardance properties and have a great potential in fire safety applications. Yet, the enhancements are far from what is needed for various fire safety applications. In this manner, numerous technical barriers such as dispersion of fillers within their matrix, structural control, contact between individual fillers, and interfacial interaction filler/matrix should be considered to realize the wide applications of these advanced composites [129].

The poor interfacial adhesion between metal filler and polymeric matrix has been the essential factors in designing and fabricating high flame retardance performance composites. This is due to most metallic derivatives that are incompatible with most organic polymers and the number of residual groups on the filler surface is still insufficient to produce strong adhesion with the polymer matrix. Metal compounds as the inorganic components show catalytic effect on reducing smoke emission and promote the char forming process. Thus, the hybridizing of metal/polymer composites in a feasible manner would result in significant advantages to promote these characteristics, which represent intriguing characteristics superior to normal FRs [130]. For instance, the good compatibility of metal oxides reinforced polyurethane composite exhibited superior pHRR reduction and LOI value as compared to normal FRs [90].

Good dispersion and orientation of metal derivative fillers in polymers for fire retardant applications are also fundamental challenges. In some cases, the metal fillers that are in uniaxial alignment inside polymer matrix would reduce dispersivity of the fillers well. Besides that, the application of metallic fillers in polymer composites would limit their dispersivity, due to the presence of abundant impurities and residual groups on the fillers' surface. This will increase the interlayer spacing and decrease the van der Waals interaction between the metallic fillers and polymer resins [131].

Whether metal derivative fillers can have a better impact on thermal and flame retardance properties than common FRs remains an open question for polymer-based composites. Nevertheless, common FRs may not compete with metal derivative fillers in terms of cost since a huge amount of filler volume is needed for composite applications. The structure of FRs composites and appropriate manufacturing technique has to be developed to ensure an optimized used of metal fillers and to obtain high performance flame retardance composites.

7. Conclusions and Future Outlooks

This review discusses various types of metal components of FRs with different types of polymer, such as LDPE, PP, rubber, and PA. The incorporation of metal to polymer composite systems has ushered in a modern age in polymer composites for a variety of applications. Different types of metal components with distinct and special properties have promising strong electrical and thermal conductivity features as polymer matrix reinforcements. One of the most notable discoveries of this review is that metal components efficiently enhance the flame retardancy of polymer composites. The use of qualitative case studies is a well-established approach for determining the chemical nature of FRs, which includes metal oxides and hydroxides, boron-based, halogen-based, phosphorus-based, synergistic, and intumescent. Metal components are combined with FR's chemical nature to strengthen the thermal properties and flammability of polymers. Moreover, different compositions of metal components and chemical nature of FRs prove that self-reinforced composite properties can be modified to achieve better properties. The incorporation of metal components into polymer composites has been shown to significantly improve the fire resistance, providing an insulation layer, decreased flammability, and increased tensile strength. Initially, several metal components added to FR's chemical nature do not exhibit any improvement during the CONE test, TGA, SEM, and EDS. However, with the right ratio of material compositions, the properties of the reinforced composites exhibit better improvement than the original composites. Therefore, finding the compatible ratio of the components is significant in this experiment. Most metal components used to improve fire retardancy positively affect the reinforced composites. According to the articles reviewed in this study, a positive potential outlook for research on the integration and functionalization of polymer with metal components for a new generation of high-performance composites can be expected and could have a bright future. Although there is no doubt that metal filled polymer composites can promote pioneering science and lead to industrial advancements. In recommendation, additional fundamental studies are needed to gain a deeper understanding of the relationship between metal components in this rapidly growing class of polymer composite materials. The crucial understanding and characterization of each metal in these advanced polymer composites, as well as experimental and theoretical proofs, is needed to make a prediction on the overall metal-polymer interaction. Based on the latest applications, metal-polymer hybridization has the ability to be used in a variety of fields, including aerospace, sport, electronic, and computer applications.

Author Contributions: Conceptualization, R.A.I. and S.M.S.; validation, M.N.F.N. and M.R.M.A.; resources, D.A.Z.N.D. and J.J.N.A.; writing—original draft preparation, R.A.I., S.M.S., D.A.Z.N.D., and J.J.N.A.; writing—review and editing, H.A.A., M.R.M.A., D.A.Z.N.D., J.J.N.A., M.R.I., M.S.A.R., M.N.F.N., N.M.N., M.R., S.S., and M.R.R.; supervision, R.A.I. and S.M.S.; project administration, R.A.I. and S.M.S.; funding acquisition, M.R.R. All authors have read and agreed to the published version of the manuscript.

Funding: This research was funded by Universiti Teknologi Malaysia, project CRG 30.3: Retardant coating using graphene/bamboo aerogel mixtures on SAR robotics system, grant number PY/2020/03495—R.J130000.7351.4B534". The authors would like to express gratitude for the financial support received from PP/LESTARI/2021 and XX-2018-008 by Universiti Kebangsaan Malaysia, Malaysia.

Institutional Review Board Statement: Not applicable.

Informed Consent Statement: Not applicable.

Data Availability Statement: No new data were created or analyzed in this study. Data sharing is not applicable to this article.

Acknowledgments: The authors would like express gratitude for the financial support received from Universiti Teknologi Malaysia, project CRG 30.3: Retardant coating using graphene/bamboo aerogel mixtures on SAR robotics system, grant number PY/2020/03495—R.J130000.7351.4B534". The authors would like to express gratitude for the financial support received from PP/LESTARI/2021 and XX-2018-008 by Universiti Kebangsaan Malaysia, Malaysia.

Conflicts of Interest: The authors declare no conflict of interest.

References

1. Ye, T.-P.; Liao, S.-F.; Zhang, Y.; Chen, M.-J.; Xiao, Y.; Liu, X.-Y.; Liu, Z.-G.; Wang, D.-Y. Cu(0) and Cu(II) decorated graphene hybrid on improving fireproof efficiency of intumescent flame-retardant epoxy resins. *Compos. Part B Eng.* **2019**, *175*, 107189. [CrossRef]
2. Mamunya, Y.P.; Davydenko, V.V.; Pissis, P.; Lebedev, E.V. Electrical and thermal conductivity of polymers filled with metal powders. *Eur. Polym. J.* **2002**, *38*, 1887–1897. [CrossRef]
3. Suriani, M.J.; Radzi, F.S.M.; Ilyas, R.A.; Petrů, M.; Sapuan, S.M.; Ruzaidi, C.M. Flammability, Tensile, and Morphological Properties of Oil Palm Empty Fruit Bunches Fiber/Pet Yarn-Reinforced Epoxy Fire Retardant Hybrid Polymer Composites. *Polymers* **2021**, *13*, 1282. [CrossRef] [PubMed]
4. Suriani, M.J.; Sapuan, S.M.; Ruzaidi, C.M.; Nair, D.S.; Ilyas, R.A. Flammability, morphological and mechanical properties of sugar palm fiber/polyester yarn-reinforced epoxy hybrid biocomposites with magnesium hydroxide flame retardant filler. *Text. Res. J.* **2021**, 1–12. [CrossRef]
5. Suriani, M.J.; Rapi, H.Z.; Ilyas, R.A.; Petrů, M.; Sapuan, S.M. Delamination and Manufacturing Defects in Natural Fiber-Reinforced Hybrid Composite: A Review. *Polymers* **2021**, *13*, 1323. [CrossRef] [PubMed]
6. Rusu, M.; Sofian, N.; Rusu, D. Mechanical and thermal properties of zinc powder filled high density polyethylene composites. *Polym. Test.* **2001**, *20*, 409–417. [CrossRef]
7. Nurazreena; Hussain, L.B.; Ismail, H.; Mariatti, M. Metal filled high density polyethylene composites—Electrical and tensile properties. *J. Thermoplast. Compos. Mater.* **2006**, *19*, 413–425. [CrossRef]
8. Omran, A.A.B.; Mohammed, A.A.B.A.; Sapuan, S.M.; Ilyas, R.A.; Asyraf, M.R.M.; Koloor, S.S.R.; Petrů, M. Micro- and Nanocellulose in Polymer Composite Materials: A Review. *Polymers* **2021**, *13*, 231. [CrossRef]
9. Asyraf, M.R.M.; Ishak, M.R.; Sapuan, S.M.; Yidris, N. Utilization of Bracing Arms as Additional Reinforcement in Pultruded Glass Fiber-Reinforced Polymer Composite Cross-Arms: Creep Experimental and Numerical Analyses. *Polymers* **2021**, *13*, 620. [CrossRef]
10. Gonzalez-Gutierrez, J.; Cano, S.; Schuschnigg, S.; Kukla, C.; Sapkota, J.; Holzer, C. Additive manufacturing of metallic and ceramic components by the material extrusion of highly-filled polymers: A review and future perspectives. *Materials* **2018**, *11*, 840. [CrossRef]
11. Asyraf, M.R.M.; Ishak, M.R.; Sapuan, S.M.; Yidris, N.; Rafidah, M.; Ilyas, R.A.; Razman, M.R. Potential application of green composites for cross arm component in transmission tower: A brief review. *Int. J. Polym. Sci.* 2020. [CrossRef]

12. Amir, A.L.; Ishak, M.R.; Yidris, N.; Zuhri, M.Y.M.; Asyraf, M.R.M. Potential of Honeycomb-Filled Composite Structure in Composite Cross-Arm Component: A Review on Recent Progress and Its Mechanical Properties. *Polymers* **2021**, *13*, 1341. [CrossRef] [PubMed]
13. Alsubari, S.; Zuhri, M.Y.M.; Sapuan, S.M.; Ishak, M.R.; Ilyas, R.A.; Asyraf, M.R.M. Potential of Natural Fiber Reinforced Polymer Composites in Sandwich Structures: A Review on Its Mechanical Properties. *Polymers* **2021**, *13*, 423. [CrossRef]
14. Ilyas, R.A.; Sapuan, S.M.; Harussani, M.M.; Hakimi, M.Y.A.Y.; Haziq, M.Z.M.; Atikah, M.S.N.; Asyraf, M.R.M.; Ishak, M.R.; Razman, M.R.; Nurazzi, N.M.; et al. Polylactic Acid (PLA) Biocomposite: Processing, Additive Manufacturing and Advanced Applications. *Polymers* **2021**, *13*, 1326. [CrossRef] [PubMed]
15. Özgür, S.M.; Amar, M.K.; Manjusri, M. *Fiber Technology for Fiber-Reinforced Composites*, 1st ed.; Özgür, S.M., Amar, M.K., Manjusri, M., Eds.; Elsevier: Duxford, UK, 2017; ISBN 9780081018712.
16. Ilyas, R.A.; Sapuan, M.S.; Norizan, M.N.; Norrrahim, M.N.F.; Ibrahim, R.; Atikah, M.S.N.; Huzaifah, M.R.M.; Radzi, A.M.; Izwan, S.; Azammi, A.M.N.; et al. Macro to nanoscale natural fiber composites for automotive components: Research, development, and application. In *Biocomposite and Synthetic Composites for Automotive Applications*; Sapuan, M.S., Ilyas, R.A., Eds.; Woodhead Publishing Series: Amsterdam, The Netherland, 2020.
17. Asyraf, M.R.M.; Ishak, M.R.; Sapuan, S.M.; Yidris, N.; Ilyas, R.A. Woods and composites cantilever beam: A comprehensive review of experimental and numerical creep methodologies. *J. Mater. Res. Technol.* **2020**, *9*, 6759–6776. [CrossRef]
18. Mohd Nurazzi, N.; Asyraf, M.R.M.; Khalina, A.; Abdullah, N.; Sabaruddin, F.A.; Kamarudin, S.H.; Ahmad, S.; Mahat, A.M.; Lee, C.L.; Aisyah, H.A.; et al. Fabrication, Functionalization, and Application of Carbon Nanotube-Reinforced Polymer Composite: An Overview. *Polymers* **2021**, *13*, 1047. [CrossRef] [PubMed]
19. Asyraf, M.R.M.; Ishak, M.R.; Sapuan, S.M.; Yidris, N.; Ilyas, R.A.; Rafidah, M.; Razman, M.R. Evaluation of design and simulation of creep test rig for full-scale cross arm structure. *Adv. Civ. Eng.* **2020**. [CrossRef]
20. Asyraf, M.R.M.; Rafidah, M.; Ishak, M.R.; Sapuan, S.M.; Ilyas, R.A.; Razman, M.R. Integration of TRIZ, Morphological Chart and ANP method for development of FRP composite portable fire extinguisher. *Polym. Compos.* **2020**, 1–6. [CrossRef]
21. Asyraf, M.R.M.; Ishak, M.R.; Sapuan, S.M.; Yidris, N. Comparison of Static and Long-term Creep Behaviors between Balau Wood and Glass Fiber Reinforced Polymer Composite for Cross-arm Application. *Fibers Polym.* **2021**, *22*. [CrossRef]
22. Asyraf, M.R.M.; Ishak, M.R.; Sapuan, S.M.; Yidris, N. Conceptual design of creep testing rig for full-scale cross arm using TRIZ-Morphological chart-analytic network process technique. *J. Mater. Res. Technol.* **2019**, *8*, 5647–5658. [CrossRef]
23. Asyraf, M.R.M.; Ishak, M.R.; Sapuan, S.M.; Yidris, N. Conceptual design of multi-operation outdoor flexural creep test rig using hybrid concurrent engineering approach. *J. Mater. Res. Technol.* **2020**, *9*, 2357–2368. [CrossRef]
24. Johari, A.N.; Ishak, M.R.; Leman, Z.; Yusoff, M.Z.M.; Asyraf, M.R.M. Influence of CaCO3 in pultruded glass fibre/unsaturated polyester composite on flexural creep behaviour using conventional and TTSP methods. *Polimery* **2020**, *65*, 46–54. [CrossRef]
25. Nurazzi, N.M.; Asyraf, M.R.M.; Khalina, A.; Abdullah, N.; Aisyah, H.A.; Rafiqah, S.A.; Sabaruddin, F.A.; Kamarudin, S.H.; Norrrahim, M.N.F.; Ilyas, R.A.; et al. A Review on Natural Fiber Reinforced Polymer Composite for Bullet Proof and Ballistic Applications. *Polymers* **2021**, *13*, 646. [CrossRef] [PubMed]
26. Ilyas, R.A.; Sapuan, S.M.; Atikah, M.S.N.; Asyraf, M.R.M.; Rafiqah, S.A.; Aisyah, H.A.; Nurazzi, N.M.; Norrrahim, M.N.F. Effect of hydrolysis time on the morphological, physical, chemical, and thermal behavior of sugar palm nanocrystalline cellulose (Arenga pinnata (Wurmb.) Merr). *Text. Res. J.* **2021**, *91*, 152–167. [CrossRef]
27. Woodrow, B. "Fire as Vulnerability": The Value Added from Adopting a Vulnerability Approach. *World Fire Stat. Bull.* **2012**, *28*, 1–26.
28. Sari, N.H.; Pruncu, C.I.; Sapuan, S.M.; Ilyas, R.A.; Catur, A.D.; Suteja, S.; Sutaryono, Y.A.; Pullen, G. The effect of water immersion and fibre content on properties of corn husk fibres reinforced thermoset polyester composite. *Polym. Test.* **2020**, *91*, 106751. [CrossRef]
29. Syafri, E.; Sudirman; Mashadi; Yulianti, E.; Deswita; Asrofi, M.; Abral, H.; Sapuan, S.M.; Ilyas, R.A.; Fudholi, A. Effect of sonication time on the thermal stability, moisture absorption, and biodegradation of water hyacinth (*Eichhornia crassipes*) nanocellulose-filled bengkuang (*Pachyrhizus erosus*) starch biocomposites. *J. Mater. Res. Technol.* **2019**, *8*, 6223–6231. [CrossRef]
30. Siakeng, R.; Jawaid, M.; Asim, M.; Saba, N.; Sanjay, M.R.; Siengchin, S.; Fouad, H. Alkali treated coir/pineapple leaf fibres reinforced PLA hybrid composites: Evaluation of mechanical, morphological, thermal and physical properties. *Polym. Lett.* **2020**, *14*, 717–730. [CrossRef]
31. Abral, H.; Ariksa, J.; Mahardika, M.; Handayani, D.; Aminah, I.; Sandrawati, N.; Sapuan, S.M.; Ilyas, R.A. Highly transparent and antimicrobial PVA based bionanocomposites reinforced by ginger nanofiber. *Polym. Test.* **2019**, 106186. [CrossRef]
32. Abral, H.; Ariksa, J.; Mahardika, M.; Handayani, D.; Aminah, I.; Sandrawati, N.; Pratama, A.B.; Fajri, N.; Sapuan, S.M.; Ilyas, R.A. Transparent and antimicrobial cellulose film from ginger nanofiber. *Food Hydrocoll.* **2020**, *98*, 105266. [CrossRef]
33. Prachayawarakorn, J.; Limsiriwong, N.; Kongjindamunee, R.; Surakit, S. Effect of Agar and Cotton Fiber on Properties of Thermoplastic Waxy Rice Starch Composites. *J. Polym. Environ.* **2012**, *20*, 88–95. [CrossRef]
34. Kumar, T.S.M.; Chandrasekar, M.; Senthilkumar, K.; Ilyas, R.A.; Sapuan, S.M.; Hariram, N.; Rajulu, A.V.; Rajini, N.; Siengchin, S. Characterization, Thermal and Antimicrobial Properties of Hybrid Cellulose Nanocomposite Films with in-Situ Generated Copper Nanoparticles in *Tamarindus indica* Nut Powder. *J. Polym. Environ.* **2020**, 1–10. [CrossRef]

35. Aisyah, H.A.; Paridah, M.T.; Sapuan, S.M.; Khalina, A.; Berkalp, O.B.; Lee, S.H.; Lee, C.H.; Nurazzi, N.M.; Ramli, N.; Wahab, M.S.; et al. Thermal Properties of Woven Kenaf/Carbon Fibre-Reinforced Epoxy Hybrid Composite Panels. *Int. J. Polym. Sci.* **2019**, *2019*, 1–8. [CrossRef]
36. Mazani, N.; Sapuan, S.M.; Sanyang, M.L.; Atiqah, A.; Ilyas, R.A. Design and Fabrication of a Shoe Shelf from Kenaf Fiber Reinforced Unsaturated Polyester Composites. In *Lignocellulose for Future Bioeconomy*; Ariffin, H., Sapuan, S.M., Hassan, M.A., Eds.; Elsevier: Amsterdam, The Netherlands, 2019; pp. 315–332. ISBN 9780128163542.
37. Aiza Jaafar, C.N.; Zainol, I.; Ishak, N.S.; Ilyas, R.A.; Sapuan, S.M. Effects of the Liquid Natural Rubber (LNR) on Mechanical Properties and Microstructure of Epoxy/Silica/Kenaf Hybrid Composite for Potential Automotive Applications. *J. Mater. Res. Technol.* **2021**, *12*, 1026–1038. [CrossRef]
38. Sabaruddin, F.A.; Paridah, M.T.; Sapuan, S.M.; Ilyas, R.A.; Lee, S.H.; Abdan, K.; Mazlan, N.; Roseley, A.S.M.; Abdul Khalil, H.P.S. The effects of unbleached and bleached nanocellulose on the thermal and flammability of polypropylene-reinforced kenaf core hybrid polymer bionanocomposites. *Polymers* **2020**, *13*, 116. [CrossRef] [PubMed]
39. Suriani, M.J.; Zainudin, H.A.; Ilyas, R.A.; Petrů, M.; Sapuan, S.M.; Ruzaidi, C.M.; Mustapha, R. Kenaf Fiber/Pet Yarn Reinforced Epoxy Hybrid Polymer Composites: Morphological, Tensile, and Flammability Properties. *Polymers* **2021**, *13*, 1532. [CrossRef]
40. Jumaidin, R.; Adam, N.W.; Ilyas, R.A.; Hussin, M.S.F.; Taha, M.M.; Mansor, M.R.; Azlan, U.A.-A.; Yob, M.S. Water Transport and Physical Properties of Sugarcane Bagasse Fibre Reinforced Thermoplastic Potato Starch Biocomposite. *J. Adv. Res. Fluid Mech. Therm. Sci.* **2019**, *61*, 273–281.
41. Asrofi, M.; Sujito; Syafri, E.; Sapuan, S.M.; Ilyas, R.A. Improvement of Biocomposite Properties Based Tapioca Starch and Sugarcane Bagasse Cellulose Nanofibers. *Key Eng. Mater.* **2020**, *849*, 96–101. [CrossRef]
42. Asrofi, M.; Sapuan, S.M.; Ilyas, R.A.; Ramesh, M. Characteristic of composite bioplastics from tapioca starch and sugarcane bagasse fiber: Effect of time duration of ultrasonication (Bath-Type). *Mater. Today Proc.* **2020**. [CrossRef]
43. Correa-Aguirre, J.P.; Luna-Vera, F.; Caicedo, C.; Vera-Mondragón, B.; Hidalgo-Salazar, M.A. The Effects of Reprocessing and Fiber Treatments on the Properties of Polypropylene-Sugarcane Bagasse Biocomposites. *Polymers* **2020**, *12*, 1440. [CrossRef] [PubMed]
44. Nassiopoulos, E.; Njuguna, J. Thermo-mechanical performance of poly(lactic acid)/flax fibre-reinforced biocomposites. *Mater. Des.* **2015**, *66*, 473–485. [CrossRef]
45. Syafri, E.; Kasim, A.; Abral, H.; Asben, A. Cellulose nanofibers isolation and characterization from ramie using a chemical-ultrasonic treatment. *J. Nat. Fibers* **2018**, 1–11. [CrossRef]
46. Battegazzore, D.; Noori, A.; Frache, A. Hemp hurd and alfalfa as particle filler to improve the thermo-mechanical and fire retardant properties of poly (3-hydroxybutyrate-co-3-hydroxyhexanoate). *Polym. Compos.* **2019**, *40*. [CrossRef]
47. Prachayawarakorn, J.; Chaiwatyothin, S.; Mueangta, S.; Hanchana, A. Effect of jute and kapok fibers on properties of thermoplastic cassava starch composites. *Mater. Des.* **2013**, *47*, 309–315. [CrossRef]
48. Gupta, M.; Singh, R. PLA-coated sisal fibre-reinforced polyester composite: Water absorption, static and dynamic mechanical properties. *J. Compos. Mater.* **2019**, *53*, 65–72. [CrossRef]
49. Ayu, R.S.; Khalina, A.; Harmaen, A.S.; Zaman, K.; Isma, T.; Liu, Q.; Ilyas, R.A.; Lee, C.H. Characterization Study of Empty Fruit Bunch (EFB) Fibers Reinforcement in Poly(Butylene) Succinate (PBS)/Starch/Glycerol Composite Sheet. *Polymers* **2020**, *12*, 1571. [CrossRef] [PubMed]
50. Jumaidin, R.; Diah, N.A.; Ilyas, R.A.; Alamjuri, R.H.; Yusof, F.A.M. Processing and Characterisation of Banana Leaf Fibre Reinforced Thermoplastic Cassava Starch Composites. *Polymers* **2021**, *13*, 1420. [CrossRef]
51. Ilyas, R.A.; Sapuan, S.M.; Atiqah, A.; Ibrahim, R.; Abral, H.; Ishak, M.R.; Zainudin, E.S.; Nurazzi, N.M.; Atikah, M.S.N.; Ansari, M.N.M.; et al. Sugar palm (*Arenga pinnata* [Wurmb.] Merr) starch films containing sugar palm nanofibrillated cellulose as reinforcement: Water barrier properties. *Polym. Compos.* **2020**, *41*, 459–467. [CrossRef]
52. Rozilah, A.; Jaafar, C.N.A.; Sapuan, S.M.; Zainol, I.; Ilyas, R.A. The Effects of Silver Nanoparticles Compositions on the Mechanical, Physiochemical, Antibacterial, and Morphology Properties of Sugar Palm Starch Biocomposites for Antibacterial Coating. *Polymers* **2020**, *12*, 2605. [CrossRef]
53. Atiqah, A.; Jawaid, M.; Sapuan, S.M.; Ishak, M.R.; Ansari, M.N.M.; Ilyas, R.A. Physical and thermal properties of treated sugar palm/glass fibre reinforced thermoplastic polyurethane hybrid composites. *J. Mater. Res. Technol.* **2019**, *8*, 3726–3732. [CrossRef]
54. Ilyas, R.A.; Sapuan, S.M.; Ishak, M.R. Isolation and characterization of nanocrystalline cellulose from sugar palm fibres (Arenga Pinnata). *Carbohydr. Polym.* **2018**, *181*, 1038–1051. [CrossRef] [PubMed]
55. Ilyas, R.A.; Sapuan, S.M.; Ishak, M.R.; Zainudin, E.S. Development and characterization of sugar palm nanocrystalline cellulose reinforced sugar palm starch bionanocomposites. *Carbohydr. Polym.* **2018**, *202*, 186–202. [CrossRef] [PubMed]
56. Ilyas, R.A.; Sapuan, S.M.; Ibrahim, R.; Abral, H.; Ishak, M.R.; Zainudin, E.S.; Atiqah, A.; Atikah, M.S.N.; Syafri, E.; Asrofi, M.; et al. Thermal, Biodegradability and Water Barrier Properties of Bio-Nanocomposites Based on Plasticised Sugar Palm Starch and Nanofibrillated Celluloses from Sugar Palm Fibres. *J. Biobased Mater. Bioenergy* **2020**, *14*, 234–248. [CrossRef]
57. Atikah, M.S.N.; Ilyas, R.A.; Sapuan, S.M.; Ishak, M.R.; Zainudin, E.S.; Ibrahim, R.; Atiqah, A.; Ansari, M.N.M.; Jumaidin, R. Degradation and physical properties of sugar palm starch/sugar palm nanofibrillated cellulose bionanocomposite. *Polimery* **2019**, *64*, 680–689. [CrossRef]
58. Ilyas, R.A.; Sapuan, S.M.; Ibrahim, R.; Abral, H.; Ishak, M.R.; Zainudin, E.S.; Atikah, M.S.N.; Mohd Nurazzi, N.; Atiqah, A.; Ansari, M.N.M.; et al. Effect of sugar palm nanofibrillated cellulose concentrations on morphological, mechanical and physical

properties of biodegradable films based on agro-waste sugar palm (*Arenga pinnata* (Wurmb.) Merr) starch. *J. Mater. Res. Technol.* **2019**, *8*, 4819–4830. [CrossRef]
59. Hazrol, M.D.; Sapuan, S.M.; Ilyas, R.A.; Othman, M.L.; Sherwani, S.F.K. Electrical properties of sugar palm nanocrystalline cellulose reinforced sugar palm starch nanocomposites. *Polimery* **2020**, *65*, 363–370. [CrossRef]
60. Ilyas, R.A.; Sapuan, S.M.; Ishak, M.R.; Zainudin, E.S. Sugar palm nanofibrillated cellulose (Arenga pinnata (Wurmb.) Merr): Effect of cycles on their yield, physic-chemical, morphological and thermal behavior. *Int. J. Biol. Macromol.* **2019**, *123*, 379–388. [CrossRef]
61. Asyraf, M.R.M.; Rafidah, M.; Azrina, A.; Razman, M.R. Dynamic mechanical behaviour of kenaf cellulosic fibre biocomposites: A comprehensive review on chemical treatments. *Cellulose* **2021**. [CrossRef]
62. Zhang, H.; Zhang, X.; Fang, Z.; Huang, Y.; Xu, H.; Liu, Y.; Wu, D.; Zhuang, J.; Sun, J. Recent Advances in Preparation, Mechanisms, and Applications of Thermally Conductive Polymer Composites: A Review. *J. Compos. Sci.* **2020**, *4*, 180. [CrossRef]
63. Shrigondekar, H.; Chowdhury, A.; Prabhu, S.V. Performance of water mist system with base injection in extinguishing small container fires. *J. Loss Prev. Process Ind.* **2021**, *71*, 104448. [CrossRef]
64. Kalyon, D.M.; Birinci, E.; Yazici, R.; Karuv, B.; Walsh, S. Electrical properties of composites as affected by the degree of mixedness of the conductive filler in the polymer matrix. *Polym. Eng. Sci.* **2002**, *42*, 1609–1617. [CrossRef]
65. Feng, N.L.; Malingam, S.D.; Ping, C.W. Mechanical characterisation of kenaf/PALF reinforced composite-metal laminates: Effects of hybridisation and weaving architectures. *J. Reinf. Plast. Compos.* **2021**, *40*, 193–205. [CrossRef]
66. Chandrasekar, M.; Ishak, M.R.; Sapuan, S.M.; Leman, Z.; Jawaid, M.; Shahroze, R.M. Fabrication of Fibre Metal Laminate with Flax and Sugar Palm Fibre based Epoxy Composite and Evaluation of their Fatigue Properties. *J. Polym. Mater.* **2019**, *35*, 463–473. [CrossRef]
67. Krishnasamy, P.; Rajamurugan, G.; Thirumurugan, M. Performance of fiber metal laminate composites embedded with AL and CU wire mesh. *J. Ind. Text.* **2020**. [CrossRef]
68. El-Sabbagh, A.; Steuernagel, L.; Ziegmann, G. Low combustible polypropylene/flax/magnesium hydroxide composites: Mechanical, flame retardation characterization and recycling effect. *J. Reinf. Plast. Compos.* **2013**, *32*, 1030–1043. [CrossRef]
69. Bar, M.; Alagirusamy, R.; Das, A. Flame retardant polymer composites. *Fibers Polym.* **2015**, *16*, 705–717. [CrossRef]
70. Marosi, G.; Szolnoki, B.; Bocz, K.; Toldy, A. Fire-retardant recyclable and biobased polymer composites. In *Novel Fire Retardant Polymers and Composite Materials*; Wang, D.-Y., Ed.; Elsevier: Duxford, UK, 2017; pp. 117–146.
71. Kim, N.K.; Dutta, S.; Bhattacharyya, D. A review of flammability of natural fibre reinforced polymeric composites. *Compos. Sci. Technol.* **2018**, *162*, 64–78. [CrossRef]
72. Akash; Girisha, K.G.; Venkatesha Gupta, N.S.; Sreenivas Rao, K.V. A study on flammability and moisture absorption behavior of sisal/coir fiber reinforced hybrid composites. *IOP Conf. Ser. Mater. Sci. Eng.* **2017**, *191*. [CrossRef]
73. Zhao, C.-S.; Huang, F.-L.; Xiong, W.-C.; Wang, Y.-Z. A novel halogen-free flame retardant for glass-fiber-reinforced poly(ethylene terephthalate). *Polym. Degrad. Stab.* **2008**, *93*, 1188–1193. [CrossRef]
74. Zadeh, K.M.; Ponnamma, D.; Al Ali Al-Maadeed, M. Date palm fibre filled recycled ternary polymer blend composites with enhanced flame retardancy. *Polym. Test.* **2017**, *61*, 341–348. [CrossRef]
75. Yuan, B.; Sheng, H.; Mu, X.; Song, L.; Tai, Q.; Shi, Y.; Liew, K.M.; Hu, Y. Enhanced flame retardancy of polypropylene by melamine-modified graphene oxide. *J. Mater. Sci.* **2015**, *50*, 5389–5401. [CrossRef]
76. Vieira, L.M.G.; dos Santos, J.C.; Panzera, T.H.; Rubio, J.C.C.; Scarpa, F. Novel fibre metal laminate sandwich composite structure with sisal woven core. *Ind. Crops Prod.* **2017**, *99*, 189–195. [CrossRef]
77. Ishak, N.M.; Malingam, S.D.; Mansor, M.R.; Razali, N.; Mustafa, Z.; Ghani, A.F.A. Investigation of natural fibre metal laminate as car front hood. *Mater. Res. Express* **2021**, *8*. [CrossRef]
78. Levchik, S.V. Introduction to Flame Retardancy and Polymer Flammability. In *Flame Retardant Polymer Nanocomposites*; Morgan, A.B., Wilkie, C.A., Eds.; John Wiley & Sons, Inc.: Hoboken, NJ, USA, 2006; pp. 1–29. ISBN 9780471734260.
79. Zhang, P.; Song, L.; Lu, H.; Wang, J.; Hu, Y. The thermal property and flame retardant mechanism of intumescent flame retardant paraffin system with metal. *Ind. Eng. Chem. Res.* **2010**. [CrossRef]
80. Gao, W.; Yu, Y.; Chen, T.; Zhang, Q.; Chen, Z.; Chen, Z.; Jiang, J. Enhanced flame retardancy of unsaturated polyester resin composites containing ammonium polyphosphate and metal oxides. *J. Appl. Polym. Sci.* **2020**, *137*, 49148. [CrossRef]
81. Jia, Y.-W.; Zhao, X.; Fu, T.; Li, D.-F.; Guo, Y.; Wang, X.-L.; Wang, Y.-Z. Synergy effect between quaternary phosphonium ionic liquid and ammonium polyphosphate toward flame retardant PLA with improved toughness. *Compos. Part B Eng.* **2020**, *197*, 108192. [CrossRef]
82. Guo, B.; Zhang, T.; Zhang, W.; Dou, Y. Influence of surface flame-retardant layer containing ammonium polyphosphate and expandable graphite on the performance of jute/polypropylene composites. *J. Therm. Anal. Calorim.* **2019**, *135*, 2367–2375. [CrossRef]
83. Pang, X.; Xin, Y.; Shi, X.; Xu, J. Effect of different size-modified expandable graphite and ammonium polyphosphate on the flame retardancy, thermal stability, physical, and mechanical properties of rigid polyurethane foam. *Polym. Eng. Sci.* **2019**, *59*, 1381–1394. [CrossRef]
84. 8Shi, X.; Pan, Y.; Wang, Y.; Jia, Z.; Chen, T.; Gong, J.; Jiang, J. Synergistic Effects of Graphene and Ammonium Polyphosphate Modified with Vinyltrimethoxysilane on the Properties of High-Impact Polystyrene Composites. *Polymers* **2021**, *13*, 881. [CrossRef]

85. Zhou, R.; Ming, Z.; He, J.; Ding, Y.; Jiang, J. Effect of Magnesium Hydroxide and Aluminum Hydroxide on the Thermal Stability, Latent Heat and Flammability Properties of Paraffin/HDPE Phase Change Blends. *Polymers* **2020**, *12*, 180. [CrossRef] [PubMed]
86. He, W.; Song, P.; Yu, B.; Fang, Z.; Wang, H. Flame retardant polymeric nanocomposites through the combination of nanomaterials and conventional flame retardants. *Prog. Mater. Sci.* **2020**, *114*, 100687. [CrossRef]
87. Kalali, E.N.; Zhang, L.; Shabestari, M.E.; Croyal, J.; Wang, D.-Y. Flame-retardant wood polymer composites (WPCs) as potential fire safe bio-based materials for building products: Preparation, flammability and mechanical properties. *Fire Saf. J.* **2019**, *107*, 210–216. [CrossRef]
88. Yen, Y.Y.; Wang, H.T.; Guo, W.J. Synergistic flame retardant effect of metal hydroxide and nanoclay in EVA composites. *Polym. Degrad. Stab.* **2012**, *97*, 863–869. [CrossRef]
89. Zhang, S.; Horrocks, A.R. A review of flame retardant polypropylene fibres. *Prog. Polym. Sci.* **2003**, *28*, 1517–1538. [CrossRef]
90. Lin, M.; Li, B.; Li, Q.; Li, S.; Zhang, S. Synergistic effect of metal oxides on the flame retardancy and thermal degradation of novel intumescent flame-retardant thermoplastic polyurethanes. *J. Appl. Polym. Sci.* **2011**, *121*, 1951–1960. [CrossRef]
91. Sain, M.; Park, S.H.; Suhara, F.; Law, S. Flame retardant and mechanical properties of natural fibre-PP composites containing magnesium hydroxide. *Polym. Degrad. Stab.* **2004**, *83*, 363–367. [CrossRef]
92. Jeencham, R.; Suppakarn, N.; Jarukumjorn, K. Effect of flame retardants on flame retardant, mechanical, and thermal properties of sisal fiber/polypropylene composites. *Compos. Part B Eng.* **2014**. [CrossRef]
93. Wang, B.; Sheng, H.; Shi, Y.; Hu, W.; Hong, N.; Zeng, W.; Ge, H.; Yu, X.; Song, L.; Hu, Y. Recent advances for microencapsulation of flame retardant. *Polym. Degrad. Stab.* **2015**, *113*, 96–109. [CrossRef]
94. Wang, D.Y.; Liu, Y.; Wang, Y.Z.; Artiles, C.P.; Hull, T.R.; Price, D. Fire retardancy of a reactively extruded intumescent flame retardant polyethylene system enhanced by metal chelates. *Polym. Degrad. Stab.* **2007**, *92*, 1592–1598. [CrossRef]
95. Wang, D.L.; Liu, Y.; Wang, D.Y.; Zhao, C.X.; Mou, Y.R.; Wang, Y.Z. A novel intumescent flame-retardant system containing metal chelates for polyvinyl alcohol. *Polym. Degrad. Stab.* **2007**. [CrossRef]
96. Rothon, R.; Hornsby, P. Fire Retardant Fillers for Polymers. In *Polymer Green Flame Retardants*; Elsevier: Amsterdam, The Netherlands, 2014; pp. 289–321. ISBN 9780444538093.
97. Braun, U.; Bahr, H.; Sturm, H.; Schartel, B. Flame retardancy mechanisms of metal phosphinates and metal phosphinates in combination with melamine cyanurate in glass-fiber reinforced poly(1,4-butylene terephthalate): The influence of metal cation. *Polym. Adv. Technol.* **2008**, *19*, 680–692. [CrossRef]
98. Wu, N.; Yang, R. Effects of metal oxides on intumescent flame-retardant polypropylene. *Polym. Adv. Technol.* **2011**. [CrossRef]
99. Lujan-Acosta, R.; Sánchez-Valdes, S.; Ramírez-Vargas, E.; Ramos-DeValle, L.F.; Espinoza-Martinez, A.B.; Rodriguez-Fernandez, O.S.; Lozano-Ramirez, T.; Lafleur, P.G. Effect of Amino alcohol functionalized polyethylene as compatibilizer for LDPE/EVA/clay/flame-retardant nanocomposites. *Mater. Chem. Phys.* **2014**, *146*, 437–445. [CrossRef]
100. Wang, Z.; Han, E.; Ke, W. Effect of nanoparticles on the improvement in fire-resistant and anti-ageing properties of flame-retardant coating. *Surf. Coat. Technol.* **2006**. [CrossRef]
101. Xiao, F.; Wu, K.; Luo, F.; Yao, S.; Lv, M.; Zou, H.; Lu, M. Influence of Ionic Liquid-Based Metal–Organic Hybrid on Thermal Degradation, Flame Retardancy, and Smoke Suppression Properties of Epoxy Resin Composites. *J. Mater. Sci.* **2018**, *53*, 10135–10146. [CrossRef]
102. Gallo, E.; Schartel, B.; Acierno, D.; Russo, P. Flame retardant biocomposites: Synergism between phosphinate and nanometric metal oxides. *Eur. Polym. J.* **2011**. [CrossRef]
103. Beyer, G. Flame retardant properties of EVA-nanocomposites and improvements by combination of nanofillers with aluminium trihydrate. *Fire Mater.* **2001**, *25*, 193–197. [CrossRef]
104. Murphy, J. Flame retardants: Trends and new developments. *Reinf. Plast.* **2001**, *45*, 42–46. [CrossRef]
105. Sharma, S.K.; Saxena, N.K. Flame Retardant Smoke Suppressant Protection for Poly Vinylchloride. *Fire Technol.* **2004**, *40*, 385–398. [CrossRef]
106. Beyer, G. Flame retardancy of nanocomposites from research to technical products. *J. Fire Sci.* **2005**. [CrossRef]
107. Chen, L.; Wang, Y.Z. A review on flame retardant technology in China. Part I: Development of flame retardants. *Polym. Adv. Technol.* **2010**, *21*, 1–26. [CrossRef]
108. Dogan, M.; Murat Unlu, S. Flame retardant effect of boron compounds on red phosphorus containing epoxy resins. *Polym. Degrad. Stab.* **2014**, *99*, 12–17. [CrossRef]
109. Topfer, O.; Clauss, M.; Futterer, T.; Schmitt, E. Flame retardants for engineering thermoplastics used in electric and electronic equipment like connectors. In *Proceedings of the Electronics Goes Green 2012+, ECG 2012—Joint International Conference and Exhibition, Proceedings*; Lang, K.-D., Nissen, N.F., Middendorf, A., Chancerel, P., Eds.; Fraunhofer Verlag: Berlin, Germany, 2012.
110. Kusakli, S.; Kocaman, S.; Ceyhan, A.A.; Ahmetli, G. Improving the flame retardancy and mechanical properties of epoxy composites using flame retardants with red mud waste. *J. Appl. Polym. Sci.* **2020**, 1–15. [CrossRef]
111. Song, P.; Fang, Z.; Tong, L.; Jin, Y.; Lu, F. Effects of metal chelates on a novel oligomeric intumescent flame retardant system for polypropylene. *J. Anal. Appl. Pyrolysis* **2008**, *82*, 286–291. [CrossRef]
112. Malkapuram, R.; Kumar, V.; Singh Negi, Y. Recent development in natural fiber reinforced polypropylene composites. *J. Reinf. Plast. Compos.* **2009**, *28*, 1169–1189. [CrossRef]
113. Müller, P.; Morys, M.; Sut, A.; Jäger, C.; Illerhaus, B.; Schartel, B. Melamine poly(zinc phosphate) as flame retardant in epoxy resin: Decomposition pathways, molecular mechanisms and morphology of fire residues. *Polym. Degrad. Stab.* **2016**. [CrossRef]

114. Hu, C.; Xue, J.; Dong, L.; Jiang, Y.; Wang, X.; Qu, L.; Dai, L. Scalable preparation of multifunctional fire-retardant ultralight graphene foams. *ACS Nano* **2016**. [CrossRef] [PubMed]
115. Davies, P.J.; Horrocks, A.R.; Alderson, A. The sensitisation of thermal decomposition of ammonium polyphosphate by selected metal ions and their potential for improved cotton fabric flame retardancy. *Polym. Degrad. Stab.* **2005**. [CrossRef]
116. Srivastava, S.K.; Kuila, T. Fire Retardancy of Elastomers and Elastomer Nanocomposites. In *Polymer Green Flame Retardants*; Papaspyrides, C.D., Kiliaris, P., Eds.; Elsevier: Oxford, UK, 2014; pp. 597–651. ISBN 9780444538093.
117. Holdsworth, A.F.; Horrocks, A.R.; Kandola, B.K.; Price, D. The potential of metal oxalates as novel flame retardants and synergists for engineering polymers. *Polym. Degrad. Stab.* **2014**. [CrossRef]
118. Otsuki, M.; Ogino, T. Flame-retardant additives for lithium-ion batteries. In *Lithium-Ion Batteries: Science and Technologies*; Yoshio, M., Brodd, R.J., Kozawa, A., Eds.; Springer-Verlag New York: New York, NY, USA, 2009; pp. 275–290. ISBN 9780387344447.
119. Abd El-Wahab, H.; Abd El-Fattah, M.; El-alfy, H.M.Z.; Owda, M.E.; Lin, L.; Hamdy, I. Synthesis and characterisation of sulphonamide (Schiff base) ligand and its copper metal complex and their efficiency in polyurethane varnish as flame retardant and antimicrobial surface coating additives. *Prog. Org. Coat.* **2020**. [CrossRef]
120. Levchik, S.V.; Weil, E.D. A review of recent progress in phosphorus-based flame retardants. *J. Fire Sci.* **2006**. [CrossRef]
121. Chang, M.K.; Hwang, S.S.; Liu, S.P. Flame retardancy and thermal stability of ethylene-vinyl acetate copolymer nanocomposites with alumina trihydrate and montmorillonite. *J. Ind. Eng. Chem.* **2014**. [CrossRef]
122. Wang, L.; Zhang, T.; Yan, H.; Peng, M.; Fang, Z. Modification of ramie fabric with a metal-ion-doped flame-retardant coating. *J. Appl. Polym. Sci.* **2013**. [CrossRef]
123. Üreyen, M.E.; Kaynak, E. Effect of Zinc Borate on Flammability of PET Woven Fabrics. *Adv. Polym. Technol.* **2019**, *2019*, 1–13. [CrossRef]
124. Suppakarn, N.; Jarukumjorn, K. Mechanical properties and flammability of sisal/PP composites: Effect of flame retardant type and content. *Compos. Part B Eng.* **2009**, *40*, 613–618. [CrossRef]
125. Li, H.; Ning, N.; Zhang, L.; Wang, Y.; Liang, W.; Tian, M. Different flame retardancy effects and mechanisms of aluminium phosphinate in PPO, TPU and PP. *Polym. Degrad. Stab.* **2014**, *105*, 86–95. [CrossRef]
126. Bulei, C.; Kiss, I.; Alexa, V. Development of metal matrix composites using recycled secondary raw materials from aluminium wastes. *Mater. Today Proc.* **2021**. [CrossRef]
127. Vo Dong, P.A.; Azzaro-Pantel, C.; Boix, M.; Jacquemin, L.; Domenech, S. *Modelling of Environmental Impacts and Economic Benefits of Fibre Reinforced Polymers Composite Recycling Pathways*; Elsevier: Amsterdam, The Netherlands, 2015; Volume 37, ISBN 9780444635761.
128. Tahir, F.; Mabrouk, A.; Al-Ghamdi, S.G.; Krupa, I.; Sedlacek, T.; Abdala, A.; Koc, M. Sustainability Assessment and Techno-Economic Analysis of Thermally Enhanced Polymer Tube for Multi-Effect Distillation (MED) Technology. *Polymers* **2021**, *13*, 681. [CrossRef] [PubMed]
129. Du, J.; Cheng, H.M. The fabrication, properties, and uses of graphene/polymer composites. *Macromol. Chem. Phys.* **2012**, *213*, 1060–1077. [CrossRef]
130. Pan, Y.T.; Zhang, Z.; Yang, R. The rise of MOFs and their derivatives for flame retardant polymeric materials: A critical review. *Compos. Part B Eng.* **2020**, *199*. [CrossRef]
131. Su, Z.; Wang, H.; Ye, X.; Tian, K.; Huang, W.; He, J.; Guo, Y.; Tian, X. Anisotropic thermally conductive flexible polymer composites filled with hexagonal born nitride (h-BN) platelets and ammine carbon nanotubes (CNT-NH2): Effects of the filler distribution and orientation. *Compos. Part A Appl. Sci. Manuf.* **2018**, *109*, 402–412. [CrossRef]

polymers

MDPI

Article

Kenaf Fiber/Pet Yarn Reinforced Epoxy Hybrid Polymer Composites: Morphological, Tensile, and Flammability Properties

M. J. Suriani [1,2,*], Hasliana Asyikin Zainudin [1], R. A. Ilyas [3,4,*], Michal Petrů [5], S. M. Sapuan [6,7], C. M. Ruzaidi [1,2] and Rohani Mustapha [1]

1 Faculty of Ocean Engineering Technology and Informatics, Universiti Malaysia Terengganu, Kuala Nerus 21030, Terengganu, Malaysia; hasliana.zainudin@yahoo.com (H.A.Z.); ruzaidi@umt.edu.my (C.M.R.); rohani.m@umt.edu.my (R.M.)
2 Marine Materials Research Group, Faculty of Ocean Engineering Technology and Informatics, Universiti Malaysia Terengganu, Kuala Nerus 21030, Terengganu, Malaysia
3 School of Chemical and Energy Engineering, Faculty of Engineering, Universiti Teknologi Malaysia, UTM, Johor Bahru 81310, Johor, Malaysia
4 Centre for Advanced Composite Materials, Universiti Teknologi Malaysia, UTM, Johor Bahru 81310, Johor, Malaysia
5 Faculty of Mechanical Engineering, Technical University of Liberec, Studentská 2, 461 17 Liberec, Czech Republic; michal.petru@tul.cz
6 Laboratory of Biocomposite Technology, Institute of Tropical Forestry and Forest Products (INTROP), Universiti Putra Malaysia, UPM, Serdang 43400, Selangor, Malaysia; sapuan@upm.edu.my
7 Advanced Engineering Materials and Composites Research Centre (AEMC), Department of Mechanical and Manufacturing Engineering, Faculty of Engineering, Universiti Putra Malaysia, UPM, Serdang 43400, Selangor, Malaysia
* Correspondence: surianimatjusoh@umt.edu.my (M.J.S.); ahmadilyas@utm.my (R.A.I.)

Citation: Suriani, M.J.; Zainudin, H.A.; Ilyas, R.A.; Petrů, M.; Sapuan, S.M.; Ruzaidi, C.M.; Mustapha, R. Kenaf Fiber/Pet Yarn Reinforced Epoxy Hybrid Polymer Composites: Morphological, Tensile, and Flammability Properties. *Polymers* **2021**, *13*, 1532. https://doi.org/10.3390/polym13091532

Academic Editor: Nadka Tz. Dintcheva

Received: 17 March 2021
Accepted: 6 May 2021
Published: 10 May 2021

Abstract: The application of natural fibers is rapidly growing in many sectors, such as construction, automobile, and furniture. Kenaf fiber (KF) is a natural fiber that is in demand owing to its eco-friendly and renewable nature. Nowadays, there are various new applications for kenaf, such as in absorbents and building materials. It also has commercial applications, such as in the automotive industry. Magnesium hydroxide ($Mg(OH)_2$) is used as a fire retardant as it is low in cost and has good flame retardancy, while polyester yarn (PET) has high tensile strength. The aim of this study was to determine the horizontal burning rate, tensile strength, and surface morphology of kenaf fiber/PET yarn reinforced epoxy fire retardant composites. The composites were prepared by hybridized epoxy and $Mg(OH)_2$ PET with different amounts of KF content (0%, 20%, 35%, and 50%) using the cold press method. The specimen with 35% KF (epoxy/PET/KF-35) displayed better flammability properties and had the lowest average burning rate of 14.55 mm/min, while epoxy/PET/KF-50 with 50% KF had the highest tensile strength of all the samples. This was due to fewer defects being detected on the surface morphology of epoxy/PET/KF-35 compared to the other samples, which influenced the mechanical properties of the composites.

Keywords: kenaf composite; flammability; fire retardant; hybrid composite; tensile; morphology; pet yarn; epoxy

1. Introduction

In the last few decades, advanced application of composite materials has gained a lot of interest from various sectors, especially transmission towers, defense, aircrafts, marine, and the automotive industry [1–4]. Industries have sought environmentally friendly materials for their products due to increasing global awareness about the environmental impact [5,6]. Natural fibers are renewable sources that have emerged as a new generation of reinforcement and supplement for polymers [7,8]. Growing industrial activities have prompted continuous demand for better materials that satisfy various requirements, such as higher strength, higher modulus, reduced cost, etc. These requirements often involve a combination of many properties that are difficult to attain. The use of a composite material whose constituents will synergize to solve the needs of the application is therefore necessary. Many researchers in the past have developed composites using natural fibers, such as sugar palm [9–15], oil palm [16], sugarcane [17–19], water hyacinth [20], kenaf [21–23], corn husk [24], bamboo [25], coir [26], sisal [27], cogon [28], and banana [29,30]. These natural fibers possess good reinforcing capability when properly combined with polymers [31].

Nowadays, hybrid composites are widely used in industrial and automobile applications. A hybrid composite can be defined as a combination of two or more different types of fibers in which one fiber balances the deficiency of another fiber [32–35]. They are composites with two or more reinforcing materials incorporated in a mixture and filling materials in a single matrix [36]. The properties of hybrid composite are influenced by the fiber content, length, and orientation. The selection of the fiber constituent for hybrid composites affects the hybridization and the requirement of the material being constructed. The selection of compatible fibers and fiber properties is therefore a critical aspect in designing hybrid composites [37]. This step is crucial in order to achieve hybrid composites with the best performance [38]. A previous study found that sisal and oil palm fibers were compatible combinations for hybrid composites due to the high tensile strength and toughness of the sisal and oil palm fibers, respectively [39]. In addition, there are some specific advantages of hybrid composites compared to conventional composites. For instance, hybrid composites can balance strength and stiffness, improve fatigue resistance, balance bending and membrane mechanical properties, increase fracture toughness, and improve impact resistance [40]. The advantages of hybrid composites could complement other fiber constituents. The hybridization of kenaf with polyester (PET) and polyoxymethylene (POM) composites improved the mechanical performance of the composites due to the high strength of PET fiber [41]. Ramesh et al. [42] studied the effect of hybridization on mechanical properties for sisal, jute natural fiber, and glass fiber woven mat. They found that a sisal/jute/glass composite enhanced the mechanical properties compared to a composite made of only sisal and jute fibers. Polymer composites can be hybridized with different combinations, such as cellulosic/cellulosic, synthetic/synthetic, and cellulosic/synthetic fibers [43]. As for cost reduction of fibers such as carbon and glass, an attempt was made by combining jute, a cost-effective and eco-friendly natural fiber, with fiber metal laminate (FML) [44]. The use of natural resources to make high-performance materials is increasing [45]. Fiber hybridization with synthetic fibers results in improvement of the properties of hybrid composites by increasing their strength [46,47]. Synthetic fibers have good potential for hybridization with natural fibers due to their low cost and high strength. PET fibers have good mechanical properties and hydrophobic characteristics, which means hybridization of kenaf fiber with PET yarn could improve the strength of the composite material.

Natural fibers contain hydrogen bonds and other linkages that give them strength and stiffness. In Western Europe, natural fibers are being used in the automotive industry and for packaging materials. Natural fibers are cellulosic materials, which have several fibrils that run all along the length of the fiber [48]. The demand for natural fiber composites has grown rapidly for various applications, such as automotive components, building materials, and the aerospace industry, due to their ecological and economic advantages compared to conventional composites [49]. The use of sisal, bamboo, sugar palm, kenaf, cotton, and jute from natural resources has helped engineers develop high-performance and environmentally safe engineering products.

One of the potential constituents of natural fiber reinforced plastic composites in Malaysia is kenaf fiber (KF). Research on kenaf has grown tremendously. Kenaf long fibers could be used for a wide variety of applications if the properties were found to be comparable to existing synthesis composites. Nowadays, there are various new applications for kenaf, such as in furniture, composite boards, automotive panels, insulation mats, geotextiles, packaging, absorbents, building materials, and paper products. Kenaf also has commercial applications, such as in the automotive industry [37,50,51]. Large global corporations, such as Toyota Motor Corporation, have led the world in the use of kenaf. Biocomposites can result in a 25% reduction of vehicle weight, consequently contributing to saving 39.45 trillion of crude oil [52]. Moreover, this material can be used for composite frames for electric vehicles as it will reduce energy consumption. Because kenaf is always available in long fiber form, the mechanical properties could be of use in many industrial applications, such as insulator seals. In addition, kenaf fibers have the advantage of being biodegradable, low density, nonabrasive during processing, and environmentally safe [53]. The attractive features of kenaf fibers are low cost, lightweight, renewability, biodegradability, and high specific mechanical properties. Kenaf has a bast fiber that contains 75% cellulose and 15% lignin and offers the advantage of being biodegradable and environmentally safe [54]. Kenaf, like all agro (lignocellulosic) fibers or cell wall polymers, is a three-dimensional polymeric composite primarily made up of cellulose (56–64%), hemicelluloses (29–35%), lignin (11–14%), and small amounts of extractives and ash.

When using natural fibers as reinforcement, the composite is prone to fire damage, and it is therefore essential to find additives that have low total emission of heat [55]. Therefore, it is really important to improve the flame retardancy of composite materials [49]. With natural fibers being increasingly used, the development of safe and environmentally friendly flame retardant polymer composites is very important. Previous studies have used magnesium hydroxide and zinc borate as flame retardants to enhance the mechanical and flammability properties of sisal/polypropylene (PP) composites. They showed that addition of flame retardant to sisal/PP composites increased the thermal stability of the composites and reduced burning. Just like other organic materials, polymers and natural fibers are very sensitive to flame. In order to minimize the fire risk and meet fire safety requirements, solutions to prevent the ignition of these materials have been developed. Thus, flame retardant composites were produced [56]. Flame retardants are chemical substances that are added to materials both to constrain and delay the spread of fire after ignition [57]. Magnesium hydroxide is low in cost and has good flame retardancy for use as a fire retardant. It is an environmentally friendly, inorganic flame retardant with low smoke and nontoxic characteristics [58]. Besides examining the addition of flame retardant material in natural fiber/polymer composites, there is also research being undertaken for treatment of natural fibers with fire retardant materials. The preparation and characterization of thermosetting and thermoplastic composite materials reinforced with the most commonly used fibers with and without treatment using different methods have been conducted in the past. Treatment using alkaline on the surface of the fiber changes the surface wettability, altering the mechanical and physical properties of the natural fibers. Previous studies have also shown that, as the fiber content increases in accordance with the mixture rule, the strength and modulus of longitudinal composites in tensile and flexural loading also increases [59].

Compared to other natural fibers, kenaf is eco-friendly and has a high growth rate. Therefore, in the present work, an investigation was carried out to study the horizontal burning rate, with magnesium hydroxide used to determine its capability as a fire retardant composite. Different percentages of kenaf fibers were added, with PET yarn and magnesium hydroxide used as controls. The objective was to determine the horizontal burning rate, tensile strength, and surface morphology of kenaf fiber/PET yarn reinforced epoxy fire retardant composites.

2. Materials and Methods

2.1. Materials

The main materials used in the study were kenaf fiber and PET yarn. Kenaf fiber was ordered from Innovative Pultrusion Sdn. Bhd., Negeri Sembilan. First, the kenaf fibers were combed to clean and untangle the strong bonds between individual fibers. Combed fibers exhibit stronger mechanical properties compared to uncombed ones [60]. Then, the kenaf fibers were cut to 22 cm pieces using scissors. PET yarn was bought from Wellward

Sdn. Bhd., Johor, and were also cut into 22 cm pieces using scissors. The matrix used in this research was epoxy resin and hardener supplied by HR Team Enterprise, Kuala Terengganu, Terengganu. Magnesium hydroxide was ordered from HR Team Enterprise, Kuala Terengganu, Terengganu. Table 1 shows the properties of the epoxy resin and hardener. The densities of the materials used in this research are shown in Table 2.

Table 1. Properties of epoxy resin and hardener.

Property	Epoxy Resin	Hardener
Form	Liquid	Liquid
Density (g/cm^3)	1.21	1.03
Curing time (hours)	24	24
Ratio	2	1

Table 2. Densities of materials used.

Material	Density (g/cm^3)
Kenaf	0.37
PET yarn	1.38
Epoxy	1.21
Magnesium hydroxide ($Mg(OH_2)$) powder	1.20

2.2. Research Methodology

Figure 1 shows the research methodology of the experiment. The experiment started with the preparation of materials, followed by fabrication of hybrid composites, and then tests carried out to determine burning, tensile, and surface morphology properties of the hybrid composites. The data obtained were analyzed and are critically discussed.

2.3. Preparation of Materials

The composite comprised a hybrid of kenaf fiber/PET yarn as reinforcement, while epoxy resin was used as the matrix and magnesium hydroxide as the fire retardant [61]. Epoxy with a density of 1.21 g/cm^3 was used as a binding material to fabricate the composite specimen. The recommended mixing ratio for epoxy resin and hardener is 2:1. The volume of epoxy used depended on the different ratios of kenaf fiber/epoxy, which determined the weight of epoxy needed. In this study, magnesium hydroxide was used as the control and a high filled-type inorganic flame retardant, with 5% magnesium hydroxide mixed with the epoxy resin to enhance the flame retardancy of the composite material.

2.4. Fabrication Process

The composite samples were prepared using a hybrid of kenaf fibers/PET yarn with the epoxy resin as binder and magnesium hydroxide as fire retardant through a random hand lay-up technique. The composition of the samples of kenaf fiber reinforced epoxy fire retardant composites is shown in Table 3.

Table 3. Composition of samples.

Sample Designation	Percentage (%)			
	KF	PET Yarn	$Mg(OH)_2$	Epoxy + Hardener
Epoxy/Mg/PET	0	5	5	90
Epoxy/PET/KF-20	20	5	5	70
Epoxy/PET/KF-35	35	5	5	55
Epoxy/PET/KF-50	50	5	5	40

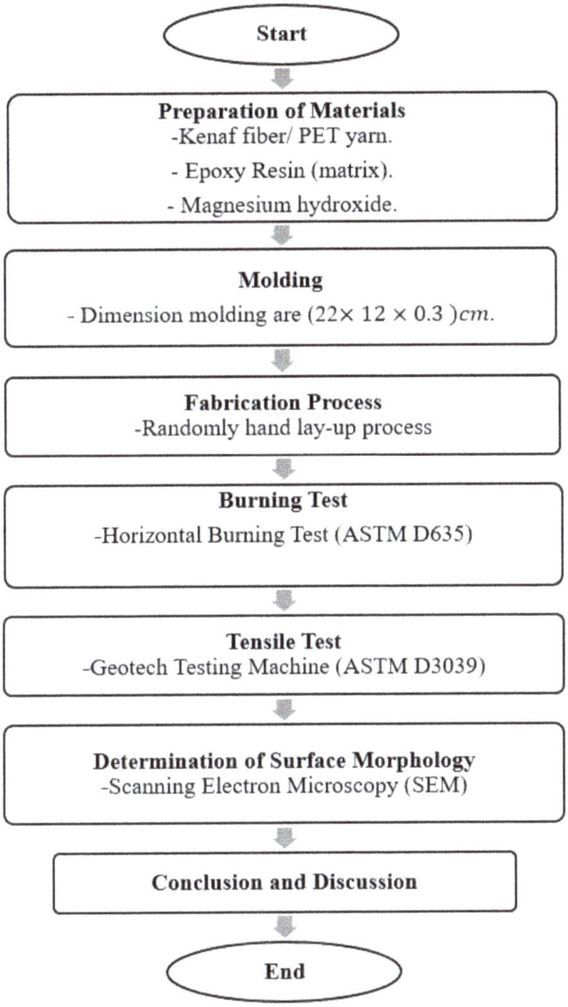

Figure 1. Research methodology.

Before fabricating, kenaf fibers were combed and then cut into 22 cm pieces. After that, the fibers were weighed based on the density of the fiber to determine the exact portion of fiber that should be used in the fiber polymer composite samples to obtain the designated fiber volume content. One of the problems with kenaf fiber reinforced epoxy that needs to be addressed is uneven fiber distribution [62]. In addition, it is difficult to manually and visually separate kenaf fibers during manufacturing [63].

The fabricating process of the composites in this study was as follows. The molds were polished using grease oil (releasing agent) before starting fabrication. Epoxy resin was mixed with a ratio of 2:1 (resin: hardener by weight) according to the manufacturer's specifications. The mixture was stirred slowly for 10 minutes to ensure that both epoxy and hardener were uniformly mixed. Then, magnesium hydroxide was poured into the mixture of epoxy and hardener. The mixture of magnesium hydroxide was then stirred slowly for 5–10 min to ensure the mixtures were uniformly mixed.

A transparent plastic sheet was placed on the surface of the mold. Then, the mixture of magnesium hydroxide with epoxy and hardener was poured slowly into the mold. Next, the fibers and PET yarn were completely immersed in resin by portion and then distributed randomly in the mold. The composite was covered by a transparent plastic sheet to avoid formation of air bubbles inside the mold. The plastic cover was used to prevent the fibers from attaching to the steel plate during curing. The composite was pressed using a flat plate to remove the remaining bubbles inside the mold. The mold was then compressed by another steel plate on top of the transparent plastic sheet. Demoulding of specimens was carried out after 24 h of fabrication. Figure 2 depicts the arrangement of KF and PET yarn with epoxy in the metal mold. Finally, the mold was cleaned to be used for the next specimen.

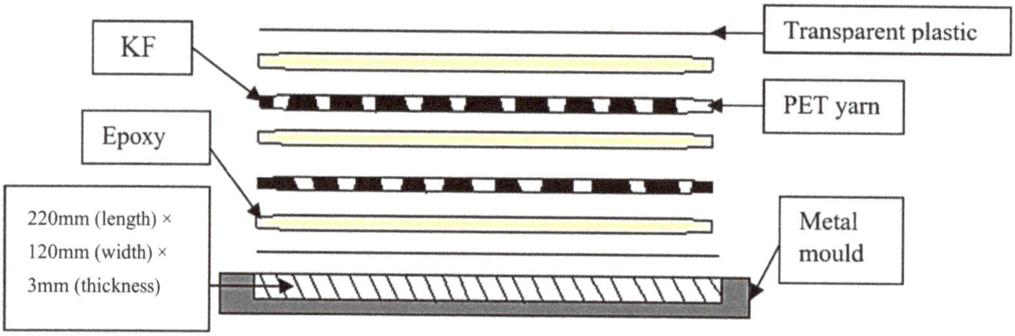

Figure 2. Arrangement of kenaf fiber and PET yarn with epoxy in the metal mold.

For each composition, the fiber and its binder were calculated to determine the weight of the fiber and epoxy needed to fill the mold. After fabrication of the composite was complete, the specimen of each composition was cut into nine pieces of 1.3 cm width for use in the horizontal burning test and eight pieces of 1.5 cm width for use in the tensile test. The images of these composites were also captured after the tensile test. Table 4 shows the results for all samples. Figure 3 presents the samples of each composition that were completely cured.

Table 4. Results of the calculation for all samples.

Descriptions	Samples			
	Epoxy/Mg/PET	Epoxy/PET/KF-20	Epoxy/PET/KF-35	Epoxy/PET/KF-50
Volume of specimen	79.2 cm^3	79.2 cm^3	79.2 cm^3	79.2 cm^3
Volume of fiber	0 cm^3	15.84 cm^3	27.72 cm^3	39.6 cm^3
Weight of fiber, W_f	0 g	18.37 g	32.16 g	45.94 g
Volume of PET yarn	3.96 cm^3	3.96 cm^3	3.96 cm^3	3.96 cm^3
Weight of PET yarn	5.46 g	5.46 g	5.46 g	5.46 g
Volume of fire retardant	3.96 cm^3	3.96 cm^3	3.96 cm^3	3.96 cm^3
Weight of fire retardant	9.28 g	9.28 g	9.28 g	9.28 g
Volume of matrix	71.28 cm^3	55.44 cm^3	43.56 cm^3	31.68 cm^3
Weight of matrix, W_m	85.54 g	66.58 g	52.27 g	38.02 g
Weight of epoxy	57.02 g	44.38 g	34.85 g	25.34 g
Weight of hardener	28.51 g	22.19 g	17.42 g	12.67 g

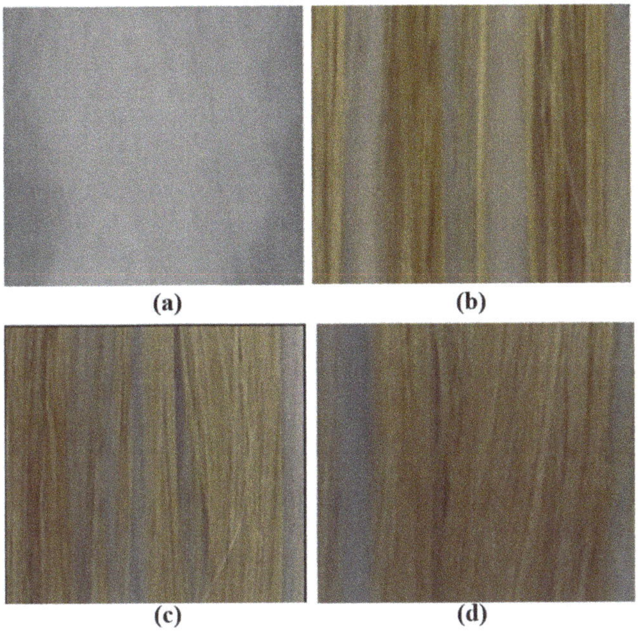

Figure 3. Observation of sample: (**a**) epoxy/Mg/PET, (**b**) epoxy/PET/KF-20, (**c**) epoxy/PET/KF-35, and (**d**) epoxy/PET/KF-50.

2.5. Testing and Experiment

2.5.1. Horizontal Burning Test

The burning test was conducted to determine the burning rate of the samples. The samples were set up according to ASTM standard D635. For these points, all specimens were cut to exact length with dimension of 12.5 cm length × 1.3 cm width × 0.3 cm thickness. The test was carried out in triplicate. The test samples were supported horizontally at one end. Next, the free end was exposed to a specified gas flame for 30 s. The time and extent of burning were measured using a stopwatch, and it was noted if the specimen did not burn. An average burning rate was noted for the material if it was burned to the 100 mm mark from the ignited end. The formula below was used to determine the average burning rate of each sample:

$$\text{Average burning rate (mm/min)} = \frac{\text{Damaged length, } L \text{ (mm)}}{\text{Time, } t \text{ (min)}} \tag{1}$$

2.5.2. Tensile Test

The fiber tensile test was used to determine the strength of the composite materials. The samples were set up according to ASTM standard D3039M. First, all of the samples were cut into the same size of 20 cm height × 1.5 cm width × 0.3 cm thickness. Then, four aluminum plates with a dimension of 25 mm × 15 mm × 1 mm were attached to the two sides of both ends of the samples using the epoxy resin. The test was carried out in five replications using the Universal Testing Machine (INSTRON 5556) (Norwood, MA, USA) with a 5 kN load cell; the crosshead speed was maintained at 2 mm/min.

2.5.3. Scanning Electron Microscopy

SEM is capable of performing analyses of selected point locations on a sample, which is especially useful for qualitatively or semiquantitatively determining chemical composition (using energy-dispersive (EDS)), crystalline structure, and crystal orientations (using electron backscatter diffraction (EBSD)). The design and function of the SEM are very similar to EPMA, and considerable overlap in capabilities exists between the two instruments. From the tensile test fracture samples, the samples at the edge of the fracture side were cut using a jigsaw machine. Then, the samples were coated with gold using a vacuum sputter coater to make them conductive prior to SEM observation. The cross section of samples was coated with gold as shown in Figure 4. The samples underwent observation to detect defects on the surface using SEM.

Figure 4. Cross section of sample after being coated with gold.

3. Results and Discussion

3.1. Flammability Properties

From the results of horizontal burning test for every sample, the average horizontal burning rate was determined. Figure 5 shows samples of each composition after the burning test. The average burning rate for each sample was converted to a graph so that the trend of this average can be clearly seen (Figure 6).

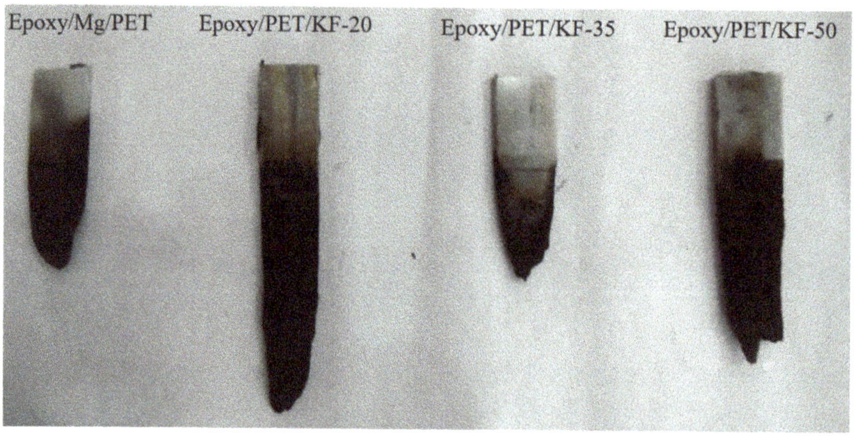

Figure 5. Samples of each composition after the burning test.

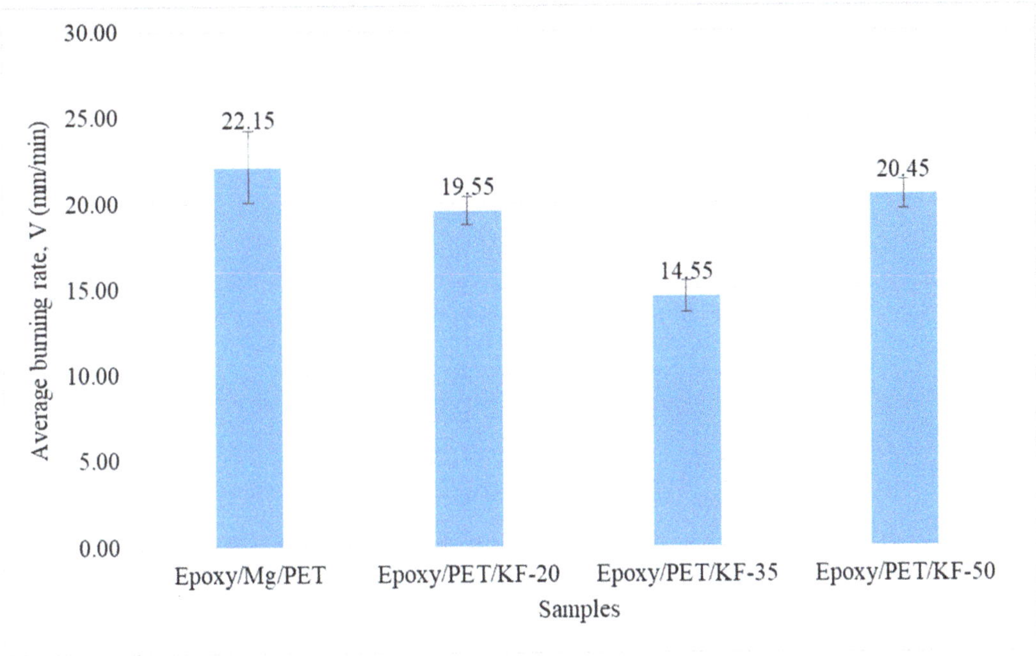

Figure 6. Graph of horizontal burning test for each composition.

From Figure 6, it can be clearly seen that the epoxy/Mg/PET with 0% KF showed the highest average burning rate of 22.151 mm/min compared to the other samples. Epoxy/Mg/PET ignited earlier, released more heat overall, and took shorter average time to reach its burning point. The ignition time is related to the fiber volume fraction of resin exposed to the flame on the composite surface [64]. Polymeric matrices have poor flammability behavior. The polymer matrix depends on the reinforcements and fillers, and it has no prominent role in improving the flame resistance of composites. The PET yarn was melted and moved away from the flame and burned with smoky flame during the burning test.

The KF content for epoxy/PET/KF-20, epoxy/PET/KF-35, and epoxy/PET/KF-50 was 20%, 35%, and 50%, respectively. The average burning rate of epoxy/PET/KF-35 was the lowest at 14.553 mm/min compared to the other samples. As a result, the composite with 35% KF and 5% magnesium hydroxide showed better result due to the compatible combination of fiber and fire retardant compared to 20% and 50% KF. The fiber volume content in epoxy/PET/KF-35 was the best compared to epoxy/PET/KF-20 and epoxy/PET/KF-50 samples because of the lower burning rate and higher fire retardant effectiveness. When the fiber content was increased above 20%, the characteristics of the composite became more similar to lignocellulosic materials [65]. A recent study showed that the presence of kenaf fiber increased the smoke density rating due to the characteristics of kenaf fiber that promote smoke production through char formation when exposed to flame [66].

The flame took a long time to propagate along epoxy/PET/KF-35 and also produced char at the same time. The production of char was the reaction of Mg in front of the flame. Charring is known as a chemical process of incomplete combustion of certain solids when subjected to high heat. By the action of heat, charring removes hydrogen and oxygen, so only carbon remains in the char. In addition, as acid and halogen-free flame retardant, Mg releases the water of hydration when it decomposes endothermically, which helps the flame retarding action [67]. The energy absorption mechanism then occurs because the magnesium compound undergoes a highly endothermic decomposition reaction that slows the heating rate of most materials in fire [68]. Epoxy/PET/KF-50 showed the highest average burning rate among the samples that contained the natural fiber. This might be due to possible environmental conditions that promote increased burning rate [66].

3.2. Tensile Properties

The ultimate tensile strength, elongation break, and elastic modulus were determined from the stress–strain curve of hybrid kenaf fiber/PET yarn reinforced epoxy fire retardant composite after the test. Statistical analysis was conducted based on the data of five specimens for each composite sample. Figure 7 shows the stress–strain curve of hybrid kenaf fiber/PET yarn reinforced epoxy fire retardant composites with different fiber volume contents (epoxy/Mg/PET, epoxy/PET/KF-20, epoxy/PET/KF-35, and epoxy/PET/KF-50), which was linear and followed the Hooke's law. It can be seen that epoxy/PET/KF-50 had better tensile properties compared to the other samples. Its ultimate tensile stress was 32.02 MPa, which was higher than the those of epoxy/Mg/PET, epoxy/PET/KF-20, and epoxy/PET/KF-35, which were 10.87, 19.96, and 25.24 MPa, respectively.

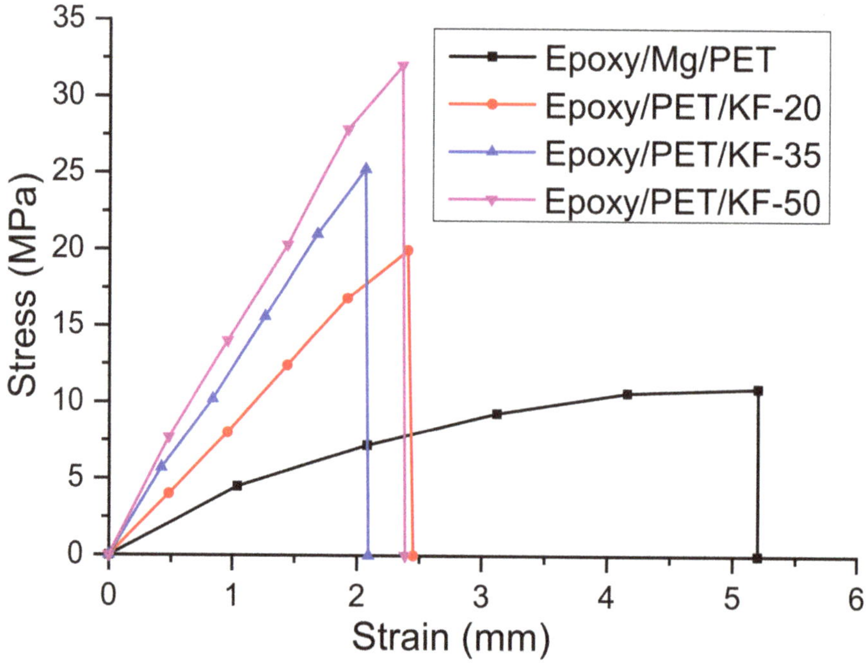

Figure 7. The stress–strain diagram of hybrid kenaf fiber/PET yarn reinforced epoxy fire retardant composites.

The tensile strength of hybrid kenaf fiber/PET yarn increased with the increase of fiber volume content, as shown in Figure 8. The tensile strength of the composites is tabulated in Table 5. The combination of kenaf fiber and PET yarn resulted in an enhancement of tensile properties of the composites. The ultimate strain of the epoxy/Mg/PET composite was the lowest compared to the other samples. The tensile strength is greatly dependent on effective and uniform stress distribution [69–71]. The increased strength of the hybrid composite was primarily due to the high strength of the hemp fiber rather than the low strength of the PET fiber [72].

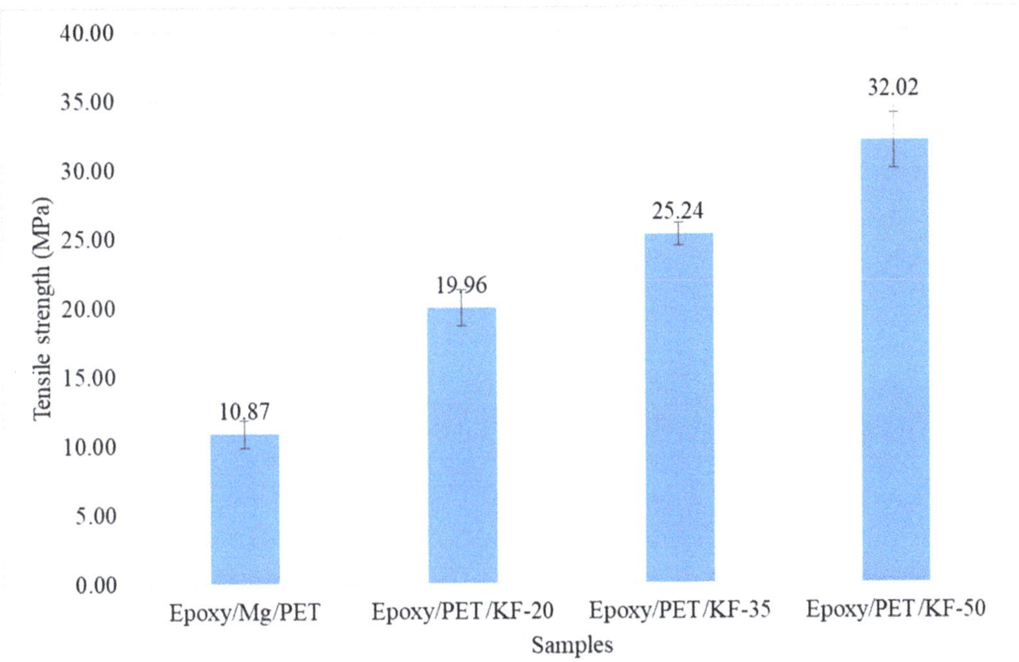

Figure 8. Effect of kenaf fiber loading on the tensile strength of epoxy/Mg/PET, epoxy/PET/KF-20, epoxy/PET/KF-35, and epoxy/PET/KF-50.

Table 5. Mechanical properties of hybrid kenaf fiber/PET yarn for all samples.

Samples Designation	Tensile Strength (MPa)	Elastic Modulus (MPa)
Epoxy/Mg/PET	10.87 ± 1.92	326.16 ± 20.13
Epoxy/PET/KF-20	19.96 ± 2.49	598.66 ± 15.12
Epoxy/PET/KF-35	25.24 ± 1.56	757.49 ± 25.95
Epoxy/PET/KF-50	32.02 ± 3.91	960.80 ± 30.87

According to the study by Mobedi et al. [73], the addition of magnesium hydroxide has a negative impact on the tensile properties, with magnesium hydroxide found to be a cursor for controlling the polymer degradation rate and accelerating the rate of degradation of composites. A previous study revealed that the deterioration of strength was insignificant with addition of 15% and 20% magnesium hydroxide [74]. However, in this study, the epoxy/Mg/PET composite had the lowest tensile strength. This might be due to the defects that occurred in the composites during fabrication. It might have also been caused by the addition of only 5% (control) magnesium hydroxide, which did not have much influence on the tensile strength compared to the increase in percentage of KF.

3.3. Surface Morphology

The surface morphology of all the specimens after the tensile test was studied by capturing images of the fractured area. From this analysis, it was concluded that several types of defects occurred on the specimens. Defects typically take place during the production of composites and sometimes by manual construction known as "lay-up". Variations in the amount of defects in composites during manufacturing will increase the likelihood of composite failure.

Two problems are critical to the material response: defects (built-in at manufacture) and damage (changes due to use). The effects of damage have been extensively covered in the open literature. Heslehurst and Scott [75] found that the "level of structural degradation in engineering properties varied" with defect severity, defect location and orientation, frequency of defect occurrence, component load path criticality and stress state, and defect idealization. As shown in Figure 9, the presence of agglomeration could be seen in the SEM micrograph of the epoxy/Mg/PET sample. It is known that $Mg(OH)_2$ has a great tendency to form agglomerates. In fact, agglomeration is a well-known phenomenon, and its probability increases with decreasing particle size. These agglomerated particles could be stress concentrator points and could affect the final performance of the composites. This might result from the defects that occur in the composites during fabrication. Moreover, it is possible to identify the needled structure of $Mg(OH)_2$ particles in images of higher magnification, e.g., 950×. This behavior might be an indication of poor interaction between the matrix and $Mg(OH)_2$ [76].

Figure 9. Agglomeration of $Mg(OH)_2$ in the epoxy/Mg/PET sample.

As shown in Figure 10, there was good interfacial bonding of matrix and fiber in the epoxy/PET/KF-20 sample. The effective stress transfer between the fiber and the matrix was due to good interfacial bonding and was further promoted by maximizing the fiber strength [43]. The small particles of magnesium hydroxide possessed good dispersity and compatibility in the matrix, which provided a strengthening and toughening effect. Therefore, the tensile strength and elongation at break were substantially increased. Figure 11 presents the SEM micrograph of the tensile-fractured surface for debonding of the fiber within the epoxy/PET/KF-35 sample. The poor compactness of the composite caused weaker bonding strength between the matrix and the fiber.

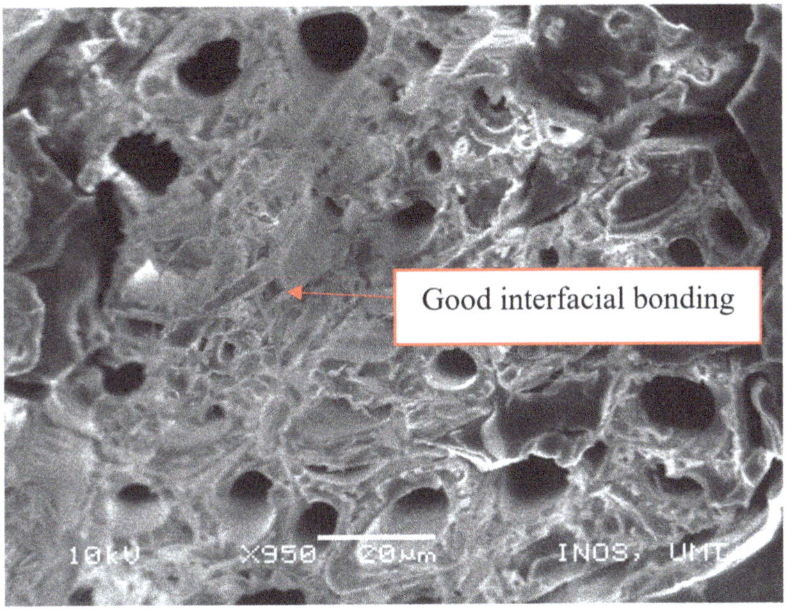

Figure 10. Good interfacial bonding in the epoxy/PET/KF-20 sample.

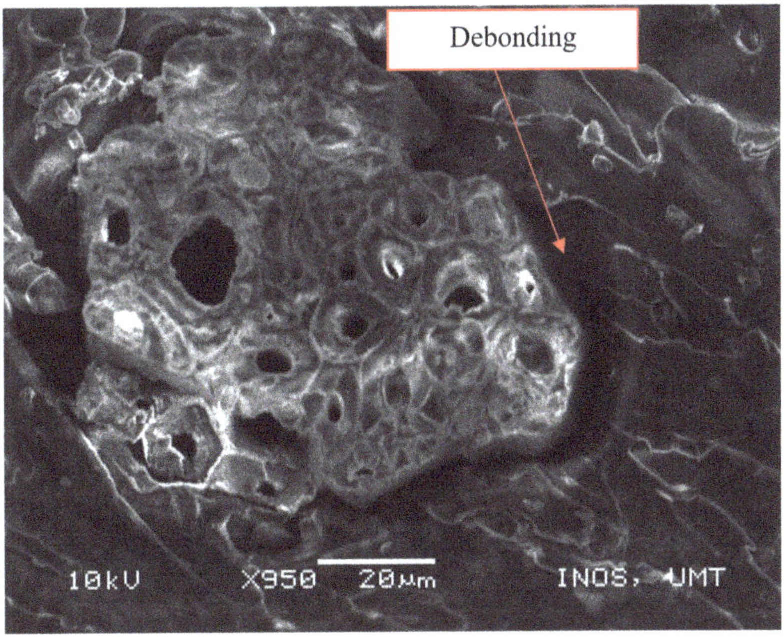

Figure 11. Debonding of fiber and polymer in the epoxy/PET/KF-35 sample.

Figure 12 shows the tensile damage that occurred on the composite, which consisted of fiber pull-outs and ensured enhanced interfacial adhesion in the epoxy/PET/KF-50 sample. As reported by Sanadi et al. [77], the primary disadvantage of natural fibers is the poor interfacial adhesion due to the hydrophilic character of cellulose, which is commonly mismatched with the hydrophobic matrix material [78,79]. Therefore, this phenomenon leads to poor fiber dispersion and fiber–matrix interfacial adhesion. This might also be due to the high fiber content, with the quantity of matrix not enough to cover the fiber [80–82].

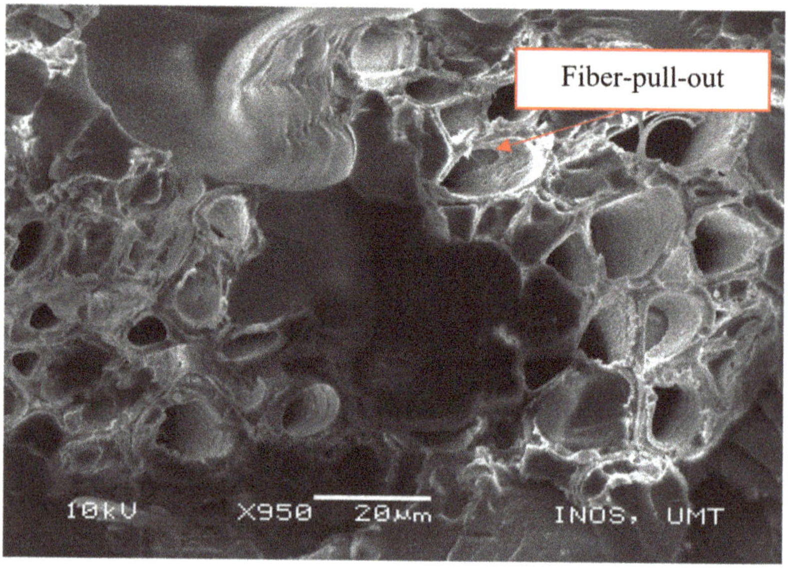

Figure 12. Fiber pull-outs in the epoxy/PET/KF-50 sample.

4. Conclusions

In order to reduce cost and provide a pollution-free environment, the flammability and mechanical properties of hybrid kenaf fiber/PET yarn reinforced epoxy fire retardant composites was analyzed. The horizontal burning test showed that epoxy/Mg/PET with 0% KF had the highest average burning rate of 22.151 mm/min compared to epoxy/PET/KF-20, epoxy/PET/KF-35, and epoxy/PET/KF-50, which contained 20, 35, and 50% KF, respectively. The average burning rate of epoxy/PET/KF-35 was the lowest at 14.553 mm/min compared to the other samples. This suggests that epoxy/PET/KF-35 would be better for application as a fire retardant composite because of the optimum composition. Epoxy/PET/KF-50 demonstrated the highest tensile strength of 32.022 MPa compared to the other samples. The tensile strengths of epoxy/Mg/PET, epoxy/PET/KF-20, and epoxy/PET/KF-35 were 10.867, 19.956, and 25.244 MPa, respectively. The tensile strength of the hybrid kenaf fiber/PET yarn increased with the increase of fiber volume content. The combination of kenaf fiber and PET yarn enhanced the tensile properties of the composites. In addition, too many defects result in poor mechanical properties of composites. Epoxy/PET/KF-50 showed the fewest defects. In general, the surface morphology of the hybrid kenaf fiber/PET yarn possessed defects such as fiber pull-out and debonding.

Author Contributions: Conceptualization, M.J.S.; methodology, M.J.S. and H.A.Z.; validation, M.J.S.; formal analysis, M.J.S. and H.A.Z.; investigation, M.J.S. and H.A.Z.; resources, M.J.S.; writing—original draft preparation, M.J.S. and H.A.Z.; writing—review and editing, C.M.R., M.P., R.M., R.A.I. and S.M.S.; supervision, M.J.S.; project administration, M.J.S.; funding acquisition, M.J.S. and M.P. All authors have read and agreed to the published version of the manuscript.

Funding: The result was obtained through financial support of the Ministry of Education, Youth, and Sports of the Czech Republic and the European Union (European Structural and Investment Funds—Operational Program Research, Development, and Education) in the frames of the project "Modular platform for autonomous chassis of specialized electric vehicles for freight and equipment transportation", Reg. No. CZ.02.1.01/0.0/0.0/16_025/0007293.

Institutional Review Board Statement: Not applicable.

Informed Consent Statement: Not applicable.

Data Availability Statement: The data presented in this study are available on request from the corresponding author.

Acknowledgments: The authors would like to thank all the members of CoREG (Composites Research Group) UMT, technicians, and support staff from the Maritime Technology Lab, UMT, Biocomposite Lab, INTROP, and Advanced Engineering Materials and Composites Research Centre, UPM, for their great contribution and help.

Conflicts of Interest: The authors declare no conflict of interest.

References

1. Mohd Nurazzi, N.; Asyraf, M.R.M.; Khalina, A.; Abdullah, N.; Sabaruddin, F.A.; Kamarudin, S.H.; Ahmad, S.; Mahat, A.M.; Lee, C.L.; Aisyah, H.A.; et al. Fabrication, Functionalization, and Application of Carbon Nanotube-Reinforced Polymer Composite: An Overview. *Polymers* **2021**, *13*, 1047. [CrossRef]
2. Suriani, M.J.; Rapi, H.Z.; Ilyas, R.A.; Petrů, M.; Sapuan, S.M. Delamination and Manufacturing Defects in Natural Fiber-Reinforced Hybrid Composite: A Review. *Polymers* **2021**, *13*, 1323. [CrossRef]
3. Nurazzi, N.M.; Asyraf, M.R.M.; Khalina, A.; Abdullah, N.; Aisyah, H.A.; Rafiqah, S.A.; Sabaruddin, F.A.; Kamarudin, S.H.; Norrrahim, M.N.F.; Ilyas, R.A.; et al. A Review on Natural Fiber Reinforced Polymer Composite for Bullet Proof and Ballistic Applications. *Polymers* **2021**, *13*, 646. [CrossRef]
4. Asyraf, M.R.M.; Ishak, M.R.; Sapuan, S.M.; Yidris, N.; Ilyas, R.A. Woods and composites cantilever beam: A comprehensive review of experimental and numerical creep methodologies. *J. Mater. Res. Technol.* **2020**, *9*, 6759–6776. [CrossRef]
5. Ilyas, R.A.; Sapuan, S.M. Biopolymers and Biocomposites: Chemistry and Technology. *Curr. Anal. Chem.* **2020**, *16*, 500–503. [CrossRef]
6. Ilyas, R.A.; Sapuan, S.M. The Preparation Methods and Processing of Natural Fibre Bio-polymer Composites. *Curr. Org. Synth.* **2020**, *16*, 1068–1070. [CrossRef] [PubMed]
7. Diyana, Z.N.; Jumaidin, R.; Selamat, M.Z.; Ghazali, I.; Julmohammad, N.; Huda, N.; Ilyas, R.A. Physical Properties of Thermoplastic Starch Derived from Natural Resources and Its Blends: A Review. *Polymers* **2021**, *13*, 1396. [CrossRef] [PubMed]
8. Ilyas, R.A.; Sapuan, S.M.; Harussani, M.M.; Hakimi, M.Y.A.Y.; Haziq, M.Z.M.; Atikah, M.S.N.; Asyraf, M.R.M.; Ishak, M.R.; Razman, M.R.; Nurazzi, N.M.; et al. Polylactic Acid (PLA) Biocomposite: Processing, Additive Manufacturing and Advanced Applications. *Polymers* **2021**, *13*, 1326. [CrossRef]
9. Ilyas, R.A.; Sapuan, S.M.; Atikah, M.S.N.; Asyraf, M.R.M.; Rafiqah, S.A.; Aisyah, H.A.; Nurazzi, N.M.; Norrrahim, M.N.F. Effect of hydrolysis time on the morphological, physical, chemical, and thermal behavior of sugar palm nanocrystalline cellulose (*Arenga pinnata* (*Wurmb.*) *Merr*). *Text. Res. J.* **2021**, *91*, 152–167. [CrossRef]
10. Ilyas, R.A.; Sapuan, S.M.; Ibrahim, R.; Abral, H.; Ishak, M.R.; Zainudin, E.S.; Atiqah, A.; Atikah, M.S.N.; Syafri, E.; Asrofi, M.; et al. Thermal, Biodegradability and Water Barrier Properties of Bio-Nanocomposites Based on Plasticised Sugar Palm Starch and Nanofibrillated Celluloses from Sugar Palm Fibres. *J. Biobased Mater. Bioenergy* **2020**, *14*, 234–248. [CrossRef]
11. Ilyas, R.A.; Sapuan, S.M.; Ishak, M.R. Isolation and characterization of nanocrystalline cellulose from sugar palm fibres (*Arenga pinnata*). *Carbohydr. Polym.* **2018**, *181*, 1038–1051. [CrossRef]
12. Ilyas, R.A.; Sapuan, S.M.; Ibrahim, R.; Abral, H.; Ishak, M.R.; Zainudin, E.S.; Atikah, M.S.N.; Mohd Nurazzi, N.; Atiqah, A.; Ansari, M.N.M.; et al. Effect of sugar palm nanofibrillated celluloseconcentrations on morphological, mechanical andphysical properties of biodegradable films basedon agro-waste sugar palm (*Arenga pinnata*(*Wurmb.*) *Merr*) starch. *J. Mater. Res. Technol.* **2019**, *8*, 4819–4830. [CrossRef]
13. Ilyas, R.A.; Sapuan, S.M.; Ishak, M.R.; Zainudin, E.S. Sugar palm nanofibrillated cellulose (*Arenga pinnata* (*Wurmb.*) *Merr*): Effect of cycles on their yield, physic-chemical, morphological and thermal behavior. *Int. J. Biol. Macromol.* **2019**, *123*, 379–388. [CrossRef]
14. Ilyas, R.A.; Sapuan, S.M.; Atiqah, A.; Ibrahim, R.; Abral, H.; Ishak, M.R.; Zainudin, E.S.; Nurazzi, N.M.; Atikah, M.S.N.; Ansari, M.N.M.; et al. Sugar palm (*Arenga pinnata* [*Wurmb.*] *Merr*) starch films containing sugar palm nanofibrillated cellulose as reinforcement: Water barrier properties. *Polym. Compos.* **2020**, *41*, 459–467. [CrossRef]
15. Ilyas, R.A.; Sapuan, S.M.; Ibrahim, R.; Abral, H.; Ishak, M.R.; Zainudin, E.S.; Asrofi, M.; Atikah, M.S.N.; Huzaifah, M.R.M.; Radzi, A.M.; et al. Sugar palm (*Arenga pinnata* (*Wurmb.*) *Merr*) cellulosic fibre hierarchy: A comprehensive approach from macro to nano scale. *J. Mater. Res. Technol.* **2019**, *8*, 2753–2766. [CrossRef]
16. Suriani, M.J.; Radzi, F.S.M.; Ilyas, R.A.; Petrů, M.; Sapuan, S.M.; Ruzaidi, C.M. Flammability, Tensile, and Morphological Properties of Oil Palm Empty Fruit Bunches Fiber/Pet Yarn-Reinforced Epoxy Fire Retardant Hybrid Polymer Composites. *Polymers* **2021**, *13*, 1282. [CrossRef] [PubMed]
17. Jumaidin, R.; Adam, N.W.; Ilyas, R.A.; Hussin, M.S.F.; Taha, M.M.; Mansor, M.R.; Azlan, U.A.-A.; Yob, M.S. Water transport and physical properties of sugarcane bagasse fibre reinforced thermoplastic potato starch biocomposite. *J. Adv. Res. Fluid Mech. Therm. Sci.* **2019**, *61*, 273–281.

18. Asrofi, M.; Syafri, S.E.; Sapuan, S.M.; Ilyas, R.A. Improvement of Biocomposite Properties Based Tapioca Starch and Sugarcane Bagasse Cellulose Nanofibers. *Key Eng. Mater.* **2020**, *849*, 96–101. [CrossRef]
19. Asrofi, M.; Sapuan, S.M.; Ilyas, R.A.; Ramesh, M. Characteristic of composite bioplastics from tapioca starch and sugarcane bagasse fiber: Effect of time duration of ultrasonication (Bath-Type). *Mater. Today Proc.* **2020**. [CrossRef]
20. Syafri, E.; Sudirman; Mashadi; Yulianti, E.; Deswita; Asrofi, M.; Abral, H.; Sapuan, S.M.; Ilyas, R.A.; Fudholi, A. Effect of sonication time on the thermal stability, moisture absorption, and biodegradation of water hyacinth (*Eichhornia crassipes*) nanocellulose-filled bengkuang (*Pachyrhizus erosus*) starch biocomposites. *J. Mater. Res. Technol.* **2019**, *8*, 6223–6231. [CrossRef]
21. Sabaruddin, F.A.; Paridah, M.T.; Sapuan, S.M.; Ilyas, R.A.; Lee, S.H.; Abdan, K.; Mazlan, N.; Roseley, A.S.M.; Abdul Khalil, H.P.S. The effects of unbleached and bleached nanocellulose on the thermal and flammability of polypropylene-reinforced kenaf core hybrid polymer bionanocomposites. *Polymers* **2020**, *13*, 116. [CrossRef] [PubMed]
22. Aisyah, H.A.; Paridah, M.T.; Sapuan, S.M.; Khalina, A.; Berkalp, O.B.; Lee, S.H.; Lee, C.H.; Nurazzi, N.M.; Ramli, N.; Wahab, M.S.; et al. Thermal Properties of Woven Kenaf/Carbon Fibre-Reinforced Epoxy Hybrid Composite Panels. *Int. J. Polym. Sci.* **2019**, *2019*, 5258621. [CrossRef]
23. Aiza Jaafar, C.N.; Zainol, I.; Ishak, N.S.; Ilyas, R.A.; Sapuan, S.M. Effects of the Liquid Natural Rubber (LNR) on Mechanical Properties and Microstructure of Epoxy/Silica/Kenaf Hybrid Composite for Potential Automotive Applications. *J. Mater. Res. Technol.* **2021**, *12*, 1026–1038. [CrossRef]
24. Sari, N.H.; Pruncu, C.I.; Sapuan, S.M.; Ilyas, R.A.; Catur, A.D.; Suteja, S.; Sutaryono, Y.A.; Pullen, G. The effect of water immersion and fibre content on properties of corn husk fibres reinforced thermoset polyester composite. *Polym. Test.* **2020**, *91*, 106751. [CrossRef]
25. Aruchamy, K.; Pavayee Subramani, S.; Palaniappan, S.K.; Sethuraman, B.; Velu Kaliyannan, G. Study on mechanical characteristics of woven cotton/bamboo hybrid reinforced composite laminates. *J. Mater. Res. Technol.* **2020**, *9*, 718–726. [CrossRef]
26. Siakeng, R.; Jawaid, M.; Asim, M.; Saba, N.; Sanjay, M.R.; Siengchin, S.; Fouad, H. Alkali treated coir/pineapple leaf fibres reinforced PLA hybrid composites: Evaluation of mechanical, morphological, thermal and physical properties. *Express Polym. Lett.* **2020**, *14*, 717–730. [CrossRef]
27. Kandola, B.K.; Mistik, S.I.; Pornwannachai, W.; Anand, S.C. Natural fibre-reinforced thermoplastic composites from woven-nonwoven textile preforms: Mechanical and fire performance study. *Compos. Part B Eng.* **2018**, *153*, 456–464. [CrossRef]
28. Jumaidin, R.; Khiruddin, M.A.A.; Asyul Sutan Saidi, Z.; Salit, M.S.; Ilyas, R.A. Effect of cogon grass fibre on the thermal, mechanical and biodegradation properties of thermoplastic cassava starch biocomposite. *Int. J. Biol. Macromol.* **2020**, *146*, 746–755. [CrossRef] [PubMed]
29. Hossain, M.; Mobarak, M.B.; Rony, F.K.; Sultana, S.; Mahmud, M.; Ahmed, S. Fabrication and Characterization of Banana Fiber Reinforced Unsaturated Polyester Resin Based Composites. *Nano Hybrids Compos.* **2020**, *29*, 84–92. [CrossRef]
30. Jumaidin, R.; Diah, N.A.; Ilyas, R.A.; Alamjuri, R.H.; Yusof, F.A.M. Processing and Characterisation of Banana Leaf Fibre Reinforced Thermoplastic Cassava Starch Composites. *Polymers* **2021**, *13*, 1420. [CrossRef] [PubMed]
31. Azammi, A.M.N.; Ilyas, R.A.; Sapuan, S.M.; Ibrahim, R.; Atikah, M.S.N.; Asrofi, M.; Atiqah, A. Characterization studies of biopolymeric matrix and cellulose fibres based composites related to functionalized fibre-matrix interface. In *Interfaces in Particle and Fibre Reinforced Composites*; Elsevier: London, UK, 2020; pp. 29–93, ISBN 9780081026656.
32. Aisyah, H.A.; Paridah, M.T.; Sapuan, S.M.; Ilyas, R.A.; Khalina, A.; Nurazzi, N.M.; Lee, S.H.; Lee, C.H. A Comprehensive Review on Advanced Sustainable Woven Natural Fibre Polymer Composites. *Polymers* **2021**, *13*, 471. [CrossRef]
33. Alsubari, S.; Zuhri, M.Y.M.; Sapuan, S.M.; Ishak, M.R.; Ilyas, R.A.; Asyraf, M.R.M. Potential of natural fiber reinforced polymer composites in sandwich structures: A review on its mechanical properties. *Polymers* **2021**, *13*, 423. [CrossRef]
34. Nurazzi, N.M.; Khalina, A.; Sapuan, S.M.; Ilyas, R.A.; Rafiqah, S.A.; Hanafee, Z.M. Thermal properties of treated sugar palm yarn/glass fiber reinforced unsaturated polyester hybrid composites. *J. Mater. Res. Technol.* **2020**, *9*, 1606–1618. [CrossRef]
35. Atiqah, A.; Jawaid, M.; Sapuan, S.M.; Ishak, M.R.; Ansari, M.N.M.; Ilyas, R.A. Physical and thermal properties of treated sugar palm/glass fibre reinforced thermoplastic polyurethane hybrid composites. *J. Mater. Res. Technol.* **2019**, *8*, 3726–3732. [CrossRef]
36. Fu, S.Y.; Xu, G.; Mai, Y.W. On the elastic modulus of hybrid particle/short-fiber/polymer composites. *Compos. Part B Eng.* **2002**, *33*, 291–299. [CrossRef]
37. Mansor, M.R.; Sapuan, S.M.; Zainudin, E.S.; Nuraini, A.A.; Hambali, A. Hybrid natural and glass fibers reinforced polymer composites material selection using Analytical Hierarchy Process for automotive brake lever design. *Mater. Des.* **2013**, *51*, 484–492. [CrossRef]
38. Rozilah, A.; Jaafar, C.N.A.; Sapuan, S.M.; Zainol, I.; Ilyas, R.A. The Effects of Silver Nanoparticles Compositions on the Mechanical, Physiochemical, Antibacterial, and Morphology Properties of Sugar Palm Starch Biocomposites for Antibacterial Coating. *Polymers* **2020**, *12*, 2605. [CrossRef]
39. Jacob, M.; Thomas, S.; Varughese, K.T. Mechanical properties of sisal/oil palm hybrid fiber reinforced natural rubber composites. *Compos. Sci. Technol.* **2004**, *64*, 955–965. [CrossRef]
40. Banerjee, S.; Sankar, B.V. Mechanical properties of hybrid composites using finite element method based micromechanics. *Compos. Part B Eng.* **2014**, *58*, 318–327. [CrossRef]
41. Dan-mallam, Y.; Hong, T.W.; Abdul Majid, M.S. Mechanical Characterization and Water Absorption Behaviour of Interwoven Kenaf/PET Fibre Reinforced Epoxy Hybrid Composite. *Int. J. Polym. Sci.* **2015**, *2015*, 1–13. [CrossRef]

42. Ramesh, M.; Palanikumar, K.; Reddy, K.H. Comparative Evaluation on Properties of Hybrid Glass Fiber- Sisal/Jute Reinforced Epoxy Composites. *Procedia Eng.* **2013**, *51*, 745–750. [CrossRef]
43. Jawaid, M.; Abdul Khalil, H.P.S. Cellulosic/synthetic fibre reinforced polymer hybrid composites: A review. *Carbohydr. Polym.* **2011**, *86*, 1–18. [CrossRef]
44. Vasumathi, M.; Murali, V. Effect of Alternate Metals for use in Natural Fibre Reinforced Fibre Metal Laminates under Bending, Impact and Axial Loadings. *Procedia Eng.* **2013**, *64*, 562–570. [CrossRef]
45. Faruk, O.; Bledzki, A.K.; Fink, H.-P.; Sain, M. Biocomposites reinforced with natural fibers: 2000–2010. *Prog. Polym. Sci.* **2012**, *37*, 1552–1596. [CrossRef]
46. Sanjay, M.R.; Siengchin, S. Exploring the applicability of natural fibers for the development of biocomposites. *Express Polym. Lett.* **2021**, *15*, 193. [CrossRef]
47. Rangappa, S.M.; Siengchin, S.; Dhakal, H.N. Green-composites: Ecofriendly and Sustainability. *Appl. Sci. Eng. Prog.* **2020**, *13*, 183–184. [CrossRef]
48. Amir, N.; Abidin, K.A.Z.; Shiri, F.B.M. Effects of Fibre Configuration on Mechanical Properties of Banana Fibre/PP/MAPP Natural Fibre Reinforced Polymer Composite. *Procedia Eng.* **2017**, *184*, 573–580. [CrossRef]
49. Sain, M.; Park, S.H.; Suhara, F.; Law, S. Flame retardant and mechanical properties of natural fibre-PP composites containing magnesium hydroxide. *Polym. Degrad. Stab.* **2004**, *83*, 363–367. [CrossRef]
50. Mazani, N.; Sapuan, S.M.; Sanyang, M.L.; Atiqah, A.; Ilyas, R.A. Design and Fabrication of a Shoe Shelf From Kenaf Fiber Reinforced Unsaturated Polyester Composites. In *Lignocellulose for Future Bioeconomy*; Ariffin, H., Sapuan, S.M., Hassan, M.A., Eds.; Elsevier: Amsterdam, The Netherlands, 2019; pp. 315–332, ISBN 9780128163542.
51. Sapiai, N.; Jumahat, A.; Jawaid, M.; Midani, M.; Khan, A. Tensile and Flexural Properties of Silica Nanoparticles Modified Unidirectional Kenaf and Hybrid Glass/Kenaf Epoxy Composites. *Polymers* **2020**, *12*, 2733. [CrossRef]
52. Hassan, F.; Zulkifli, R.; Ghazali, M.J.; Azhari, C.H. Kenaf Fiber Composite in Automotive Industry: An Overview. *Int. J. Adv. Sci. Eng. Inf. Technol.* **2017**, *7*, 315. [CrossRef]
53. Nishino, T.; Hirao, K.; Kotera, M.; Nakamae, K.; Inagaki, H. Kenaf reinforced biodegradable composite. *Compos. Sci. Technol.* **2003**, *63*, 1281–1286. [CrossRef]
54. Kian, L.K.; Saba, N.; Jawaid, M.; Sultan, M.T.H. A review on processing techniques of bast fibers nanocellulose and its polylactic acid (PLA) nanocomposites. *Int. J. Biol. Macromol.* **2019**, *121*, 1314–1328. [CrossRef]
55. Shih, Y.-F.; Wang, Y.-T.; Jeng, R.-J.; Wei, K.-M. Expandable graphite systems for phosphorus-containing unsaturated polyesters. I. Enhanced thermal properties and flame retardancy. *Polym. Degrad. Stab.* **2004**, *86*, 339–348. [CrossRef]
56. Costes, L.; Laoutid, F.; Brohez, S.; Dubois, P. Bio-based flame retardants: When nature meets fire protection. *Mater. Sci. Eng. R Rep.* **2017**, *117*, 1–25. [CrossRef]
57. Saba, N.; Safwan, A.; Sanyang, M.L.; Mohammad, F.; Pervaiz, M.; Jawaid, M.; Alothman, O.Y.; Sain, M. Thermal and dynamic mechanical properties of cellulose nanofibers reinforced epoxy composites. *Int. J. Biol. Macromol.* **2017**, *102*, 822–828. [CrossRef]
58. Hao, A.; Zhao, H.; Chen, J.Y. Kenaf/polypropylene nonwoven composites: The influence of manufacturing conditions on mechanical, thermal, and acoustical performance. *Compos. Part B Eng.* **2013**, *54*, 44–51. [CrossRef]
59. Ratna Prasad, A.V.; Mohana Rao, K. Mechanical properties of natural fibre reinforced polyester composites: Jowar, sisal and bamboo. *Mater. Des.* **2011**, *32*, 4658–4663. [CrossRef]
60. Bernard, M.; Khalina, A.; Ali, A.; Janius, R.; Faizal, M.; Hasnah, K.S.; Sanuddin, A.B. The effect of processing parameters on the mechanical properties of kenaf fibre plastic composite. *Mater. Des.* **2011**, *32*, 1039–1043. [CrossRef]
61. Zhang, H.; Wang, H.; Wang, H. Flame retardant mechanism and surface modification of magnesium hydroxide flame retardant. *IOP Conf. Ser. Earth Environ. Sci.* **2018**, *170*, 032028. [CrossRef]
62. Joseph, K.; Tolêdo Filho, R.D.; James, B.; Thomas, S.; de Carvalho, L.H. A Review on Sisal Fiber Reinforced Polymer Composites. *Rev. Bras. Eng. Agrícola Ambient.* **1999**, *3*, 367–379. [CrossRef]
63. Zampaloni, M.; Pourboghrat, F.; Yankovich, S.A.; Rodgers, B.N.; Moore, J.; Drzal, L.T.; Mohanty, A.K.; Misra, M. Kenaf natural fiber reinforced polypropylene composites: A discussion on manufacturing problems and solutions. *Compos. Part A Appl. Sci. Manuf.* **2007**, *38*, 1569–1580. [CrossRef]
64. Bar, M.; Alagirusamy, R.; Das, A. Flame retardant polymer composites. *Fibers Polym.* **2015**, *16*, 705–717. [CrossRef]
65. Chapple, S.; Anandjiwala, R. Flammability of natural fiber-reinforced composites and strategies for fire retardancy: A review. *J. Thermoplast. Compos. Mater.* **2010**, *23*, 871–893. [CrossRef]
66. Concrete, F.R.; Muda, Z.C.; Mohd, N.L.; Composites, P.; Ismail, A.E. Flammability of self-extinguishing kenaf/ABS nanoclays composite for aircraft secondary structure. *IOP Conf. Ser. Mater. Sci. Eng.* **2016**, *152*, 012068. [CrossRef]
67. Prabhakar, M.N.; Shah AU, R.; Song, J.I. A Review on the Flammability and Flame Retardant Properties of Natural Fibers and Polymer Matrix Based Composites. *Compos. Res.* **2015**, *28*, 28–39. [CrossRef]
68. Martins, L.A.L.; Bastian, F.L.; Netto, T.A. Reviewing some design issues for filament wound composite tubes. *Mater. Des.* **2014**, *55*, 242–249. [CrossRef]
69. El-Shekeil, Y.A.; Sapuan, S.M.; Jawaid, M.; Al-Shuja'a, O.M. Influence of fiber content on mechanical, morphological and thermal properties of kenaf fibers reinforced poly(vinyl chloride)/thermoplastic polyurethane poly-blend composites. *Mater. Des.* **2014**, *58*, 130–135. [CrossRef]

70. Jagadeesh, P.; Thyavihalli Girijappa, Y.G.; Puttegowda, M.; Rangappa, S.M.; Siengchin, S. Effect of natural filler materials on fiber reinforced hybrid polymer composites: An Overview. *J. Nat. Fibers* **2020**, 1–16. [CrossRef]
71. Sanjay, M.R.; Siengchin, S.; Parameswaranpillai, J.; Jawaid, M.; Pruncu, C.I.; Khan, A. A comprehensive review of techniques for natural fibers as reinforcement in composites: Preparation, processing and characterization. *Carbohydr. Polym.* **2019**, *207*, 108–121. [CrossRef]
72. Hanifawati, I.N.; Azmah Hanim, M.A.; Sapuan, S.M.; Zainuddin, E.S. Tensile and Flexural Behavior of Hybrid Banana Pseudostem/Glass Fibre Reinforced Polyester Composites. *Key Eng. Mater.* **2011**, *471–472*, 686–691. [CrossRef]
73. Mobedi, H.; Nekoomanesh, M.; Orafaei, H.; Mivechi, H. Studying the Degradation of Poly(L-Lactide) in Presence of Magnesium Hydroxide. *Iran. Polym. J.* **2006**, *15*, 31–39.
74. Datta, J.; Kopczyńska, P. Effect of kenaf fibre modification on morphology and mechanical properties of thermoplastic polyurethane materials. *Ind. Crops Prod.* **2015**, *74*, 566–576. [CrossRef]
75. Heslehurst, R.B.; Scott, M. Review of defects and damage pertaining to composite aircraft components. *Compos. Polym.* **1990**, *3*, 103–133.
76. El-Sabbagh, A.; Steuernagel, L.; Ziegmann, G. Low combustible polypropylene/flax/magnesium hydroxide composites: Mechanical, flame retardation characterization and recycling effect. *J. Reinf. Plast. Compos.* **2013**, *32*, 1030–1043. [CrossRef]
77. Sanadi, A.R.; Hunt, J.F.; Caulfield, D.F.; Kovacsvolgyi, G.; Destree, B. High fiber-low matrix composites: Kenaf fiber/polypropylene. In Proceedings of the Sixth International Conference on Woodfiber-Plastic Composites, Madison, WI, USA, 15–16 May 2001; Forest Research Society: Madison, WI, USA, 2002; pp. 121–124.
78. Nurazzi, N.M.; Khalina, A.; Chandrasekar, M.; Aisyah, H.A.; Rafiqah, S.A.; Ilyas, R.A.; Hanafee, Z.M. Effect of fiber orientation and fiber loading on the mechanical and thermal properties of sugar palm yarn fiber reinforced unsaturated polyester resin composites. *Polimery* **2020**, *65*, 115–124. [CrossRef]
79. Baihaqi, N.M.Z.N.; Khalina, A.; Nurazzi, N.M.; Aisyah, H.A.; Sapuan, S.M.; Ilyas, R.A. Effect of fiber content and their hybridization on bending and torsional strength of hybrid epoxy composites reinforced with carbon and sugar palm fibers. *Polimery* **2021**, *66*, 36–43. [CrossRef]
80. Suriani, M.J.; Sapuan, S.M.; Ruzaidi, C.M.; Nair, D.S.; Ilyas, R.A. Flammability, morphological and mechanical properties of sugar palm fiber/polyester yarn-reinforced epoxy hybrid biocomposites with magnesium hydroxide flame retardant filler. *Text. Res. J.* **2021**, 1–12. [CrossRef]
81. Sapuan, S.M.; Aulia, H.S.; Ilyas, R.A.; Atiqah, A.; Dele-Afolabi, T.T.; Nurazzi, M.N.; Supian, A.B.M.; Atikah, M.S.N. Mechanical properties of longitudinal basalt/woven-glass-fiber-reinforced unsaturated polyester-resin hybrid composites. *Polymers* **2020**, *12*, 2211. [CrossRef] [PubMed]
82. Ayu, R.S.; Khalina, A.; Harmaen, A.S.; Zaman, K.; Isma, T.; Liu, Q.; Ilyas, R.A.; Lee, C.H. Characterization study of empty fruit bunch (EFB) fibers reinforcement in poly(butylene) succinate (PBS)/starch/glycerol composite sheet. *Polymers* **2020**, *12*, 1571. [CrossRef] [PubMed]

polymers

MDPI

Review

Wheat Biocomposite Extraction, Structure, Properties and Characterization: A Review

Abdulrahman A. B. A. Mohammed [1], Abdoulhdi A. Borhana Omran [1,2,*], Zaimah Hasan [1], R. A. Ilyas [3,4,*] and S. M. Sapuan [5,6,*]

1. Department of Mechanical Engineering, College of Engineering, Universiti Tenaga Nasional, Jalan Ikram-Uniten, Kajang 43000, Selangor, Malaysia; rahman.aziz@uniten.edu.my (A.A.B.A.M.); zaimah@uniten.edu.my (Z.H.)
2. Department of Mechanical Engineering, College of Engineering Science & Technology, Sebha University, Sabha 00218, Libya
3. School of Chemical and Energy Engineering, Faculty of Engineering, Universiti Teknologi Malaysia, Johor Bahru 81310, Johor, Malaysia
4. Centre for Advanced Composite Materials (CACM), Universiti Teknologi Malaysia, Johor Bahru 81310, Johor, Malaysia
5. Laboratory of Biocomposite Technology, Institute of Tropical Forestry and Forest Products (INTROP), Universiti Putra Malaysia, Serdang 43400, Selangor, Malaysia
6. Advanced Engineering Materials and Composites Research Centre (AEMC), Department of Mechanical and Manufacturing Engineering, Universiti Putra Malaysia, Serdang 43400, Selangor, Malaysia

* Correspondence: amhmad@uniten.edu.my (A.A.B.O.); ahmadilyas@utm.my (R.A.I.); sapuan@upm.edu.my (S.M.S.)

Abstract: Biocomposite materials create a huge opportunity for a healthy and safe environment by replacing artificial plastic and materials with natural ingredients in a variety of applications. Furniture, construction materials, insulation, and packaging, as well as medical devices, can all benefit from biocomposite materials. Wheat is one of the world's most widely cultivated crops. Due to its mechanical and physical properties, wheat starch, gluten, and fiber are vital in the biopolymer industry. Glycerol as a plasticizer considerably increased the elongation and water vapor permeability of wheat films. Wheat fiber developed mechanical and thermal properties as a result of various matrices; wheat gluten is water insoluble, elastic, non-toxic, and biodegradable, making it useful in biocomposite materials. This study looked at the feasibility of using wheat plant components such as wheat, gluten, and fiber in the biocomposite material industry.

Keywords: wheat biocomposite; wheat starch; wheat gluten; wheat fiber; antioxidant; antimicrobial

Citation: Mohammed, A.A.B.A.; Omran, A.A.B.; Hasan, Z.; Ilyas, R.A.; Sapuan, S.M. Wheat Biocomposite Extraction, Structure, Properties and Characterization: A Review. *Polymers* **2021**, *13*, 3624. https://doi.org/10.3390/polym13213624

Academic Editors: Adriana Kovalcik and Carlo Santulli

Received: 20 August 2021
Accepted: 7 October 2021
Published: 21 October 2021

1. Introduction

Plastic materials cause significant environmental damage and are one of humanity's greatest issues. Petroleum-based plastics are non-biodegradable, even after a hundred years. Plastic polymers, which are created from non-renewable elements, are one of the primary causes of global warming. Biocomposite materials are the ideal choice for possibly replacing fossil-based polymers. However, biocomposite materials require further development in terms of their characteristics [1].

Improving the properties of biocomposite material is still being investigated by researchers [2–7]. There is an abundance of research on wood and non-wood plants to extract starch, gluten and fiber in order to produce bio-composite materials. The ingredients of biocomposite materials are extracted from various types of agricultural crops, such as wheat, corn, cassava, hemp, jute, kenaf and other crops [8]. The advantages that make plants more useful than other sources for biopolymers are their availability, quality and quantity. In addition, plants offer variation in physical properties such as thickness, density, water content, water absorption and water solubility. There exists a variation in chemical constituents such as cellulose, hemicellulose, lignin and protein content in fiber, amylose and amylopectin ratio in starch [9]. Furthermore, their diversity in degree of polymerization, degree of crystallinity, water-vapor permeability and porosity make a difference in the biocomposite properties.

Wheat is a non-wood plants based fiber [10], which is planted in many countries and produces a lot of waste. Starch is the primary component of wheat, having a number of food and industrial applications [11]. In biocomposite application, wheat starch is used as biopolymer film with or without filler. Wheat fiber can be extracted from different parts of the plant to be used as reinforcement filler for either natural or synthesis matrix. Surface treatment is a method that is commonly used to clean, modify and improve the fiber surface to decrease surface tension and to improve the interaction between the fiber filler and the starch film matrix or synthesis matrix [12–16]. Several publications have addressed the effects of sodium hydroxide treatment on the structure and properties of natural fibers such as kenaf, flax, jute, hemp, sugar palm and wheat fiber [17–22].

Straws such as wheat, rice and rapeseed straws, which known as cereal straws, are not only highly abundant but they are also a low-cost, potential candidate to be utilized in the development of green composites [23]. Wheat is one of the crops that is most sought after, and it is widely cultivated. The source of it comes from a grass named (*Triticum*) that is grown in countless countries around the entire globe. The total production of wheat in 2019–2020 was 763.9 million metric tons [24] and this percentage increases yearly.

One of the co-products from the starch and bioethanol industry is wheat gluten, which is utilized in many food and non-food application. It is widely used to develop films and other Bioplastics [25–29]. In 36 days, the decomposition of wheat gluten takes place in aerobic fermentation and takes 50 days in farmland soil without releasing any toxic residues into the environment [30]. Wheat gluten protein has a high decomposition rate, even when it is subjected to chemical and physical treatments. Therefore, wheat gluten polymer is a perfect alternative for the development of new biodegradable polymers, because of its decomposition properties and its unique viscoelastic and gas barrier properties [31]. Furthermore, wheat gluten has been explored as a raw material for non-food applications such as biopolymers [32–34]. In order to develop the eco-industry on our planet, biodegradable materials such as wheat-based biocomposites, which are distinguished with unique advantages such as, renewability, availability and low-cost raw materials.

Plasticizers used with starch to create the polymeric entangled phase, by reducing intramolecular hydrogen bonding [35–37]. Adding plasticizer to wheat starch improves the physical and mechanical properties because plasticizer increases the flexibility of the material. There are many types of plasticizer such as, fructose, sorbitol, urea and glycerol used to improve physical and mechanical properties. Similarly, to enhance mechanical and physical properties, plasticizers have been applied in many biocomposite materials, such as corn [38–40], sugar palm (*Arenga pinnata*) starch [41], cassava [42] and rice starch [43,44].

In this work, we conduct a comprehensive study on wheat fiber, as well as wheat starch biopolymers. This review paper will reveal the improvement of the properties in terms of mechanical response, thermal behavior, antioxidant, antimicrobial, and morphological properties of different parts of the wheat plant that can be used as a bioplastic material.

2. Wheat Plant

Wheat is a grass plant of the *Poaceae* plants family; the scientific name of wheat plant is *Triticum*. Wheat is one of the world's most ancient and essential cereal crops, which is grown across a wide range of climates and types of soils [45].

The main parts of the wheat plant are head spike, stem, leaves and roots. Wheat plants grow up to 2–4 feet tall. Figure 1 shows wheat plants' main parts. The kernel of the wheat (also called the wheat berry) is the seed of the wheat plant [46], while the part that covers the kernel and protects it is called the beard; similar to all the grass plants, wheat plants stand on the stem. The leaves of wheat plants are long and comparatively thin; flog leaves are in the top of the leaves, which are responsible for the protection of the leaves. The nourishment from the soil to the plant comes through roots in the bottom of the plant [47].

Figure 1. Wheat plant main parts [48].

3. Film Preparation and Properties Characterization of Films Based Wheat Starch

There are many factors that affect biopolymer properties, including: starch type, treatment temperature, additions such as plasticizer and co-biopolymers [35]. In this section, we will discuss properties of film-based wheat starch.

3.1. Physical and Chemical Properties of Wheat Starch

Wheat is one of the most widely farmed crops worldwide; the type of the soil and soil-dryness conditions affects the quality of the starch and other plant parts. The gelatinization enthalpy and swelling power of moderate soil-dryness treated starch are low. When compared to well-watered conditions, however, a greater gelatinization temperature, retrogradation enthalpy, and retrogradation percentage are found. According to Weiyang Zhang et al. [49], soil dryness affects amylose structure more than amylopectin structure in wheat grains. Furthermore, moderate soil dryness improves molecular structure and functional properties of the starch. Table 1 shows a comparison between the chemical and physical structure of wheat, corn, rice and potato starches. There is no significant difference between the chemical composition of various starches.

The starch basically contains Amylose and Amylopectin. In biocomposite materials, it is important to identify the percentage of Amylose and Amylopectin, which directly affect the properties of the film or the matrix of the bio-polymer [50]. Amylose has a lower molecular weight than amylopectin; however, the high relative weight of Amylopectin reduces the mobility of polymer chains, resulting in high viscosity, whereas the linear structure of Amylose has demonstrated behavior more similar to that of conventional synthetic polymers [51]. The majority of natural starches are semicrystalline. Depending on the resource of the starch, the crystallinity of starch is around 20–45% percent. The short-branched chains in Amylopectin are mostly responsible for crystalline regain and appear as double helices with a length of around 5 nm. In the crystalline areas, the Amylopectin segments are all parallel to the big helix's axis [52]. Since proteins and polysaccharides are the primary components of natural polymers, the structure–property relationships in these materials are determined by their interactions with water and with each other in an aquatic medium [53].

Thianming Zhu et al. [54] applied different techniques to determine the percentage of Amylose in the starch; techniques included Differential Scanning Calorimetry (DSC), High-Performance Size-Exclusion Chromatography (HPSEC), iodine binding, and Megazyme amylose/amylopectin. Michael Ronoubigouwa Ambouroue Avaro [55] developed a method that used Tristimulus CIE Lab Values and developed a specific color board of Starch-iodine complex solution, the conversion of the regression values L*a*b* to Red, Green, Blue (RGB) values and to color hexadecimal codes. This method used a colorimeter device. A spectrophotometer is another device that can be used to detect the percentage of the Amylose by calculating the absorbent light that gets through the mixture of the starch and iodine solution [56–58].

Table 1. A comparison between the chemical composition and physical properties of wheat, corn, rice and potato starches [59–72].

	Type of Starch			
Parameter	**Wheat Starch**	**Corn Starch**	**Rice Starch**	**Potato Starch**
Amylose (%)	16.0–31.5	20.0–28	20–28	25–31
Amylopectin (%)	68.5–75	75–83	65–85	76–83
Ash (%)	0.20–0.29	0.32–0.62	0.17–0.19	15.95–16.05
Proteins (%)	0.40–0.46	0.38–7.7	0.33–0.38	4.26–4.82
Density (g/cm^3)	1.5	1.356–1.4029	1.282	0.763
Moisture content (%)	10.65–13.3	10.45–10.82	3.60	15.98 ± 0.36

3.2. *Production of Films Based Wheat Starch*

In order to produce starch-based films, starch should be isolated from granules [73], then the isolated starch is mixed with distilled water and plasticizer to prepare the slurry. Subsequently, casting and drying processes takes place.

3.2.1. Wheat Starch Isolation

Zuosheng Zhang et al. [74] discussed different methods of starch isolation, including Alkaline Washing (ALW), Ultrasonic Assist Ethanol Soaking (UAES), Hot Water Soaking (HWS) and Ultrasonic Assist Hot Water Soaking (UAWS). A similar crystalline pattern of C-type was found for all the isolated starch samples; starch isolated using the ALW and UAES methods shows a greater degree of crystallinity than the other isolation methods. FT-IR spectra analysis shows similar chemical interactions with different isolation methods. Starch isolated using the UAES method exhibited the highest water solubility. The HWS and ALW methods resulted in greater enthalpy and gelatinization temperatures, while the UAES and UAWS isolation methods resulted in greater peak viscosity.

According to Ali et al. [75], starch can be isolated from the kernel by soaking 1 kg of flour in four liter of distilled water and keeping the mixture at 4 °C for 12 h. Then the slurry mixture is diluted 10 times (volume/volume) with distilled water. Then a 20 g of sodium hydroxide is dissolved in 1000 mL of water to make 0.5 M of NaOH. The solution of NaOH is then added to the diluted slurry. The diluted slurry is then mixed by continuously stirring for one hour. The starch is filtered and centrifuged for 30 min at 10 °C. The sediment gained from the surface is scraped and the lower white portion recovered as starch and, subsequently, dried at 40 °C in a hot air oven.

3.2.2. Wheat Starch Film Preparation

The process of preparing wheat-starch film starts with mixing the pure starch that has been isolated from other kernel ingredients with distilled water. Then the mixture is put on a hot plate mechanical stirrer for full dispersion in a temperature around 80 °C to 100 °C [76,77]. If the process contains the addition of a plasticizer, to ensure high homogeneity in the film, the plasticizer addition is recommended to take place after the starch is dispersed in the distilled water [39]. Once the slurry is cooled to room temperature, the slurry is casted on a petri dish or Teflon mold. However, Teflon mold prevents sticking of the film that happens with petri dishes [78,79]. Subsequently, the slurry is put into a drying oven at 45 °C with air circulation to remove water and moisture [76]. After the film is fully dried, it is peeled off carefully as to avoid rupture. Figure 2 shows the casting method of film formation.

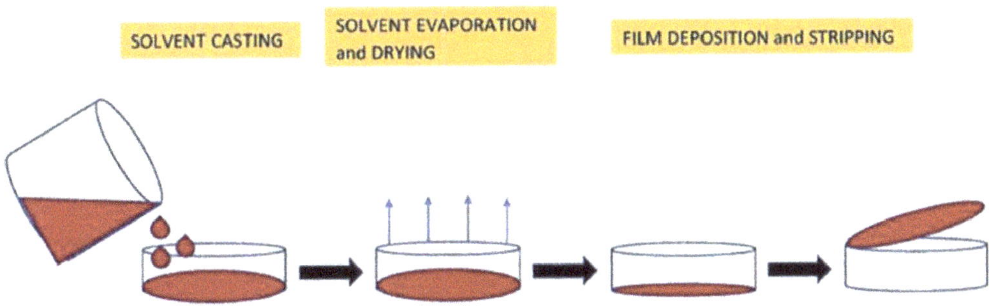

Figure 2. Casting method of film formation [80].

3.3. Properties Characteristics of Wheat Starch-Based Film

3.3.1. Pasting Properties

The pasting properties of starch samples can be determined using a Rapid Visco-Analyzer RVA; the properties of pasting viscosity profiles are shown in Figure 3. The process of testing the pasting properties can be undertaken by following H. Liu method, where weighed starch and distilled water is mixed and stirred in the aluminum Rapid Visco-Analyzer RVA sample canister to obtain a 10.0% starch suspension. A programmed cooling and heating is used to record the amylograms of the pastes [81]. Studies by Huanxin Zhang et al. [77] show that the paste viscosity of the waxy wheat-starch was gradually enhanced and reached a peak at 73.6 °C, while the normal wheat starch peak temperature was 94.7 °C.

3.3.2. Morphological Properties

Morphological properties in biocomposite are extremely important in order to see how homogeneous the composite is to get through. The scanning electronic microscope test also gives a structural explanation for other properties. For example, if the film surface is homogenous, this would indicate integrity of other properties. Wheat starch has a bimodal size distribution, with small, round B granules (2–10 mm) and large, lenticular (20–32 mm) [82–84]. Figure 4 shows the morphology of the wheat starch.

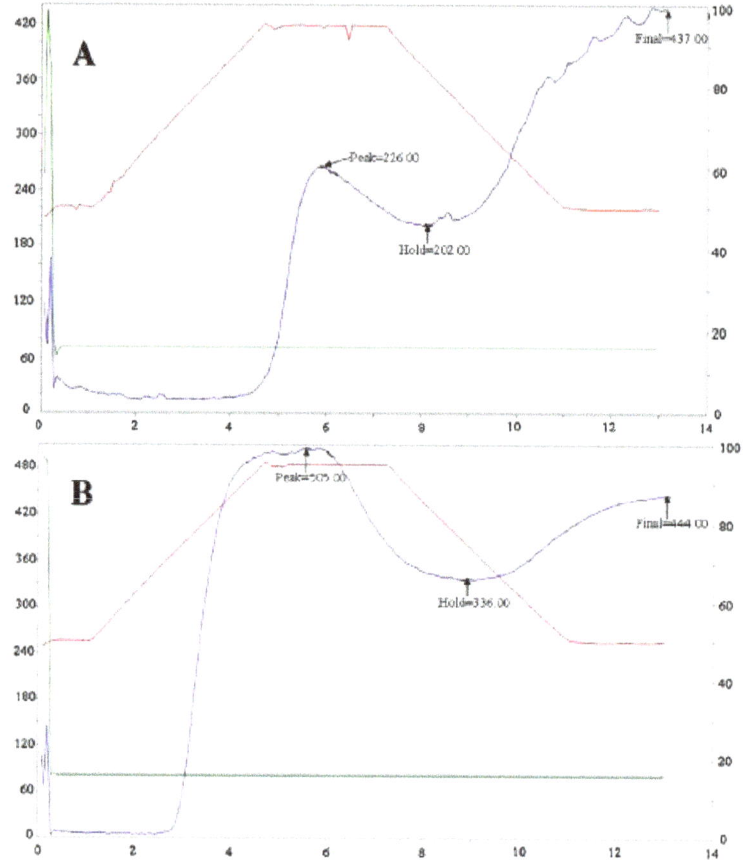

Figure 3. Rapid Visco-Analyzer pasting profiles of (**A**) normal wheat and (**B**) waxy wheat starch [77].

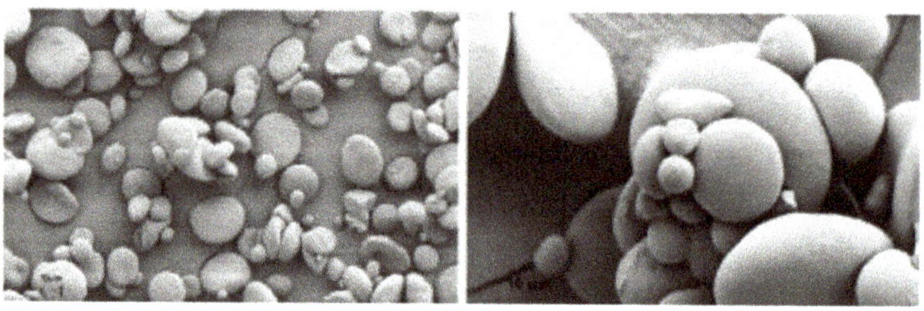

Figure 4. Morphology of native wheat starch [85].

Non-plasticized films usually have cracks or pores and some undissolved particles, which could make it easier for water vapor to pass through the film. Plasticizing wheat films with Glycerol reduced those cracks and pores. Plasticizing also improves the adhesion between the particles of the material, as shown in Figure 5. Similar results were reported for plasticized starch-based films such as corn starch-based films [86], cassava starch-based films [42], sago starch-based films [87], rice starch-based films and potato starch-based films [88].

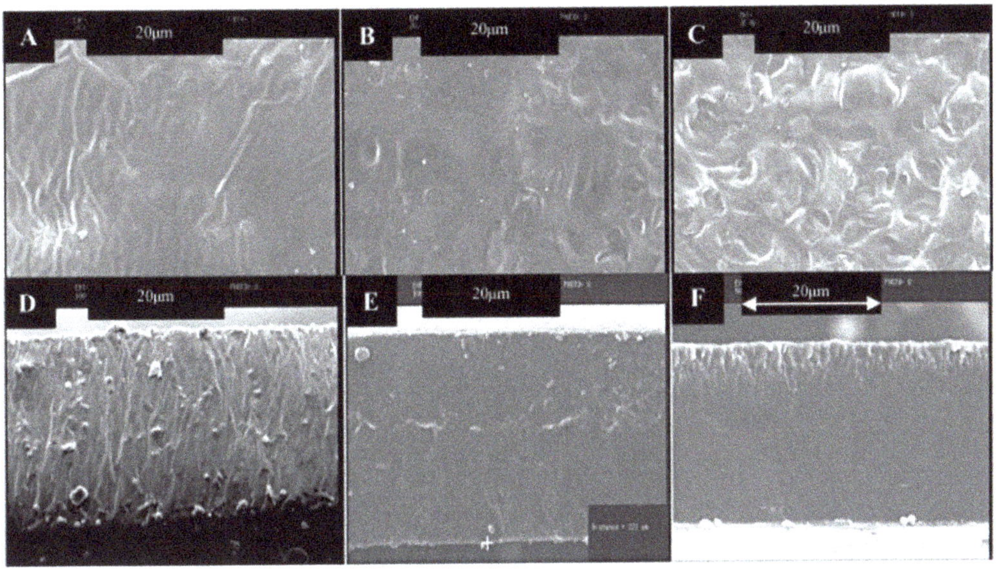

Figure 5. Morphology of biopolymer-based wheat starch (**A–C**) represent wheat starch biopolymer surface with 0, 20 and 50% of Glycerol, respectively, while (**D–F**) represent wheat starch biopolymer cross-section with 0, 20 and 50% of Glycerol, respectively [76].

3.3.3. Film Transparency

The film transparency (reverse of opacity) is used to manifest the ability of light to pass through the film. Films that have a low degree of opacity are usually referred to as being acceptable as packaging material because they offer the better visual view of the food [89]. However, the variation of transparency gives more options in different applications. The opacity of the film can be calculated with the equation below:

$$\text{Opacity} = \frac{\text{Abs600}}{\text{x}} \quad (1)$$

where: x represents the thickness (in mm) of the film, and Abs600 is the absorbance of light measured at 600 nm [90]. Lower values of the opacity value mean greater transparency.

The bioplastic wheat-starch-based films have lower opacity than that of the corn-starch-based films. This indicates that wheat-starch-based films exhibit higher transparency compared to corn-starch-based films. However, the highest transparency has been found in the potato-starch-based films [88,91]. The high opacity of corn-based film can be attributed to high lipid content of corn-starch film [92], while the addition of protein in starch films also improves the transparency [93]. However, the Amylopectin in potato starch contains a high number of long chains, which contribute to the formation of the compact structure that leads to a more transparent starch matrix [94].

3.3.4. Thermal Properties

Thermal tests are important to gauge the information about the thermal behavior of the biocomposite film. Thermogravimetric Analysis (TGA) is used to measure the temperature change over the time, while Derivative Thermogravimetry (DTG) is used to show the phases degradation of the material [95]. Differential Scanning Calorimetry (DSC) is used to measure the thermal properties of starch such as peak gelatinization temperature (Tp), gelatinization onset temperature (To), gelatinization conclusion temperature (Tc), and enthalpy of gelatinization (ΔH) [96,97]. Wheat starch begins to breakdown at nearly 275 °C, according to research. The temperature at which wheat nanofibers films degraded was roughly 296 °C [98]. Jie Zeng et al. [99] found that the gelatinization onset temperature of wheat starch is 59.43 °C, the peak gelatinization temperature is 64.23 °C, the gelatinization conclusion temperature is 78.02 °C, and the enthalpy of gelatinization is 2.915 J g^{-1}. Sorghum starch (Broom-corn) thermal properties show a little more peak gelatinization temperatures (Tp) and enthalpy compared to wheat starch, peak gelatinization temperatures (Tp) was reported for sorghum ranging from 68.2 °C to 77.8 °C, while the enthalpy values ranged from 8.2–16.4 J g^{-1} [100]. The onset temperature in biocomposites based starch and plasticizer is around 300 °C, the elimination of hydrogen functional groups, degradation, and depolymerization of the starch carbon chains polymer happened at this stage [101], while creating strong bonds by adding additives such as fiber and cross-linkers delays degradation temperature [102].

3.3.5. Water Vapor Permeability (WVP)

Water vapor transmission rate (WVTR) or moisture vapor transmission rate (MVTR) is a measure of the passing of water vapor through the substance. It is a measure of the permeability for vapor barriers. According to ASTM E96-00 standard, the films should be placed in the dryer oven for 48 h under 25 °C and 67% relative humidity before starting the test [103]. WVP is calculated from the transmission of the vapor across the films due to the difference in the partial pressure [104–106].

X. Guo et al. [107] tested the ratio of zein to wheat gluten. The researchers found that when the ratio of zein to wheat gluten is increased, the WVP decreases. WVP is related to the protein's characteristics. Gluten is made up of multiple proteins with more polar residues in the gluten molecules. Zein, on the other hand, contains a higher proportion of hydrophobic residues. For this reason, when the ratio of zein to wheat gluten is increased, WVP decreases.

Plasticizers such as glycerol have a great effect on water vapor permeability. Wheat-starch films without plasticizer have higher WVP compared to plasticized wheat films with 20% and 30% glycerol. However, the WVP of the starch film with 50% glycerol was greater, which can be attributed to micro cracks in the film [76]. The addition of hydroxypropylation, cross-linkers and antioxidants to starch bio-polymers such as corn, rice and wheat starch improves water barrier resistance [108], because the addition of those additives reduce polymer polarity, which results in low hydrogen bonding [109]. Film thickness does not affect the WVP, since the amount of casted solution does not molecularly rearranged during the drying process [110].

3.3.6. Crystallinity

Crystalline substance, in most conditions, exhibits a polycrystalline structure. Each grain being separated from the next one by a boundary, along which the atomic configurations are heavily distorted [111]. X-ray diffraction (XRD) is a technique that is used to analyze and measure the crystalline phases of a different types of substances, basically for mineralogical analysis and identification of unknown substances. Powder diffraction data are fundamentally derived X-ray Diffraction by the atomic and molecular arrangements explained by the physics of crystallography. One advantage of using X-ray diffraction is its ability to characterize crystal index with high-accuracy [112].

Granular starches are partially crystalline because they contains of approximately 25% w/w of the linear polysaccharide amylose and 75% w/w of the branched polysaccharide amylopectin [113]. Starches from various sources have close crystalline index. Corn starch, rice starch, and potato starch have crystalline indexes of 43–48%, 38% and 23–53%, respectively, while wheat starch crystalline index is 36–39% [114]. The relative crystallinity of wheat starch decreased with heat moisture treatment, because the heat moisture treatment disrupt helical structures in the amorphous and crystalline region [115]. Amylose has 33.3% crystallinity index while Amylopectin has 0% crystallinity index, preparing films by blending Amylose and Amylopectin shows co-crystallization between them. Starch-based film shows higher crystallinity than expected, which refers to crystallization of Amylopectin [116].

4. Wheat Gluten-Based Film; Preparation and Characterization

Wheat Gluten (WG) is the primary protein in wheat grains [117]. Films that are made from wheat gluten have potential to develop an edible film, adhesives, binders, and biomedical substances. The main advantages of wheat gluten films include being insoluble in water, elastic in nature, and non-toxic. Gluten matrix is biodegradable and glassy, with characteristics similar to epoxy resin [118–120].

4.1. Production of Wheat Gluten-Based Film

Wheat-gluten based films can be produced via two common methods:

4.1.1. Wet Method

Wet-type mechanical milling is a common approach for producing nanoparticles for a variety of bio-materials, including starch and gluten [121]. For gluten, a milling process is used to obtain gluten powder. The wheat gluten suspension solution is made by mixing the gluten powder with ethanol (70% aqueous ethanol). Then fibers are immersed in gluten suspension-solution. After the mixture is homogenized, the composite is dried in a vacuum air oven to allow the solvents (water and ethanol) to evaporate more quickly [97].

4.1.2. Dry Method

This method can be performed by either; (1) spreading dry powder with dry fibers in the mold, where the gluten powder will be first distributed in the mold. Next, the dry fiber preforms will be placed into the mold. Subsequently, another gluten powder layer would be added through a sieve. These steps will be repeated until the desired thickness is achieved (2), by spreading dry powder on wet fiber in the mold. In this method, fiber must be wetted again (after combing and drying), as the water will be a processing aid, after casting the gluten powder and wet fiber on the mold, the drying process needs to be conducted in dryer oven [122].

4.2. Properties Characterization of Films Based Wheat Gluten

Due to the fact that polar amino acids such as glutamic acid, aspartic acid, lysine, arginine, serine, threonine, and tyrosine are present in proteins, the addition of protein in biocomposite films improves the mechanical properties. Amino acids contain reactive groups that can be useful in cross-linking and creating covalent connections, improving the mechanical characteristics of biocomposites [123]. It has been found that proteins rich in sulfur amino acids, particularly rapeseed proteins when combined with rubber, cause a substantial enhancement of the cross-linking process. Protein-rich composites have a higher thermal resistance due to the high number of nitrogen atoms in a single polypeptide molecule [124].

Wheat-gluten films revealed lower water absorption (settled on 80% after 4000 min), this amount of water absorption is a response for (C=O, C=C) bonds existence in gluten film [125]. While the starch-based films revealed higher water absorption, which reached approximately 520% after 210 min on cassava-starch-based films [42] and 295% after 240 min on corn-starch-based films [126]. All starch-based films showed very strong water absorption capacity. However, the amount of absorbed water is different from one starch to another. This behavior is attributed to the size of starch particles, the smaller the particles the earlier and higher water absorption. Also FTIR analysis shows hydrogen bonded hydroxyl group peak more intensely with small-particle content compared to the larger particles, this explains the increase in water absorption capacity [127]. Wheat-gluten-based films, plasticized with glycerol show elongation at break in the range from 320.5–474.5%, 6.33 MPa tensile strength, while the moisture content was just about 5% [128]; the addition of a plasticizer reduces hydrogen bonding, which allows molecules to move and increase the elongation, while the high tensile appears when starch-starch hydrogen bonds overcomes starch-plasticizer bonds in a low amount of plasticizer [129]. Reinforcing wheat-gluten with flax fiber improves the tensile strength and the elastic modulus, because of the hydrogen bonding between the fiber and the protein [122,130–133]. Heat treatment of wheat gluten at temperature higher than 100 °C reduces the effect of the reinforcing filler which reflected as reduction in the Young's modulus. This explains the reduction of wheat gluten adhesion when it is heat treated [134]. However, treating the filler with alkaline and/or silane improves adhesion between wheat gluten and filler. This surface treatment increases the mechanical properties by reducing the fiber pullout length [135] As confirmed by FTIR results, fiber chemical treatment removes lignin and hemicellulose and reduces the hydrophilic nature of the fiber and, hence, improves the interfacial adhesion between fiber and matrix [136,137]. Natural structures of bio-polymers have relatively low degradation temperatures [138]. This refers to the low energy level required to break the weak interactions between the polymer chains. To avoid undesirable decomposition of wheat-gluten-based bioplastics, hydrophobic liquids, e.g., castor or silicone oil are used [139,140]. Blending gluten with hydrophobic polymers, such as polyvinylalcohol improves the degradation temperature [25,141]. The addition of hydrophobic polymers widens the gap between the energy required to break bond interactions and the energy required to cause chains breakdown. Although wheat-gluten-based films also prepared with solution cast method, compression molding have given better properties [142]. The wheat-gluten films reinforced with fiber filler can be prepared either by wet or dry method:

Tensile strength increased when drying temperature increased at 35% RH, while it decreased when temperature increased at 70% RH [143]. N. Vo Hong et al. [120] used water as a processing aid together with the use of unidirectional flax fibers to obtain the strongest properties in the fiber direction. Pakanita Muensri et al. [144] found that lignin content in the fibers does not affect the fiber/matrix adhesion. The type of wheat proteins and compression molding conditions controls the properties of wheat-protein films [145]. To make edible films out of wheat gluten, Francisco Zubeldía et al. [146] employed the dry process. They observed that molding temperature has a greater impact on the films' ultimate mechanical and physical properties than mixing time. This was due to increased disulfide bonding during heating, resulting in a more cross-linked polymeric network, according to the study. Further work needs to be undertaken to understand the mechanism of cross-linking wheat gluten with fillers [147].

5. Wheat Fiber

Wheat fiber is an isolated dietary fiber made from the wheat plant. This fiber goes through a special thermo-physical process followed by milling, sieving, and standardizing into application specific grades. Wheat fiber is a white to light beige, fibrous, and odorless powder [148]. Wheat plant is a good source of fiber from different parts, most fibers extracted from wheat husk, straw, and barn. Tables 2 and 3 show comparison of wheat, corn and rice fibers from husk and straw based on their properties, while Table 4 shows wheat bran properties. Wheat straw and husk show high amounts of cellulose, therefore, they consider as a good source for nano and microfibers.

Jing Huang et al. [149] illustrated the relationship between the data from the chemical method and Near-Infrared (NIR) to identify the fiber chemical composition. The analytical methods that are used to analyse the NIR results were the partial least squares (PLS) and principal component regression (PCR). PLS is proved to be a better quantitative method than PCR [150]. The fiber composition can also be identified through Neutral Detergent Fiber (NDF), Acid Detergent Fiber (ADF), and Acid Detergent Lignin (ADL) [151,152].

Wheat straw has a high amount of cellulose and offers several advantages over the other types of reinforcement fillers. The advantages include being low in density, nonabrasive nature, low cost and having accessibility and renewability [153]. Wheat straw fibers were utilized by Beatriz Montano-Leyva et al. [119] to modify the mechanical characteristics of wheat gluten-based film. By adopting a solvent-free method, the ultimate cost of the materials was lowered. Increases in fiber content of up to 11.1% result in increases in Young's modulus and stress at break, as well as a reduction in strain at break [154].

Yi Zou et al. [155] used long and untreated wheat straw fiber (WS) (10 cm) with polypropylene (PP) webs to develop a lightweight and cost-effective thermos-plastic composite. In this study, whole straw and split straw have been compared. Split WS–PP composites have improved over whole straw composite by 39% in modulus of elasticity, 69% enhancement in tensile strength and 18% improvement in impact resistance properties, 26% enhancement Young's modulus, 69% improvement in flexural strength. Comparing lightweight WS–PP composites with Jute–PP composites of the same density, showed that, mechanically split WS–PP composites have 114% improvement in flexural strength, 38% improvement in modulus of elasticity, 140% improvement in Young's modulus, 10% enhancement in tensile strength, better sound absorption properties and 50% lower impact resistance.

Other applications of wheat straw include extracting off hemicellulose from wheat straw (WS) and it is used as reinforcing filler for kappa carrageenan-locust bean gum polymeric blend films [156]. Wheat straw is also used as reinforcement fiber and injected with polylactic acid (PLA), PLA–WS (70:30) [157]. Additionally, wheat straw is also used with thermoplastic resins to improve their properties [158,159] and used in thermosetting resins-straw boards [160,161].

Wheat bran is the hard outer layer of cereal grains [162]. Lucia Fama et al. [163] reinforced cassava matrix with wheat bran; they found that the interaction between starch and fillers increased with the availability of hydroxyl groups in the film, which involved in a dynamic exchange with water. Zong-qiang Fu et al. [164] used wheat bran as a filler with corn starch matrix. WVP is poor in starch-based films that are not supplemented with wheat bran fiber. By increasing the wheat bran fiber content, the elongation at break of films is decreased. The tensile strength increased up to 10% w/w (up to 5.07 MPa) with the addition of wheat bran fiber content, then declined when the wheat bran fiber content was increased.

Lucia Fama et al. [163] reinforced cassava matrix with wheat bran, they found that the interaction between starch and fillers increased with the availability of hydroxyl groups in the film, which involved in a dynamic exchange with water.

Due to the strong mechanical properties and biocompatibility of isolated cellulose, it is gaining a lot of interest as a reinforcing material [165]. However, in comparison to all-cellulose composites (ACCs), where the reinforcement and matrix are both cellulose, reinforcing polymers with cellulose gives relative poor dispersion of cellulose with synthesis and bio-matrix resulted in reduced interfacial affinity [166].

Table 2. Chemical structure and physical properties of wheat, corn and rice husk [167–174].

	Type of Husk		
Parameter	**Wheat Husk**	**Corn Husk**	**Rice Husk**
Density (g/cm^3)	0.75	1.49–1.18	0.1214
Moisture content (%)	6–6.05	7.6–8.7	9
Cellulose (%)	36–39.2	31.3–47	34.34–43.80
Hemicellulose (%)	18–26.4	34–43.91	19–25
Protein (%)	6	7	1.70–7.26
Fats (%)	5	17.2	0.38–2.98
Lignin (%)	6.8–16	1.5–14.3	16

Table 3. Chemical composition and physical properties of wheat straw [175–183].

	Type of Straw		
Parameter	**Wheat Straw**	**Corn Straw**	**Rice Straw**
Density (g/cm^3)	0.3231–0.871	0.033–0.069	0.194
Moisture content (%)	8–60	25–30	6.58–18
Cellulose (%)	28.8–51.5	28–44	29.2–38
Hemicellulose (%)	10.5–39.1	36.05–36.83	12.0–29.3
Protein (%)	3–6.3	4–9	3–7
Lignin (%)	5.4–30	7–29	12–19.0

Table 4. Chemical composition and physical properties of wheat bran [184–188].

Wheat Bran	
Parameter	**Amount**
Density (g/cm^3)	0.17–0.25
Water holding capacity (g/g)	3.39–6.49
Water retention capacity (g/g)	2.17–5.76
Moisture content (%)	8.2
Cellulose (%)	11.65–13.15
Hemicellulose (%)	49.7
Starch (%)	55.9–70.53
Protein (%)	15.8–16.88
Lipid (%)	3.8–4.13
Lignin (%)	5.3

6. Antioxidant Properties of Wheat Based Film

The inhibition of oxidation improves the stability of polymers to be effective in more applications [189]. The addition of antioxidant into films can change the structure of the film [190], where the reduction in the antioxidant impairs the resistance to degradation [191]. Antioxidant materials are added to prolong the useful life of the constituents of polymers [117,192], the polymer type and the compound formulation and the end use application are governing the selection of the correct combination of antioxidants [193]. Wheat starch–chitosan films show the highest antioxidant (α-tocopherol) capacity. However, the addition of α-tocopherol led to more heterogeneous film structure [194]. Feruloylated arabinoxylans extracted from wheat bran show high antioxidant activity in the presence of bound ferulic acid [195].

7. Antimicrobial Properties of Wheat Based Film

Antimicrobial property has received more attention recently, especially in the biopackaging food industry [196]. It has been found that composite wheat gluten-chitosan-based films can prevent microbial growth in intermediate-moisture conditions [197], where gluten is thought to act as an antimicrobial agents carrier [198–200]. Organic acids, enzymes, various plant extracts, bacteriocins, and essential oils have been integrated into biopolymers as antimicrobial agents [201–203]. Essential oils (EOs) used in food packaging films to inhibit the growth of bacteria and fungi [204–206]. Essential oils are natural, volatile, complex compounds with a strong odor extracted from plants [207]. They have health benefits, antimicrobial and antioxidant properties [208,209]. (EOs) used to reinforce bio-matrix composites [210], such as reinforcing corn wheat starch matrix with lemon oil, and the addition of lemon oil, significantly increased antimicrobial activity [211]. However, the addition of (EOs) concentration reduced the tensile strength, while the elongation at break does not change [212]. Potassium Sorbate (PS) has been used as an antimicrobial agent for wheat gluten films. (PS) shows antimicrobial activity, but it has been found that when the film is exposed to an absorbing medium, most of the PS is released [213]. Thymol has been added as an antimicrobial to hydroxyethyl cellulose wheat-starch-based films and the results show the film kept the same chemical properties, whereas mechanical properties improved [214].

8. Wheat Biocomposite

8.1. Wheat Biocomposite Advantages and Applications

One of the significant advantages of agriculture-based biocomposites' resources such as wheat, is the renewability of agriculture crops; this advantage is limited in forest-based biocomposite plants, unless the green cover of forest is constantly replaced and renewed.

In many countries around the world, wheat is considered the main ingredient of their diet. In the recent year, wheat consumption has increased at a faster rate than all other cereals, which generates enormous amounts of waste [204]. The waste is increasing with the wheat consumption and production [12]. Fibrous tissue in wheat straw reach 67%, which can be considered as a high percentage among cereal plants [205]. Furthermore, wheat has the highest amount of proteins amongst other cereals.

Additional to its application in bioplastics, wheat gluten can be used as a binder with fibers [206]. As the mechanical and physical properties of starch and wheat-gluten-based biocomposites improved with fiber reinforcement [207], these improvements in the properties with the reduction of moisture content due to the addition of wheat gluten make the wheat based biocomposites a good choice in various applications, such as food packaging and drug delivery systems [208–210]. Furthermore, starch-based biocomposites foam is used to produce ecofriendly food containers and bioplastic sheets [37]. The abundance of wheat fiber make it a good choice to be included in various applications including printed circuit boards (PCB), cars, interior components, mobile phone casing and other various fields. Besides, wheat straw has been used as a filler in biodegradable matrices to make different products such as diches and trays; it can also be used with other different types of matrices such as thermosetting matrices, and thermoplastic matrices. Wheat biocomposites are found useful for indoor building insulation applications [211–213] as they are proven to be environmentally friendly and contribute to cost and energy savings [144,214,215]. Producing micro- and nano-composites separating from wheat wastes, would be one of the conceivable advancement in biocomposites-based wheat such as reinforcing thermoplastic starch polymer with wheat straw nanofibers [85], while the effect of agronomical aspects in micro- and nano-biocomposites needs more investigation [216].

8.2. Wheat Biocomposite Fabrication

The fabrication of biocomposite materials by reinforcing natural lignocellulosic fibers (e.g., sugar palm, water hyacinth, sisal, ginger, cotton, sugarcane bagasse, flax, jute, hemp, arrowroot, banana etc.) with polymer composite is frequently advocated to enhance agricultural materials [215–225]. Natural fibers have key advantages such low price, fully biodegradability, high tensile strength and stiffness, and non-abrasive behavior during processing and high availability with worldwide existence of sources [6,226] According to Azammi et al. [227], the mechanical properties of fiber reinforced polymer composite are depend on 4 factors such (1) fiber type, (2) content/loading of fiber, (3) the orientation and dispersion of fibers within the polymer matrix, and (4) the adhesion at the interface between the polymer matrix and fibers. Suitable type of fiber and optimum fiber loading, as well as good orientation and dispersion of fiber within polymer would result in good adhesion, in which ensures a good stress transfer from the matrix to the filler.

Ecological concerns in recent years have been directed at encouraging the development non-food sources for a new materials from renewable sources. Wheat gluten was effectively employed as a by-product of the starch industry for the manufacturing of environmentally friendly agricultural materials. This is due to its biodegradability [228], non-ecotoxicity [228], high availability at a reasonable price (1.4–1.8 USD/kg), as well as intriguing practical features including adhesion characteristics and efficient lipid barrier properties [229], gases [230], and aroma compounds [231]. Besides that, due to its excellent film and thermoplastic qualities, wheat gluten-based products can be produced through either compression molding [232,233], extrusion [234,235] or casting [98,236]. However, due to the high glass transition temperature of wheat gluts, the inclusion of hydrophilic plasticizers is frequently essential for thermal processing and film flexibility. The inclusion of plasticizer within the wheat gluten would resulting in changes of mechanical properties. Various researchers [237–239] found that the inclusion of plasticizer within the wheat would improve the elongation at break and reduce the strength at break and Young's modulus. Therefore, in order to improve the mechanical, water-barrier, thermal and physical properties of plasticized wheat-gluten-based materials, natural and synthetic fibers were introduced. This is undertaken in order to find new balances between process needs and material stiffness preferences [222,240,241] The addition of protein in biocomposite as a component improves their mechanical properties [124].

Table 5 displays the fabrication, filler loading and optimum mechanical properties of wheat biocomposite. From Table 5, it can be observed that many studies have been conducted on wheat biocomposite. Various techniques have been utilized to fabricate wheat composite such as solution-casting, mixing and compression molding, extrusion and compression molding, and extrusion and injection molding. Usually, the selection of method is based on the final product of the composite, such as film or mold composites. Besides, various polymers had been used to be reinforced with wheat fiber such as Modified potato starch, natural rubber, Polyethylene, Ecovio, PHBV, PLA, Polyester resin, and Polypropylene. Reinforcement of wheat with polylactic acid (PLA) shows the highest tensile modulus and tensile strength, with value of 3450 and 61.2 MPa, respectively. Moreover, many researchers also used wheat biopolymer to be reinforced with various filler such as coconut coir, eucalyptus, wheat straw fibers, hydroxyethyl cellulose, chemlali olive pomace, CNCs rice, CNCs oat, and CNCs. Monta et al. [154] conducted study on wheat straw fiber reinforced with wheat gluten. According to Monta et al. [154], this is the first experiment that had been conducted focusing on incorporating processed wheat fiber into wheat gluten. The wheat straw fibers were prepared using impact milling (IM), cut milling (CM) and ball milling (BM) processes. The result shows that the incorporation of 11.1% of IM or CM wheat straw, or 1.2% of BM wheat straw fiber, increased the mechanical performance of the biocomposite. Additionally, wheat nanocellulose reinforced polymer nanocomposite had also been studied by Alemdar [98]. Alemdar [98] conducted a study on the effect of wheat straw nanofibers reinforced with modified potato starch on morphology, thermal and mechanical properties of bio-nanocomposites. The morphological image of wheat straw can be observed in Figure 6a. The diameter of the wheat fiber decreases as it underwent chemical treatment, as displayed in Figure 6b. Subsequently, the isolation process using chemical treatment resulted in the nano-sized diameters of the nanofibers, which are within the range of 10–80 nm with lengths of a few thousand nanometers, as shown in Figure 6c. The wheat straw nanofibers are well distributed in the modified potato-starch biomatrix, according to scanning electron microscopy (SEM) tests. Furthermore, tensile testing revealed that nanocomposites exhibited a 145% increase in tensile strength and modulus compared to pure thermoplastic modified potato starch.

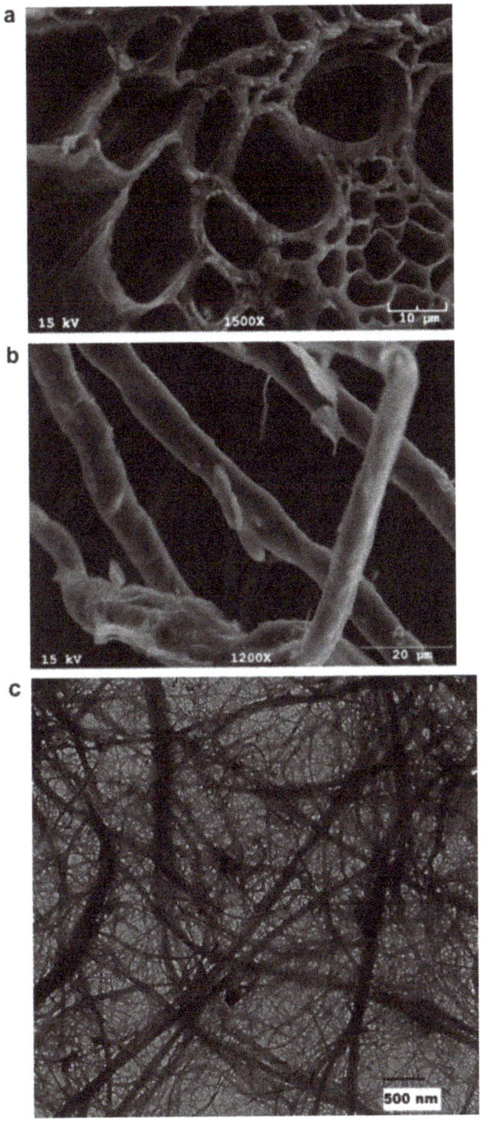

Figure 6. SEM images of the wheat straw cross-section (**a**), microfibers (**b**), and TEM images (magnification ×15,000) of the wheat straw nanofibers (**c**) [98].

In the second section of Table 5, examples of corn biocomposites have been added to compare it with wheat biocomposites. Reinforcing Polylactic acid (PLA) with corn cob exhibited lower mechanical properties compared with reinforcing PLA with wheat straw, while corn husk shows higher mechanical properties with natural matrix compared with wheat fiber.

Moazzen, N. et al. [242] developed biocomposite by Blending PVA with starch plasticized with glycerol and reinforced with carboxy methyl cellulous (CMC); the results reveal OC=O stretching and C=O stretching functional groups, PLA, CMC and Glycerol have improved the tensile strength and the hydrogen-bonded hydroxyl group of starch has reduced it. The increase of C-O and C-H stretching vibration intensity confirms the addition of cellulose concentration. Thus, the increase in the intensity of these groups indicates the increase in the crystallinity and mechanical properties of the biocomposite [243].

For hybrid wheat composites, Reddy et al. [244] had conducted experimentation on the preparation and characterization of wheat straw/clay reinforced polypropylene hybrid biocomposite. The hybrid biocomposite samples were fabricated through a melt-blending method using a co-rotating twin-screw extruder, and injection molding. The result shows that the increase in wheat straw loading would reduce the resistance for water absorption and increased the flexural modulus. Additionally, the hybridization of wheat straw (30 wt%) and organo-clay (5%) resulted in the increase in flexural modulus of hybrid composite.

Table 5. Fabrication, filler loading and optimum mechanical properties of wheat biocomposite.

Polymer	Filler	Fabrication Process	Filler Loading (%)	Optimum Tensile Modulus (MPa)	Optimum Yield Strength (MPa)	Ref.
Modified potato starch	Wheat straw nanofiber	Solution casting	2–10	271 ± 27.4	7.71 ± 0.67	[98]
Wheat gluten	Coconut coir	Mixing and compression molding	10	2.29 ± 0.47	123.2 ± 34.7	[144]
Natural rubber	Wheat bran	Mixing and compression molding	10–50 phr	-	22	[245]
Wheat gluten	Wheat straw fibers	Mixing and compression molding	0–11.1	18.4 ± 2.3	41.7 ± 3.4	[154]
Polyethylene	Wheat Bran	Extrusion	10–50	371	11.5	[246]
Wheat gluten	Hydroxyethyl cellulose	Mixing and compression molding	0–35	70	2.4	[247]
Ecovio	Wheat husk	Mixing and compression molding	13.5	Flexural: 60	Flexural: 0.75	[248]
Wheat gluten	Chemlal olive pomace	Mixing and compression molding	0–20	40	3.5	[249]
Native Wheat	CNCs rice	Solution casting	0.18 g	34.86 ± 3.3	3.64 ± 0.18	[236]
	CNCs oat			56.58 ± 9.06	5.07 ± 0.33	
	CNCs eucalyptus			70.81 ± 8.22	4.32 ± 0.13	
Phosphorylated Wheat	CNCs rice			31.94 ± 1.38	3.78 ± 0.08	
	CNCs oat			24.37 ± 1.5	3.52 ± 0.14	
	CNCs eucalyptus			30.12 ± 0.35	3.08 ± 0.02	

Table 5. *Cont.*

Polymer	Filler	Fabrication Process	Filler Loading (%)	Optimum Tensile Modulus (MPa)	Optimum Yield Strength (MPa)	Ref.
PHBV	Wheat straw fibers	Extrusion and compression molding	20	3100 ± 200	21 ± 2	[235]
PLA	Wheat straw fibers	Extrusion and injection molding	0–40	3450	61.2	[234]
Polyester resin	Wheat straw strands	Mixing and compression molding	25	Flexural: 2427.2	Flexural: 28.21	[250]
Polypropylene	Wheat straw/Clay	Extrusion and injection molding	Wheat: 0–50Clay: 0–5	Flexural: 2400	-	[245]
Fabrication, filler loading and optimum mechanical properties of corn biocomposite						
PLA	Corn Cob	Mixing and compression molding	0–40	3.7	53	[251]
Corn starch	Corn husk	Solution casting	0–8	620	13	[40]
Polypropylene (PP)	Corn stalk	Mixing and injection molding	40	4.3	34.1	[252]

CNCs = cellulose nanocrystals; MPa = MegaPascal; PHBV = Poly(3-hydroxybutyrate-co-3-hydroxyvalerate); PLA = polylactic acid.

9. Conclusions

Conventional plastic-based petroleum materials cause significant environmental damage and are one of humanity's greatest issues. Using wheat starch and wheat residues to produce biocomposite materials is a promising alternative for plastic-based petroleum. Wheat-starch biopolymer and fiber needs more concern and study, due to their availability, highly abundant, renewability, low cost, good properties and the possibility of using many parts of the wheat plant, which makes wheat plants a good resource for different kinds of biocomposite. Therefore, biocomposite-based wheat represents a good opportunity for biocomposites production in the future. Wheat bran and wheat straw are good sources of fiber to reinforce synthetic polymers and biopolymers. Films that are made from wheat gluten provide the potential to develop an edible film, adhesives, binders, and biomedical substances. The main advantages of wheat-gluten films include that they are insoluble in water, elastic in nature, and non-toxic. The production of wheat-starch-based films can be made via two common methods: (1) wet method and (2) dry method, whereas wheat biocomposite can be fabricated using several techniques such as solution casting, mixing and compression, and extrusion and compression. Plasticizer such as glycerol improves the flexibility and physical properties of wheat-starch-based films. Furthermore, the study of the mechanical properties of wheat biocomposite revealed that biocomposites exhibited increase in tensile strength and modulus when incorporated with fiber. Additionally, the influence of different types of plasticizers, chemical treatment, and addition of cross-linking agents for wheat-starch-based composite is not thoroughly evaluated in the literature. Hence, wheat-gluten-based biocomposites need more research work to better understand functionality and mechanical response of wheat-gluten-based biocomposite films. Molecular weight distribution and the mechanism of cross-linking between the proteins and any other additions such as plasticizers and fibers need further investigation.

Author Contributions: Conceptualization, A.A.B.O., R.A.I., A.A.B.A.M.; Methodology, A.A.B.O., S.M.S., A.A.B.A.M.; Formal analysis, A.A.B.O., S.M.S., Z.H., A.A.B.A.M.; Investigation, A.A.B.O., S.M.S.; Resources, A.A.B.O., A.A.B.A.M., R.A.I.; Writing—original draft preparation, A.A.B.A.M., A.A.B.O., Z.H., R.A.I.; Writing—review and editing, A.A.B.O., S.M.S., A.A.B.O., Z.H., R.A.I.; Supervision, A.A.B.O., Z.H.; Project administration, A.A.B.O., Z.H., S.M.S.; Funding acquisition, A.A.B.O. All authors have read and agreed to the published version of the manuscript.

Funding: The APC was covered by Universiti Tenaga Nasional (UNITEN), Malaysia, through BOLD re-fresh publication fund 2021 (J510050002-BOLDRefresh2025-Centre of Excellence).

Data Availability Statement: No data were used to support this study.

Acknowledgments: The authors would like express gratitude for the financial support received from BOLD2021 research grant (J510050002/2021073) by Innovation & Research Management Center (iRMC), Universiti Tenaga Nasional, Malaysia.

Conflicts of Interest: The authors declare no conflict of interest.

References

1. Omran, A.A.B.; Mohammed, A.A.B.A.; Sapuan, S.M.; Ilyas, R.A.; Asyraf, M.R.M.; Koloor, S.S.R.; Petrů, M. Micro- and Nanocellulose in Polymer Composite Materials: A Review. *Polymers* **2021**, *13*, 231. [CrossRef]
2. Nurazzi, N.M.; Asyraf, M.R.M.; Khalina, A.; Abdullah, N.; Aisyah, H.A.; Rafiqah, S.A.; Sabaruddin, F.A.; Kamarudin, S.H.; Norrrahim, M.N.F.; Ilyas, R.A.; et al. A Review on Natural Fiber Reinforced Polymer Composite for Bullet Proof and Ballistic Applications. *Polymers* **2021**, *13*, 646. [CrossRef]
3. Alsubari, S.; Zuhri, M.Y.M.; Sapuan, S.M.; Ishak, M.R.; Ilyas, R.A.; Asyraf, M.R.M. Potential of natural fiber reinforced polymer composites in sandwich structures: A review on its mechanical properties. *Polymers* **2021**, *13*, 423. [CrossRef]
4. Diyana, Z.N.; Jumaidin, R.; Selamat, M.Z.; Ghazali, I.; Julmohammad, N.; Huda, N.; Ilyas, R.A. Physical Properties of Thermoplastic Starch Derived from Natural Resources and Its Blends: A Review. *Polymers* **2021**, *13*, 1396. [CrossRef]
5. Mohd Nurazzi, N.; Asyraf, M.R.M.; Khalina, A.; Abdullah, N.; Sabaruddin, F.A.; Kamarudin, S.H.; Ahmad, S.; Mahat, A.M.; Lee, C.L.; Aisyah, H.A.; et al. Fabrication, Functionalization, and Application of Carbon Nanotube-Reinforced Polymer Composite: An Overview. *Polymers* **2021**, *13*, 1047. [CrossRef] [PubMed]
6. Ilyas, R.A.; Sapuan, S.M.; Harussani, M.M.; Hakimi, M.Y.A.Y.; Haziq, M.Z.M.; Atikah, M.S.N.; Asyraf, M.R.M.; Ishak, M.R.; Razman, M.R.; Nurazzi, N.M.; et al. Polylactic Acid (PLA) Biocomposite: Processing, Additive Manufacturing and Advanced Applications. *Polymers* **2021**, *13*, 1326. [CrossRef] [PubMed]
7. Supian, A.B.M.; Sapuan, S.M.; Jawaid, M.; Zuhri, M.Y.M.; Ilyas, R.A.; Syamsir, A. Crashworthiness Response of Filament Wound Kenaf/Glass Fibre-reinforced Epoxy Composite Tubes with Influence of Stacking Sequence under Intermediate-velocity Impact Load. *Fibers Polym.* **2021**, 1–12. [CrossRef]
8. Simon, F.; Loussert-Ajaka, I.; Damond, F.; Saragosti, S.; Barin, F.; Brun-Vézinet, F. A REVIEW ON NATURAL FIBERS. *AIDS Res. Hum. Retrovir.* **1996**, *12*, 1427–1433. [CrossRef] [PubMed]
9. Nam, S.; Netravali, A.N. Green composites. I. physical properties of ramie fibers for environment-friendly green composites. *Fibers Polym.* **2006**, *7*, 372–379. [CrossRef]
10. Ramesh, M.; Palanikumar, K.; Reddy, K.H. Plant fibre based bio-composites: Sustainable and renewable green materials. *Renew. Sustain. Energy Rev.* **2017**, *79*, 558–584. [CrossRef]
11. Shevkani, K.; Singh, N.; Bajaj, R.; Kaur, A. Wheat starch production, structure, functionality and applications—a review. *Int. J. Food Sci. Technol.* **2017**, *52*, 38–58. [CrossRef]
12. Liu, W.; Mohanty, A.K.; Askeland, P.; Drzal, L.T.; Misra, M. Influence of fiber surface treatment on properties of Indian grass fiber reinforced soy protein based biocomposites. *Polymer* **2004**, *45*, 7589–7596. [CrossRef]
13. Bledzki, A.K.; Gassan, J. Composites reinforced with cellulose based fibers. *Prog. Polym. Sci.* **1999**, *24*, 221–274. [CrossRef]
14. Norrrahim, M.N.F.; Huzaifah, M.R.M.; Farid, M.A.A.; Shazleen, S.S.; Misenan, M.S.M.; Yasim-Anuar, T.A.T.; Naveen, J.; Nurazzi, N.M.; Rani, M.S.A.; Hakimi, M.I.; et al. Greener Pretreatment Approaches for the Valorisation of Natural Fibre Biomass into Bioproducts. *Polymers* **2021**, *13*, 2971. [CrossRef] [PubMed]
15. Nurazzi, N.M.; Sabaruddin, F.A.; Harussani, M.M.; Kamarudin, S.H.; Rayung, M.; Asyraf, M.R.M.; Aisyah, H.A.; Norrrahim, M.N.F.; Ilyas, R.A.; Abdullah, N.; et al. Mechanical Performance and Applications of CNTs Reinforced Polymer Composites—A Review. *Nanomaterials* **2021**, *11*, 2186. [CrossRef]
16. Chan, J.X.; Wong, J.F.; Petrů, M.; Hassan, A.; Nirmal, U.; Othman, N.; Ilyas, R.A. Effect of Nanofillers on Tribological Properties of Polymer Nanocomposites: A Review on Recent Development. *Polymers* **2021**, *13*, 2867. [CrossRef] [PubMed]
17. Sreenivasan, S.; Iyer, P.B.; Iyer, K.R.K. Influence of delignification and alkali treatment on the fine structure of coir fibres (Cocos Nucifera). *J. Mater. Sci.* **1996**, *31*, 721–726. [CrossRef]
18. Gassan, J.; Bledzki, A.K. Alkali Treatment of Jute Fibers: Relationship Between. *J. Appl. Polym. Sci.* **1998**, *71*, 623–629. [CrossRef]
19. Janker-Obermeier, I.; Sieber, V.; Faulstich, M.; Schieder, D. Solubilization of hemicellulose and lignin from wheat straw through microwave-assisted alkali treatment. *Ind. Crops Prod.* **2012**, *39*, 198–203. [CrossRef]
20. Ilyas, R.A.; Sapuan, S.M.; Atikah, M.S.N.; Asyraf, M.R.M.; Rafiqah, S.A.; Aisyah, H.A.; Nurazzi, N.M.; Norrrahim, M.N.F. Effect of hydrolysis time on the morphological, physical, chemical, and thermal behavior of sugar palm nanocrystalline cellulose (Arenga pinnata (Wurmb.) Merr). *Text. Res. J.* **2021**, *91*, 152–167. [CrossRef]
21. Ilyas, R.A.; Sapuan, S.M.; Ishak, M.R.; Zainudin, E.S. Sugar palm nanofibrillated cellulose (Arenga pinnata (Wurmb.) Merr): Effect of cycles on their yield, physic-chemical, morphological and thermal behavior. *Int. J. Biol. Macromol.* **2019**, *123*, 379–388. [CrossRef]

22. Ahmad Ilyas, R.; Mohd Sapuan, S.; Ibrahim, R.; Abral, H.; Ishak, M.R.; Zainudin, E.S.; Asrofi, M.; Siti Nur Atikah, M.; Muhammad Huzaifah, M.R.; Radzi, M.A.; et al. Sugar palm (Arenga pinnata (Wurmb.) Merr)cellulosic fibre hierarchy: A comprehensiveapproach from macro to nano scale. *J. Mater. Res. Technol.* **2019**, *8*, 2753–2766. [CrossRef]
23. Zhao, L.; Xia, W.; Tarverdi, K.; Song, J. Biocomposite boards from wheat straw without addition of bonding agent. *Mater. Sci. Technol.* **2014**, *30*, 603–610. [CrossRef]
24. USDA. World agricultural production. *Ekon. APK* **2021**. Available online: https://www.fas.usda.gov/data/world-agricultural-production (accessed on 14 August 2021).
25. Dicharry, R.M.; Ye, P.; Saha, G.; Waxman, E.; Asandei, A.D.; Parnas, R.S. Wheat gluten-thiolated poly(vinyl alcohol) blends with improved mechanical properties. *Biomacromolecules* **2006**, *7*, 2837–2844. [CrossRef]
26. Yang, Z.; Peng, H.; Wang, W.; Liu, T. Crystallization behavior of poly(ε-caprolactone)/layered double hydroxide nanocomposites. *J. Appl. Polym. Sci.* **2010**, *116*, 2658–2667. [CrossRef]
27. Gällstedt, M.; Mattozzi, A.; Johansson, E.; Hedenqvist, M.S. Transport and Tensile Properties of Compression-Molded Wheat Gluten Films. *Biomacromolecules* **2004**, *5*, 2020–2028. [CrossRef] [PubMed]
28. Kunanopparat, T.; Menut, P.; Morel, M.-H.; Guilbert, S. Reinforcement of plasticized wheat gluten with natural fibers: From mechanical improvement to deplasticizing effect. *Compos. Part A Appl. Sci. Manuf.* **2008**, *39*, 777–785. [CrossRef]
29. Zárate-Ramírez, L.S.; Martínez, I.; Romero, A.; Partal, P.; Guerrero, A. Wheat gluten-based materials plasticised with glycerol and water by thermoplastic mixing and thermomoulding. *J. Sci. Food Agric.* **2011**, *91*, 625–633. [CrossRef]
30. Domenek, S.; Feuilloley, P.; Gratraud, J.; Morel, M.-H.; Guilbert, S. Biodegradability of wheat gluten based bioplastics. *Chemosphere* **2004**, *54*, 551–559. [CrossRef]
31. Edwin, A.; Habeych, N. *Development of Starch-Based Materials*; Wageningen University: Wageningen, The Netherlands, 2009; ISBN 9789085854333.
32. Woerdeman, D.L.; Veraverbeke, W.S.; Parnas, R.S.; Johnson, D.; Delcour, J.A.; Verpoest, I.; Plummer, C.J.G. Designing New Materials from Wheat Protein. *Biomacromolecules* **2004**, *5*, 1262–1269. [CrossRef]
33. Lagrain, B.; Goderis, B.; Brijs, K.; Delcour, J.A. Molecular Basis of Processing Wheat Gluten toward Biobased Materials. *Biomacromolecules* **2010**, *11*, 533–541. [CrossRef]
34. Jansens, K.J.A.; Vo, N.; Telen, L.; Brijs, K.; Lagrain, B.; Willem, A.; Vuure, V.; Van Acker, K.; Verpoest, I.; Van Puyvelde, P.; et al. Effect of molding conditions and moisture content on the mechanical properties of compression molded glassy, wheat gluten bioplastics. *Ind. Crops Prod.* **2013**, *44*, 480–487. [CrossRef]
35. Thakur, R.; Pristijono, P.; Scarlett, C.J.; Bowyer, M.; Singh, S.P.; Vuong, Q.V. Starch-based films: Major factors affecting their properties. *Int. J. Biol. Macromol.* **2019**, *132*, 1079–1089. [CrossRef]
36. Punia Bangar, S.; Nehra, M.; Siroha, A.K.; Petrů, M.; Ilyas, R.A.; Devi, U.; Devi, P. Development and Characterization of Physical Modified Pearl Millet Starch-Based Films. *Foods* **2021**, *10*, 1609. [CrossRef]
37. Kumari, N.; Bangar, S.P.; Petrů, M.; Ilyas, R.A.; Singh, A.; Kumar, P. Development and Characterization of Fenugreek Protein-Based Edible Film. *Foods* **2021**, *10*, 1976. [CrossRef] [PubMed]
38. Isotton, F.S.; Bernardo, G.L.; Baldasso, C.; Rosa, L.M.; Zeni, M. The plasticizer effect on preparation and properties of etherified corn starchs films. *Ind. Crops Prod.* **2015**, *76*, 717–724. [CrossRef]
39. Ibrahim, M.I.J.; Sapuan, S.M.; Zainudin, E.S.; Zuhri, M.Y.M. Physical, thermal, morphological, and tensile properties of cornstarch-based films as affected by different plasticizers. *Int. J. Food Prop.* **2019**, *22*, 925–941. [CrossRef]
40. Ibrahim, M.I.J.; Sapuan, S.M.; Zainudin, E.S.; Zuhri, M.Y.M. Preparation and characterization of cornhusk/sugar palm fiber reinforced Cornstarch-based hybrid composites. *J. Mater. Res. Technol.* **2020**, *9*, 200–211. [CrossRef]
41. Sanyang, M.; Sapuan, S.; Jawaid, M.; Ishak, M.; Sahari, J. Effect of Plasticizer Type and Concentration on Tensile, Thermal and Barrier Properties of Biodegradable Films Based on Sugar Palm (Arenga pinnata) Starch. *Polymers* **2015**, *7*, 1106. [CrossRef]
42. Edhirej, A.; Sapuan, S.M.; Jawaid, M.; Zahari, N.I. Effect of various plasticizers and concentration on the physical, thermal, mechanical, and structural properties of cassava-starch-based films. *Starch/Staerke* **2017**, *69*, 1–11. [CrossRef]
43. Laohakunjit, N.; Noomhorm, A. Effect of Plasticizers on Mechanical and Barrier Properties of Rice Starch Film. *Starch* **2004**, *56*, 348–356. [CrossRef]
44. Hong-rui, C.; Hai-tao, C.; Shuang, L.; Guo-qiang, D.; Ying, Z. ScienceDirect Effect of Plasticizers on Properties of Rice Straw Fiber Film. *J. Northeast Agric. Univ.* **2014**, *21*, 67–72. [CrossRef]
45. Wheat | Production, Types, Nutrition, Uses, & Facts | Britannica. Available online: https://www.britannica.com/plant/wheat (accessed on 14 August 2021).
46. A Kernel of Wheat | National Festival of Breads. Available online: https://nationalfestivalofbreads.com/nutrition-education/a-kernel-of-wheat (accessed on 14 August 2021).
47. The Parts of a Wheat Plant. Available online: https://sciencing.com/the-parts-of-a-wheat-plant-12211988.html (accessed on 14 August 2021).
48. Wheat — Louisiana Ag in the Classroom, Reproduced with the permission of Louisiana Agriculture in the Classroom. Available online: https://aitcla.org/wheat (accessed on 14 August 2021).
49. Zhang, W.; Gu, J.; Wang, Z.; Wei, C.; Yang, J.; Zhang, J. Comparison of Structural and Functional Properties of Wheat Starch under Different Soil Drought Conditions. *Sci. Rep.* **2017**, *7*, 1–18. [CrossRef]

50. Jha, P.; Dharmalingam, K.; Nishizu, T.; Katsuno, N.; Anandalakshmi, R. Effect of Amylose–Amylopectin Ratios on Physical, Mechanical, and Thermal Properties of Starch-Based Bionanocomposite Films Incorporated with CMC and Nanoclay. *Starch/Staerke* **2020**, *72*, 1–9. [CrossRef]
51. Jiang, T.; Duan, Q.; Zhu, J.; Liu, H.; Yu, L. Starch-based biodegradable materials: Challenges and opportunities. *Adv. Ind. Eng. Polym. Res.* **2020**, *3*, 8–18. [CrossRef]
52. Liu, H.; Yu, L.; Simon, G.; Zhang, X.; Dean, K.; Chen, L. Effect of annealing and pressure on microstructure of cornstarches with different amylose/amylopectin ratios. *Carbohydr. Res.* **2009**, *344*, 350–354. [CrossRef] [PubMed]
53. Yu, L.; Dean, K.; Li, L. Polymer blends and composites from renewable resources. *Prog. Polym. Sci.* **2006**, *31*, 576–602. [CrossRef]
54. Zhu, T.; Jackson, D.S.; Wehling, R.L.; Geera, B. Comparison of amylose determination methods and the development of a dual wavelength iodine binding technique. *Cereal Chem.* **2008**, *85*, 51–58. [CrossRef]
55. Avaro, M.R.A.; Pan, Z.; Yoshida, T.; Wada, Y. Two Alternative Methods to Predict Amylose Content of Rice Grain by Using Tristimulus CIE Lab Values and Developing a Specific Color Board Of Starch-Iodine Complex Solution. *Plant Prod. Sci.* **2011**, *14*, 164–168. [CrossRef]
56. Jian, Y.; Xiaorong, Y.; Zhaoci, W.; Xiarong, Y.; Zhaoci, W. Research on method for determination of amylose content in rice. *Proc. 7th Int. Work. Conf. Stored-Product Prot.* **1998**, *2*, 1710–1714.
57. Boonpo, S.; Kungwankunakorn, S. Study on Amylose Iodine Complex from Cassava Starch by Colorimetric Method. *J. Adv. Agric. Technol.* **2017**, *4*, 345–349. [CrossRef]
58. Landers, P.S.; Gbur, E.E.; Sharp, R.N. Comparison of Two Models to Predict Amylose Concentration in Rice Flours as Determined by Spectrophotometric Assay. *Cereal Chem.* **1991**, *68*, 545–548.
59. Cauvain, S.P. *Bread Making: Improving Quality*; CRC Press: Boca Raton, FL, USA, 2003; ISBN 9781855735538.
60. Chen, X.; He, X.; Fu, X.; Huang, Q. In vitro digestion and physicochemical properties of wheat starch/flour modified by heat-moisture treatment. *J. Cereal Sci.* **2015**, *63*, 109–115. [CrossRef]
61. Chen, G.X.; Zhou, J.W.; Liu, Y.L.; Lu, X.B.; Han, C.X.; Zhang, W.Y.; Xu, Y.H.; Yan, Y.M. Biosynthesis and Regulation of Wheat Amylose and Amylopectin from Proteomic and Phosphoproteomic Characterization of Granule-binding Proteins. *Sci. Rep.* **2016**, *6*, 33111. [CrossRef]
62. Qiu, S.; Yadav, M.P.; Liu, Y.; Chen, H.; Tatsumi, E.; Yin, L. Effects of corn fiber gum with different molecular weights on the gelatinization behaviors of corn and wheat starch. *Food Hydrocoll.* **2016**, *53*, 180–186. [CrossRef]
63. De Pilli, T.; Legrand, J.; Derossi, A.; Severini, C. Effect of proteins on the formation of starch-lipid complexes during extrusion cooking of wheat flour with the addition of oleic acid. *Int. J. Food Sci. Technol.* **2015**, *50*, 515–521. [CrossRef]
64. Wang, S.; Luo, H.; Zhang, J.; Zhang, Y.; He, Z.; Wang, S. Alkali-induced changes in functional properties and in vitro digestibility of wheat starch: The role of surface proteins and lipids. *J. Agric. Food Chem.* **2014**, *62*, 3636–3643. [CrossRef] [PubMed]
65. Dengate, H.N.; Baruch, D.W.; Meredith, P. The Density of Wheat Starch Granules: A Tracer Dilution Procedure for Determining the Density of an Immiscible Dispersed Phase. *Starch-Stärke* **1978**, *30*, 80–84. [CrossRef]
66. Jang, J.K.; Pyun, Y.R. Effect of moisture content on the melting of wheat starch. *Starch/Staerke* **1996**, *48*, 48–51. [CrossRef]
67. Bertoft, E. Understanding starch structure: Recent progress. *Agronomy* **2017**, *7*, 56. [CrossRef]
68. Zakaria, N.H.; Muhammad, N.; Abdullah, M.M.A.B. Potential of Starch Nanocomposites for Biomedical Applications. *IOP Conf. Ser. Mater. Sci. Eng.* **2017**, *209*, 012087. [CrossRef]
69. Yoo, S.H.; Chang, Y.H. Effect of tara gum addition on steady and dynamic shear rheological properties of rice starch isolated from the Korean rice variety "Boramchan". *Prev. Nutr. Food Sci.* **2018**, *23*, 254–259. [CrossRef] [PubMed]
70. Puncha-arnon, S.; Uttapap, D. Rice starch vs. rice flour: Differences in their properties when modified by heat-moisture treatment. *Carbohydr. Polym.* **2013**, *91*, 85–91. [CrossRef] [PubMed]
71. Tharise, N.; Julianti, E.; Nurminah, M. Evaluation of physico-chemical and functional properties of composite flour from cassava, rice, potato, soybean and xanthan gum as alternative of wheat flour. *Int. Food Res. J.* **2014**, *21*, 1641–1649.
72. Marichelvam, M.K.; Jawaid, M.; Asim, M. Corn and Rice Starch-Based Bio-Plastics as Alternative Packaging Materials. *Fibers* **2019**, *7*, 32. [CrossRef]
73. Gifuni, I.; Olivieri, G.; Krauss, I.R.; D'Errico, G.; Pollio, A.; Marzocchella, A. Microalgae as new sources of starch: Isolation and characterization of microalgal starch granules. *Chem. Eng. Trans.* **2017**, *57*, 1423–1428. [CrossRef]
74. Zhang, Z.; Saleh, A.S.M.; Wu, H.; Gou, M.; Liu, Y.; Jing, L.; Zhao, K.; Su, C.; Zhang, B.; Li, W. Effect of Starch Isolation Method on Structural and Physicochemical Properties of Acorn Kernel Starch. *Starch/Staerke* **2020**, *72*, 1900122. [CrossRef]
75. Ali, A.; Wani, T.A.; Wani, I.A.; Masoodi, F.A. Comparative study of the physico-chemical properties of rice and corn starches grown in Indian temperate climate. *J. Saudi Soc. Agric. Sci.* **2016**, *15*, 75–82. [CrossRef]
76. Farahnaky, A.; Saberi, B.; Majzoobi, M. Effect of glycerol on physical and mechanical properties of wheat starch edible films. *J. Texture Stud.* **2013**, *44*, 176–186. [CrossRef]
77. Zhang, H.; Zhang, W.; Xu, C.; Zhou, X. Morphological features and physicochemical properties of waxy wheat starch. *Int. J. Biol. Macromol.* **2013**, *62*, 304–309. [CrossRef]
78. Zelaziński, T.; Ekielski, A.; Tulska, E.; Vladut, V.; Durczak, K. Wood dust application for improvement of selected properties of thermoplastic starch. *INMATEH-Agric. Eng.* **2019**, *58*, 37–44. [CrossRef]
79. Fahma, F.; Sunarti, T.C.; Indriyani, S.M.; Lisdayana, N. Thermoplastic Cassava Starch-PVA Composite Films with Cellulose Nanofibers from Oil Palm Empty Fruit Bunches as Reinforcement Agent. *Int. J. Polym. Sci.* **2017**, *2017*, 2745721. [CrossRef]

80. Suhag, R.; Kumar, N.; Petkoska, A.T.; Upadhyay, A. Film formation and deposition methods of edible coating on food products: A review. *Food Res. Int.* **2020**, *136*, 109582. [CrossRef]
81. Liu, H.; Corke, H.; Ramsden, L. Functional properties and enzymatic digestibility of cationic and cross-linked cationic ae, wx, and normal maize starch. *J. Agric. Food Chem.* **1999**, *47*, 2523–2528. [CrossRef]
82. Jane, J.-L.; Kasemsuwan, T.; Leas, S.; Zobel, H.; Robyt, J.F. Anthology of starch granule morphology by scanning electron microscopy. *Starch-Stärke* **1994**, *46*, 121. [CrossRef]
83. Kim, H.S.; Huber, K.C. Physicochemical properties and amylopectin fine structures of A- and B-type granules of waxy and normal soft wheat starch. *J. Cereal Sci.* **2010**, *51*, 256–264. [CrossRef]
84. Kim, H.S.; Huber, K.C. Channels within soft wheat starch A- and B-type granules. *J. Cereal Sci.* **2008**, *48*, 159–172. [CrossRef]
85. Properties and Biodegradation Nature of Thermoplastic Starch. Available online: https://books.google.com.hk/books?hl=zh-TW&lr=&id=e8qgDwAAQBAJ&oi=fnd&pg=PA57&dq=Properties+and+Biodegradation+Nature+of+Thermoplastic+Starch&ots=zPn6_EwbF6&sig=qfWSktlyiYCQanAQG5YwYYZgjjg&redir_esc=y#v=onepage&q=Properties%20and%20Biodegradation%20Nature%20of%20Thermoplastic%20Starch&f=false (accessed on 15 August 2021).
86. Oromiehie, A.R.; Taherzadeh, T.; Rabiee, A. Physical and Thermal Mechanical Properties of Corn Starch/LDPE Composites. *J. Appl. Polym. Sci.* **2013**, *127*, 1128–1134. [CrossRef]
87. Sondari, D.; Falah, F.; Suryaningrum, R.; Sari, F.P.; Sari, F.P.; Septefani, A.A.; Septefani, A.A.; Restu, W.K.; Restu, W.K.; Sampora, Y.; et al. Biofilm Based on Modified Sago Starch: Preparation and Characterization. *Reaktor* **2019**, *19*, 125–130. [CrossRef]
88. Domene-López, D.; García-Quesada, J.C.; Martin-Gullon, I.; Montalbán, M.G. Influence of starch composition and molecular weight on physicochemical properties of biodegradable films. *Polymers* **2019**, *11*, 1084. [CrossRef] [PubMed]
89. Zhong, Y.; Li, Y.; Liang, W.; Liu, L.; Li, S.; Xue, J.; Guo, D. Comparison of gelatinization method, starch concentration, and plasticizer on physical properties of high-amylose starch films. *J. Food Process Eng.* **2017**, *41*, e12645. [CrossRef]
90. Luchese, C.; Garrido, T.; Spada, J.; Tessaro, I.; De la Caba, K. Development and characterization of cassava starch films incorporated with blueberry pomace. *Int. J. Biol. Macromol.* **2017**, *106*, 834–839. [CrossRef] [PubMed]
91. Dai, L.; Zhang, J.; Cheng, F. Effects of starches from different botanical sources and modification methods on physicochemical properties of starch-based edible films. *Int. J. Biol. Macromol.* **2019**, *132*, 897–905. [CrossRef] [PubMed]
92. Basiak, E.; Lenart, A.; Debeaufort, F. Effect of starch type on the physico-chemical properties of edible films. *Int. J. Biol. Macromol.* **2017**, *98*, 348–356. [CrossRef] [PubMed]
93. Zuo, G.; Song, X.; Chen, F.; Shen, Z. Physical and structural characterization of edible bilayer films made with zein and corn-wheat starch. *J. Saudi Soc. Agric. Sci.* **2019**, *18*, 324–331. [CrossRef]
94. Mali, S.; Karam, L.B.; Ramos, L.P.; Grossmann, M.V.E. Relationships among the composition and physicochemical properties of starches with the characteristics of their films. *J. Agric. Food Chem.* **2004**, *52*, 7720–7725. [CrossRef] [PubMed]
95. Karak, N. 10—Vegetable oil-based polymer composites. In *Vegetable Oil-Based Polymers: Properties, Processing and Applications*; Woodhead Publishing: Cambridge, UK, 2012; pp. 247–270. ISBN 978-0-85709-710-1.
96. Sui, Z.; Yao, T.; Zhao, Y.; Ye, X.; Kong, X.; Ai, L. Effects of heat-moisture treatment reaction conditions on the physicochemical and structural properties of maize starch: Moisture and length of heating. *Food Chem.* **2015**, *173*, 1125–1132. [CrossRef]
97. Krieger, K.M.; Duvick, S.A.; Pollak, L.M.; White, P.J. Thermal properties of corn starch extracted with different blending methods: Microblender and homogenizer. *Cereal Chem.* **1997**, *74*, 553–555. [CrossRef]
98. Alemdar, A.; Sain, M. Biocomposites from wheat straw nanofibers: Morphology, thermal and mechanical properties. *Compos. Sci. Technol.* **2008**, *68*, 557–565. [CrossRef]
99. Zeng, J.; Gao, H.; Li, G. Functional properties of wheat starch with different particle size distribution. *J. Sci. Food Agric.* **2014**, *94*, 57–62. [CrossRef]
100. Taylor, J.R.N. Chapter 1—Sorghum and Millets: Taxonomy, History, Distribution, and Production. In *Sorghum and Millets: Chemistry, Technology, and Nutritional Attributes*, 2nd ed.; Taylor, J.R.N., Duodu, K.G., Eds.; AACC International Press: Duxford, UK, 2019; pp. 1–21. ISBN 978-0-12-811527-5.
101. Nascimento, T.A.; Calado, V.; Carvalho, C.W.P. Development and characterization of flexible film based on starch and passion fruit mesocarp flour with nanoparticles. *Food Res. Int.* **2012**, *49*, 588–595. [CrossRef]
102. Zhang, R.; Wang, X.; Cheng, M. Preparation and characterization of potato starch film with various size of Nano-SiO_2. *Polymers* **2018**, *10*, 1172. [CrossRef]
103. Sun, Q.; Sun, C.; Xiong, L. Mechanical, barrier and morphological properties of pea starch and peanut protein isolate blend films. *Carbohydr. Polym.* **2013**, *98*, 630–637. [CrossRef]
104. Ghanbarzadeh, B.; Almasi, H.; Entezami, A.A. Improving the barrier and mechanical properties of corn starch-based edible films: Effect of citric acid and carboxymethyl cellulose. *Ind. Crops Prod.* **2011**, *33*, 229–235. [CrossRef]
105. Patricia Miranda, S.; Garnica, O.; Lara-Sagahon, V.; Cárdenas, G. Water Vapor Permeability and Mechanical Properties of Chitosan Composite Films. *J. Chil. Chem. Soc.* **2004**, *49*, 173–178. [CrossRef]
106. Bertuzzi, M.A.; Castro Vidaurre, E.F.; Armada, M.; Gottifredi, J.C. Water vapor permeability of edible starch based films. *J. Food Eng.* **2007**, *80*, 972–978. [CrossRef]
107. Guo, X.; Lu, Y.; Cui, H.; Jia, X.; Bai, H.; Ma, Y. Factors Affecting the Physical Properties of Edible Composite Film Prepared from Zein and Wheat Gluten. *Molecules* **2012**, *17*, 3794. [CrossRef] [PubMed]

108. Shah, U.; Naqash, F.; Gani, A.; Masoodi, F.A. Art and Science behind Modified Starch Edible Films and Coatings: A Review. *Compr. Rev. Food Sci. Food Saf.* **2016**, *15*, 568–580. [CrossRef] [PubMed]
109. Kester, J.J.; Fennema, O.R. Edible films and coatings: A review. *Food Technol.* **1986**, *40*, 47–59.
110. Henrique, C.M.; Teófilo, R.F.; Sabino, L.; Ferreira, M.M.C.; Cereda, M.P. Classification of Cassava Starch Films by Physicochemical Properties and Water Vapor Permeability Quantification by FTIR and PLS. *J. Food Sci.* **2007**, *72*, E184–E189. [CrossRef]
111. Bénière, F.; Bocquet, J.L.; Brébec, G.; Limoge, Y. Diffusion in Solids. In *Encyclopedia of Materials: Science and Technology*; Buschow, K.H.J., Ed.; Elsevier: Oxford, UK, 2001; pp. 2159–2170. ISBN 978-0-08-043152-9.
112. Bunaciu, A.A.; Udriştioiu, E.G.; Aboul-Enein, H.Y. X-Ray Diffraction: Instrumentation and Applications. *Crit. Rev. Anal. Chem.* **2015**, *45*, 289–299. [CrossRef] [PubMed]
113. Jenkins, P.J.; Donald, A.M. Gelatinisation of starch: A combined SAXS/WAXS/DSC and SANS study. *Carbohydr. Res.* **1998**, *308*, 133–147. [CrossRef]
114. Ogunsona, E.; Ojogbo, E.; Mekonnen, T. Advanced material applications of starch and its derivatives. *Eur. Polym. J.* **2018**, *108*, 570–581. [CrossRef]
115. Ruiz, E.; Srikaeo, K.; la Revilla, L.S. Effects of Heat Moisture Treatment on Physicochemical Properties and Starch Digestibility of Rice Flours Differing in Amylose Content. *Food Appl. Biosci. J.* **2018**, *6*, 140–153.
116. Rindlav-Westling, Å.; Stading, M.; Gatenholm, P. Crystallinity and morphology in films of starch, amylose and amylopectin blends. *Biomacromolecules* **2002**, *3*, 84–91. [CrossRef]
117. Pouplin, M.; Redl, A.; Gontard, N. Glass transition of wheat gluten plasticized with water, glycerol, or sorbitol. *J. Agric. Food Chem.* **1999**, *47*, 538–543. [CrossRef] [PubMed]
118. Patni, N.; Yadava, P.; Agarwal, A.; Maroo, V. Study on Wheat Gluten Biopolymer: A Novel Way to Eradicate Plastic Waste. *Indian J. Appl. Res.* **2011**, *3*, 253–255. [CrossRef]
119. ROY, S.B.; SHIT, D.S.C.; SEN GUPTA, D.R.A.; SHUKLA, D.P.R. A Review on Bio-Composites: Fabrication, Properties and Applications. *Int. J. Innov. Res. Sci. Eng. Technol.* **2014**, *03*, 16814–16824. [CrossRef]
120. Vo Hong, N.; Van Puyvelde, P.; Van Vuure, A.W.; Verpoest, I. Preparation of biocomposites based on gluten resin and unidirectional flax fibers. In Proceedings of the 15th European Conference on Composite Materials (ECCM 2012), Venice, Italy, 24–28 June 2012.
121. Hou, T.H.; Su, C.H.; Liu, W.L. Parameters optimization of a nano-particle wet milling process using the Taguchi method, response surface method and genetic algorithm. *Powder Technol.* **2007**, *173*, 153–162. [CrossRef]
122. Hong, N.V.; Pyka, G.; Wevers, M.; Goderis, B.; Van Puyvelde, P.; Verpoest, I.; Van Vuure, A.W. Processing rigid wheat gluten biocomposites for high mechanical performance. *Compos. Part A* **2015**, *79*, 74–81. [CrossRef]
123. Kim, J.T.; Netravali, A.N. Mechanical, thermal, and interfacial properties of green composites with ramie fiber and soy resins. *J. Agric. Food Chem.* **2010**, *58*, 5400–5407. [CrossRef] [PubMed]
124. Zelaziński, T.; Słoma, J.; Skudlarski, J.; Ekielski, A. The rape pomace and microcrystalline cellulose composites made by press processing. *Sustainability* **2020**, *12*, 1311. [CrossRef]
125. Hemsri, S.; Thongpin, C.; Supatti, N.; Manomai, P.; Socharoentham, A. Bio-based Blends of Wheat Gluten and Maleated Natural Rubber: Morphology, Mechanical Properties and Water Absorption. *Energy Procedia* **2016**, *89*, 264–273. [CrossRef]
126. Edhirej, A.; Sapuan, S.M.; Jawaid, M.; Zahari, N.I. Cassava/sugar palm fiber reinforced cassava starch hybrid composites: Physical, thermal and structural properties. *Int. J. Biol. Macromol.* **2017**, *101*, 75–83. [CrossRef]
127. Lee, J.; Cousineau, A. Production and Characterization of Wheat Gluten Films. Master's Thesis, University of Waterloo, Waterloo, ON, Canada, 2012.
128. Mojumdar, S.C.; Moresoli, C.; Simon, L.C.; Legge, R.L. Edible wheat gluten (WG) protein films. *J. Therm. Anal. Calorim. J. Therm. Anal. Calorim.* **2011**, *104*, 929–936. [CrossRef]
129. Tarique, J.; Sapuan, S.M.; Khalina, A. Effect of glycerol plasticizer loading on the physical, mechanical, thermal, and barrier properties of arrowroot (Maranta arundinacea) starch biopolymers. *Sci. Rep.* **2021**, *11*, 1–17. [CrossRef]
130. Dou, Y.; Zhang, L.; Zhang, B.; He, M.; Shi, W.; Yang, S.; Cui, Y.; Yin, G. Preparation and characterization of edible dialdehyde carboxymethyl cellulose crosslinked feather keratin films for food packaging. *Polymers* **2020**, *12*, 158. [CrossRef] [PubMed]
131. Ye, P.; Reitz, L.; Horan, C.; Parnas, R. Manufacture and biodegradation of wheat gluten/basalt composite material. *J. Polym. Environ.* **2006**, *14*, 1–7. [CrossRef]
132. Reddy, N.; Yang, Y. Biocomposites developed using water-plasticized wheat gluten as matrix and jute fibers as reinforcement. *Polym. Int.* **2011**, *60*, 711–716. [CrossRef]
133. Muneer, F.; Johansson, E.; Hedenqvist, M.S.; Gällstedt, M.; Newson, W.R. Preparation, properties, protein cross-linking and biodegradability of plasticizer-solvent free hemp fibre reinforced wheat gluten, glutenin, and gliadin composites. *BioResources* **2014**, *9*, 5246–5261. [CrossRef]
134. Kunanopparat, T.; Menut, P.; Morel, M.H.; Guilbert, S. Plasticized wheat gluten reinforcement with natural fibers: Effect of thermal treatment on the fiber/matrix adhesion. *Compos. Part A Appl. Sci. Manuf.* **2008**, *39*, 1787–1792. [CrossRef]
135. Hemsri, S.; Grieco, K.; Asandei, A.D.; Parnas, R.S. Wheat gluten composites reinforced with coconut fiber. *Compos. Part A Appl. Sci. Manuf.* **2012**, *43*, 1160–1168. [CrossRef]
136. Nguyen, H.D.; Thuy Mai, T.T.; Nguyen, N.B.; Dang, T.D.; Phung Le, M.L.; Dang, T.T. A novel method for preparing microfibrillated cellulose from bamboo fibers. *Adv. Nat. Sci. Nanosci. Nanotechnol.* **2013**, *4*, 015016. [CrossRef]

137. Tan, M.Y.; Nicholas Kuan, H.T.; Lee, M.C. Characterization of Alkaline Treatment and Fiber Content on the Physical, Thermal, and Mechanical Properties of Ground Coffee Waste/Oxobiodegradable HDPE Biocomposites. *Int. J. Polym. Sci.* **2017**, *2017*, 6258151. [CrossRef]
138. Harwalkar, C.Y.M. Thermal analysis of food carbohydrates FCA. Elsevier Applied Science: London, UK, 1990; pp. 168–222.
139. Rouilly, A.; Rigal, L. Agro-materials: A bibliographic review. *J. Macromol. Sci.-Polym. Rev.* **2002**, *42*, 441–479. [CrossRef]
140. Song, Y.; Zheng, Q. Improved tensile strength of glycerol-plasticized gluten bioplastic containing hydrophobic liquids. *Bioresour. Technol.* **2008**, *99*, 7665–7671. [CrossRef] [PubMed]
141. Dong, J.; Dicharry, R.; Waxman, E.; Parnas, R.S.; Asandei, A.D. Imaging and thermal studies of wheat gluten/poly(vinyl alcohol) and wheat gluten/thiolated poly(vinyl alcohol) blends. *Biomacromolecules* **2008**, *9*, 568–573. [CrossRef]
142. Mangavel, C.; Rossignol, N.; Perronnet, A.; Barbot, J.; Popineau, Y.; Guéguen, J. Properties and microstructure of thermo-pressed wheat gluten films: A comparison with cast films. *Biomacromolecules* **2004**, *5*, 1596–1601. [CrossRef]
143. Kayserilioğlu, B.Ş.; Bakir, U.; Yilmaz, L.; Akkaş, N. Drying temperature and relative humidity effects on wheat gluten film properties. *J. Agric. Food Chem.* **2003**, *51*, 964–968. [CrossRef]
144. Muensri, P.; Kunanopparat, T.; Menut, P.; Siriwattanayotin, S. Composites: Part A Effect of lignin removal on the properties of coconut coir fiber / wheat gluten biocomposite. *Compos. Part A* **2011**, *42*, 173–179. [CrossRef]
145. Chen, L.; Reddy, N.; Wu, X.; Yang, Y. Thermoplastic films from wheat proteins. *Ind. Crops Prod.* **2012**, *35*, 70–76. [CrossRef]
146. Zubeldía, F.; Ansorena, M.R.; Marcovich, N.E. Wheat gluten films obtained by compression molding. *Polym. Test.* **2015**, *43*, 68–77. [CrossRef]
147. Gianibelli, M.C.; Larroque, O.R.; MacRitchie, F.; Wrigley, C.W. Biochemical, genetic, and molecular characterization of wheat glutenin and its component subunits. *Cereal Chem.* **2001**, *78*, 635–646. [CrossRef]
148. "RightFiber Wheat Fiber | The Ingredient House.". Available online: https://theingredienthouse.com/product/wheat-fiber/ (accessed on 27 October 2020).
149. Huang, J.; Yu, C. Determination of cellulose, hemicellulose and lignin content using near-infrared spectroscopy in flax fiber. *Text. Res. J.* **2019**, *89*, 4875–4883. [CrossRef]
150. The Canadian Society for Bioengineering Qualitative and Quantitative Analysis of Lignocellulosic Biomass using Infrared Spectroscopy. Available online: https://library.csbe-scgab.ca/docs/meetings/2009/CSBE09307.pdf (accessed on 14 August 2021).
151. Wolfrum, E.J.; Lorenz, A.J.; deLeon, N. Correlating detergent fiber analysis and dietary fiber analysis data for corn stover collected by NIRS. *Cellulose* **2009**, *16*, 577–585. [CrossRef]
152. Hindrichsen, I.K.; Kreuzer, M.; Madsen, J.; Bach Knudsen, K.E. Fiber and lignin analysis in concentrate, forage, and feces: Detergent versus enzymatic-chemical method. *J. Dairy Sci.* **2006**, *89*, 2168–2176. [CrossRef]
153. Sain, M.; Panthapulakkal, S. Bioprocess preparation of wheat straw fibers and their characterization. *Ind. Crops Prod.* **2006**, *23*, 1–8. [CrossRef]
154. Monta, B.; Ghizzi, G.; Silva, D.; Gastaldi, E.; Torres-chávez, P.; Gontard, N.; Angellier-coussy, H. Biocomposites from wheat proteins and fibers: Structure/mechanical properties relationships. *Ind. Crops Prod.* **2013**, *43*, 545–555. [CrossRef]
155. Zou, Y.; Huda, S.; Yang, Y. Lightweight composites from long wheat straw and polypropylene web. *Bioresour. Technol.* **2010**, *101*, 2026–2033. [CrossRef]
156. Ruiz, H.A.; Rodríguez-Jasso, R.M.; Fernandes, B.D.; Vicente, A.A.; Teixeira, J.A. Hydrothermal processing, as an alternative for upgrading agriculture residues and marine biomass according to the biorefinery concept: A review. *Renew. Sustain. Energy Rev.* **2013**, *21*, 35–51. [CrossRef]
157. Pradhan, R.; Misra, M.; Erickson, L.; Mohanty, A. Compostability and biodegradation study of PLA-wheat straw and PLA-soy straw based green composites in simulated composting bioreactor. *Bioresour. Technol.* **2010**, *101*, 8489–8491. [CrossRef]
158. Le Digabel, F.; Boquillon, N.; Dole, P.; Monties, B.; Averous, L. Properties of thermoplastic composites based on wheat-straw lignocellulosic fillers. *J. Appl. Polym. Sci.* **2004**, *93*, 428–436. [CrossRef]
159. Mengeloglu, F.; Karakus, K. Thermal degradation, mechanical properties and morphology of wheat straw flour filled recycled thermoplastic composites. *Sensors* **2008**, *8*, 500. [CrossRef]
160. Karr, G.S.; Cheng, E.; Sun, X.S. Physical properties of strawboard as affected by processing parameters. *Ind. Crops Prod.* **2000**, *12*, 19–24. [CrossRef]
161. Zhao, L. Novel Bio-Composites Based on Whole Utilisation of Wheat Straw. Ph.D. Thesis, Brunel University, London, UK, 2013.
162. A Biocomposite Made with Wheat Bran | JEC Group. Available online: http://www.jeccomposites.com/knowledge/international-composites-news/biocomposite-made-wheat-bran (accessed on 28 October 2020).
163. Fama, L.; Bittante, A.M.B.Q.; Sobral, P.J.A.; Goyanes, S.; Gerschenson, L.N. Garlic powder and wheat bran as fillers: Their effect on the physicochemical properties of edible biocomposites. *Mater. Sci. Eng. C* **2010**, *30*, 853–859. [CrossRef]
164. Fu, Z.Q.; Wu, H.J.; Wu, M.; Huang, Z.G.; Zhang, M. Effect of Wheat Bran Fiber on the Behaviors of Maize Starch Based Films. *Starch/Staerke* **2020**, *72*, 1900319. [CrossRef]
165. Oishi, Y.; Nakaya, M.; Matsui, E.; Hotta, A. Structural and mechanical properties of cellulose composites made of isolated cellulose nanofibers and poly(vinyl alcohol). *Compos. Part A Appl. Sci. Manuf.* **2015**, *73*, 72–79. [CrossRef]
166. Wei, X.; Wei, W.; Cui, Y.H.; Lu, T.J.; Jiang, M.; Zhou, Z.W.; Wang, Y. All-cellulose composites with ultra-high mechanical properties prepared through using straw cellulose fiber. *RSC Adv.* **2016**, *6*, 93428–93435. [CrossRef]

167. Bledzki, A.K.; Mamun, A.A.; Volk, J. Physical, chemical and surface properties of wheat husk, rye husk and soft wood and their polypropylene composites. *Compos. Part A Appl. Sci. Manuf.* **2010**, *41*, 480–488. [CrossRef]
168. Jagwani, D.; Joshi, P. Deportation of Toxic Phenol From Aqueous System by Wheat Husk. *Int. J. Plant Anim. Environ. Sci.* **2014**, *4*, 58–64.
169. Mendes, C.A.D.C.; Adnet, F.A.D.O.; Leite, M.C.A.M.; Furtado, C.R.G.; Sousa, A.M.F. De Chemical, physical, mechanical, thermal and morphological characterization of corn husk residue. *Cellul. Chem. Technol.* **2014**, *49*, 727–735.
170. Pandecha, K.; Pongtornkulpanich, A.; Sukchai, S. Thermal properties of corn husk fiber as insulation for flat plate solar collector. *J. Renew. Energy Smart Grid Technol.* **2015**, *10*, 27–36.
171. Norashikin, M.Z.; Ibrahim, M.Z. The potential of natural waste (corn husk) for production of environmental friendly biodegradable film for seedling. *World Acad. Sci. Eng. Technol.* **2009**, *58*, 176–180. [CrossRef]
172. Table 1 | Physical and Acoustical Properties of Corn Husk Fiber Panels. Available online: https://www.hindawi.com/journals/aav/2016/5971814/tab1/ (accessed on 8 June 2020).
173. Nordin, R.; Ismail, H. Properties of Rice Husk Powder/Natural Rubber Composite. *Solid State Sci. Technol.* **2007**, *15*, 83–91.
174. Faisal Bukhori, M.S.D. Effect of Rice Husk Waste and Rice Husk Ash Composition as Filler in Plastic Bottle Drink Waste Composites on Water Absorption Properties. *Int. J. Sci. Res.* **2015**, *4*, 2146–2148.
175. Gummert, M.; Van Hung, N.; Chivenge, P.; Douthwaite, B. Sustainable Rice Straw Management. 2020. Springer International Publishing: Midtown Manhattan, NY, USA, 2020; pp. 1–13. [CrossRef]
176. Kapoor, M.; Panwar, D.; Kaira, G.S. *Bioprocesses for Enzyme Production Using Agro-Industrial Wastes: Technical Challenges and Commercialization Potential*; Elsevier Inc.: Amsterdam, The Netherlands, 2016; ISBN 9780128026120.
177. Galanakis, C.M. *Food Waste Recovery: Processing Technologies and Industrial Techniques*; Elsevier Inc.: Amsterdam, The Netherlands, 2015; ISBN 9780128004197.
178. Lam, P.S.; Sokhansanj, S.; Bi, X.; Mani, S.; Lim, C.J.; Womac, A.R.; Hoque, M.; Peng, J.; Jayashankar, T.; Nalmi, L.J.; et al. Physical characterization of wet and dry wheat straw and switchgrass—Bulk and specific density. In Proceedings of the 2007 ASAE Annual Meeting, Minneapolis, MN, USA, 17–20 June 2007. [CrossRef]
179. Jiang, D.; An, P.; Cui, S.; Sun, S.; Zhang, J.; Tuo, T. Effect of Modification Methods of Wheat Straw Fibers on Water Absorbency and Mechanical Properties of Wheat Straw Fiber Cement-Based Composites. *Adv. Mater. Sci. Eng.* **2020**, *2020*, 5031025. [CrossRef]
180. Bouasker, M.; Belayachi, N.; Hoxha, D.; Al-Mukhtar, M. Physical characterization of natural straw fibers as aggregates for construction materials applications. *Materials* **2014**, *7*, 3034. [CrossRef]
181. Passoth, V.; Sandgren, M. Biofuel production from straw hydrolysates: Current achievements and perspectives. *Appl. Microbiol. Biotechnol.* **2019**, *103*, 5105–5116. [CrossRef]
182. Rehman, N.; de Miranda, M.I.G.; Rosa, S.M.L.; Pimentel, D.M.; Nachtigall, S.M.B.; Bica, C.I.D. Cellulose and Nanocellulose from Maize Straw: An Insight on the Crystal Properties. *J. Polym. Environ.* **2014**, *22*, 252–259. [CrossRef]
183. Li, Y.; Yan, F.; Li, T.; Zhou, Y.; Jiang, H.; Qian, M.; Xu, Q. High-solid anaerobic digestion of corn straw for methane production and pretreatment of bio-briquette. *Bioresour. Technol.* **2018**, *250*, 741–749. [CrossRef]
184. Appendix XI: Bulk Density, Pelletability and Particle Size. Available online: http://www.fao.org/3/S4314E/s4314e0q.htm (accessed on 28 October 2020).
185. Onipe, O.O.; Beswa, D.; Jideani, A.I.O. Effect of size reduction on colour, hydration and rheological properties of wheat bran. *Food Sci. Technol.* **2017**, *37*, 389–396. [CrossRef]
186. Mayo, B. The proteolytic system of lactic acid bacteria. *Microbiologia* **1993**, *9*, 90–106. [CrossRef] [PubMed]
187. Liu, L.; Song, J.; Li, Y.; Li, P.; Wang, H. Robust and cost-saving static solid cultivation method for lipid production using the chlamydospores of Phanerochaete chrysosporium. *Biotechnol. Biofuels* **2019**, *12*, 1–14. [CrossRef] [PubMed]
188. Matavire, T.O. Bran to produce entrapment materials for the controlled by Master of Engineering (Chemical Engineering). 2018. Available online: https://www.semanticscholar.org/paper/Extraction-and-modification-of-hemicellulose-from_Matavire/09d110349e4453a83d864a6c73183a5b7e2c9556?sort=is-influential&pdf=true (accessed on 8 June 2020).
189. Nagarajan, S.; Nagarajan, R.; Kumar, J.; Salemme, A.; Togna, A.R.; Saso, L.; Bruno, F. Antioxidant activity of synthetic polymers of phenolic compounds. *Polymers* **2020**, *12*, 1646. [CrossRef]
190. Eça, K.S.; Sartori, T.; Menegalli, F.C. Films and edible coatings containing antioxidants—A review. *Braz. J. Food Technol.* **2014**, *17*, 98–112. [CrossRef]
191. Abd El-Ghaffar, M.A.; Shaffei, K.A.; Abdelwahab, N. Evaluation of some conducting polymers as novel antioxidants for rubber vulcanizates. *Int. J. Polym. Sci.* **2014**, *2014*, 893542. [CrossRef]
192. Jacob, J.; Thomas, S.; Loganathan, S.; Valapa, R.B. Chapter 10—Antioxidant incorporated biopolymer composites for active packaging. In *Processing and Development of Polysaccharide-Based Biopolymers for Packaging Applications*; Zhang, Y., Ed.; Elsevier: Amsterdam, The Netherlands, 2020; pp. 239–260. ISBN 978-0-12-818795-1.
193. Cheremisinoff, N.P. Condensed Encyclopedia of polymer Engineering Terms. *Choice Rev. Online* **2001**, *39*, 1288.
194. Bonilla, J.; Talón, E.; Atarés, L.; Vargas, M.; Chiralt, A. Effect of the incorporation of antioxidants on physicochemical and antioxidant properties of wheat starch-chitosan films. *J. Food Eng.* **2013**, *118*, 271–278. [CrossRef]
195. Yilmaz-Turan, S.; Jiménez-Quero, A.; Menzel, C.; de Carvalho, D.M.; Lindström, M.E.; Sevastyanova, O.; Moriana, R.; Vilaplana, F. Bio-based films from wheat bran feruloylated arabinoxylan: Effect of extraction technique, acetylation and feruloylation. *Carbohydr. Polym.* **2020**, *250*, 116916. [CrossRef]

196. Díez-Pascual, A.M. Antimicrobial polymer-based materials for food packaging applications. *Polymers* **2020**, *12*, 731. [CrossRef]
197. Park, S.-K.; Bae, D.H. Antimicrobial properties of wheat gluten-chitosan composite film in intermediate-moisture food systems. *Food Sci. Biotechnol.* **2006**, *15*, 133–137.
198. Güçbilmez, Ç.M.; Yemenicioğlu, A.; Arslanoğlu, A. Antimicrobial and antioxidant activity of edible zein films incorporated with lysozyme, albumin proteins and disodium EDTA. *Food Res. Int.* **2007**, *40*, 80–91. [CrossRef]
199. Pintado, C.; Ferreira, S.; Sousa, I. Properties of Whey Protein-Based Films Containing Organic Acids and Nisin To Control Listeria monocytogenes. *J. Food Prot.* **2009**, *72*, 1891–1896. [CrossRef]
200. Türe, H.; Eroglu, E.; Soyer, F.; Özen, B. Antifungal activity of biopolymers containing natamycin and rosemary extract against Aspergillus niger and Penicillium roquefortii. *Int. J. Food Sci. Technol.* **2008**, *43*, 2026–2032. [CrossRef]
201. Pranoto, Y.; Rakshit, S.K.; Salokhe, V.M. Enhancing antimicrobial activity of chitosan films by incorporating garlic oil, potassium sorbate and nisin. *LWT-Food Sci. Technol.* **2005**, *38*, 859–865. [CrossRef]
202. Seydim, A.C.; Sarikus, G. Antimicrobial activity of whey protein based edible films incorporated with oregano, rosemary and garlic essential oils. *Food Res. Int.* **2006**, *39*, 639–644. [CrossRef]
203. Sivarooban, T.; Hettiarachchy, N.S.; Johnson, M.G. Physical and antimicrobial properties of grape seed extract, nisin, and EDTA incorporated soy protein edible films. *Food Res. Int.* **2008**, *41*, 781–785. [CrossRef]
204. Iamareerat, B.; Singh, M.; Sadiq, M.B.; Anal, A.K. Reinforced cassava starch based edible film incorporated with essential oil and sodium bentonite nanoclay as food packaging material. *J. Food Sci. Technol.* **2018**, *55*, 1953–1959. [CrossRef]
205. Acevedo-Fani, A.; Salvia-Trujillo, L.; Rojas-Graü, M.A.; Martín-Belloso, O. Edible films from essential-oil-loaded nanoemulsions: Physicochemical characterization and antimicrobial properties. *Food Hydrocoll.* **2015**, *47*, 168–177. [CrossRef]
206. Ribeiro-Santos, R.; Andrade, M.; Sanches-Silva, A. Application of encapsulated essential oils as antimicrobial agents in food packaging. *Curr. Opin. Food Sci.* **2017**, *14*, 78–84. [CrossRef]
207. Kan, Y.; Chen, T.; Wu, Y.; Wu, J.; Wu, J. Antioxidant activity of polysaccharide extracted from Ganoderma lucidum using response surface methodology. *Int. J. Biol. Macromol.* **2015**, *72*, 151–157. [CrossRef]
208. Atarés, L.; Chiralt, A. Essential oils as additives in biodegradable films and coatings for active food packaging. *Trends Food Sci. Technol.* **2016**, *48*, 51–62. [CrossRef]
209. Viuda-Martos, M.; El Gendy, A.E.N.G.S.; Sendra, E.; Fernández-López, J.; El Razik, K.A.A.; Omer, E.A.; Pérez-Alvarezj, J.A. Chemical composition and antioxidant and anti-Listeria activities of essential oils obtained from some Egyptian plants. *J. Agric. Food Chem.* **2010**, *58*, 9063–9070. [CrossRef] [PubMed]
210. Syafiq, R.; Sapuan, S.M.; Zuhri, M.Y.M.; Ilyas, R.A.; Nazrin, A.; Sherwani, S.F.K.; Khalina, A. Antimicrobial activities of starch-based biopolymers and biocomposites incorporated with plant essential oils: A review. *Polymers* **2020**, *12*, 2403. [CrossRef] [PubMed]
211. Song, X.; Zuo, G.; Chen, F. Effect of essential oil and surfactant on the physical and antimicrobial properties of corn and wheat starch films. *Int. J. Biol. Macromol.* **2018**, *107*, 1302–1309. [CrossRef] [PubMed]
212. Jamróz, E.; Juszczak, L.; Kucharek, M. Investigation of the physical properties, antioxidant and antimicrobial activity of ternary potato starch-furcellaran-gelatin films incorporated with lavender essential oil. *Int. J. Biol. Macromol.* **2018**, *114*, 1094–1101. [CrossRef]
213. Türe, H.; Gällstedt, M.; Hedenqvist, M.S. Antimicrobial compression-moulded wheat gluten films containing potassium sorbate. *Food Res. Int.* **2012**, *45*, 109–115. [CrossRef]
214. Khairuddin, N.; Muhamad, I.I.; Abd Rahman, W.A.W.; Siddique, B.M. Physicochemical and thermal characterization of hydroxyethyl cellulose—Wheat starch based films incorporated thymol intended for active packaging. *Sains Malays.* **2020**, *49*, 323–333. [CrossRef]
215. Rozilah, A.; Jaafar, C.N.A.; Sapuan, S.M.; Zainol, I.; Ilyas, R.A. The Effects of Silver Nanoparticles Compositions on the Mechanical, Physiochemical, Antibacterial, and Morphology Properties of Sugar Palm Starch Biocomposites for Antibacterial Coating. *Polymers* **2020**, *12*, 2605. [CrossRef]
216. Ilyas, R.A.; Sapuan, S.M.; Ibrahim, R.; Abral, H.; Ishak, M.R.; Zainudin, E.S.; Atiqah, A.; Atikah, M.S.N.; Syafri, E.; Asrofi, M.; et al. Thermal, Biodegradability and Water Barrier Properties of Bio-Nanocomposites Based on Plasticised Sugar Palm Starch and Nanofibrillated Celluloses from Sugar Palm Fibres. *J. Biobased Mater. Bioenergy* **2020**, *14*, 234–248. [CrossRef]
217. Ilyas, R.A.; Sapuan, S.M.; Ishak, M.R.; Zainudin, E.S. Development and characterization of sugar palm nanocrystalline cellulose reinforced sugar palm starch bionanocomposites. *Carbohydr. Polym.* **2018**, *202*, 186–202. [CrossRef]
218. Ilyas, R.A.; Sapuan, S.M.; Ibrahim, R.; Abral, H.; Ishak, M.R.; Zainudin, E.S.; Atikah, M.S.N.; Mohd Nurazzi, N.; Atiqah, A.; Ansari, M.N.M.; et al. Effect of sugar palm nanofibrillated celluloseconcentrations on morphological, mechanical andphysical properties of biodegradable films basedon agro-waste sugar palm (Arenga pinnata(Wurmb.) Merr) starch. *J. Mater. Res. Technol.* **2019**, *8*, 4819–4830. [CrossRef]
219. Syafri, E.; Sudirman; Mashadi; Yulianti, E.; Deswita; Asrofi, M.; Abral, H.; Sapuan, S.M.; Ilyas, R.A.; Fudholi, A. Effect of sonication time on the thermal stability, moisture absorption, and biodegradation of water hyacinth (Eichhornia crassipes) nanocellulose-filled bengkuang (Pachyrhizus erosus) starch biocomposites. *J. Mater. Res. Technol.* **2019**, *8*, 6223–6231. [CrossRef]
220. Abral, H.; Ariksa, J.; Mahardika, M.; Handayani, D.; Aminah, I.; Sandrawati, N.; Pratama, A.B.; Fajri, N.; Sapuan, S.M.; Ilyas, R.A. Transparent and antimicrobial cellulose film from ginger nanofiber. *Food Hydrocoll.* **2020**, *98*, 105266. [CrossRef]

221. Abral, H.; Ariksa, J.; Mahardika, M.; Handayani, D.; Aminah, I.; Sandrawati, N.; Sapuan, S.M.; Ilyas, R.A. Highly transparent and antimicrobial PVA based bionanocomposites reinforced by ginger nanofiber. *Polym. Test.* **2020**, *81*, 106186. [CrossRef]
222. Kumar, T.S.M.; Chandrasekar, M.; Senthilkumar, K.; Ilyas, R.A.; Sapuan, S.M.; Hariram, N.; Rajulu, A.V.; Rajini, N.; Siengchin, S. Characterization, Thermal and Antimicrobial Properties of Hybrid Cellulose Nanocomposite Films with in-Situ Generated Copper Nanoparticles in Tamarindus indica Nut Powder. *J. Polym. Environ.* **2021**, *29*, 1134–1142. [CrossRef]
223. Asrofi, M.; Sapuan, S.M.; Ilyas, R.A.; Ramesh, M. Characteristic of composite bioplastics from tapioca starch and sugarcane bagasse fiber: Effect of time duration of ultrasonication (Bath-Type). *Mater. Today Proc.* **2021**, *46*, 1626–1630. [CrossRef]
224. Tarique, J.; Sapuan, S.M.; Khalina, A.; Sherwani, S.F.K.; Yusuf, J.; Ilyas, R.A. Recent developments in sustainable arrowroot (Maranta arundinacea Linn) starch biopolymers, fibres, biopolymer composites and their potential industrial applications: A review. *J. Mater. Res. Technol.* **2021**, *13*, 1191–1219. [CrossRef]
225. Jumaidin, R.; Diah, N.A.; Ilyas, R.A.; Alamjuri, R.H.; Yusof, F.A.M. Processing and Characterisation of Banana Leaf Fibre Reinforced Thermoplastic Cassava Starch Composites. *Polymers* **2021**, *13*, 1420. [CrossRef]
226. Aisyah, H.A.; Paridah, M.T.; Sapuan, S.M.; Ilyas, R.A.; Khalina, A.; Nurazzi, N.M.; Lee, S.H.; Lee, C.H. A comprehensive review on advanced sustainable woven natural fibre polymer composites. *Polymers* **2021**, *13*, 471. [CrossRef]
227. Azammi, A.M.N.; Ilyas, R.A.; Sapuan, S.M.; Ibrahim, R.; Atikah, M.S.N.; Asrofi, M.; Atiqah, A. Characterization studies of biopolymeric matrix and cellulose fibres based composites related to functionalized fibre-matrix interface. In *Interfaces in Particle and Fibre Reinforced Composites—From Macro to Nano Scales*; Woodhead Publishing: London, UK, 2019; pp. 1–68. ISBN 9780081026656.
228. Domenek, S.; Brendel, L.; Morel, M.-H.; Guilbert, S. Influence of Degree of Protein Aggregation on Mass Transport Through Wheat Gluten Membranes and Their Digestibility—An In Vitro Study. *Cereal Chem. J.* **2004**, *81*, 423–428. [CrossRef]
229. Guillaume, C.; Pinte, J.; Gontard, N.; Gastaldi, E. Wheat gluten-coated papers for bio-based food packaging: Structure, surface and transfer properties. *Food Res. Int.* **2010**, *43*, 1395–1401. [CrossRef]
230. Gontard, N.; Thibault, R.; Cuq, B.; Guilbert, S. Influence of Relative Humidity and Film Composition on Oxygen and Carbon Dioxide Permeabilities of Edible Films. *J. Agric. Food Chem.* **1996**, *44*, 1064–1069. [CrossRef]
231. Chalier, P.; Peyches-Bach, A.; Gastaldi, E.; Gontard, N. Effect of Concentration and Relative Humidity on the Transfer of Alkan-2-ones through Paper Coated with Wheat Gluten. *J. Agric. Food Chem.* **2007**, *55*, 867–875. [CrossRef]
232. Angellier-Coussy, H.; Torres-Giner, S.; Morel, M.-H.; Gontard, N.; Gastaldi, E. Functional properties of thermoformed wheat gluten/montmorillonite materials with respect to formulation and processing conditions. *J. Appl. Polym. Sci.* **2008**, *107*, 487–496. [CrossRef]
233. Sun, S.; Song, Y.; Zheng, Q. Thermo-molded wheat gluten plastics plasticized with glycerol: Effect of molding temperature. *Food Hydrocoll.* **2008**, *22*, 1006–1013. [CrossRef]
234. Yang, S.; Bai, S.; Wang, Q. Sustainable packaging biocomposites from polylactic acid and wheat straw: Enhanced physical performance by solid state shear milling process. *Compos. Sci. Technol.* **2018**, *158*, 34–42. [CrossRef]
235. Berthet, M.-A.; Gontard, N.; Angellier-Coussy, H. Impact of fibre moisture content on the structure/mechanical properties relationships of PHBV/wheat straw fibres biocomposites. *Compos. Sci. Technol.* **2015**, *117*, 386–391. [CrossRef]
236. Bruni, G.P.; Oliveira, J.P.; Fonseca, L.M.; Silva, F.T.; Dias, A.R.G.; da Rosa Zavareze, E. Biocomposite Films Based on Phosphorylated Wheat Starch and Cellulose Nanocrystals from Rice, Oat, and Eucalyptus Husks. *Starch-Stärke* **2020**, *72*, 1900051. [CrossRef]
237. Nurazzi, N.M.; Khalina, A.; Sapuan, S.M.; Ilyas, R.A. Mechanical properties of sugar palm yarn / woven glass fiber reinforced unsaturated polyester composites: Effect of fiber loadings and alkaline treatment. *Polimery* **2019**, *64*, 12–22. [CrossRef]
238. Suriani, M.J.; Radzi, F.S.M.; Ilyas, R.A.; Petrů, M.; Sapuan, S.M.; Ruzaidi, C.M. Flammability, Tensile, and Morphological Properties of Oil Palm Empty Fruit Bunches Fiber/Pet Yarn-Reinforced Epoxy Fire Retardant Hybrid Polymer Composites. *Polymers* **2021**, *13*, 1282. [CrossRef]
239. Suriani, M.J.; Rapi, H.Z.; Ilyas, R.A.; Petrů, M.; Sapuan, S.M. Delamination and Manufacturing Defects in Natural Fiber-Reinforced Hybrid Composite: A Review. *Polymers* **2021**, *13*, 1323. [CrossRef] [PubMed]
240. Suriani, M.J.; Sapuan, S.M.; Ruzaidi, C.M.; Nair, D.S.; Ilyas, R.A. Flammability, morphological and mechanical properties of sugar palm fiber/polyester yarn-reinforced epoxy hybrid biocomposites with magnesium hydroxide flame retardant filler. *Text. Res. J.* **2021**, 004051752110086. [CrossRef]
241. Nurazzi, N.M.; Asyraf, M.R.M.; Fatimah Athiyah, S.; Shazleen, S.S.; Rafiqah, S.A.; Harussani, M.M.; Kamarudin, S.H.; Razman, M.R.; Rahmah, M.; Zainudin, E.S.; et al. A Review on Mechanical Performance of Hybrid Natural Fiber Polymer Composites for Structural Applications. *Polymers* **2021**, *13*, 2170. [CrossRef] [PubMed]
242. Moazzen, N.; Khanmohammadi, M.; Bagheri Garmarudi, A.; Kazemipour, M.; Ansari Dogaheh, M. Optimization and infrared spectrometric evaluation of the mechanical properties of PLA-based biocomposites. *J. Macromol. Sci. Part A Pure Appl. Chem.* **2019**, *56*, 17–25. [CrossRef]
243. Mat Zain, N.F. Preparation and Characterization of Cellulose and Nanocellulose From Pomelo (Citrus grandis) Albedo. *J. Nutr. Food Sci.* **2014**, *5*, 10–13. [CrossRef]
244. Reddy, C.R.; Sardashti, A.P.; Simon, L.C. Preparation and characterization of polypropylene–wheat straw–clay composites. *Compos. Sci. Technol.* **2010**, *70*, 1674–1680. [CrossRef]

245. Formela, K.; Hejna, A.; Piszczyk, Ł.; Saeb, M.R.; Colom, X. Processing and structure–property relationships of natural rubber/wheat bran biocomposites. *Cellulose* **2016**, *23*, 3157–3175. [CrossRef]
246. Sasimowski, E.; Majewski, Ł.; Grochowicz, M. Influence of the Design Solutions of Extruder Screw Mixing Tip on Selected Properties of Wheat Bran-Polyethylene Biocomposite. *Polymers* **2019**, *11*, 2120. [CrossRef]
247. Song, Y.; Zheng, Q.; Liu, C. Green biocomposites from wheat gluten and hydroxyethyl cellulose: Processing and properties. *Ind. Crops Prod.* **2008**, *28*, 56–62. [CrossRef]
248. Muthuraj, R.; Lacoste, C.; Lacroix, P.; Bergeret, A. Sustainable thermal insulation biocomposites from rice husk, wheat husk, wood fibers and textile waste fibers: Elaboration and performances evaluation. *Ind. Crops Prod.* **2019**, *135*, 238–245. [CrossRef]
249. Boudria, A.; Hammoui, Y.; Adjeroud, N.; Djerrada, N.; Madani, K. Effect of filler load and high-energy ball milling process on properties of plasticized wheat gluten/olive pomace biocomposite. *Adv. Powder Technol.* **2018**, *29*, 1230–1238. [CrossRef]
250. Mahmood, H.; Mehmood, S.; Shakeel, A.; Iqbal, T.; Kazmi, M.A.; Khurram, A.R.; Moniruzzaman, M. Glycerol Assisted Pretreatment of Lignocellulose Wheat Straw Materials as a Promising Approach for Fabrication of Sustainable Fibrous Filler for Biocomposites. *Polymers* **2021**, *13*, 388. [CrossRef] [PubMed]
251. Chun, K.S.; Husseinsyah, S. Polylactic acid/corn cob eco-composites: Effect of new organic coupling agent. *J. Thermoplast. Compos. Mater.* **2014**, *27*, 1667–1678. [CrossRef]
252. Rodriguez, M.; Rodriguez, A.; R, J.B.; Vilaseca, F.; Girones, J.; Mutje, P. Determination of corn stalk fibers' strength through modeling of the mechanical properties of its composites. *BioResources* **2010**, *5*, 2535–2546. [CrossRef]